Accession no.
36124136

KU-295-410

Atmosphere, Weather and Climate

WITHDRAWN

This book presents a comprehensive introduction to weather processes and climatic conditions around the world, their observed variability and changes, and projected future trends. Extensively revised and updated, this ninth edition retains its tried and tested structure while incorporating recent advances in the field. From clear explanations of the basic physical and chemical principles of the atmosphere, to descriptions of regional climates and their changes, the book presents a comprehensive coverage of global meteorology and climatology. In this new edition the latest scientific ideas are again expressed in a clear, non-mathematical manner.

New features include:

- Extended and updated treatment of atmospheric models
- Final chapter on climate variability and change has been completely rewritten to take account of the IPCC 2007 scientific assessment.
- New full colour text design featuring over 30 colour plates
- Over 360 diagrams have been redrawn in full colour to improve clarity and aid understanding
- A new companion website can be found at: www.routledge.com/eresources/9780415465694

Atmosphere, Weather and Climate continues to be an indispensable source for all those studying the earth's atmosphere and world climate, whether from environmental and earth sciences, geography, ecology, agriculture, hydrology, or related disciplinary perspectives. Its pedagogic value is enhanced by several features: learning points at the opening of each chapter and discussion topics at their ending, boxes on topical subjects and on twentieth century advances in the field.

Roger G. Barry is Distinguished Professor of Geography, University of Colorado at Boulder, Director of the World Data Center for Glaciology and the National Snow and Ice Data center, and a Fellow of the Cooperative Institute for Research in Environmental Sciences.

The late **Richard J. Chorley** was Professor of Geography at the University of Cambridge.

Atmosphere, Weather and Climate

Roger G. Barry and
Richard J. Chorley

LIS LIBRARY
Date 29.7.10 Fund 9
Order No 2134950
University of Chester

LONDON AND NEW YORK

First published 1968 by Methuen & Co. Ltd
Second edition 1971
Third edition 1976
Fourth edition 1982
Fifth edition 1987
Reprinted by Routledge 1989, 1990
Sixth edition 1992
Reprinted 1995
Seventh edition 1998 by Routledge
Eighth edition 2003 by Routledge

Ninth edition 2010
by Routledge
2 Park Square, Milton Park, Abingdon, Oxon, OX14 4RN

Simultaneously published in the USA and Canada
by Routledge
270 Madison Avenue, New York, NY 10016

Routledge is an imprint of the Taylor & Francis Group and informa business

© 1968, 1971, 1976, 1982, 1987, 1992, 1998, 2003, 2010 Roger G. Barry and Richard J. Chorley

Typeset in Minion and Univers by
Florence Production Ltd, Stoodleigh, Devon
Printed and bound in India by
Replika Press Pvt. Ltd.

All rights reserved. No part of this book may be reprinted, or reproduced or utilized in any form or by any electronic, mechanical, or other means, now known or hereafter invented, including photocopying and recording, or in any information storage or retrieval system, without permission in writing from the publishers.

British Library Cataloguing in Publication Data
A catalogue record for this book is available from the British Library

Library of Congress Cataloging in Publication Data
Barry, Roger Graham.
Atmosphere, weather, and climate/Roger G. Barry and Richard J. Chorley. – 9th ed.
p. cm.
1. Meteorology. 2. Atmospheric physics. 3. Climatology.
I. Chorley, Richard J. II. Title.
QC861.3.B37 2009
551.5 – dc22 2009009579

ISBN10: 0–415–46569–9 (hbk)
ISBN10: 0–415–46570–2 (pbk)
ISBN10: 0–203–87102–2 (ebk)

ISBN13: 978–0–415–46569–4 (hbk)
ISBN13: 978–0–415–46570–0 (pbk)
ISBN13: 978–0–203–87102–7 (ebk)

Contents

Preface to the ninth edition

This revised ninth edition of *Atmosphere, Weather and Climate* will prove invaluable to all those studying the earth's atmosphere and world climate, whether from environmental, atmospheric and earth sciences, geography, ecology, agriculture, hydrology or related disciplinary perspectives.

Atmosphere, Weather and Climate provides a comprehensive introduction to weather processes and climatic conditions. Since the last edition, we have added an introductory overview of the historical development of the field and its major components. Following this there is an extended treatment of atmospheric composition and energy, stressing the heat budget of the earth and the causes of the greenhouse effect. We then turn to the manifestations and circulation of atmospheric moisture, including atmospheric stability and precipitation patterns in space and time. A consideration of atmospheric and oceanic motion on small to large scales leads on to a chapter on modelling of the atmospheric circulation and climate that also presents weather forecasting on different time scales. This was revised by my colleague Dr Tom Chase of CIRES and Geography at the University of Colorado, Boulder. This is followed by a discussion of the structure of air masses, the development of frontal and non-frontal cyclones and of mesoscale convective systems in mid-latitudes. The treatment of weather and climate in temperate latitudes begins with studies of Europe and America, extending to the conditions of their subtropical and high-latitude margins and includes the Mediterranean, Australasia, North Africa, the southern westerlies, and the subarctic and polar regions. Tropical weather and climate are also described through an analysis of the climatic mechanisms of monsoon Asia, Africa, Australia and Amazonia, together with the tropical margins of Africa and Australia and the effects of ocean movement and the El Niño–Southern Oscillation and teleconnections. Small-scale climates – including urban climates – are considered from the perspective of energy budgets. The final chapter revised by Dr Mark Serreze of CIRES stresses the structure and operation of the atmosphere–earth–ocean system and the causes of its climate changes. Since the previous edition appeared in 2003, the pace of research on the climate system and attention to global climate change has accelerated. A discussion of the various modeling strategies adopted for the prediction of climate change is undertaken, in particular relating to the IPCC 1990–2007 models. A consideration of other environmental impacts of climate change is also included.

The new information age and wide use of the World Wide Web has led to significant changes in presentation. Apart from the two revised chapters 8 and 13, new features include all figures redrawn in color. Wherever possible, the criticisms and suggestions of colleagues and reviewers have been taken into account in preparing this latest edition.

This edition benefited greatly from the ideas and work of my long-time friend and co-author Professor Richard J. Chorley, who sadly passed away on 12 May 2002. His knowledge, enthusiasm and inspiration are sorely missed.

Roger G. Barry
CIRES and Department of Geography
University of Colorado, Boulder

Acknowledgments

The authors are very much indebted to Mr A. J. Dunn for his considerable contribution to the first edition; to the late Professor F. Kenneth Hare of the University of Toronto, Ontario, for his thorough and authoritative criticism of the preliminary text and his valuable suggestions for its improvement; also to Mr Alan Johnson of Barton Peveril College, Eastleigh, Hampshire, for helpful comments on chapters 1 to 3; and to Dr C. Desmond Walshaw, formerly of the Cavendish Laboratory, Cambridge, and Mr R. H. A. Stewart of the Nautical College, Pangbourne, for offering valuable criticisms and suggestions at an early stage in the preparation of the original manuscript. Gratitude is also expressed to the following persons for their helpful comments with respect to the fourth edition: Dr Brian Knapp of Leighton Park School, Reading; Dr L. F. Musk of the University of Manchester; Dr A. H. Perry of University College, Swansea; Dr R. Reynolds of the University of Reading; and Dr P. Smithson of the University of Sheffield. Dr C. Ramage, of the University of Hawaii, made numerous helpful suggestions on the revision of Chapter 6 for the fifth edition. Dr Z. Toth and Dr D. Gilman of the National Meteorological Center, Washington, DC, kindly helped in the updating of Chapter 4 I and Dr M. Tolbert of the University of Colorado assisted with the environmental chemistry in the seventh edition. Thanks are also due to Dr N. Cox of Durham University for his many textual suggestions which contributed significantly to the improvement of the seventh edition. The authors accept complete responsibility for any remaining textual errors.

The redrawn color figures were prepared by Mr Paul Coles in the Geography Department at Sheffield University, UK, building on the illustrative imagination and cartographic expertise of Mr M. Young of the Department of Geography, Cambridge University. The authors owe a considerable debt of gratitude to both individuals.

Thanks are also due to Natasha Vizcarra, Jessica Erven, Jody Hoon-Starr, Sam Massey and Mike Laxer, NSIDC, for office support for the ninth edition; and to Dr Eileen McKim for preparing the index.

The authors would like to thank the following learned societies, editors, publishers, organizations and individuals for permission to reproduce figures, tables and plates.

LEARNED SOCIETIES

American Association for the Advancement of Science for Figure 7.32 from *Science*.

American Geographical Society for Figure 2.17 from the *Geographical Review*.

American Geophysical Union for Figure 13.3 from the *Review of Geophysics and Space Physics*; for Figures 2.4, 2.12 and 5.20 from the *Journal of Geophysical Research*; for Figure 13.6 from *Geophysical Research Letters* and for Figure 10.39 from *Arctic Meteorology and Climatology* by D. H. Bromwich and C. R. Stearns (eds).

American Meteorological Society for Figures 2.2, 3.22A, 3.26C, 5.11, 7.21C, 9.29 and 10.34 from the *Bulletin*; for figure 9.8 and 10.38 from the *Journal of Applied Meteorology*; for Figures 7.21A, B and 9.33 from the *Journal of Atmospheric* Sciences; for Figure 10.24 from the *Journal pf Climate;* for Figure 4.12 from the *Journal of Hydrometeorology*, Figures 7.8, 7.24, 7.25, 8.1, 9.2B, 9.6, 9.10, 11.5, 11.11 and 11.33 from the *Monthly Weather Review*; for Figure 7.28 from the *Journal of Physical Oceanography* and for Figures 9.9, 9.15 and 9.17 from *Extratropical Cyclones* by C. W. Newton and E. D. Holopainen (eds).

American Planning Association for Figure 12.30 from the *Journal*.

Association of American Geographers for Figure 4.21 from the *Annals*.

Geographical Association for Figure 10.4 from *Geography*.

Geographical Society of China for Figures 11.34 and 11.37.

Institute of British Geographers for Figures 4.11 and 4.14 from the *Transactions*; and for figure 4.21 from the *Atlas of Drought in Britain 1975–76* by J. C. Doornkamp and K. J. Gregory (eds).

Institution of Civil Engineers for Figure 4.15 from the *Proceedings*.

International Glaciological Society for Figure 12.6.

Royal Meteorological Society for Figures 9.12, 10.7, 10.8, 11.3 and 12.14 from the *Quarterly Journal*; for Figures 5.17 and 10.9 from the *Journal of Climatology*; and for Figures 4.7, 4.8, 5.9, 5.13, 5.15, 9.30, 10.5, 10.12, 11.55 and 12.20 and for Figure 10.5, 10.12 from *Weather*.

Royal Society of Canada for Figure 3.15 from *Special Publication* 5.

Royal Society of London for Figure 9.27 from the *Proceedings, Section A*.

US National Academy of Sciences for Figures 13.4 and 13.5 from *Natural Climate Variability on Decade-to-century Time Scales* by P. Grootes.

EDITORS

Advances in Space Research for Figures 3.8 and 5.12.

American Scientist for Figure 11.49.

Climate Monitor for Figure 13.13.

Endeavour for Figure 5.18.

Erdkunde for Figures 11.21, 112.31 and A1.2B.

Geographical Reports of Tokyo Metropolitan University for Figure 11.36.

International Journal of Climatology (John Wiley & Sons, Chichester) for Figures 4.16, 10.33 and A1.1.

Japanese Progress in Climatology for Figure 12.28.

Meteorological Magazine for Figures 9.11, 10.6, 10.35 and 11.31.

Meteorological Monographs for Figures 9.2 and 9.4.

Meteorologische Rundschau for Figure 12.9.

Meteorologiya Gidrologiya (Moscow) for Figure 11.17.

New Scientist for Figures 9.25 and 9.28.

Science for Figure 7.32 .

Tellus for Figures 10.10, 10.11 and 11.25.

Zeitschrift für Geomorphologie for Figure 12.4 from *Supplement 21*.

PUBLISHERS

Academic Press, New York, for Figures 9.13, 9.14, 9.31 and 11.10 from *Advances in Geophysics*; for Figure 11.15 from *Monsoon Meteorology* by C. S. Ramage.

Allen and Unwin, London, for Figures 3.14 and 3.16B from *Oceanography for Meteorologists* by H. V. Sverdrup.

Butterworth-Heinemann for Figure 7.27 from *Ocean Circulation* by G. Beerman.

Cambridge University Press for Figure 5.8 from *Clouds, Rain and Rainmaking* by B. J. Mason; for Figure 7.7 from *World Weather and Climate* by D. Riley and L. Spalton; for Figure 10.30 from *The Warm Desert Environment* by A. Goudie and J. Wilkinson; for Figure 12.21 from *Air: Composition and Chemistry* by P. Brimblecombe (ed.); for Figures 2.4, 2.5, 2.8, 13.15 and 13.16 from *Climate Change: The IPCC Scientific Assessment 2001*; for Figure 13.10 from *Climate Change 1995: The Science of Climate Change., IPCC 1996;* for Figures 13.1, 13.14, 13.17, 13.18, 13.19, 13.20, 13.21 and 13.22 from *Climate Change 2007: The Physical Science Basis, IPCC 2007;* for Figure 8.2 from *Climate System Modelling* by K. E. Trenberth; Figures 3.19, 4.20 and 11.44 from *Mountain, weather and climate* by Roger Barry; and for Figure 11.52 from *Teleconnections Linking Worldwide Climate Anomalies* by M. H. Glantz *et al.* (eds).

Chapman and Hall, New York, for Figure 7.30 from *Elements of Dynamic Oceanography* by D. Tolmazin; for Figure 10.40 from *Encyclopedia of Climatology* by J. Oliver and R. W. Fairbridge (eds) and for Figure 9.22 from *Weather Systems* by L.F. Musk.

The Controller, Her Majesty's Stationery Office (Crown Copyright Reserved) for Figure 11.31 from *Geophysical Memoir 115* by J. Findlater; for Figures 9.11, 10.6 and 11.31 from the *Meteorological Magazine*; for Figure 9.11 from *A Course in Elementary Meteorology* by D. E. Pedgley; for Figures 10.26 and 10.27 from *Weather in the Mediterranean 1*, 2nd edn (1962); for the tephigram base of Figure 5.1 from *RAF Form 2810;* and for Figure 7.33 from *Global Ocean Surface Temperature Atlas* by M. Bottomley *et al.*

CRC Press, Florida, for Figure 3.6 from *Meteorology Theoretical and Applied,* by E. Hewson and R. Longley.

Elsevier Science, Amsterdam, for Figure 11.38 from *Palaeogeography, Palaeoclimatology, Palaeoecology*; and for Figure 11.47 from *Climates of Central and South America* by W. Schwerdtfeger (ed.); for Figure 10.29 from *Climates of the World* by D. Martyn; for Figure 10.29 from *Climates of the Soviet Union* by P. E. Lydolph, and for Figures 11.11 and 11.12 from *Advances in Geophysics.*

Generalstabens Litografiska Anstalt, Stockholm, for Figure 9.18 from *Klimatologi* by G. H. Liljequist.

Hutchinson, London, for Figure 12.20A and 12.27 from the *Climate of London* by T. J. Chandler; and for Figures 11.41 and 11.42 from *The Climatology of West Africa* by D. F. Hayward and J. S. Oguntoyinbo.

Kluwer Academic Publishers, Dordrecht, Holland, for Figure 2.1 from *Air–Sea Exchange of Gases and Particles* by P. S. Liss and W. G. N. Slinn (eds); for Figures 4.5 and 4.17 from *Variations in the Global Water Budget* by A. Street-Perrott *et al.* (eds).

Longman, London, for Figure 7.17 from *Contemporary Climatology* by A. Henderson-Sellers and P. J. Robinson.

McGraw-Hill Book Company. New York, for Figure 7.23 from *Dynamical and Physical Meteorology* by G. J. Haltiner and F. L. Martin; for Figures 12.12A and 12.13B from *Forest Influences* by J. Kittredge; for Figures 4.9 and 5.17 from *Introduction to Meteorology* by S. Petterssen; for Figures 11.1 and 11.6 from *Tropical Meteorology* by H. Riehl; and for Figure 11.21 from *The Earth's Problem Climates* by G. T. Trewartha.

National Academy Press, Washington, DC, for Figure 13.5.

North-Holland Publishing Company, Amsterdam, for Figure 4.18 from the *Journal of Hydrology.*

Plenum Publishing Corp., New York, for Figure 10.35B from *The Geophysics of Sea Ice* by N. Untersteiner (ed.).

D. Reidel, Dordrecht, for Figure 10.31 from *Climatic Change*; for Figure 12.26 from

Interactions of Energy and Climate by W. Bach, J. Pankrath and J. Williams (eds).

Rowman & Littlefield, Lanham, MS, for Figures 3.20, 12.11, 12.12 and 12.13 from *The Climate near the Ground* by R. Geiger (1965). For Figures 11.41 and 11.42 from *The Climatology of West Africa* by D. F. Hayward and J. S. Oguntoyinbo.

Routledge, London, for Figure 7.20 from *Models in Geography* by R. J. Chorley and P. Haggett (eds); for Figures 12.2, 12.5, 12.7, 12.15, 12.19, 12.23, 12.24, 12.25 and 12.29 from *Boundary Layer Climates* by T. R. Oke; for Figure 11.51 from *Climate Since AD 1500* by R. S. Bradley and P. D. Jones (eds); and for Figure 13.12 from *Climate of the British Isles* by P. D. Jones *et al.*

Scientific American Inc., New York, for Figure 3.25 by R. E. Newell and for Figure 2.12B by M. R. Rapino and S. Self.

Springer-Verlag, Heidelberg, for Figures 11.22 and 11.24.

Springer-Verlag, Vienna, for Figure 6.10 from *Archiv für Meteorologie, Geophysik und Bioklimatologie.*

University of California Press, Berkeley, for Figure 11.7 from *Cloud Structure and Distributions over the Tropical Pacific Ocean* by J. S. Malkus and H. Riehl.

University of Chicago Press for Figures 3.1, 3.5, 3.20, 3.27, 4.4B, 4.5, 12.8 and 12.10 from *Physical Climatology* by W. D. Sellers.

University of Wisconsin Press for Figure 10.20 from *The Earth's Problem Climates* by G. Trewartha.

Van Nostrand Reinhold Company, New York, for Figure 11.56 from *The Encyclopedia of Atmospheric Sciences and Astrogeology* by R. W. Fairbridge (ed.).

Walter De Gruyter, Berlin, for Figure 10.2 from *Allgemeine Klimageographie* by J. Blüthgen.

John Wiley, Chichester, for Figures 10.9, 11.30. 11.43 and A1.1 from the *International Journal of Climatology*; for Figures 2.7 and 2.10 from *The Greenhouse Effect, Climatic Change, and Ecosystems* by G. Bolin *et al.* for Figure 3.6 from *Meteorology, Theoretical and Applied* by E. W. Hewson and R. W. Longley; for Figures 11.16, 11.28, 11.29, 11.32 and 11.34 from *Monsoons* by J. S. Fein and P. L. Stephens (eds) and for Figure 7.31 from *Ocean Science* by K. Stowe.

ORGANIZATIONS

Climate Diagnostics Center, NOAA for Plate 7.3.

Deutscher Wetterdienst, Zentralamt, Offenbach am Main, for Figure 11.27.

Directorate-General Science, Research and Development, European Commission, Brussels, for Figure 10.25.

Geographical Branch, Department of Energy, Mines and Resources, Ottawa, for Figure 10.15 from *Geographical Bulletin.*

Laboratory of Climatology, Centerton, New Jersey, for Figure 10.22.

Goddard Institute for space sciences, NASA, for figures 13.7, 13.8 and 13.9.

National Academy of Sciences, Washington, DC, for Figure 13.4.

National Aeronautics and Space Administration (NASA) for Figures 2.16 and 7.26, and for Plates 3.2, 5.2, 5.15, 5.17, 9.1, 9.3 and 11.3.

National Environmental Research Council, UK, for Figures 2.7 and 4.4A from NERC News, July 1993 by K. A. Browning.

National Geophysical Data Center, NOAA, Boulder, for Figure 3.2 and Plate 3.1.

National Oceanic and Atmospheric Administration (NOAA), United States Department of Commerce, Washington, DC, for Figures 7.3, 7.4, 7.9, 7.10, 7.12, 7.15, 8.5, 8.6, 8.7, 8.8, 10.13, and for Plates 5.1, 5.6, 5.9, 5.16, 9.4, 11.1, 11.2 and 11.4.

National Snow and Ice Data Center, Boulder, for Plate 3.3.

New Zealand Alpine Club for Figure 5.15.

New Zealand Meteorological Service, Wellington, New Zealand, for Figures 11.26 and 11.57 from the *Proceedings of the Symposium on Tropical Meteorology* by J. W. Hutchings (ed.).

Nigerian Meteorological Service for Figure 11.39 from *Technical Note 5.*

Quartermaster Research and Engineering Command, Natick, MA., for Figure 10.17 by J. N. Rayner.

Risø National Laboratory, Roskilde, Denmark, for Figures 6.24 and 10.1 from *European Wind Atlas* by I. Troen and E. L. Petersen.

Smithsonian Institution, Washington, DC, for Figure 2.12A.

United Nations Food and Agriculture Organization, Rome, for Figure 12.17 from *Forest Influences.*

United States Department of Health, Education and Welfare for Figure 12.22.

United States Department of Agriculture, Washington, DC, for Figure 12.16 from *Climate and Man.*

United States Environmental Data Service for Figure 4.10.

United States Geological Survey, Washington, DC, for Figures 10.19, 10.21 and 10.23 from *Professional Paper 1052* and for Figure 10.23 mostly from *Circular 1120-A.*

United States Naval Oceanographic Office for Figure 7.29.

United States Weather Bureau for Figure 9.21 from *Research Paper 40.*

University of Tokyo for Figure 11.35 from *Bulletin of the Department of Geography.*

World Meteorological Organization for Figure 11.50 from *The Global Climate System 1982–84*; for Figure 3.24 from *GARP Publications Series, Rept No. 16;* and for Figure 13.2 from WMO *Publication No. 537.*

INDIVIDUALS

Dr Mark Anderson for Plates 5.14 and 9.3.

Dr R.M. Banta for Figure 6.12.

The late Dr R. P. Beckinsale, of Oxford University, for suggested modification to Figure 9.7.

Dr Otis B. Brown, of the University of Miami, for Plate 7.4.

The late Dr R. A. Bryson for Figure 10.15.

The late Dr M. I. Budyko for Figure 4.6.

Dr N. Caine for Plate 5.13.

Dr T. J. Chinn, of the Institute of Geological Sciences, Dunedin, for Figure 5.16.

Dr G. C. Evans, of the University of Cambridge, for Figure 12.18A.

The late Professor H. Flohn, of the University of Bonn, for Figures 7.14 and 11.14.

Dr S. Gregory, of the University of Sheffield, for Figures 11.13 and 11.53B.

Dr S. L. Hastenrath, of the University of Wisconsin, for Figure 4.19.

Dr R. A. Houze, Jr., of the University of Washington, for Figures 9.13, 9.14, 11.11 and 11.12.

Dr Patrick Koch for Plate 7.2.

Dr V. E. Kousky, of São Paulo, for Figure 11.48.

Dr Y. Kurihara, of Princeton University, for Figure 11.10.

Dr Kiuo Maejima, of Tokyo Metropolitan University, for Figure 11.36.

Dr J. Maley, of the Université des Sciences et des Techniques du Languedoc, for Figure 11.40.

Dr. M. E. Manu of Pennsylvania State University for Figure 13.6.

The late Dr J. R. Mather, of the University of Delaware, for Figure 10.22.

Dr Yale Mintz, of the University of California, for Figure 7.17.

The late Dr L. F. Musk, of the University of Manchester, for Figures 9.22 and 11.9.

Dr T. R. Oke, of the University of British Columbia, for Figures 6.11, 12.2, 12.3, 12.7, 12.15. 12.19, 12.23, 12.24, 12.25 and 12.29.

Dr W. Palz for Figure 11.25.

Dr L. R. Ratisbona, of the Servicio Meteorologico Nacional, Rio de Janeiro, for Figures 11.46 and 11.47.

Mr D. A. Richter, of Analysis and Forecast Division, National Meteorological Center, Washington, DC, for Figure 9.24.

Dr J. C. Sadler, of the University of Hawaii, for Figure 11.19.

The late Dr B. Saltzman, of Yale University, for Figure 8.4.

Dr Glenn E. Shaw, of the University of Alaska, for Figure 2.1A.

Dr Tao Shi-yan, of the Chinese Meteorological Society, for Figures 11.24 and 11.34.

Dr W. G. N. Slinn for Figure 2.1B.
Dr K. Stowe, of the California State Polytechnic College, for Figure 7.31.
The late Dr A. N. Strahler, of Santa Barbara, California, for Figures 3.3C and 5.10.

The publishers would be grateful to hear from any copyright holder who is not here acknowledged and will undertake to rectify any errors or omissions in future editions.

CHAPTER ONE

Introduction and history of meteorology and climatology

1

LEARNING OBJECTIVES

When you have read this chapter you will:

- be familiar with key concepts in meteorology and climatology
- know how these fields of study evolved and the contributions of leading individuals.

A THE ATMOSPHERE

The atmosphere, vital to terrestrial life, envelops the earth to a thickness of only 1 percent of the earth's radius. It had evolved to its present form and composition at least 400 million years ago by which time a considerable vegetation cover had developed on land. At its base, the atmosphere rests on the land and ocean surface, the latter which, at present, covers some 71 percent of the surface of the globe. Although air and water share somewhat similar physical properties, they differ in one important respect – air is compressible, while water is largely incompressible. In other words, in contrast to water, if one were to 'squeeze' a given sample of air, its volume would decrease. Study of the atmosphere has a long history involving observations, theory, and, since the 1960s, numerical modeling. Like most scientific fields, incremental progress has been interspesed by moments of great insight and rapid advance.

Scientific measurements only became possible with the invention of appropriate instruments; most had a long and complex evolution. A thermometer was invented by Galileo in the early 1600s, but accurate liquid-in-glass thermometers with calibrated scales were not available until the early 1700s (Fahrenheit), or 1740s (Celsius). In 1643 Torricelli invented the barometer, and demonstrated that the weight of the atmosphere at sea level would support a 10m column of water or a 760mm column of liquid mercury. Pascal used a barometer of Torricelli to show that pressure decreases with altitude, by taking one up the Puy de Dome in France. This paved the way for Boyle (1660) to demonstrate the compressibility of air by propounding his law that volume is inversely proportional to pressure. It was not until 1802 that Charles made the discovery that air volume is also directly proportional to its temperature. Combining Boyle's and Charles' laws yields the ideal gas law relating pressure, volume and temperature, one of the most important relationships in atmospheric science. By the end of the nineteenth century the four major constituents of the dry atmosphere (nitrogen 78.08 percent, oxygen 20.98 percent, argon 0.93 percent and carbon dioxide 0.035 percent) had been identified. It had been long suspected that human activities could have the potential to alter climate. While the atmospheric 'greenhouse effect' was discovered in 1824 by Joseph Fourier, the first serious consideration of a link between climate change, the greenhouse effect and changes in atmospheric carbon dioxide also emerged in the late nineteenth century through the insights of Swedish scientist Svante Arthenius. His expectation that carbon dioxide levels and temperature would rise due to fossil fuel burning has sadly turned out to be correct.

The hair hygrograph, designed to measure relative humidity (the amount of water vapor in the atmosphere relative to how much it can hold at saturation, expressed as a percent), was invented in 1780 by de Saussure. Rainfall records exist from the late seventeenth century in England, although early measurements are described from India in the fourth century BC, Palestine about AD 100 and Korea in the 1440s. A cloud classification scheme was devised by Luke Howard in 1803, but was not fully developed and implemented in observational practice until the 1920s. Equally vital was the establishment of networks of observing stations, following a standardized set of procedures for observing the weather and its elements, and a rapid means of exchanging the data (the telegraph). These two developments went hand-in-hand in Europe and North America in the 1850s–1860s.

The greater density of water compared with that of air (a factor of about 1000 at mean sea level pressure) gives water a higher specific heat. In other words, much more heat is required to raise the temperature of a cubic meter of water by 1°C than to raise the temperature of an equal volume of air by the same amount. It is interesting to note that just the top 10–15cm of ocean waters contain as much heat as does the total atmosphere; the total heat in the ocean in turn dwarfs that of the atmosphere. As is now known, this tremendous reservoir of heat in the upper ocean and its exchanges with the atmosphere is key to understanding climate variability. Another important feature of the behavior of air and water appears during the process of evaporation or condensation. As Black showed in 1760, during evaporation, heat energy of water is translated into kinetic energy of water vapor molecules (i.e., latent heat), whereas subsequent condensation in a cloud or as fog releases kinetic energy which returns as heat energy. The amount of water which can be stored in water vapor depends on the temperature of the air. This is why the condensation of warm, moist tropical air releases large amounts of latent heat increasing the instability of tropical air masses. This may be considered as part of the process of convection in which heated air expands, decreases in density and rises, perhaps resulting in precipitation, whereas cooling air contracts, increases in density and subsides.

The combined use of the barometer and thermometer allowed the vertical structure of the atmosphere to be investigated. While it is common experience to the aviator and mountain traveler that temperature tends to decrease with height, the reverse pattern of temperature increasing with height, known as an inversion, is also quite common, and in fact dominates in certain regions and atmospheric levels. A low-level (i.e., near-surface) temperature inversion was discovered in 1856 at a height of about 1km on a mountain in Tenerife. Later investigations revealed that this so-called Trade Wind Inversion is found over the eastern subtropical oceans where subsiding dry high pressure air overlies cool, moist maritime air close to the ocean surface. Such inversions inhibit vertical (convective) air movements and, consequently, form a lid to some atmospheric activity. The Trade Wind Inversion was shown in the 1920s to differ in elevation between some 500m and 2km in different parts of the Atlantic Ocean in the belt 30°N to 30°S. Around 1900 a more important continuous and widespread temperature inversion was revealed by balloon flights to exist at about 10km at the equator and 8km at high latitudes. This inversion level (the tropopause) was recognized to mark the top of the so-called troposphere within which most weather systems form and decay. By 1930 balloons equipped with an array of instruments to measure pressure, temperature and humidity, and report them back to earth by radio (radiosonde), were routinely investigating the atmosphere. Observations from both kites and balloons also revealed that strong inversions extending up to about 1000m are a near ubiquitous feature of the Arctic in winter.

B SOLAR ENERGY

Differential solar heating of low and high latitudes is the mechanism which drives the earth's large-scale atmospheric and oceanic circulations. Most of the energy from the sun entering the atmosphere as short-wave radiation (or insolation) reaches the earth's surface. Some is reflected back to space. The remainder is absorbed by the surface which then warms the atmosphere above it. The atmosphere and surface together radiate long-wave (thermal) radiation back to space. Although the land and ocean parts of the surface absorb different amounts of solar radiation and have different thermal characteristics, the differential solar heating between low and high latitudes dominates, fostering an equator-to-pole gradient in atmospheric and upper ocean temperatures.

Although increased solar heating of the tropical regions compared with the higher latitudes had long been apparent, it was not until 1830 that Schmidt made a key calculation, namely heat gains and losses for each latitude by incoming solar radiation and by outgoing longwave radiation from the earth. This showed that equatorward of about latitudes 35° there is an excess of incoming solar over outgoing longwave energy, while poleward of those latitudes the longwave loss exceeds solar input. If, at each latitude, the longwave loss to space equaled the solar radiation input (termed radiative equilibrium), this pattern would not be seen. That it exists is direct evidence that there must be an overall tranasfer of energy from lower to higher latitudes via the atmospheric and oceanic circulations. Put differently, while the differential solar heating gives rise to the equator-to-pole temperature gradient, the poleward energy transports work to reduce this gradient. Later and more refined calculations showed that the poleward flow (or flux) of atmospheric energy reaches a maximum around latitudes 30° and 40°, with the maximum ocean transport occurring at lower latitudes. The total poleward transport in both hemispheres is in turn dominated by the atmosphere. The amount of solar energy being received and re-radiated from the earth's surface can be computed theoretically by mathematicians and astronomers. Following Schmidt, many such calculations were made, notably by Meech (1857), Wiener (1877) and Angot (1883) who calculated the amount of extraterrestrial insolation received at the outer limits of the atmosphere at all latitudes. Theoretical calculations of insolation in the past by Milankovitch (1920, 1930), and Simpson's (1928–1929) calculated values of the insolation balance over the earth's surface, were important contributions to understanding astronomic controls of climate. Nevertheless, the solar radiation received by the earth was only accurately determined by satellites in the 1990s.

C GLOBAL CIRCULATION

While differential solar heating of the surface and the atmospheric temperature gradient that it generates fosters the large-scale transport of energy from equatorial to polar regions, what are the mechanisms by which this atmospheric transport is accomplished? While we now know that the transport is accomplished by the Hadley circulation in lower latitudes and in higher latitudes through disturbances in the basic westerly (west to east) flow in the form of transient cyclones and anticyclones, it is fascinating to briefly outline how our modern view of the global circulation emerged.

The first attempt to explain the global atmospheric circulation was based on a simple convectional concept. In 1686 Halley associated the easterly Trade Winds with low-level convergence on the equatorial belt of greatest heating (i.e., the thermal equator). These flows are compensated at high levels by return flows aloft. Poleward of these convectional regions, the air cools and subsides to feed the northeasterly and southeasterly Trade Winds at the surface. This simple mechanism, however, presented two significant problems: what mechanism produced the observed high pressure in the subtropics and what was responsible for the belts of dominantly westerly winds poleward of this high pressure zone? It is interesting to note that it was not until 1883 that Teisserenc de Bort produced the first global mean sea-level map showing the main zones of high and low pressure. The climatic significance of Halley's work rests also in his thermal convectional theory for the origin of the Asiatic monsoon which was based on the differential thermal behavior of land and sea; i.e., the land reflects more and stores less of the incoming solar radiation and therefore heats and cools faster. This heating causes continental surface pressures to be generally lower than oceanic ones in summer and higher in winter, causing seasonal wind reversals. The role of seasonal movements of the thermal equator in monsoon systems was only recognized much later. Some of the difficulties faced by Halley's simplistic large-scale circulation theory began to be addressed by Hadley in 1735, who was particularly concerned with the deflection of winds on a rotating globe, to the right (left) in the Northern (Southern) Hemisphere. Like Halley, he advocated a thermal circulatory mechanism, but was perplexed by the existence of the westerlies. Following the mathematical analysis of moving bodies on a rotating earth by Coriolis (1831), Ferrel (1856) developed a three-cell model of hemispherical atmospheric circulation by suggesting a mechanism for the production of high pressure in the subtropics (i.e., 35°N and S latitude). The tendency for cold upper air to subside in the subtropics, together with the latitudinal increase in the deflective force (the Coriolis force, the product of wind speed and the the Coriolis parameter which increases with latitude) applied by terrestrial rotation to upper air moving poleward above the Trade Wind Belt, would cause a buildup of air (and therefore of pressure) in the subtropics. Equatorward of these subtropical highs the thermally direct Hadley cells dominate the Trade Wind Belt but poleward of them air tends to flow towards higher latitudes at the surface. This airflow, increasingly deflected with latitude, constitutes the westerly winds in both hemispheres. In the Northern Hemisphere, the highly variable northern margin of the westerlies is situated where the westerlies are undercut by polar air moving equatorward. This margin was compared with a battlefield front by Bergeron who, in 1922, termed it the Polar Front. Thus, Ferrel's three cells consisted of two thermally direct Hadley cells (where warm air rises and cool air sinks), separated by a weak, indirect Ferrel cell in mid-latitudes. The relation between pressure distribution and wind speed and direction was demonstrated by Buys-Ballot in 1860.

D CLIMATOLOGY

During the nineteenth century it became possible to assemble a large body of global climatic data and to use it to make useful regional generalizations. In 1817 Alexander von Humboldt produced his valuable treatise on global temperatures containing a map of mean annual isotherms (lines of equal temperature) for the Northern Hemisphere but it was not until 1848 that Dove published the first world maps of monthly mean temperature. An early world map of precipitation was produced by Berghaus in 1845; in 1882 Loomis produced the first world map of precipitation employing mean annual isohyets (lines of equal precipitation); and in 1886 de Bort published the first world maps of annual and monthly cloudiness.

These generalizations allowed, in the later decades of the century, attempts to classify climates regionally. In the 1870s Wladimir Koeppen, a St Petersburg-trained biologist, began producing maps of climate based on plant geography, as did de Candolle (1875) and Drude (1887). In 1883 Hann's massive, three-volume *Handbook of Climatology* appeared, which remained a standard until 1930–1940 when the five-volume work of the same title by Koeppen and Geiger replaced it. At the end of World War I Koeppen (1918) produced the first detailed classification of world climates based on terrestrial vegetation cover. This was followed by Thornthwaite's (1931–1933) classification of climates employing evaporation and precipitation amounts, which he made more widely applicable in 1948 by the use of the theoretical concept of potential evapo-transpiration. The Inter-War period was particularly notable for the appearance of a number of climatic ideas which were not brought to fruition until the 1950s. These included the use of frequencies of various weather types (Federov 1921), the concepts of variability of temperature and rainfall (Gorczynski 1942 and 1945) and microclimatology, the study of the fine climate structure near the surface (Geiger 1927).

Despite the problems of obtaining detailed measurements over the large ocean areas, the later nineteenth century saw much climatic research which was concerned with pressure and wind distributions. In 1868 Buchan produced the first world maps of monthly mean pressure; eight years later Coffin composed the first world wind charts for land and sea areas, and in 1883 L. Teisserenc de Bort produced the first mean global pressure maps showing the cyclonic and anticyclonic 'centers of action' on which the general circulation is based. In 1887 de Bort began producing maps of upper-air pressure distributions and in 1889 his world map of January mean pressures in the lowest 4km of the atmosphere was particularly effective in depicting the great belt of the westerlies between 30° and 50° north latitudes.

E MID-LATITUDE DISTURBANCES

Theoretical ideas about the atmosphere and its weather systems evolved in part through the needs of nineteenth-century mariners for information about winds and storms, especially predictions of future behavior. At low levels in the westerly belt (approximately 40° to 70° latitude) there is a complex pattern of moving high and low pressure systems, while between 6000m and 20,000m there is a coherent westerly airflow. Dove (1827 and 1828) and Fitz Roy (1863) supported the 'opposing current' theory of cyclone (i.e., depression) formation, where the energy for the systems was produced by converging airflow. Espy (1841) set out more clearly a convection theory of energy production in cyclones with the release of latent heat (condensation of water vapor) as the main source. In 1861 Jinman held that storms develop where opposing air currents form lines of confluence (later termed 'fronts'). Ley (1878) gave a three-dimensial picture of a low pressure system with a cold air wedge behind a sharp temperature discontinuity cutting into warmer air, and Abercromby (1883) described storm systems in terms of a pattern of closed isobars (lines of equal pressure) with typical associated weather types. By this time, although the energetics were far from clear, a picture, correct in is basics, was emerging of mid-latitude storms being generated by the mixing of warm tropical and cool polar air as a fundamental result of the latitudinal temperature gradients created by the patterns of incoming solar radiation and of outgoing terrestrial radiation. Towards the end of the nineteenth century two important European research groups were dealing with storm formation: the Vienna Group under Margules, including Exner and Schmidt; and the Swedish Group led by Vilhelm Bjerknes. The former workers were concerned with the origins of cyclone kinetic energy (energy of motion) which was thought to be due to differences in the potential energy of opposing air masses of different temperature. Potential energy is energy associated with the height of air parcels above the surface. Gradients in potential energy on a pressure surface provide conditions to convert potential to kinetic energy. This was set forth in the work of Margules (1901) who showed that the potential energy of a typical depression is less than 10 percent of the kinetic energy of its constituent winds. In Stockholm V. Bjerknes' group concentrated on frontal development (Bjerknes, 1897 and 1902) but its researches were particularly important during the period 1917–1929 after J. Bjerknes moved to Bergen and worked with Bergeron. In 1918 the warm front was identified, the occlusion process was described in 1919, and the full Polar Front Theory of cyclone development was presented in 1922 (J. Bjerknes and Solberg). After about 1930, meteorological research concentrated increasingly on the importance of mid- and upper-tropospheric influences for global weather phenomena. This was led by Sir Napier Shaw in Britain and by Rossby, with Namias and others, in the USA. The airflow in the 3–10km high layer of the polar vortex of the Northern Hemisphere westerlies was shown to form large-scale horizontal (Rossby) waves due to latitudunal gradients in the Coriolis parameter, the influence of which was simulated by rotating 'dish pan' experiments in the 1940s and 1950s. The number and amplitude of these waves appear to depend on the hemispheric energy gradient, or 'index'. At times of high index, especially in winter, there may be as few as three Rossby waves of small amplitude giving a strong zonal (i.e., west to east) flow. A weaker hemispheric energy gradient (i.e., low index) is characterized by four to six Rossby waves of larger amplitude. As with most broad, fluid-like flows in nature, the upper westerlies were shown by observations in the 1920s and 1930s, and particularly by aircraft observations in World War II, to possess narrow high-velocity threads, termed 'jet streams' by Seilkopf in 1939. The higher and more important jet streams approximately lie along the Rossby waves. The most important jet stream, located at 10km, clearly affects surface weather by guiding the low pressure systems which tend to form beneath it. In addition, air subsiding beneath the jet streams strengthens the subtropical high pressure cells.

F THE POLAR REGIONS

The earliest view of the Arctic's atmospheric circulation can be traced to the late nineteenth-century work of von Helmholtz, who argued that the region was dominated by a more or less permanent surface high pressure cell, a view developed in the earlier part of the twentieth century by Hobbs in his 'glacial anticyclone' theory. In 1945, Hobbs elaborated further on this basic idea, advocating the existence of a persistent anticyclone over the Greenland ice sheet, having strong impacts on middle latitudes. Given the general lack of data until the 1940s and 1950s, such a misconception is not surprising. Sea level pressure analyses in the US Historical Weather Map Series produced during World War II contained strong positive biases prior to the 1930s away from the North Atlantic sector. Part of the problem, as noted by Jones, was that these maps were prepared by relatively untrained analysts, who tended to extrapolate into the data-poor Arctic with the prevailing mindset of an Arctic high pressure cell. Even by the early 1950s, some studies erroneously depicted traveling cyclones as largely restricted to the periphery of the Arctic Ocean. The emergence in North America of more modern views of the Arctic circulation in the late 1950s and 1960s, fostered by the growing database of upper-air data and surface observations, appeared in the work of research groups at McGill University led by F. K. Hare and the University of Washington led by R. J. Reed. R. G. Barry participated in the work at McGill, and made many contributions. Interestingly, in the Soviet Union, a relatively modern view of the summer circulation had already been formulated in 1945 by B. L. Dzerdzeevskii.

Knowledge of Antarctica lagged behind that of the Arctic. The remoteness and extremely harsh conditions of this continent were barriers to progress. Furthermore, while the Arctic was a key strategic region during the Cold War, leading to extensive research and the rapid establishment of observing networks, the Antarctic did not benefit from Cold War activity to the same degree. Some features had long been recognized, such as the existence of a trough of low pressure surrounding the continent, and persistent strong katabatic (downslope) winds. Much progress was made following extensive observations made during the International Geophysical Year (IGY) of 1957–1958, which was modeled on the International Polar Years of 1882–1883 and 1932–1933 (Box 1.1). A preliminary survey of the Southern Hemisphere westerlies, based in part on upper-air observations during the IGY, was published by H.H. Lamb in 1959. Even today, direct observations are much more sparse over the Antarctic than over the Arctic. Weather forecasts in this region rely especially strongly on data collected by earth-orbiting satellites.

G TROPICAL WEATHER

The success in modeling the life cycle of the mid-latitude frontal depression, and its value as a forecasting tool, naturally led to attempts in the immediate pre-World War II period to apply it to the atmospheric conditions which dominate the tropics (i.e., 30°N–30°S), comprising half the surface area of the globe. This attempt was largely doomed to failure, as observations made during the air war in the Pacific soon demonstrated. This failure was due to the lack of frontal temperature discontinuities between air masses and the absence of a strong Coriolis effect and thus of Rossby-like waves. Tropical airmass discontinuities are based on moisture differences. Tropical weather results mainly from strong convectional features such as heat lows, tropical cyclones (hurricanes and typhoons) and the Inter-Tropical Convergence Zone (ITCZ), the axis of which represents the dividing line between the southeast and northeast Trade Winds of the Northern and Southern Hemisphere. The huge instability of tropical airmasses means that even mild convergence in the Trade Winds gives rise to atmospheric waves travelling westward with characteristic weather patterns.

Above the Pacific and Atlantic Oceans the ITCZ is quasi-stationary with a latitudinal displacement annually of 5° or less, but elsewhere it varies between latitudes 17°S and 8°N in January and between 2°N and 27°N in July – i.e., during the southern and northern summer monsoon seasons, respectively. The seasonal movement of the ITCZ and the existence of other convective influences make the south and east Asian monsoon the most significant seasonal global weather phenomenon.

Investigations of weather conditions over the broad expanses of the tropical oceans were assisted by satellite observations after about 1960. Observations of waves in the tropical easterlies began in the Caribbean during the mid-1940s, but the structure of meso-scale cloud clusters and associated storms was only recognized in the 1970s. Satellite observations also proved very valuable in detecting the generation of hurricanes over the great expanses of the tropical oceans.

In the late 1940s and subsequently, important work was conducted on the relations between the south Asian monsoon mechanism concerning the westerly subtropical jet stream, the Himalayan mountain barrier and the displacement of the ITCZ. The very significant failure of the Indian summer monsoon in 1877 had led Blanford (1860) in India, Todd (1888) in Australia, and others, to seek correlations between Indian monsoon rainfall and other climatic phenomena such as the amount of Himalayan snowfall (which influences large-scale differential heating between the land and ocean) and the strength of the southern Indian Ocean high pressure center. Such correlations were studied intensively by Sir Gilbert Walker and his co-workers in India between about 1909 and the late 1930s. In 1924 a major advance was made when Walker identified the 'Southern Oscillation' – an east–west seesaw of atmospheric pressure and resulting rainfall (i.e., negative correlation) between Indonesia and the eastern Pacific. Other north–south climatic oscillations were identified in the North Atlantic (Azores vs. Iceland, known as the North Atlantic Oscillation) and the North Pacific (Alaska vs. Hawaii). In the phase of the Southern Oscillation when there is high pressure over the eastern Pacific, westward-flowing central Pacific surface waters, with a consequent upwelling of cold water, plankton-rich, off the coast of South America, are associated with ascending air giving heavy summer rains over Indonesia. Periodically, weakening and breakup of the eastern Pacific high pressure cell leads to important consequences. The chief among these are subsiding air and drought over India and Indonesia and the removal of the mechanism of the cold coastal upwelling off the South American coast with the consequent failure of the fisheries there. The presence of warm coastal water is termed 'El Niño'. Although the central role played by lower latitude high pressure systems over the global circulations of atmosphere and oceans is well recognized, the cause of the east Pacific pressure change which gives rise to El Niño is not yet fully understood. There was a waning of interest in the Southern Oscillation and associated phenomena during the 1940s to mid-1960s, but the work of Berlage (1957), the increase in the number of Indian droughts during the period 1965–1990, and especially the strong El Niño which caused immense economic hardship in 1972, led to a revival of interest and research. One feature of this research has been the thorough study of the 'teleconnections' (correlations between climatic conditions in widely separated regions of the earth) pointed out by Sir Gilbert Walker.

H PALEOCLIMATES

Before the middle of the twentieth century thirty years of records was generally regarded as sufficient in order to define a given climate. By the 1960s the idea of a static climate was recognized as being untenable. New approaches to paleoclimatology, the study of past climates, were developed in the 1960s to 1970s. The astronomical theory to explain the great ice ages of the Pleistocene proposed by Croll (1867), and developed mathematically by Milankovitch (1920), seemed to conflict with evidence for dated climate changes. However, in 1976, Hays, Imbrie and Shackleton recalculated Milankovitch's chronology using powerful new statistical techniques and showed that it correlated well with past temperature records, especially for ocean paleotemperatures derived from isotopic ($^{18}0/^{16}0$) ratios in marine organisms recorded in ocean cores. The idea behind Milankovitch forcings is that periodic changes in the eccentricity of the earth's orbit, the tilt of the earth's axis and the timing of the equinoxes cause variations in the amount of solar radiation received at different times of the year over different parts of the surface. As is now widely accepted, the major ice ages as well as intervening interglacials over about the past two million years reflect influences of these Milankovitch cycles and attendant climate feeedbacks that amplify change. Paleoclimate information from ocean cores and terrestrial sources is complemented by ice cores obtained from the Greenland and Antarctic ice sheets, ice caps in Canada and elsewhere. As well as documenting climate links with Milankovitch cycles, these records provide evidence of rapid, large-scale shifts in climate. The longest ice core record currently available from Dome C in the eastern Antarctic spans 800,000 years and shows that interglacials before 450,000 years ago were weaker (less warm) than subsequently. Reconstructed temperature records from ice cores are obtained from oxygen isotope ratios ($\partial^{18}O$). Samples of past atmospheres trapped as bubbles in ice cores also document close links between climate and atmospheric carbon dioxide concentrations, and show convincingly that present-day concentrations of this greenhouse gas are higher than at any time during at least the past 800,000 years.

Other paleoclinatic information is obtained from annual tree rings that reflect growing season temperature and moisture, lake and peat bog sediments that contain pollen records of regional vegetation, reconstructed temperature records from oxygen isotope ratios in cave stalagmite and annual growth rings in ocean corals.

Major advances have been made in paleoclimate reconstruction through the use of general circulation models with past boundary conditions (paleogeography, paleovegetation) and changed earth orbital characteristics.

I THE GLOBAL CLIMATE SYSTEM

Undoubtedly the most important outcome of work in the second half of the twentieth century was the recognition of the existence of the global climate system (see Box 1.1). The climate system involves not just the atmosphere elements, but the five major subsystems: the atmosphere (the most unstable and rapidly changing); the ocean (very sluggish in terms of its thermal inertia and therefore important in regulating atmospheric variations); the snow and ice cover (the cryosphere); and the land surface with its vegetation cover (the lithosphere and biosphere). Physical, chemical and biological processes take place in and among these complex subsystems. The most important interaction takes place between the highly dynamic atmosphere, through which solar energy is input into the system, and the oceans which store and transport large amounts of energy (especially thermal), thereby acting as a regulator to more rapid atmospheric changes. A further complication is provided by the living matter of the biosphere, which influences the incoming radiation and outgoing re-radiation and affects the atmospheric composition via greenhouse gases. In the oceans, marine biota play a major role in the dissolution and storage of CO_2. All subsystems are linked by fluxes of mass, heat and momentum into a very complex whole. The coupled climate system always has and always will be characterized by variability on numerous time and space scales. However, the introduction of humans into the system has added a new dimension. Indeed, by the dawn of the twenty-first century, overwhelming evidence had amassed of a discernible and growing human impact upon global climate.

The driving mechanism of global climate change is referred to as 'radiative forcing'. In an equilibrium climate state, globally averaged solar energy absorbed by the earth system is balanced by the globally averaged longwave radiation emitted to space. In other words, there is radiative equilibrium at the top of the atmosphere. An imbalance, or radiative forcing, is defined as positive when less energy is emitted than absorbed and negative for the reverse. In response to radiative forcing, the system tries to come back to a new equilibrium, attended by, respectively warming or cooling at the surface. Radiation imbalances arise from both natural processes (e.g., astronomical effects on incoming shortwave solar radiation, changes in total solar output and volcanic eruptions, the latter of which load the atmosphere with aerosols, tiny particles suspended in the air) and human influences (e.g., changes in greenhouse gas and aerosol concentrations owing to fossil fuel burning and a suite of other activities, such as deforestation and agriculture). Direct solar radiation measurements have been made via satellites since about 1980, but the correlation between small changes in solar radiation and in the thermal economy of the global climate system is still somewhat unclear. However, observed human-induced increases in the greenhouse gas content of the atmosphere (0.1 percent of which is composed of the trace gases carbon dioxide, methane, nitrous oxide and ozone) appear to have been very significant in increasing the proportion of terrestrial longwave radiation trapped by the atmosphere (a positive radiative forcing), thereby raising surface air temperature over the past 100 years.

Adjustments to a radiative forcing take place in a matter of months in the surface and tropospheric subsystems but are slower (centuries or longer) in the ocean. In turn, the amount of surface warming

SIGNIFICANT 20TH-CENTURY ADVANCE

1.1 Global Atmospheric Research Programme (GARP) and the World Climate Research Programme (WCRP)

The idea of studying global climate through coordinated intensive programs of observations emerged through the World Meteorological Organization (WMO:http://www.wmo.ch/) and the International Council on Science (ICSU: http://www.icsu.org) in the 1970s. Three 'streams' of activity were planned: a physical basis for long-range weather forecasting; interannual climate variability; and long-term climatic trends and climate sensitivity. Global meteorological observations became a major concern and this led to a series of observational programs. The earliest was the Global Atmospheric Research Programme (GARP). This had a number of related but semi-independent components. One of the earliest was the GARP Atlantic Tropical Experiment (GATE) in the eastern North Atlantic, off West Africa, in 1974–1975. The objectives were to examine the structure of the Trade Wind inversion and to identify the conditions associated with the development of tropical disturbances. There was a series of monsoon experiments in West Africa and the Indian Ocean in the late 1970s to early 1980s and also an Alpine Experiment. The First GARP Global Experiment (FGGE), in November 1978 to March 1979, assembled global weather observations. Coupled with these observational programs, there was also a coordinated effort to improve numerical modeling of global climate processes.

The World Climate Research Programme (WCRP: http://www.wmo.ch/web/wcrp/prgs.htm), established in 1980, is sponsored by the WMO, ICSU and the International Ocean Commission (IOC). The first major global effort was the World Ocean Circulation Experiment (WOCE) which provided detailed understanding of ocean currents and the global thermohaline circulation. This was followed in the 1980s by the Tropical Ocean Global Atmosphere (TOGA).

Current major WCRP projects are Climate Variability and Predictability (CLIVAR: http://www.clivar.org/), the Global Energy and Water Cycle Experiment (GEWEX), Stratospheric Processes and their Role in Climate (SPARC), and Climate and Cryosphere (CliC; http://clic.npolar.no/) . Under GEWEX are the International Satellite Cloud Climatology Project (ISCCP) and the International Land Surface Climatology Project (ISLSCP) which provide valuable data sets for analysis and model validation. CliC, which addresses all major components of the earth's cryosphere (glaciers, ice caps and ice sheets, sea ice, snow cover, seasonally frozen ground and permafrost, the latter representing perennially frozen ground) developed from the earlier Arctic Climate System (ACSYS) effort. The WCRP has also been activity involved in the planning and implementation of the third International Polar Year (IPY), a large international scientific program focused on the Arctic and Antarctic from March 2007 to March 2009.

Reference

Houghton, J. D. and Morel, P. (1984) The World Climate Research Programme. In J. D. Houghton (ed.) *The Global Climate*, Cambridge University Press, Cambridge, pp. 1–11.

to a given radiative forcing (termed climate sensitivity) depends critically on feedbacks that amplify or dampen the climate response to the forcing. In the case of greenhouse gases, the issue is further complicated in that the radiative forcing is itself changing. Major feedbacks involve the role of snow and ice reflecting incoming solar radiation and atmospheric water vapor absorbing terrestrial re-radiation, and are positive in character. For example: the earth warms; atmospheric water vapor increases; this, in turn, increases the greenhouse effect; the result being that the earth warms further. Similar warming occurs as higher temperatures reduce snow and ice cover allowing the land or ocean to absorb more radiation. Clouds play a more complex and still incompletely understood role by reflecting solar (shortwave radiation) but also by trapping terrestrial outgoing radiation. Negative feedback, when the effect of change is damped down, is a much less important feature of the operation of the climate system, which partly explains the tendency to recent global warming. The impact of aerosols is one of the biggest areas of uncertainty. While the cooling effect of aerosols through scattering solar radiation back to space is well known, and in part masks the warming effect of greenhouse gases, some aerosols, such as soot, absorb solar radiation. Aerosols also affect the number and density of cloud droplets, changing the optical properties of clouds.

An important factor in weather and climate processes is unpredictabiliy. Weather systems display sensitivity to their initial conditions, meaning that a very small change in the initial state of a weather system can have a large, disproportionate effect on the whole system. This was first recognized by E. Lorenz (1963), who pointed out that a butterfly flapping its wings in Beijing could affect the weather thousands of miles away some days later. This sensitivity is now called the 'butterfly effect'. It is addressed in numerical model experiments by running many simulations with minute variations in the initial conditions and then examining the results of an ensemble of projections.

The Intergovernmental Panel on Climate Change (IPCC), jointly established in 1988 by the WMO and the United Nations Environmental Programme (UNEP), has served as a focal point for climate change research, and released its Fourth Assessment Report in 2007. One of the most important tools of the IPCC is numerical models of the climate system. Since the initial development of atmospheric general circulation models (GCMs) in the 1960s, the current models have become very sophisticated, and are essential for untangling the complexities of radiative forcing, feedbacks and climate response. They now incorporate coupled ocean, land and biosphere submodels. The emerging picture that these models paint is of a much wamer and different world by the end of this century, posing challenges for society that include, but are not limited to, higher sea levels and shifts in agricultural zones. Major uncertainties nevertheless remain, particularly of climate change on regional scales.

The first edition of *Atmosphere, Weather and Climate* appeared in 1968 before many of the advances described in later editions were even conceived. However, our continuous aim in writing it is to provide a largely non-technical account of how the atmosphere works, thereby helping the understanding of both weather phenomena and of global climates. As noted in the eighth edition, greater explanation inevitably results in an increase in the range of phenomena requiring explanation. As a result, this book continues to thicken with time.

DISCUSSION TOPICS

- How have technological advances contributed to the evolution of meteorology and climatology?
- Consider the relative contributions of observation, theory, and modeling to our knowledge of atmospheric processes.

REFERENCES AND FURTHER READING

Books

Allen, R., Lindsay, J. and Parker, D. (1996) *El Niño Southern Oscillations and Climatic Variability*, CSIRO, Australia, 405pp. [Modern account of ENSO and its global influences]

Fleming, J. R. (ed.) (1998) *Historical Essays in Meteorology, 1919–1995,* American Meteorological Society, Boston, MA, 617pp. [Valuable accounts of the evolution of meteorological observations, theory and modeling and of climatology]

Houghton, J.T., Ding, Y., *et al.* (eds) (2001) *Climate Change 2001; The Scientific Basis; The Climate System: An Overview*, Cambridge University Press, Cambridge, 881pp. [Working Group I contribution to The Third Assessment Report of the Intergovernmental Panel on Climate Change (IPCC); a comprehensive assessment from observations and models of past, present and future climatic variability and change. It includes a technical summary and one for policy makers]

Peterssen, S. (1969) *Introduction to Meteorology,* 3rd edn., McGraw Hill, New York, 333pp. [Classic introductory text, including world climates]

Stringer, E.T. (1972) *Foundations of Climatology: An Introduction to Physical, Dynamic, Synoptic, and Geographical Climatology* Freeman and Co., San Francisco, CA, 586pp. [Detailed and advanced survey with numerous references to key ideas; equations are in Appendices]

Van Andel, T.H. (1994) *New Views on an Old Planet,* 2nd edn, Cambridge University Press, Cambridge, 439pp. [Readable introduction to earth history and changes in the oceans, continents and climate]

Journal articles

Browning, K.A. (1996) Current research in atmospheric sciences. *Weather* 51, 167–72.

Grahame, N. S. (2000) The development of meteorology over the last 150 year as illustrated by historical weather charts. *Weather* 55(4),108–16.

Hare, F.K. (1951) Climatic classification. In: *London Essays in Geography*, L. D. Stamp and S. W. Wooldridge (eds), Longmans, London, pp. 111–34.

CHAPTER TWO

Atmospheric composition, mass and structure

2

LEARNING OBJECTIVES

When you have read this chapter you will:

- be familiar with the composition of the atmosphere – its gases and other constituents
- understand how and why the distribution of trace gases and aerosols varies with height, latitude and time
- know how atmospheric pressure, density and water vapor pressure vary with altitude
- be familiar with the vertical layers of the atmosphere, their terminology and significance.

This chapter describes the composition of the atmosphere – its major gases and impurities, their vertical distribution, and variations through time. The various greenhouse gases and their significance are discussed. It also examines the vertical distribution of atmospheric mass and the structure of the atmosphere, particularly the vertical variation of temperature.

A COMPOSITION OF THE ATMOSPHERE

1 Primary gases

Air is a mechanical mixture of gases, not a chemical compound. Dry air, by volume, is more than 99 percent composed of nitrogen and oxygen (Table 2.1). Rocket observations show that these gases are mixed in remarkably constant proportions up to about 100km altitude. Yet, despite

Table 2.1 Average composition of the dry atmosphere below 25km

Component	Symbol	Volume % (dry air)	Molecular weight
Nitrogen	N_2	78.08	28.02
Oxygen	O_2	20.95	32.00
*‡Argon	Ar	0.93	39.88
Carbon dioxide	CO_2	0.037	44.00
‡Neon	Ne	0.0018	20.18
*‡Helium	He	0.0005	4.00
†Ozone	O_3	0.00006	48.00
Hydrogen	H	0.00005	2.02
‡Krypton	Kr	0.00011	
‡Xenon	Xe	0.00009	
§Methane	CH_4	0.00017	

Notes: *Decay products of potassium and uranium. †Recombination of oxygen. ‡Inert gases. §At surface.

their predominance, these gases are of little climatic importance.

2 Greenhouse gases

In spite of their relative scarcity, the so-called *greenhouse gases* play a crucial role in the thermodynamics of the atmosphere (see Box 2.1). They trap radiation emitted by the earth, thereby producing the *greenhouse effect* (see Chapter 3C). Moreover, the concentrations of these trace gases are strongly affected by human (i.e., anthropogenic) activities:

1 Carbon dioxide (CO_2) is involved in a complex global cycle (see 2A.7). It is released from the earth's interior and produced by respiration of biota, soil microbia, fuel combustion and oceanic evaporation. Conversely, it is dissolved in the oceans and consumed by plant photosynthesis. The imbalance between emissions and uptake by the oceans and terrestrial biosphere leads to the net increase in the atmosphere
2 Methane (CH_4) is produced primarily through anaerobic (i.e., oxygen-deficient) processes by natural wetlands and rice paddies (together about 40 percent of the total), as well as by enteric fermentation in animals, by termites, through coal and oil extraction, biomass burning and from landfills.

$$CO_2 + 4H_2 \rightarrow CH_4 + 2H_2O$$

Almost two-thirds of the total production is related to anthropogenic activity.

Methane is oxidized to CO_2 and H_2O by a complex photochemical reaction system.

$$CH_4 + O_2 + 2x \rightarrow CO_2 + 2x\,H2$$

where x denotes any specific methane destroying species (e.g.H, OH, NO, Cl or Br).
3 Nitrous oxide (N_2O) is produced primarily by nitrogen fertilizers (50–75 percent) and industrial processes. Other sources are transportation, biomass burning, cattle feed lots and biological mechanisms in the oceans and soils. It is destroyed by photochemical reactions in the stratosphere involving the production of nitrogen oxides (NOx).
4 Ozone (O_3) is produced through the break up of oxygen molecules in the upper atmosphere by solar ultraviolet radiation and is destroyed by reactions involving nitrogen oxides (NO_x) and chlorine (Cl) (the latter generated by CFCs, volcanic eruptions and vegetation burning) in the middle and upper stratosphere.
5 Chlorofluorocarbons (CFCs: chiefly $CFCl_3$ (F-12) and CF_2Cl_2 (F-12)) are entirely anthropogenically produced by aerosol propellants, refrigerator coolants (e.g., 'freon'), cleansers and air-conditioners, and were not present in the atmosphere until the 1930s. CFC molecules rise slowly into the stratosphere and then move poleward, being decomposed by photochemical processes into chlorine after an estimated average lifetime of some 65–130 years.
6 Hydrogenated halocarbons (HFCs and HCFCs) are also entirely anthropogenic gases. They have increased sharply in the atmosphere over the past few decades, following their use as substitutes for CFCs. Trichloroethane ($C_2H_3Cl_3$), for example, which is used in dry-cleaning and degreasing agents, increased fourfold in the 1980s and has a seven-year residence time in the atmosphere. They generally have lifetimes of a few years, but still have substantial greenhouse effects. The role of *halogens* of carbon (CFCs and HCFCs) in the destruction of ozone in the stratosphere is described below.

Water vapor (H_2O), the primary greenhouse gas, is a vital atmospheric constituent. It averages about 1 percent by volume but is very variable both in space and time, being involved in a complex global hydrological cycle (see Chapter 3).

3 Reactive gas species

In addition to the greenhouse gases, important *reactive gas species* are produced by the cycles of sulfur, nitrogen and chlorine. These play key roles in acid precipitation and in ozone destruction. Sources of these species are as follows:

- *Nitrogen species.* The reactive species of nitrogen are nitric oxide (NO) and nitrogen dioxide (NO_2). NO*x* refers to these and other odd nitrogen species with oxygen. Their primary significance is as a catalyst for tropospheric ozone formation. Fossil fuel combustion (approximately 40 percent for transportation and 60 percent for other energy uses) is the primary source of NO*x* (mainly NO) accounting for ~25 × 10^9kg N/year. Biomass burning and lightning activity are other important sources. NO*x* emissions increased by some 200 percent between 1940 and 1980. The total source of NO*x* is about 40 × 10^9kg N/year. About 25 percent of this enters the stratosphere, where it undergoes photochemical dissociation. It is also removed as nitric acid (HNO_3) in snowfall. Odd nitrogen is also released as NH*x* by ammonia oxidation in fertilizers and by domestic animals (6–10 × 10^9kg N/year).
- *Sulfur species.* Reactive species are sulfur dioxide (SO_2) and reduced sulfur (H_2S, DMS). Atmospheric sulfur is almost entirely anthropogenic in origin: 90 percent from coal and oil combustion, and much of the remainder from copper smelting. The major sources are sulfur dioxide (80–100 × 10^9kg S/year), hydrogen sulfide (H_2S) (20–40 × 10^9g S/year) and dimethyl sulfide (DMS) (35–55 × 10^9kg S/year). DMS is primarily produced by biological productivity near the ocean surface. SO_2 emissions increased by about 50 percent between 1940 and 1980, but declined in the 1990s. China is the largest source of emissions although the USA makes the largest per capita contribution. Volcanic activity releases approximately 10^9kg S/year as sulfur dioxide. Because the lifetime of SO_2 and H_2S in the atmosphere is only about one day, atmospheric sulfur occurs largely as carbonyl sulfur (COS), which has a lifetime of about one year. The conversion of H_2S gas to sulfur particles is an important source of atmospheric aerosols.
  Despite its short lifetime, sulfur dioxide is readily transported over long distances. It is removed from the atmosphere when condensation nuclei of SO_2 are precipitated as acid rain containing sulfuric acid (H_2SO_4). The acidity of fog deposition can be more serious because up to 90 percent of the fog droplets may be deposited.
- *Acid deposition* includes both acid rain and snow (wet deposition) and dry deposition of particulates. Acidity of precipitation represents an excess of positive hydrogen ions [H^+] in a water solution. Acidity is measured on the pH scale ($1 - \log[H^+]$) ranging from 1 (most acid) to 14 (most alkaline), 7 is neutral (i.e., the hydrogen cations are balanced by anions of sulphate, nitrate and chloride). Peak pH readings in the eastern United States and Europe are $\leqslant 4.3$.
  Over the oceans, the main anions are Cl– and SO_4^{2-} from sea salt. The background level of acidity in rainfall is about pH 4.8 to 5.6, because atmospheric CO_2 reacts with water to form carbonic acid. Acid solutions in rainwater are enhanced by reactions involving both gas-phase and aqueous-phase chemistry with sulfur dioxide and nitrogen dioxide. For sulfur dioxide, rapid pathways are provided by:

$$HOSO_2 + O_2 \rightarrow HO_2 + SO_3$$

$$H_2O + SO_3 \rightarrow H_2SO_4 \text{ (gas phase)}$$

and $$H_2O + HSO_3 \rightarrow H+ + SO_4^{2-} + H_2O \text{ (aqueous phase)}$$

The OH radical is an important catalyst in gas-phase reaction and hydrogen peroxide (H_2O_2) in the aqueous phase.

Acid deposition depends on emission concentrations, atmospheric transport and chemical activity, cloud type, cloud microphysical processes, and type of precipitation. Observations in northern Europe and eastern

North America in the mid-1970s, compared with the mid-1950s, showed a twofold to threefold increase in hydrogen ion deposition and rainfall acidity. Sulfate concentrations in rainwater in Europe increased over this 20-year period by 50 percent in southern Europe and 100 percent in Scandinavia, although there has been a subsequent decrease, apparently associated with reduced sulfur emissions in both Europe and North America. The emissions from coal and fuel oil in these regions have high sulfur contents (2–3 percent) and, since major SO_2 emissions occur from elevated stacks, SO_2 is readily transported by the low-level winds. NO*x* emissions, by contrast, are primarily from automobiles and thus NO_3– is mainly deposited locally. SO_2 and NO*x* have atmospheric resident times of one to three days. SO_2 is not readily dissolved in cloud or raindrops unless oxidized by OH or H_2O_2, but dry deposition is quite rapid. NO is insoluble in water, but it is oxidized to NO_2 by reaction with ozone, and ultimately to HNO_3 (nitric acid), which readily dissolves.

In the western United States where there are fewer major sources of emission, H+ ion concentrations in rainwater are only 15–20 percent of levels in the east, while sulfate and nitrate anion concentrations are one-third to one-half of those in the east. In China, high-sulfur coal is the main energy source and rainwater sulfate concentrations are high; observations in southwest China show levels six times those in New York City. In winter, in Canada, snow has been found to contain more nitrate and less sulfate than rain, apparently because falling snow scavenges nitrate faster and more effectively. Consequently, nitrate accounts for about half of the snowpack acidity. In spring, snowmelt runoff causes an acid flush that may be harmful to fish populations in rivers and lakes, especially at the egg or larval stages.

In areas with frequent fog, or hill cloud, acidity may be greater than with rainfall; North American data indicate pH values averaging 3.4 in fog. This is a result of several factors. Small fog or cloud droplets have a large surface area, higher levels of pollutants provide more time for aqueous-phase chemical reactions, and the pollutants may act as nuclei for fog droplet condensation. In California, pH values as low as 2.0–2.5 are not uncommon in coastal fogs. Fog water in Los Angeles usually has high nitrate concentrations due to automobile traffic during the morning rush-hour.

The impact of acid precipitation depends on the vegetation cover, soil and bedrock type. Neutralization may occur by addition of cations in the vegetation canopy or on the surface. Such buffering is greatest if there are carbonate rocks (Ca, Mg cations); otherwise the increased acidity augments normal leaching of bases from the soil.

4 Aerosols

There are significant quantities of *aerosols* in the atmosphere. These are suspended particles of sulfate, sea salt, mineral dust (particularly silicates), organic matter and black carbon. Aerosols enter the atmosphere from a variety of natural and anthropogenic sources (Table 2.2). Some originate as particles that are emitted directly into the atmosphere – mineral dust particles from dry surfaces, carbon soot from coal fires and biomass burning, and volcanic dust. Figure 2.1B shows their size distributions. Others are formed in the atmosphere by gas-to-particle conversion processes (sulfur from anthropogenic SO_2 and natural H_2S; ammonium salts from NH_3; nitrogen from NO*x*). Sulfate aerosols, two-thirds of which come from coal-fired power station emissions, played an important role in countering global warming effects by reflecting incoming solar radiation during the 1960s–1980s, but that so-called 'global dimming' has subsequently been reversed ('global brightening') (see Chapter 13). Other aerosol sources are sea salts and organic matter (plant hydrocarbons and anthro-

Table 2.2 Aerosol production estimates, less than 5μm radius (10^9kg/year) and typical concentrations near the surface (μg m^{-3}) in remote and urban areas

	Production	Concentration	
		Remote	Urban
Natural			
Primary production:			
Sea salt	2300	5–10	
Mineral particles		900–1500	0.5–5*
Volcanic	20		
Forest fires and biological debris	50		
Secondary production (gas → particle):			
Sulphates from H_2S	70	1–2	
Nitrates from NO*x*	22		
Converted plant hydrocarbons	25		
Total natural	3600		
Anthropogenic			
Primary production:			
Mineral particles	0–600		
Industrial dust	50		
Combustion (black carbon)	10	100–500†	
(organic carbon)	50		
Secondary production (gas → particle):			
Sulphate from SO2	140	0.5–1.5	10–20
Nitrates from NO*x*	30	0.2	0.5
Biomass combustion (organics)	20		
Total anthropogenic	290–890		

Sources: Ramanthan *et al.* 2001; Schimel *et al.* 1996, Bridgman 1990.

Notes: *10–60μg m^{-3} during dust episodes from the Sahara over the Atlantic.
†Total suspended particles. 10^9kg = 1Tg.

pogenically derived). Natural sources are several times larger than anthropogenic sources on a global scale, but the estimates are wide-ranging. Mineral dust is particularly hard to estimate due to the episodic nature of wind events and the considerable spatial variability. For example, the wind picks up some 1500Tg (10^{12}g) of crustal material annually, about one-half from the Sahara and the Arabian Peninsula (see Plate 2.1). Most of this is deposited downwind over the Atlantic. There is similar transport from western China and Mongolia eastward over the North Pacific Ocean. Large size particles originate from mineral dust, sea salt spray, fires and plant spores (Figure 2.1A); these sink rapidly back to the surface or are washed out (scavenged) by rain after a few days. Fine particles from volcanic eruptions may reside in the upper stratosphere for one to three years.

Small (Aitken) particles form by the condensation of gas-phase reaction products and from organic molecules and polymers (natural and synthetic fibers, plastics, rubber and vinyl). There are 500–1000 Aitken particles per cm^3 in air over Europe. Intermediate-sized (accumulation mode) particles originate from natural sources such as soil surfaces, from combustion, or they accumulate by random coagulation and by repeated cycles of condensation and evaporation (Figure 2.1A). Over Europe, 2000–3500 such particles per cm^3 are measured. Particles with diameters < 2.5μm ($PM_{2.5}$) – that can cause adverse health problems – are now often documented separately. Particles with diameters of 0.1–1.0μm are highly effective in scattering solar radiation (Chapter 3B.2), and those of about 0.1μm diameter are important in cloud condensation. The climatological effects of aerosols on precipitation are complex and the overall impact is uncertain (see p.117).

Having made these generalizations about the atmosphere, we now examine the variations that occur in composition with height, latitude and time.

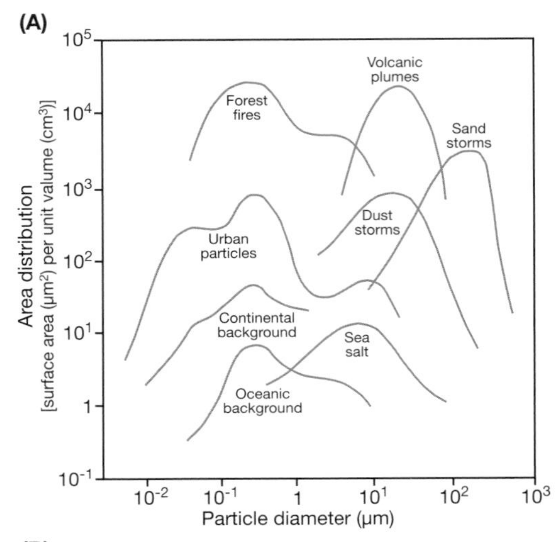

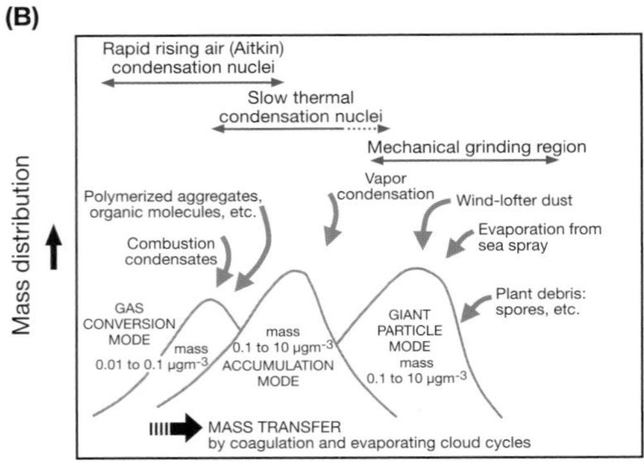

Figure 2.1 Atmospheric particles. A: Mass distribution, together with a depiction of the surface–atmosphere processes that create and modify atmospheric aerosols, illustrating the three size modes. Aitken nuclei are solid and liquid particles that act as condensation nuclei and capture ions, thus playing a role in cloud electrification. B: Distribution of surface area per unit volume.

Sources: A: After Glenn E. Shaw, University of Alaska, Geophysics Institute. B: After Slinn (1983).

5 Variations with height

The light gases (hydrogen and helium especially) might be expected to become more abundant in the upper atmosphere, but large-scale turbulent mixing of the atmosphere prevents such diffusive separation up to at least 100km above the surface. The height variations that do occur are related to the source locations of the two major non-permanent gases – water vapor and ozone. Since both absorb some solar and terrestrial radiation, the heat budget and vertical temperature structure of the atmosphere are considerably affected by the distribution of these two gases.

Water vapor comprises up to 4 percent of the atmosphere by volume (about 3 percent by weight) near the surface, but only 3–6ppmv (parts per million by volume) above 10 to 12km. It is supplied to the atmosphere by evaporation from surface water or by transpiration from plants and is transferred upwards by atmospheric turbulence. Turbulence is most effective below about 10–15km and, as the maximum possible water vapor density of cold air is very low anyway (see B.2, this chapter), there is little water vapor in the upper layers of the atmosphere.

Ozone (O_3) is concentrated mainly between 15 and 35km. The upper layers of the atmosphere are irradiated by ultraviolet radiation from the sun (see C.1, this chapter), which causes the breakup of oxygen molecules at altitudes above 30km (i.e., $O_2 \rightarrow O + O$). These separated atoms (O + O) may then combine individually with

Plate 2.1 Dust plumes over the Red Sea from MODIS on 15 January 2009. Courtesy NASA

other oxygen molecules to create ozone, as illustrated by the simple photochemical scheme:

$$O_2 + O + M \rightarrow O_3 + M$$

where M represents the energy and momentum balance provided by collision with a third atom or molecule; this Chapman cycle is shown schematically in Figure 2.2A. Such three-body collisions are rare at 80 to 100km because of the very low density of the atmosphere, while below about 35km most of the incoming ultraviolet radiation has already been absorbed at higher levels. Therefore ozone is mainly formed between 30 and 60km, where collisions between O and O_2 are more likely. Ozone itself is unstable; its abundance is determined by three different photochemical interactions. Above 40km odd oxygen is destroyed primarily by a cycle involving molecular oxygen; between 20 and 40km NO*x* cycles are dominant; while below 20km a hydrogen–oxygen radical (HO_2) is responsible.

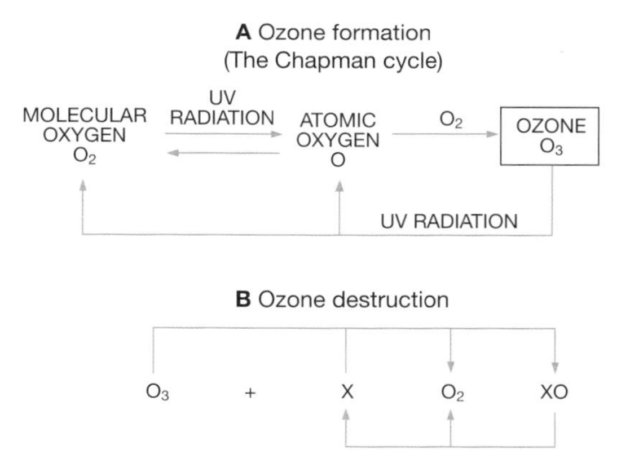

Figure 2.2 Schematic illustrations of (A) the Chapman cycle of ozone formation and (B) ozone destruction. X is any ozone-destroying species (e.g. H, OH, NO, CR, Br).

Source: After Hales (1996) from *Bulletin of the American Meteorological Society*, by permission of the American Meteorological Society.

Additional important cycles involve chlorine (ClO) and bromine (BrO) chains at various altitudes. Collisions with monatomic oxygen may re-create oxygen (see Figure 2.2B), but ozone is mainly destroyed through cycles involving catalytic reactions, some of which are photochemical associated with longer wavelength ultraviolet radiation (2.3–2.9μm). The destruction of ozone involves a recombination with atomic oxygen, causing a net loss of the odd oxygen. This takes place through the catalytic effect of a radical such as OH (hydroxyl):

$$\left.\begin{array}{l} H + O \rightarrow HO_2 \\ HO_2 + O \rightarrow OH + O_2 \\ OH + O \rightarrow H + O_2 \end{array}\right] \quad \text{net: } 2O \rightarrow O_2$$

The odd hydrogen atoms and OH result from the dissociation of water vapor, molecular hydrogen and methane (CH_4).

Stratospheric ozone is similarly destroyed in the presence of nitrogen oxides (NO*x*, i.e., NO_2 and NO) and chlorine radicals (Cl, ClO). The source gas of the NO*x* is nitrous oxide (N_2O), which is produced by combustion and fertilizer use, while chlorofluorocarbons (CFCs), manufactured for 'freon', give rise to the chlorines. These source gases are transported up to the stratosphere from the surface and are converted by oxidation into NO*x*, and by UV photodecomposition into chlorine radicals, respectively.

The chlorine chain involves:

$$2\,(Cl + O_3 \rightarrow ClO + O_2)$$
$$ClO + ClO \rightarrow Cl2O_2$$

and

$$Cl + O_3 \rightarrow ClO + O_2$$
$$OH + O_3 \rightarrow HO_3 + 2O_2$$

Both reactions result in a conversion of O_3 to O_2 and the removal of all odd oxygens. Another cycle may involve an interaction of the oxides of chlorine and bromine (Br). It appears that the increases of Cl and Br species during the decades 1970–1990 are sufficient to explain the observed decrease of stratospheric ozone over Antarctica (see Box 2.1). A mechanism that may enhance the catalytic process involves polar stratospheric clouds. These can form readily during the austral spring (October), when temperatures decrease to 185–195K, permitting the formation of particles of nitric acid (HNO_3) ice and water ice. It is apparent, however, that anthropogenic sources of the trace gases are the primary factor in the ozone decline. Conditions in the Arctic are somewhat different, as the stratosphere is warmer and there is more mixing of air from lower latitudes. Nevertheless, ozone decreases are now observed in the boreal spring in the Arctic stratosphere.

The constant metamorphosis of oxygen to ozone and from ozone back to oxygen involves a very complex set of photochemical processes, which tend to maintain an approximate equilibrium above about 40km. However, the ozone mixing ratio is at its maximum at about 35km, whereas maximum ozone concentration (see Note 1) occurs lower down, between 20 and 25km in low latitudes and between 10 and 20km in high latitudes. This is the result of a circulation mechanism transporting ozone downward to levels where its destruction is less likely, allowing an accumulation of the gas to occur. Despite the importance of the ozone layer, it is essential to realize that if the atmosphere were compressed to sea level (at normal sea-level temperature and pressure) ozone would contribute only about 3mm to the total atmospheric thickness of 8km (Figure 2.3).

6 Variations with latitude and season

Variations of atmospheric composition with latitude and season are particularly important in the case of water vapor and stratospheric ozone.

Ozone content is low over the equator and high in subpolar latitudes in spring (see Figure 2.3). If the distribution were solely the result of photochemical processes, the maximum would occur in June near the equator, so the anomalous

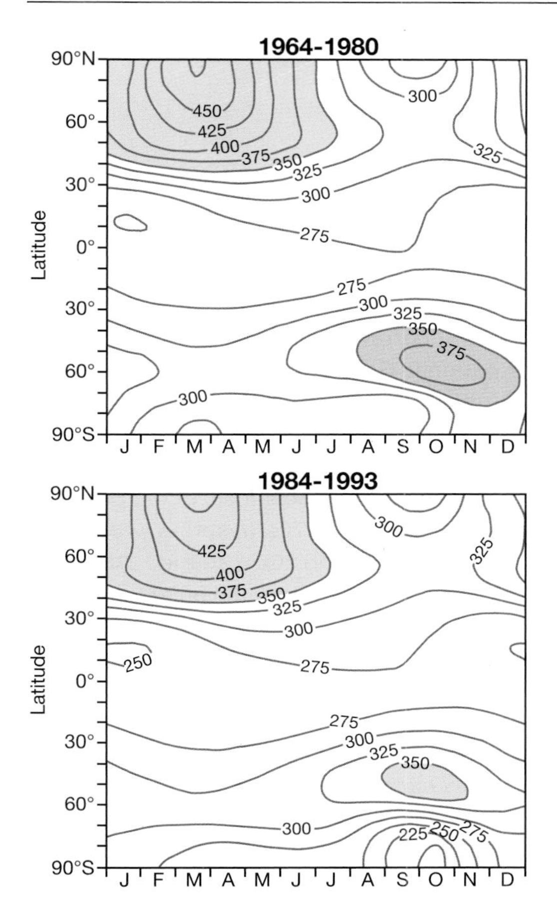

Figure 2.3 Variation of total ozone with latitude and season in Dobson units (milliatmosphere centimeters) for two time intervals: (top) 1964–1980 and (bottom) 1984–1993. Values over 350 units are stippled.

Source: From Bojkov and Fioletov (1995) from *Journal of Geophysical Research*, 100(D), Fig. 15, p. 16, 548. Courtesy American Geophysical Union.

pattern must result from a poleward transport of ozone. Apparently, ozone moves from higher levels (30–40km) in low latitudes towards lower levels (20–25km) in high latitudes during the winter months. Here the ozone is stored during the *polar night*, giving rise to an ozone-rich layer in early spring under natural conditions. It is this feature that has been disrupted by the stratospheric ozone 'hole' that now forms each spring in the Antarctic and in some recent years in the Arctic also (see Box 2.1). The type of circulation responsible for this transfer is not yet known with certainty, although it does not seem to be a simple direct one.

The water vapor content of the atmosphere is closely related to air temperature (see B.2, this chapter, and Chapter 4B and C) and is therefore greatest in summer and in low latitudes. There are, however, obvious exceptions to this generalization, such as the tropical desert areas of the world.

The carbon dioxide content of the air, which averaged 387 parts per million (ppm) in 2007, has a large seasonal range in higher latitudes in the northern hemisphere associated with photosynthesis and decay in the biosphere. At 50°N, the concentration ranges from about 380ppm in autumn to 393ppm in spring. The low summer values are related to the assimilation of CO_2 by the cold polar seas. Over the year, a small net transfer of CO_2 from low to high altitudes takes place to maintain an equilibrium content in the air.

7 Variations with time

The quantities of carbon dioxide, other greenhouse gases and particles in the atmosphere undergo long-term variations that may play an important role in the earth's radiation budget. Measurements of atmospheric trace gases show increases in nearly all of them since the Industrial Revolution began around 1750 (Table 2.3). The burning of fossil fuels is the primary source of these increasing trace-gas concentrations. Heating, transportation and industrial activities generate almost 5 × 10^{20} J/year of energy. Oil and natural gas consumption account for 60 percent of global energy and coal about 25 percent. Natural gas is almost 90 percent methane (CH_4), whereas the burning of coal and oil releases not only CO_2 but also odd nitrogen (NOx), sulfur and carbon monoxide (CO). Other factors relating to agricultural practices (land clearance, farming, paddy cultivation and cattle raising) also contribute to modifying the atmospheric

SIGNIFICANT 20TH-CENTURY ADVANCE

2.1 Ozone in the stratosphere

Ozone measurements were first made in the 1930s. Two properties are of interest: (1) the total ozone in an atmospheric column. This is measured with the Dobson spectrophotometer by comparing the solar radiation at a wavelength where ozone absorption occurs with that in another wavelength where such effects are absent; (2) the vertical distribution of ozone. This can be measured by chemical soundings of the stratosphere, or calculated at the surface using the *Umkehr* method; here the effect of solar elevation angle on the scattering of solar radiation is measured. Ozone measurements, begun in the Antarctic during the International Geophysical Year 1957–1958, showed a regular annual cycle with an austral spring (October to November) peak as ozone-rich air from mid-latitudes was transported poleward as the winter polar vortex in the stratosphere broke down. Values declined seasonally from around 450 Dobson units (DU) in spring to about 300DU in summer and continued at about this level through the autumn and winter. Scientists of the British Antarctic Survey noted a different pattern at Halley Base beginning in the 1970s. In spring, with the return of sunlight, values began to decrease steadily between about 12 and 20km altitude. Also in the 1970s, satellite sounders began mapping the spatial distribution of ozone over the polar regions. These revealed that low values formed a central core and the term 'Antarctic ozone hole' came into use. Since the mid-1970s, values start decreasing in late winter and now reach minima of 95–100DU in the austral spring.

Using a boundary of 220DU (corresponding to a thin, 2.2mm ozone layer, if all the gas were brought to sea-level temperature and pressure), the extent of the Antarctic ozone hole at the end of September averaged 21 million km^2 during 1990–1999. This expanded to cover 27 million km^2 by early September in 1999 and 2000 and continued at this level through spring 2006.

In the Arctic, temperatures in the stratosphere are not as low as over the Antarctic, but in recent years ozone depletion has been large when temperatures fall well below normal in the winter stratosphere. In February 1996, for example, column totals averaging 330DU for the Arctic vortex were recorded compared with 360DU, or higher, in other years. A series of mini-holes was observed over Greenland, the northern North Atlantic and northern Europe with an absolute low, over Greenland below 180DU. An extensive ozone hole is less likely to develop in the Arctic because the more dynamic stratospheric circulation, compared with the Antarctic, transports ozone poleward from mid-latitudes.

To combat the ozone decreases, the Montreal Protocol was agreed internationally in 1987 to phase out the production of substances thought to be responsible for ozone depletion. Subsequently, the concentrations of the most important chlorofluorocarbons (CFCs) have either leveled off or decreased, but the size of the ozone hole has not yet responded.

composition. The concentrations and sources of the most important greenhouse gases are considered in turn.

- *Carbon dioxide* (CO_2). The major reservoirs of carbon are in limestone sediments and fossil fuels. The atmosphere contains about 800 × 10^{12}kg of carbon (C), corresponding to a CO_2 concentration of 387ppm (Figure 2.4). The major fluxes of CO_2 are a result of solution/dissolution in the ocean and photosynthesis/respiration and decomposition by biota. The

Table 2.3 Anthropogenically induced changes in concentration of atmospheric trace gases

Gas	Concentration 1850*	2008	Annual increase (%) 1990s	Sources
Carbon dioxide	280ppm	385ppm	0.4	Fossil fuels
Methane	800 ppbv	1775ppbv	0.3	Rice paddies, cows, wetlands
Nitrous oxide	280ppbv	320ppbv	0.25	Microbiological activity, fertilizer, fossil fuel
CFC-11	0	0.27ppbv	≈0	Freon†
HCFC-22	0	0.11ppbv	5	CFC substitute
Ozone (troposphere)	?	10–50ppbv	≈0	Photochemical reactions

Sources: Updated from Schimel *et al.* (1996), in Houghton *et al.* (1996).

Notes: *Pre-industrial levels are primarily derived from measurements in ice cores where air bubbles are trapped as snow accumulates on polar ice sheets. †Production began in the 1930s.

ppm = parts per million; ppbv = parts per billion by volume

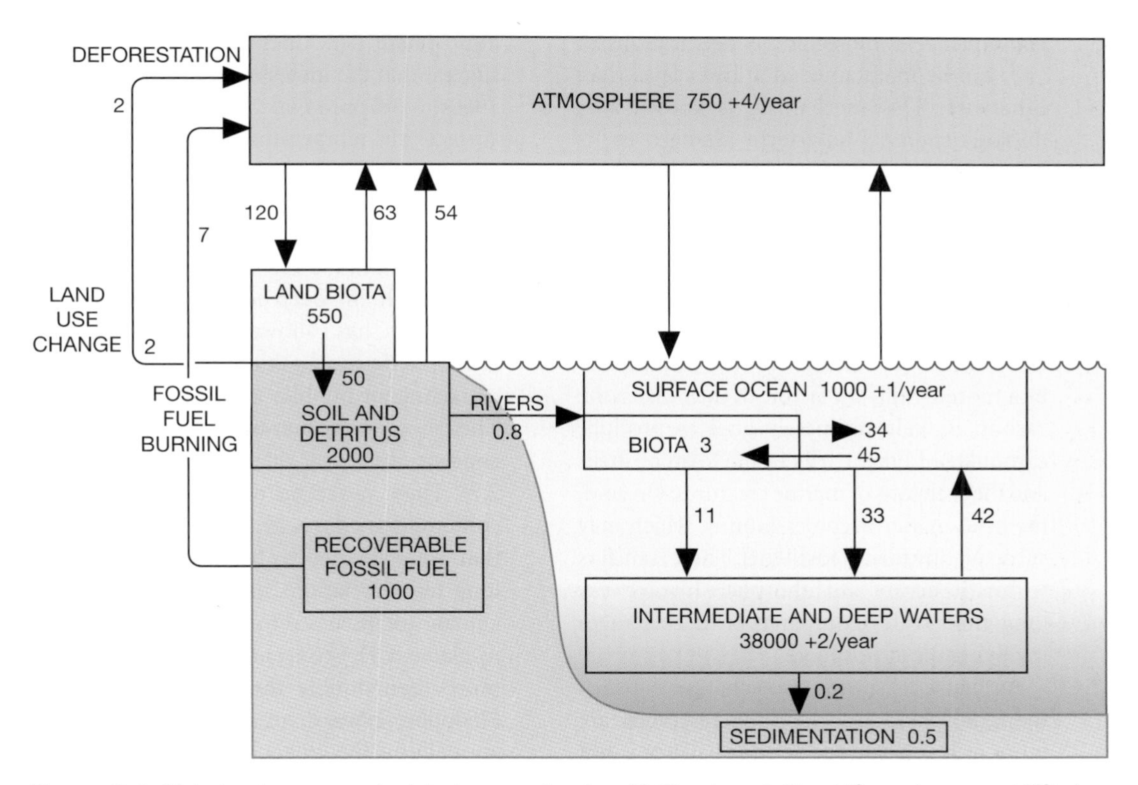

Figure 2.4 Global carbon reservoirs (gigatonnes of carbon (GtC): where 1 Gt = 10^9 metric tons = 10^{12}kg) and gross annual fluxes (GtC yr^{-1}). Numbers italicized in the reservoirs suggest the net annual accumulation due to anthropogenic causes.

Source: Based on Sundquist, Trabalka, Bolin and Siegenthaler; after IPCC (1990 and 2001).

average time for a CO_2 molecule to be dissolved in the ocean or taken up by plants is about four years. Photosynthetic activity leading to primary production on land involves 50×10^{12}kg of carbon annually, representing 7 percent of atmospheric carbon; this accounts for the annual oscillation in CO_2 observed in the Northern Hemisphere due to its extensive land biosphere.

The oceans play a key role in the global carbon cycle. Photosynthesis by phytoplankton generates organic compounds of aqueous carbon dioxide. Eventually, some of the biogenic matter sinks into deeper water, where it undergoes decomposition and oxidation back into carbon dioxide. This process transfers carbon dioxide from the surface water and sequesters it in the ocean deep water. As a consequence, atmospheric concentrations of CO_2 can be maintained at a lower level than otherwise. This mechanism is known as a 'biologic pump'; long-term changes in its operation may have caused the rise in atmospheric CO_2 at the end of the last glaciation. Ocean biomass productivity is limited by the availability of nutrients and by light. Hence, unlike the land biosphere, increasing CO_2 levels will not necessarily affect ocean productivity; inputs of fertilizers in river runoff may be a more significant factor. In the oceans, the carbon dioxide ultimately goes to produce carbonate of lime, partly in the form of shells and the skeletons of marine creatures. On land, the dead matter becomes humus, which may subsequently form a fossil fuel. These transfers within the oceans and lithosphere involve very long time scales compared with exchanges involving the atmosphere.

As Figure 2.4 shows, the exchanges between the atmosphere and the other reservoirs are more or less balanced. Yet this balance is not an absolute one; between AD 1750 and 2008 the concentration of atmospheric CO_2 is estimated to have increased by 38 percent, from 280 to 387ppm, the highest value for 650,000 years! (Figure 2.5). Half of this increase has taken place since the mid-1960s; currently, atmospheric CO_2 levels are increasing by 1.5–2ppmv per year. The primary net source is fossil fuel combustion, now accounting for 6.55×10^{12}kg C/year. Tropical deforestation and fires may contribute a further 2×10^{12}kg C/year; the figure is still uncertain. Fires destroy only above-ground biomass, and a large fraction of the carbon is stored as charcoal in the soil. The consumption of fossil fuels should actually have produced an increase almost twice as great as is observed. Uptake and dissolution in the oceans and the terrestrial biosphere primarily account for the difference.

Carbon dioxide has a significant impact on global temperature through its absorption and re-emission of radiation from the earth and atmosphere (see Chapter 3C). Calculations suggest that the increase from 320ppm in the 1960s to 387ppm (AD 2008) raised the mean surface air temperature by 0.6°C (in the absence of other factors). The rate of increase of CO_2 since AD 2000 has been about 2ppm/yr compared with less than 1ppm in the 1960s and 1.5ppm in the 1980s.

Research on deep ice cores taken from Antarctica has allowed changes in past atmospheric composition to be calculated by extracting air bubbles trapped in the old ice. This shows large natural variations in CO_2 concentration over the ice age cycles (Figure 2.7). These variations of up to 100ppm were contemporaneous with temperature changes that are estimated to be about 10°C. These long-term variations in carbon dioxide and climate are discussed further in Chapter 13.

- *Methane* (CH_4) concentration (1,775ppbv) is more than double the pre-industrial level (750ppbv). It was increasing by about 4–5ppbv annually in the 1990s but this has dropped to near zero since 1999–2000 (Figure 2.7). For unknown reasons, concentrations increased again in 2008. Methane has an atmospheric lifetime of about nine years and is responsible

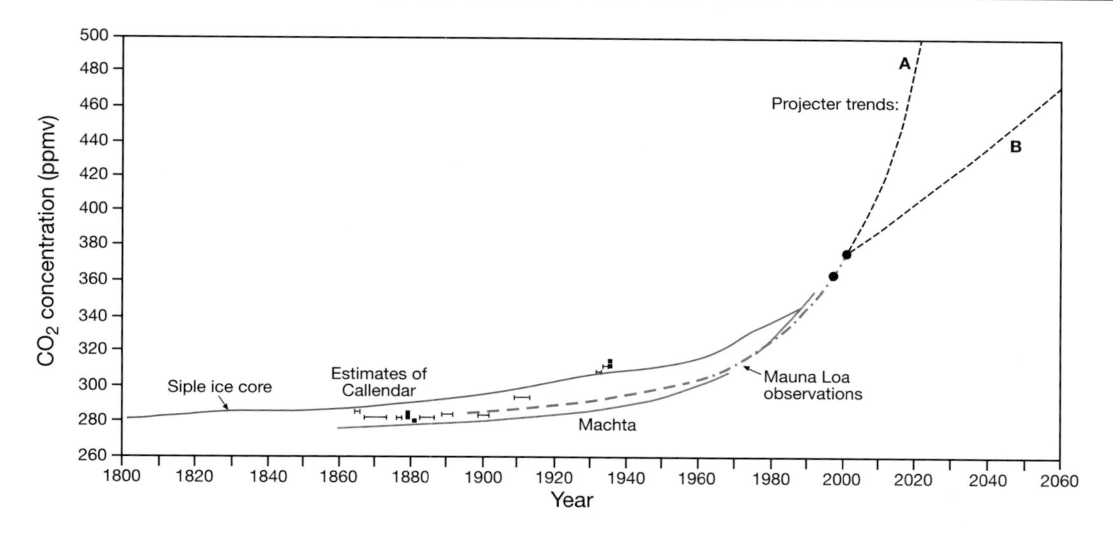

Figure 2.5 Estimated carbon dioxide concentration: since AD 1800 from air bubbles in an Antarctic ice core, early measurements from 1860–1960; observations at Mauna Loa, Hawaii, since 1957; and projected trends for this century.

Source: After Keeling, Callendar, Machta, Broecker and others.

Note: (a) and (b) indicate different scenarios of global fossil fuel use (IPCC, 2001).

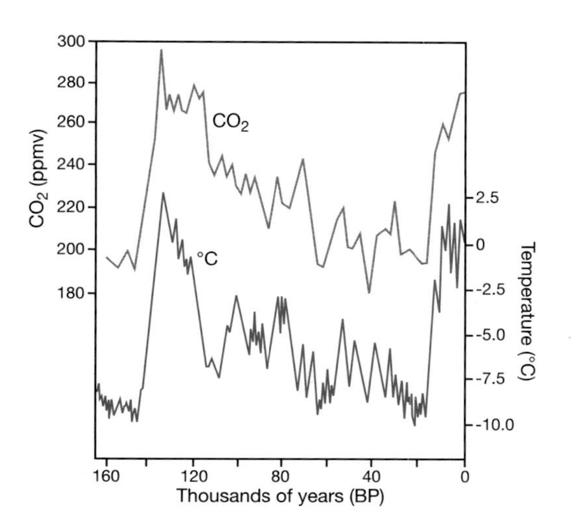

Figure 2.6 Changes in atmospheric CO_2 (ppmv: parts per million by volume) and estimates of the resulting global temperature deviations from the present value obtained from air trapped in ice bubbles in cores spanning 160,000 years at Vostok, Antarctica.

Source: *Our Future World*, Natural Environment Research Council (NERC) 1989.

for some 18 percent of the greenhouse effect. Cattle populations have increased by 5 percent/year over 30 years and paddy rice area by 7 percent/year, although it is uncertain whether these account quantitatively for the annual increase of 120ppbv in methane over the past decade. Table 2.4, showing the mean annual release and consumption, indicates the uncertainties in our knowledge of its sources and sinks.

- *Nitrous oxide* (N_2O), which is relatively inert, originates primarily from microbial activity (nitrification) in soils and in the oceans (4 to 8 × 10^9kg N/year), with about 1.0 × 10^9kg N/year from industrial processes. Other major anthropogenic sources are nitrogen fertilizers and biomass burning. The concentration of N_2O has increased from a pre-industrial level of about 285ppbv to 320ppbv (in clean air). Its increase began in around 1940 and is about 0.8ppbv/year (Figure 2.8A). The major sink of N_2O is in the stratosphere, where it is oxidized into NO*x*.

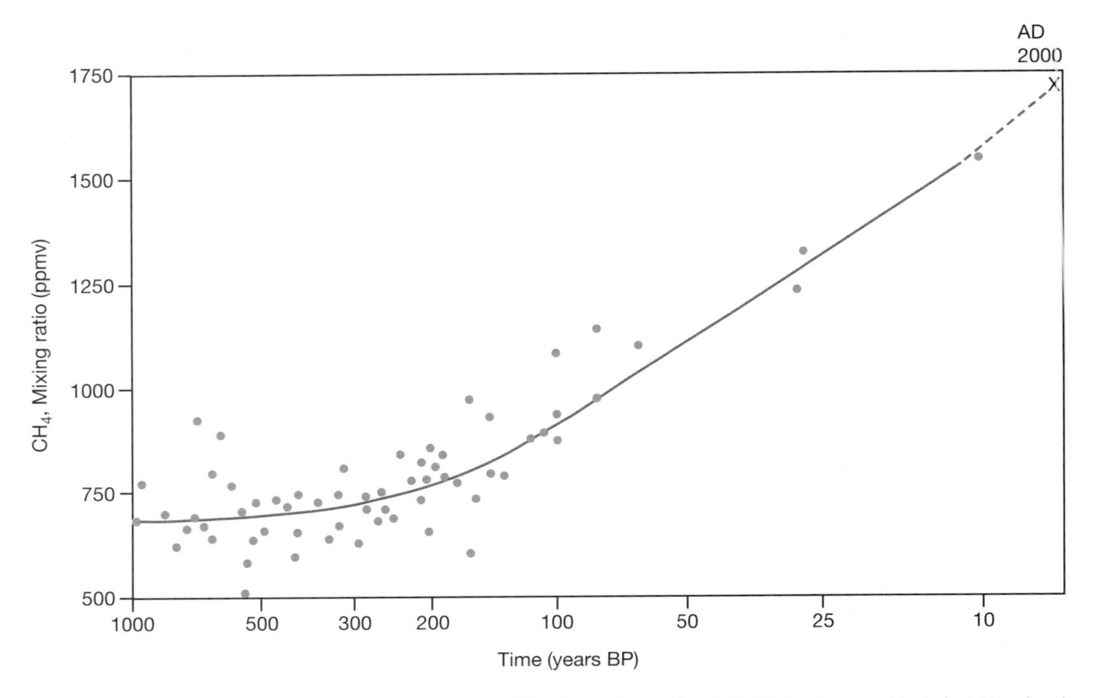

Figure 2.7 Methane concentration (parts per million by volume) in air bubbles trapped in ice dating back to 1000 years BP obtained from ice cores in Greenland and Antarctica and the global average for AD 2000 (X).

Source: Data from Rasmussen and Khalil, Craig and Chou, and Robbins; adapted from Bolin *et al.* (eds) *The Greenhouse Effect, Climatic Change, and Ecosystems* (SCOPE 29). Copyright © 1986. Reprinted by permission of John Wiley & Sons, Inc.

Table 2.4 Mean annual release and consumption of CH_4 ($T_g = 10^{12}_g$)

	Mean	**Range**
A *Release*		
Natural wetlands	115	100–200
Rice paddies	110	25–170
Enteric fermentation (mammals)	80	65–110
Gas drilling	45	25–50
Biomass burning	40	20–80
Termites	40	10–100
Landfills	40	20–70
Total	*c.* 530	
B *Consumption*		
Soils	30	15–30
Reaction with OH	500	400–600
Total	*c.* 530	

Source: Tetlow-Smith (1995).

- *Chlorofluorocarbons* (CF_2Cl_2 and $CFCl_3$), better known as 'freons' CFC-11 and CFC-12, respectively, were first produced in the 1930s and now have a total atmospheric burden of 10^{10}kg. They increased at 4–5 percent per year up to 1990, but CFC-11 has slowly declined since the mid-1990s and CFC-12 is nearly static, after peaking in 2003, as a result of the Montreal Protocol agreements to curtail production and use substitutes (see Figure 2.9B). Although their concentration is <1ppbv, CFCs account for nearly 10 percent of the greenhouse effect. They have a residence time of 55–130 years in the atmosphere. However, while the replacement of CFCs by hydrohalocarbons (HCFCs) can significantly reduce the depletion of stratospheric ozone, HCFCs still have a large greenhouse potential.

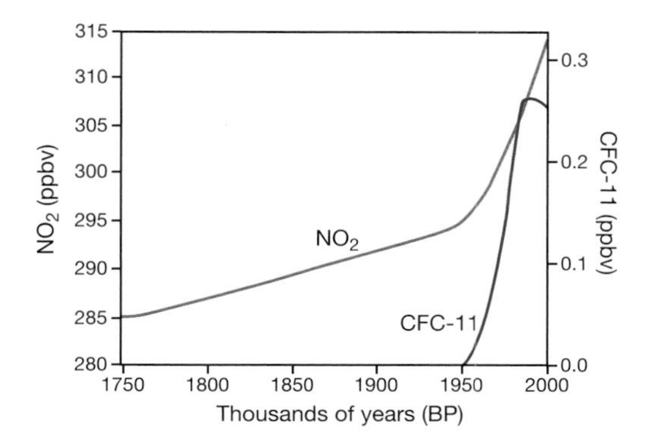

Figure 2.8 Concentration of (A) nitrous oxide, N_2O (left scale), which has increased since the mid-eighteenth century and especially since 1950; and of (B) CFC-11 since 1950 (right scale). Both in parts per billion by volume (ppbv).

Source: After IPCC (1990 and 2001).

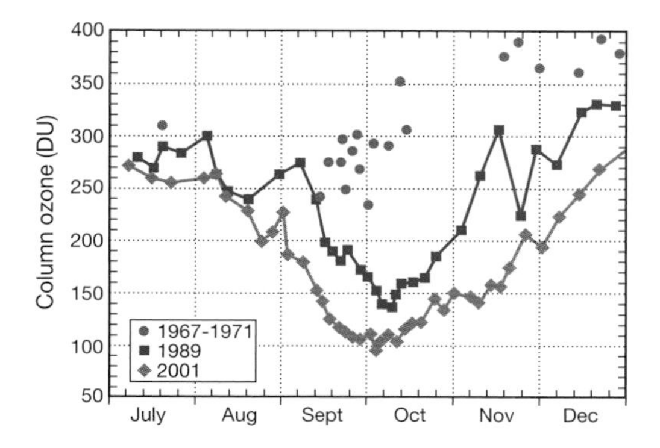

Figure 2.9 Total ozone measurements from ozonesondes over South Pole for 1967–1971, 1989 and 2001, showing deepening of the Antarctic ozone hole.

Source: Based on Climate Monitoring and Diagnostics Laboratory, NOAA.

- *Ozone* (O_3) is distributed very unevenly with height and latitude (see Figure 2.4) as a result of the complex photochemistry involved in its production (A.2, this chapter). Since the late 1970s, dramatic declines in springtime total ozone have been detected over high southern latitudes. The normal increase in stratospheric ozone associated with increasing solar radiation in spring apparently failed to develop. Observations in Antarctica show a decrease in total ozone in September to October from 320DU (10^{-3}cm at standard atmospheric temperature and pressure) in the 1960s to around 100 in the 1990s. Satellite measurements of stratospheric ozone (Figure 2.9) illustrate the presence of an 'ozone hole' over the south polar region (see Box 2.1). Similar reductions are also evident in the Arctic and at lower latitudes. Between 1979 and 1986 there was a 30 percent decrease in ozone at 30–40km altitude between latitudes 20 and 50°N and S (Figure 2.10); along with this there has been an increase in ozone in the lowest 10km as a result of anthropogenic activities. Tropospheric ozone represents about 34DU compared with 25 pre-industrially. These changes in the vertical distribution of ozone concentration are likely to lead to changes in atmospheric heating (Chapter 2C), with implications for future climate trends (see Chapter 11). The global mean column total decreased from 306DU for 1964–1980 to 297 for 1984–1993 (see Figure 2.4). The decline over the past 25

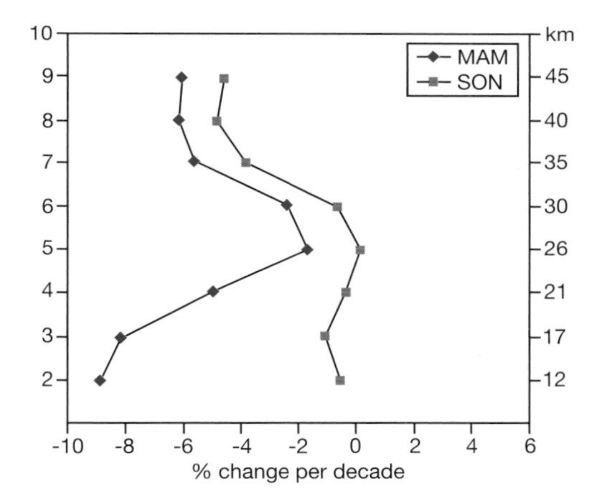

Figure 2.10 Changes in stratospheric ozone content (percent per decade) during March to May and September to November 1978–1997 over Europe (composite of Belsk, Poland, Arosa, Switzerland and Observatoire de Haute Provence, France) based on umkehr measurements.

Source: Adapted from Bojkov *et al.* (2002) *Meteorology and Atmospheric Physics*, 79, p. 148, Fig. 14a.

years has exceeded 7 percent in middle and high latitudes.

The effects of reduced stratospheric ozone are particularly important for their potential biological damage to living cells and human skin. It is estimated that a 1 percent reduction in total ozone will increase ultraviolet-B radiation by 2 percent, for example, and ultraviolet radiation at 0.30μm is a thousand times more damaging to the skin than at 0.33μm (see Chapter 3A). The ozone decrease would also be greater in higher latitudes. However, the mean latitudinal and altitudinal gradients of radiation imply that the effects of a 2 percent UV-B increase in mid-latitudes could be offset by moving poleward 20km or 100m lower in altitude! Recent polar observations suggest dramatic changes. Stratospheric ozone totals in the 1990s over Palmer Station, Antarctica (65°S), now maintain low levels from September until early December, instead of recovering in November. Hence, the altitude of the sun has been higher and the incoming radiation much greater than in previous years, especially at wavelengths ≤ 0.30μm. However, the possible effects of increased UV radiation on biota remain to be determined.

- *Aerosol* loading may change due to natural and human-induced processes. Atmospheric particle concentrations derived from volcanic dust are extremely irregular (see Figure 2.11), but individual volcanic emissions are rapidly diffused geographically. As shown in Figure 2.12, a strong westerly wind circulation carried the El Chichón dust cloud at an average velocity of 20m s^{-1} so that it encircled the globe

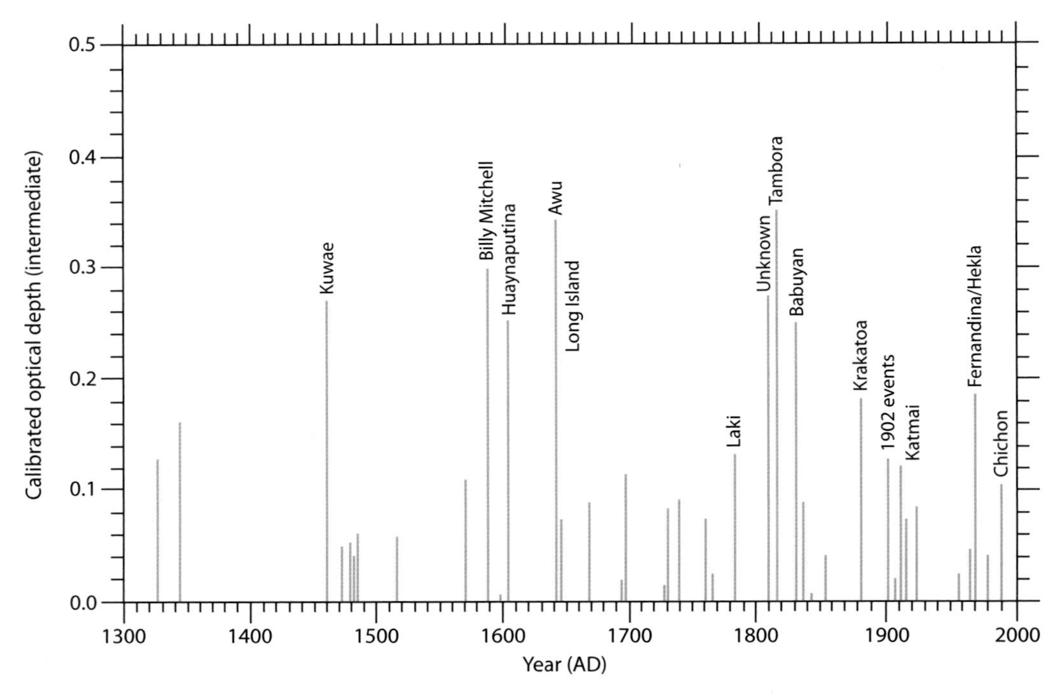

Figure 2.11 Record of volcanic eruptions in the GISP 2 ice core and calibrated visible optical depth for AD 1300–2000, together with the names of major volcanic eruptions.

Note that the record reflects eruptions in the Northern Hemisphere and equatorial region only; optical depth estimates depend on the latitude and the technique used for calibration.

Source: Updated after Zielinski (1995) *Journal of Geophysical Research*, 100(D10), p. 20,950, Fig. 6.

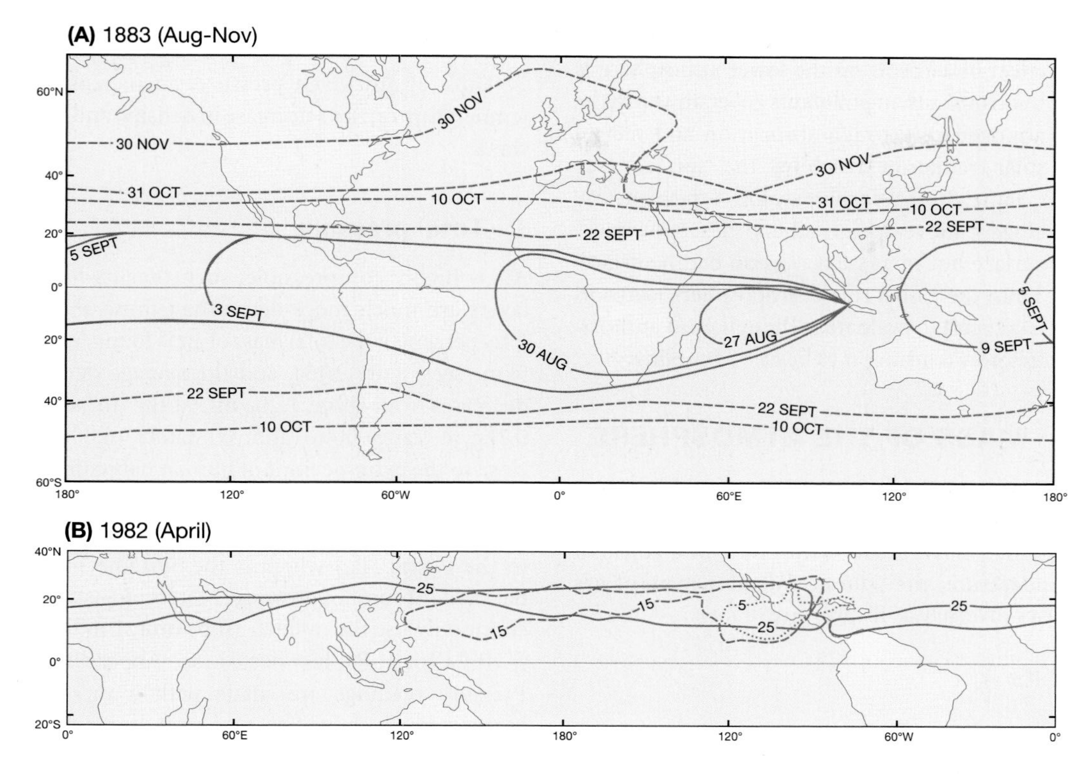

Figure 2.12 The spread of volcanic material in the atmosphere following major eruptions: A Approximate distributions of observed optical sky phenomena associated with the spread of Krakatoa volcanic dust between the eruption of 26 August and 30 November 1883. B The spread of the volcanic dust cloud following the main eruption of the El Chichón volcano in Mexico on 3 April 1982. Distributions on 5, 15 and 25 April are shown.

Sources: Russell and Archibald (1888), Simkin and Fiske (1983), Rampino and Self (1984), Robock and Matson (1983). A: by permission Smithsonian Institution.

in less than three weeks. The spread of the Krakatoa dust in 1883 was more rapid and extensive due to the greater amount of fine dust that was blasted into the stratosphere. In June 1991, the eruption of Mount Pinatubo in the Philippines injected 20 megatons of SO_2 into the stratosphere. However, only about twelve eruptions have produced measurable dust veils in the past 120 years. They occurred mainly between 1883 and 1912, and 1982 and 1992.

Volcanic eruptions, which inject dust and sulfur dioxide high into the stratosphere, are known to cause a small deficit in surface heating with a global effect of –0.1° to –0.2°C, but the effect is short-lived, lasting only a year or so after the event (see Box 13.1). In addition, unless the eruption is in low latitudes, the dust and sulfate aerosols remain in one hemisphere and do not cross the equator.

The contribution of man-made particles (particularly sulfates and mineral dust) has been progressively increasing, and now accounts for about 30 percent of the total tropospheric aerosol load. Sulfate emissions have decreased in Europe and North America since the 1990s, but increased in southern and eastern Asia; global sulfate emissions have decreased overall since the 1980s. The overall effect of aerosols on the lower atmosphere is uncertain; urban pollutants generally warm the atmosphere through absorption and reduce solar radiation reaching the surface (see Chapter 3C). Aerosols may lower the planetary albedo above a high-albedo desert or snow surface but increase it over an ocean surface. Thus, the global role of tropospheric aerosols is difficult to evaluate, although most authorities now consider it to be one of cooling.

B MASS OF THE ATMOSPHERE

Atmospheric gases obey a few simple laws in response to changes in pressure and temperature. The first, Boyle's Law, states that, at a constant temperature, the volume (V) of a mass of gas varies inversely as its pressure (P), i.e.,

$$P = \frac{k_1}{V}$$

(k_1 is a constant); and the second, Charles's Law, that, at a constant pressure, volume varies directly with absolute temperature (T) measured in degrees Kelvin (see Note 2):

$$V = k_2 T$$

These laws imply that the three qualities of pressure, temperature and volume are completely interdependent, such that any change in one of them will cause a compensating change to occur in one, or both, of the remainder. The gas laws may be combined to give the following relationship:

$$PV = RmT$$

where m = mass of air, and R = a gas constant for dry air (287 J kg^{-1} K^{-1}) (see Note 3).

If m and T are held fixed, we obtain Boyle's Law; if m and P are held fixed, we obtain Charles's Law. Since it is convenient to use density, ρ (= mass/volume), rather than volume when studying the atmosphere, we can rewrite the equation in the form known as the equation of state:

$$P = R\rho T$$

Thus, at any given pressure, an increase in temperature causes a decrease in density, and vice versa.

1 Total pressure

Air is highly compressible, such that its lower layers are much more dense than those above. Fifty percent of the total mass of air is found below 5km (see Figure 2.13), and the average density decreases from about 1.2kg m^{-3} at the surface to 0.7kg m^{-3} at 5000m (approximately 16,000ft), close to the extreme limit of human habitation.

Pressure is measured as a force per unit area. A force of 10^5 newtons acting on 1m^2, corresponds to the Pascal (Pa) which is the Système International (SI) unit of pressure. Meteorologists still commonly use the millibar (mb) unit; 1 millibar = 10^2Pa (or 1hPa; h = hecto) (see Appendix 2). Pressure readings are made with a mercury barometer, which in effect measures the height of the column of mercury that the atmosphere is able to support in a vertical glass tube. The closed upper end of the tube has a vacuum space and its open lower end is immersed in a cistern of

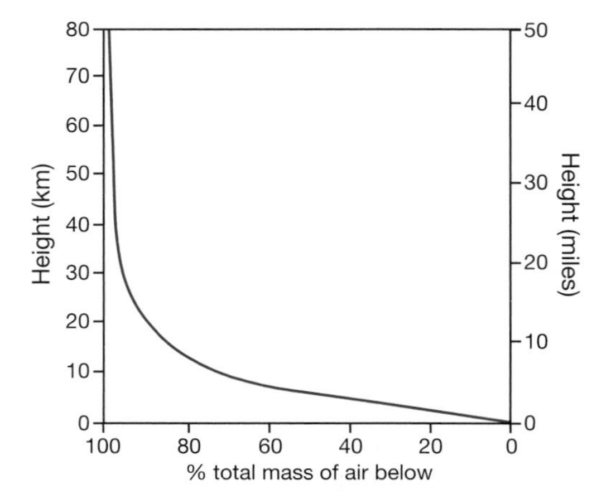

Figure 2.13 The percentage of the total mass of the atmosphere lying below elevations up to 80km (50 miles). This illustrates the shallow character of the earth's atmosphere.

mercury. By exerting pressure downward on the surface of mercury in the cistern, the atmosphere is able to support a mercury column in the tube of about 760mm (29.9in or approximately 1013mb). The weight of air on a surface at sea level is about 10,000kg per square meter.

Pressures are standardized in three ways. The readings from a mercury barometer are adjusted to correspond to those for a standard temperature of 0°C (to allow for the thermal expansion of mercury); they are referred to a standard gravity value of $9.81ms^{-2}$ at 45° latitude (to allow for the slight latitudinal variation in g from $9.78ms^{-2}$ at the equator to $9.83ms^{-2}$ at the poles); and they are calculated for mean sea level to eliminate the effect of station elevation. This third correction is the most significant, because near sea-level pressure decreases with height by about 1mb per 8m. A fictitious temperature between the station and sea level has to be assumed and in mountain areas this commonly causes bias in the calculated mean sea-level pressure (see Note 4).

The mean sea-level pressure (p_0) can be estimated from the total mass of the atmosphere (M, the mean acceleration due to gravity (g_0) and the mean earth radius (R)):

$$P_0 = g_0 (M/4 \pi R_E^2)$$

where the denominator is the surface area of a spherical earth. Substituting appropriate values into this expression: M = 5.14 × 1018kg, g_0 = $9.8ms^{-2}$, $R_E = 6.36 \times 10^6m$, we find $p_0 = 10^5kg\ ms^{-2}$ = $105Nm^{-2}$, or 10^5 Pascals. Hence the mean sea-level pressure is approximately 10^5 Pa or 1000mb. The global mean value is 1013.25mb. On average, nitrogen contributes about 760mb, oxygen 240mb and water vapor 10mb. In other words, each gas exerts a partial pressure independent of the others.

Atmospheric pressure, depending as it does on the weight of the overlying atmosphere, decreases logarithmically with height. This relationship is expressed by the *hydrostatic equation*:

$$\frac{\partial p}{\partial z} = -g\rho$$

i.e., the rate of change of pressure (p) with height (z) is dependent on gravity (g) multiplied by the air density (ρ). With increasing height, the drop in air density causes a decline in this rate of pressure decrease. The temperature of the air also affects this rate, which is greater for cold dense air (see Chapter 7A.1). The relationship between pressure and height is so significant that meteorologists often express elevations in millibars: 1000mb represents sea level, 500mb about 5500m and 300mb about 9000m. A conversion nomogram for an idealized (standard) atmosphere is given in Appendix 2.

2 Vapor pressure

At any given temperature, there is a limit to the density of water vapor in the air, with a consequent upper limit to the vapor pressure, termed the *saturation vapor pressure* (e_s). Figure 2.14A illustrates how e_s increases with temperature (the Clausius–Clapeyron relationship), reaching a maximum of 1013mb (1 atmosphere) at boiling point. Attempts to introduce more vapor into the air when the vapor pressure is at saturation produce condensation of an equivalent amount of vapor. Figure 2.14B shows that whereas the saturation vapor pressure has a single value at any temperature above freezing point, below 0°C the saturation vapor pressure above an ice surface is lower than that above a supercooled water surface. The significance of this will be discussed in Chapter 5D.1.

Vapor pressure (e) varies with latitude and season from about 0.2mb over northern Siberia in January to over 30mb in the tropics in July, but this is not reflected in the pattern of surface pressure. Pressure decreases at the surface when some of the overlying air is displaced horizontally, and in fact the air in high pressure areas is generally dry owing to dynamic factors, particularly vertical air motion (see Chapter 7A.1), whereas air in low pressure areas is usually moist.

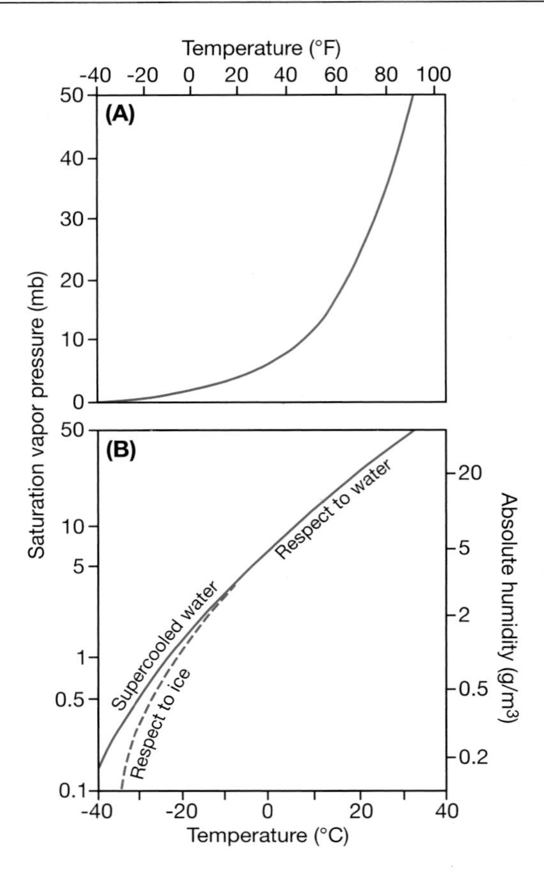

Figure 2.14 Plots of saturation vapor pressure as a function of temperature (i.e., the dew-point curve). A The semi-logarithmic plot. B shows that below 0°C the atmospheric saturation vapor pressure is less with respect to an ice surface than with respect to a water drop. Thus, condensation may take place on an ice crystal at lower air humidity than is necessary for the growth of water drops.

C THE LAYERING OF THE ATMOSPHERE

The atmosphere can be divided conveniently into a number of rather well-marked horizontal layers, mainly on the basis of temperature (Figure 2.15). The evidence for this structure comes from regular rawinsonde (radar wind-sounding) balloons, radio wave investigations, and, more recently, from rocket flights and satellite sounding systems. There are three relatively warm layers (near the surface; between 50 and 60km; and above about 120km) separated by two relatively cold layers (between 10 and 30km; and 80–100km). Mean January and July temperature sections illustrate the considerable latitudinal variations and seasonal trends that complicate the scheme (see Figure 2.15).

1 Troposphere

The lowest layer of the atmosphere is called the troposphere. It is the zone where weather phenomena and atmospheric turbulence are most marked, and it contains 75 percent of the total molecular or gaseous mass of the atmosphere and virtually all the water vapor and aerosols. Throughout this layer, there is a general decrease of temperature with height at a mean rate of about 6.5°C/km. The decrease occurs because air is compressible and its density decreases with height, allowing rising air to expand and thereby cool. In addition, turbulent heat transfer from the surface mainly heats the lower atmosphere, not direct absorption of radiation. The troposphere is capped in most places by a temperature inversion level (i.e., a layer of relatively warm air above a colder layer) and in others by a zone that is isothermal with height. The troposphere thus remains to a large extent self-contained, because the inversion acts as a 'lid' that effectively limits convection (see Chapter 4E). This inversion level or weather ceiling is called the *tropopause* (see Note 5 and Box 2.2). Its height is not constant in either space or time. It seems that the height of the tropopause at any point is correlated with sea-level temperature and pressure, which are in turn related to the factors of latitude, season and daily changes in surface pressure. There are marked variations in the altitude of the tropopause with latitude (Figure 2.16), from about 16km at the equator, where there is strong heating and vertical convective turbulence, to only 8km at the poles.

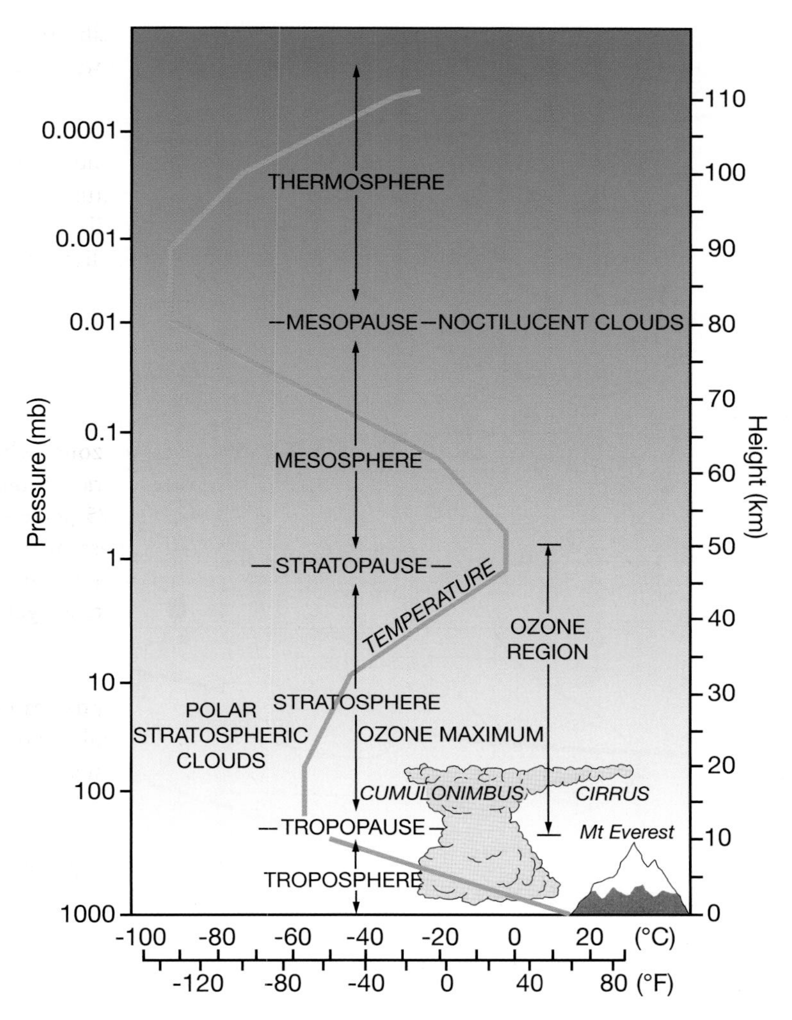

Figure 2.15 The generalized vertical distribution of temperature and pressure up to about 110km. Note particularly the tropopause and the zone of maximum ozone concentration with the warm layer above. The typical altitudes of polar stratospheric and noctilucent clouds are indicated.

Source: After NASA (n.d.). Courtesy NASA.

The equator–pole (meridional) temperature gradients in the troposphere in summer and winter are roughly parallel, as are the tropopauses (see Figure 2.16), and the strong lower mid-latitude temperature gradient in the troposphere is reflected in the tropopause breaks (see also Figure 7.8). In these zones, important interchange may occur between the troposphere and stratosphere, and vice versa. Traces of water vapor can penetrate into the stratosphere by this means, while dry, ozone-rich stratospheric air may be brought down into the mid-latitude troposphere. Thus above-average concentrations of ozone are observed in the rear of mid-latitude low pressure systems where the tropopause elevation tends to be low. Both facts are probably the result of stratospheric subsidence, which warms the lower stratosphere and causes downward transfer of the ozone.

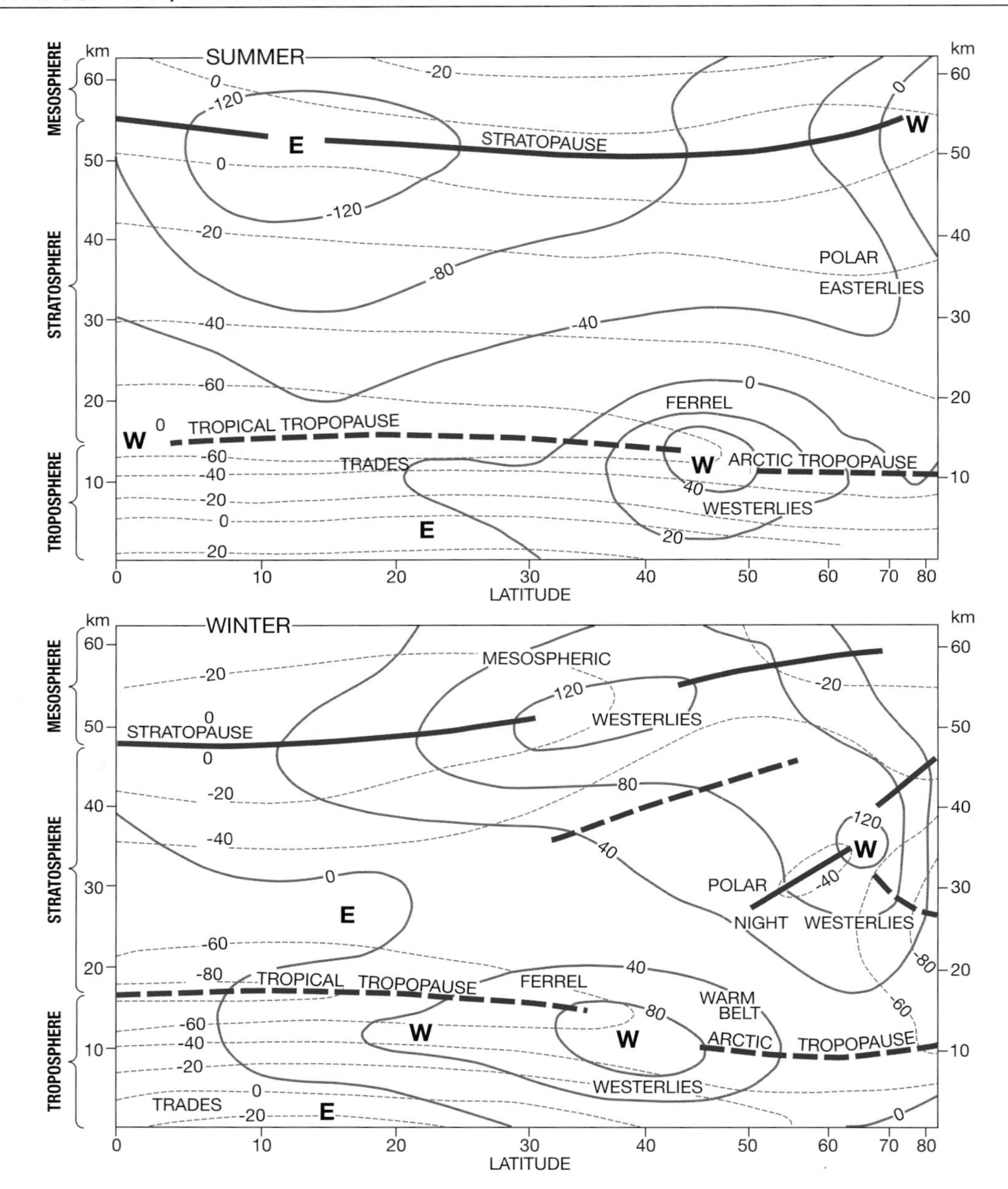

Figure 2.16 Mean zonal (westerly) winds (solid isolines, in knots; negative values from the east) and temperatures (in °C, dashed isolines), showing the broken tropopause near the mean Ferrel jet stream.

Source: After Boville (from Hare 1962).

Notes: The term 'Ferrel Westerlies' was proposed by F. K. Hare in honor of W. Ferrel (see p. 179). The heavy black lines denote reversals of the vertfical temperature gradient of the tropopause and stratopause. Summer and winter refer to the Northern Hemisphere.

SIGNIFICANT 20TH-CENTURY ADVANCE

2.2 Discovery of the tropopause and stratosphere

Early scientific exploration of the upper atmosphere began with manned balloon flights in the mid-nineteenth century. Notable among these was the ascent by Glaisher and Cox in 1862. Glaisher lost consciousness due to lack of oxygen at about 8800m altitude and they barely survived the hypoxia. In 1902 Teisserenc de Bort in France reported a totally unexpected finding; that temperatures ceased decreasing at altitudes of around 12km. Indeed, at higher elevations temperatures were commonly observed to begin increasing with altitude. This means structure is shown in Figure 2.13.

The terms troposphere (turbulent sphere) and stratosphere (stratified sphere) were proposed by Teisserenc de Bort in 1908; the use of 'tropopause' to denote the inversion or isothermal layer separating them was introduced in Great Britain during World War I. The distinctive features of the stratosphere are its stability compared with the troposphere, its dryness, and its high concentration of ozone.

2 Stratosphere

The stratosphere extends upward from the tropopause to about 50km and accounts for about 10 percent of the atmospheric mass. Although the stratosphere contains much of the total atmospheric ozone (it reaches a peak density at approximately 22km), maximum temperatures associated with the absorption of the sun's ultraviolet radiation by ozone occur at the *stratopause* where they may exceed 0°C (see Figure 2.15). The air density is much less here, so even limited absorption produces a large temperature rise. Temperatures increase fairly generally with height in summer, with the coldest air at the equatorial tropopause. In winter, the structure is more complex, with very low temperatures, averaging –80°C, at the equatorial tropopause, which is highest at this season. Similar low temperatures are found in the middle stratosphere at high latitudes, whereas over 50–60°N there is a marked warm region with nearly isothermal conditions at about –45 to –50°C. In the circumpolar low pressure vortex over both polar regions, polar stratospheric clouds (PSCs) are sometimes present at 20–30km altitude. These have a nacreous ('mother-of-pearl') appearance. They can absorb odd nitrogen and thereby cause catalytic destruction of ozone.

Marked seasonal changes of temperature affect the stratosphere. The cold 'polar night' winter stratosphere in the Arctic often undergoes dramatic *sudden warmings* associated with subsidence due to circulation changes in late winter or early spring, when temperatures at about 25km may jump from –80 to –40°C over a two-day period. The autumn cooling is a more gradual process. In the tropical stratosphere there is a quasi-biennial (26-month) wind regime, with easterlies in the layer 18 to 30km for 12 to 13 months, followed by westerlies for a similar period. The reversal begins first at high levels and takes approximately 12 months to descend from 30 to 18km (10 to 60mb).

How far events in the stratosphere are linked with temperature and circulation changes in the troposphere remains a topic of meteorological research. Any such interactions are undoubtedly complex

3 Mesosphere

Above the stratopause, average temperatures decrease to a minimum of about –133°C (140K)

or around 90km (Figure 2.15). This layer is commonly termed the mesosphere, although as yet there is no universal terminology for the upper atmospheric layers. Pressure is very low in the mesosphere, decreasing from about 1mb at 50km to 0.01mb at 90km. Above 80km, temperatures again begin rising with height and this inversion is referred to as the 'mesopause'. Molecular oxygen and ozone absorption bands contribute to heating around 85km altitude. It is in this region that *noctilucent clouds* are observed on summer 'nights', particularly over high latitudes at 80–90km altitude. Their presence appears to be due to meteoric dust particles, which act as ice crystal nuclei when traces of water vapor are carried upward by high-level convection caused by the vertical decrease of temperature in the mesosphere. However, their formation may also be related to the production of water vapor through the oxidation of atmospheric methane, since apparently they were not observed prior to the Industrial Revolution. The layers between the tropopause and the lower thermosphere are commonly referred to as the *middle atmosphere*, with the upper atmosphere designating the regions above about 100km altitude.

4 Thermosphere

Above the mesopause, atmospheric densities are extremely low, although the tenuous atmosphere still effects drag on space vehicles above 250km. The lower portion of the thermosphere is composed mainly of nitrogen (N_2) and oxygen in molecular (O_2) and atomic (O) forms, whereas above 200km atomic oxygen predominates over nitrogen (N_2 and N). Temperatures rise with height, owing to the absorption of extreme ultraviolet radiation (0.125–0.205μm) by molecular and atomic oxygen, probably approaching 800–1200K at 350km, but these temperatures are essentially theoretical. For example, artificial satellites do not acquire such temperatures owing to the rarefied air. 'Temperatures' in the upper thermosphere and exosphere undergo wide diurnal and seasonal variations. They are higher by day and are also higher during a sunspot maximum, although the changes are only represented in varying velocities of the sparse air molecules.

Above 100km, cosmic radiation, solar X-rays and ultraviolet radiation increasingly affect the atmosphere, which cause *ionization*, or electrical charging, by separating negatively charged electrons from neutral oxygen atoms and nitrogen molecules, leaving the atom or molecule with a net positive charge (an *ion*). The term *ionosphere* is commonly applied to the layers above 80km. The Aurora Borealis and Aurora Australis are produced by the penetration of ionizing particles through the atmosphere from about 300km to 80km, particularly in zones about 10–20° latitude from the earth's magnetic poles. On occasion, however, aurora may appear at heights up to 1000km, demonstrating the immense extension of a rarefied atmosphere.

5 Exosphere and magnetosphere

The base of the exosphere is between about 500km and 750km. Here atoms of oxygen, hydrogen and helium (about 1 percent of which are ionized) form the tenuous atmosphere, and the gas laws (see B, this chapter) cease to be valid. Neutral helium and hydrogen atoms, which have low atomic weights, can escape into space since the chance of molecular collisions deflecting them downward becomes less with increasing height. Hydrogen is replaced by the breakdown of water vapor and methane (CH_4) near the mesopause, while helium is produced by the action of cosmic radiation on nitrogen and from the slow but steady breakdown of radioactive elements in the earth's crust.

Ionized particles increase in frequency through the exosphere and, beyond about 200km, in the magnetosphere there are only electrons (negative) and protons (positive) derived from the solar wind – which is a plasma of electrically conducting gas.

SUMMARY

The atmosphere is a mixture of gases with constant proportions of up to 80km or more. The exceptions are ozone, which is concentrated in the lower stratosphere, and water vapor in the lower troposphere. The principal greenhouse gas is water vapor. Carbon dioxide, methane and other trace gases have increased significantly since the Industrial Revolution, especially in the twentieth century, due to the combustion of fossil fuels, industrial processes, and other anthropogenic effects, but larger natural fluctuations occurred during the geologic past.

Reactive gases include nitrogen and sulfur and chlorine species. These play important roles in acid precipitation and ozone destruction. Acid precipitation (by wet or dry deposition) results from the reaction of cloud droplets with emissions of SO_2 and NO*x*. There are large geographical variations in acid deposition. The processes leading to the destruction of stratospheric ozone are complex, but the roles of nitrogen oxides and chlorine radicals are highly important in causing polar ozone holes. Aerosols in the atmosphere originate from natural and anthropogenic sources and they play an important but complex role in climate.

Air is highly compressible, so that half of its mass occurs in the lowest 5km, and pressure decreases logarithmically with height from an average sea-level value of 1013mb. The vertical structure of the atmosphere comprises three relatively warm layers – the lower troposphere, the stratopause and the upper thermosphere – separated by a cold layer above the tropopause (in the lower stratosphere), and the mesopause. The temperature profile is determined by atmospheric absorption of solar radiation, and the decrease of density with height.

DISCUSSION TOPICS

- What properties distinguish the different layers of the atmosphere?
- What differences would exist in a dry atmosphere compared with the real atmosphere?
- What role is played by water vapor, ozone, carbon dioxide, methane and CFCs in the radiation balance of the atmosphere?
- Given the strong pressure gradient upward from the surface, why is there no large-scale upward flow of air?

REFERENCES AND FURTHER READING

Books

Andreae, M. O. and Schimel, D. S. (1989) *Exchange of Trace Gases Between Terrestrial Ecosystems and the Atmosphere*, J. Wiley & Sons, Chichester, 347pp. [Detailed technical treatment]

Bolin, B., Degens, E. T., Kempe, S. and Ketner, P. (eds) (1979) *The Global Carbon Cycle* (SCOPE 13), J. Wiley & Sons, Chichester, 528pp. [Important early overview]

Bolin, B., Döös, B. R., Jäger, J. and Warrick, R. A. (eds) (1986) *The Greenhouse Effect, Climatic Change, and Ecosystems* (SCOPE 29), J. Wiley & Sons, Chichester, 541pp.

Bridgman, H. A. (1990) *Global Air Pollution: Problems for the 1990s*, Belhaven Press, London, 201pp. [Broad survey of air pollution causes and processes by a geographer; includes greenhouse gases]

Brimblecombe, P. (1986) *Air: Composition and Chemistry*, Cambridge University Press, Cambridge, 224pp.

Craig, R. A. (1965) *The Upper Atmosphere: Meteorology and Physics*, Academic Press, New York, 509pp. [Classic text on the upper atmosphere, prior to the recognition of the ozone problem]

Crowley, T. J. and North, G. R. (1991) *Paleoclimatology*, Oxford University Press, New York and Oxford, 339pp. [Thorough, modern overview of climate history]

Houghton, J. T., Jenkins, G. J. and Ephraums, J. J. (eds) (1990) *Climate Change: The IPCC Scientific Assessment*, Cambridge University Press, Cambridge, 365pp. [The first comprehensive assessment of global climate change]

Kellogg, W. W. and Schware, R. (1981) *Climate Change and Society*, Westview Press, Boulder, CO., 178pp. [Early coverage of the societal implications of climate change]

NERC (1989) *Our Future World: Global Environmental Research*, NERC, London, 28pp. [Brief overview of major issues]

Rex, D. F. (ed.) (1969) *Climate of the Free Atmosphere, Vol.1: World Survey of Climatology*, Elsevier, Amsterdam, 450pp. [Useful reference on atmospheric structure and the stratosphere]

Roland, F. S. and Isaksen, I. S. A. (eds) (1988) *The Changing Atmosphere*, J. Wiley & Sons, Chichester, 296pp. [Treats atmospheric chemistry, especially trace gases, aerosols, tropospheric pollution and acidification]

Russell, F. A. R. and Archibold, E. D.(1888) *The Eruption of Krakatoa and Subsequent Phenomena*, Report of the Krakatoa Committee of the Royal Society, London.

Simkin, T. and Fiske, R. S. (1983*) Krakatau 1883: The Volcanic Eruption and Its Effects*, Washington, DC: Smithsonian Institution Press, 464pp.

Journal articles

Bach, W. (1976) Global air pollution and climatic change. *Rev. Geophys. Space Phys.* 14, 429–74.

Bojkov, R. D. and Fioletov, V. E. (1995) Estimating the global ozone characteristics during the last 30 years. *J. Geophys. Res.* 100(D8), 16537–551.

Bojkov, R. D. *et al.* (2002) Vertical ozone distribution characteristics deduced from ~ 44,000 re-evaluated Umkehr profiles (1957–2000). *Met. Atmos. Phys.* 79(3–4), 1217–58.

Bolle, H-J., Seiler, W. and Bolin, B. (1986) Other greenhouse gases and aerosols, in Bolin, B. *et al.* (eds) *The Greenhouse Effect, Climatic Change, and Ecosystems*, J. Wiley & Sons, Chichester, 157–203.

Brugge, R. (1996) Back to basics: atmospheric stability. Part I – Basic concepts. *Weather* 51(4), 134–40.

Defant, F. R. and Taba, H. (1957) The threefold structure of the atmosphere and the characteristics of the tropopause. *Tellus* 9, 259–74.

Ghan, S. J. and Schwartz, S. E. (2007) Aerosol properties and processes: a path from field and laboratory measurements to global climate models. *Bull. Amer. Met. Soc.* 88(7): 1059–83.

Hales, J. (1996) Scientific background for AMS policy statement on atmospheric ozone. *Bull. Amer. Met. Soc.* 77(6), 1249–53.

Hare, F. K. (1962) The stratosphere. *Geog. Rev.* 52, 525–47.

Hastenrath, S. L. (1968) Der regionale und jahrzeithliche Wandel des vertikalen Temperaturgradienten und seine Behandlung als Wärmhaushaltsproblem. *Meteorologische Rundschau* 1, 46–51.

Husar, R. B. *et al.* (2001), Asian dust events of April 1998. *J. Geophys. Res.* 106(D16), 18317–330.

Jiang, Y. B., Yung, Y. L. and Zurek, R. W. (1996) Decadal evolution of the Antarctic ozone hole. *J. Geophys. Res.* 101(D4), 8985–9000.

Kondratyev, K. Y. and Moskalenko, N. I. (1984) The role of carbon dioxide and other minor gaseous components and aerosols in the radiation budget, in Houghton, J. T. (ed.) *The Global Climate*, Cambridge University Press, Cambridge, 225–33.

LaMarche, V. C., Jr. and Hirschboeck, K. K. (1984) Frost rings in trees as records of major volcanic eruptions. *Nature* 307, 121–6.

Lashof, D. A. and Ahnja, D. R. (1990) Relative contributions of greenhouse gas emissions to global warming. *Nature* 344, 529–31.

London, J. (1985) The observed distribution of atmospheric ozone and its variations, in Whitten, R. C. and Prasad, T. S. (eds) *Ozone in the Free Atmosphere*, Van Nostrand Reinhold, New York, 11–80.

Mason, B. J. (1990) Acid rain – cause and consequence. *Weather* 45 , 70–9.

McElroy, M. B. and Salawitch, R. J. (1989) Changing composition of the global stratosphere. *Science*, 243, 763–70.

Machta, L. (1972) The role of the oceans and biosphere in the carbon dioxide cycle, in

Dyrssen, D. and Jagner, D. (eds) *The Changing Chemistry of the Oceans*, Nobel Symposium 20, Wiley, New York, 121–45.

Neuendorffer, A. C. (1996) Ozone monitoring with the TIROS-N operational vertical sounders. *J. Geophys. Res.* 101(D13), 8807–28.

Paffen, K. (1967) Das Verhältniss der Tages – zur Jahreszeitlichen Temperaturschwankung. *Erdkunde* 21, 94–111.

Pearce, F. (1989) Methane: the hidden greenhouse gas. *New Scientist* 122, 37–41.

Plass, G. M. (1959) Carbon dioxide and climate. *Sci. American* 201, 41–7.

Prather, M. and Enhalt, D. (eds) (2001) Atmospheric chemistry and greenhouse gases, in Houghton, J.T. *et al. Climate Change 2001: The Scientific Basis,* Cambridge University Press, Cambridge, pp.239–87.

Prospero, J. M. (2001) African dust in America. *Geotimes* 46(11), 24–7.

Ramanathan, V., Cicerone, R. J., Singh, H. B. and Kiehl, J. T. (1985) Trace gas trends and their potential role in climatic change, *J. Geophys. Res.* 90(D3), 5547–66.

Ramanathan, V., Crutzen, P. J., Kiehl, J. T. and Rosenfeld, D. (2001). Aerosols, climate, and the hydrologic cycle. *Science* 294 (5549): 2119 -24.

Rampino, M. R. and Self, S. (1984) The atmospheric effects of El Chichón. *Sci. American* 250(1), 34–43.

Raval, A. and Ramanathan, V. (1989) Observational determination of the greenhouse effect. *Nature* 342, 758–61.

Robock, A. and Matson, M. (1983) Circumglobal transport of the El Chichón volcanic dust cloud. *Science* 221, 195–7.

Rodhe, H. (1990) A comparison of the contribution of various gases to the greenhouse effect. *Science* 244, 763–70.

Schimel, D. *et al.* (1996) Radiative forcing of climate change, in Houghton, J. T. *et al.* (eds) *Climate Change 1995: The Science of Climate Change,* Cambridge University Press, Cambridge, 65–131.

Shanklin, J. D. (2001). Back to basics: the ozone hole. *Weather,* 56, 222–30.

Shine, K. (1990) Effects of CFC substitutes. *Nature* 344, 492–3.

Slinn, W. G. N. (1983) Air-to-sea transfer of particles, in Liss, P. S. and Slinn, W. G. N. (eds) *Air–Sea Exchange of Gases and Particles,* D. Reidel, Dordrecht, 299–407.

Solomon, S. (1988) The mystery of the Antarctic ozone hole. *Rev. Geophys.* 26, 131–48.

Staehelin, J., Harris, N. R. P., Appenzeller, C. and Eberhard, J. (2001). Ozone trends: a review. *Rev. Geophys.* 39(2), 231–90.

Strangeways, I. (2002) Back to basics: the 'met. enclosure': Part 8(a) – barometric pressure, mercury barometers. *Weather* 57(4), 132–9.

Tetlow-Smith, A. (1995) Environmental factors affecting global atmospheric methane concentrations. *Prog. Phys. Geog.* 19, 322–35.

Thompson, R. D. (1995) The impact of atmospheric aerosols on global climate. *Prog. Phys. Geog.* 19, 336–50.

Trenberth, K. E., Houghton, J. T. and Meira Filho, L. G. (1996) The climate system: an overview, in Houghton, J. T. *et al.* (eds) *Climate Change 1995: The Science of Climate Change,* Cambridge University Press, Cambridge, 51–64.

Webb, A. R. (1995) To burn or not to burn. *Weather* 50(5), 150–4.

World Meteorological Organization (1964) Regional basic networks. *WMO Bulletin* 13, 146–7.

Zielinski, G. A. (1995) Stratospheric loading and optical depth estimates of explosive volcanism over the last 2100 years derived from the Greenland Ice Sheet Project 2 ice core. *J. Geophys. Res.* 100(20),937–955.

CHAPTER THREE

Solar radiation and the global energy budget

3

LEARNING OBJECTIVES

When you have read this chapter you will:

- know the characteristics of solar radiation and the electromagnetic spectrum
- know the effects of the atmosphere on solar and terrestrial radiation
- understand the cause of the atmospheric greenhouse effect
- understand the earth's heat budget and the importance of horizontal transfers of energy as sensible and latent heat.

This chapter describes how radiation from the sun enters the atmosphere and reaches the surface. The effects on solar radiation of absorbing gases and the scattering effects of aerosols are examined. Then terrestrial longwave (infrared) radiation is discussed in order to explain the radiation balance. At the surface, an energy balance exists due to the additional transfers of sensible and latent heat to the atmosphere. The effects of heating on surface temperature characteristics are then presented.

A SOLAR RADIATION

The source of the energy injected into our atmosphere is the sun, which is continually shedding part of its mass by radiating waves of electromagnetic energy and high-energy particles into space. This constant emission represents all the energy available to the earth (except for a small amount emanating from the radioactive decay of earth minerals). The amount of energy received at the top of the atmosphere is affected by four factors: solar output, the sun–earth distance, the altitude of the sun, and day length.

1 Solar output

Solar energy originates from nuclear reactions within the sun's hot core (16×10^6K), and is transmitted to the sun's surface by radiation and hydrogen convection. Visible solar radiation (light) comes from a 'cool' (~ 6000K) outer surface layer called the *photosphere*. Temperatures rise again in the outer chromosphere (10,000K) and corona (10^6K), which is continually expanding into space. The outflowing hot gases (plasma) from the sun, referred to as the *solar wind* (with a speed of 1.5×10^6km hr^{-1}), interact with the earth's magnetic field and upper atmosphere. The earth intercepts both the normal electromagnetic radiation and energetic particles emitted during solar flares.

The sun behaves virtually as a *black body;* i.e., it absorbs all energy received and in turn radiates energy at the maximum rate possible for a given temperature. The energy emitted at a particular wavelength by a perfect radiator of given temperature is described by a relationship due to Max Planck. The black-body curves in Figure 3.1 illustrate this relationship. The area under each curve gives the total energy emitted by a black body (F); its value is found by integration of Planck's equation, known as Stefan's Law:

$$F = \sigma T^4$$

where $\sigma = 5.67 \times 10^{-8}$W m^{-2} K^{-4} (the Stefan–Boltzmann constant), i.e., the energy emitted is proportional to the fourth power of the absolute temperature of the body (T).

The total solar output to space, assuming a temperature of 5760K for the sun, is 3.84×10^{26}W, but only a tiny fraction of this is intercepted by the earth, because the energy received is inversely proportional to the square of the solar distance (150 million km). The energy received at the top of the atmosphere on a surface perpendicular to the solar beam for mean solar distance is termed the *solar constant* (see Note 1). Satellite measurements since 1980 indicate a value of about 1366W m^{-2}, with an absolute uncertainty of about ±2W m^{-2}. Figure 3.1 shows the wavelength range of solar (shortwave) radiation and the infrared (longwave) radiation emitted by the earth and atmosphere. For solar radiation, about 8 percent is ultraviolet (0.2–0.4μm), 40 percent visible light (0.4–0.7μm) and 52 percent near-infrared (>0.7μm); (1μm = 1 micrometer = 10^{-6}m). The figure illustrates the black-body radiation curves for 6000K at the top of the atmosphere (which slightly exceeds the observed extraterrestrial radiation), for 300K, and for 263K. The mean temperature of the earth's surface is about 288K (15°C) and of the atmosphere about 250K (–23°C). Gases do not behave as black bodies, and Figure 3.1 shows the absorption bands in the atmosphere, which cause its emission to be much less than that from an equivalent black body. The wavelength of maximum emission (λmax) varies inversely with the absolute temperature of the radiating body:

$$\lambda\text{max} = \frac{2897}{T} 10^{-6}\text{m (Wien's Law)}$$

Thus solar radiation is very intense and is mainly shortwave between about 0.2 and 4.0μm, with a maximum (per unit wavelength) at 0.5μm because $T \sim 6000$K. The much weaker terrestrial radiation with $T \approx 280$K has a peak intensity at about 10μm and a range from about 4 to 100μm.

The solar constant undergoes small periodic variations of just over 1W m^{-2} related to sunspot activity. Sunspot number and positions change in a regular manner, known as sunspot cycles. Satellite measurements during the latest cycle show a small decrease in solar output as sunspot number approached its *minimum*, and a subsequent recovery. *Sunspots* are dark (i.e., cooler) areas visible on the sun's surface. Although sunspots are cool, bright areas of activity known as *faculae* (or *plages*), that have higher temperatures, surround them (Plate 3.1). The net effect is for solar output to vary in parallel with the number of sunspots. Thus, the solar 'irradiance' decreases by about 1.1W m^{-2} from sunspot maximum to minimum. Sunspot cycles have wavelengths averaging 11 years (the Schwabe cycle, varying between 8 and 13 years), the 22-year (Hale) magnetic cycle, much less importantly 37.2 years (18.6 years – the luni–solar oscillation), and 88 years (Gleissberg). Figure 3.2 shows the estimated variation of sunspot activity since 1610. Between the thirteenth and eighteenth centuries sunspot activity was generally low, except during AD 1350–1400 and 1600–1645. Output within the ultraviolet part of the spectrum shows considerable variability, with up to 20 times more ultraviolet radiation emitted at certain wavelengths during a sunspot maximum than a minimum.

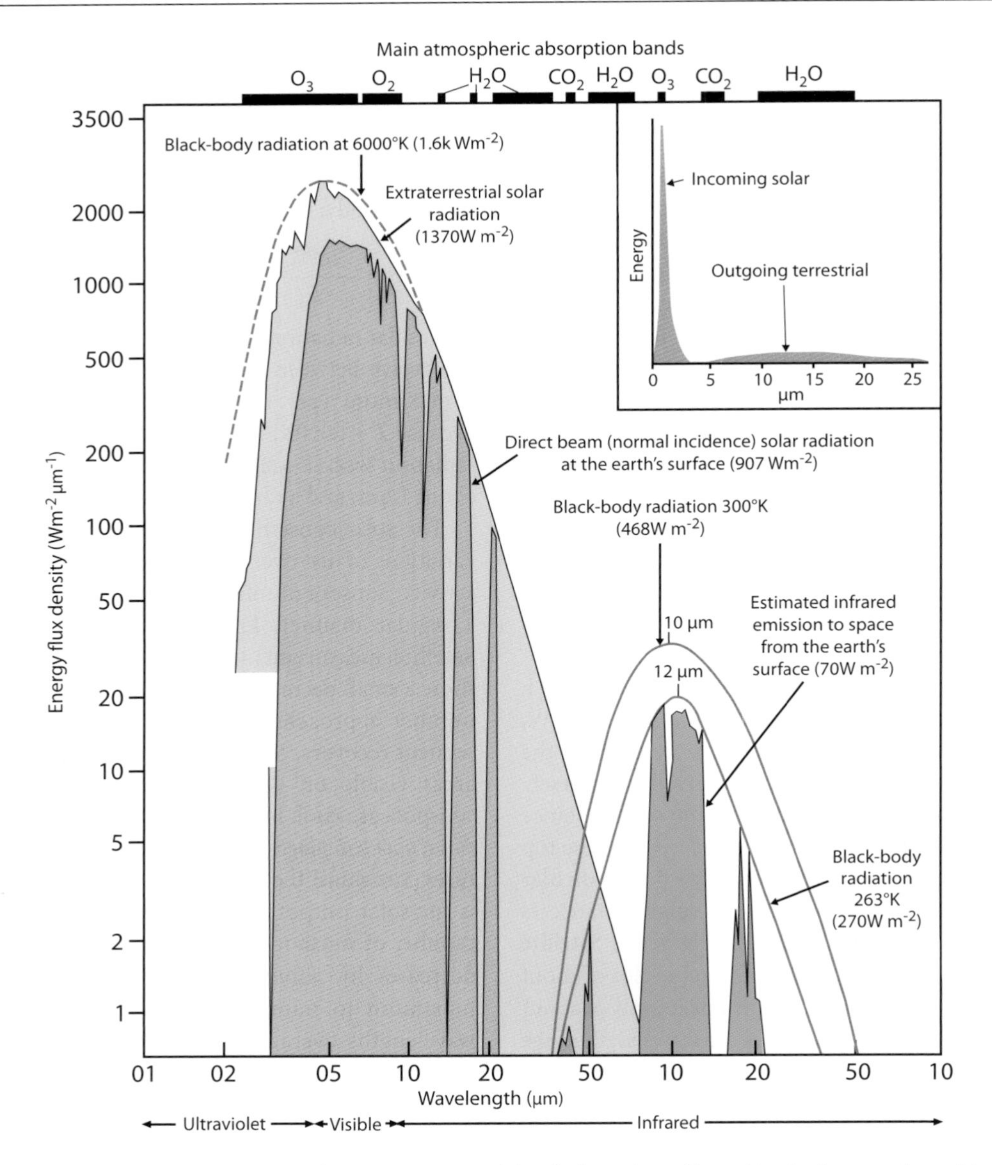

Figure 3.1 Spectral distribution of solar and terrestrial radiation, plotted logarithmically, together with the main atmospheric absorption bands due to trace gases (top). The cross-hatched areas in the infrared spectrum indicate the 'atmospheric windows' where radiation escapes to space. The black-body radiation at 6000K is that proportion of the flux which would be incident on the top of the atmosphere. The inset shows the same curves for incoming and outgoing radiation with the wavelength plotted arithmetically on an arbitrary vertical scale.

Source: Mostly after Sellers (1965). Courtesy of University of Chicago Press.

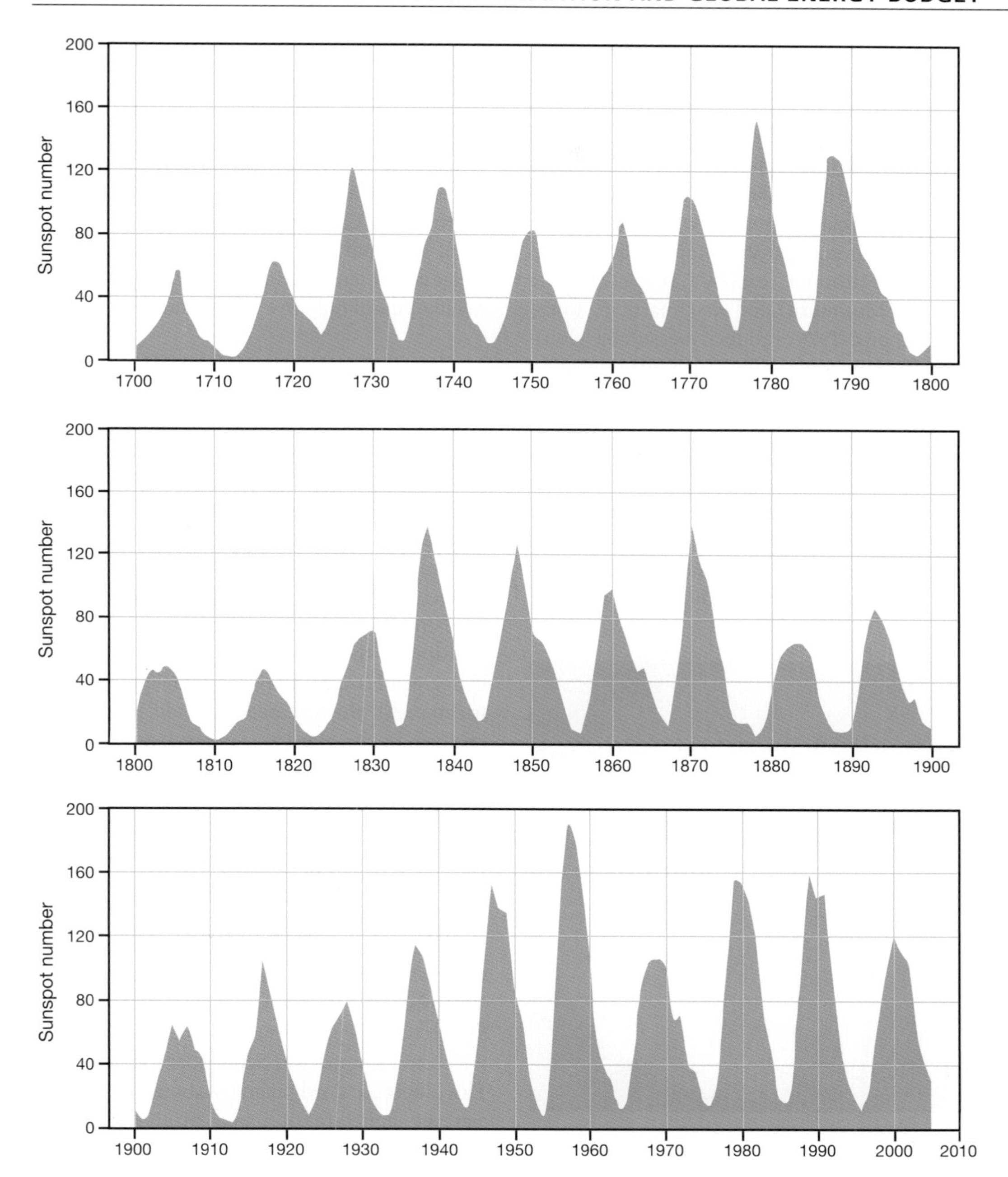

Figure 3.2 Yearly sunspot numbers for the sun's visible surface for the period 1700–2005.

Sources: Reproduced by courtesy of the National Geophysical Data Center, NOAA, Boulder, Colorado.

How to translate sunspot activity into solar radiation and terrestrial temperatures is a matter of some dispute. It has been suggested that the sun is more active when the sunspot cycle length is short, but this is disputed. However, anomalies of temperature over Northern Hemisphere land areas do correlate inversely with cycle length between 1860 and 1985. Prolonged time-spans of sunspot minima (e.g., AD 1645–1715, the Maunder Minimum) and maxima (e.g., 1895–1940 and post-1970) produce measurable global cooling and warming, respectively. Solar radiation may have been reduced by 0.25 percent during the Maunder Minimum. It is suggested that almost three-quarters of the variations in global temperature between 1610 and 1800 were attributable to fluctuations in solar radiation and during the twentieth century there is evidence for a modest contribution from solar forcing. Shorter term relationships are more difficult to support, but mean annual temperatures have been correlated with the combined 10–11 and 18.6-year solar cycles. Assuming that the earth behaves as a black body, a persistent anomaly of 1 percent in the solar constant could change the effective mean temperature of the earth's surface by as much as 0.6°C. However, the observed fluctuations of about 0.1 percent would change the mean global temperature by $\leq$0.06°C (based on calculations of radiative equilibrium).

Plate 3.1 A Hydrogen-Alpha image of the sun on 22 January 1990 from the spectroheliograph at the Observatoire de Paris. The bright, active areas are termed plages and the dark bands are filaments related to magnetic fields (sunspots are apparent only in visible light).

Source: Courtesy of the National Geophysical Data Center, NOAA.

2 Distance from the sun

The annually changing distance of the earth from the sun produces seasonal variations in solar energy received by the earth. Owing to the eccentricity of the earth's orbit around the sun, the receipt of solar energy on a surface normal to the beam is 7 percent more on 3 January at the perihelion than on 4 July at the aphelion (Figure 3.3). In theory (that is, discounting the interposition of the atmosphere and the difference in degree of conductivity between large land and sea masses), this difference should produce an increase in the effective January world surface temperatures of about 4°C over those of July. It should also make northern winters warmer than those in the Southern Hemisphere, and southern summers warmer than those in the Northern Hemisphere. In practice, atmospheric heat circulation and the effects of continentality mask this global tendency, and the actual seasonal contrast between the hemispheres is reversed. Moreover, the northern summer half-year (21 March to 22 September) is five days longer than the austral summer (22 September to 21 March). This difference slowly changes; about 10,000 years ago the aphelion occurred in the Northern Hemisphere winter, and northern summers received 3–4 percent more radiation than today (Figure 3.3B). This same pattern will return about 10,000 years from now.

Figure 3.4 graphically illustrates the seasonal variations of energy receipt with latitude. Actual amounts of radiation received on a horizontal

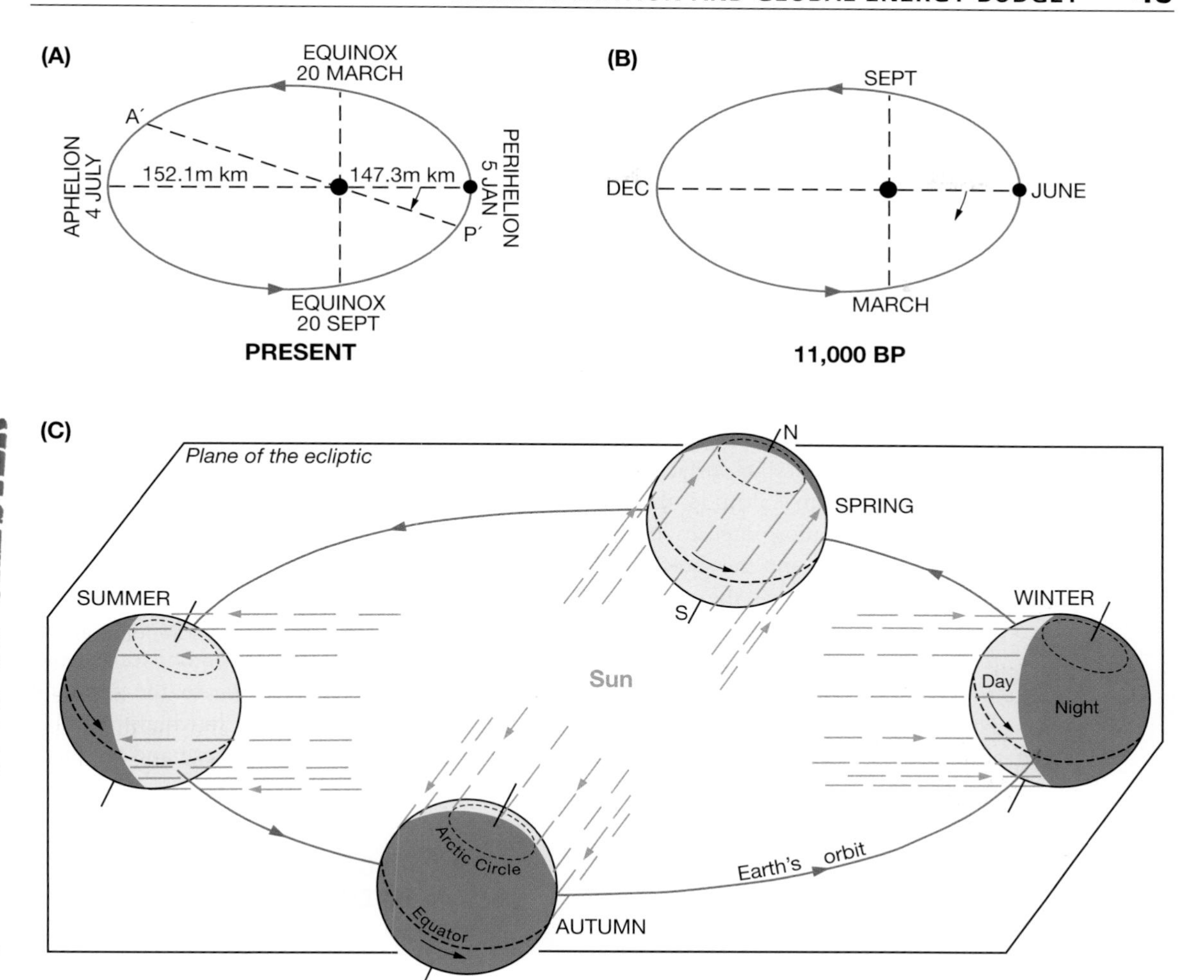

Figure 3.3 Perihelion shifts. A The present timing of perihelion; B The direction of its shift and the situation at 11,000 years BP; C The geometry of the present seasons (Northern Hemisphere).
Source: Partly after Strahler (1965).

LIBRARY, UNIVERSITY OF CHESTER

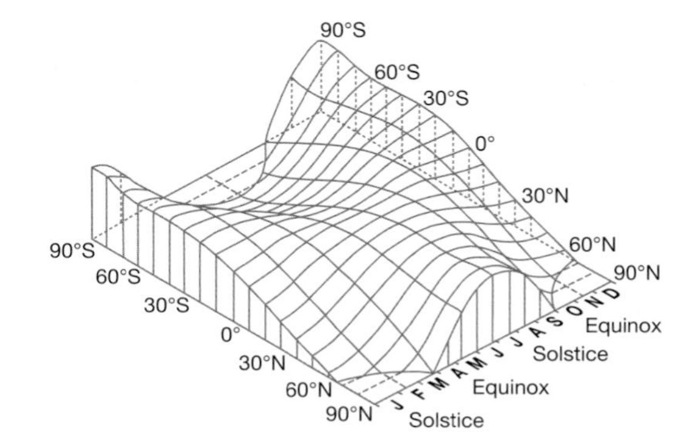

Figure 3.4 The variations of solar radiation with latitude and season for the whole globe, assuming no atmosphere. This assumption explains the abnormally high amounts of radiation received at the poles in summer, when daylight lasts for 24 hours each day.
Source: After W. M. Davis.

surface outside the atmosphere are given in Table 3.1. The intensity on a horizontal surface (I_h) is determined from:

$$I_h = I_0 \sin d$$

where I_0 = the solar constant and d = the angle between the surface and the solar beam.

3 Altitude of the sun

The altitude of the sun (i.e., the angle between its rays and a tangent to the earth's surface at the point of observation) also affects the amount of solar radiation received at the surface of the earth. The greater the sun's altitude, the more concentrated is the radiation intensity per unit area at the earth's surface and the shorter is the path length of the beam through the atmosphere, which decreases the atmospheric absorption. There are, in addition, important variations with solar altitude of the proportion of radiation reflected by the surface, particularly in the case of a water surface (see B.5, this chapter). The principal factors that determine the sun's altitude are, of course, the latitude of the site, the time of day and the season (see Figure 3.3). At the June solstice, the sun's altitude is a constant 23 1⁄2° throughout the day at the North Pole and the sun is directly overhead at noon at the Tropic of Cancer (23 1⁄2°N).

4 Length of day

The length of daylight also affects the amount of radiation that is received. Obviously, the longer the time that the sun shines the greater is the quantity of radiation that a given portion of the earth will receive. At the equator, for example, the day length is close to 12 hours in all months, whereas at the poles it varies between 0 and 24 hours from winter (polar night) to summer (see Figure 3.3).

The combination of all of these factors produces the pattern of receipt of solar energy at the top of the atmosphere shown in Figure 3.4. The polar regions receive their maximum amounts of solar radiation during their summer solstices, which is the period of continuous day. The amount received during the December solstice in the Southern Hemisphere is theoretically greater than that received by the Northern Hemisphere during the June solstice, due to the previously mentioned elliptical path of the earth around the sun (see Table 3.1). The equator has two radiation maxima at the equinoxes and two minima at the solstices, due to the apparent passage of the sun during its double annual movement between the Northern and Southern Hemispheres.

B SURFACE RECEIPT OF SOLAR RADIATION AND ITS EFFECTS

1 Energy transfer within the earth–atmosphere system

So far, we have described the distribution of solar radiation as if it were all available at the earth's surface. This is, of course, unrealistic owing to the

Table 3.1 Daily solar radiation on a horizontal surface outside the atmosphere: W m^{-2}.

Date	90°N	70	50	30	0	30	50	70	90°S
Dec 21	0	0	86	227	410	507	514	526	559
Mar 21	0	149	280	378	436	378	280	149	0
June 22	524	492	482	474	384	213	80	0	0
Sept 23	0	147	276	373	430	372	276	147	0

Source: After Berger (1996).

effect of the atmosphere on energy transfer. Heat energy can be transferred by three mechanisms:

1 *Radiation*: Electromagnetic waves transfer energy (both heat and light) between two bodies, without the necessary aid of an intervening material medium, at a speed of 300×10^6m s^{-1} (i.e., the speed of light). This is so with solar energy through space, whereas the earth's atmosphere allows the passage of radiation only at certain wavelengths and restricts that at others.

   Radiation entering the atmosphere may be absorbed in certain wavelengths by atmospheric gases but, as shown in Figure 3.1, most shortwave radiation is transmitted without absorption. Scattering occurs if the direction of a photon of radiation is changed by interaction with atmospheric gases and aerosols. Two types of scattering are distinguished. For gas molecules smaller than the radiation wavelength (λ), *Rayleigh scattering* occurs in all directions (i.e., it is *isotropic*) and is proportional to ($1/\lambda^4$). As a result, the scattering of blue light ($\lambda \sim 0.4\mu$m) is an order of magnitude (i.e., × 10) greater than that of red light ($\lambda \sim 0.7\mu$m), thus creating the daytime blue sky. However, when water droplets or aerosol particles, with similar sizes (0.1–0.5μm radius) to the radiation wavelength, are present, most of the light is scattered forward. This *Mie scattering* gives the greyish appearance of polluted atmospheres.

   Within a cloud, or between low clouds and a snow-covered surface, radiation undergoes multiple scattering. In the latter case, the 'white-out' conditions typical of polar regions in summer (and mid-latitude snowstorms) are experienced, when surface features and the horizon become indistinguishable.

2 *Conduction*: By this mechanism, heat passes through a substance from a warmer to a colder part through the transfer of adjacent molecular vibrations. Air is a poor conductor so this type of heat transfer is negligible in the atmosphere, but it is important in the ground. The thermal conductivity increases as the water content of a given soil increases and is greater in frozen than in unfrozen soil.

3 *Convection*: This occurs in fluids (including gases) that are able to circulate internally and distribute heated parts of the mass. It is the chief means of atmospheric heat transfer due to the low viscosity of air and its almost continual motion. *Forced convection* (mechanical turbulence) occurs when eddies form in airflow over uneven surfaces. In the presence of surface heating, *free* (thermal) *convection* develops.

Convection transfers energy in two forms. The first is the *sensible heat* content of the air (called enthalpy by physicists), which is transferred directly by the rising and mixing of warmed air. It is defined as c_pT, where T is the temperature and c_p (= 1004J kg^{-1} K^{-1}) is the specific heat at constant pressure (the heat absorbed by unit mass for unit temperature increase). Sensible heat is also transferred by conduction. The second form of energy transfer by convection is indirect, involving *latent heat*. Here, there is a phase change but no temperature change. Whenever water is converted into water vapor by evaporation (or boiling), heat is required. This is referred to as the latent heat of vaporization (L). At 0°C, L is 2.50×10^6J kg^{-1} of water. More generally,

$$L\,(10^6 J\,kg^{-1}) = (2.5 - 0.00235T)$$

where T is in °C. When water condenses in the atmosphere (see Chapter 4D), the same amount of latent heat is given off as is used for evaporation *at the same temperature*. Similarly, for melting ice at 0°C, the latent heat of fusion is required, which is 0.335×10^6J kg^{-1}. If ice evaporates without melting, the latent heat of this sublimation process is 2.83×10^6J kg^{-1} at 0°C (i.e., the sum of the latent heats of melting and vaporization). In all of these phase changes of water there is an energy transfer. We discuss other aspects of these processes in Chapter 4.

2 Effect of the atmosphere

Solar radiation is virtually all in the short wavelength range, less than 4μm (see Figure 3.1). About 18 percent of the incoming energy is absorbed directly by ozone and water vapor. Ozone absorption is concentrated in three solar spectral bands (0.20–0.31, 0.31–0.35 and 0.45–0.85μm), while water vapor absorbs to a lesser degree in several bands between 0.9 and 2.1μm (see Figure 3.1). Solar wavelengths shorter than 0.285μm scarcely penetrate below 20km altitude, whereas those >0.295μm reach the surface. Thus the 3mm (equivalent) column of stratospheric ozone attenuates ultraviolet radiation almost entirely, except for a partial window around 0.20μm, where radiation reaches the lower stratosphere. About 30 percent of incoming solar radiation is immediately reflected back into space from the atmosphere, clouds and the earth's surface, leaving approximately 70 percent to heat the earth and its atmosphere. The surface absorbs almost half of the incoming energy available at the top of the atmosphere and re-radiates it outwards as long (infrared) waves of greater than 3μm (see Figure 2.1). Much of this re-radiated longwave energy is then absorbed by the water vapor, carbon dioxide and ozone in the atmosphere, the rest escaping through atmospheric *windows* back into outer space, principally between 8 and 13μm (see Figure 3.1). This retention of energy by the atmosphere is vital to most lifeforms, because otherwise the average temperature of the earth's surface would fall by some 40°C!

The atmospheric scattering, noted above, gives rise to *diffuse* (or sky) radiation and this is sometimes measured separately from the direct beam radiation. On average, under cloud-free conditions the ratio of diffuse to total (or global) solar radiation is about 0.15–0.20 at the surface. For average cloudiness, the ratio is about 0.5 at the surface, decreasing to around 0.1 at 4km, as a result of the decrease in cloud droplets and aerosols with altitude. During a total solar eclipse experienced over much of western Europe in August 1999, the elimination of direct beam radiation caused diffuse radiation to drop from 680W m^{-2} at 10.30 a.m. to only 14W m^{-2} at 11.00 a.m.at Bracknell in southern England.

Between 1961 and the 1990, solar radiation receipts globally decreased by some 4 percent, a phenomenon termed 'global dimming'; amounts then recovered again in the 1990s ('brightening'). The reason for these trends appears to have been increased aerosol absorption (by black carbon) and backscatter (by sulfates, nitrate and dust) during the first period and a decrease in aerosol loading subsequently. Sulfate aerosols have a direct radiative forcing globally of – 0.4W m^{-2}, fossil fuel black carbon + 0.2W m^{-2}, and mineral dust – 0.1W m^{-2}, out of a total aerosol direct effect of – 0.5W m^{-2}. There is also an indirect effect on clouds whereby aerosols increase the

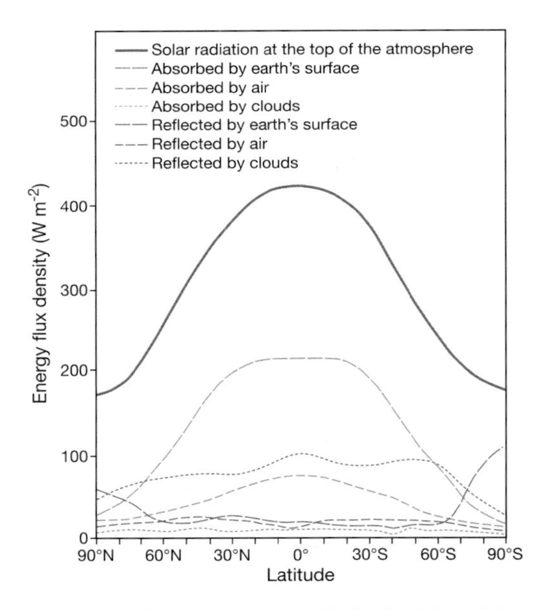

Figure 3.5 The average annual latitudinal disposition of solar radiation in W m^{-2}. Of 100 percent radiation entering the top of the atmosphere, about 20 percent is reflected back to space by clouds, 3 percent by air (plus dust and water vapor), and 8 percent by the earth's surface. Three percent is absorbed by clouds, 18 percent by the air and 48 percent by the earth.

Source: After Sellers (1965). Courtesy of University of Chicago Press.

number of cloud water droplets and increase the cloud albedo, giving a cooling effect of about $-0.7 W m^{-2}$.

Figure 3.5 illustrates the relative roles of the atmosphere, clouds and the earth's surface in reflecting and absorbing solar radiation at different latitudes. (A more complete analysis of the heat budget of the earth–atmosphere system is given in D, this chapter.)

3 Effect of cloud cover

Thick and continuous cloud cover forms a significant barrier to the penetration of radiation. The drop in surface temperature often experienced on a sunny day when a cloud temporarily cuts off the direct solar radiation illustrates our reliance upon the sun's radiant energy. How much radiation is actually reflected by clouds depends on the amount of cloud cover and its thickness (Figure 3.6). The proportion of incident radiation that is reflected is termed the *albedo*, or reflection coefficient (expressed as a fraction or percentage). Cloud type affects the albedo. Aircraft measurements show that the albedo of a complete overcast ranges from 44 to 50 percent for cirrostratus to 90 percent for cumulonimbus. Average albedo values, as determined by satellites, aircraft and surface measurements, are summarized in Table 3.2 (see Note 2).

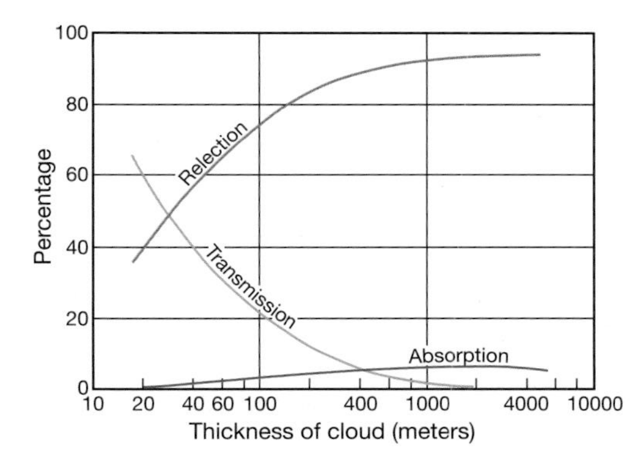

Figure 3.6 Percentage of reflection, absorption and transmission of solar radiation by cloud layers of different thickness.

Source: From Hewson and Longley (1944).

It should be noted that the albedo (α) is defined by the ratio of reflected radiation ($S\uparrow$) to the incoming radiation ($S\downarrow$) received at the top of the atmosphere, at the cloud top, at the top of the vegetation canopy, or at the ground surface:

$\alpha = S\uparrow/S\downarrow$, expressed as a fraction or percent.

The total (or global) solar radiation received at the surface on cloudy days is

$$S = S_0 [b + (1 - b)(1 - c)]$$

where S_0 = global solar radiation for clear skies;
c = cloudiness (fraction of sky covered);
b = a coefficient depending on cloud type and thickness; and the depth of atmosphere through which the radiation must pass.

For mean monthly values for the United States, $b \approx 0.35$, so that

$$S \approx S_0 [1 - 0.65c]$$

Table 3.2 The average (integrated) fractional albedo of various surfaces

Planet earth	0.31
Global surface	0.14–0.16
Global cloud	0.23
Cumulonimbus	0.9
Stratocumulus	0.6
Cirrus	0.4–0.5
Fresh snow	0.8–0.9
Melting snow	0.4–0.6
Sand	0.30–0.35
Grass, cereal crops	0.18–0.25
Deciduous forest	0.15–0.18
Coniferous forest	0.09–0.15
Tropical rainforest	0.07–0.15
Water bodies*	0.06–0.10

Note: *Increases sharply at low solar angles.

The effect of cloud cover also operates in reverse, since it serves to retain much of the heat that would otherwise be lost from the earth by longwave radiation throughout the day and night. In this way, cloud cover appreciably lessens the daily temperature range by preventing high maxima by day and low minima by night. As well as interfering with the transmission of radiation, clouds act as temporary thermal reservoirs because they absorb a certain proportion of the energy they intercept. The modest effects of cloud reflection and absorption of solar radiation are illustrated in figures 3.5 to 3.7.

Global cloudiness is not yet accurately known. Ground-based observations are mostly at land stations and refer to a small (~ 250km^2) area. Satellite estimates are derived from the reflected shortwave radiation and infrared irradiance measurements, with various threshold assumptions for cloud presence/absence; typically they refer to a grid area of 2500km^2 to 37,500km^2. Surface-based observations tend to be about 10 percent greater than satellite estimates due to the observer's perspective. Average winter and summer distributions of total cloud amount from surface observations are shown in Figure 3.8. The cloudiest areas are the Southern Ocean and the mid- to high-latitude North Pacific and North Atlantic storm tracks. Lowest amounts are over the Saharan–Arabian desert area. Total global cloud cover is just over 60 percent in January and July. The low altitude cloud fraction is shown in Plate 3.2.

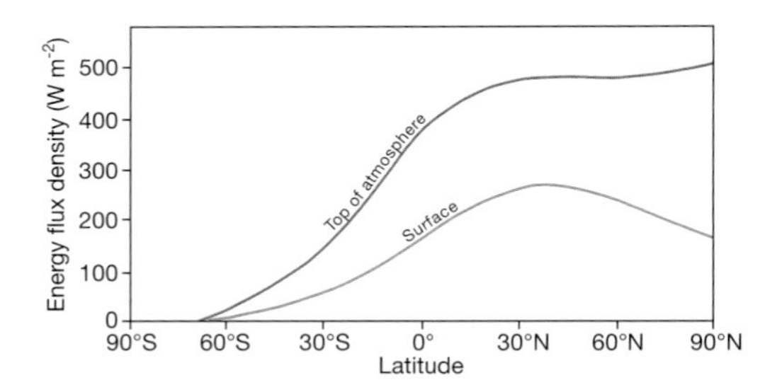

Figure 3.7 The average receipt of solar radiation with latitude at the top of the atmosphere and at the earth's surface during the June solstice.

4 Effect of latitude

As Figure 3.4 has already shown, different parts of the earth's surface receive different amounts of solar radiation. The time of the year is one factor controlling this, more radiation being received in summer than in winter owing to the higher altitude of the sun and the longer days. Latitude is a very important control because this determines the duration of daylight and the distance travelled through the atmosphere by the oblique rays of the sun. However, actual calculations show the effect of the latter to be negligible near the poles, apparently due to the low vapor content of the air limiting tropospheric absorption. Figure 3.7 shows that in the upper atmosphere over the North Pole there is a marked maximum of solar radiation at the June solstice, yet only about 30 percent is absorbed at the surface. This may be compared with the global average of 48 percent of solar radiation being absorbed at the surface. The explanation lies in the high average cloudiness over the Arctic in summer and also in the high reflectivity of the snow and ice surfaces. This example illustrates the complexity of the radiation budget and the need to take into account the interaction of several factors.

A special feature of the latitudinal receipt of radiation is that the maximum temperatures experienced at the earth's surface do not occur at the equator, as one might expect, but at the tropics. A number of factors need to be taken into account. The apparent migration of the vertical sun is relatively rapid during its passage over the equator, but its rate slows down as it reaches the tropics. Between 6°N and 6°S the sun's rays remain almost vertically overhead for only 30 days during each of the spring and autumn equinoxes, allowing little time for any large buildup of surface heat and high temperatures. On the other hand, between 17.5° and 23.5° latitude the sun's rays shine down almost vertically for 86 consecutive days during the period of the solstice. This longer interval, combined with the fact that the tropics experience longer days than at the equator, makes the maximum zones of heating occur nearer

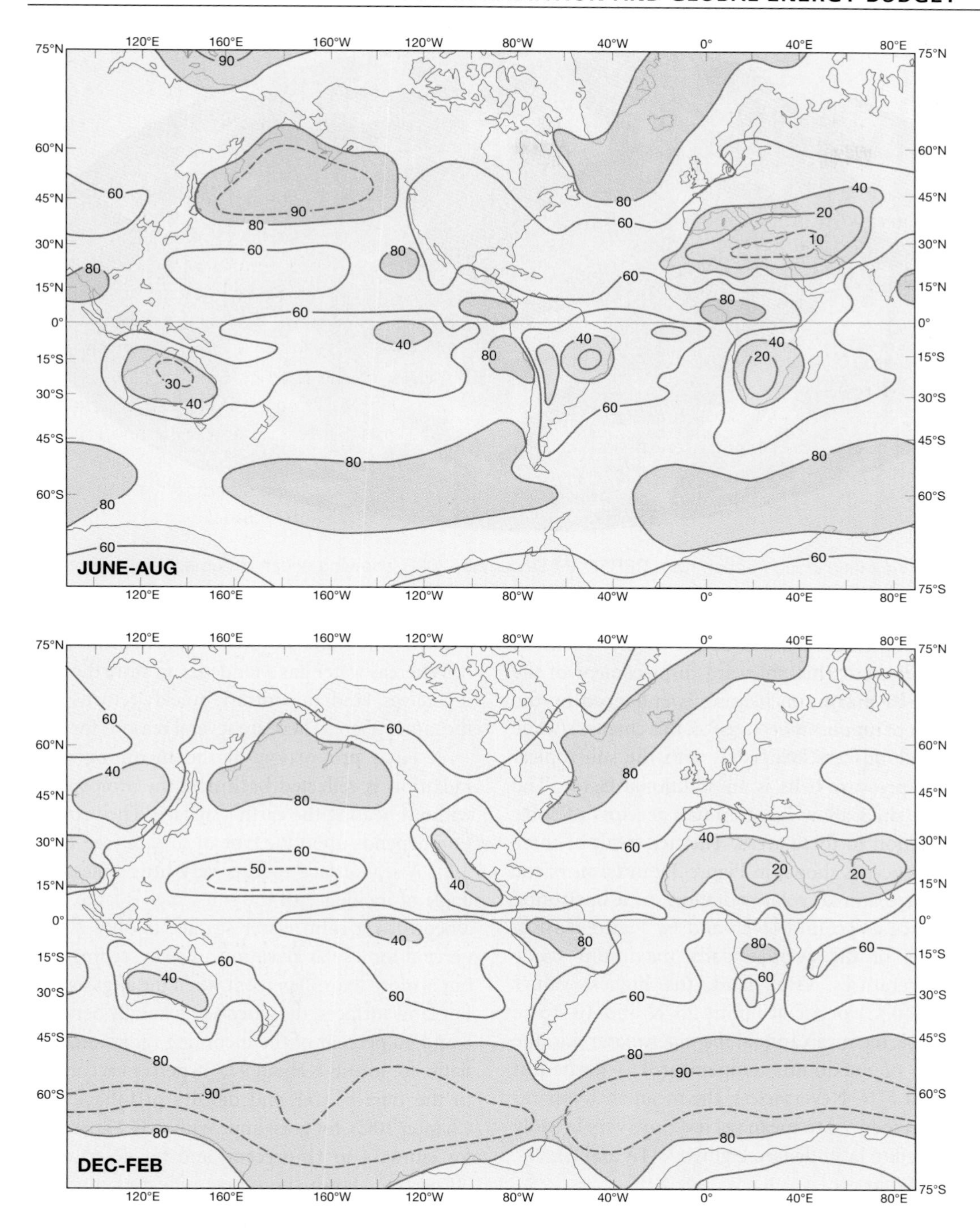

Figure 3.8 The global distribution of total cloud amount (percent) derived from surface-based observations during the period 1971–1981, averaged for the months June to August (above) and December to February (below). High percentages are shaded and low percentages stippled.

Source: From London *et al*. (1989).

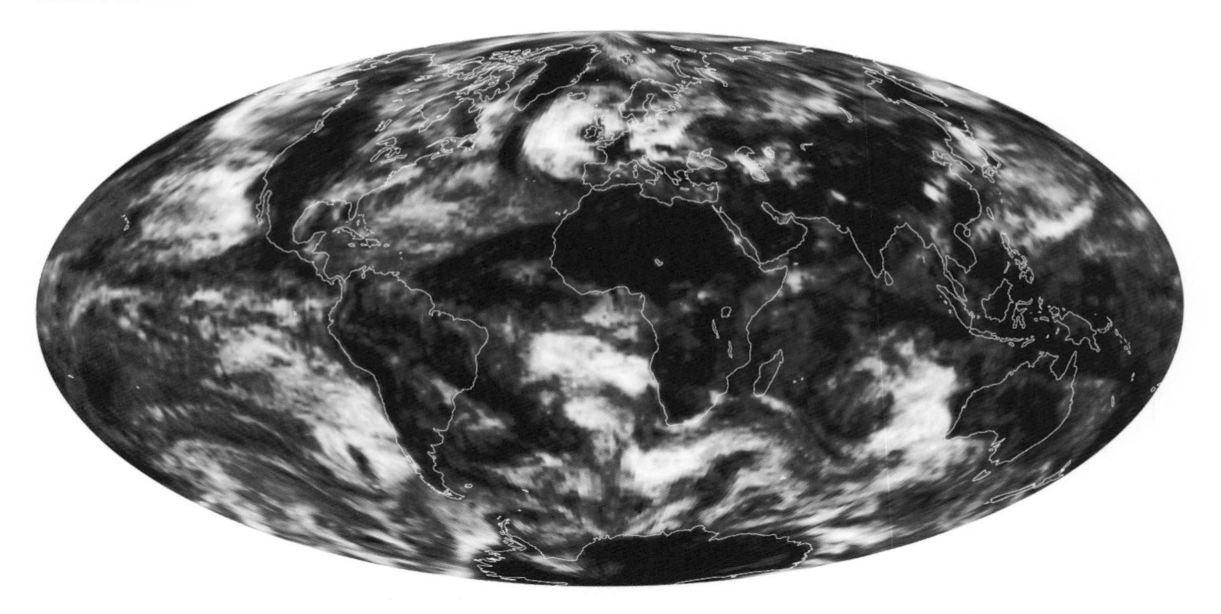

Plate 3.2 Low-altitude cloud fraction from CERES, 27 December 2008, showing ocean maxima.
Source: Jesse Allen, NASA.

the tropics than the equator. In the Northern Hemisphere, this poleward displacement of the zone of maximum heating is enhanced by the effect of *continentality* (see B.5, this chapter), while low cloudiness associated with the subtropical high pressure belts is an additional factor. The clear skies allow large annual receipts of solar radiation in these areas. The net result of these influences is shown in Figure 3.9 in terms of the average annual solar radiation on a horizontal surface at ground level, and by Figure 3.10 in terms of the average daily maximum shade temperatures. Over land, the highest values (38–40°C) occur at about 23°N and 10–15°S. Hence, the mean annual *thermal equator* (i.e., the zone of maximum temperature) is located at about 5°N. Nevertheless, the mean air temperatures, reduced to mean sea level, are very broadly related to latitude (see Figures 3.11A and B).

5 Effect of land and sea

Another important control on the effect of incoming solar radiation stems from the different ways in which land and sea are able to profit from it. Whereas water has a tendency to store the heat it receives, land, in contrast, quickly returns it to the atmosphere. There are several reasons for this.

A large proportion of the incoming solar radiation is reflected back into the atmosphere without heating the earth's surface. The proportion depends upon the type of surface (see Table 3.2). A sea surface reflects very little unless the angle of incidence of the sun's rays is large. The albedo for a calm water surface is only 2 to 3 percent for a solar elevation angle exceeding 60°, but is more than 50 percent when the angle is 15°. For land surfaces, the albedo is generally between 8 and 40 percent of the incoming radiation. The figure for forests is about 9 to 18 percent according to the type of tree and density of foliage (see Chapter 10C), for grass approximately 25 percent, for cities 14 to 18 percent, and for desert sand 30 percent. Fresh snow may reflect as much as 90 percent of solar radiation, but snow cover on vegetated, especially forested, surfaces is much less reflective (30–50 percent). The long duration of snow cover on the northern continents (see

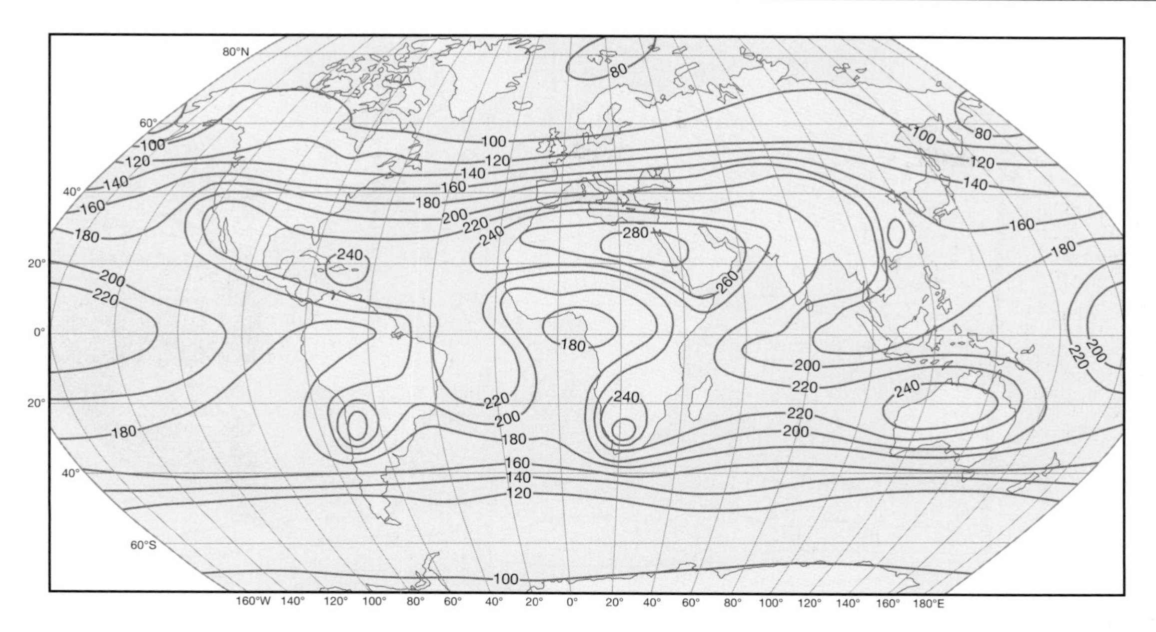

Figure 3.9 The mean annual global solar radiation $(Q + q)(\mathrm{W\ m^{-2}})$(i.e., on a horizontal surface at ground level). Maxima are found in the world's hot deserts, where as much as 80 percent of the solar radiation annually incident on the top of the unusually cloud-free atmosphere reaches the ground.

Source: After Budyko *et al.* (1962).

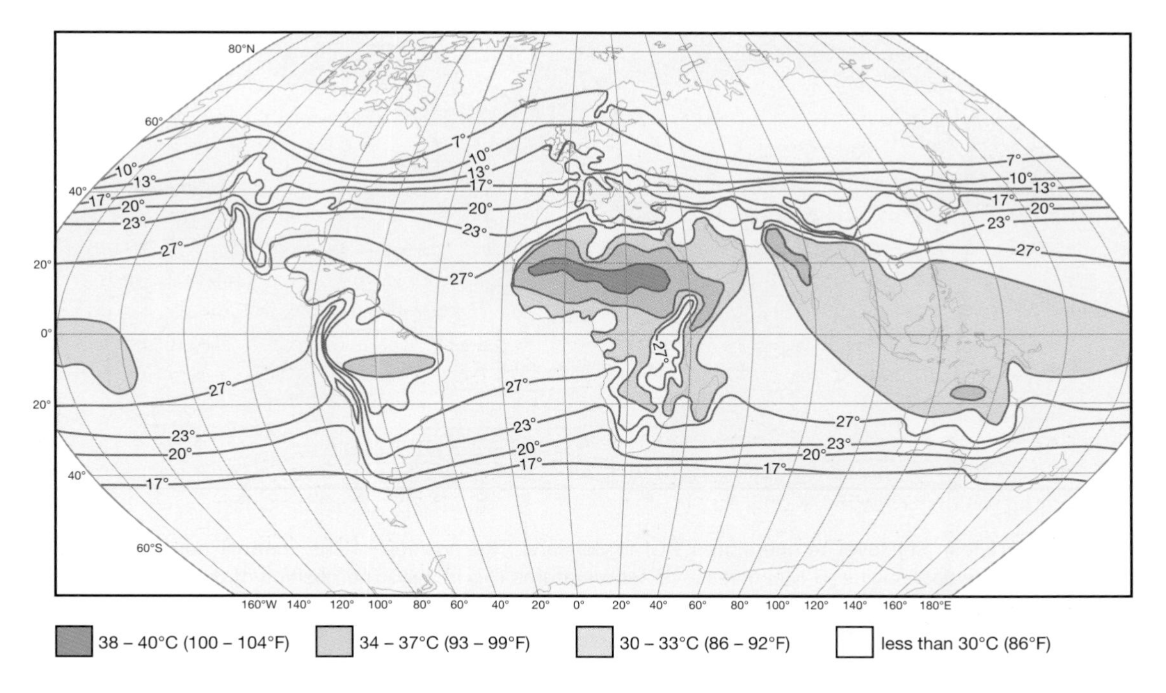

Figure 3.10 Mean daily maximum shade air temperature (°C).

Source: After Ransom (1963). Courtesy of the Royal Meteorological Society.

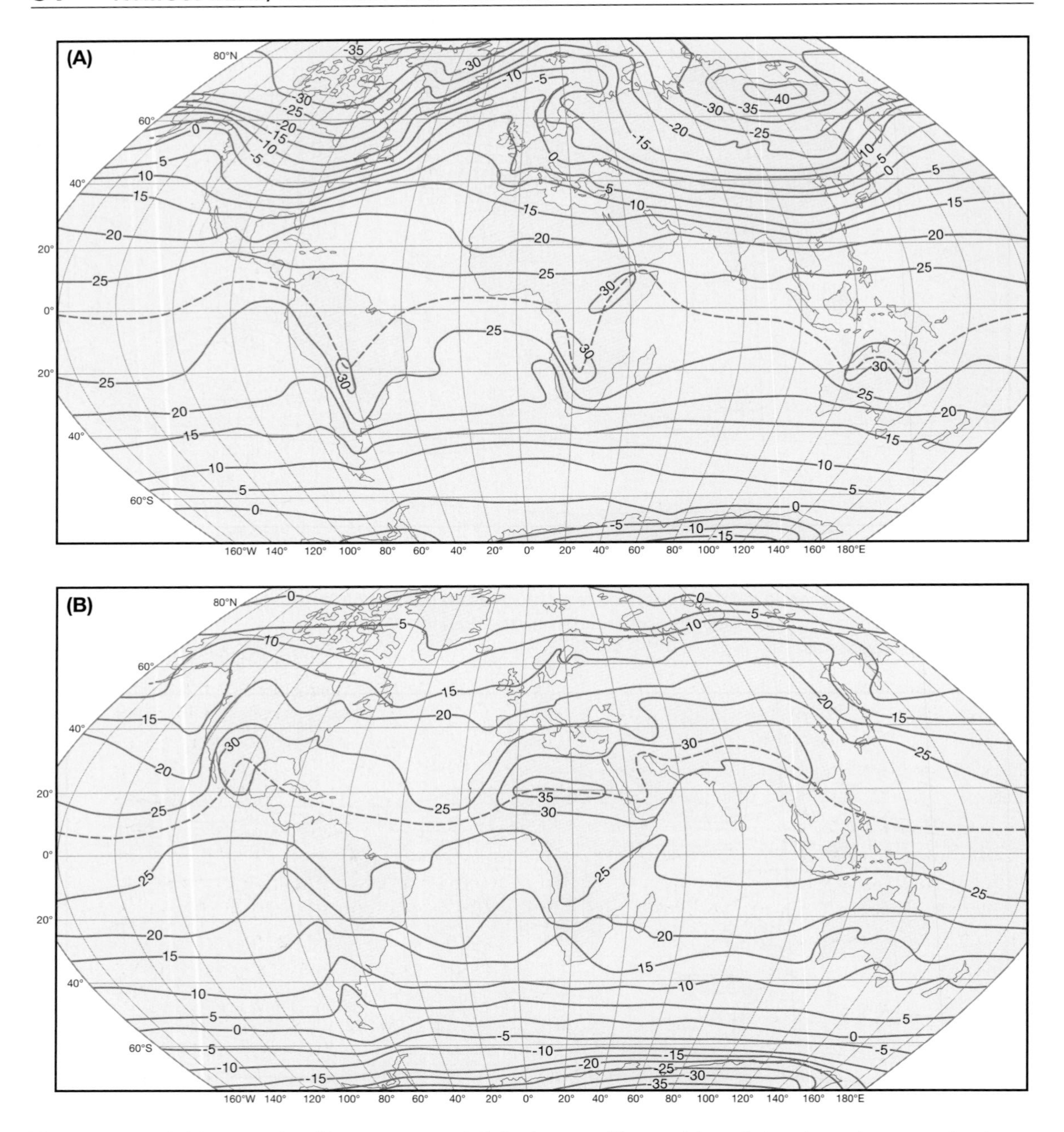

Figure 3.11 A Mean sea-level temperatures (°C) in January. The position of the thermal equator is shown approximately by the dashed line. B Mean sea-level temperatures (°C) in July. The position of the thermal equator is shown approximately by the dashed line.

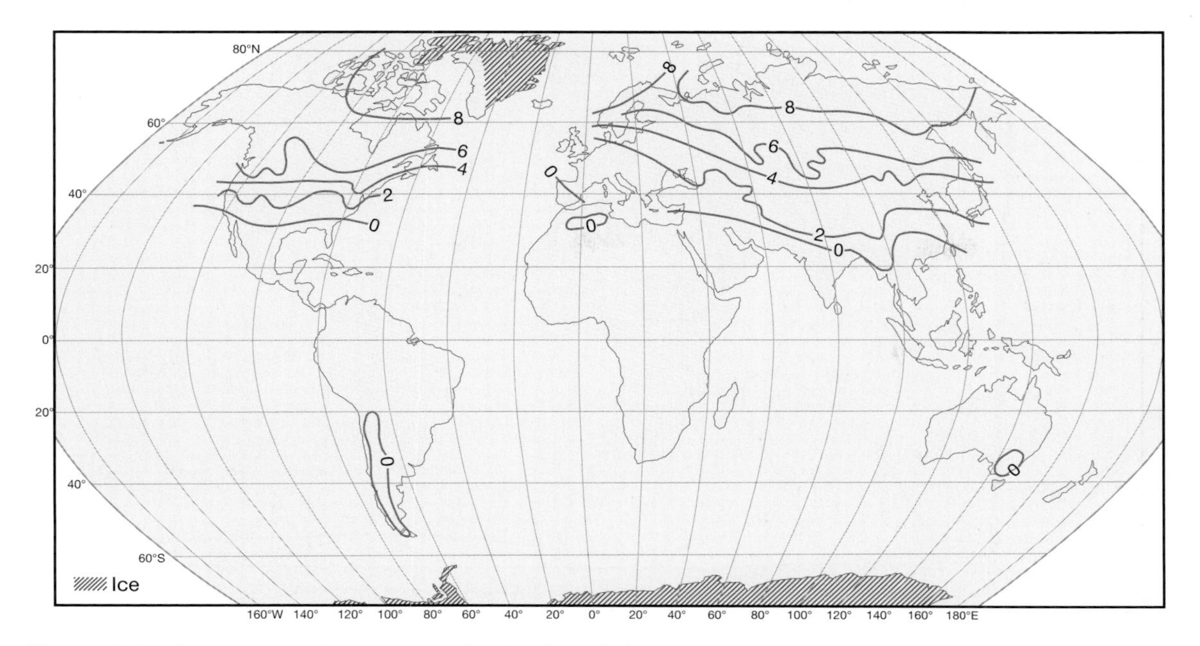

Figure 3.12 Average annual snow cover duration (months).
Source: Henderson-Sellers and Wilson (1983). Courtesy of the American Geophysical Union.

Figure 3.12 and Plate 3.3) causes much of the incoming radiation in winter to spring to be reflected. However, the global distribution of annual average surface albedo (Figure 3.13A) shows mainly the influence of the snow-covered Arctic sea ice and Antarctic ice sheet (cf. Figure 3.13B for planetary albedo).

The global solar radiation absorbed at the surface is determined from measurements of radiation incident on the surface and its albedo (α). It may be expressed as

$$S\downarrow(100 - \alpha)$$

where the albedo is a percentage. A snow cover will absorb only about 15 percent of the incident radiation, whereas for the sea the figure generally exceeds 90 percent. The ability of the sea to absorb the heat received also depends upon its transparency. As much as 20 percent of the radiation penetrates as far down as 9m (30ft). Figure 3.14 illustrates how much energy is absorbed by the sea at different depths. However, the heat absorbed by the sea is carried down to considerable depths by the turbulent mixing of water masses by the action of waves and currents. Figure 3.15, for example, illustrates the mean monthly variations with depth in the upper 100 meters of the waters of the eastern North Pacific (around 50°N, 145°W; it shows the development of the seasonal thermocline under the influences of surface heating, vertical mixing and surface conduction.

A measure of the difference between the subsurfaces of land and sea is given in Figure 3.16, which shows ground temperatures at Kaliningrad (Königsberg) and sea temperature deviations from the annual mean at various depths in the Bay of Biscay. Heat transmission in the soil is carried out almost wholly by conduction, and the degree of conductivity varies with the moisture content and porosity of each particular soil.

Air is an extremely poor conductor, and for this reason a loose, sandy soil surface heats up rapidly by day, as the heat is not conducted away. Increased soil moisture tends to raise the conductivity by filling the soil pores, but too much moisture increases the soil's heat capacity, thereby

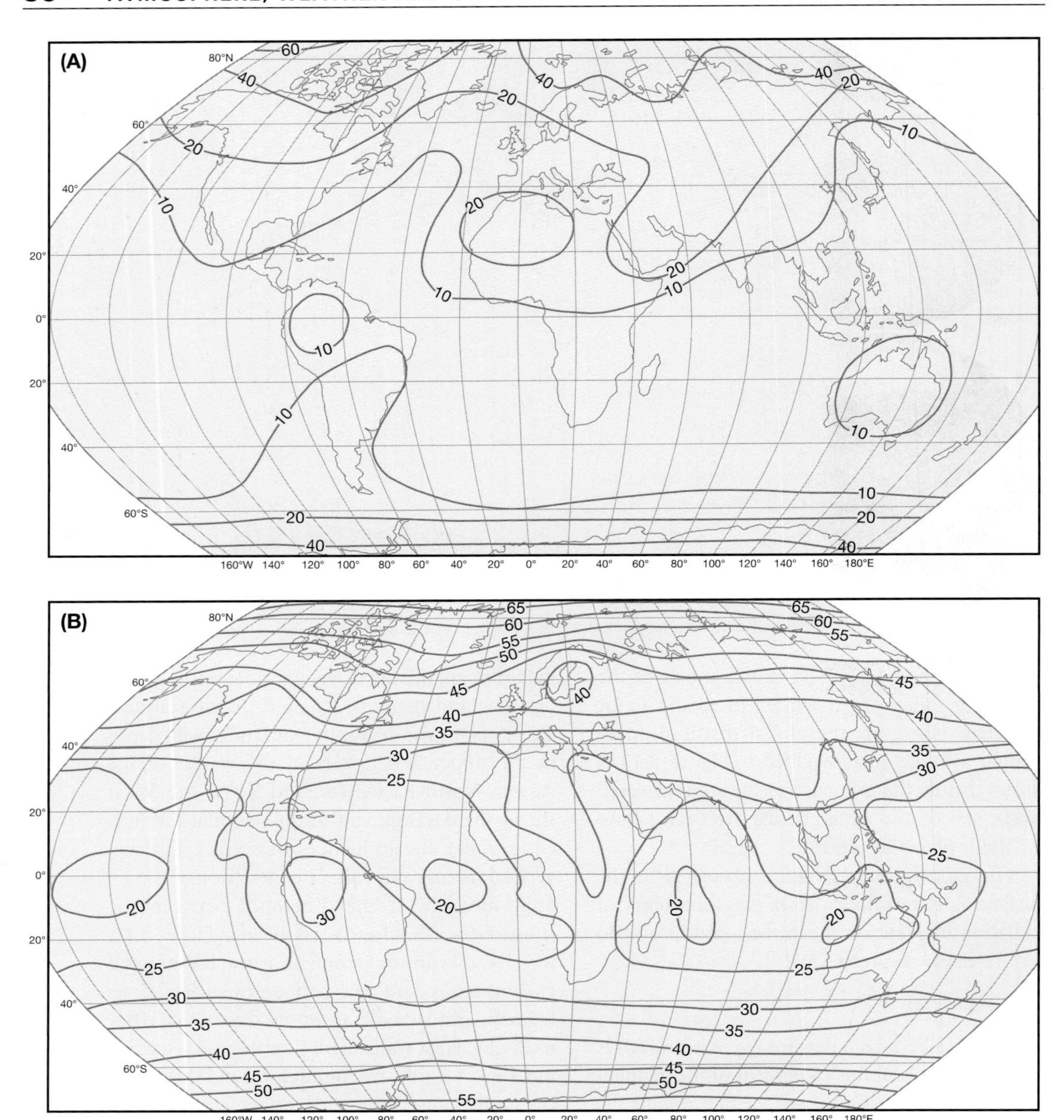

Figure 3.13 Mean annual albedos (percent): A At the earth's surface; B On a horizontal surface at the top of the atmosphere.

Source: After Hummel and Reck; from Henderson-Sellers and Wilson (1983) and Stephens *et al.* (1981).

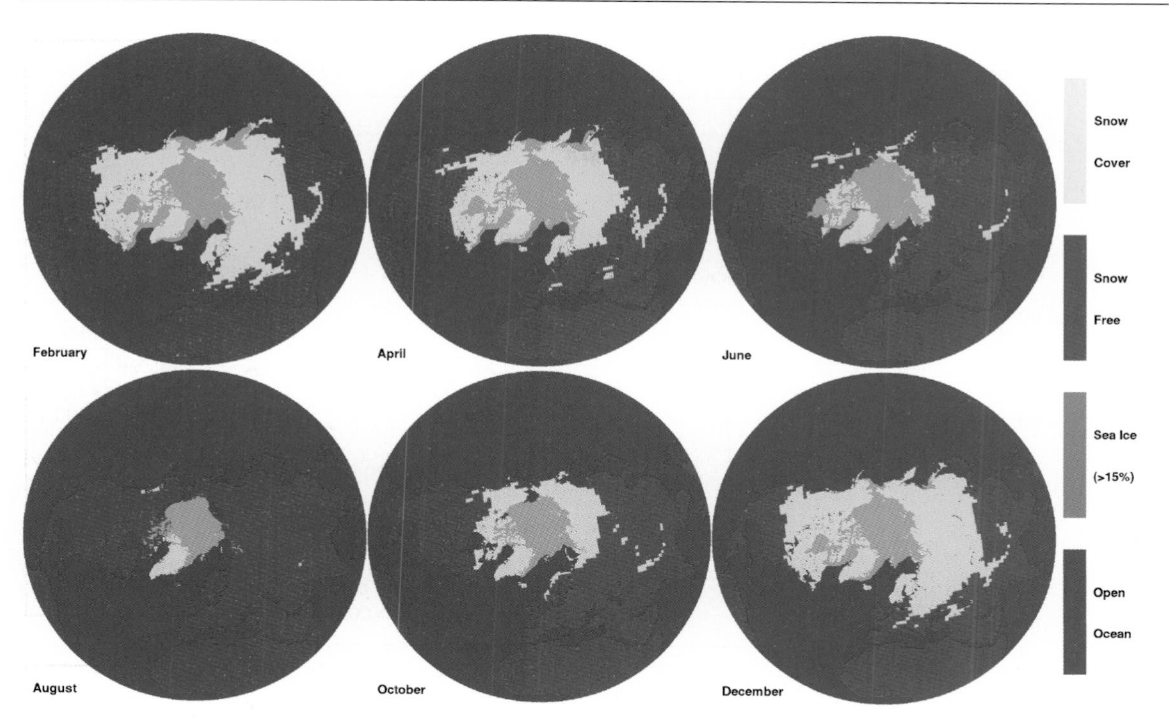

Plate 3.3 Average snow cover and sea ice extent in the Northern Hemisphere for the months of February, April, June, August, October and December derived from weekly data for the period 1978–2007.

Source: National Snow and Ice Data Center (NSIDC), Boulder, Colorado (courtesy Mary Jo Brodzik, NSIDC).

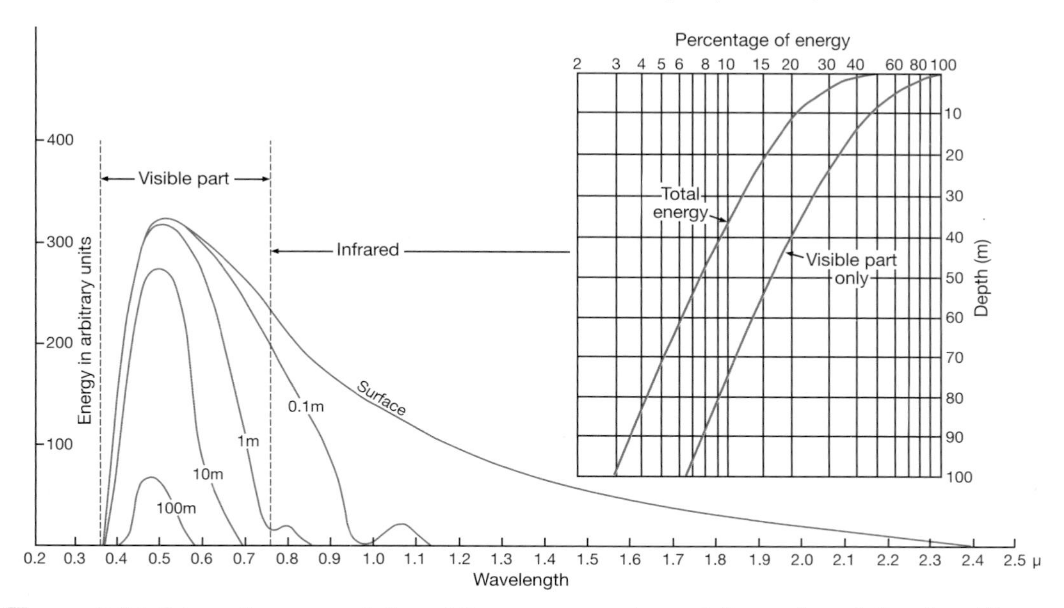

Figure 3.14 Schematic representation of the energy spectrum of the sun's radiation (in arbitrary units) that penetrates the sea surface to depths of 0.1, 1, 10 and 100m. This illustrates the absorption of infrared radiation by water, and also shows the depths to which visible (light) radiation penetrates.

Source: From Sverdrup (1945). Courtesy of Allen & Unwin.

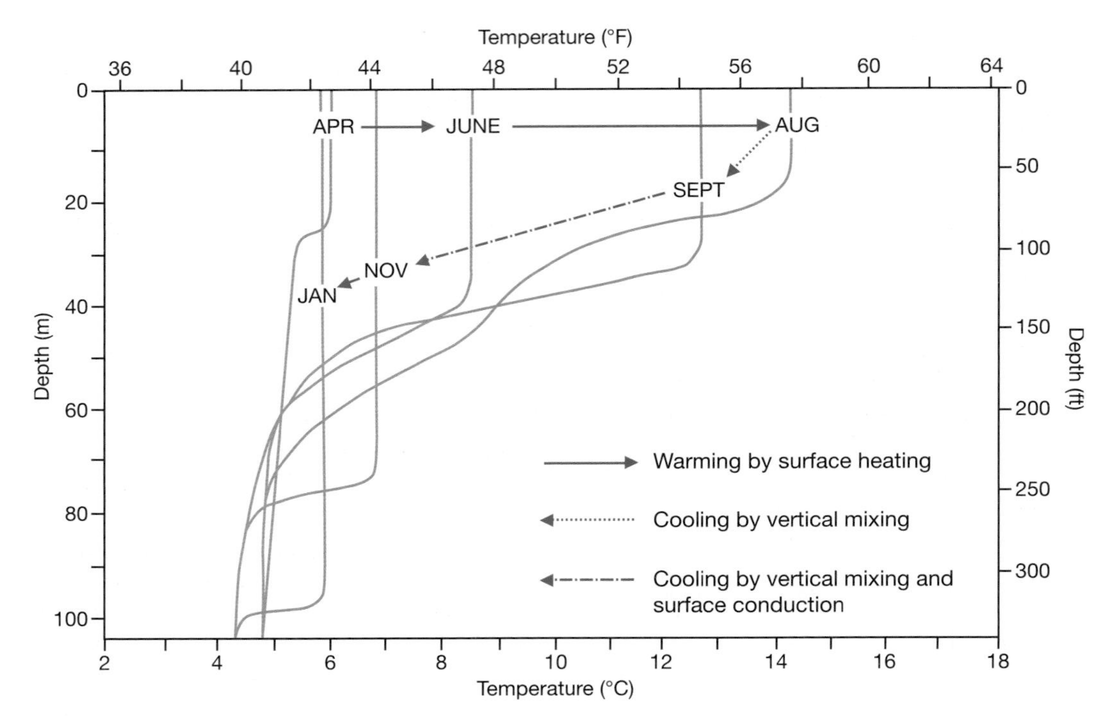

Figure 3.15 Mean monthly variations of temperature with depth in the surface waters of the eastern North Pacific. The layer of rapid temperature change is termed the thermocline.
Source: After Tully and Giovando (1963, p. 13, Fig. 4). Reproduced by permission of the Royal Society of Canada.

reducing the temperature response. The relative depths over which the annual and diurnal temperature variations are effective in wet and dry soils are roughly as follows:

	Diurnal variation	Annual variation
Wet soil	0.5m	9m
Dry sand	0.2m	3m

However, the *actual* temperature change is greater in dry soils. For example, the following values of diurnal temperature range have been observed during clear summer days at Sapporo, Japan:

	Sand	Loam	Peat	Clay
Surface	40°C	33°C	23°C	21°C
5cm	20	19	14	14
10cm	7	6	2	4

The different heating qualities of land and water are also partly accounted for by their different *specific heats*. The specific heat (c) of a substance may be represented by the number of thermal units required to raise a unit mass of it through 1°C (4184J kg^{-1} K^{-1}). The specific heat of water is much greater than for most other common substances, and water must absorb five times as much heat energy to raise its temperature by the same amount as a comparable mass of dry soil. Thus for dry sand, $c = 840$ J kg^{-1} K^{-1}.

If unit volumes of water and soil are considered, the heat capacity, ρc, of the water, where ρ = density ($\rho c = 4.18 \times 10^6$J m^{-3} K^{-1}), exceeds that of the sand approximately threefold ($\rho c = 1.3 \times 1.6$ J m^{-3} K^{-1}) if the sand is dry and twofold if it is wet. When this water is cooled the situation is reversed, for then a large quantity of heat is released. A meter-thick layer of sea water

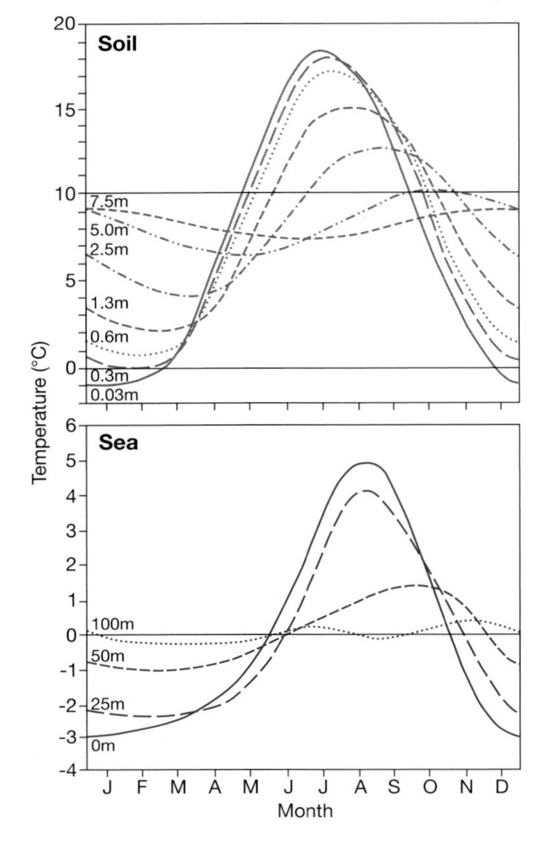

Figure 3.16 Annual variation of temperature at different depths in soil at Kaliningrad, European Russia (above) and in the water of the Bay of Biscay (at approximately 47°N, 12 (W) (below), illustrating the relatively deep penetration of solar energy into the oceans as distinct from that into land surfaces. The bottom figure shows the temperature deviations from the annual mean for each depth.
Sources: Geiger (1965) and Sverdrup (1945).

being cooled by as little as 0.1°C will release enough heat to raise the temperature of an approximately 30m-thick air layer by 10°C. In this way, the oceans act as a very effective reservoir for much of the world's heat. Similarly, evaporation of sea water causes large heat expenditure because a great amount of energy is needed to evaporate even a small quantity of water (see Chapter 3C).

The thermal role of the ocean is an important and complex one (see Chapter 7D). The ocean has three thermal layers:

1 A seasonal boundary, or upper mixed, layer, lying above the thermocline. This is less than 100m deep in the tropics but is hundreds of meters deep in the subpolar seas. It is subject to annual thermal mixing from the surface.
2 A warm water sphere or lower mixed layer. This underlies layer 1 and slowly exchanges heat with it down to many hundreds of meters.
3 The deep ocean. This contains some 80 percent of the total oceanic water volume and exchanges heat with layer 1 in the polar seas.

This vertical thermal circulation allows global heat to be conserved in the oceans, thus damping down the global effects of climatic change produced by thermal forcing (see Chapter 13). The time for heat energy to diffuse within the upper mixed layer is two to seven months, within the lower mixed layer seven years, and within the deep ocean upward of 300 years. The comparative figure for the outer thermal layer of the solid earth is only 11 days.

These differences between land and sea help to produce what is termed *continentality*. Continentality implies, first, that a land surface heats and cools much quicker than that of an ocean. Over the land, the lag between maximum (minimum) periods of radiation and the maximum (minimum) surface temperature is only one month, but over the ocean and at coastal stations the lag is up to two months. Second, the annual and diurnal ranges of temperature are greater in continental than in coastal locations. Figure 3.17 illustrates the annual variation of temperature at Toronto, Canada and Valentia, southwestern Ireland, while diurnal temperature ranges experienced in continental and maritime areas are described below (see pp. 68–69). The third effect of continentality results from the global distribution of the landmasses. The smaller ocean area of the Northern Hemisphere causes the

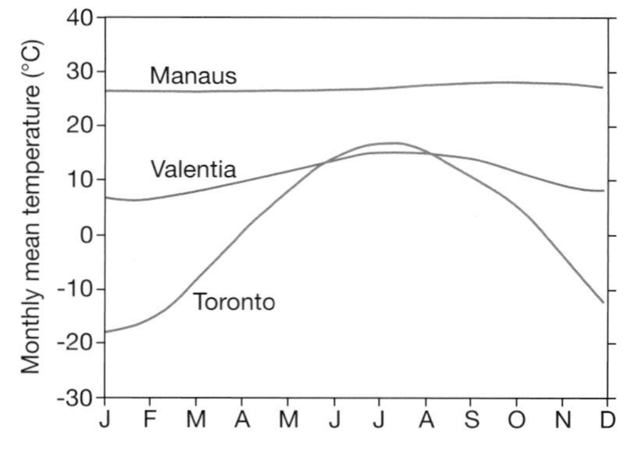

Figure 3.17 Mean annual temperature regimes in various climates: Manaus, Braziil (equatorial), Valentia, Ireland (temperate maritime) and Toronto, Canada (temperate continental).

boreal summer to be warmer but its winters to be colder on the average than the austral equivalents of the Southern Hemisphere (summer, 22.4°C versus 17.1°C; winter, 8.1°C versus 9.7°C). Heat storage in the oceans causes them to be warmer in winter and cooler in summer than land in the same latitude, although ocean currents give rise to some local departures from this rule. The distribution of temperature anomalies for the latitude in January and July (Figure 3.18) illustrates the significance of continentality and the influence of the warm currents in the North Atlantic and the North Pacific in winter.

Sea-surface temperatures can now be estimated by the use of infrared satellite imagery (see C, this chapter). Plate 7.4 shows a false-color satellite thermal image of the western North Atlantic with the relatively warm, meandering Gulf Stream. Maps of sea-surface temperatures are now routinely constructed from such images.

6 Effect of elevation and aspect

When we come down to the local scale, differences in the elevation of the land and its *aspect* (that is, the direction in which the surface faces) strongly control the amount of solar radiation received.

High elevations that have a much smaller mass of air above them (see Figure 2.13) receive considerably more direct solar radiation under clear skies than locations near sea level due to the concentration of water vapor in the lower troposphere (Figure 3.19). On the average in mid latitudes the intensity of incident solar radiation increases by 5–15 percent for each 1000m increase in elevation in the lower troposphere. The difference between sites at 200 and 3000m in the Alps, for instance, can amount to 70W m^{-2} on cloudless summer days. However, there is also a correspondingly greater net loss of terrestrial radiation at higher elevations because the low density of the overlying air results in a smaller fraction of the outgoing radiation being absorbed. The overall effect is invariably complicated by the greater cloudiness associated with most mountain ranges, and it is therefore impossible to generalize from the limited available data.

Figure 3.20 illustrates the effect of aspect and slope angle on theoretical maximum solar radiation receipts at two locations in the Northern Hemisphere. The general effect of latitude on insolation amounts is clearly shown, but it is also apparent that increasing latitude causes a relatively greater radiation loss for north-facing slopes, as distinct from south-facing slopes. The radiation intensity on a sloping surface (Is) is

$$Is = Io \cos i$$

where i = the angle between the solar beam and a beam normal to the sloping surface. Relief may also affect the quantity of insolation and the duration of direct sunlight when a mountain barrier screens the sun from valley floors and sides at certain times of the day. In many Alpine valleys, settlement and cultivation are noticeably concentrated on southward-facing slopes (the adret or sunny side), whereas northward slopes (ubac or shaded side) remain forested.

7 Variation of free air temperature with height

Chapter 2C described the gross characteristics of the vertical temperature profile in the atmosphere.

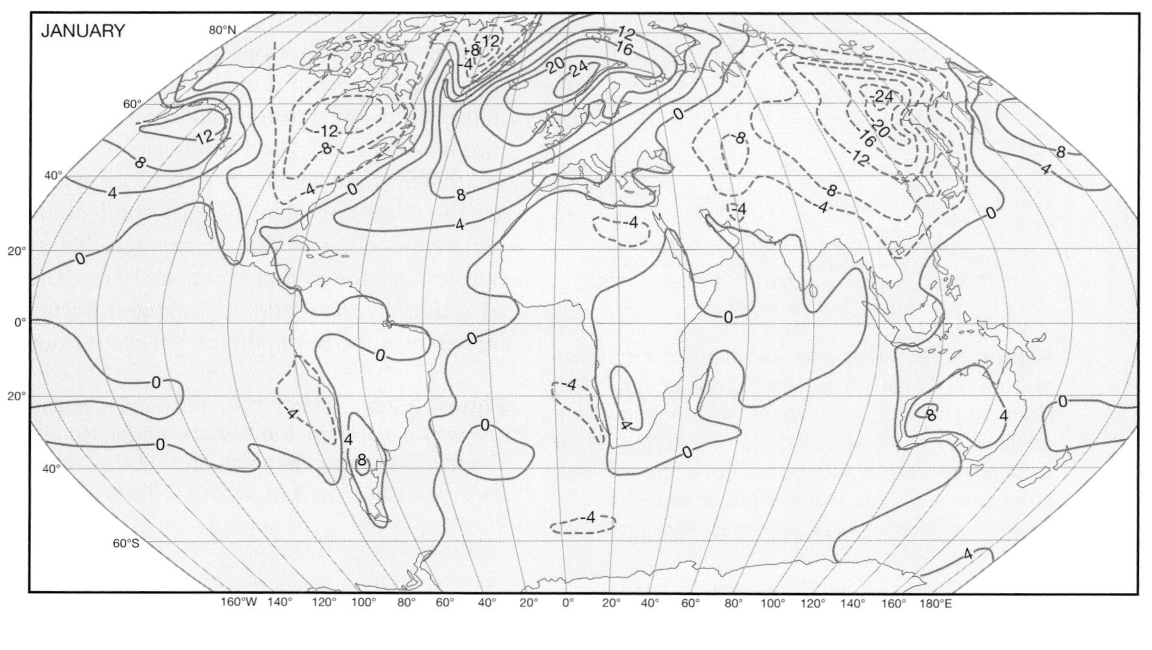

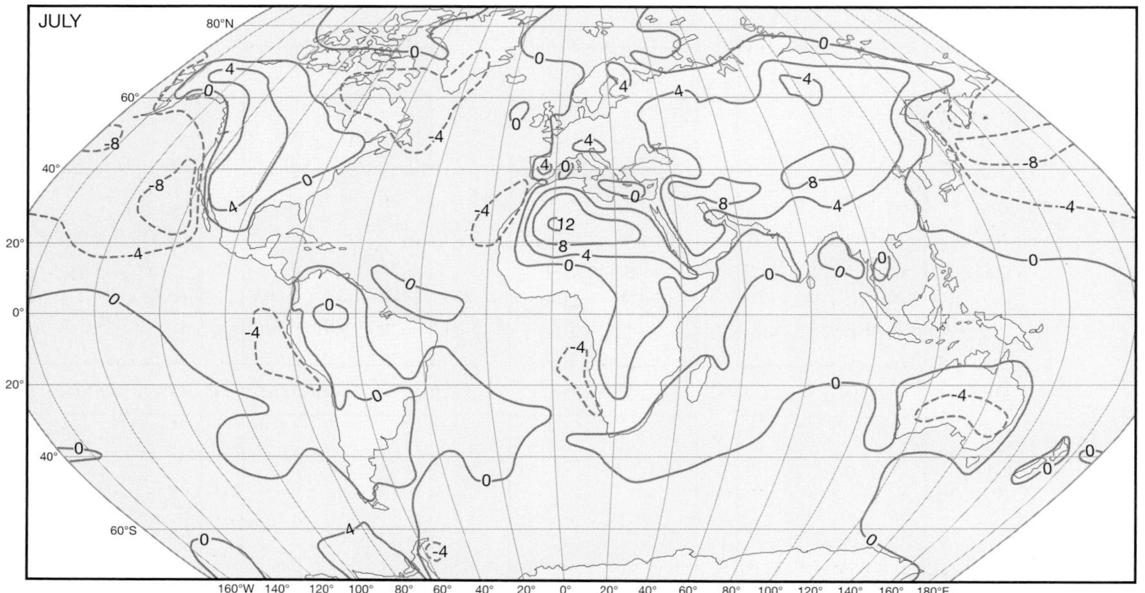

Figure 3.18 World temperature anomalies (i.e., the difference between recorded temperatures °C and the mean for that latitude) for January and July. Solid lines indicate positive, and dashed lines negative, anomalies.

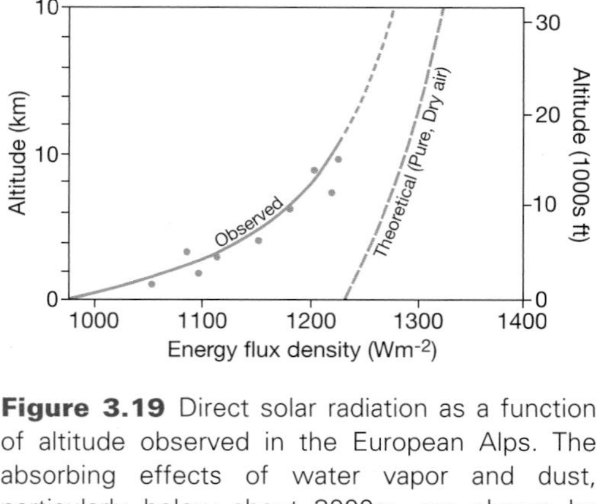

Figure 3.19 Direct solar radiation as a function of altitude observed in the European Alps. The absorbing effects of water vapor and dust, particularly below about 3000m, are shown by comparison with a theoretical curve for an ideal atmosphere without water vapor or aerosols.

Source: After Albetti, Kastrov, Kimball and Pope; from Barry (2008).

Now we examine the vertical temperature gradient in the lower troposphere.

Vertical temperature gradients are determined in part by energy transfers and in part by vertical motion of the air. The various factors interact in a highly complex manner. The energy terms are the release of latent heat by condensation, radiative cooling of the air and sensible heat transfer from the ground. Horizontal temperature advection, by the motion of cold and warm air masses, may also be important. Vertical motion is

Figure 3.20 Average direct beam solar radiation (W m^{-2}) incident at the surface under cloudless skies at Trier, West Germany, and Tucson, Arizona, as a function of slope, aspect, time of day and season of year.

Source: After Geiger (1965) and Sellers (1965).

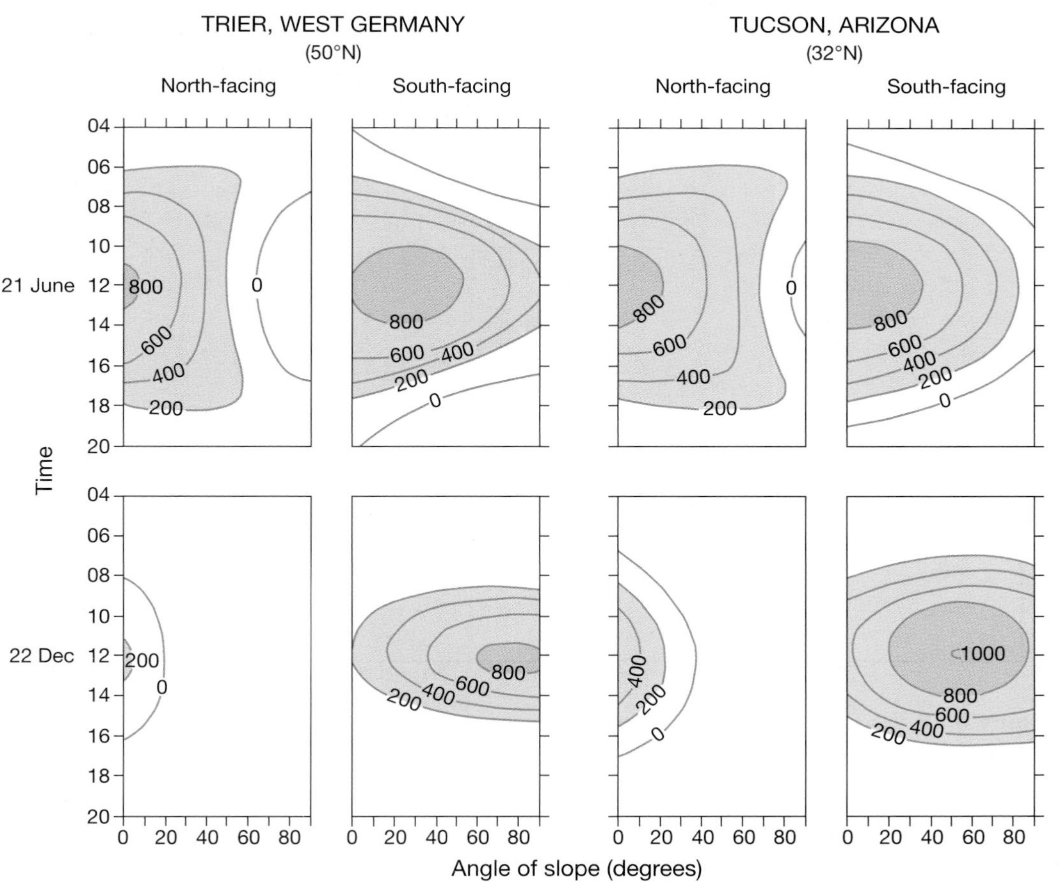

dependent on the type of pressure system. High pressure areas are generally associated with descent and warming of deep layers of air, hence decreasing the temperature gradient and frequently causing temperature inversions in the lower troposphere. In contrast, low pressure systems are associated with rising air, which cools upon expansion and increases the vertical temperature gradient. Moisture is an additional complicating factor (see Chapter 3E), but it remains true that the middle and upper troposphere is relatively cold above a surface low pressure area, leading to a steeper temperature gradient.

The overall vertical decrease of temperature, or *lapse rate*, in the troposphere is about 6.5°C/km. However, this is by no means constant with height, season or location. Average global values calculated by C. E. P. Brooks for July show increasing lapse rate with height: about 5°C/km in the lowest 2km, 6°C/km between 4 and 5km, and 7°C/km between 6 and 8km. The seasonal regime is very pronounced in continental regions with cold winters. Winter lapse rates are generally small and, in areas such as central Canada or eastern Siberia, may even be negative (i.e., temperatures increase with height in the lowest layer) as a result of excessive radiational cooling over a snow surface. A similar situation occurs when dense, cold air accumulates in mountain basins on calm, clear nights. On such occasions, mountain summits may be many degrees warmer than the valley floor below (see Chapter 5C.1). For this reason, the adjustment of average temperature of upland stations to mean sea level may produce misleading results. Observations in Colorado at Pike's Peak (4301m) and Colorado Springs (1859m) show the mean lapse rate to be 4.1°C/km in winter and 6.2°C/km in summer. It should be noted that such topographic lapse rates may bear little relation to free air lapse rates in nocturnal radiation conditions, and the two must be carefully distinguished.

In the Arctic and over Antarctica, surface temperature inversions persist for much of the year. In winter, these inversions are due to intense radiational cooling at the snow/ice surface which cools the air layer above up to a height of about 1km; in summer they are the result of the surface cooling, by conduction, of warmer air that is advected (transported horizontally) over the ice surfaces of the polar regions. The inversions persist owing to the prevailing high pressure situations that prevent the spread of cloud cover associated with storm systems. The tropical and subtropical deserts have very steep lapse rates in summer causing considerable heat transfer from the surface and generally ascending motion; subsidence associated with high pressure cells is predominant in the desert zones in winter. Over the subtropical oceans, sinking air leads to warming and a subsidence inversion near the surface (see Chapter 13).

C TERRESTRIAL INFRARED RADIATION AND THE GREENHOUSE EFFECT

Radiation from the sun is predominantly shortwave, whereas that leaving the earth is longwave, or infrared, radiation (see Figure 3.1). The infrared emission from the surface is slightly less than that from a black body at the same temperature and, accordingly, Stefan's equation (see p. 41) is modified by an emissivity coefficient (ϵ), which is generally between 0.90 and 0.95, i.e., $F = \epsilon\sigma\ T^4$. Figure 3.1 shows that the atmosphere is highly absorbent to infrared radiation (due to the effects of water vapour, carbon dioxide and other trace gases), except between about 8.5 and 13.0μm – the 'atmospheric window'. The opaqueness of the atmosphere to infrared radiation, relative to its transparency to shortwave radiation, is commonly referred to as the *greenhouse effect*. However, in the case of an actual greenhouse, the effect of the glass roof is probably as significant in reducing cooling by restricting the turbulent heat loss as it is in retaining the infrared radiation.

The total 'greenhouse' effect results from the net infrared absorption capacity of water vapor, carbon dioxide and other trace gases – methane

(CH_4), nitrous oxide (N_2O) and tropospheric ozone (O_3). These gases absorb strongly at wavelengths within the atmospheric window region, in addition to their other absorbing bands (see Figure 3.1 and Table 3.3). Moreover, because concentrations of these trace gases are low, their radiative effects increase approximately linearly with concentration, whereas the effect of CO_2 is related to the logarithm of the concentration. In addition, owing to the long atmospheric residence time of nitrous oxide (132 years) and CFCs (65–140 years), the cumulative effects of human activities will be substantial. It is estimated that between 1765 and 2000, the radiative effect of increased CO_2 concentration was 1.5W m^{-2}, and of all trace gases about 2.5W m^{-2} (*cf.* the solar constant value of 1366W m^{-2}).

The net warming contribution of the natural (non-anthropogenic) greenhouse gases to the mean 'effective' planetary temperature of 255K (corresponding to the emitted infrared radiation) is approximately 33K. Water vapor accounts for 21K of this amount, carbon dioxide 7K, ozone 2K, and other trace gases (nitrous oxide, methane) about 3K. The present global mean surface temperature is 288K, but the surface was considerably warmer during the early evolution of the earth, when the atmosphere contained large quantities of methane, water vapor and ammonia. The largely carbon dioxide atmosphere of Venus creates a 500K greenhouse effect on that planet.

Stratospheric ozone absorbs significant amounts of both incoming ultraviolet radiation, harmful to life, and outgoing terrestrial longwave re-radiation, so that its overall thermal role is a complex one. Its net effect on earth surface temperatures depends on the elevation at which the absorption occurs, being to some extent a trade-off between shortwave and longwave absorption in that:

1. An increase of ozone above about 30km absorbs relatively more incoming shortwave radiation, causing a net *decrease* of surface temperatures.
2. An increase of ozone below about 25km absorbs relatively more outgoing longwave radiation, causing a net *increase* of surface temperatures.

Longwave radiation is not merely terrestrial in the narrow sense. The atmosphere radiates to space, and clouds are particularly effective since

Table 3.3 Influence of greenhouse gases on atmospheric temperature

Gas	Centres of main absorption bands (μm)	Temperature increase (K) for × 2 present concentration	Global warming potential on a weight basis (kg^{-1} of air)†
Water vapor (H_2O)	6.3–8.0, >15		
	(8.3–12.5)*		
Carbon dioxide (CO_2)	(5.2), (10), 14.7	3.0 ± 1.5	1
Methane (CH_4)	6.52, 7.66	0.3–0.4	11
Ozone (O_3)	4.7, 9.6, (14.3)	0.9	
Nitrous oxide (N_2O)	7.78, 8.56, 17.0	0.3	270
Chlorofluoromethanes			
($CFCl_3$)	4.66, 9.22, 11.82	0.1	3400
(CF_2Cl_2)	8.68, 9.13, 10.93		7100

Sources: After Campbell; Ramanathan; Lashof and Ahuja; Luther and Ellingson; IPCC (1992).

Notes: *Important in moist atmospheres.
†Refers to direct annual radiative forcing for the surface-troposphere system.

SIGNIFICANT 20TH-CENTURY ADVANCE

3.1 The greenhouse effect

The natural greenhouse effect of the earth's atmosphere is attributable primarily to water vapor. It accounts for 21K of the 33K difference between the effective temperature of a dry atmosphere and the real atmosphere through the trapping of infrared radiation. Water vapor is strongly absorptive around 2.4–3.1μm, 4.5–6.5μm and above 16μm. The concept of greenhouse gas-induced warming is commonly applied to the effects of the increases in atmospheric carbon dioxide concentrations resulting from anthropogenic activities, principally the burning of fossil fuels. Sverre Arrhenius in Sweden drew attention to this possibility in 1896, but observational evidence was only forthcoming some 40 years later (Callendar, 1938, 1961). However, a careful record of of atmospheric concentrations of carbon dioxide was lacking until Charles Keeling installed calibrated instruments at the Mauna Loa Observatory, Hawaii, in 1957. Within a decade, these observations became the global benchmark. They showed an annual cycle of about 5ppm at the Observatory, caused by the biospheric uptake and release, and the ca. 0.4 percent annual increase in CO_2 from 315ppm in 1957 to 383ppm in 2007, due to fossil fuel burning. The annual increase is about half of the total emission due to CO_2 uptake by the oceans and the land biosphere. The principal absorption band for radiation by carbon dioxide is around 14–16μm, but there are others at 2.6 and 4.2μm. Most of the effect of increasing CO_2 concentration is by enhanced absorption in the latter, as the main band is almost saturated. The sensitivity of mean global air temperature to a doubling of CO_2 is in the range 2–5°C, while a removal of all atmospheric CO_2 might lower the mean surface temperature by more than 10°C.

The important role of other trace greenhouse gases (methane, nitrous oxide, fluorocarbons) was recognized in the 1980s and many additional trace gases began to be monitored. The latest is nitrogen trifluoride used during the manufacture of liquid crystal flat-panel displays, thin-film solar cells and microcircuits. Although concentrations of the gas are currently only 0.454 parts per trillion, it is 17,000 times more potent as a global warming agent than a similar mass of carbon dioxide.

The past histories of greenhouse gases, reconstructed from ice core records, show that the pre-industrial level of CO_2 was 280ppm and methane 750ppb compared with 383ppm and 1790ppb, respectively, today. Their concentrations decreased to about 180 ppm and 350ppb, respectively, during the maximum phases of continental glaciation in the Pleistocene Ice Age.

The positive feedback effect of CO_2, which involves greenhouse gas-induced warming leading to an enhanced hydrological cycle with a larger atmospheric vapor content and therefore further warming, is still not well resolved quantitatively.

they act as black bodies. For this reason, cloudiness and cloud-top temperature can be mapped from satellites by day and night using infrared sensors. Radiative cooling of cloud layers averages about 1.5°C per day.

For the globe as a whole, satellite measurements show that in cloud-free conditions the mean absorbed solar radiation is approximately 285W m^{-2}, whereas the emitted terrestrial radiation is 265Wm^{-2}. Including cloud-covered areas, the corresponding global values are 235W m^{-2} for both terms. Clouds reduce the absorbed solar radiation by 50W m^{-2}, but reduce the emitted radiation by only 30W m^{-2}. Hence, global cloud cover causes a net radiative loss of about 20W m^{-2}, due to the dominance of cloud albedo reducing shortwave radiation absorption. In lower latitudes, this effect is much larger (up to –50 to –100W m^{-2}) whereas in high latitudes the two factors are close to balance, or the increased infrared absorption by clouds may lead to a small positive value. These results are important in terms of changing concentrations of greenhouse gases, since the net radiative forcing by cloud cover is four times that expected from CO_2 doubling (see Chapter 13).

D HEAT BUDGET OF THE EARTH

We can now summarize the net effect of the transfers of energy in the earth–atmosphere system averaged over the globe and over an annual period.

The incident solar radiation averaged over the globe is

$$\text{Solar constant} \times \pi r^2 / 4\pi r^2$$

where r = radius of the earth and $4\pi r^2$ is the surface area of a sphere. This figure is approximately 342W m^{-2}, or 11 × 10^9J m^{-2} yr^{-1} (10^9J = 1GJ); for convenience we will regard it as 100 units. Referring to Figure 3.21, incoming radiation is absorbed in the stratosphere (3 units), by ozone mainly, and 20 units are absorbed in the troposphere by carbon dioxide (1), water vapour (13), dust (3) and water droplets in clouds (3). Twenty units are reflected back to space from clouds, which cover about 62 percent of the earth's surface, on average. A further 9 units are similarly reflected from the surface and 3 units are returned by atmospheric scattering. The total reflected radiation is the *planetary albedo* (31 percent or 0.31). The remaining 49 units reach the earth

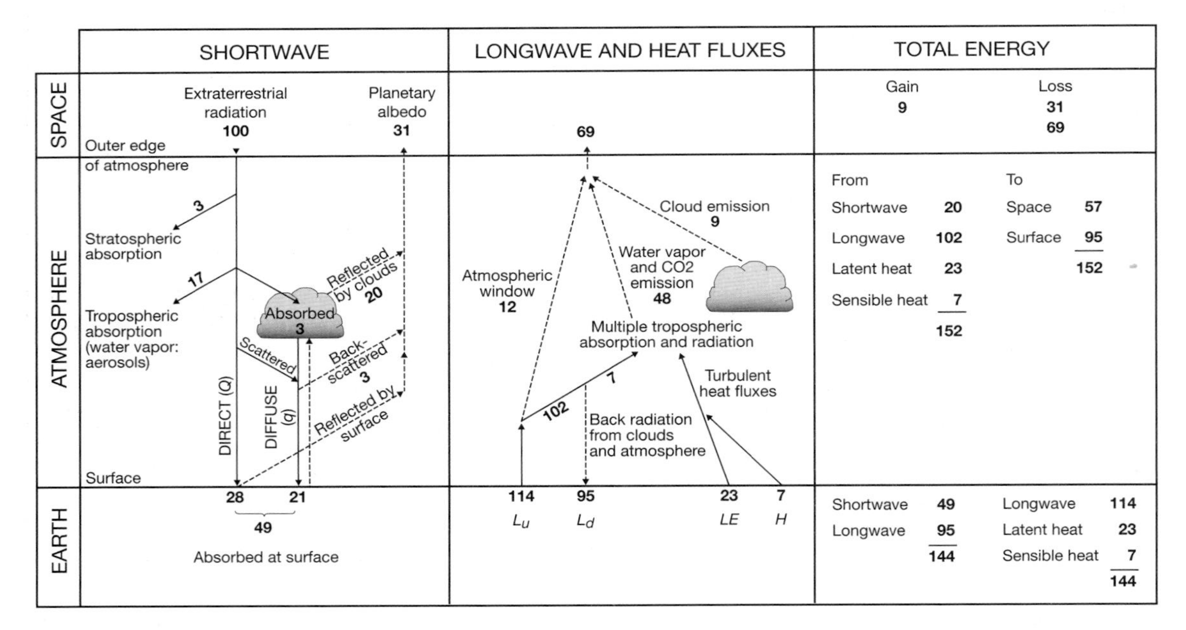

Figure 3.21 The balance of the atmospheric energy budget. The transfers are explained in the text. Solid lines indicate energy gains by the atmosphere and surface in the left-hand diagram and the troposphere in the right-hand diagram. The exchanges are referred to 100 units of incoming solar radiation at the top of the atmosphere (equal to 342W m^{-2}).

Source: After Kiehl and Trenberth (1997) *Bulletin of the American Meteorological Society*, by permission of the American Meteorological Society.

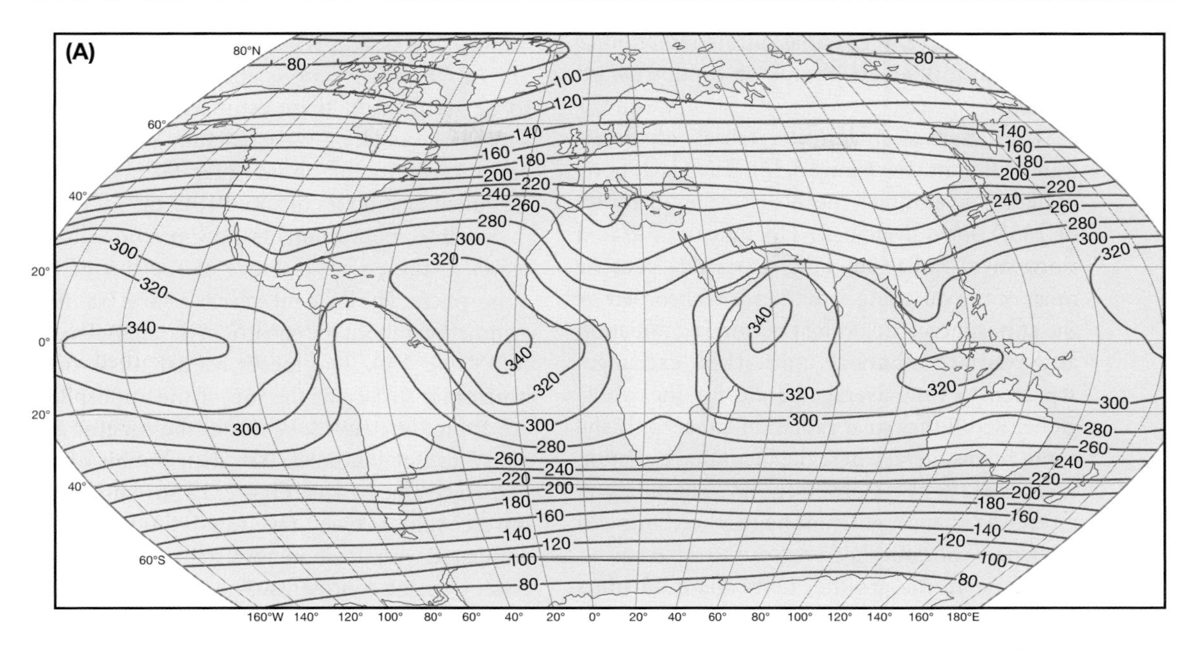

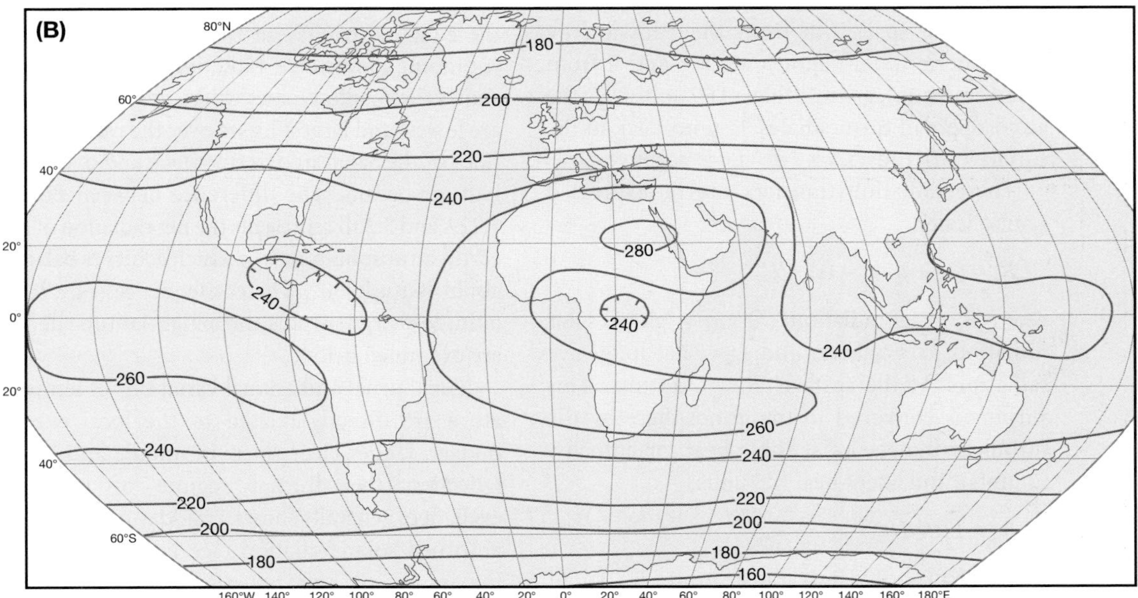

Figure 3.22 Planetary short and longwave radiation (W m^{-2}): (A) Mean annual absorbed shortwave radiation for the period April 1979 to March 1987; B Mean annual net planetary longwave radiation (L_n) on a horizontal surface at the top of the atmosphere.

Sources: A Ardanuy *et al.* (1992) and Kyle *et al.* (1993) *Bulletin of the American Meteorological Society*, by permission of the American Meteorological Society. B Stephens *et al.* (1981).

either directly ($Q = 28$) or as diffuse radiation ($q = 21$) transmitted via clouds or by downward scattering.

The pattern of outgoing terrestrial radiation is quite different (see Figure 3.22). The black-body radiation, assuming a mean surface temperature of 288K, is equivalent to 114 units of infrared (longwave) radiation. This is possible because most of the outgoing radiation is reabsorbed by the atmosphere; the *net* loss of infrared radiation at the surface is only 19 units. These exchanges represent a time-averaged state for the whole globe. Recall that solar radiation affects only the sunlit hemisphere, where the incoming radiation exceeds 342W m^{-2}. Conversely, no solar radiation is received by the night-time hemisphere. Infrared exchanges continue, however, due to the accumulated heat in the ground. Only about 12 units escape through the atmospheric window directly from the surface. The atmosphere itself radiates 57 units to space (48 from the emission by atmospheric water vapor and CO_2 and 9 from cloud emission), giving a total of 69 units (L_u); the atmosphere in turn radiates 95 units back to the surface (Ld); thus, $Lu + Ld = Ln$ is negative.

These radiation transfers can be expressed symbolically:

$$R_n = (Q + q)\,(1 - \alpha) + L_n$$

where R_n = net radiation, $(Q + q)$ = global solar radiation, α = albedo and L_n = net longwave radiation. At the surface, R_n = 30 units. This surplus is conveyed to the atmosphere by the turbulent transfer of sensible heat, or enthalpy (7 units), and latent heat (23 units),

$$R_n = LE + H$$

where H = sensible heat transfer and LE = latent heat transfer. There is also a flux of heat into the ground (B.5, this chapter), but for annual averages this is approximately zero.

Figure 3.22 summarizes the total balances at the surface (± 144 units) and for the atmosphere (± 152 units). The total absorbed solar radiation and emitted radiation for the entire earth–atmosphere system is estimated to be ±7GJ m^{-2} yr^{-1} (± 69 units). Various uncertainties have still to be resolved in these estimates. The surface short-wave and long-wave radiation budgets have an uncertainty of about 20W m^{-2}, and the turbulent heat fluxes of about 10W m^{-2}.

Satellite measurements now provide global views of the energy balance at the top of the atmosphere. The incident solar radiation is almost symmetrical about the equator in the annual mean (cf. Table 3.1). The mean annual totals on a horizontal surface at the top of the atmosphere are approximately 420W m^{-2} at the equator and 180W m^{-2} at the poles. The distribution of the planetary albedo (see Figure 3.13B) shows the lowest values over the low-latitude oceans compared with the more persistent areas of cloud cover over the continents. The highest values are over the polar ice caps. The resulting planetary shortwave radiation ranges from 340W m^{-2} at the equator to 80W m^{-2} at the poles. The net (outgoing) longwave radiation (Figure 3.22B) shows the smallest losses where the temperatures are lowest and largest losses over the mainly clear skies of the Saharan desert surface and over low-latitude oceans. The difference between Figure 3.22A and 3.23B represents the net radiation of the earth–atmosphere system which achieves balance about latitude 30°. The consequences of a low-latitude energy surplus and a high-latitude deficit are examined below.

The diurnal and annual variations of temperature are directly related to the local energy budget. Under clear skies, in middle and lower latitudes, the diurnal regime of radiative exchanges generally shows a midday maximum of absorbed solar radiation (see Figure 3.23A). A maximum of infrared (longwave) radiation (see Figure 3.1) is also emitted by the heated ground surface at midday, when it is warmest. The atmosphere re-radiates infrared radiation downward, but there is a net loss at the surface (L_n). The difference between the absorbed solar radiation and L_n is the net radiation, R_n; this is generally positive between about an hour after

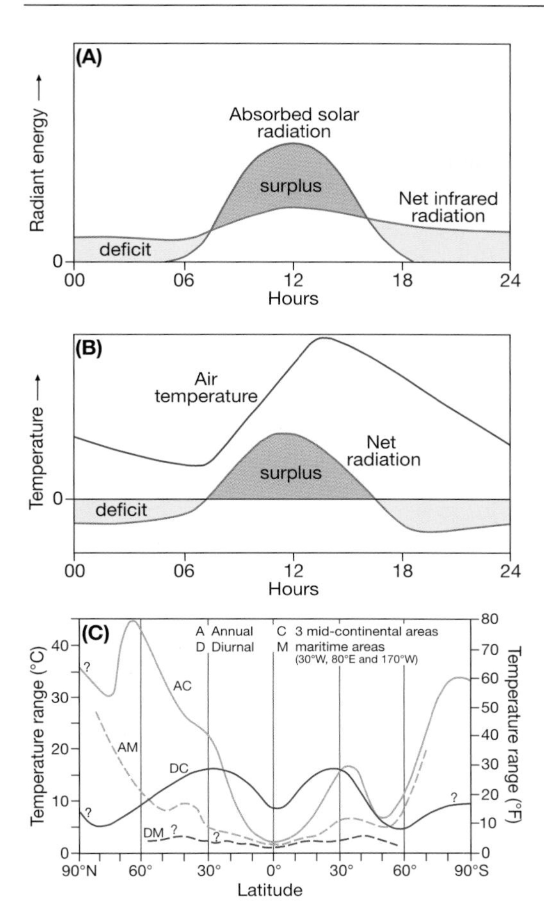

Figure 3.23 Curves showing diurnal variations of radiant energy and temperature. A Diurnal variations in absorbed solar radiation and infrared radiation in the middle and low latitudes. B Diurnal variations in net radiation and air temperature in the middle and low latitudes. C Annual (A) and diurnal (D) temperature ranges as a function of latitude and of continental (C) or maritime (M) location.
Source: From Paffen (1967). Courtesy of *Erdkunde*.

sunrise and an hour or so before sunset, with a midday maximum. The delay in the occurrence of the maximum air temperature until about 14:00 hours local time (Figure 3.23B) is caused by the gradual heating of the air by convective transfer from the ground. Minimum R_n occurs in the early evening, when the ground is still warm; there is a slight increase thereafter. The temperature decrease after midday is slowed by heat supplied from the ground. Minimum air temperature occurs shortly after sunrise due to the lag in the transfer of heat from the surface to the air. The annual pattern of the net radiation budget and temperature regime is closely analogous to the diurnal pattern, with a seasonal lag in the temperature curve relative to the radiation cycle, as noted above (p. 59).

There are marked latitudinal variations in the diurnal and annual ranges of temperature. Broadly, the annual range is a maximum in higher latitudes, with extreme values about 65°N related to the effects of continentality and distance to the ocean in interior Asia and North America (Figure 3.24). In contrast, in low latitudes the annual range differs little between land and sea owing to the thermal similarity between tropical rainforests and tropical oceans. The diurnal range is a maximum over tropical land areas, but it is in the equatorial zone that the diurnal variation of heating and cooling exceeds the annual variation (Figure 3.23C), due to the small seasonal change in solar elevation angle at the equator.

E ATMOSPHERIC ENERGY AND HORIZONTAL HEAT TRANSPORT

Thus far, we have given an account of the earth's heat budget and its components. We have already referred to two forms of energy: internal (or heat) energy, due to the motion of individual air molecules, and latent energy, which is released by condensation of water vapor. Two other forms of energy are important: geopotential energy due to gravity and height above the surface, and kinetic energy associated with air motion.

Geopotential and internal energy are interrelated, since the addition of heat to an air column not only increases its internal energy but also adds to its geopotential as a result of the vertical expansion of the air column. In a column extending to the top of the atmosphere, the geopotential is approximately 40 percent of the

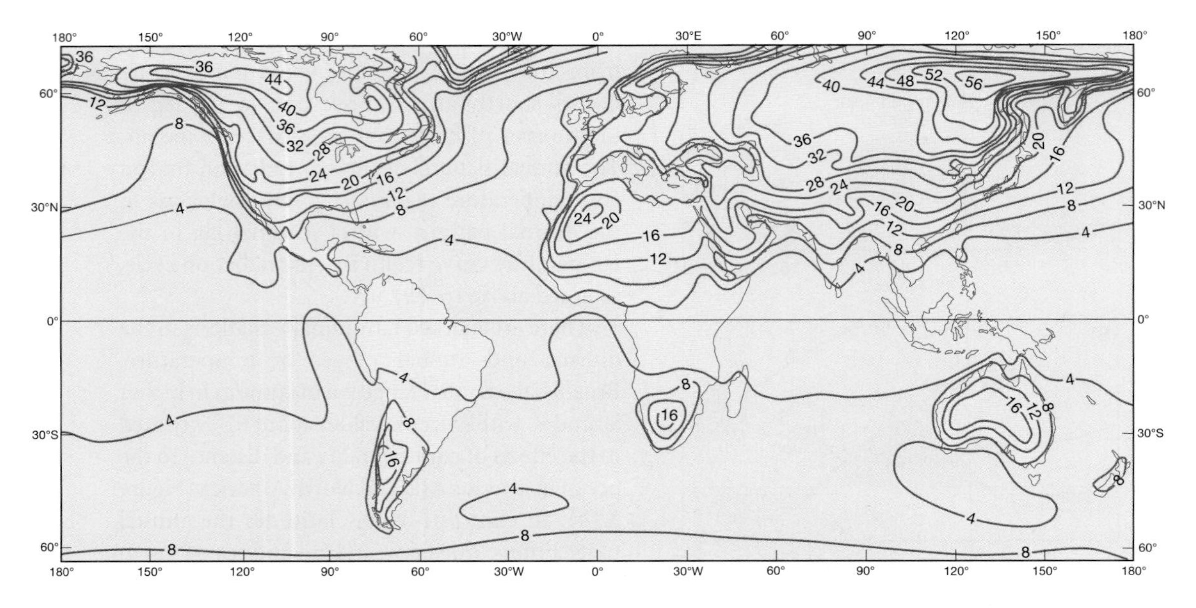

Figure 3.24 The mean annual temperature range (°C) at the earth's surface.
Source: Monin (1975). Courtesy World Meteorological Organization.

internal energy. These two are therefore usually considered together and termed the total potential energy (*PE*). For the whole atmosphere

potential energy ≈ 10^{24}J
kinetic energy ≈ 10^{10}J

In a later section (Chapter 6C), we shall see how energy is transferred from one form to another, but here we consider only heat energy. It is apparent that the receipt of heat energy is very unequal geographically and that this must lead to great lateral transfers of energy across the surface of the earth. In turn, these transfers give rise, at least indirectly, to the observed patterns of global weather and climate.

The amounts of energy received at different latitudes vary substantially, the equator on the average receiving 2.5 times as much annual solar energy as the poles. Clearly, if this process were not modified in some way the variations in receipt would cause a massive accumulation of heat within the tropics (associated with gradual increases of temperature) and a corresponding deficiency at the poles. Yet this does not happen,

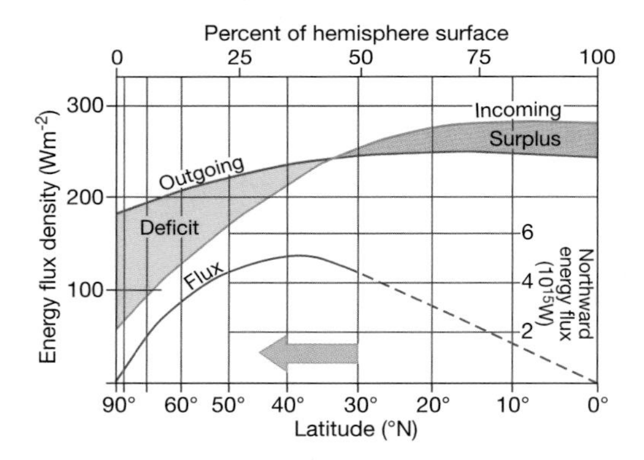

Figure 3.25 A meridional illustration of the balance between incoming solar radiation and outgoing radiation from the earth and atmosphere* in which the zones of permanent surplus and deficit are maintained in equilibrium by a poleward energy transfer.†
Sources: *Data from Houghton; after Newell (1964). †After Gabites.

and the earth as a whole is roughly in a state of thermal equilibrium. One explanation of this equilibrium could be that for each region of the world there is equalization between the amount of

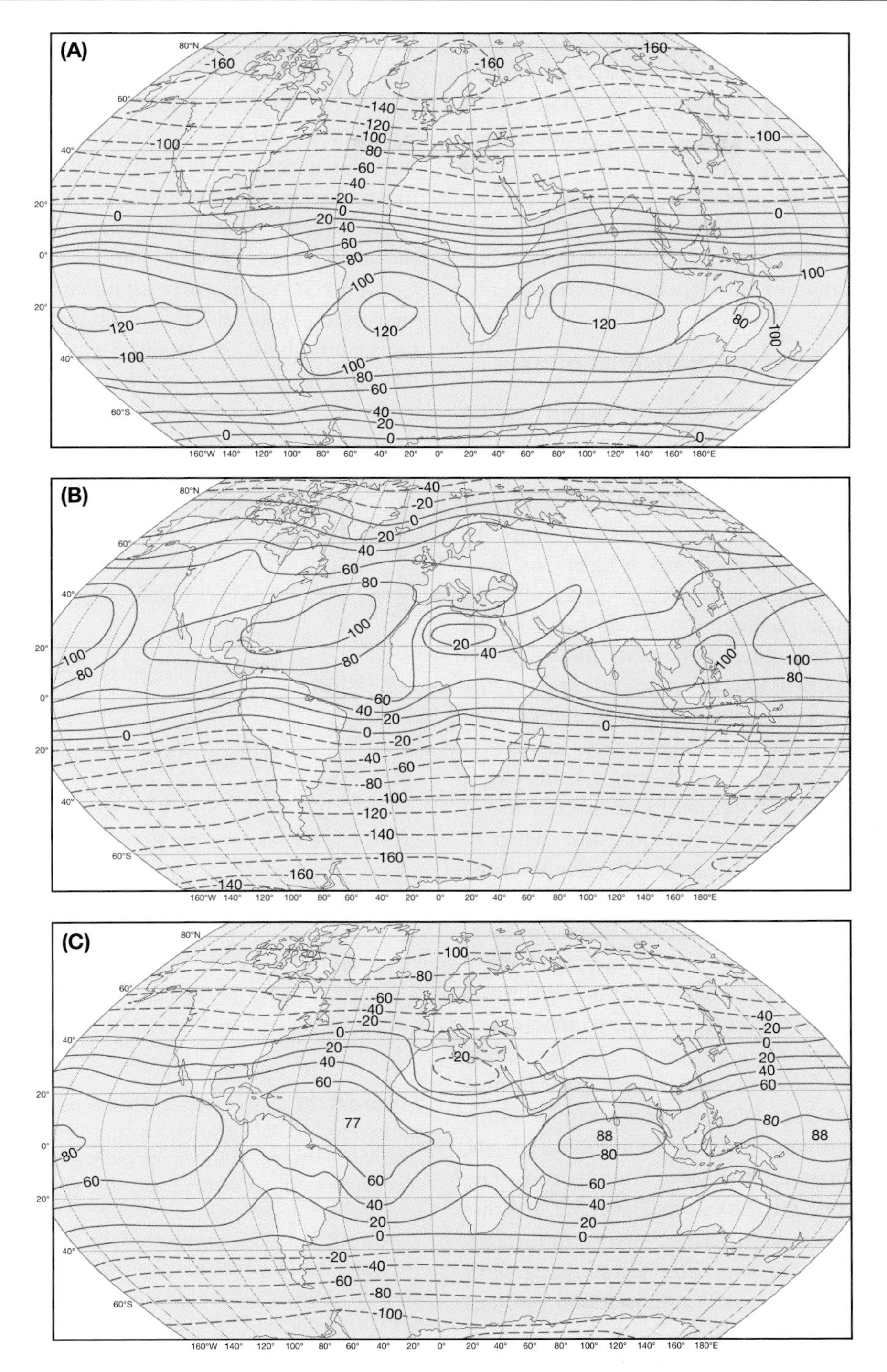

Figure 3.26 Mean net planetary radiation budget (R_n) (W m^{-2}) for a horizontal surface at the top of the atmosphere (i.e., for the earth–atmosphere system). A January, B July, C Annual.

Sources: Ardanuy *et al.* (1992), Kyle *et al.* (1993) and Stephens *et al.* (1981). C: From *Bulletin of the American Meteorological Society*, by permission of the American Meteorological Society.

incoming and outgoing radiation. However, observation shows that this is not so (Figure 3.25), because, whereas incoming radiation varies appreciably with changes in latitude, being highest at the equator and declining to a minimum at the poles, outgoing radiation has a more even latitudinal distribution owing to the rather small variations in atmospheric temperature. Some other explanation therefore becomes necessary.

1 The horizontal transport of heat

If the net radiation for the whole earth–atmosphere system is calculated, it is found that there is a positive budget between 35°S and 40°N, as shown in Figure 3.26C. The latitudinal belts in each hemisphere separating the zones of positive and negative net radiation budgets oscillate dramatically with season (Figure 3.26A and B). As the tropics do not get progressively hotter or the high latitudes colder, a redistribution of world heat energy must occur constantly, taking the form of a continuous movement of energy from the tropics to the poles. In this way, the tropics shed their excess heat and the poles, being global heat sinks, are not allowed to reach extremes of cold. If there were no meridional interchange of heat, a radiation balance at each latitude would be achieved only if the equator were 14°C warmer and the North Pole 25°C colder than at present. This poleward heat transport takes place within the atmosphere and oceans, and it is estimated that the former accounts for approximately two-thirds of the required total. The horizontal transport (*advection* of heat) occurs in the form of both latent heat (that is, water vapor, which subsequently condenses) and sensible heat (that is, warm air masses). It varies in intensity according to the latitude and the season. Figure 3.27B shows the mean annual pattern of energy transfer by the three mechanisms. The latitudinal zone of maximum total transfer rate is found between latitudes 35° and 45° in both hemispheres, although the patterns for the individual components are quite different from one another. The latent heat transport, which occurs almost wholly in the lowest 2 or 3km, reflects the global wind belts on either side of the subtropical high pressure zones (see Chapter 8B). The more important meridional transfer of sensible heat has a double maximum not only latitudinally but also in the vertical plane, where there are maxima near the surface and at about 200mb. The high-level transport is particularly significant over the subtropics, whereas the primary latitudinal maximum about 50° to 60°N is related to the travelling low pressure systems of the westerlies.

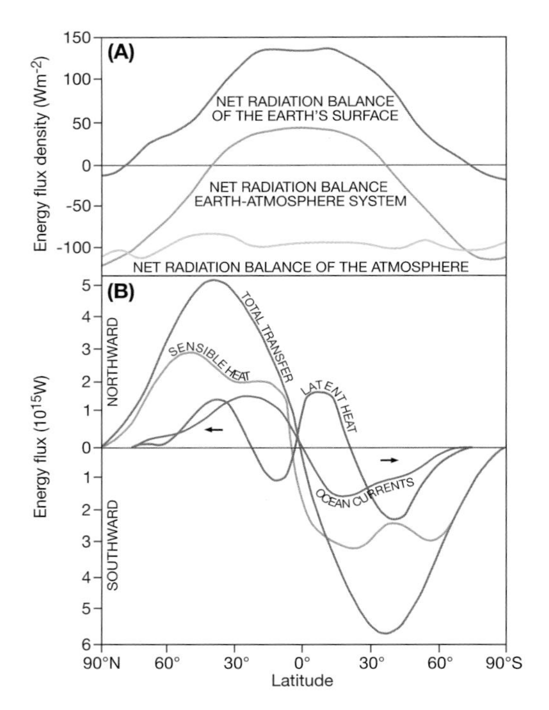

Figure 3.27 A: Net radiation balance for the earth's surface of 101W m^{-2} (incoming solar radiation of 156W m^{-2}, minus outgoing longwave energy to the atmosphere of 55W m^{-2}); for the atmosphere of –101W m^{-2} (incoming solar radiation of 84W m^{-2}, minus outgoing longwave energy to space of 185W m^{-2}); and for the whole earth–atmosphere system of zero. B: The average annual latitudinal distribution of the components of the poleward energy transfer (in 10^{15}W) in the earth–atmosphere system.

Source: From Sellers (1965). Courtesy of University of Chicago Press.

The intensity of the poleward energy flow is closely related to the meridional (that is, north–south) temperature gradient. In winter, this temperature gradient is at a maximum and in consequence the hemispheric air circulation is most intense. The nature of the complex transport mechanisms will be discussed in Chapter 8C.

As shown in Figure 3.27B, ocean currents account for a significant proportion of the poleward heat transfer in low latitudes. Indeed, recent satellite estimates of the required total poleward energy transport indicate that the previous figures are too low. The ocean transport may be 47 percent of the total at 30–35°N and as much as 74 percent at 20°N; the Gulf Stream and Kuro Shio currents are particularly important. In the Southern Hemisphere, poleward transport is mainly in the Pacific and Indian Oceans (see Figure 8.30). The energy budget equation for an ocean area must be expressed as

$$R_n = LE + H + G + \Delta A$$

where ΔA = horizontal advection of heat by currents and G = the heat transferred into or out of storage in the water. The storage is more or less zero for annual averages.

2 Spatial pattern of the heat budget components

The mean latitudinal values of the heat budget components discussed above conceal great spatial variations. Figure 3.28 shows the global distribution of the annual net radiation at the surface. Broadly, its magnitude decreases poleward from about 25° latitude. However, as a result of the high absorption of solar radiation by the sea, net radiation is greater over the oceans – exceeding 160W m^{-2} in latitudes 15–20° – than over land areas, where it is about 80–105W m^{-2} in the same latitudes. Net radiation is also lower in arid continental areas than in humid areas, because in spite of the increased insolation

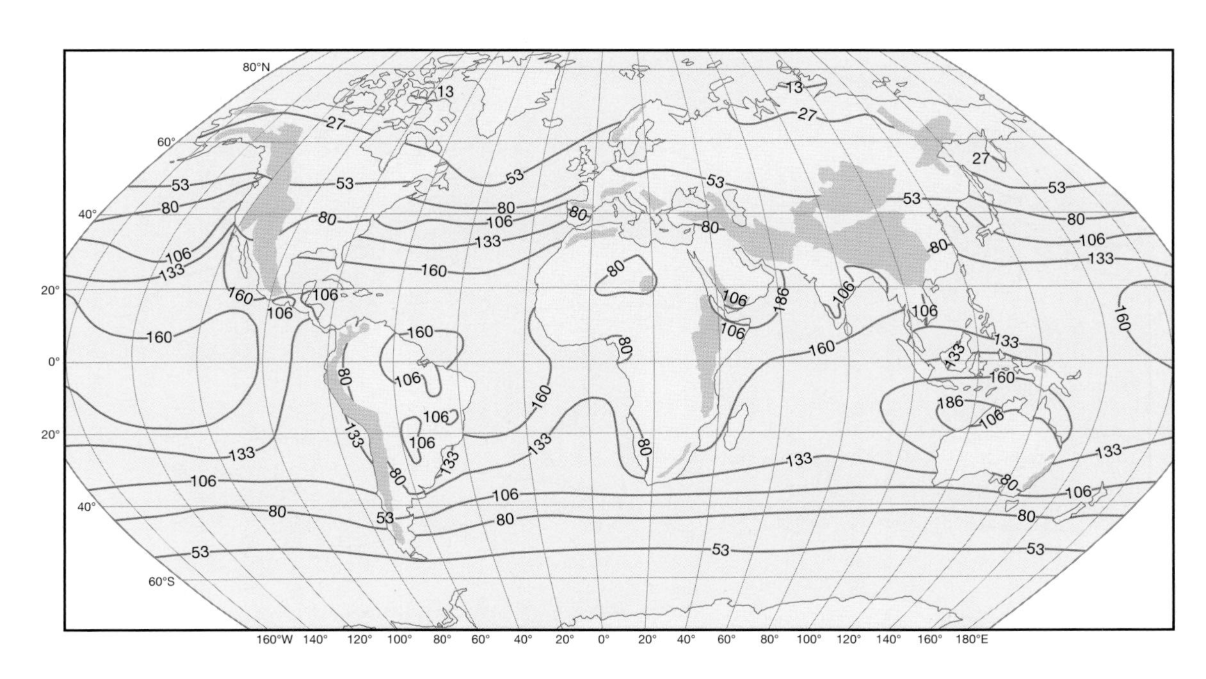

Figure 3.28 Global distribution of the annual net radiation at the surface, in W m^{-2}.
Source: After Budyko *et al.* (1962).

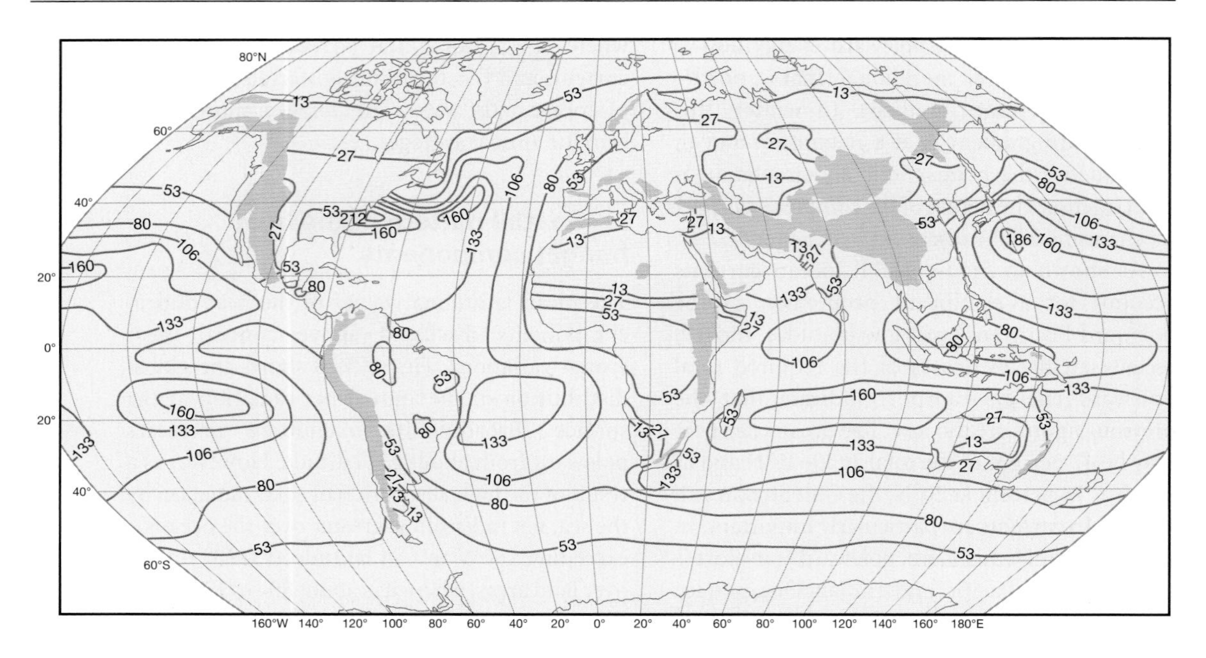

Figure 3.29 Global distribution of the vertical transfer of latent heat, in W m^{-2}.
Source:After Budyko *et al.* (1962).

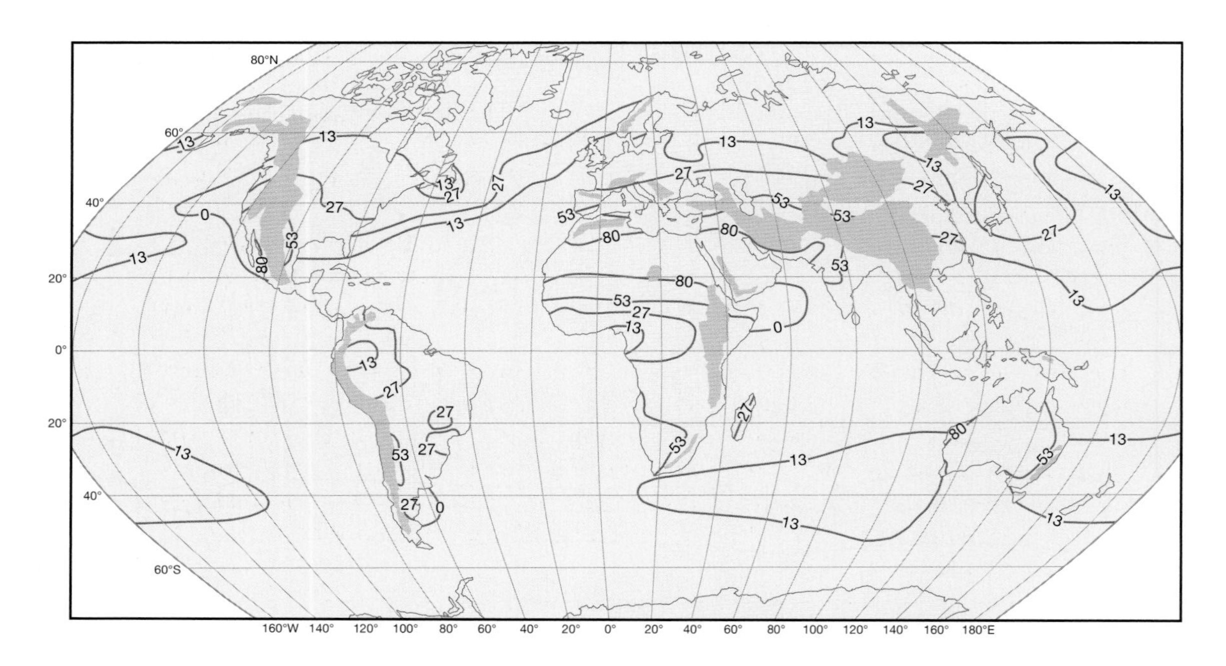

Figure 3.30 Global distribution of the vertical transfer of sensible heat, in W m^{-2}.
Source:After Budyko *et al.* (1962).

receipts under clear skies there is at the same time greater net loss of terrestrial radiation.

Figures 3.29 and 3.30 show the annual vertical transfers of latent and sensible heat to the atmosphere. Both fluxes are distributed very differently over land and seas. Heat expenditure for evaporation is at a maximum in tropical and subtropical ocean areas, where it exceeds 160W m^{-2}. It is less near the equator, where wind speeds are somewhat lower and the air has a vapor pressure close to the saturation value (see Chapter 3A). It is clear from Figure 3.29 that the major warm currents greatly increase the evaporation rate. On land, the latent heat transfer is largest in hot, humid regions. It is least in arid areas with low precipitation and in high latitudes, where there is little available energy or moisture.

The largest exchange of sensible heat occurs over tropical deserts, where more than 80W m^{-2} is transferred to the atmosphere (see Figure 3.30). In contrast to latent heat, the sensible heat flux is generally small over the oceans, only reaching 25–40W m^{-2} in areas of warm currents. Indeed, negative values occur (transfer *to* the ocean) where warm continental air masses move offshore over cold currents.

SUMMARY

Almost all energy affecting the earth is derived from solar radiation, which is of short wavelength (<4μm) due to the high temperature of the sun (6000K) (i.e., Wien's Law). The solar constant has a value of approximately 1366W m^{-2}. The sun and the earth radiate almost as black bodies (Stefan's Law, $F = \sigma T^4$), whereas the atmospheric gases do not. Terrestrial radiation, from an equivalent black body, amounts to only about 270W m^{-2} due to its low radiating temperature (263K); this is infrared (longwave) radiation between 4 and 100μm. Water vapor and carbon dioxide are the major absorbing gases for infrared radiation, whereas the atmosphere is largely transparent to solar radiation (the greenhouse effect). Trace-gas increases are now augmenting the 'natural' greenhouse effect (33K). Solar radiation is lost by reflection, mainly from clouds, and by absorption (largely by water vapor). The planetary albedo is 31 percent; 49 percent of the extraterrestrial radiation reaches the surface. The atmosphere is heated primarily from the surface by the absorption of terrestrial infrared radiation and by turbulent heat transfer. Temperature usually decreases with height at an average rate of about 6.5°C/km in the troposphere. In the stratosphere and thermosphere, it increases with height due to the presence of radiation absorbing gases.

The excess of net radiation in lower latitudes leads to a poleward energy transport from tropical latitudes by ocean currents and by the atmosphere. This is in the form of sensible heat (warm air masses/ocean water) and latent heat (atmospheric water vapor). Air temperature at any point is affected by the incoming solar radiation and other vertical energy exchanges, surface properties (slope, albedo, heat capacity), land and sea distribution and elevation, and also by horizontal advection due to air mass movements and ocean currents.

DISCUSSION TOPICS

- Explain the respective roles of the earth's orbit around the sun and the tilt of the axis of rotation for global climate.
- Explain the differences between the transmission of solar and terrestrial radiation by the atmosphere.
- What is the relative importance of incoming solar radiation, turbulent energy exchanges and other factors in determining local daytime temperatures?
- Consider the role of clouds in global climate from a radiative perspective.
- What effects do ocean currents have on regional climates? Consider the mechanisms involved for both warm and cold currents.
- Explain the concept of 'continentality'. What climatic processes are involved and how do they operate?

REFERENCES AND FURTHER READING

Books

Barry, R. G. (2008) *Mountain Weather and Climate,* 3rd edn, Cambridge University Press, Cambridge, 506pp. [comprehensive survey]

Budyko, M. I. (1974) *Climate and Life,* Academic Press, New York, 508pp. [Provides ready access to the work of a pre-eminent Russian climatologist]

Campbell, I. M. (1986) *Energy and the Atmosphere: A Physical–Chemical Approach* (2nd edn), John Wiley & Sons, Chichester, 337 pp.

Davis, W. M. (1894) *Elementary Meteorology,* Ginn & Co., Boston, MA.

Essenwanger, O. M. (1985) *General Climatology, Vol. 1A: Heat Balance Climatology. World Survey of Climatology.* Elsevier, Amsterdam, 224pp. [Comprehensive overview of net radiation, latent, sensible and ground heat fluxes; units are calories]

Fröhlich, C. and London, J. (1985) *Radiation Manual,* World Meteorological Organization, Geneva. [standard handbook]

Geiger, R, (1965) *The Climate Near the Ground,* Harvard University Press, Cambridge, MA, 611pp.

Herman, J. R. and Goldberg, R. A. (1985) *Sun, Weather and Climate,* Dover, New York, 360pp. [Useful survey of solar variability (sunspots, electromagnetic and corpuscular radiation, cosmic rays and geomagetic sector structure), long- and short-term relations with weather and climate, and design of experiments]

Hewson,E. W. and Longley, R. W. (1944) *Meteorology: Theoretical and Applied,* Wiley, New York, 468pp.

Miller, D. H. (1981) *Energy at the Surface of the Earth.* Academic Press, New York, 516pp. [Comprehensive treatment of radiation and energy fluxes in ecosystems and fluxes of carbon; many original illustrations, tables and references]

NASA (n.d.) *From Pattern to Process: The Strategy of the Earth Observing System* (Vol. III), EOS Science Steering Committee Report, NASA, Houston, TX.

Sellers, W. D. (1965) *Physical Climatology,* University of Chicago Press, Chicago, IL, 272pp. [Classic treatment of the physical mechanisms of radiation, the budgets of energy, momentum and moisture, turbulent transfer and diffusion]

Simpkin, T. and Fiske, R. S. (1983) *Krakatau 1883,* Smithsonian Institution Press, Washington, DC, 464pp.

Strahler, A. N. (1965) *Introduction to Physical Geography,* Wiley, New York, 455pp.

Sverdrup, H.V. (1945) *Oceanography for Meteorologists,* Allen & Unwin, London, 235pp.

Journal articles

Ahmad, S. A. and Lockwood, J. G. (1979) Albedo. *Prog. Phys. Geog.* 3, 520–43.

Ardanuy, P. E., Kyle, H. L. and Hoyt, D. (1992) Global relationships among the earth's radiation budget, cloudiness, volcanic aerosols and surface temperature. *J. Climate* 5(10), 1120–39.

Barry, R. G. (1985) The cryosphere and climatic change, in MacCracken, M. C. and Luther, F. M.

(eds) *Detecting the Climatic Effects of Increasing Carbon Dioxide*, DOE/ER-0235, US Department of Energy, Washington, DC, 109–48.

Barry, R. G. and Chambers, R. E. (1966) A preliminary map of summer albedo over England and Wales. *Quart. J. Roy. Met. Soc.* 92, 543–8.

Beckinsale, R. P. (1945) The altitude of the zenithal sun: a geographical approach. *Geog. Rev.* 35, 596–600.

Berger, A. (1996) Orbital parameters and equations, in Schneider, S. H. (ed.) *Encyclopedia of Climate and Weather*, Vol. 2, Oxford University Press, New York, 552–7.

Budyko, M. I., Nayefimova, N. A., Zubenok, L. I. and Strokhina, L. A. (1962) The heat balance of the surface of the earth. *Soviet Geography* 3(5), 3–16.

Callendar, G. S. (1938) The artificial production of carbon dioxide and its influence on climate. *Quart. J. Roy. Met. Soc.* 64, 223–40.

Callendar, G. S. (1961) Temperature fluctuations and trends over the earth. *Quart. J. Roy. Met.* Soc. 87, 1–12.

Currie, R. G. (1993) Luni–solar 18.6 and solar cycle 10–11 year signals in U.S.A. air temperature records. *Int. J. Climatology* 13, 31–50.

Foukal, P. V. (1990) The variable sun. *Sci. American* 262(2), 34–41.

Garnett, A. (1937) Insolation and relief. *Trans. Inst. Brit. Geog.* 5 (71pp.).

Henderson-Sellers, A. and Wilson, M. F. (1983) Surface albedo data for climate modeling. *Rev. Geophys. Space Phys.* 21(1),743–78.

Kiehl, J. T. and Trenberth, K. E. (1997) Earth's annual global mean energy budget. *Bull. Amer. Met. Soc.* 78, 197–208.

Kraus, H. and Alkhalaf, A. (1995) Characteristic surface energy balances for different climate types. *Int. J. Climatology* 15, 275–84.

Kung, E. C., Bryson, R. A. and Lenschow, D. H. (1964) Study of a continental surface albedo on the basis of flight measurements and structure of the earth's surface cover over North America. *Mon. Weather Rev.* 92, 543–64.

Kyle, H. L. *et al.* (1993) The Nimbus Earth Radiation Budget (ERB) experiment: 1975–1992. *Bull. Amer. Met. Soc.* 74, 815–30.

Lean, J. (1991) Variations in the sun's radiative output. *Rev. Geophys.* 29, 505–35.

Lean, J. and Rind, D. (1994) Solar variability: implications for global change. *EOS* 75(1), 1 and 5–7.

London, J., Warren, S. G. and Hahn, C. J. (1989) The global distribution of observed cloudiness – a contribution to the ISCCP. *Adv. Space Res.* 9, 161–5.

Lumb, F. E. (1961) Seasonal variation of the sea surface temperature in coastal waters of the British Isles, Sci. Paper No. 6, Meteorological Office, HMSO, London (21pp.).

McFadden, J. D. and Ragotzkie, R. A. (1967) Climatological significance of albedo in central Canada. *J. Geophys. Res.* 72(1),135–43.

Minami, K. and Neue, H-U. (1994) Rice paddies as a methane source. *Climatic Change* 27, 13–26.

Monin, A. S. (1975) The role of the oceans in climatic models, in *The Physical Basis of Climate and Climate Modelling*, GARP Publishing, Series No. 16, World Meteorological Organization, Geneva, 201–5.

Newell, R. E. (1964) The circulation of the upper atmosphere. *Sci. Amer.* 210, 62–74.

Paffen, K. (1967) Das Verhaeltniss der Tages zur Jahrzeitlichen Temperaturschwankung. *Erdkunde* 21, 94–111.

Ramanathan, V., Barkstrom, B. R. and Harrison, E. F. (1990) Climate and the earth's radiation budget. *Physics Today* 42, 22–32.

Ramanathan, V., Cess, R. D., Harrison, E. F., Minnis, P., Barkstrom, B. R., Ahmad, E. and Hartmann, D. (1989) Cloud-radiative forcing and climate: results from the Earth Radiation Budget Experiment. *Science* 243, 57–63.

Ransom, W. H. (1963) Solar radiation and temperature. *Weather* 8, 18–23.

Sellers, W. D. (1980) A comment on the cause of the diurnal and annual temperature cycles. *Bull. Amer. Met. Soc.* 61, 741–55.

Stephens, G. L., Campbell, G. G. and Vonder Haar, T. H. (1981) Earth radiation budgets. *J. Geophys. Res.* 86(C10), 9739–60.

Stone, R. (1955) Solar heating of land and sea. *J. Geography* 40, 288.

Strangeways, I. (1998) Back to basics: the 'met. enclosure'. Part 3: Radiation. *Weather* 53, 43–9.

Tully, J. P. and Giovando, L. F. (1963) Seasonal temperature structure in the eastern subarctic Pacific Ocean, in *Maritime Distributions, Roy. Soc. Canada, Spec.Pub.* 5, Dunbar, M. J. (ed.), 10–36.

Weller, G. and Wendler, G. (1990) Energy budgets over various types of terrain in polar regions. *Ann. Glac.* 14, 311–14.

Wilson, R. C. and Hudson, H. S. (1991) The sun's luminosity over a complete solar cycle. *Nature* 351, 42–3.

CHAPTER FOUR

Atmospheric moisture budget

4

LEARNING OBJECTIVES

When you have read this chapter you will:

- be familiar with the major atmospheric components of the hydrological cycle
- know the main controls of evaporation and condensation
- be aware of the spatial and temporal characteristics of moisture in the atmosphere, evaporation and precipitation
- know the different forms of precipitation and typical statistical characteristics
- know the major geographical and altitudinal patterns of precipitation and their basic causes
- understand the nature and characteristics of droughts.

This chapter considers the role of water in its various phases (solid, liquid and vapor) in the climate system and the transfers (or cycling) of water between the major reservoirs – the oceans, the land surface and the atmosphere. We discuss measures of humidity, large-scale moisture transport, moisture balance, evaporation and condensation.

A THE GLOBAL HYDROLOGICAL CYCLE

The global hydrosphere consists of a series of reservoirs interconnected by water cycling in various phases. These reservoirs are the oceans; ice sheets and glaciers; terrestrial water (rivers, soil moisture, lakes and ground water); the biosphere (water in plants and animals); and the atmosphere. The oceans, with a mean depth of 3.8km and covering 71 percent of the earth's surface, hold 97 percent of *all* the earth's water (23.4×10^6km^3). Approximately 70 percent of the total *fresh* water is locked up in ice sheets and glaciers, while almost all of the remainder is ground water. It is an astonishing fact that rivers and lakes hold only 0.3 percent of all fresh water and the atmosphere a mere 0.04 percent (Figure 4.1). The average residence time of water within these reservoirs varies from hundreds or thousands of years for the oceans and polar ice to only about 10 days for the atmosphere. Water cycling involves evaporation, the transport of water vapor in the atmosphere, condensation, precipitation and terrestrial runoff. The equations of the water budget for the atmosphere and for the surface are respectively:

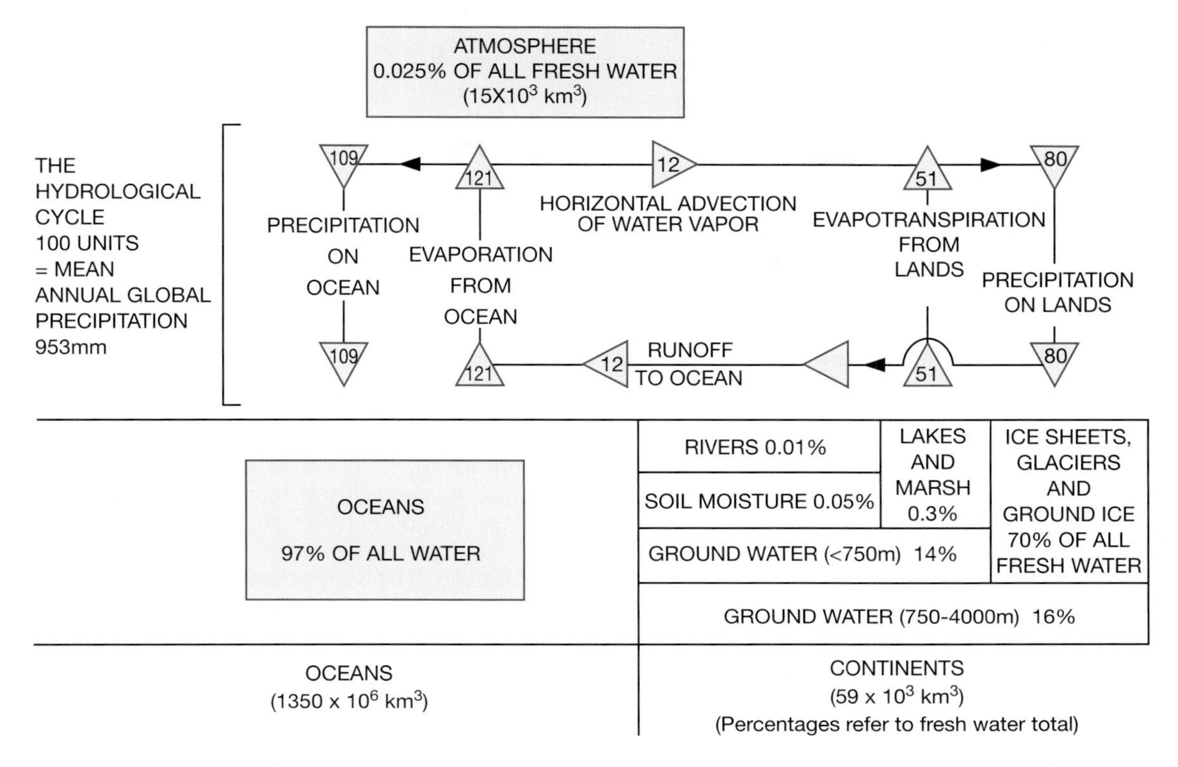

Figure 4.1 The hydrological cycle and water storage of the globe. The exchanges in the cycle are referred to 100 units, which equal the mean annual global precipitation of 953mm. The percentage storage figures for atmospheric and continental water are percentages of all *fresh* water. The saline ocean waters make up 97 percent of *all* water. The horizontal advection of water vapor indicates the *net* transfer from oceans to land. The land runoff of 29 units corresponds to 12 units over the oceans – an area ratio of 0.42.

Source: Based on data in Rudolfl and Rubel (2005).

$$\Delta Q = E - P + D_Q$$

and $\Delta S = P - E - r$

where ΔQ is the time change of moisture in an atmospheric column, E = evaporation, P = precipitation, D_Q = moisture divergence out of the column, ΔS = surface storage of water and r = runoff. For short-term processes, the water balance of the atmosphere may be assumed to be in equilibrium; however, over periods of tens of years, global warming may increase its water storage capacity.

Because of its large heat capacity, the global occurrence and transport of water is closely linked to global energy. Atmospheric water vapor is responsible for the bulk of total global energy lost into space by infrared radiation. Over 75 percent of the energy input from the surface into the atmosphere is a result of the liberation by condensation of latent heat (that is generated during evaporation) and, principally, the

Table 4.1 Mean water content of the atmosphere (in mm of rainfall equivalent)

	Northern Hemisphere	Southern Hemisphere	World
January	19	25	22
July	34	20	27

Source: After Sutcliffe (1956).

formation of clouds and the production of rainfall.

The average storage of water vapor in the atmosphere (Table 4.1), termed the precipitable water content (about 25mm), is sufficient for only about 10 days, supply of rainfall over the earth as a whole. However, intense (horizontal) influx of moisture into the air over a given region makes possible short-term rainfall totals greatly in excess of 30mm. The phenomenal record total of 1870mm fell on the island of Réunion, off Madagascar, during 24 hours in March 1952, and much greater intensities have been observed over shorter periods (see E.2a, this chapter).

B HUMIDITY

1 Moisture content

Atmospheric moisture comprises water vapor, and water droplets and ice crystals in clouds. Moisture content is determined by local evaporation, air temperature and the horizontal atmospheric transport of moisture. Cloud water, on average, amounts to only 4 percent of atmospheric moisture. The moisture content of the atmosphere can be expressed in several ways, apart from the vapor pressure (p. 31), depending on which aspect the user wishes to emphasize. The total mass of water in a given volume of air (i.e., the density of the water vapor) is one such measure. This is termed the *absolute humidity* (r_w) and is measured in grams per cubic metre (g m^{-3}). Volumetric measurements are seldom used in meteorology and more convenient is the *mass mixing ratio* (x). This is the mass of water vapor in grams per kilogram of dry air. For most practical purposes, the *specific humidity* (q) is identical, being the mass of vapor per kilogram of air, including its moisture.

More than 50 percent of the atmospheric moisture content is located below 850mb (approximately 1450m) and more than 90 percent below 500mb (5575m). Figure 4.2 illustrates typical vertical distributions in spring in mid-

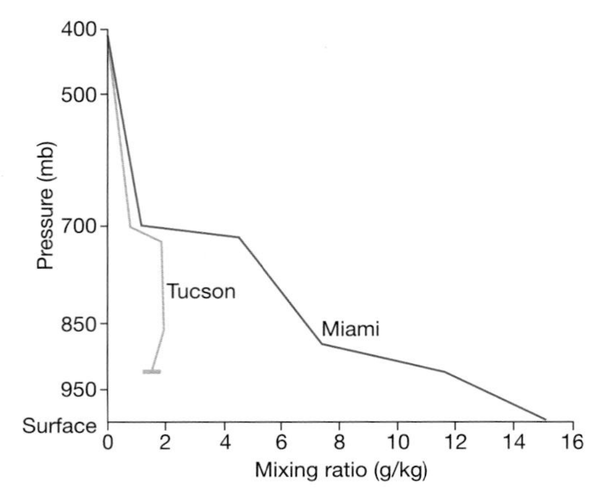

Figure 4.2 The vertical variation of atmospheric vapor content (g/kg) at Tucson, AZ, and Miami, FL, at 12 UTC on 27 March 2002.

latitudes. It is also apparent that the seasonal effect is most marked in the lowest 3000m (i.e., below 700mb). Air temperature sets an upper limit to water vapor pressure – the saturation value (i.e., 100 percent relative humidity); consequently we may expect the distribution of mean vapor content to reflect this control. In January, minimum values of 1–2mm (equivalent depth of water) occur in northern continental interiors and high latitudes, with secondary minima of 5–10mm in tropical desert areas, where there is subsiding air (Figure 4.3). Maximum vapor contents of 50–60mm are over southern Asia during the summer monsoon and over equatorial latitudes of Africa and South America.

Another important measure is *relative humidity* (r), which expresses the actual moisture content of a sample of air as a percentage of that contained in the same volume of saturated air at the same temperature. The relative humidity is defined with reference to the mixing ratio, but it can be determined approximately in several ways:

$$r = \frac{x}{x_s} \times 100 < \frac{q}{q_s} \times 100 < \frac{e}{e_s} \times 100$$

where the subscript s refers to the respective saturation values at the same temperature; e denotes vapor pressure.

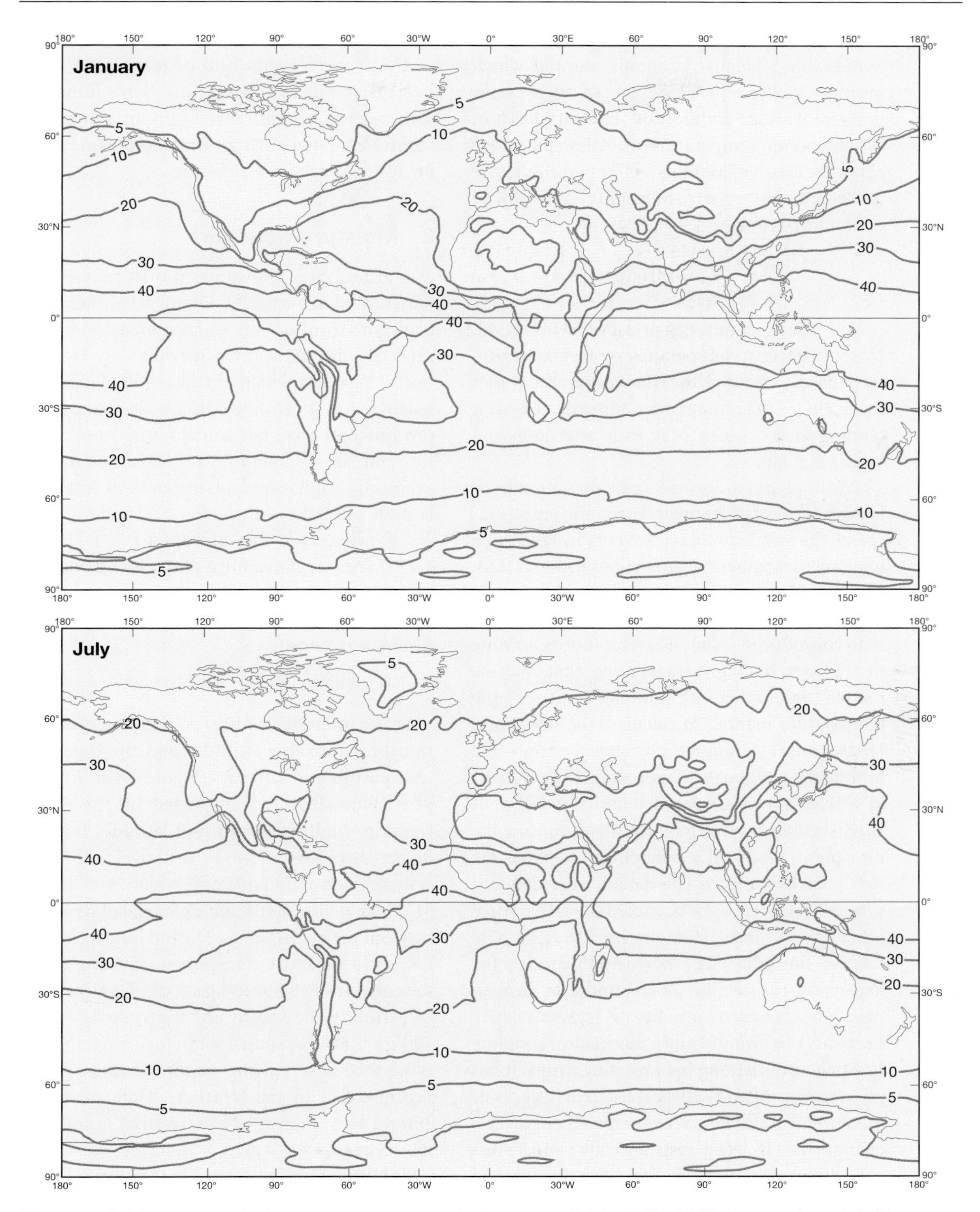

Figure 4.3 Mean atmospheric water vapor content in January and July, 1970–1999, in mm of precipitable water.
Source: Climate Diagnostics Center, CIRES-NOAA, Boulder, CO.

A further index of humidity is the dew-point temperature. This is the temperature at which saturation occurs if air is cooled at constant pressure without addition or removal of vapor. When the air temperature and dew-point are equal the relative humidity is 100 percent, and it is evident that relative humidity can also be determined from

$$\frac{e_s \text{ at dew-point}}{e_s \text{ at air temperature}} \times 100$$

The relative humidity of a parcel of air will change if either its temperature or its mixing ratio is changed. In general, the relative humidity varies inversely with temperature during the day, tending to be lower in the early afternoon and higher at night.

Atmospheric moisture can be measured by at least six types of instrument. For routine measurements the *wet-bulb thermometer* is installed in a louvered instrument shelter (Stevenson screen). The bulb of the standard thermometer is wrapped in muslin, which is kept moist by a wick from a reservoir of pure water. The evaporative cooling of this wet bulb gives a reading that may be used in conjunction with a simultaneous dry-bulb temperature reading to calculate the dew-point temperature. A similar portable device – the aspirated *psychrometer* – uses a forced flow of air at a fixed rate over the dry and wet bulbs. A sophisticated instrument for determining the dew-point, based on a different principle, is the *dew-point hygrometer*. This detects when condensation first occurs on a cooled surface. Three other types of instrument are used to determine relative humidity. The *hygrograph* utilizes the expansion/contraction of a bundle of human hair, in response to humidity, to register relative humidity continuously by a mechanical coupling to a pen arm marking on a rotating drum. It has an accuracy of ± 5–10 percent. For upper air measurements, a *lithium chloride* element detects changes in electrical resistance to vapor pressure differences. Relative humidity changes are accurate within ± 3 percent. Automatic weather stations (AWS) commonly employ an electrical method where a thin film of material changes its capacitance in relation to relative humidity The material is a thin metal film on a thin glass substrate coated with an organic polymer that forms the capacitor's dielectric.

2 Moisture transport

The atmosphere transports moisture horizontally as well as vertically. Figure 4.1 shows a net transport from oceans to land areas. Moisture must also be transported meridionally (south–north) in order to maintain the required moisture balance at a given latitude; i.e., evaporation – precipitation = net horizontal transport of moisture into an air column. Comparison of annual average precipitation and evaporation totals for latitude zones shows that in low and mid-latitudes $P > E$, whereas in the subtropics $P < E$ (Figure 4.4A). These regional imbalances are maintained by net moisture transport into (convergence) and out of (divergence) the respective zones (D_Q, where divergence is positive).

$$E - P = D_Q$$

A prominent feature is the equatorward transport into low latitudes and the poleward transport in mid-latitudes (Figure 4.4B). Atmospheric moisture is transported by the global westerly wind systems of mid-latitudes towards higher latitudes and by the easterly Trade Wind systems towards the equatorial region (see Chapter 8). There is also significant exchange of moisture between the hemispheres. During June to August there is a moisture transport northward across the equator of 18.8×10^8kg s^{-1}; during December to February the southward transport is 13.6×10^8kg s^{-1}. The net annual south to north transport is 3.2×10^8kg s^{-1}, giving an annual excess of net precipitation in the Northern Hemisphere of 39mm. This is returned by terrestrial runoff into the oceans.

It is important to stress that local evaporation is, in general, not the major source of local

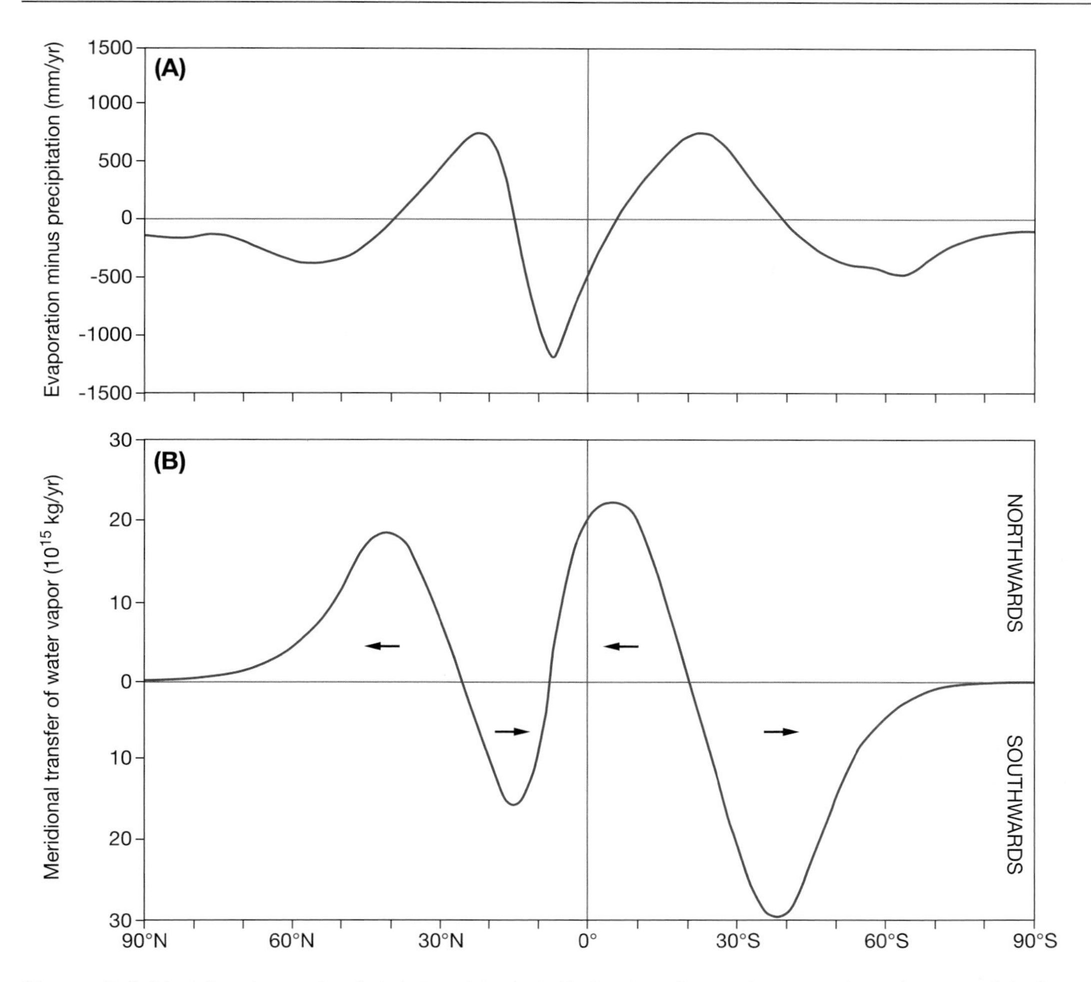

Figure 4.4 Meridional aspects of global moisture. A: Estimates of annual evaporation minus precipitation (in cm) as a function of latitude; B: Annual meridional transfer of water vapor (in 10^{15}kg).

Source: A: After J. Dodd. From Browning 1993. By permission NERC. B: From Sellers 1965. Courtesy of University of Chicago Press.

precipitation. For example, 32 percent of the summer season precipitation over the Mississippi River basin and between 25 and 35 percent of that over the Amazon basin is of 'local' origin, the remainder being transported into these basins by moisture advection. Even when moisture is available in the atmosphere over a region, only a small portion of it is usually precipitated. This depends on the efficiency of the condensation and precipitation mechanisms, both microphysical and large scale.

Using atmospheric sounding data on winds and moisture content, global patterns of average water vapor flux divergence (i.e., $E - P > 0$) or convergence (i.e., $E - P < 0$) can be determined. The distribution of atmospheric moisture 'sources' (i.e., $P < E$) and 'sinks' (i.e., $P > E$) form an important basis for understanding global climates. Strong divergence (outflow) of moisture occurs over the northern Indian Ocean in summer, providing moisture for the monsoon. Subtropical divergence zones are associated with the high pressure areas. The oceanic subtropical highs are evaporation sources; divergence over land may reflect underground water supply or may be artifacts of sparse data.

C EVAPORATION

Evaporation (including transpiration from plants) provides the moisture input into the atmosphere; the oceans provide 87 percent and the continents 13 percent.

The highest annual values (1500mm), averaged zonally around the globe, occur over the tropical oceans, associated with Trade Wind belts, and over equatorial land areas in response to high solar radiation receipts and luxuriant vegetation growth (Figure 4.5A). The larger oceanic evaporative losses in winter, for each hemisphere (Figure 4.5B), represent the effect of outflows of cold continental air over warm ocean currents in the western North Pacific and North Atlantic (Figure 4.6) and stronger Trade Winds in the cold season of the Southern Hemisphere.

Evaporation requires an energy source at a surface that is supplied with moisture; the vapor pressure in the air must be below the saturated value (*es*); and air motion removes the moisture transferred into the surface layer of air. As illustrated in Figure 2.16, the saturation vapor pressure increases with temperature. The change in state from liquid to vapor requires energy to be expended in overcoming the intermolecular attractions of the water particles. This energy is often acquired by the removal of heat from the immediate surroundings, causing an apparent heat loss (*latent heat*), as discussed on p. 72, and a consequent drop in temperature. The latent heat of vaporization needed to evaporate 1kg of water at 0°C is 2.5×10^6J. Conversely, condensation releases this heat, and the temperature of an air mass in which condensation is occurring is increased as the water vapor reverts to the liquid

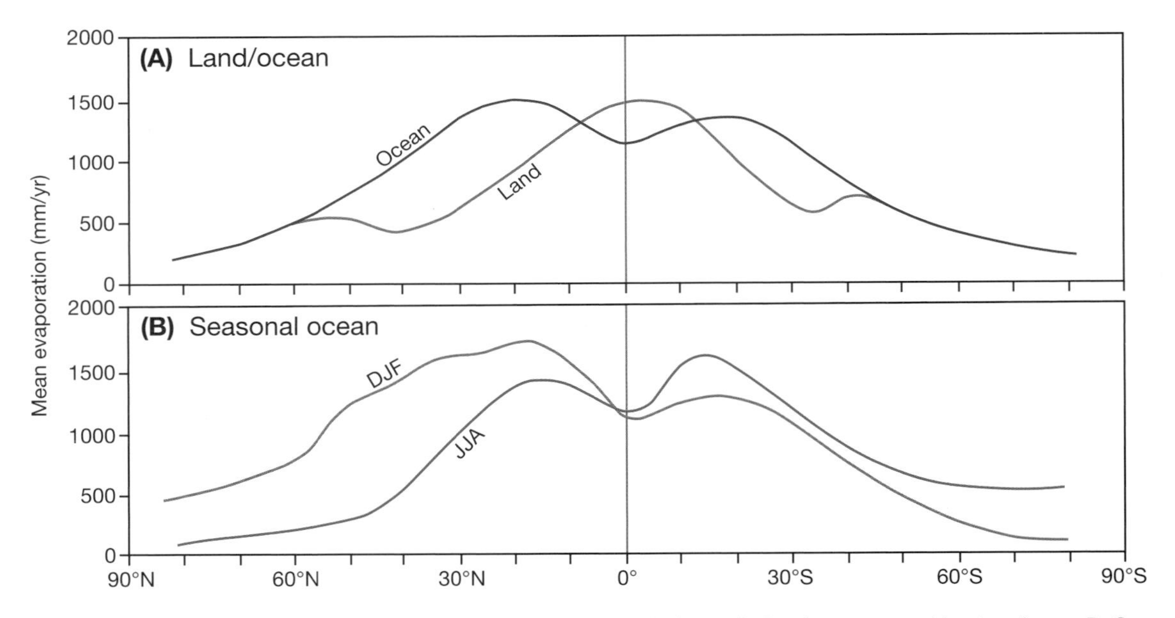

Figure 4.5 Zonal distribution of mean evaporation (mm/year). A: Annually for the ocean and land surfaces; B: Over the oceans for December to February and June to August.

Sources: After Peixoto and Oort (1983), Fig 22. Copyright (c) D. Reidel, Dordrecht, by kind permission of Kluwer Academic Publishers. Also partly from Sellers (1965).

state. In the case of ice, the latent heat of fusion ($0.33 \times 10^6 Jkg^{-1}$) is needed to melt the ice to water at 0°C. The same amount of heat is released during freezing. Sublimation/deposition of ice directly to vapor, or vice versa, involves the sum of the two latent heats (i.e., $2.83 \times 10^6 Jkg^{-1}$) and therefore sublimation is less common than evaporation. Nevertheless, in dry windy climates 15–30 percent of the annual snow pack may be lost through sublimation in situ combined with the more important sublimation of blowing snow.

The diurnal range of temperature can be moderated by humid air, when evaporation takes place during the day and condensation at night. The relationship of saturation vapor pressure to temperature (Figure 2.14) means that evaporation processes limit low-latitude ocean surface temperature (i.e., where evaporation is at a maximum) to values of about 30°C. This plays an important role in regulating the temperature of ocean surfaces and overlying air in the tropics.

The rate of evaporation depends on a number of factors. The two most important are the difference between the saturation vapor pressure at the water surface and the vapor pressure of the air, and the existence of a continual supply of energy to the surface. Wind velocity also affects the evaporation rate, because the wind is generally associated with the advection of unsaturated air, which will absorb the available moisture.

Water loss from plant surfaces, chiefly leaves, is a complex process termed *transpiration*. It occurs when the vapor pressure in the leaf cells is greater than the atmospheric vapor pressure. It is vital as a life function in that it causes a rise of plant nutrients from the soil and cools the leaves. The cells of the plant roots can exert an osmotic tension of up to about 15 atmospheres upon the water films between the adjacent soil particles. As these soil water films shrink, however, the tension within them increases. If the tension of the soil films exceeds the osmotic root tension, the continuity of the plant's water uptake is broken, and wilting occurs. Transpiration is controlled by the atmospheric factors that determine evaporation as well as by plant factors such as the stage of plant growth, leaf area and leaf temperature, and also by the amount of soil moisture (see Chapter 12C). It occurs mainly during the day, when the *stomata* (small pores in the leaves), through which transpiration takes place, are open. This opening is determined primarily by light intensity. Transpiration naturally varies greatly with season, and during the winter months in mid-latitudes conifers lose only 10–18 percent of their total annual transpiration losses and deciduous trees less than 4 percent.

In practice, it is difficult to separate water evaporated from the soil, *intercepted* (liquid or solid), remaining on vegetation surfaces after precipitation and subsequently evaporated or sublimated, and transpiration. For this reason, evaporation, or the compound term *evapotranspiration,* may be used to refer to the total loss. Over land, annual evaporation is 52 percent due to transpiration, 28 percent soil evaporation and 20 percent interception.

Evapotranspiration losses from natural surfaces cannot be measured directly. There are, however, various indirect methods of assessment, as well as theoretical formulae. One method of estimation is based on the moisture balance equation at the surface:

$$P - E = r + \Delta S$$

ΔS is the storage change in the block of soil and this term may also include the storage of water in the snow pack. This equation may be applied to a gauged river catchment, where precipitation and runoff (r) are measured, or to a block of soil. In the latter case we measure the percolation through an enclosed block of soil with a vegetation cover (usually grass, but occasionally a large area with tree cover) and record the rainfall upon it. The block, termed a *lysimeter*, is weighed regularly so that weight changes unaccounted for by rainfall or runoff may be ascribed to evapotranspiration losses, provided the grass is kept short! The technique allows the determination of daily evapotranspiration amounts. If the soil block is regularly 'irrigated' so that the vegetation cover is always yielding the maximum possible evapotranspiration, the water loss is called *potential evapotranspiration* (or PE). More generally, PE may be defined as the water loss corresponding to the available energy. Potential evapotranspiration forms the basis for the climate classification developed by C. W. Thornthwaite (see Appendix 1).

In regions where snow cover is long-lasting, evaporation/sublimation from the snow pack can be estimated by lysimeters (pans or box containers) sunk into the snow that are weighed regularly.

A meteorological solution to the calculation of evaporation uses sensitive instruments to measure the net effect of eddies of air transporting moisture upward and downward near the surface. In this 'eddy correlation' (or eddy covariance) technique, the vertical component of wind and the atmospheric moisture content are measured simultaneously at the same level (say, 1.5m) every 10^{-1} s (10Hz). The product of each pair of measurements is then averaged over a time interval of 15–60 minutes to determine the evaporation (or condensation). This method requires delicate rapid-response instruments, so it cannot be used in very windy conditions. Sonic anemometers are used to measure the vertical and horizontal wind components. These use sound pulses to measure the difference in time that it takes for sound to travel with and against the wind, allowing the wind speed to be calculated. The humidity is determined by measuring the absorption of infrared radiation by water vapor in the air.

Theoretical methods for determining evaporation rates have followed two lines of approach. The first relates average monthly evaporation (E) from large water bodies to the mean wind speed (u) and the mean vapor pressure difference between the water surface and the air ($e_w - e_d$) in the form:

$$E = Ku(e_w - e_d)$$

where K is an empirical constant. This is termed the aerodynamic (or bulk) approach because it takes account of the factors responsible for removing vapor from the water surface. The second method is based on the energy budget. The *net balance* of solar and terrestrial radiation at the surface (R_n) is used for evaporation (E) and the transfer of heat to the atmosphere (H). A small proportion also heats the soil by day, but since nearly all of this is lost at night it can be disregarded. Thus:

$$R_n = LE + H$$

where L is the latent heat of evaporation ($2.5 \times 10^6 \mathrm{J\ kg^{-1}}$). Rn can be measured with a net radiometer and the ratio $H/LE = \beta$, referred to as Bowen's ratio, can be estimated from measurements of air temperature and vapor content (dew-point) at two levels near the surface; the levels are typically at about 0.5 and 2m. β ranges from <0.1 for water to $\geqslant 10$ for a desert surface. The use of this ratio assumes that the vertical transfers of heat and water vapor by turbulence take place with equal efficiency. Evaporation is then determined from an expression of the form:

$$E = \frac{R_n}{L(1 + \beta)}$$

The conversion of evaporation to energy units is 1mm evaporation = $2.5 \times 10^6 \mathrm{J\ m^{-2}}$.

The most satisfactory climatological method so far devised combines the energy budget and aerodynamic approaches. In this way, H. L. Penman succeeded in expressing evaporation losses in terms of four meteorological elements that are regularly measured, at least in Europe and North America. These are net radiation (or an estimate based on sunshine duration), mean air temperature, mean air humidity and mean wind speed (which limit the losses of heat and vapor from the surface).

The relative roles of these factors are illustrated by the global pattern of evaporation (see Figure 4.6). Losses decrease sharply in high latitudes, where there is little available energy. In middle and lower latitudes, there are appreciable differences between land and sea. Rates are naturally high

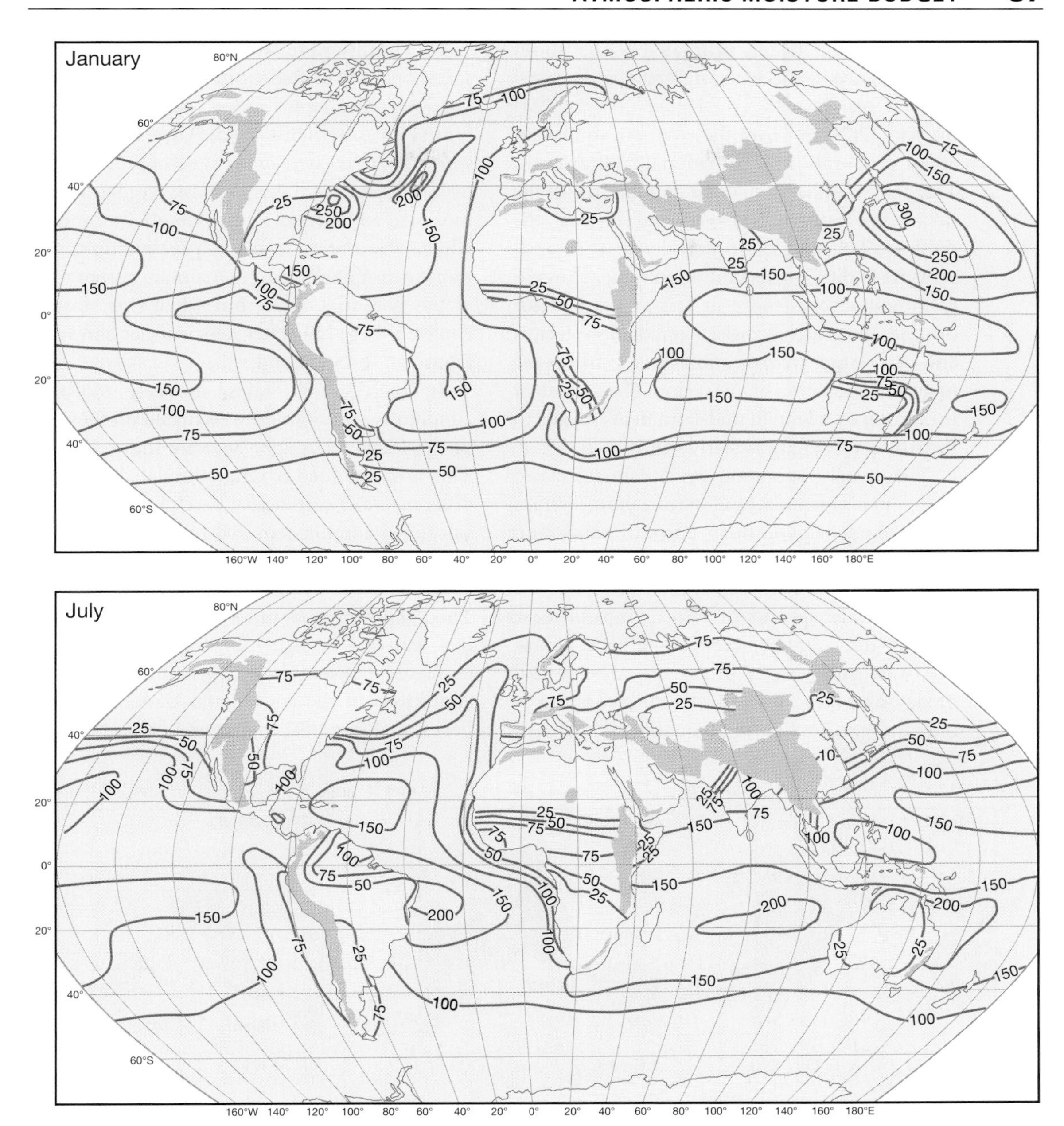

Figure 4.6 Mean evaporation (mm) for January and July (after M. I. Budyko, *Heat Budget Atlas of the Earth*, 1958).

over the oceans in view of the unlimited availability of water, and on a seasonal basis the maximum rates occur in January over the western Pacific and Atlantic, where cold continental air blows across warm ocean currents. On an annual basis, maximum oceanic losses occur about 15–20°N and 10–20°S, in the belts of the constant Trade Winds (see Figures 4.5B and 4.6). The highest annual losses, estimated to be about 2000mm, are in the western Pacific and central Indian Ocean near 15°S (cf. Figure 3.30); 2460MJ m^{-2} yr^{-1} (78W m^{-2} over the year) are equivalent to an evaporation of 1000mm of water/cm^2). There is a subsidiary equatorial minimum over the oceans as a result of the lower wind speeds in the doldrum belt and the proximity of the vapor pressure in the air to its saturation value. The land maximum occurs more or less at the equator owing to the relatively high solar radiation receipts and the large transpiration losses from the luxuriant vegetation of this region. The secondary maximum over land in mid-latitudes is related to the strong prevailing westerly winds.

The annual evaporation over Britain, calculated by Penman's formula, ranges from about 380mm in Scotland to 500mm in parts of south and southeast England. Since this loss is concentrated in the period May to September, there may be seasonal water deficits of 120–150mm in these parts of the country, necessitating considerable use of irrigation water by farmers. The annual moisture budget can also be determined approximately by a bookkeeping method devised by C. W. Thornthwaite, where potential evapotranspiration is estimated from mean temperature. Figure 4.7 illustrates this for stations in western, central and eastern Britain (cf. Figure 10.25). In the winter months, there is an excess of precipitation over evaporation; this goes to recharging the soil moisture, and further surplus runoff. In summer, when evaporation exceeds precipitation, soil moisture is initially used to maintain evaporation at the potential value, but when this store is depleted there is a water deficiency as shown in Figure 4.7 for Southend.

In the United States, monthly moisture conditions are commonly evaluated on the basis of the Palmer Drought Severity Index (PDSI). This is determined from accumulated weighted differences between actual precipitation and the calculated amount required for evapotranspiration, soil recharge and runoff. Accordingly, it takes account of the persistence effects of droughts. The PDSI ranges from ≥4 (extremely moist) to ≤–4 (extreme drought). Figure 4.8 indicates an oscillation between drought and unusually moist conditions in the continental USA during the period October 1992 to August 1993.

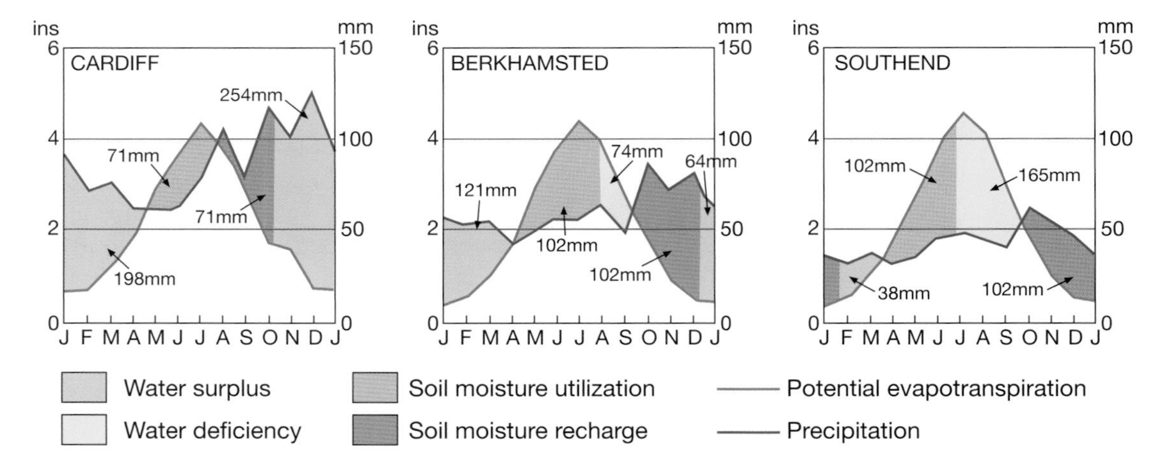

Figure 4.7 The average annual moisture budget for stations in western, central and eastern Britain determined by Thornthwaite's method. When potential evaporation exceeds precipitation, soil moisture is used; at Berkhamsted in central England and Southend on the east coast, this is depleted by July to August. Autumn precipitation excess over potential evaporation goes into replenishing the soil moisture until field capacity is reached.

Source: After Howe (1956). Courtesy of the Royal Meteorological Society.

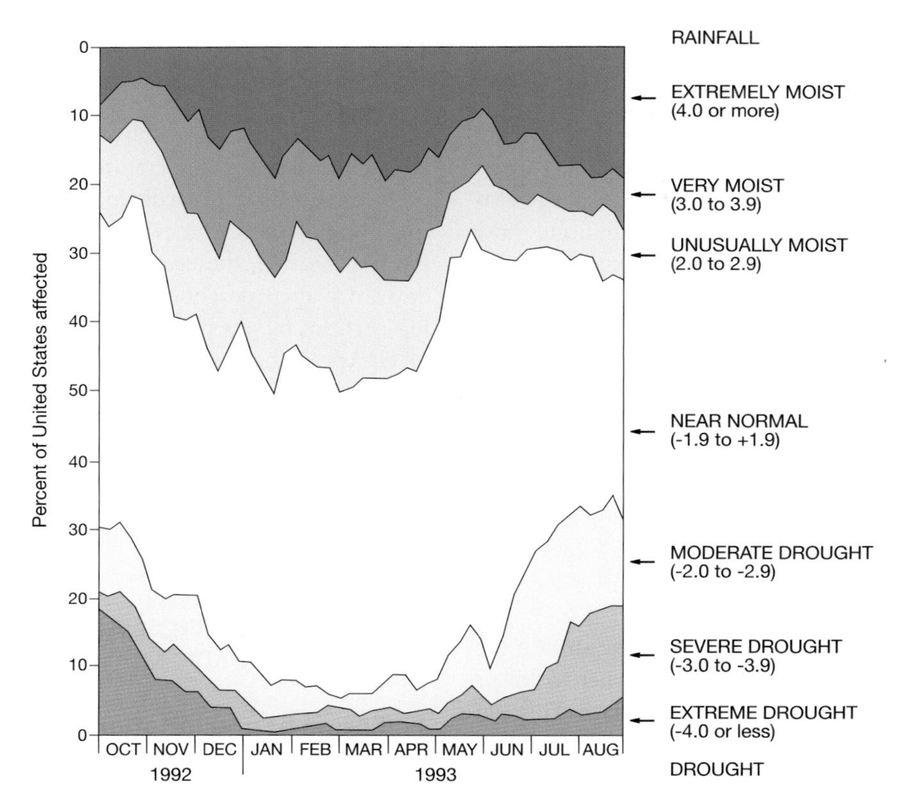

Figure 4.8 Percentage of the continental USA affected by wet spells or drought, based on the Palmer Index (see scale on right), during the period October 1992 to August 1993.

Sources: US Climate Analysis Center and Lott (1994). Reprinted from *Weather* by permission of the Royal Meteorological Society. Crown copyright ©.

D CONDENSATION

Condensation is the direct cause of all the various forms of precipitation. It occurs as a result of changes in air volume, temperature, pressure or humidity. Four mechanisms may lead to condensation: (1) the air is cooled to dew-point but its volume remains constant; (2) the volume of the air is increased without addition of heat; this cooling occurs because adiabatic expansion causes energy to be consumed through work (see Chapter 5); (3) a joint change of temperature and volume reduces the moisture-holding capacity of the air below its existing moisture content; or (4) evaporation adds moisture to the air. The key to understanding condensation lies in the fine balance that exists between these variables. Whenever the balance between one or more of them is disturbed beyond a certain limit, condensation may result.

The most common circumstances favoring condensation are those producing a drop in air temperature; namely contact cooling, radiative cooling, mixing of air masses of different temperatures and dynamic cooling of the atmosphere. Contact cooling occurs within warm, moist air passing over a cold land surface. On a clear winter's night, strong radiation will cool the surface very quickly. This surface cooling gradually extends to the moist lower air, reducing the temperature to a point where condensation occurs in the form of dew, fog or frost, depending on the amount of moisture involved, the thickness of the cooling air layer and the dew-point value. When the latter is below 0°C, it is referred to as the hoar frost point if the air is saturated with respect to ice.

The mixing of contrasting layers within a single air mass, or of two different air masses, can also produce condensation. Figure 4.9 indicates how the horizontal mixing of two air masses (A and B), of given temperature and moisture characteristics, may produce an air mass (C) that is supersaturated at the intermediate temperature and consequently forms cloud. Vertical mixing of an air layer, discussed in Chapter 5 (see Figure 5.7), can have the same effect. Fog, or low stratus, with drizzle – known as 'crachin' – is common along the coasts of south China and the Gulf of Tonkin in February to April. It develops either through air mass mixing or warm advection over a colder surface.

The addition of moisture into the air near the surface by evaporation occurs when cold air moves out over a warm water surface. This can produce steam fog, which is common in Arctic regions. Attempts at fog dispersal are one area where some progress has been made in local weather modification. Cold fogs can be dissipated locally by the use of dry ice (frozen CO_2) or the release of propane gas through expansion nozzles to produce freezing and the subsequent fallout of ice crystals (cf. p. 127). Warm fogs (i.e., having drops above freezing temperatures) present bigger problems, but attempts at dissipation have shown some limited success in evaporating droplets by artificial heating, the use of large fans to draw down dry air from above, the sweeping out of fog particles by jets of water, and the injection of electrical charges into the fog to produce coagulation.

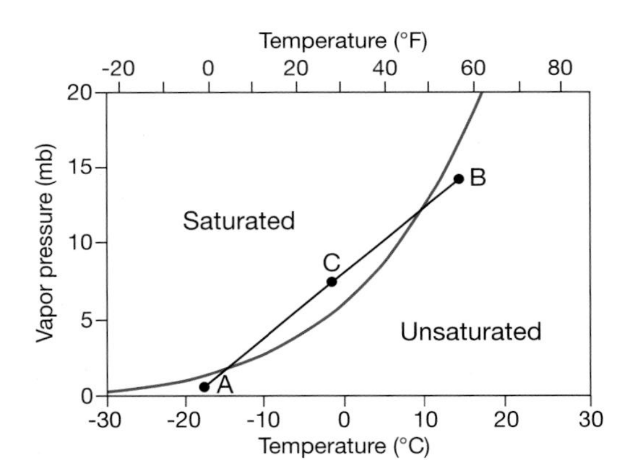

Figure 4.9 The effect of air-mass mixing. The horizontal mixing of two unsaturated air masses A and B will produce one supersaturated air mass C. The saturation vapor pressure curve is shown (cf. Figure 2.13B, which is a semi-logarithmic plot).
Source: After Petterssen (1969).

The most effective cause of condensation is undoubtedly the dynamic process of adiabatic cooling associated with instability. This is discussed in the next chapter.

E PRECIPITATION CHARACTERISTICS AND MEASUREMENT

1 Forms of precipitation

Strictly, *precipitation* refers to all liquid and frozen forms of water. The primary ones are:

- *Rain* – falling water drops with a diameter of at least 0.5mm and typically 2mm; droplets of less than 0.5mm are termed *drizzle*. Rainfall has an accumulation rate of ⩾1mm/hour. Rain (or drizzle) that falls on a surface at subzero temperature forms a glazed ice layer and is termed *freezing rain*. During the protracted 'ice storm' of 5–9 January 1998 in the northeastern United States and eastern Canada, some areas received up to 100mm of freezing rain.
- *Snow* – ice crystals falling in branched clusters as snowflakes. Wet snow has crystals bonded by liquid water in interior pores and crevices. Individual crystals have a hexagonal form (needles or platelets). At low temperatures (–40°C), crystals may float in the air, forming 'diamond dust'.
- *Hail* – hard pellets, balls or irregular lumps of ice, at least 5mm across, formed of alternating

shells of opaque and clear ice. The core of a hailstone is a frozen water drop (ice pellet) or an ice particle (graupel).

- *Graupel* – snow pellets, opaque conical or rounded ice particles 2–5mm in diameter formed by aggregation ofice crystals.
- *Sleet* – refers in the UK to a rain–snow mixture; in North America, to small translucent ice pellets (frozen raindrops) or snowflakes that have melted and refrozen.
- *Dew* – condensation droplets on the ground surface or grass, deposited when the surface temperature is below the air's dew-point temperature. *Hoar frost* is the frozen form, when ice crystals are deposited on a surface.
- *Rime* – clear crystalline or granular ice deposited when supercooled fog or cloud droplets encounter a vertical structure, trees, or suspended cable. The rime deposit grows into the wind in a triangular form related to the wind speed. It is common in cold, maritime climates and on mid-latitude mountains in winter.

In general, only rain and snow make significant contributions to precipitation totals. In many parts of the world, the term *rainfall* can be used interchangeably with precipitation. Precipitation is measured in a rain gauge, a cylindrical container capped by a funnel to reduce evaporative losses, which most commonly stands on the ground. Its height is about 60cm and its diameter about 20cm. More than 50 types of rain gauge are in use by national meteorological services around the world! In windy and snowy regions they are often equipped with a wind shield to increase the catch efficiency. It must be emphasized that precipitation records are only *estimates.* Factors of gauge location, its height above ground, turbulence in the airflow, splash-in and evaporation all introduce errors in the catch. Gauge design differences affect the airflow over the gauge aperture, retention by wetting, and the evaporation losses from the container. Falling snow is particularly subject to wind effects, which can result in underrepresentation of the true amount by 50 percent or more. It has been shown that a double snow fence around the gauge installation greatly improves the measured catch. Corrections to gauge data need to take account of the proportion of precipitation falling in liquid and solid form, wind speeds during precipitation events, and precipitation intensity. Studies in Switzerland suggest that observed totals underestimate the true amounts by 7 percent in summer and 11 percent in winter below 2000m, but by as much as 15 percent in summer and 35 percent in winter in the Alps between 2000 and 3000m.

The density of gauge networks limits the accuracy of areal precipitation estimates. The number of gauges per 10,000km^2 area ranges from 245 gauges in Britain to ten in the United States and only three in Canada and Asia. The coverage is particularly sparse in mountain and polar regions. In many land areas, weather radar provides unique information on storm systems and quantitaive estimates of area-averaged precipitation (see Box 4.1). Ocean data come from island stations and ship observations of precipitation frequency and relative intensity. Satellite remote sensing, using infrared and passive microwave data, provides independent estimates of large-scale ocean rainfall.

2 Precipitation characteristics

The climatological characteristics of precipitation may be described in terms of mean annual precipitation, the annual cycle, annual variability and decadal trends. However, hydrologists are interested in the properties of individual rainstorms. Weather observations usually indicate the amount, duration and frequency of precipitation, and these enable other derived characteristics to be determined. Three of these are discussed below.

Rainfall intensity

The intensity (= amount/duration) of rainfall during an individual storm, or a still shorter period, is of vital interest to hydrologists and water

SIGNIFICANT 20TH-CENTURY ADVANCE

4.1 Radar meteorology

Radio detection and ranging (radar), developed for aircraft detection during World War II, was swiftly applied to tracking precipitation areas from the radar echoes. Radio waves transmitted by an antenna in the cm wavelength range (typically 3 and 10cm) are back scattered by raindrops and ice particles, as well as by cloud droplets, particulates, insect swarms and flocks of birds. The return signal and its time delay provide information on the objects in the path of the beam and their direction, distance and altitude. The need to detect tropical rainstorms led to the first training programs in radar interpretation in 1944. In 1946–1947, the Thunderstorm Project led by H. R. Byers used radar to track the growth and organization of thunderstorms in Florida and Ohio. Gradually, indicators of storm severity were devised based on the shape and arrangement of echoes, their vertical extent and the strength of the back scatter measured in decibels (dB). Much of this process is now automated. Specifically designed weather radars for the U.S. Weather Bureau became available only in 1957. In the 1970s, the Doppler radar, which uses the frequency shift produced by a moving target to determine the horizontal motion relative to the radar location, began to be used for research on hail and tornadoes. Dual Doppler systems are used to calculate the horizontal wind vector. The Next Generation Weather Radar (NEXRAD) deployed in the 1990s in the United States, and similar systems in Canada and European countries, are modern Doppler instruments. The vertical profile of winds in the atmosphere can be determined with vertically pointing Doppler radar operating in the VHF (30MHz) to UHF (3GHz) ranges. The wind velocity is calculated from variations in the clear air refractive index caused by turbulence. Beginning in the 1980s, but more particularly over the past ten years, millimeter wavelength (35 and 94GHz) radars have come into use to study small cloud droplets and ice crystals in clouds. In 2006 a 94GHz radar was launched on CloudSat.

A major application of radar is in estimating precipitation intensity. R. Wexler and J.S. Marshall and colleagues first established a relationship between radar reflectivity and rain rate in 1947. The reflectivity, Z, was found to depend on the droplet concentration (N) times the sixth power of the diameter (D^6). Estimates are generally calibrated with reference to rain gauge measurements.

Reference

Kollias, P. *et al.* (2006) Millimeter-wavelength radars. *Bull. Amer. Met. Soc.* 88(10), 1608–24.

Rogers, R. R. and Smith, P. L. (1996) A short history of radar meteorology, in Fleming, J. R. (ed.) *Historical Essays on Meteorology 1919–1995*. American Meteorological Society, Boston, MA, 57–98.

engineers concerned with flood forecasting and prevention, as well as to conservationists dealing with soil erosion. Chart records of the rate of rainfall (*hyetograms*) are necessary to assess intensity, which varies markedly with the time interval selected. Average intensities are greatest for short periods (thunderstorm-type downpours), as Figure 4.10 illustrates for Milwaukee, USA.

In the case of extreme rates at different points over the earth (Figure 4.11), the record intensity over 10 minutes is approximately three times that for 100 minutes, and the latter exceeds by as much again the record intensity over 1000 minutes (i.e., 16.5 hours). Note that many of the records for events with a duration greater than a day are from the tropics. The 24-hour and the 12 calendar month records for India are both

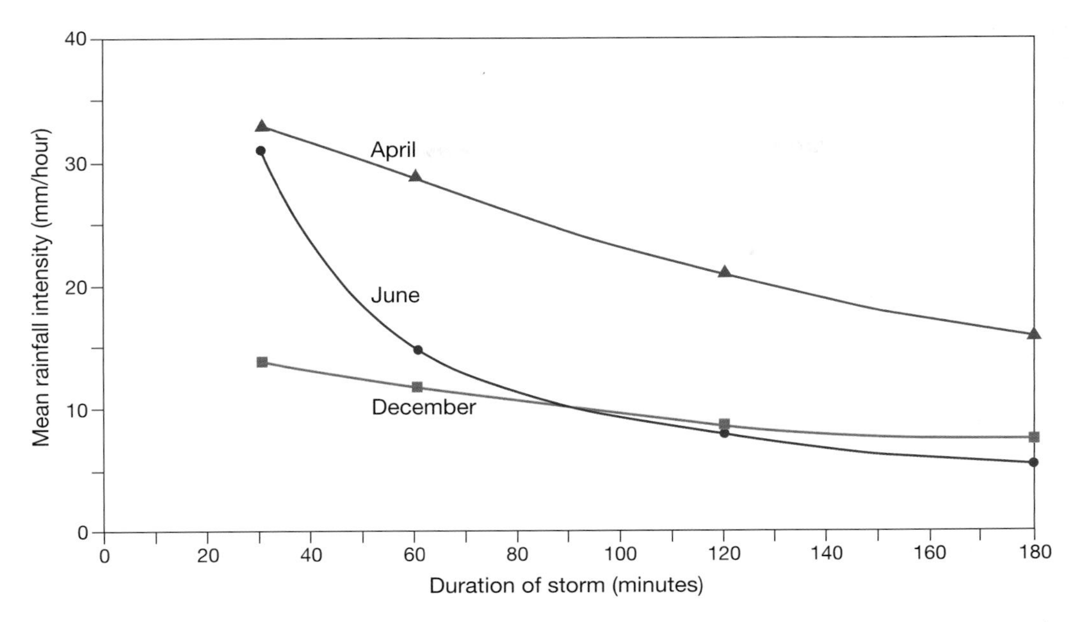

Figure 4.10 Relation between rainfall intensity and duration for Milwaukee, USA, during three months in 1973.

Source: US Environmental Data Service (1974). Courtesy US Environmental Data Service.

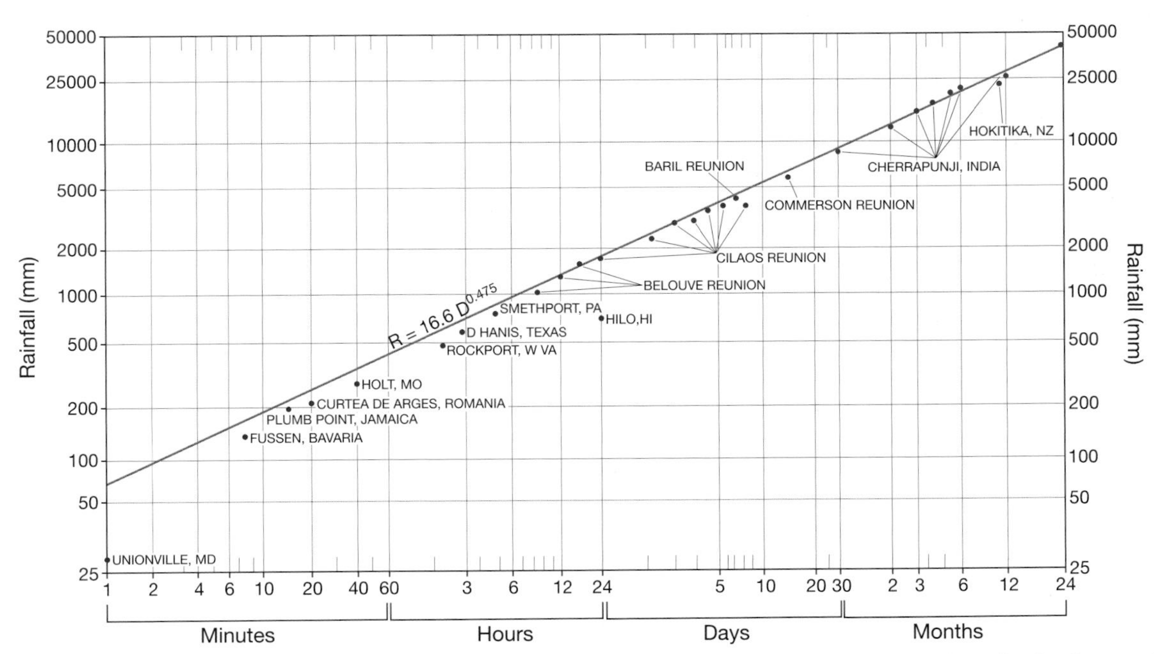

Figure 4.11 World record rainfalls (mm) with an envelope line prior to 1967. The equation of the line is given and the state or country where important records were established.

Source: Modified and updated after Rodda (1970). Courtesy of the Institute of British Geographers.

from Cherrapunji – 1563mm and 22992mm, respectively.

High-intensity rain is associated with increased drop size rather than an increased number of drops. For example, with precipitation intensities of 0.1, 1.3 and 10.2cm/hr (or. 0.05, 0.5 and 4.0in/hr), the most frequent raindrop diameters are 0.1, 0.2 and 0.3cm, respectively. Figure 4.12 shows maximum expected precipitation for storms of different duration and frequency in the USA. The maxima are along the Gulf Coast and in Florida.

Areal extent of a rainstorm

The rainfall totals received in a given time interval depend on the size of the area that is considered. Rainfall averages for a 24-hour storm covering 100,000km^2 may be only one-third to one-tenth of those for a storm over a 25km^2 area. The curvilinear relationship is similar to that for rainfall duration and intensity. Figure 4.13 illustrates the relationship between rain area and frequency of occurrence in Illinois, USA. Here a log-log plot gives a straight line fit. For 100-year, or heavier falls, the storm frequency in this region can be estimated from 0.0011 (area)$^{0.896}$ where the area is in km^2.

Frequency of rainstorms

It is useful to know the average time period within which rainfall of a specified amount or intensity may be expected to occur once. This is termed the *recurrence interval* or *return period.* Figure 4.14 gives this type of information for six contrasting stations. From this, it would appear that on average, each 20 years, a 24-hour rainfall of at least 95mm is likely to occur at Cleveland and 216mm at Lagos. However, this *average* return period does not mean that such falls necessarily

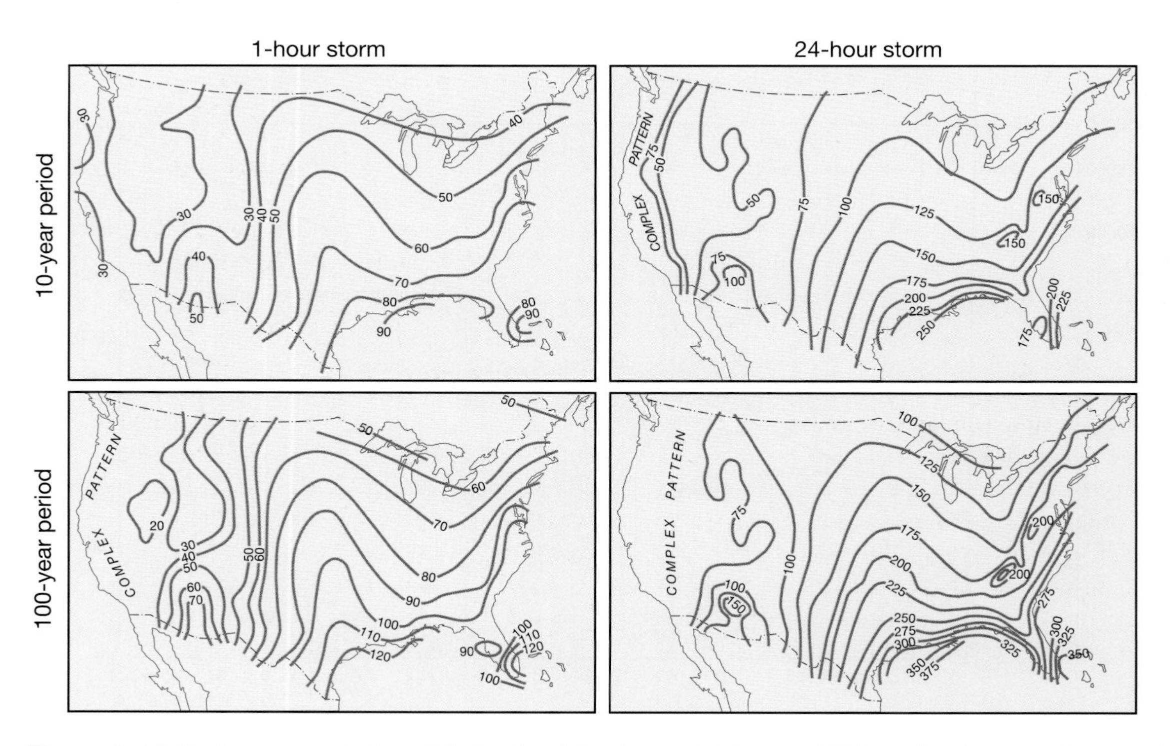

Figure 4.12 Maximum expected precipitation (mm) for storms of 1-hour and 24-hour duration occurring once in 10 years and once in 100 years over the continental United States, calculated from records prior to 1961.
Source: US National Weather Service courtesy NOAA.

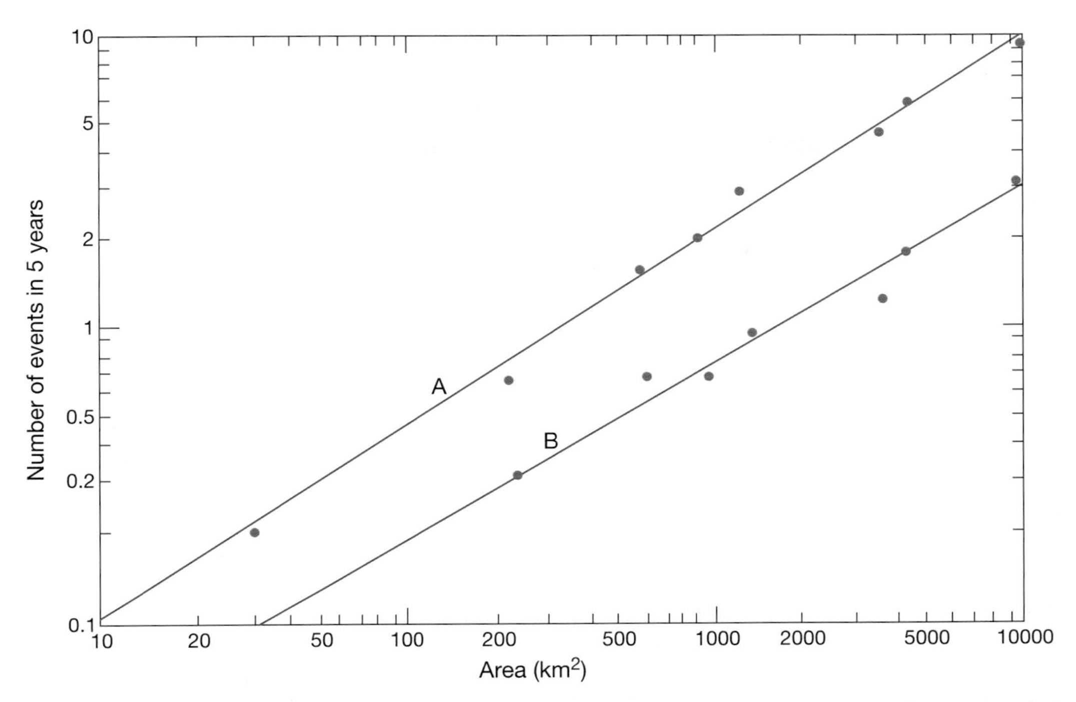

Figure 4.13 The relationship between area (km^2) and frequency of occurrence, during a five-year period, of rainstorms that produce A: 25-year and B: 100-year or heavier rain amounts for 6–12-hour periods over 50 percent or more of each area in Illinois.

Source: S.A. Chagnon (2002). From *Journal of Hydrometeorology* by permission of the American Meteorological Society.

occur in the twentieth year of a selected period. Indeed, they may occur in the first or not at all! These estimates require long periods of observational data, but the approximately linear relationships shown by such graphs are of great practical significance for the design of flood-control systems, dams and reservoirs.

Studies of rainstorm events have been carried out in many different climatic areas. An example for southwest England is shown in Figure 4.15. The 24-hour storm had an estimated 150–200-year return period. By comparison, tropical rainstorms have much higher intensities and shorter recurrence intervals for comparable totals.

3 The world pattern of precipitation

Globally, 79 percent of total precipitation falls on the oceans and 21 percent on land (Figure 4.1). A glance at the maps of precipitation amount for December to February and June to August (Figure 4.16) indicates that the distributions are considerably more complex than those, for example, of mean temperature (see Figure 3.11). Comparison of Figure 4.16 with the meridional profile of average precipitation for each latitude (Figure 4.17) brings out the marked longitudinal variations that are superimposed on the zonal pattern. The zonal pattern has several significant features:

1. The 'equatorial' maximum, which is displaced into the Northern Hemisphere. This is related primarily to the converging Trade Wind systems and monsoon regimes of the summer hemisphere, particularly in South Asia and West Africa. Annual totals over large areas are of the order of 2000–2500mm or more.

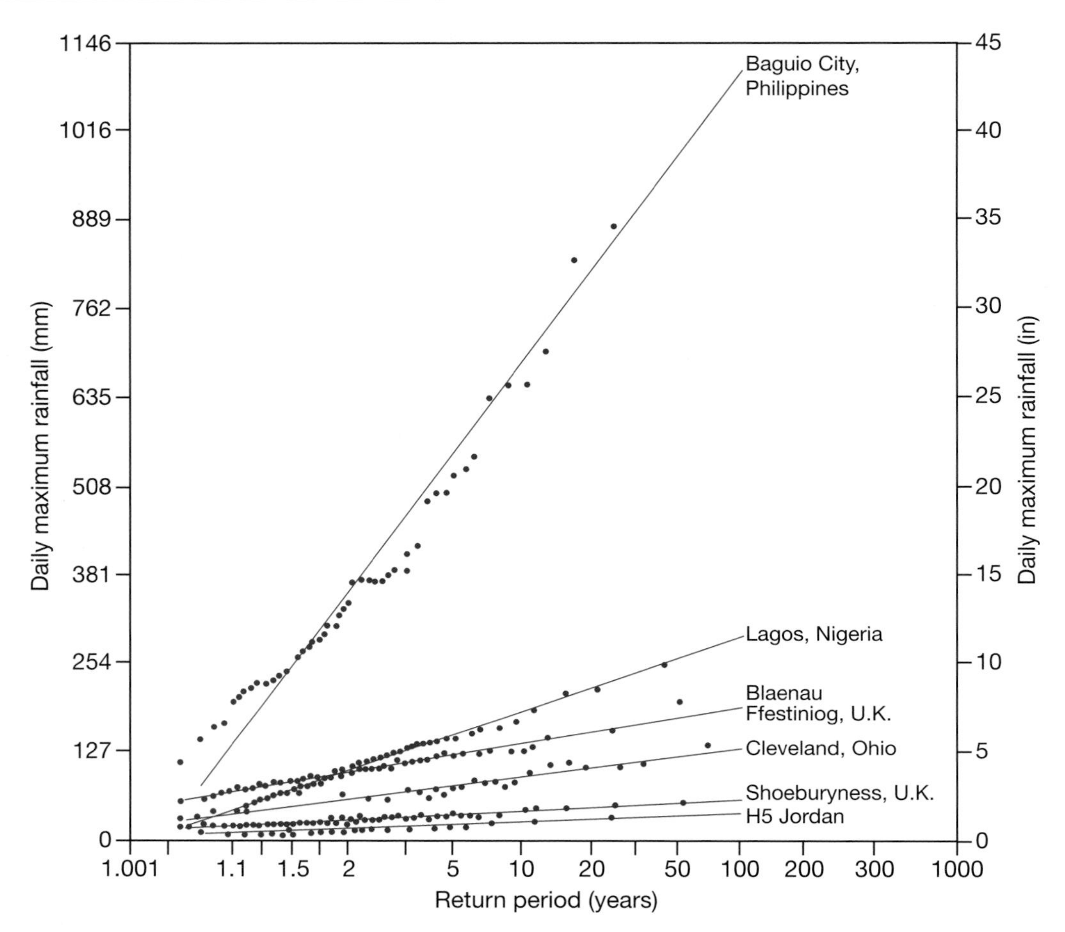

Figure 4.14 Rainfall/duration/frequency plots for daily maximum rainfalls in respect of a range of stations from the Jordan desert to an elevation of 1482m in the monsoonal Philippines.
Source: After Rodda (1970); Linsley and Franzini (1964); Ayoade (1976).

2 The west coast maxima of mid-latitudes associated with the storm tracks in the westerlies. The precipitation in these areas has a high degree of reliability.

3 The dry areas of the subtropical high pressure cells, which include not only many of the world's major deserts but also vast oceanic expanses. In the Northern Hemisphere, the remoteness of the continental interiors extends these dry conditions into mid-latitudes. In addition to very low average annual totals (less than 150mm), these regions have considerable year-to-year variability.

4 Low precipitation in high latitudes and in winter over the continental interiors of the Northern Hemisphere. This reflects the low vapor content of the extremely cold air. Most of this precipitation occurs in solid form.

Figure 4.16 demonstrates why the subtropics do not appear as particularly dry on the meridional profile in spite of the known aridity of the subtropical high pressure areas (see Chapter 10). In these latitudes, the eastern sides of the continents receive considerable rainfall in summer.

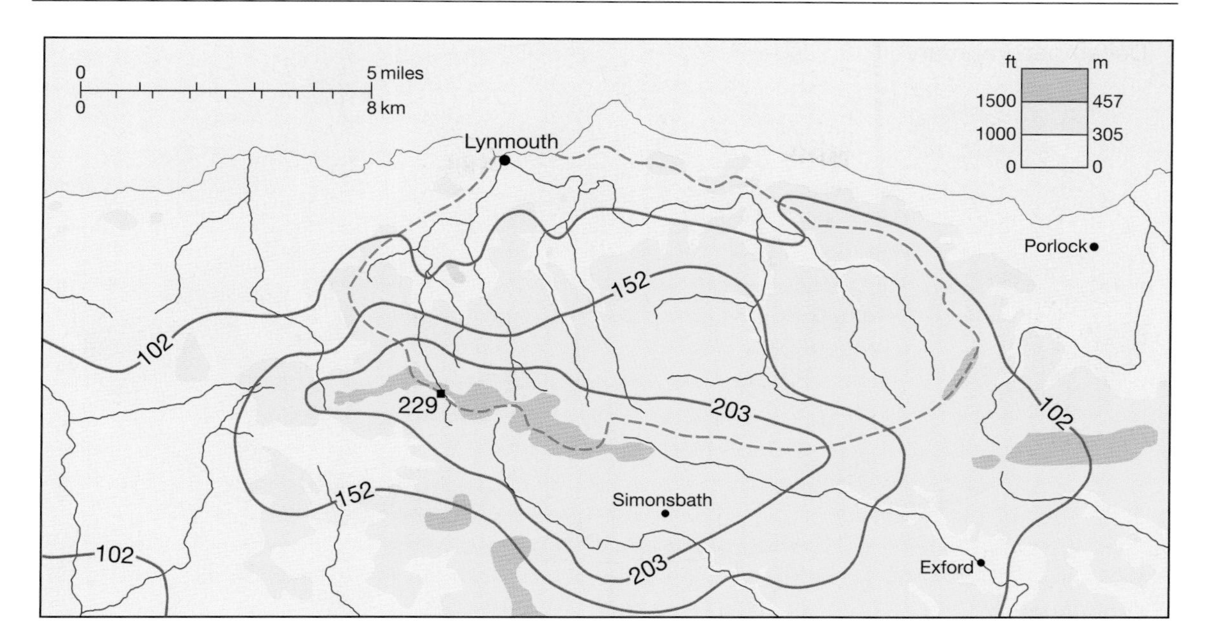

Figure 4.15 Distribution of rainfall (mm) over Exmoor, southwest England during a 24-hour period on 15 August 1952 which produced catastrophic local flooding at Lynmouth. The catchment is marked (dashed); 75 percent of the rain fell in just seven hours.

Souce: Dobbie and Wolf (1953). Courtesy of the Institute of Civil Engineers.

In view of the complex controls involved, no brief explanation of these precipitation distributions can be very satisfactory. Various aspects of selected precipitation regimes are examined in chapters 10 and 11, after consideration of the fundamental ideas about atmospheric motion and weather disturbances. Here we simply point out four factors that have to be taken into account in studying figures 4.16 and 4.17:

1 The limit imposed on the maximum moisture content of the atmosphere by air temperature. This is important in high latitudes and in winter in continental interiors.
2 The major latitudinal zones of moisture influx due to atmospheric advection. This in itself is a reflection of the global wind systems and their disturbances (i.e., the converging Trade Wind systems and the cyclonic westerlies, in particular).
3 The distribution of the land masses. The Southern Hemisphere lacks the vast, arid, mid-latitude continental interiors of the Northern. The oceanic expanses of the Southern Hemisphere allow the mid-latitude storms to increase the zonal precipitation average for 45°S by about one-third compared with that of the Northern Hemisphere for 50°N. Longitudinal irregularities are also created by the monsoon regimes, especially in Asia.
4 The orientation of mountain ranges with respect to the prevailing winds.

4 Regional variations in the altitudinal maximum of precipitation

The increase of mean precipitation with height on mountain slopes is a widespread characteristic in mid-latitudes, although actual profiles of precipitation differ regionally and seasonally. An

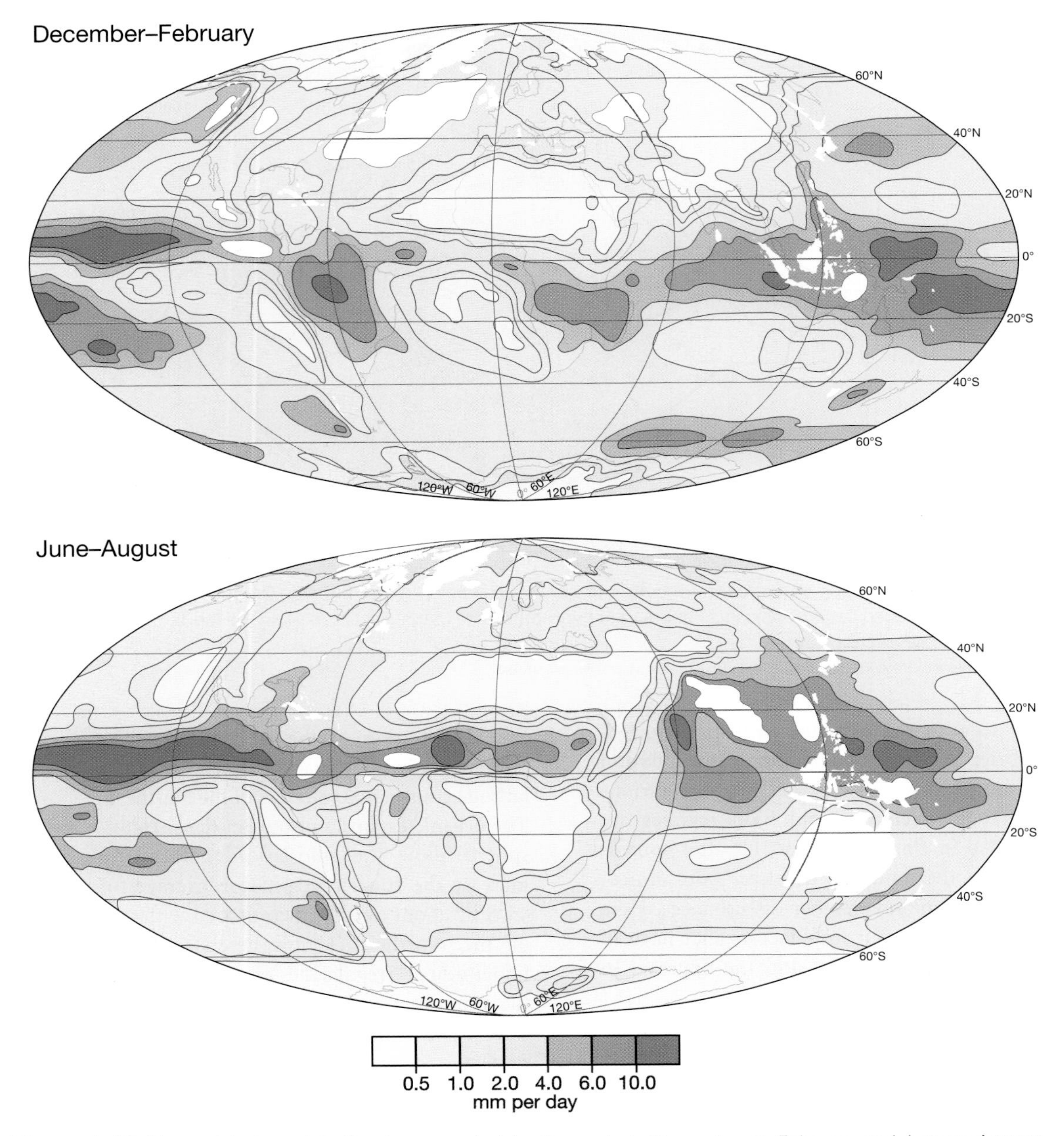

Figure 4.16 Mean global precipitation (mm per day) for the periods December to February and June to August.
Source: From Legates (1995). From *International Journal of Climatology*, copyright © John Wiley & Sons Ltd. Reproduced with permission.

increase may be observed up to at least 3000–4000m in the Rocky Mountains in Colorado. In western North America the maximum occurs on the windward slopes of the Sierra Nevada, while in western Canada there is a close association of terrain and precipitation maxima. In the Alps patterns vary with maxima at high elevations in the central Alps and at low elevations on the outer

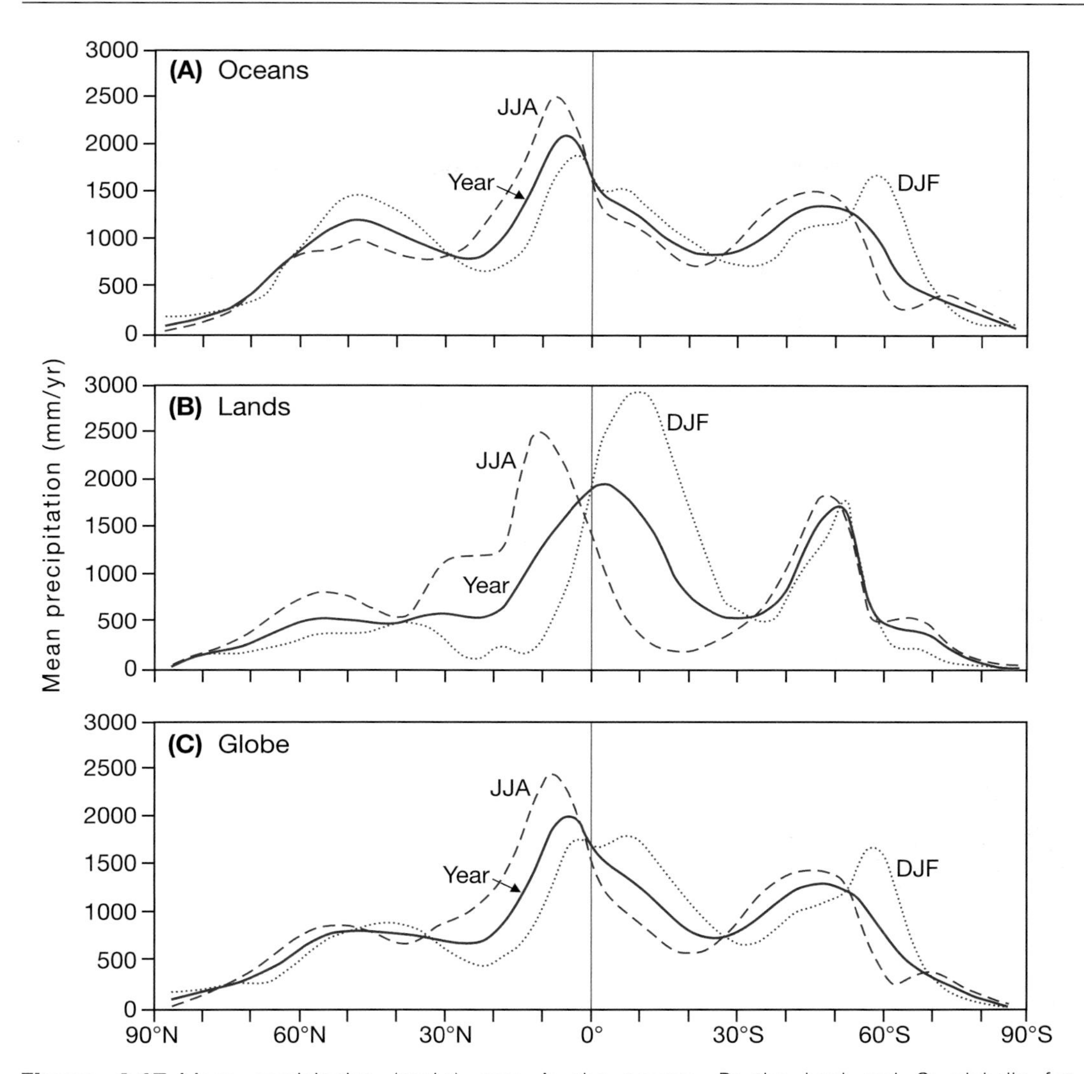

Figure 4.17 Mean precipitation (cm/yr) over A: the oceans, B: the land and C: globally for December–February, June –August and annually.

Source: Peixoto and Oort (1983), Fig. 23. Copyright © D. Reidel, Dordrecht, by kind permission of Kluwer Academic Publishers.

northern and southern ranges. In western Britain, with mountains of about 1000m, the maximum falls are recorded to leeward of the summits. This probably reflects the general tendency of air to go on rising for a while after it has crossed the crestline and the time lag involved in the precipitation process after condensation. Over narrow uplands, the horizontal distance may allow insufficient time for maximum cloud buildup and the occurrence of precipitation. However, a further factor may be the effect of eddies, set up in the airflow by the mountains, on the catch of rain gauges. Studies in Bavaria at the Hohenpeissenberg Observatory show that standard rain gauges may overestimate amounts by about 10 percent on the lee slopes and underestimate them by 14 percent on windward slopes.

In the tropics and subtropics, maximum precipitation occurs below the higher mountain summits, from which level it decreases upward towards the crest. Observations are generally sparse in the tropics, but numerous records from Java show that the average elevation of greatest precipitation is approximately 1200m. Above about 2000m, the decrease in amounts becomes quite marked. Similar features are reported from Hawaii and, at a rather higher elevation, on mountains in East Africa (see Chapter 11H.2). Figure 4.19A shows that, despite the wide range of records for individual stations, this effect is clearly apparent along the Pacific flank of the Guatemalan highlands. Further north along the coast, the occurrence of a precipitation maximum below the mountain crest is observed in the Sierra Nevada, despite some complication introduced by the shielding effect of the Coast Ranges (Figure 4.18B), but in the Olympic Mountains of Washington precipitation increases right up to the summits. Precipitation gauges on mountain crests may underestimate the actual precipitation due to the effect of eddies, and this is particularly true where much of the precipitation falls in the form of snow, which is very susceptible to blowing by the wind.

One explanation of the orographic difference between tropical and temperate rainfall is based on the concentration of moisture in a fairly shallow layer of air near the surface in the tropics (see Chapter 11). Much of the orographic precipitation seems to derive from warm clouds (particularly cumulus congestus), composed of water droplets, which commonly have an upper limit at about 3000m. It is probable that the height of the maximum precipitation zone is close to the mean cloud base, since the maximum size and number of falling drops will occur at that level. Thus, stations located above the level of mean cloud base will receive only a proportion of the orographic increment. In temperate latitudes, much of the precipitation, especially in winter, falls from stratiform cloud, which commonly extends through a considerable depth of the troposphere. In this case, there tends to be a smaller fraction of the total cloud depth below the station level. These differences according to cloud type and depth are apparent even on a day-to-day basis in mid-latitudes. Seasonal variations in the

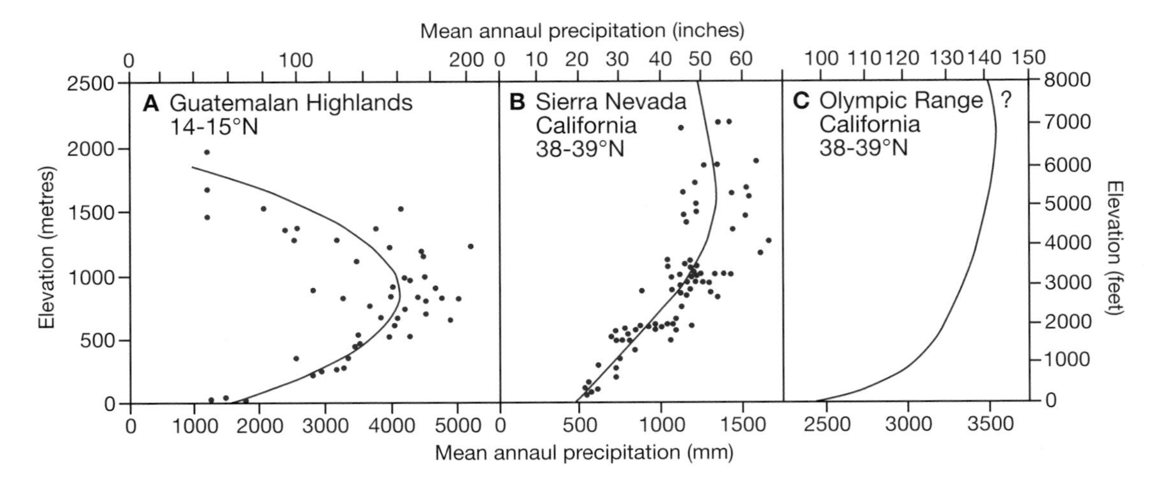

Figure 4.18 Generalized curves showing the relationship between elevation and mean annual precipitation for west-facing mountain slopes in Central and North America. The dots give some indication of the wide scatter of individual precipitation readings.

Source: Adapted from Hastenrath (1967) and Armstrong and Stidd (1967).

altitude of the mean condensation level and zone of maximum precipitation are similarly observed. In the Pamir and Tien Shan of Central Asia, for instance, the maximum is reported to occur at about 1500m in winter and at 3000m or more in summer. A further difference between orographic effects on precipitation in the tropics and the mid-latitudes relates to the great instability of many tropical air masses. Where mountains obstruct the flow of moist tropical air masses, the upwind turbulence may be sufficient to trigger convection, producing a rainfall maximum at low elevations. This is illustrated in Figure 4.19A for Papua New Guinea, where there is a seasonally alternating wind regime – northwesterly (southeasterly) in the austral summer (winter). By contrast, in more stable mid-latitude airflow, the rainfall maximum is closely related to the topography (see Figure 4.19B for the Swiss Alps).

5 Drought

The term 'drought' implies an absence of significant precipitation for a period long enough to cause moisture deficits in the soil through evapotranspiration and decreases in stream flow, so disrupting normal biological and human activities. Crop damage and water shortages are typical results of drought conditions. Thus, drought may occur after only three or four weeks without rain in parts of Britain, whereas areas of the tropics regularly experience many successive dry months. There is no universally applicable definition of drought. Specialists in meteorology, agriculture, hydrology and socio-economic studies, who have differing perspectives, have suggested at least 150 different definitions! All regions suffer the temporary but irregularly recurring condition of drought, but particularly

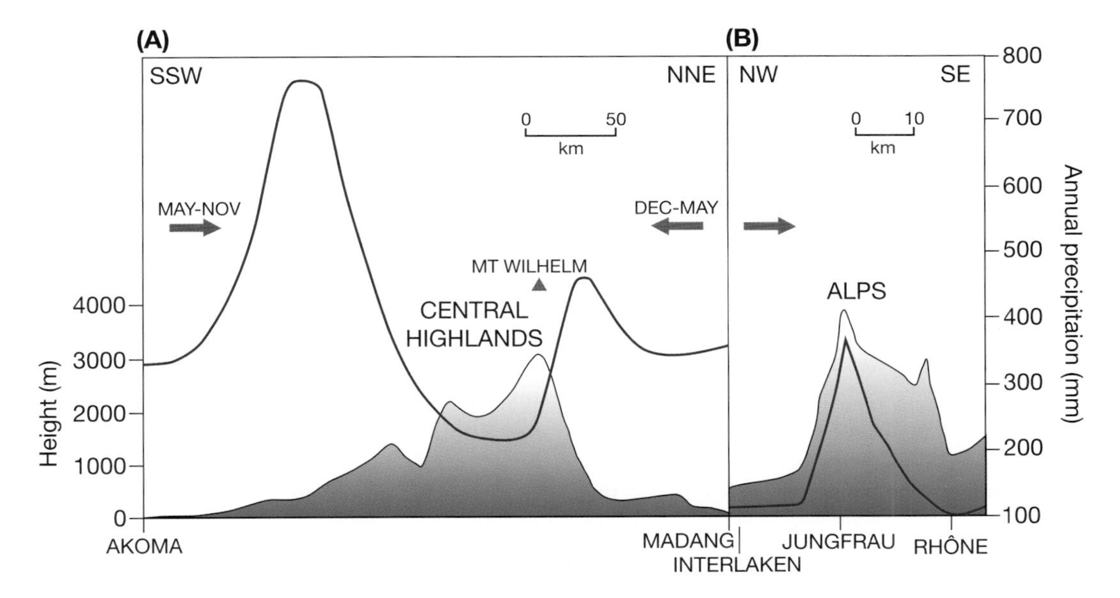

Figure 4.19 The relationship between precipitation (broken line) and relief in the tropics and mid-latitudes. A: The highly saturated air masses over the Central Highlands of Papua New Guinea give seasonal maximum precipitations on the windward slopes of the mountains with changes in the monsoonal circulation; B: Across the Jungfrau massif in the Swiss Alps the precipitation is much less than in A and is closely correlated with the topography on the windward side of the mountains. The arrows show the prevailing airflow directions.

Sources: A After Barry (2008). B After Maurer and Lütschg from Barry (2008).

those with marginal climates alternately influenced by differing climatic mechanisms. Causes of drought conditions include:

1 Increases in the size and persistence of subtropical high pressure cells. The major droughts in the African Sahel (see Figure 13.11) have been attributed to an eastward and southward expansion of the Azores anticyclone.
2 Changes in the summer monsoon circulation. This may cause a postponement or failure of moist tropical incursions in areas such as West Africa or the Punjab of India. In India, monsoon failures in the years 1965–1966, 1972 and 1987 produced the most extensive and damaging droughts in the records for 1950–2000.
3 Anomalous lower ocean-surface temperatures produced by changes in currents or increased upwelling of cold waters. Rainfall in California and Chile may be affected by such mechanisms (see p. 377), and adequate rainfall in the drought-prone region of northeast Brazil appears to be strongly dependent on high sea-surface temperatures at 0–15°S in the South Atlantic. Warm ocean waters off Peru and associated teleconnections (see pp. 375–383) caused severe drought in Australia in 1982–1983.
4 Displacement of mid-latitude storm tracks. This may be associated with an expansion of the circumpolar westerlies into lower latitudes or with the development of persistent blocking circulation patterns in mid-latitudes (see Figure 8.25). It has been suggested that droughts on the Great Plains east of the Rockies in the 1890s and 1930s were due to such changes in the general circulation. However, the droughts of the 1910s and 1950s in this area were caused by persistent high pressure in the southeast and the northward displacement of storm tracks (Figure 4.20).

From a global analysis of simulated soil moisture variations, droughts of ≤ 6 months tend to occur in the tropics and mid-latitudes where there is high interannual climate variability, while droughts of 7–12 months are more frequent in mid- to high latitudes. Droughts lasting 12 months or more are limited to the Sahel and higher northern latitudes. Severe droughts in northern Asia typically occur with persistent winter anomalies of soil moisture.

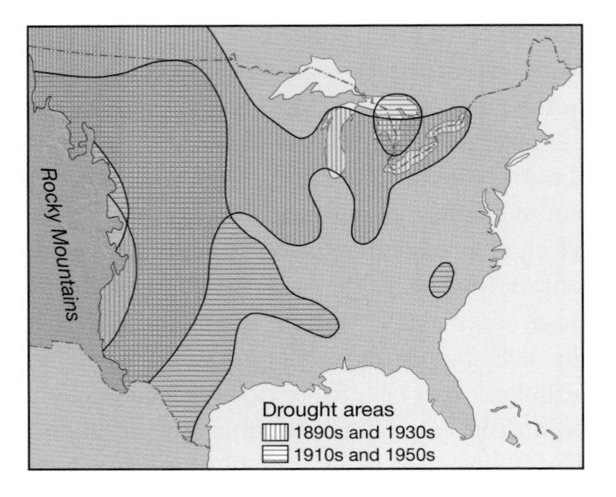

Figure 4.20 Drought areas of the central USA, based on areas receiving less than 80 percent of the normal July–August precipitation.

Source: After Borchert (1971). Courtesy of the Association of American Geographers.

From May 1975 to August 1976, parts of northwest Europe from Sweden to western France experienced severe drought conditions. Southern England received less than 50 percent of average rainfall, the most severe and prolonged drought since records began in 1727 (Figure 4.21). The immediate causes of this regime were the establishment of a persistent blocking ridge of high pressure over the area, displacing depression tracks 5–10° latitude northward over the eastern North Atlantic. Upstream, the circulation over the North Pacific had changed earlier, with the development of a stronger high pressure cell and stronger upper-level westerlies, perhaps associated with a cooler-than-average sea surface. The westerlies were displaced northward over both the Atlantic and the Pacific. Over Europe, the dry conditions at the surface increased the stability of the atmosphere, further lessening the possibility

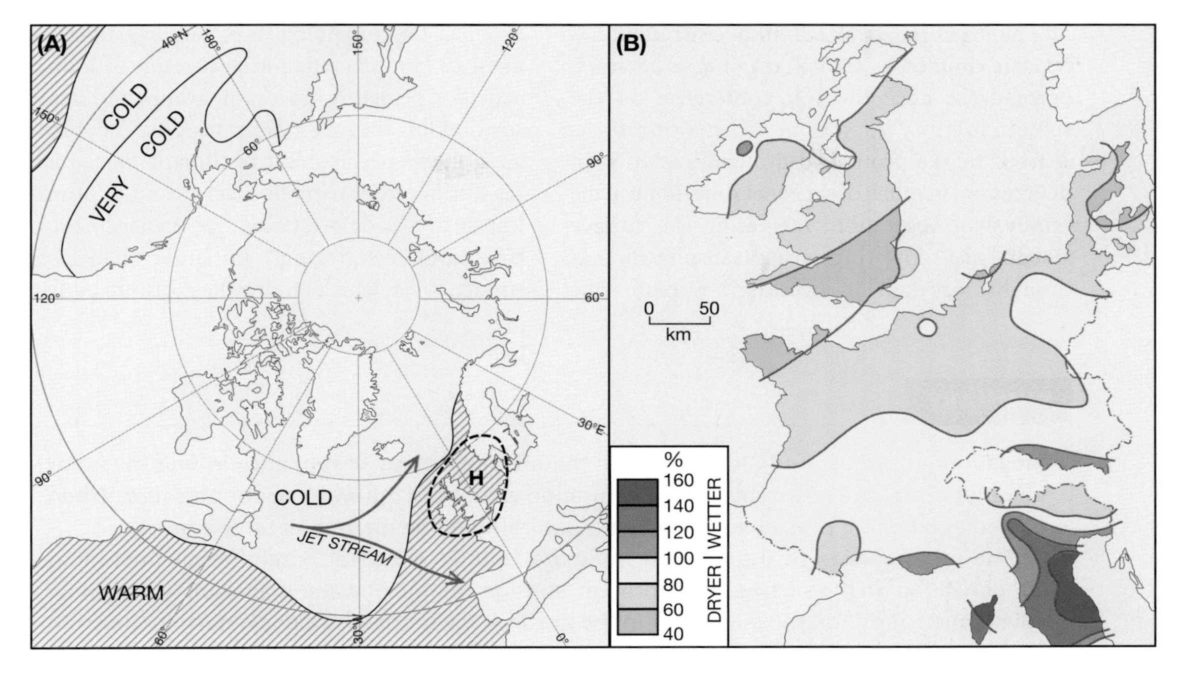

Figure 4.21 The drought of northwest Europe during May 1975 to August 1976. A: Conditions of a blocking high pressure over Britain, jet stream bifurcation and low sea-surface temperatures; B: Rainfall over Western Europe between May 1975 and August 1976 expressed as a percentage of a 30-year average.

Source: From Doornkamp *et al.* (1980). Courtesy of the Institute of British Geographers.

of precipitation. Other major droughts in England and Wales (1800–2006) occurred during spring 1990 to summer 1992 and spring 1995 to summer 1997. Rainfall for April to August 1995 over England and Wales was only 46 percent of average (compared with 47 percent in 1976), again associated with a northward extension of the Azores anticyclone. This deficit has an estimated return period in excess of 200 years! A further drought was experienced in 2004–2006 and this was especially severe in southern England. However, these droughts do not compare in length with the droughts of 1854–1860 and 1890–1910.

Persistent, severe droughts involve combinations of several mechanisms. The prolonged drought in the Sahel – a 3000 by 700km zone stretching along the southern edge of the Sahara from Mauretania to Chad – which began in 1969 and has continued with interruptions up to the present (see Figure 13.11), has been attributed to several factors. These include an expansion of the circumpolar westerly vortex, shifting of the subtropical high pressure belt towards the equator, and lower sea-surface temperatures in the eastern North Atlantic. There is no evidence that the subtropical high pressure was further south, but dry easterly airflow was stronger across Africa during drought years. Severe drought occurred in 1983–1984. In September 1984, for example, 69 percent of the Sahel (10–20°N, 20°W–20°E) experienced drought and 18/20 ranked months of spatial drought extent were reported in the 1980s.

The 1988 drought in the central United States is ranked first (1950–2000) in terms of summer drought extent and is estimated to have caused losses to agriculture of $30 billion. The causes of its development and longevity are attributed to stationary anomalies of atmospheric circulation, sea-surface temperature anomalies, and soil moisture-precipitation feedbacks.

The meteorological definition of drought has become clouded by the subject of *desertification*, particularly since the UN conference on the subject in 1977 in Nairobi. This concern was sparked by the protracted drought, resulting in desiccation, in much of the Sahel zone. In turn, the removal of vegetation, increasing the surface albedo and lowering evapotranspiration, is thought to result in decreased rainfall. The problem for climatologists is that desertification involves land degradation as a result of human activities, primarily in areas of savanna and steppe surrounding the major desert regions. These areas have always been subject to climatic *fluctuations* (as distinct from climatic *change*) and to human impacts (e.g., deforestation, mismanagement of irrigation, overgrazing.) initiating changes in surface cover, which modify the moisture budget.

SUMMARY

Measures of atmospheric humidity are: the absolute mass of moisture in unit mass (or volume) of air, as a proportion of the saturation value; and the water vapor pressure. When cooled at constant pressure, air becomes saturated at the dew-point temperature.

The components of the surface moisture budget are total precipitation (including condensation on the surface), evaporation, storage change of water in the soil or in a snow cover, and runoff (on the surface or in the ground). Evaporation rate is determined by the available energy, the surface–air difference in vapor pressure, and the wind speed, assuming the moisture supply is unlimited. If the moisture supply is limited, soil water tension and plant factors affect the evaporation rate. Evapotranspiration is best determined with a lysimeter. Otherwise, it may be calculated by formulae based on the energy budget, or on the aerodynamic profile method using the measured gradients of wind speed, temperature and moisture content near the ground.

Condensation in the atmosphere may occur by continued evaporation into the air; by mixing of air of different temperatures and vapor pressures such that the saturation point is reached; or by adiabatic cooling of the air through lifting until the condensation level is reached.

Rainfall is described statistically by the intensity, areal extent and frequency (or recurrence interval) of rainstorms. Orography intensifies the precipitation on windward slopes, but there are geographical differences in this altitudinal effect. Global patterns of precipitation amount and annual regime are determined by the regional atmospheric circulation, the proximity to ocean areas, orography, sea-surface temperatures, and the atmospheric moisture budget. Droughts may occur in many different climatic regions due to various causal factors. In mid-latitudes, blocking anticyclones are a major factor. The primary cause of protracted drought in the African Sahel seems to be climatic fluctuations.

DISCUSSION TOPICS

- Trace the possible paths of a water molecule through the hydrological cycle and consider the measurements that need to be made to determine the quantities of water involved in the various transformations.
- What processes lead to phase changes of water in the atmosphere and what are some of their consequences?
- What is the significance of clouds in the global water balance?
- Compare the moisture balance of an air column and that of a small drainage basin.
- What are the various statistics used to characterize rainfall events and for what different applications are they important?
- Consider how an annual water budget diagram might differ between a wet year and a dry year at the same location.

REFERENCES AND FURTHER READING

Books

Anderson, B. R. (1975) *Weather in the West. From the Midcontinent to the Pacific.* American West, Palo Alto, CA, 223pp. [Popular account]

Barry, R. G. (2008) *Mountain Weather and Climate*, 3rd edn, Cambridge University Press, Cambridge, 506pp. [Comprehensive survey]

Baumgartner, A. and Reichel, E. (1975) *The World Water Balance: Mean Annual Global, Continental and Maritime Precipitation, Evaporation and Runoff*, Elsevier, Amsterdam, 179 pp. [Statistical assessment of the major components of the hydrological cycle; one of the standard summaries]

Bruce, J. P. and Clark, R. H. (1966) *Introduction to Hydrometeorology*. Oxford: Pergamon, 319pp.

Brutsaert, W. (1982) *Evaporation into the Atmosphere: Theory, History and Applications*. Kluwer, Dordrecht, 279pp. [Thorough survey of evaporation processes and applications]

Doornkamp, J. C., Gregory, K. J. and Burn, A.S. (eds) (1980) *Atlas of Drought in Britain 1975–6*. Institute of British Geographers, London, 82pp. [Detailed case study of a major UK drought]

Gash, J. and Shuttleworth, J. (2007) *Evaporation*. International Association of Hydrological Sciences, Wallingford, UK, Benchmark Papers in Hydrology Series No. 2, 524pp. [Comprehensive modern survey]

Korzun, V. I. (ed.-in-chief), USSR Committee for the International Hydrological Decade (1978) *World Water Balance and Water Resources of the Earth*. UNESCO, Paris (translation of Russian edition, Leningrad, 1974), 663pp. [Comprehensive account of atmospheric and terrestrial components of the water balance for the globe and by continent; numerous figures, tables and extensive references]

Linsley, R. K. and Franzini, J. B. (1964) *Water Resources Engineering*, McGraw-Hill, New York, 654pp.

Linsley, R.K., Franzini, J. B., Freyberg. D. L. and Tchbanoglous, G. (1992) *Water-resources Engineering,* 4th edn. McGraw-Hill, New York, 841pp. [Chapters on descriptive and quantitative hydrology and ground water; water supply and engineering topics predominate]

Miller, D. H. (1977) *Water at the Surface of the Earth*. Academic Press, New York, 557pp. [Comprehensive treatment of all components of the water cycle and water in ecosystems; well illustrated with many references]

Pearl, R. T. *et al.* (1954) *The Calculation of Irrigation Need*. Tech. Bull. No. 4, Ministry of Agriculture Fish and Food, HMSO, London, 35pp. [Handbook based on the Penman formulae for the UK]

Peixoto, J. P. and Oort, A. H. (1992) *Physics of Climate*. American Institute of Physics, New York, ch. 12 [Deals with the water cycle in the atmosphere]

Penman, H. L. (1963) *Vegetation and Hydrology*. Tech. Comm. No. 53, Commonwealth Bureau of

Soils, Harpenden, 124pp. [A survey of the literature on the effects of vegetation on the hydrological cycle through interception, evapotranspiration, infiltration and runoff, and of related catchment experiments around the world]

Petterssen, S. (1969) *Introduction to Meteorology*, 3rd edn, McGraw-Hill, New York, 416pp.

Rudolf, B. and Rubel, F. (2005) Global precipitation, in Hantel, M. (ed.) *Observed Global Climate.* Springer, Berlin, pp. 11.1–11.43 [An up-to-date overview]

Sellers, W. D. (1965) *Physical Climatology*, University of Chicago Press, Chicago, IL, 272 pp.

Strangeways, I. C. (2003) *Measuring the Natural Environment*, 2nd edn. Cambridge University Press, Cambridge, 548pp. [A complete account of all kinds of instrumentation]

Sumner, G. (1988) *Precipitation. Process and Analysis.* J. Wiley and Sons, Chichester, 455 pp. [Comprehensive discussion of cloud and precipitation formation, precipitation systems, surface measurements and their analysis in time and space]

World Meteorological Organization (1972) *Distribution of Precipitation in Mountainous Areas* (2 vols). WMO No. 326, Geneva, 228 and 587pp. [Conference proceedings with many valuable papers]

Journal articles

Acreman, M. (1989) Extreme rainfall in Calderdale, 19 May 1989. *Weather* 44, 438–46.

Agnew, C. T. and Chappell, A. (2000). Desiccation in the Sahel, in McLaren, S. J. and Kniveton, D. R. (eds), *Linking Climate Change to Land Surface Changes*, Kluwer, Dordrecht, 27–48.

Armstrong, C. F. and Stidd, C. K. (1967). A moisture balance profile on the Sierra Nevada. *J. Hydrol.* 5, 258–68.

Atlas, D., Chou, S-H. and Byerly, W. P. (1983) The influence of coastal shape on winter mesoscale air-sea interactions. *Monthly Weather Review* 111, 245–52.

Ayoade, J. A. (1976) A preliminary study of the magnitude, frequency and distribution of intense rainfall in Nigeria. *Hydro. Sci. Bull.* 21(3), 419–29.

Bannon, J. K. and Steele, L. P. (1960) Average water-vapour content of the air. *Geophysical Memoirs* 102, Meteorological Office (38pp.).

Borchert, J. R. (1971) The dust bowl in the 1970s. *Ann. Assn Amer. Geogr.* 61, 1–22.

Browning, K. (1993) The global energy and water cycle. *NERC News* July, 21–3.

Bryson, R. A. (1973) Drought in the Sahel: who or what is to blame? *The Ecologist* 3(10), 366–71.

Chacon, R. E. and Fernandez, W. (1985) Temporal and spatial rainfall variability in the mountainous region of the Reventazon River Basin, Costa Rica. *J. Climatology* 5, 175–88.

Chagnon, S. A. (2002) Frequency of heavy rainstorms on areas from 10 to 10,000km^2 , defined using dense rain gauge networks. *J. Hydromet.* 3(2), 220–3.

Choudhury, B. J. (1993). Desertification, in Gurney, R. J., *et al.* (eds) *Atlas of Satellite Observations Related to Global Change,* Cambridge University Press, Cambridge, 313–25.

Deacon, E. L. (1969) Physical processes near the surface of the earth, in Flohn, H. (ed.) *General Climatology*, World Survey of Climatology 2, Elsevier, Amsterdam, 39–104.

Dobbie, C. H. and Wolf, P. O. (1953) The Lynmouth flood of August 1952. *Pro. Inst. Civ. Eng.,* Part III, 522–88.

Dorman, C. E. and Bourke, R. H. (1981) Precipitation over the Atlantic Ocean, 30°S to 70°N. *Monthly Weather Review* 109, 554–63.

Garcia-Prieto, P. R., Ludlam, F. H. and Saunders, P. M. (1960) The possibility of artificially increasing rainfall on Tenerife in the Canary Islands. *Weather* 15, 39–51.

Gilman, C. S. (1964) Rainfall, in Chow, V. T. (ed.) *Handbook of Applied Hydrology*, McGraw-Hill, New York, section 9.

Guhathakurta, P. (2007) Highest recorded point rainfall over India. *Weather* 62, 349.

Harrold, T. W. (1966) The measurement of rainfall using radar. *Weather* 21, 247–9 and 256–8.

Hastenrath, S. L. (1967) Rainfall distribution and regime in Central America. *Archiv. Met. Geophys. Biokl.* B. 15(3), 201–41.

Hershfield, D. M. (1961) Rainfall frequency atlas of the United States for durations from 30 minutes to 24 hours and return periods of 1 to 100 years. *US Weather Bureau, Tech. Rept.* 40.

Howarth, D. A. and Rayner, J. N. (1993) An analysis of the atmospheric water balance over the southern hemisphere. *Phys. Geogr.* 14, 513–35.

Howe, G. M. (1956) The moisture balance in England and Wales. *Weather* 11, 74–82.

Iesanmi, O. O. (1971) An empirical formulation of an ITD rainfall model for the tropics: a case study for Nigeria. *J. App. Met.* 10(5), 882–91.

Jaeger, L. (1976) Monatskarten des Niederschlags für die ganze Erde, *Berichte des Deutsches Wetterdienstes* 18(139), Offenbach am Main, 38pp. + plates.

Jiusto, J. E. and Weickmann, H. K. (1973) Types of snowfall. *Bull. Amer. Met. Soc.* 54, 148–62.

Kelly, P. M. and Wright, P. B. (1978) The European drought of 1975–6 and its climatic context. *Prog. Phys. Geog.* 2, 237–63.

Klemes, V. (1990). The modeling of mountain hydrology: the ultimate challenge, in Molnar, L. (ed.) *Hydrology of Mountainous Areas, Int. Assoc. Hydrol. Sci., Publ.* 190: 29–43.

Landsberg, H. E. (1974) Drought, a recurring element of climate. Graduate Program in Meteorology, University of Maryland, Contribution No. 100, 47pp.

Legates, D. R. (1995) Global and terrestrial precipitation: a comparative assessment of existing climatologies. *Int. J. Climatol.* 15, 237–58.

Legates, D. R. (1996) Precipitation, in Schneider, S. H. (ed.) *Encyclopedia of Climate and Weather,* Oxford University Press, New York, 608–12.

Lott, J. N. (1994) The U.S. summer of 1993: a sharp contrast in weather extremes. *Weather* 49, 370–83.

MacDonald, J. E. (1962) The evaporation–precipitation fallacy. *Weather* 17, 168–77.

Markham, C. G. and McLain, D. R. (1977) Sea-surface temperature related to rain in Ceará, north-eastern Brazil. *Nature* 265, 320–3.

Marsh, T., Cole, G. and Wilby, R. (2007) Major droughts in England and Wales, 1800–2006. *Weather* 62(4), 87–93.

Mather, J. R. (1985) The water budget and the distribution of climates, vegetation and soils, *Publications in Climatology* 38(2), University of Delaware, Center for Climatic Research, Newark, Del. 36 pp.

McCallum, E. and Waters, A. J. (1993) Severe thunderstorms over southeast England, 20/21 July 1992. *Weather* 48, 198–208.

Möller, F. (1951) Vierteljahrkarten des Niederschlags für die ganze Erde. *Petermanns Geographische Mitteilungen*, 95 (Jahrgang), 1–7.

More, R. J. (1967) Hydrological models and geography, in Chorley, R. J. and Haggett, P. (eds) *Models in Geography,* Methuen, London, 145–85.

Palmer, W. C. (1965) Meteorological drought, Research Paper No. 45, US Weather Bureau, Washington, DC.

Parrett, C., Melcher, N. B. and James, R. W., Jr. (1993) Flood discharges in the upper Mississippi River basin. *U.S. Geol. Sur. Circular* 1120–A (14pp.).

Paulhus, J. L. H. (1965) Indian Ocean and Taiwan rainfall set new records. *Monthly Weather Review* 93, 331–5.

Peixoto, J. P. and Oort, A. H. (1983) The atmospheric branch of the hydrological cycle and climate, in Street-Perrott, A., Beran, M. and Ratcliffe, R. (eds) *Variations in the Global Water Budget,* D. Reidel, Dordrecht, 5–65.

Pike, W. S. (1993) The heavy rainfalls of 22–23 September 1992. *Met. Mag.* 122, 201–9.

Ratcliffe, R. A. S. (1978) Meteorological aspects of the 1975–6 drought. *Proc. Roy. Soc. Lond. Sect. A* 363, 3–20.

Reitan, C. H. (1960) Mean monthly values of precipitable water over the United States, 1946–56. *Mon. Weather Rev.* 88, 25–35.

Roach, W. T. (1994) Back to basics: Fog. Part 1–Definitions and basic physics. *Weather* 49(12), 411–15.

Rodda, J. C. (1970) Rainfall excesses in the United Kingdom. *Trans. Inst. Brit. Geog.* 49, 49–60.

Rodhe, H. (1989) Acidication in a global perspective. *Ambio* 18, 155–60.

Rossow, W. B. 1993. Clouds, in Gurney, R. J., Foster, J. L. and Parkinson, C. L. (eds) *Atlas of Satellite Observations Related to Global Change,* Cambridge University Press, Cambridge., 141–63.

Schwartz, S. E. (1989) Acid deposition: unravelling a regional phenomenon. *Science* 243, 753–63.

Sevruk, B. (ed.) (1985) Correction of precipitation measurements. *Zürcher Geogr. Schriften* 23 (also appears as WMO Rep. No. 24, Instruments and Observing Methods, WMO, Geneva) (288pp.).

Sheffield, B. and Wood, E. F. (2007) Characteristics of global and regional drought, 1950–2000: analysis of soil moisture data from off-line simulation of the terrestrial hydrologic cycle. *J. Geophys. Res.* 112, D17115, 21pp.

Shuttleworth, W. J. (2008) Evapotranspiration: measurement methods. *Southwest Hydrol.* 7, 22–3.

Smith, F. B. (1991) An overview of the acid rain problem. *Met. Mag.* 120, 77–91.

So, C. L. (1971) Mass movements associated with the rainstorm of June 1966 in Hong Kong. *Trans. Inst. Brit. Geog.* 53, 55–65.

Strangeways, I. (1996) Back to basics: the 'met. enclosure'. Part 2: Raingauges. *Weather* 51, 274–9; 298–303.

Strangeways, I. (2001). Back to basics: the 'met. enclosure'. Part 7: Evaporation. *Weather* 56, 419–27.

Sutcliffe, R. C. (1956). Water balance and the general circulation of the atmosphere. *Quart. J. Roy. Met. Soc.* 82, 385–95.

Weischet, W. (1965) Der tropische-konvective und der ausser tropischeadvektive Typ der vertikalen Niederschlagsverteilung. *Erdkunde* 19, 6–14.

Wilhite, D. A. and Glantz, M. H. (1982) Understanding the drought phenomenon: the role of definitions, *Water Internat.* 10, 111–30.

Yarnell, D. L. (1935) Rainfall intensity–frequency data. US Dept. Agr., Misc. Pub. No. 204.

CHAPTER FIVE

Atmospheric instability, cloud formation and precipitation processes

5

LEARNING OBJECTIVES

When you have read this chapter you will:

- know the effects of vertical displacements on the temperature of unsaturated and saturated air parcels
- know what determines atmospheric stability/instability
- be familiar with the basic cloud types and how they form
- understand the two main mechanisms leading to precipitation formation
- know the basic features of thunderstorms and how lightning develops.

To understand how clouds form and precipitation occurs, we first discuss the change of temperature with height in a rising air parcel and temperature lapse rates. Then we consider atmospheric stability/instability and what causes air to rise and condensation to occur. Cloud mechanisms and cloud classifications are described next, followed by a discussion of the growth of raindrops and precipitation processes, and finally thunderstorms.

A ADIABATIC TEMPERATURE CHANGES

When an air parcel moves to an environment of lower pressure (without heat exchange with surrounding air) its volume increases. Volume increase involves work and the consumption of energy; this reduces the heat available per unit volume and hence the temperature falls. Such a temperature change, involving no subtraction (or addition) of heat, is termed *adiabatic*. Vertical displacements of air are the major cause of adiabatic temperature changes.

Near the earth's surface, most temperature changes are non-adiabatic (also termed *diabatic*) because of energy transfer from the surface and the tendency of air to mix and modify its characteristics by lateral movement and turbulence. When an air parcel moves vertically, the changes that take place are generally adiabatic, because air is fundamentally a poor thermal conductor, and the air parcel tends to retain its own thermal identity, which distinguishes it from the surrounding air. However, in some circumstances, mixing of air with its surroundings must be taken into account.

Consider the changes that occur when an air parcel rises: the decrease of pressure (and density) causes its volume to increase and temperature to decrease (see Chapter 2B). The rate at which temperature decreases in a rising, expanding air parcel is called the *adiabatic lapse rate*. If the upward movement of air does not produce condensation, then the energy expended by expansion will cause the temperature of the mass to fall at the constant *dry adiabatic lapse rate* or DALR (9.8°C/km). However, prolonged cooling of air invariably produces condensation, and when this happens latent heat is liberated, counteracting the dry adiabatic temperature decrease to a certain extent. Therefore, rising and saturated (or precipitating) air cools at a slower rate (the *saturated adiabatic lapse rate* or SALR) than air that is unsaturated. Another difference between the dry and saturated adiabatic rates is that whereas the DALR is constant the SALR varies with temperature. This is because air at higher temperatures is able to hold more moisture and therefore on condensation releases a greater quantity of latent heat. At high temperatures, the saturated adiabatic lapse rate may be as low as 4°C/km, but this rate increases with decreasing temperatures, approaching 9°C/km at –40°C. The DALR is reversible (i.e., subsiding air warms at 9.8°C/km); whereas in any descending cloud air saturation cannot persist because droplets evaporate.

Three different lapse rates must be distinguished: two dynamic and one static. The static, *environmental lapse rate* (ELR) is the actual temperature decrease with height on any occasion, such as an observer ascending in a balloon or climbing a mountain would record (see Chapter 2C.1). This is not an adiabatic rate, therefore, and may assume any value depending on the local vertical profile of air temperature. In contrast, the dynamic *adiabatic dry and saturated lapse rates* (or cooling rates) apply to rising parcels of air moving through their environment. Above a heated surface, the vertical temperature gradient sometimes exceeds the dry adiabatic lapse rate (i.e., it is superadiabatic). This is common in arid areas in summer. Over most ordinary dry surfaces, the lapse rate approaches the dry adiabatic value at an elevation of 100m or so.

The changing properties of rising air parcels can be determined by plotting *path curves* on suitably constructed graphs such as the skew *T*-log *p* chart and the *tephigram*, or *T*-ϕ-gram, where ϕ refers to entropy. A tephigram (Figure 5.1) displays five sets of lines representing properties of the atmosphere:

1. Isotherms – i.e., lines of constant temperature (parallel lines from bottom left to top right).
2. Dry adiabats (parallel lines from bottom right to top left).
3. Isobars – i.e., lines of constant pressure and corresponding height contours (slightly curved nearly horizontal lines).
4. Saturated adiabats (curved lines sloping up from right to left).
5. Saturation mixing ratio lines (at a slight angle to the isotherms).

Air temperature, dew-point temperature and wind velocity are determined from atmospheric soundings made by rawinsondes (radar wind soundings). Helium-filled balloons with a suspended instrument package and a radar reflector for tracking them are released at upper-air stations around the world once or twice daily, The instruments in the package are an aneroid barometer to determine altitude, a temperature sensor and a dew-point sensor. Radar is used to track the balloon as it rises and to calculate the wind speed and direction. The data are reported at standard levels (1000, 850, 700, 500, 300, 200, 100, 50, 20 and 10mb) and at intermediate levels where significant departures occur from a linear interpolation between standard levels.

Air temperature and dew-point temperature are the variables that are commonly plotted on an adiabatic chart. The dry adiabats are also lines of constant potential temperature, θ (or isentropes). Potential temperature is the temperature of an air parcel brought dry adiabatically to a pressure of 1000mb. Mathematically,

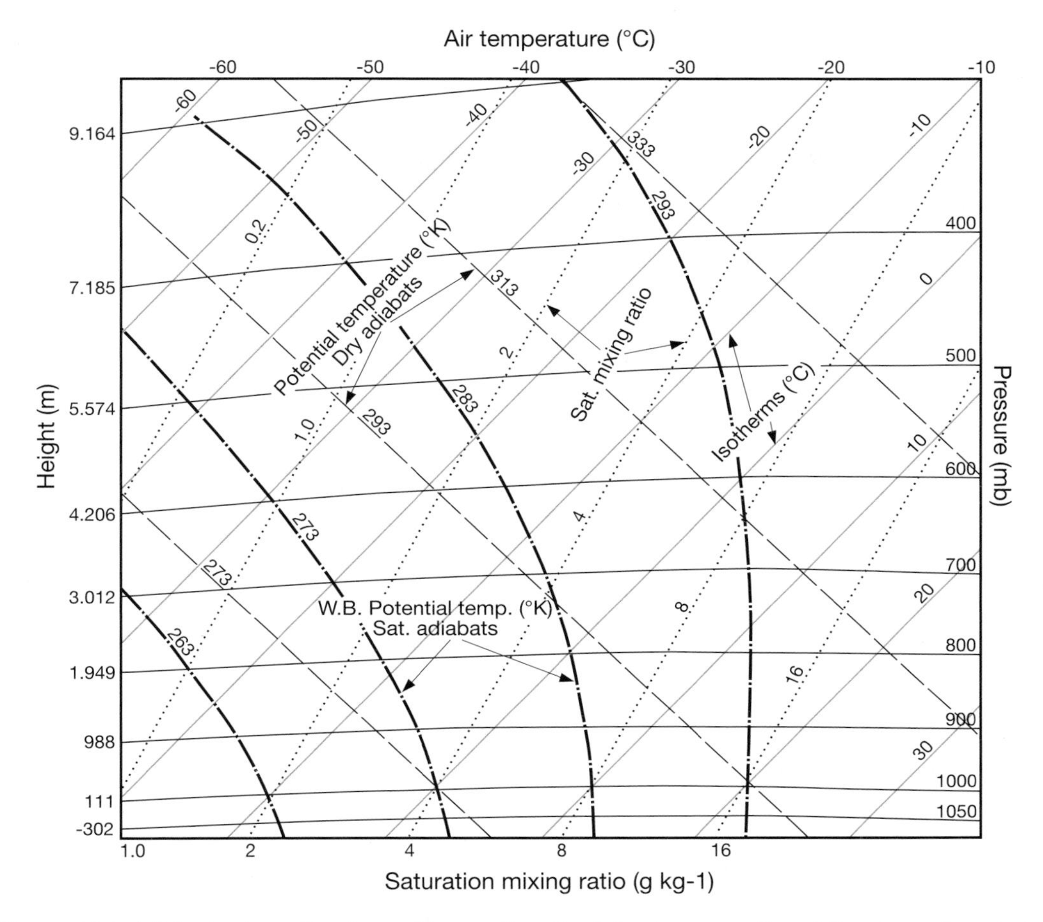

Figure 5.1 Adiabatic charts like the tephigram allow the following properties of the atmosphere to be displayed: temperature, pressure, potential temperature, wet-bulb potential temperature and saturation (humidity) mixing ratio.

$$\theta = T\left\{\frac{1000}{p}\right\}^{0.286}$$

where θ and T are in K, and p = pressure (mb).

The relationship between T and θ; also between T and θw, the wet-bulb potential temperature (where the air parcel is brought to a pressure of 1000mb by a saturated adiabatic process), is shown schematically in Figure 5.2. Potential temperature provides an important yardstick for air mass characteristics, since if the air is affected only by dry adiabatic processes the potential temperature remains constant. This helps to

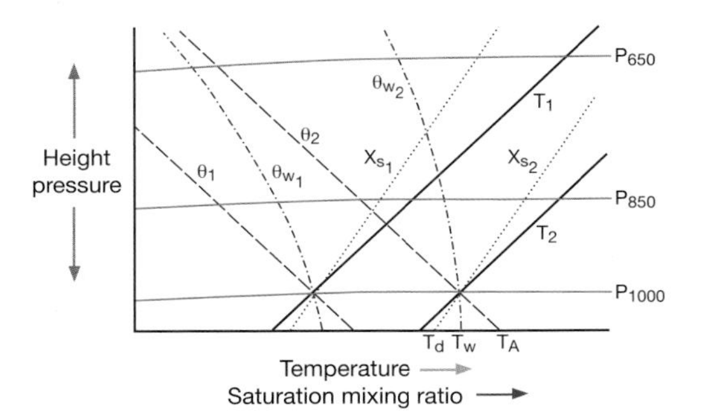

Figure 5.2 Graph showing the relationships between temperature (T), potential temperature (θ), wet-bulb potential temperature (θ_w) and saturation mixing ratio (x_s). T_d = dew-point, T_w = wet-bulb temperature and T_A = air temperature.

identify different air masses and indicates when latent heat has been released through saturation of the air mass or when non-adiabatic temperature changes have occurred.

B CONDENSATION LEVEL

Rising air cools as air parcels expand and the relative humidity level of the air increases.

After reaching saturation – 100 percent relative humidity – condensation occurs and cloud forms above the condensation level. Convection may occur as free or forced convection. *Free convection* is caused by density differences in the atmosphere that give rise to thermals – rising currents due to differential heating of the atmosphere. *Forced convection* involves uplift by mechanical forces such as flow over orographic barriers, frontal uplift, turbulence due to surface friction or ascent due to convergent windflow.

Figure 5.2 illustrates an important property of the tephigram. A line along a dry adiabat (θ) through the dry-bulb temperature of the surface air (T_A), an isopleth of saturation mixing ratio (xs) through the dew-point (Td), and a saturated adiabat (θw) through the wet-bulb temperature (Tw), all intersect at a point corresponding to saturation for the air mass. This relationship, known as Normand's theorem, is used to estimate the *lifting condensation level* (see Figure 5.3). For example, with an air temperature of 20°C and a dew-point of 10°C at 1000mb surface pressure on

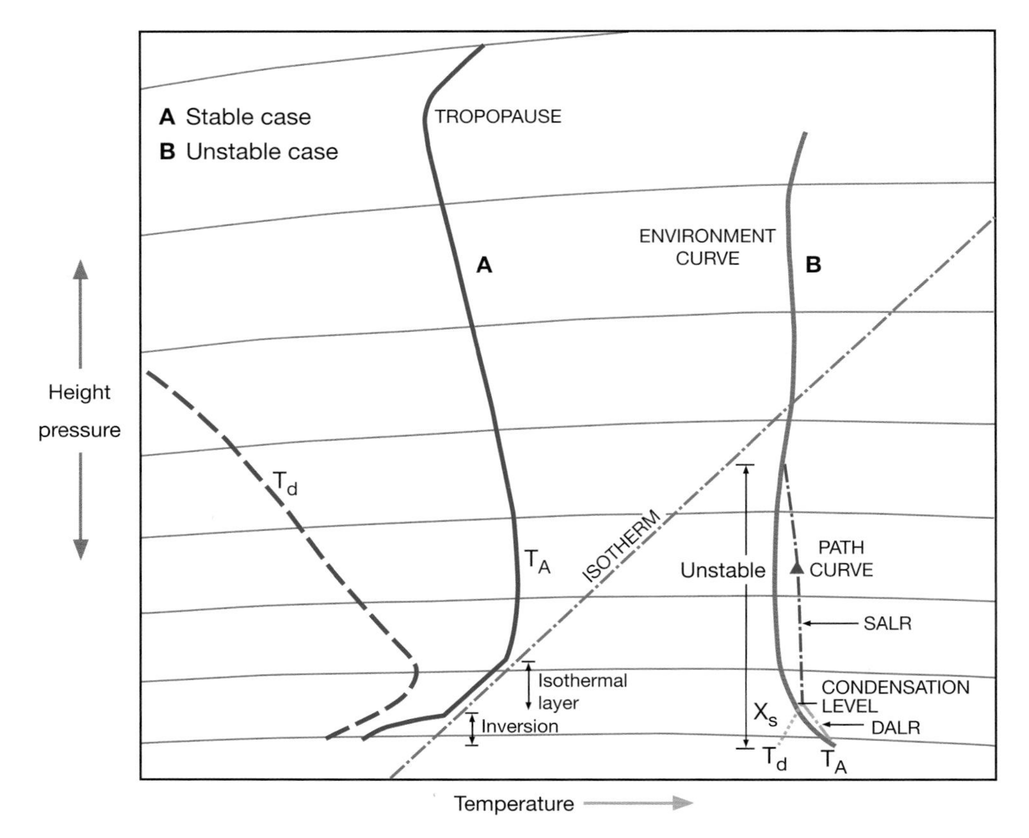

Figure 5.3 Tephigram showing (A) stable air case – T_A is the air temperature and T_d the dew-point; and (B) unstable air case. The lifting condensation level is shown, together with the path curve (arrowed) of a rising air parcel. x_s is the saturation humidity mixing ratio line through the dew-point temperature (see text).

Figure 5.1, the lifting condensation level is at 860mb with a temperature of 8°C. The height of this 'characteristic point' is approximately

$$h\,(\mathrm{m}) = 120(T - Td)$$

where T = air temperature and Td = dew-point temperature at the surface in °C.

The lifting condensation level (LCL) formulation does not take account of vertical mixing. A modified calculation defines a *convective condensation level* (CCL). In the near-ground layer, surface heating may establish a superadiabatic lapse rate, but convection modifies this to the DALR profile. Daytime heating steadily raises the surface air temperature from T_0 to T_1, T_2 and T_3 (Figure 5.4). Convection also equalizes the humidity mixing ratio, assumed equal to the value for the initial temperature. The CCL is located at the intersection of the environment temperature curve with a saturation mixing ratio line corresponding to the average mixing ratio in the surface layer (1000–1500m). Expressed in another way, the surface air temperature is the minimum that will allow cloud to form as a result of free convection. Because the air near the surface is often well mixed, the CCL and LCL, in practice, are commonly nearly identical.

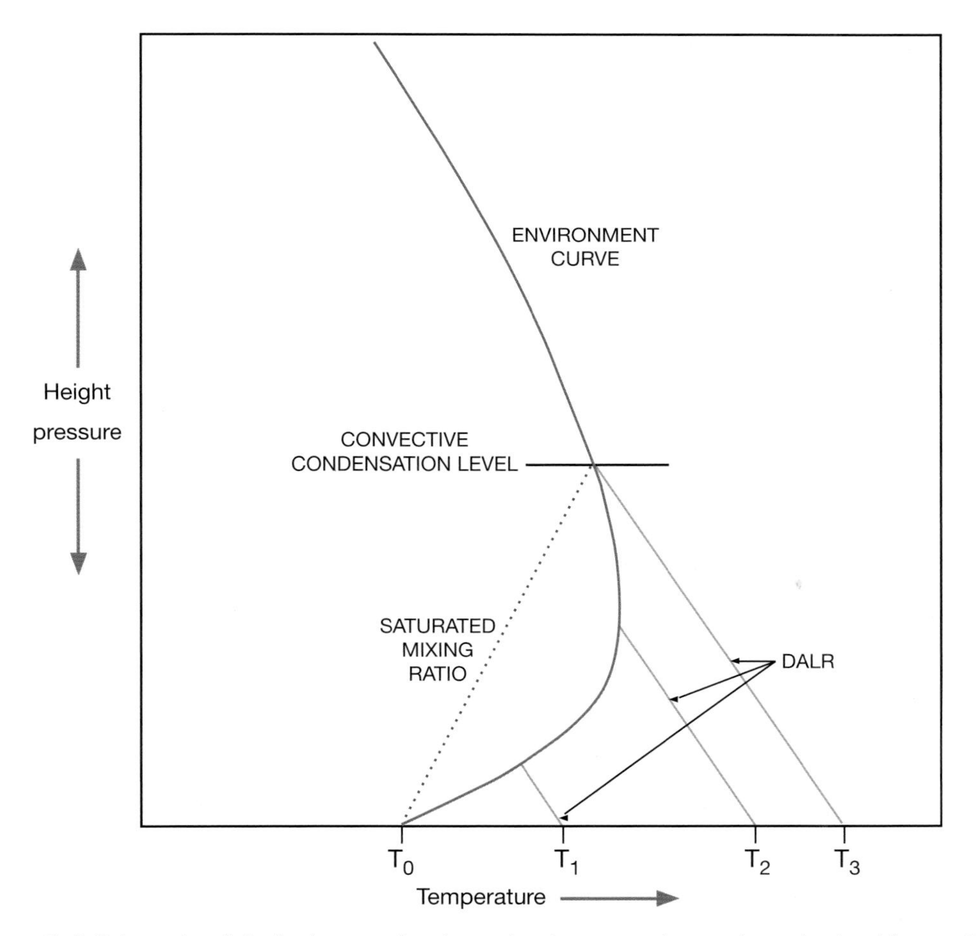

Figure 5.4 Schematic adiabatic chart used to determine the convective condensation level (see p. 113). T_0 represents the early morning temperature; T_1, T_2 and T_3 illustrate daytime heating of the surface air.

Experimentation with a tephigram shows that both the convective and the lifting condensation levels rise as the surface temperature increases, with little change of dew-point. This is commonly observed in the early afternoon, when the base of cumulus clouds tends to be at higher levels.

C AIR STABILITY AND INSTABILITY

If stable (unstable) air is forced up or down it has a tendency to return to (continue to move away from) its former position once the motivating force ceases. Figure 5.3 shows the reason for this important characteristic. The environment temperature curve (A) lies to the right of any *path curve* representing the lapse rate of an unsaturated air parcel cooling dry adiabatically when forced to rise. At any level, the rising parcel is cooler and more dense than its surroundings and therefore tends to revert to its former level. Similarly, if air is forced downward it will warm at the dry adiabatic rate; the parcel will always be warmer and less dense than the surrounding air, and tend to return to its former position (unless prevented from doing so). However, if local surface heating causes the environmental lapse rate near the surface to exceed the dry adiabatic lapse rate (B), then the adiabatic cooling of a convective air parcel allows it to remain warmer and less dense than the surrounding air, so it continues to rise through buoyancy. The characteristic of unstable air is a tendency to continue to move away from its original level when set in motion. The transition between the stable and unstable states is termed *neutral*.

We can summarize the five basic states of static stability which determine the ability of air at rest to remain laminar or become turbulent through buoyancy: The key is the temperature of a displaced air parcel relative to that in the surrounding air.

Absolutely stable: ELR < SALR
Saturated neutral: ELR = SALR
Conditionally unstable: SALR < ELR < DALR
Dry neutral: ELR = DALR
Absolutely unstable: ELR > DALR

Air that is colder than its surroundings tends to sink. Cooling in the atmosphere usually results from radiative processes, but subsidence also results from horizontal convergence of upper tropospheric air (see Chapter 6B.2). Subsiding air has a typical vertical velocity of only 1–10cm s^{-1}, unless convective downdraft conditions prevail (see below). Subsidence can produce substantial changes in the atmosphere; for instance, if a typical air mass sinks about 300m, all average-size cloud droplets will usually be evaporated through the adiabatic warming.

Figure 5.5 illustrates a common situation where the air is stable in the lower layers. If the air is forced upward by a mountain range or through local surface heating, the path curve may eventually cross to the right of the environment curve (the level of free convection). The air, now warmer than its surroundings, is buoyant and free to rise. This is termed *conditional instability*; the development of instability is dependent on the air mass becoming saturated. Since the environmental lapse rate is frequently between the dry and saturated adiabatic rates, a state of conditional instability is common. The path curve intersects the environment curve at 650mb. Above this level the atmosphere is stable, but the buoyant energy gained by the rising parcel enables it to move some distance into this region. The theoretical upper limit of cloud development can be estimated from the tephigram by determining an area (B) above the intersection of the environment and path curves equal to that between the two curves from the level of free convection to the intersection (A) in Figure 5.5. The tephigram is so constructed that equal areas represent equal energy.

These examples assume that a small air parcel is being displaced without any compensating air motion or mixing of the parcel with its surroundings. These assumptions are rather unrealistic. Dilution of an ascending air parcel by mixing of

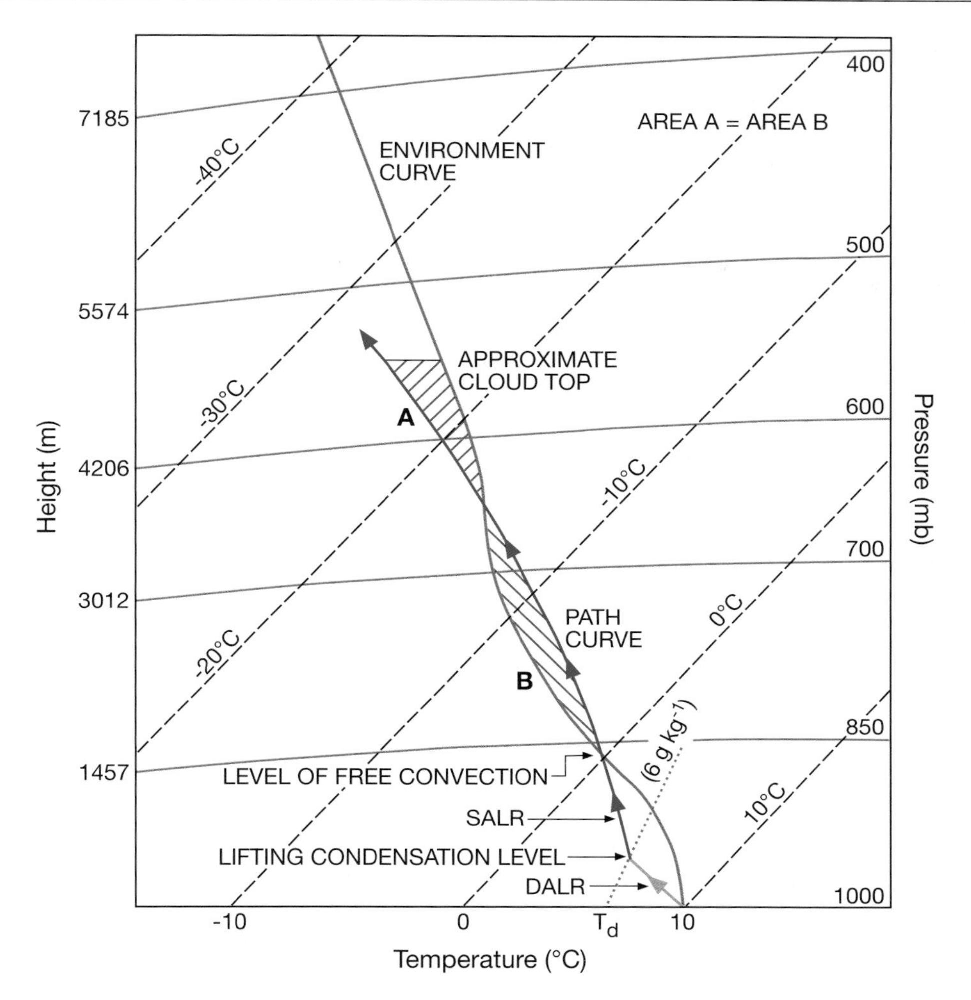

Figure 5.5 Schematic tephigram illustrating the conditions associated with the conditional instability of an air mass that is forced to rise. The saturation mixing ratio is a broken line and the lifting condensation level (cloud base) is below the level of free convection.

the surrounding air with it through *entrainment* will reduce its buoyant energy. However, the parcel method is generally satisfactory for routine forecasting because the assumptions approximate conditions in the updraft of cumulonimbus clouds.

In some situations a deep layer of air may be displaced over an extensive topographic barrier. Figure 5.6 shows a case where the air in the upper levels is less moist than that below. If the whole layer is forced upward, the drier air at B cools at the dry adiabatic rate, and so initially will the air around A. Eventually the lower air reaches condensation level, after which this layer cools at the saturated adiabatic rate. This results in an increase in the actual lapse rate of the total thickness of the raised layer, and, if this new rate exceeds the saturated adiabatic, the air layer becomes unstable and may overturn. This is termed *convective* (or *potential*) *instability*.

Vertical mixing of air was identified earlier as a possible cause of condensation. This is best

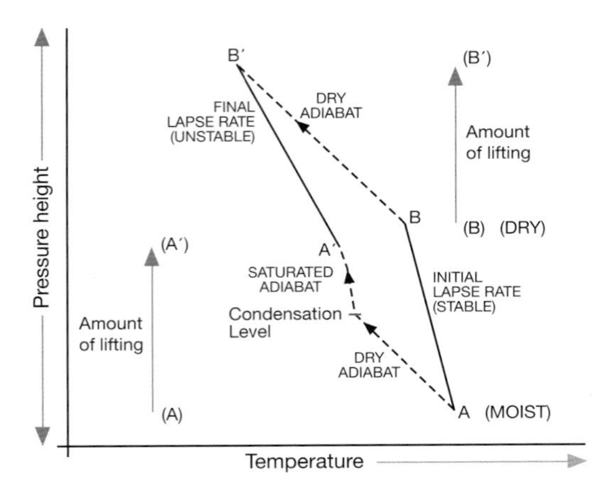

Figure 5.6 Convective instability. AB represents the initial state of an air column; moist at A, dry at B. After uplift of the whole air column the temperature gradient A′ B′ exceeds the saturated adiabatic lapse rate, so the air column is unstable.

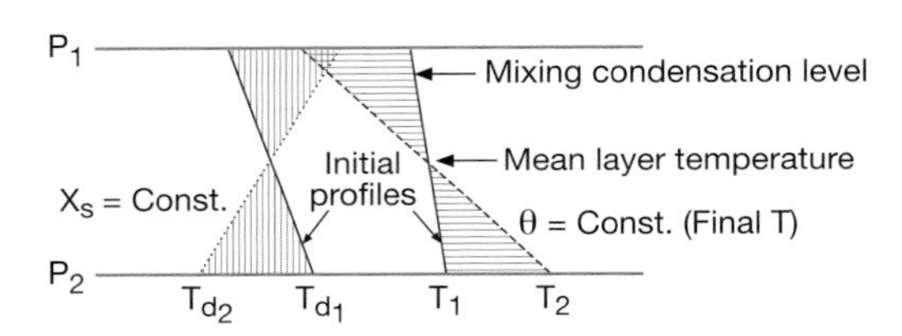

Figure 5.7 Graph illustrating the effects of vertical mixing in an air mass. The horizontal lines are pressure surfaces (P_2, P_1). The initial temperature (T_1) and dew-point temperature (T_{d1}) gradients are modified by turbulent mixing to T_2 and T_{d2}. The condensation level occurs where the dry adiabat (u) through T_1 intersects the saturation humidity mixing ratio line (x_s) through T_{d2}.

illustrated by use of a tephigram. Figure 5.7 shows an initial distribution of temperature and dew-point. Vertical mixing leads to averaging these conditions through the layer affected. Thus, the *mixing condensation level* is determined from the intersection of the average values of saturation humidity mixing ratio and potential temperature. The areas above and below the points where these average-value lines cross the initial environment curves are equal.

D CLOUD FORMATION

The formation of clouds depends on atmospheric instability and vertical motion but it also involves microscale processes. These are discussed before we examine cloud development and cloud types.

1 Condensation nuclei

Remarkably, condensation occurs with the utmost difficulty in *clean* air; moisture needs a suitable surface upon which it can condense. If clean air is cooled below its dew-point it becomes *supersaturated* (i.e., relative humidity exceeding 100 percent). To maintain a pure water drop of radius 10^{-7}cm (0.001mm) requires a relative humidity of 320 percent, and for one of radius 10^{-5}cm (0.1mm) only 101 percent.

Usually, condensation occurs on a foreign surface; this can be a land or plant surface in the case of dew or frost, while in the free air condensation begins on *hygroscopic nuclei.* These are microscopic particles – *aerosols* – the surfaces of which (like the weather enthusiast's seaweed!) have the property of *wettability.* Aerosols include dust, smoke, salts and chemical compounds. Sea salts, which are particularly hygroscopic, enter the atmosphere by the bursting of air bubbles in foam. They are a major component of the aerosol load near the ocean surface but tend to be removed rapidly due to their size. Other contributions are from fine soil particles and various natural, industrial and domestic combustion products raised by the wind. A further source is the conversion of atmospheric trace gas to particles through photochemical reactions, particularly over urban areas. Nuclei range in size from 0.001μm radius, which are ineffective owing to the high supersaturation required for their activation, to *giants* of over 10μm, which do not remain airborne for very long (see pp. 16–18). On average, oceanic air contains 1 million condensation nuclei per liter (i.e., dm^3), and land air holds some 5 or 6 million. In the marine troposphere there are fine particles, mainly ammonium sulphate. A photochemical origin associated with anthropogenic activities accounts for about half of these in the Northern Hemisphere. Dimethyl sulfide (DMS), associated with algal decomposition, also undergoes oxidation to sulfate. Over the tropical continents, aerosols are produced by forest vegetation and surface litter, and through biomass burning; particulate organic carbon predominates. In mid-latitudes, remote from anthropogenic sources, coarse particles are mostly of crustal origin (calcium, iron, potassium and aluminium) whereas crustal, organic and sulfate particles are almost equally represented in the fine aerosol load.

Aerosols have a substantial effect on cloud properties and therefore on the initiation of precipitation. Polluted atmospheres typically have aerosol concentrations x 10–100 times those of pristine oceanic air masses. The effects of aerosols on clouds involve radiative processes and microphysical effects. Aerosol layers decrease the solar radiation reaching the surface acting to lower surface temperatures and reduce evaporation and convection. Aerosols also serve as cloud condensation nuclei (CCN), which by nucleating large numbers of small droplets, slow the conversion of cloud droplets into raindrops. The net effect seems to be to decrease precipitation from shallow clouds, but to enhance precipitation from deep convective clouds with warm cloud bases. The maximum enhancement occurs for intermediate CCN values, whereas at higher concentrations both the radiative and microphysical effects work towards a lower release of convective energy.

Hygroscopic aerosols are soluble. This is very important since the saturation vapor pressure is less over a solution droplet (for example, sodium chloride or sulfuric acid) than over a pure water drop of the same size and temperature (Figure 5.8). Indeed, condensation begins on hygroscopic particles before the air is saturated; in the case of sodium chloride nuclei at 78 percent relative humidity. Figure 5.8 illustrates Kohler curves showing droplet radii for three sets of solution droplets of sodium chloride (a common sea salt) in relation to their equilibrium relative humidity. Droplets in an environment where values are below/above the appropriate curve will evaporate/grow. Each curve has a maximum beyond which the droplet can grow in air with less supersaturation.

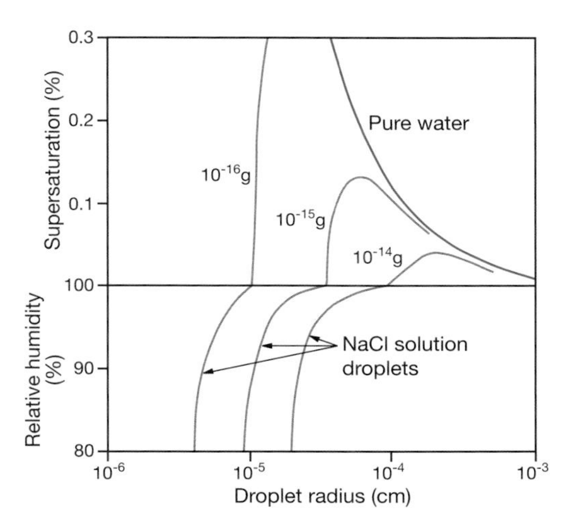

Figure 5.8 Kohler curves showing the variation of equilibrium relative humidity or supersaturation (%) with droplet radius for pure water and NaCl solution droplets. The numbers show the mass of sodium chloride (a similar family of curves is obtained for sulfate solutions). The pure water droplet line illustrates the curvature effect.

Once formed, the growth of water droplets is far from simple. In the early stages, the solution effect is predominant and small drops grow more quickly than large ones, but as the size of a droplet increases, its growth rate by condensation decreases (Figure 5.9). Radial growth rate slows down as the drop size increases, because there is a greater surface area to cover with every increment of radius. However, the condensation rate is limited by the speed with which the released latent heat can be lost from the drop by conduction to the air; this heat reduces the vapor gradient. In addition, competition between droplets for the available moisture acts to reduce the degree of supersaturation.

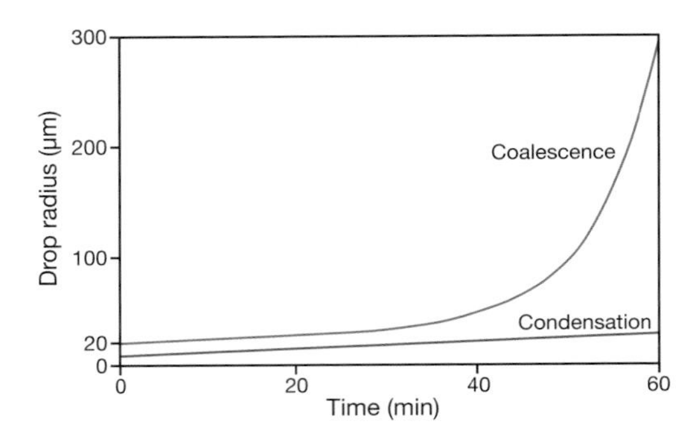

Figure 5.9 Droplet growth by condensation and coalescence.

Source: Jonas (1994a). Reprinted from *Weather* by permission of the Royal Meteorological Society. Crown copyright ©.

Supersaturation in clouds rarely exceeds 1 percent and, because the saturation vapor pressure is greater over a curved droplet surface than over a plane water surface, minute droplets (<0.1µm radius) are readily evaporated (see Figure 5.8). Initially, the nucleus size is important; for supersaturation of 0.05 percent, a droplet of 1µm radius with a salt nucleus of mass 10^{-13}g reaches 10µm in 30 minutes, whereas one with a salt nucleus of 10^{-14}g would take 45 minutes. Later, when the dissolved salt has ceased to have significant effect, the radial growth rate slows due to decreasing supersaturation.

Figure 5.9 illustrates the very slow growth of water droplets by condensation – in this case at 0.2 percent super saturation from an initial radius of 10µm. As there is an immense size difference between cloud droplets (<1 to 50µm radius) and raindrops (>1mm diameter), it is apparent that the gradual process of condensation cannot explain the rates of formation of raindrops that are often observed. For example, in most clouds precipitation develops within an hour. The alternative coalescence mechanism illustrated in Figure 5.9 is described below (p. 128). It must be remembered too that falling raindrops undergo evaporation in the unsaturated air below the cloud base. A droplet of 0.1mm radius evaporates after falling only 150m at a temperature of 5°C and 90 percent relative humidity, but a drop of 1mm radius would fall 42km before evaporating. On average, clouds contain only 4 percent of the total water in the atmosphere at any one time but they are a crucial element in the hydrological cycle.

2 Cloud types

The great variety of cloud forms necessitates a classification for purposes of weather reporting. The internationally adopted system is based upon (1) the general shape, structure and vertical extent of the clouds, and (2) their altitude. This approach was originally developed by Luke Howard in 1803.

These primary characteristics are used to define the ten basic groups (or genera) as shown in Figure 5.10. High cirriform cloud is composed of ice crystals, giving a generally fibrous appearance. Stratiform clouds are in layers, while cumuliform clouds have a heaped appearance and usually show progressive vertical development. Other prefixes are *alto-* for middle-level (medium) clouds and *nimbo-* for thick, low clouds which appear dark grey and from which continuous rain is falling.

The height of the cloud base may show a considerable range for of any of these types and varies with latitude. The approximate limits in thousands of meters for different latitudes are shown in Table 5.1.

Following taxonomic practice, the classification subdivides the major groups into species and varieties with Latin names according to their appearance. The *International Cloud Atlas* and Plates 5.1–5.15 provide illustrations.

Clouds may also be grouped according to their mode of origin. A genetic grouping can be made

Table 5.1 Cloud base height (in 000 m)

	Tropics	Mid-latitudes	High latitudes
High cloud	6–18	5–13	3–8
Medium cloud	2–8	2–7	2–4
Low cloud	Below 3	Below 2	Below 2

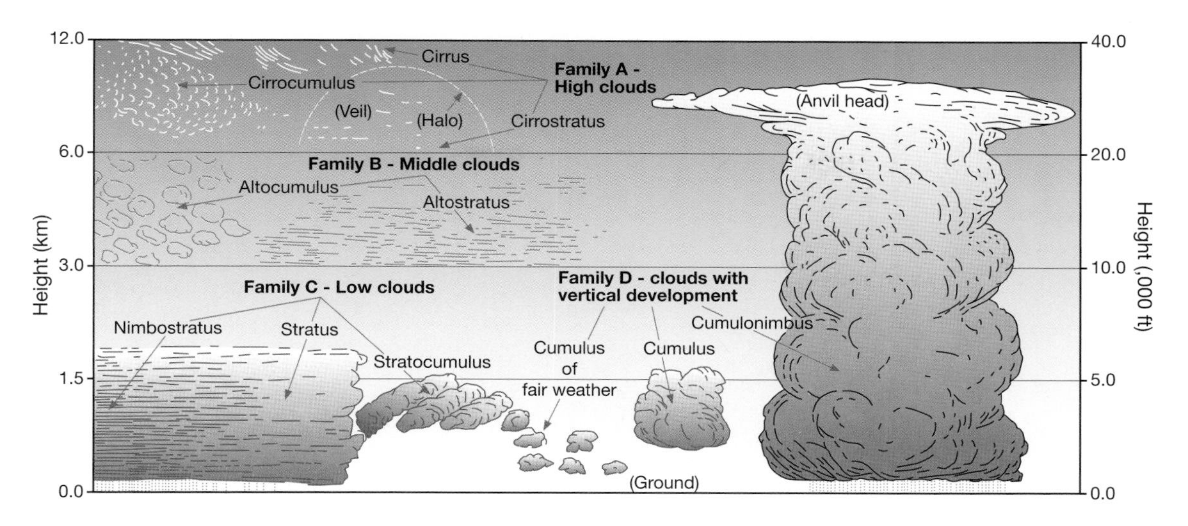

Figure 5.10 The ten basic cloud groups classified according to height and form.
Source: Modified after Strahler (1965).

based on the mechanism of vertical motion that produces condensation. Four categories are:

1. gradual uplift of air over a wide area in association with a low pressure system;
2. thermal convection (on the local cumulus scale);
3. uplift by mechanical turbulence (*forced convection*);
4. ascent over an orographic barrier.

Group 1 includes a wide range of cloud types and is discussed more fully in Chapter 9D.2. With cumuliform clouds (group 2), upward convection currents (thermals) form plumes of warm air that, as they rise, expand and are carried downwind. Towers in cumulus and other clouds are caused not by thermals of surface origin, but by those set up *within* the cloud as a result of the release of latent heat through condensation. Thermals gradually lose their impetus as mixing of cooler, drier air from the surroundings dilutes the more buoyant warm air. Cumulus towers also tend to evaporate as updrafts diminish, leaving a shallow, oval-shaped 'shelf' cloud (*stratocumulus cumulogenitus*), which may amalgamate with others to produce a high overcast. Group 3 includes fog, stratus or stratocumulus and is important whenever air near the surface is cooled to dew-point by conduction or night-time radiation and the air is stirred by irregularities of the ground. The final group (4) includes stratiform or cumulus clouds produced by forced uplift of air over mountains. Hill fog is simply stratiform cloud enveloping high ground. A special and important category is the wave (lenticular) cloud, which develops when air flows over hills, setting up a wave motion in the air current downwind of the ridge (see Chapter 6C.2). Clouds form in the crest of these waves if the air reaches its condensation level.

Operational weather satellites provide information on global cloudiness, and on cloud patterns in relation to weather systems. They supply direct-readout imagery and information not obtainable by ground observations. Special classifications of cloud elements and patterns have been devised in order to analyze satellite imagery. A common pattern seen on satellite photographs is cellular, or honeycomb-like, with a typical diameter of 30km. This develops from the movement of cold air over a warmer sea surface.

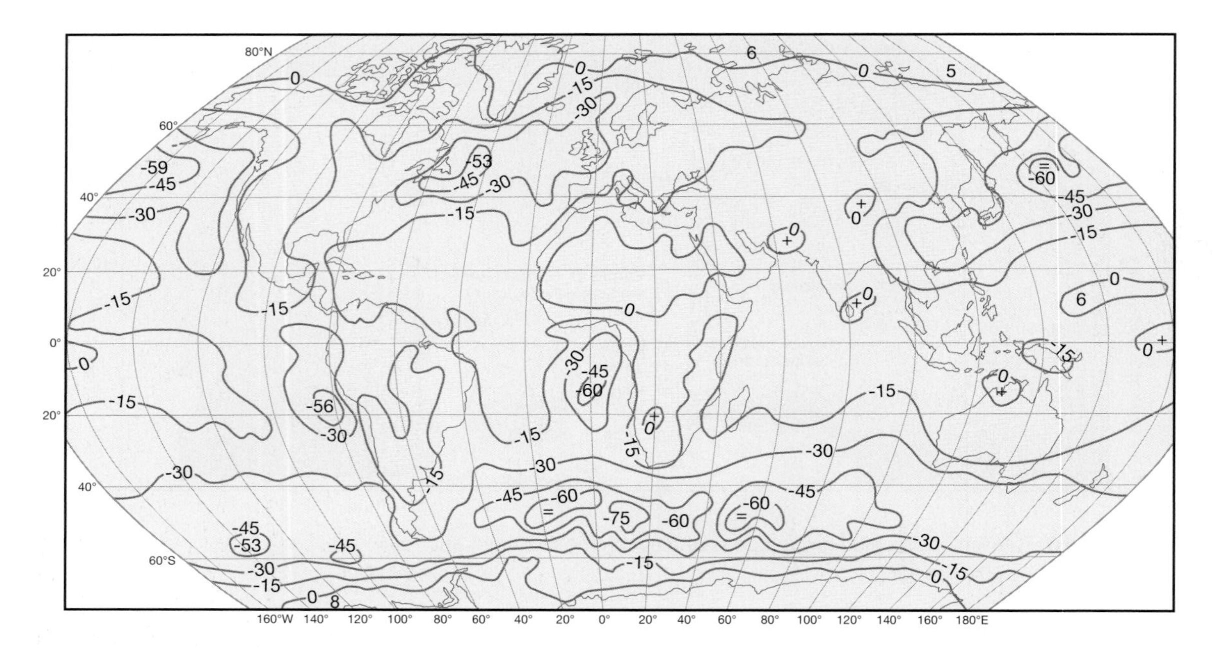

Figure 5.11 Mean annual net cloud forcing (W m^{-2}) observed by the *Nimbus-7* ERB satellite for the period June 1979 to May 1980.

Source: Kyle *et al.* (1993). From *Bulletin of the American Meteorological Society*, by permission of the American Meteorological Society.

An open cellular pattern, where cumulus clouds are along the cell sides, forms where there is a large air–sea temperature difference, whereas closed polygonal cells occur if this difference is small. In both cases there is subsidence above the cloud layer. Open (closed) cells are more common over warm (cool) ocean currents to the east (west) of the continents. The honeycomb pattern has been attributed to mesoscale convective mixing, but the cells have a width–depth ratio of about 30:1, whereas laboratory thermal convection cells have a corresponding ratio of only 3:1. Thus the true explanation may be more complicated. Less common is a radiating cellular pattern. Another common pattern over oceans and uniform terrain is provided by linear cumulus cloud 'streets'. Helical motion in these two-dimensional cloud cells develops with surface heating, particularly when outbreaks of polar air move over warm seas and there is a capping inversion.

3 Global cloud cover

There are difficulties in determining cloud cover and layer structure from both satellite and ground observations. Surface-based estimates of cloud amounts are some 10 percent greater than those derived from satellites, mainly owing to the problem of estimating gaps near the horizon. The greatest discrepancies occur in summer in the subtropics and in polar regions. Total cloud amounts show characteristic geographical, latitudinal and seasonal distributions (see Figures 3.8 and 5.11). During the northern summer there are high percentages over West Africa, northwestern South America and Southeast Asia, with minima over the Southern Hemisphere continents, southern Europe, North Africa and the Near East. During austral summer, there are high percentages over tropical land areas in the Southern Hemisphere, partly due to convection

Plate 5.1 Fair weather cumulus over Coconut Creek, Florida, December 1977.
Source: National Weather Service (NWS) Collection. Photographer: Ralph F. Kresge #0655.

Plate 5.2 Cumulus humilis over Swanage, Dorset.

Plate 5.3 Cloud streets in Davis Strait from MODIS visible imagery. In late March 2002, winter ice is still blocking much of the northern portion of the Strait. The streamers of cloud are referred to as 'cloud streets'. They are formed when cold Arctic air moves warmer, open water. Usually the streets are aligned in the direction of a low-level wind.
Source: Jacques Descloitres, MODIS Land Rapid Response Team, NASA/GSFC. Courtesy NASA Earth Observatory.

Plate 5.4 Cumulus over Mount Etna from heat and vapor supplied by the volcano.

along the Intertropical Convergence Zone, and in subpolar ocean areas due to moist air advection. Minimal cloud cover is associated with the subtropical high pressure regions throughout the year, whereas persistent maximum cloud cover occurs over the Southern Ocean storm belt at 50–70°S and over much of the ocean area north of 45°N (Figure 5.12).

Cloud acts both as an important sink for radiative energy in the earth–atmosphere system, through absorption, as well as a source due to reflection and re-radiation (see Chapter 3B, C). Globally, the mean annual net forcing effect of clouds is negative (~ $-20Wm^{-2}$) because the albedo effect on incoming solar radiation outweighs the infrared absorption. However, cloud forcing is complex; for example, more total cloud implies more absorption of outgoing terrestrial radiation (positive forcing, leading to warming) whereas more high cloud produces increased reflection of incoming solar radiation (negative forcing, leading to cooling) (Figure 5.11).

There is evidence that cloud amounts increased during the twentieth century. For example, there was a striking increase in cloud cover over the United States (especially between 1940 and 1950).

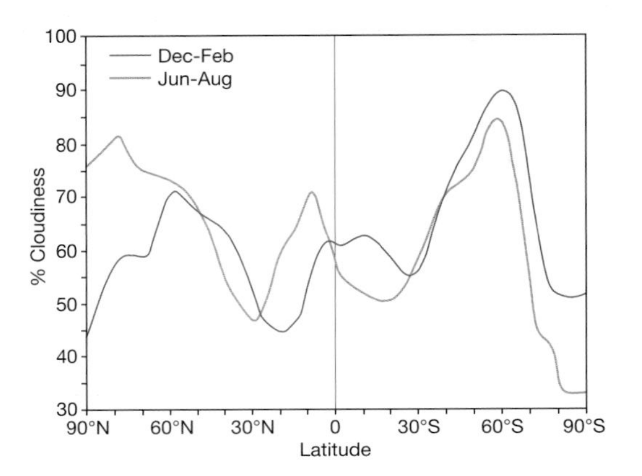

Figure 5.12 Average zonal distribution of total cloud amount (%), derived from surface observations over the total global surface (i.e., land plus water) for the months of December–February and June–August during the period 1971–1981.

Source: From London *et al.* (1989). Courtesy of Cospar and Elsevier.

This may be associated with higher atmospheric sulfate concentrations due to increased coal burning. The relationship with temperature is unclear.

E FORMATION OF PRECIPITATION

The puzzle of raindrop formation has already been noted. The simple growth of cloud droplets through condensation is apparently an inadequate mechanism and more complex processes have to be envisaged.

Various early theories of raindrop growth can be discounted. Proposals were that differently charged droplets could coalesce through electrical attraction, but it appears that distances between drops are too great and the difference between the electrical charges too small for this to happen. It was suggested that large drops might grow at the expense of small ones. However, observations show that the distribution of droplet size in a cloud tends to maintain a regular pattern; the average radius is between 10 and 15μm, and few are larger than 40μm. A further idea was that atmospheric turbulence might bring warm and cold cloud droplets into close conjunction. The supersaturation of the air with reference to the cold droplets and the undersaturation with reference to the warm droplets would cause the latter to evaporate and cold droplets to develop at their expense. However, except perhaps in some tropical clouds, the temperature of cloud droplets is too low for this differential mechanism to operate. Figure 2.4 shows that, below about –10°C, the slope of the saturation vapor pressure curve is low. Another theory was that raindrops grow around exceptionally large condensation nuclei (observed in some tropical storms). Large nuclei do experience a more rapid rate of initial condensation, but after this stage they are subject to the same limiting rates of growth that apply to all cloud drops.

Current theories for the rapid growth of raindrops involve either the growth of ice crystals at the expense of water drops, or the coalescence of small droplets by the sweeping action of falling drops.

1 Bergeron–Findeisen theory

This widely accepted theory is based on the fact that at subzero temperatures the atmospheric vapor pressure decreases more rapidly over an ice surface than over water (Figure 2.14). The saturation vapor pressure over water becomes greater than over ice, especially between temperatures of –5 and –25°C, where the difference exceeds 0.2mb. If ice crystals and supercooled water droplets exist together in a cloud, the drops tend to evaporate and direct deposition takes place from the vapor on to the ice crystals.

Freezing nuclei are necessary before ice particles can form – usually at temperatures of about –15 to –25°C. Small water droplets can, in fact, be supercooled in pure air to –40°C before spontaneous freezing occurs. But ice crystals generally predominate in clouds where temperatures are below about –22°C. Freezing nuclei are far less numerous than condensation nuclei; there may be

Plate 5.5 A spring shower over the Front Range, CO from Niwot Ridge.

Plate 5.6 Towering thunderhead with anvil top. NOAA wea00031.

Plate 5.7 Stratocumulus over Indian House Lake, Quebec.

Plate 5.8 Stratiform cloud over the High Plains from Niwot Ridge, CO.

as few as 10 per liter at –30°C and probably rarely more than 1000. However, some become active at higher temperatures. Kaolinite, a common clay mineral, initially becomes active at –9°C and on subsequent occasions at –4°C. The origin of freezing nuclei has been a subject of much debate but it is generally considered that very fine soil particles are a major source. Biogenic aerosols emitted by decaying plant litter, in the form of complex chemical compounds, also serve as freezing nuclei. In the presence of certain associated bacteria, ice nucleation can take place at only –2 to –5°C.

Tiny ice crystals grow readily by deposition from vapor, with different hexagonal forms developing at different temperature ranges. The number of ice crystals also tends to increase progressively because small splinters become detached by air currents during growth and act as fresh nuclei. The freezing of supercooled water drops may also produce ice splinters (see F, this chapter). Figure 5.13 shows that a low density of ice particles is capable of rapid growth in an environment of cloud water droplets. This results in a slower decrease in the average size of the much larger number of cloud droplets although this still takes place on a time scale of 101 minutes. Ice crystals readily aggregate upon collision due to their branched (dendritic) shape, and tens of thousands of crystals may form a single snowflake. Temperatures between about 0 and –5°C are particularly favorable to aggregation, because fine films of water on the crystal surfaces freeze when two crystals touch, binding them together. When the fall speed of the growing ice mass exceeds the existing velocities of the air up-currents the snowflake falls, melting into a raindrop if it falls about 250m below the freezing level.

This theory can account for most precipitation in middle and higher latitudes, yet it is not completely satisfactory. Cumulus clouds over tropical oceans can give rain when they are only some 2000m deep and the cloud-top temperature is 5°C or more. In mid-latitudes in summer, precipitation may fall from cumuli that have no subfreezing layer (*warm clouds*). A suggested mechanism in such cases is that of 'droplet coalescence' discussed below.

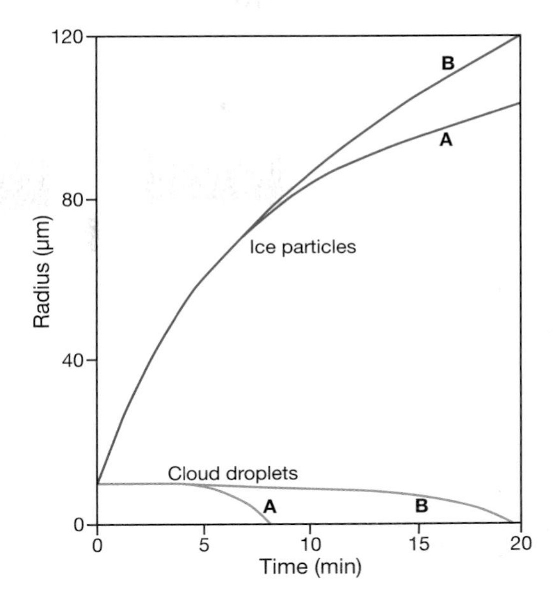

Figure 5.13 The effect of a small proportion of initially frozen droplets on the relative increase/decrease in the sizes of cloud ice and water particles. The initial droplets were at a temperature of –10°C and at water saturation. A: a density of 100 drops per cc, 1% of which were assumed to be frozen. B: a density of 1000 drops per cc, 0.1% of which were assumed to be frozen.

Source: Jonas 1994a. Reprinted from *Weather* by permission of the Royal Meteorological Society. Copyright ©.

Practical *rainmaking* has been based on the Bergeron theory with some success. The basis of such experiments is the freezing nucleus. Supercooled (water) clouds between –5 and –15°C are *seeded* with especially effective materials, such as silver iodide or 'dry ice' (CO_2) from aircraft or ground-based silver iodide generators, promoting the growth of ice crystals and encouraging precipitation. The seeding of some cumulus clouds at these temperatures probably produces a mean increase of precipitation of 10–15 percent from clouds that are already precipitating or that are 'about to precipitate'. Increases of up to 10 percent have resulted from seeding winter orographic storms. However, it appears likely that clouds with an abundance of natural ice crystals, or with above-freezing temperatures throughout, are not susceptible to rainmaking. Premature release of precipitation may destroy the updrafts and cause dissipation of the cloud. This explains why some seeding experiments have actually *decreased* the rainfall! In other instances, cloud growth and precipitation have been achieved by such methods in Australia, China and the United States. Programs aimed at increasing winter snowfall on the western slopes of the Sierra Nevada and Rocky Mountains by seeding cyclonic storms have been carried out for a number of years with mixed results. Their success depends on the presence of suitable supercooled clouds. In a recent experiment in Wyoming, ground-based-only seeding began in the winter of 2006–2007 and in 2007–2008, a plane and 25 ground stations were involved. Evaluation of the degree of success is underway. There are at least two reasons why it is difficult to establish the impact of cloud seeding: the mismatch between the scale of the impact and the scale at which seeding operates; and the large natural variability of precipitation versus the relativcly minor effect of seeding.

When several cloud layers are present in the atmosphere, natural seeding may be important. For example, if ice crystals fall from high-level cirrostratus or altostratus (a 'seeder' cloud) into nimbostratus (a 'feeder' cloud) composed of supercooled water droplets, the latter can grow rapidly through the Bergeron process and such situations may lead to extensive and prolonged precipitation. This is a frequent occurrence in cyclonic systems in winter and is important in orographic precipitation (see E3, this chapter).

2 Coalescence theories

Theories of raindrop growth use collision, coalescence and 'sweeping' as growth mechanisms. It was originally thought that cloud particle collisions due to atmospheric turbulence would cause a significant proportion to coalesce. However, particles just as easily break up if subject to collisions. Langmuir offered a variation of this simple idea. He pointed out that falling drops have terminal velocities (typically 1–10cm s^{-1}) directly related to their diameters, such that the larger drops can overtake and absorb small droplets; the latter might also be swept into the wake of larger drops and absorbed by them. Figure 5.9 gives experimental results of the rate of growth of water drops by coalescence, from an initial radius of 20mm in a cloud having a water content of 1g/m^3 (assuming maximum efficiency). Although coalescence is initially slow, droplets reach 100–200μm radius in 50 minutes. Moreover, the growth rate is rapid for drops with radii greater than 40μm. Calculations show that drops must exceed 19μm radius before they can coalesce with others; smaller droplets are swept aside without colliding. The initial presence of a few very large droplets calls for the availability of giant nuclei (e.g., salt particles) if the cloud top does not reach above the freezing level. Observations show that maritime clouds do have relatively few large condensation nuclei (10–50μm radius) and a high liquid water content, whereas continental air tends to contain many small nuclei (~ 1μm) and less liquid water. Hence, rapid onset of showers is feasible by the coalescence mechanism in maritime clouds. Alternatively, if a few ice crystals are present at higher levels in the cloud (or if seeding occurs with ice crystals falling from higher clouds) they may eventually fall through the cloud as drops and the coalescence mechanism comes into action. Turbulence in cumulus clouds serves to encourage collisions in the early stages. Thus, the coalescence process allows for more rapid growth than simple condensation, and is, in fact, common in 'warm' clouds in tropical maritime air masses, even in temperate latitudes.

Rainmaking in warm clouds has been attempted through hygroscopic seeding of clouds that makes use of compounds such as sodium, lithium and potassium salts. The idea is to generate larger droplets, either by providing larger

Plate 5.9 Altocumulus clouds.
Source: National Weather Service (NWS) Collection Photographer: Ralph F. Kresge #0863.

Plate 5.10 Lee wave clouds and 'pile of plates' over Boulder, CO.

Plate 5.11 Evening lee wave clouds over Boulder, CO.

Plate 5.12 Cirrus uncinus over Cape Dyer, Baffin Island.

nuclei to condense around or by facilitating the formation of large drops through the merging of small droplets.

3 Solid precipitation

Rain has been discussed at length because it is the most common form of precipitation. Snow occurs when the freezing level is so near the surface that aggregations of ice crystals do not have time to melt before reaching the ground. Generally, this means that the freezing level must be below 300m. Mixed snow and rain ('sleet' in British usage) is especially likely when the air temperature at the surface is about 1.5°C. Snowfall rarely occurs with a surface air temperature exceeding 4°C.

Soft hail pellets (roughly spherical, opaque grains of ice with much enclosed air) occur when the Bergeron process operates in a cloud with a small liquid water content and ice particles grow mainly by deposition of water vapor. Limited accretion of small, supercooled droplets forms an aggregate of soft, opaque ice particles 1mm or so in radius. Showers of such pellets are quite common in winter and spring from cumulonimbus clouds.

Ice pellets may develop if the soft hail falls through a region of large liquid water content above the freezing level. Accretion forms a casing of clear ice around the pellet. Alternatively, an ice pellet consisting entirely of transparent ice may result from the freezing of a raindrop or the refreezing of a melted snowflake.

True hailstones are roughly concentric accretions of clear and opaque ice. The embryo is a raindrop carried aloft in an updraft and frozen. Successive accretions of opaque ice (rime) occur due to impact of supercooled droplets, which freeze instantaneously. The clear ice (glaze) represents a wet surface layer, developed as the result of very rapid collection of supercooled drops in parts of the cloud with large liquid water content which has subsequently frozen. A major difficulty in early theories was the necessity to postulate violently fluctuating up-currents to give the observed banded hailstone structure. Modern thunderstorm models successfully account for this; the growing hailstones are recycled by a traveling storm (see Chapter 9I). On occasions, hailstones may reach giant size, weighing up to 0.76kg each (recorded in September 1970 at Coffeyville, Kansas). In view of their rapid fall speeds, hailstones may fall considerable distances with little melting. Hailstorms are a cause of severe damage to crops and property when large hail falls.

It is usual to identify three main types of precipitation – convective, cyclonic and orographic – according to the primary mode of uplift of the air. Essential to this analysis is some knowledge of storm systems. These are treated in later chapters, and the newcomer to the subject may prefer to read the following in conjunction with them (Chapter 9).

F PRECIPITATION TYPES

1 'Convective type' precipitation

This is associated with towering cumulus (cumulus congestus) and cumulonimbus clouds. Three subcategories may be distinguished according to their degree of spatial organization.

1. Scattered convective cells develop through strong heating of the land surface in summer, especially when low temperatures in the upper troposphere facilitate the release of conditional or convective instability (see B, this chapter). Precipitation, often including hail, is of the thunderstorm type, although thunder and lightning do not necessarily occur. Small areas (20 to 50km^2) are affected by individual heavy downpours, which generally last for about 30 minutes to an hour.
2. Showers of rain, snow or soft hail pellets may form in cold, moist, unstable air passing over a warmer surface. Convective cells moving with the wind can produce a streaky distribution of precipitation parallel to the wind

Plate 5.13 Valley fog from clouds spilling over the mountains north of Katmandu, Nepal, 20 November 1979.
Source: Courtesy of Nel Caine, University of Colorado.

Plate 5.14 Mammatus clouds. These form below cumulonimbus, usually presaging a thunderstorm.
Source: Courtesy of Mark Anderson, University of Nebraska.

direction. Such cells tend to occur parallel to a surface cold front in the warm sector of a depression (sometimes as a squall line), or parallel to and ahead of the warm front (see Chapter 9D). Hence the precipitation is widespread, although of limited duration at any locality.

3 In tropical cyclones, cumulonimbus cells become organized around the center in spiraling bands (see Chapter 13B.2). Particularly in the decaying stages of such cyclones, typically over land, the rainfall can be very heavy and prolonged, affecting areas of thousands of square kilometers.

2 'Cyclonic type' precipitation

Precipitation characteristics vary according to the type of low pressure system and its stage of development, but the essential mechanism is the ascent of air through horizontal convergence of airstreams in an area of low pressure (see Chapter 6B). In extra-tropical depressions, this is reinforced by uplift of warm, less dense air along an air mass boundary (see Chapter 9D.2). Such depressions give moderate and generally continuous precipitation over very extensive areas as they move, usually eastward, in the westerly wind belts between about 40 and 65° latitude. The precipitation belt in the forward sector of the storm can affect a locality in its path for 6 to 12 hours, whereas the belt in the rear gives a shorter period of thunderstorm-type precipitation. These sectors are therefore sometimes distinguished in precipitation classifications, and a more detailed breakdown is illustrated in Table 9.2. Polar lows (see Chapter 9H.3) combine the effects of airstream convergence and convective activity of category 2 (previous section), whereas troughs in the equatorial low pressure area give convective precipitation as a result of airstream convergence in the tropical easterlies (see Chapter 13B.1).

3 Orographic precipitation

Orographic precipitation is commonly regarded as a distinct type, but this requires careful qualification. Mountains are not especially efficient in causing moisture to be removed from airstreams crossing them, yet because precipitation falls repeatedly in more or less the same locations, the cumulative totals are large. An orographic barrier may produce several effects depending on its alignment and size. They include: (1) forced ascent on a smooth mountain slope, producing adiabatic cooling, condensation and precipitation; (2) triggering of conditional or convective instability by blocking of the airflow and upstream lifting; (3) triggering of convection by diurnal heating of slopes and up-slope winds; (4) precipitation from low-level cloud over the mountains by 'seeding' of ice crystals or droplets from an upper-level feeder cloud (Figure 5.14); and (5) increased frontal precipitation by retarding the movement of cyclonic systems and fronts. West coast mountains with onshore flow, such as the Western Ghats, India, during the southwest summer monsoon; the west coasts of Canada, Washington and Oregon; or coastal Norway in winter months supposedly illustrate smooth, forced ascent, yet many other processes seem to be involved. The limited width of some coastal ranges, with average wind speeds, generally allows insufficient time for the basic mechanisms of precipitation growth to operate (see Figure 4.9). In view of the complexity of

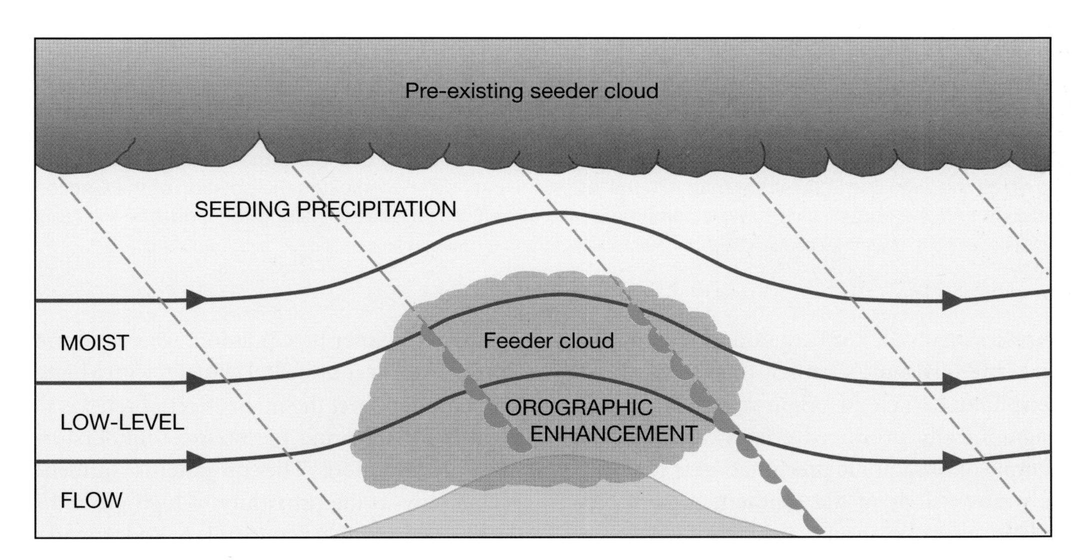

Figure 5.14 Schematic diagram of T. Bergeron's 'seeder–feeder' cloud model of orographic precipitation over hills.

Note: This process may also operate in deep nimbostratus layers.

Source: After Browning and Hill (1981), by permission of the Royal Meteorological Society.

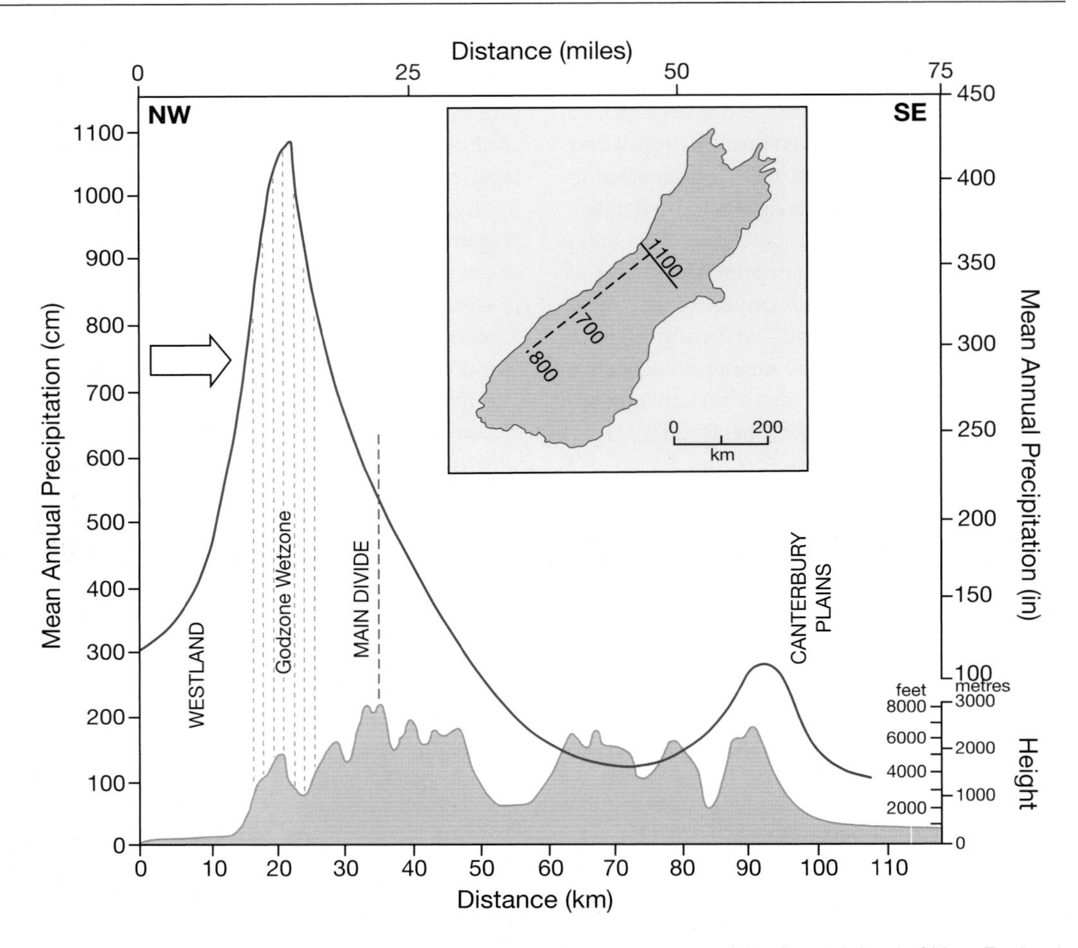

Figure 5.15 Mean annual precipitation (1951–1980) along a transect of the South Island of New Zealand, shown as the solid line in the inset map. On the latter, the dashed line indicates the position of the Godzone Wetzone and the figures give the precipitation peaks (cm) at three locations along the Godzone Wetzone.
Sources: Chinn (1979) and Henderson (1993), by permission of the New Zealand Alpine Club Inc. and T. J. Chinn.

processes involved, Tor Bergeron proposed using the term 'orogenic', rather than orographic, precipitation, (i.e., an origin related to various orographically produced effects). An extreme example of orographic precipitation is found on the western slope of the southern Alps of New Zealand, where mean annual rainfall totals exceed 10 meters! (Figure 5.15).

In mid-latitude areas where precipitation is predominantly of cyclonic origin, orographic effects tend to increase both the frequency and intensity of winter precipitation, whereas during summer and in continental climates with a higher condensation level the main effect of relief is the occasional triggering of intense thunderstorm type precipitation. The orographic influence occurs only in the proximity of high ground in the case of a stable atmosphere. Radar studies show that the main effect in this case is one of redistribution, whereas in the case of an unstable atmosphere precipitation appears to be increased, or at least redistributed on a larger scale, since the orographic effects may extend well downwind due to the activation of mesoscale rain bands (see Figure 9.13).

In tropical highland areas, there is a clearer distinction between orographic and convective contributions to total rainfall than in the mid-latitude cyclonic belt. Figure 5.16 shows that in the mountains of Costa Rica the temporal character of convective and orographic rainfalls and their seasonal occurrences are quite distinguishable. Convective rain occurs mainly in the May to November period, when 60 percent of the rain falls in the afternoons between 12:00 and 18:00 hours; orographic rain predominates between December and April, with a secondary maximum in June and July coinciding with an intensification of the Trade Winds.

Even low hills may have an orographic effect. Research in Sweden shows that wooded hills, rising only 30–50m above the surrounding lowlands, increase precipitation amounts locally by 50–80 percent during cyclonic spells. Until

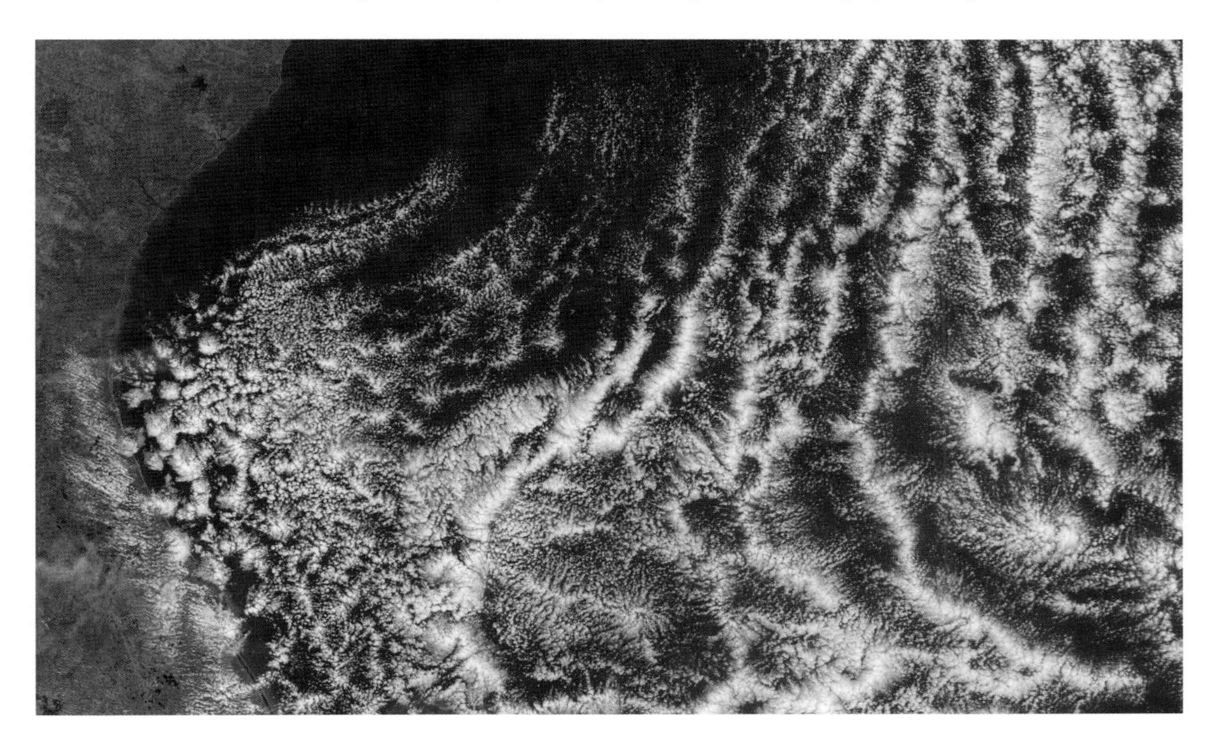

Plate 5.15 What atmospheric scientists refer to as open cell cloud formation is a regular occurrence on the back side of a low pressure system or cyclone in the mid-latitudes. In the Northern Hemisphere, a low-pressure system will draw in surrounding air and spin it counterclockwise. That means that on the back side of the low pressure center, cold air will be drawn in from the north, and on the front side, warm air will be drawn up from latitudes closer to the equator. This movement of an air mass is called advection, and when cold air advection occurs over warmer waters, open cell cloud formations often result.

This MODIS image shows open cell cloud formation over the Atlantic Ocean off the southeast coast of the United States and the Bahamas on February 19, 2002. This formation is the result of a low-pressure system in the North Atlantic Ocean a few hundred miles east of Massachusetts. Cold air is being drawn down from the north on the western side of the low and the open cell cumulus clouds being to form as the cold air passes over the warmer Caribbean waters.

Source: Jacques Descloitres, MODIS Land Rapid Response Team, NASA/GSFC.

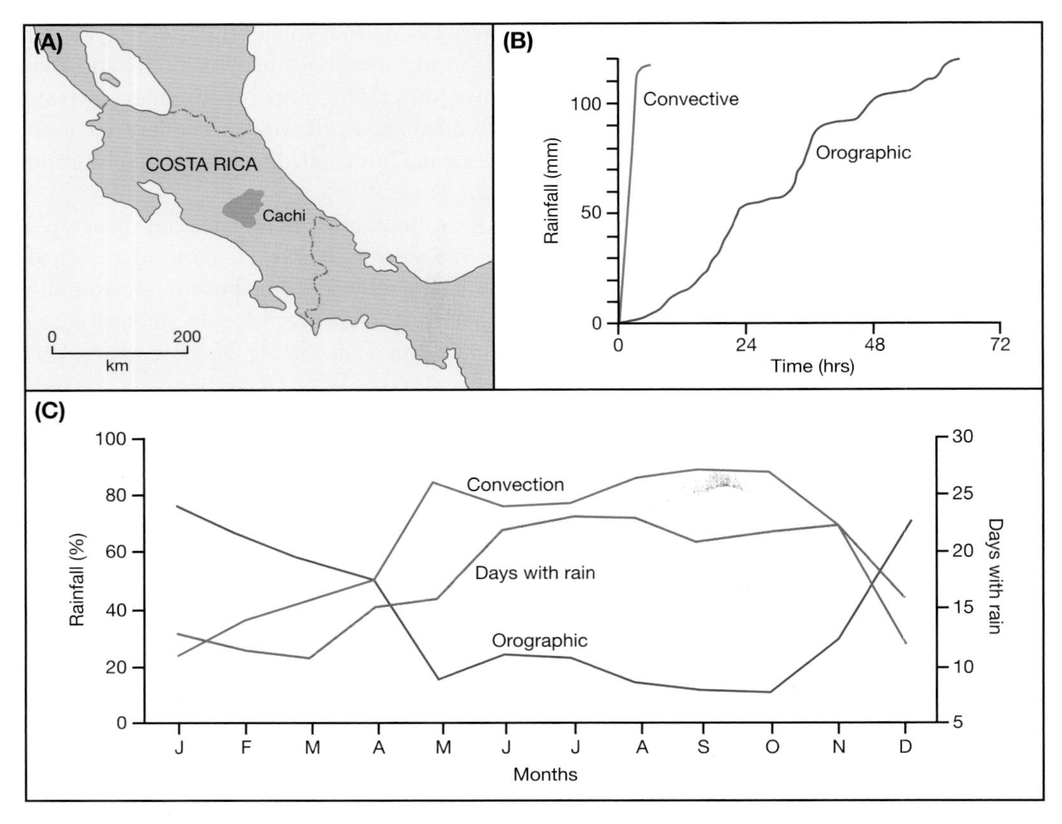

Figure 5.16 Orographic and convective rainfall in the Cachi region of Costa Rica for the period 1977–1980. A: the Cachi region, elevation 500–3000m. B: typical accumulated rainfall distributions for individual convective (duration 1–6 hours, high intensity) and orographic (1–5 days, lower intensity except during convective bursts) rainstorms. C: monthly rainfall divided into percentages of convective and orographic, plus days with rain, for Cachi (1018m).

Source: From Chacon and Fernandez (1985), by permission of the Royal Meteorological Society.

Doppler radar studies of the motion of falling raindrops became feasible, the processes responsible for such effects were unknown. A principal cause is the 'seeder–feeder' ('releaser–spender') cloud mechanism, proposed by Tor Bergeron and illustrated in Figure 5.14. In moist, stable airflow, shallow cap clouds form over hilltops. Precipitation falling from an upper layer of altostratus (the seeder cloud) grows rapidly by the washout of droplets in the lower (feeder) cloud. The seeding cloud may release ice crystals, which subsequently melt. Precipitation from the upper cloud layer alone would not give significant amounts on the ground, as the droplets would have insufficient time to grow in the airflow, which may traverse the hills in 15–30 minutes. Most of the precipitation intensification happens in the lowest kilometer-layer of moist, fast-moving airflows.

G THUNDERSTORMS

1 Development

In mid-latitudes the most spectacular example of moisture changes and associated energy releases in the atmosphere is the thunderstorm. Extreme

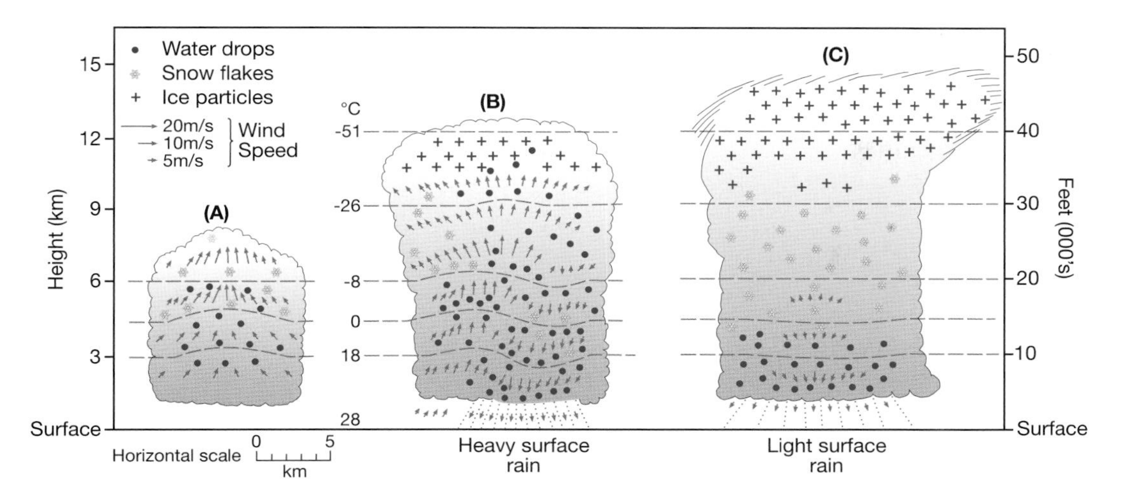

Figure 5.17 Classic view of the cycle of a local thunderstorm. The arrows indicate the direction and speed of air currents. A: the developing stage of the initial updraft. B: the mature stage with updrafts, downdrafts and heavy rainfall. C: the dissipating stage, dominated by cool downdrafts.

Source: After Byers and Braham; adapted from Petterssen (1969).

upward and downward movements of air are both the principal ingredients and motivating machinery of such storms. They occur: (1) due to rising cells of excessively heated moist air in an unstable air mass; (2) through the triggering of conditional instability by uplift over mountains; or (3) through mesoscale circulations or lifting along convergence lines (see p. 255).

The life cycle of a local storm lasts for only a few hours and begins when a parcel of air is either warmer than the air surrounding it or is actively undercut by colder encroaching air. In both instances, the air begins to rise and the embryo thunder cell forms as an unstable updraft of warm air (Figure 5.17). As condensation begins to form cloud droplets, latent heat is released and the initial upward impetus of the air parcel is augmented by an expansion and a decrease in density until the whole mass becomes completely out of thermal equilibrium with the surrounding air. At this stage, updrafts may increase from 3–5m s^{-1} at the cloud base to 8–10m s^{-1} some 2–3km higher, and they can exceed 30m s^{-1}. The constant release of latent heat continuously injects fresh supplies of energy, which accelerate the updraft. The air mass will continue to rise as long as its temperature remains greater (or, in other words, its density less) than that of the surrounding air. Cumulonimbus clouds form where the air is already moist as a result of previous penetrating towers from a cluster of clouds, and there is persistent ascent.

Raindrops begin to develop rapidly when the ice stage (or freezing stage) is reached by the vertical build up of the cell, allowing the Bergeron process to operate. They do not immediately fall to the ground, because the updrafts are able to support them. The minimum cumulus depth for showers over ocean areas seems to be between 1 and 2km, but 4–5km is more typical inland. The corresponding minimum time intervals needed for showers to fall from growing cumulus are about 15 minutes over ocean areas and ⩾30 minutes inland. Falls of hail require the special cloud processes, described in the last section, involving phases of 'dry' (rime accretion) and 'wet' growth on hail pellets. The mature stage of a storm (see Figure 5.18B) is usually associated with precipitation downpours and lightning (see Plate 5.16). The precipitation causes frictional

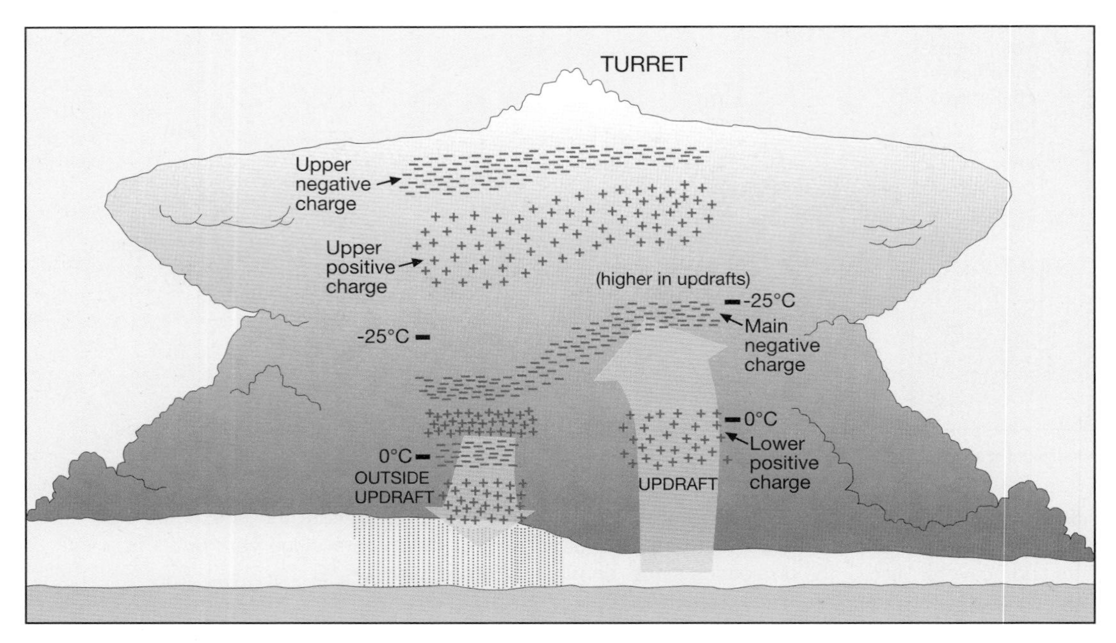

Figure 5.18 The electric charge structure in air mass storms in New Mexico, supercell storms and the convective elements of mesoscale convective systems (see Chapter 9), based on balloon soundings of the electric field – 33 in updrafts and 16 outside them. There are four vertical zones in the updraft region and six in the downdraft region, but the size, strength and relative positions of the up- and downdrafts vary, as do the heights and temperatures shown. *Source:* Stolzenburg *et al.* (1998) *J. Geophys. Res.* 103, p.14,101, Fig. 3. Courtesy of the American Geophysical Union.

downdrafts of cold air. As these gather momentum, cold air may eventually spread out below the thunder cell in a wedge. Gradually, as the moisture of the cell is expended, the supply of released latent heat energy diminishes, the downdrafts progressively gain in power over the warm updrafts, and the cell dissipates.

To simplify the explanation, a thunderstorm with only one cell was illustrated. Usually storms are far more complex in structure and consist of several cells arranged in clusters of 2–8km across, 100km or so in length and extending up to 10km altitude or more. Such systems are known as squall lines (see Chapter 9).

2 Cloud electrification and lightning

Two general hypotheses help to account for thunderstorm electrification. One involves induction in the presence of an electric field, the other is non-inductive charge transfer. The ionosphere at 30–40km altitude is positively charged (owing to the action of cosmic and solar ultraviolet radiation in ionization) and the earth's surface is negatively charged during fine weather. Thus, cloud droplets can acquire an induced positive charge on their lower side and negative charge on their upper side. Non-inductive charge transfer requires contact between cloud and precipitation particles. According to J. Latham, the major factor in cloud electrification is non-inductive charge transfer involving collisions between ice crystals growing by vapor diffusion and warmer pellets of soft hail (graupel) growing by riming. Recently, data from the TRMM satellite mission have shown that such non-inductive charging is the dominant mechanism for charge separation in all regions of the globe. The accretion of supercooled droplets (riming) on hail pellets

Plate 5.16 Time-lapse photography of cloud-to-ground lightning during a night-time thunderstorm in Norman, Oklahoma, March 1978.

Source: C. Clark; NOAA Photo Library, National Severe Storms Laboratory, Norman, OK, NOAA.

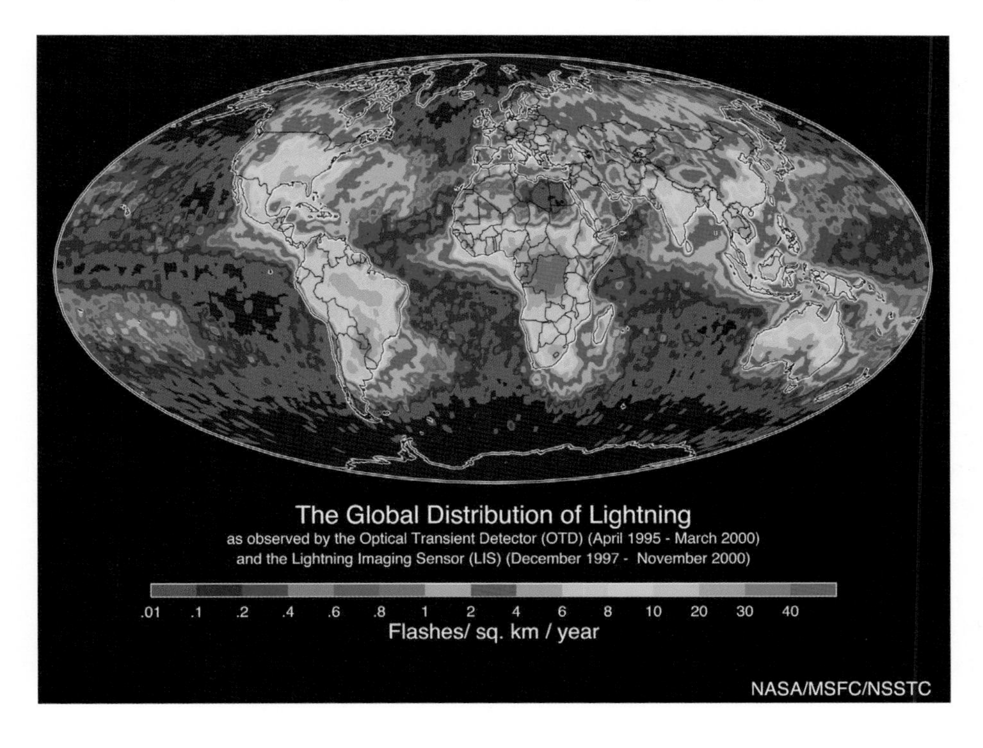

Plate 5.17 The global distribution of lightning flashes as observed by the Optical Transient Detector (April 1995 to March 2000) and the Lightning Image Sensor.

Source: courtesy of NASA Marshal Space Flight Center, NSSTC).

produces an irregular surface, which is warmed as the droplets release latent heat on freezing. The impacts of ice crystals on this irregular surface generate negative charge, while the crystals acquire positive charge. Negative charge is usually concentrated between about –10° and –25°C in a thundercloud, where ice crystal concentrations are large, and due to splintering of crystals at about the 0° to –5°C level and the ascent of the crystals in up-currents. The separation of electrical charges of opposite signs may involve several mechanisms. One is the differential movement of particles under gravity and convective updrafts. Another is the splintering of ice crystals during the freezing of cloud droplets. This operates as follows: a supercooled droplet freezes inward from its surface and this leads to a negatively charged warmer core (OH–ions) and a positively charged colder surface due to the migration of H+ ions outward down the temperature gradient. When this soft hailstone ruptures during freezing, small ice splinters carrying a positive charge are ejected by the ice shell and preferentially lifted to the upper part of the convection cell in updrafts. However, the ice-splintering mechanism appears to work only for a narrow range of temperature conditions, and the charge transfer is small.

The vertical distribution of charges in a thundercloud, based on balloon soundings, is shown in Figure 5.18. This general scheme applies to air mass thunderstorms in the southwestern USA, as well as to supercell storms and mesoscale convective systems described in Chapter 9. There are four alternating bands of positive and negative charges in the updraft and six outside the updraft area. The lower three bands of the four in the updraft are attributed to collision processes. Ice crystals carried upward may explain why the upper part of the cloud (above the –25°C isotherm) is positively charged. Negatively charged graupel accounts for the main region of negative charge. There is a temperature threshold around –10° to –20°C (depending on the cloud liquid water content and the rate of accretion on the graupel) where charge-sign reversal takes place. Above/below the altitude of this threshold, graupel pellets charge negatively/positively. The lower area of positive charge represents larger precipitation particles acquiring positive charge at temperatures higher than this threshold. The origin of the uppermost zone of negative charge is uncertain, but may involve induction (so-called 'screening layer' formation) since it is near the upper cloud boundary and the ionosphere is positively charged. The non-updraft structure may represent spatial variations or a temporal evolution of the storm system. The origin of the positive area at the cloud base outside the updraft is uncertain, but it may be a screening layer.

Radar studies show that lightning is associated with both ice particles in clouds and with rising air currents carrying small hail aloft. Lightning commonly begins more or less simultaneously with precipitation downpours and rainfall yield appearing to be correlated with flash density. The most common form of lightning (about two-thirds of all flashes) occurs within a cloud and is visible as *sheet lightning*. More significant are cloud–ground (CG) strokes. These are frequently between the lower part of the cloud and the ground which locally has an induced positive charge (Figure 5.19A). The first (leader) stage of the flash bringing down negative charge from the cloud is met about 30m above the ground by a return stroke, which rapidly takes positive charge upward along the already formed channel of ionized air. Just as the leader is neutralized by the return stroke, so the cloud neutralizes the latter in turn. Subsequent leaders and return strokes drain higher regions of the cloud until its supply of negative charge is temporarily exhausted. The total flash, with about eight return strokes, typically lasts about 0.5 seconds (Figure 5.19). The extreme heating and explosive expansion of air immediately around the path of the lightning sets up intense soundwaves, causing thunder to be heard. The sound travels at about 300m s^{-1}. Less commonly, positive CG flashes occur from the upper positive region (Figure 5.19B, case (1)), and they predominate in the stratiform cloud sector of

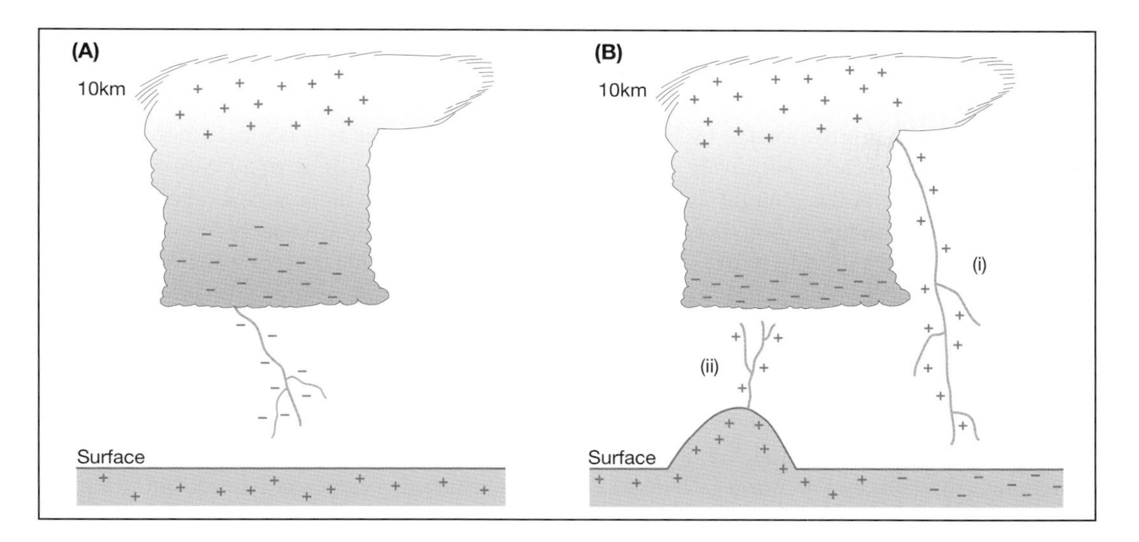

Figure 5.19 Classic view of the vertical distribution of electrostatic charges in a thundercloud and at the ground. A shows the common transfer of negative charge to the surface in a lightning stroke; B shows other cases: (1) when positive charge from the upper part of the cloud is transferred towards a locally induced area of negative charge at the surface; (2) positive charge transfer is from a summit or surface structure towards the cloud base.

a traveling convective storm (Chapter 9). Positive charge can also be transferred from a mountain top or high structure towards the cloud base (case (2)). In the United States, over 20 percent of flashes are positive in the Midwest, along the Gulf Coast and in Florida. Figure 5.19 represents a simple dipole model of cloud electricity; schemes to address the complexity shown in Figure 5.18 remain to be developed.

Lightning is only one aspect of the atmospheric electricity cycle. During fine weather, the earth's surface is negatively charged, the ionosphere positively charged. The potential gradient of this vertical electrical field in fine weather is about 100V m^{-1} near the surface, decreasing to about 1V m^{-1} at 25km, whereas beneath a thundercloud it reaches 10,000V m^{-1} immediately before a discharge. The 'breakdown potential' for lightning to occur in dry air is 3×10^6V m^{-1}, but this is ten times the largest observed potential in thunderclouds. Hence, the necessity for localized cloud droplet/ice crystal charging processes, as already described, to initiate flash leaders. Atmospheric ions conduct electricity from the ionosphere down to the earth, and hence a return supply must be forthcoming to maintain the observed electrical field. A major source is the slow *point discharge*, from objects such as buildings and trees, of ions carrying positive charge (electrons) induced by the negative thundercloud base.

Upward currents have recently been discovered high above the stratiform regions of large convective storm systems with positive CG lightning. Brief luminous emissions, due to electrical discharges, appear in the mesosphere and extend downward to 30–40km. These so-called *sprites* are red in the upper part, with blue tendrils beneath. The red color is from neutral nitrogen molecules excited by free electrons. In the ionosphere above, a luminous expanding ring (termed 'elve') may occur. High above the lightning storm, a discharge takes place because the imposed electric field of a vertical dipole exceeds the breakdown potential of the low-density air. The electrically conductive ionosphere prevents sprites extending above 90km altitude.

The other source of a return supply (estimated to be smaller in its effect over the earth as a whole than point discharges) is the instantaneous upward transfer of positive charge by lightning strokes, leaving the earth negatively charged. The joint operation of these supply currents, in approximately 1800 thunderstorms over the globe at any one instance, is thought to be sufficient to balance the air–earth leakage, and this number matches reasonably well with observations.

Globally, thunderstorms are most frequent between 12:00 and 21:00 local time, with a minimum around 03:00. An analysis of lightning on visible satellite imagery at local midnight shows a predominance of flashes over tropical land areas between 15°N and 30°S (Plate 5.17). In the austral summer, lightning signatures are along the equatorial trough and south to about 30°S over the Congo, South Africa, Brazil, Indonesia and northern Australia, with activity along cyclone paths in the Northern Hemisphere. In the boreal summer, activity is concentrated in central and northern South America, West Africa – the Congo, northern India and Southeast Asia and the southeastern United States. The North American Lightning Detection Network recorded 28–31 million flashes per year for 1998–2000. Severe storms in the USA may have peak cloud-to-ground lightning flash rates exceeding 9000 flashes per hour.. In Florida and along the Gulf Coast the mean flash density is 9 flashes/km^2. The median peak current is about16kA. Lightning is a significant environmental hazard. In the United States alone there are 100–150 deaths per year, on average, as a result of lightning accidents.

SUMMARY

Air may be lifted through instability due to surface heating or mechanical turbulence, ascent of air at a frontal zone, or forced ascent over an orographic barrier. Instability is determined by the actual rate of temperature decrease with height in the atmosphere relative to the appropriate adiabatic rate. The dry adiabatic lapse rate is 9.8°C/km; the saturated adiabatic rate is less than the DALR due to latent heat released by condensation. It is least (around 5°C/km) at high temperatures, but approaches the DALR at subzero temperatures.

Condensation requires the presence of hygroscopic nuclei such as salt particles in the air. Otherwise, supersaturation occurs. Similarly, ice crystals only form naturally in clouds containing freezing nuclei (clay mineral particles). Otherwise, water droplets may supercool to –39°C. Both supercooled droplets and ice crystals may be present at cloud temperatures of –10 to –20°C.

Clouds are classified in ten basic types according to altitude and cloud form. Satellites are providing new information on spatial patterns of cloudiness, revealing cellular (honeycomb) areas and linear cloud streets, as well as large-scale storm patterns.

Precipitation drops do not form directly by growth of cloud droplets through condensation. Two processes may be involved – coalescence of falling drops of differing sizes, and the growth of ice crystals by vapor deposition (the Bergeron–Findeisen process). Low-level cloud may be seeded naturally by ice crystals from upper cloud layers, or by introducing artificial nuclei. There is no single cause of the orographic enhancement of precipitation totals, and at least four contributing processes may be distinguished.

Thunderstorms are generated by convective uplift, which may result from daytime heating, orographic ascent or squall lines. The freezing process appears to be a major element of cloud electrification in thunderstorms. Lightning plays a key role in maintaining the electrical field between the surface and the ionosphere.

DISCUSSION TOPICS

- Account for the differences between the environmental, dry adiabatic and saturated adiabatic lapse rates.
- What processes determine the presence of stability and instability in the troposphere?
- What factors cause air to ascend/descend on small and large scales and what are the associated weather outcomes?
- Maintain a record of cloud type and amount over several days and compare what you observe with the cloud cover shown for your location on satellite imagery from an appropriate website (see Appendix 4D).
- Make a cross-section of terrain height and precipitation amounts at stations along a height transect in your own region/country. Use daily, monthly or annual data as available. In addition, note the prevailing wind direction with respect to the mountains/hills.
- From national records/websites, examine the occurrence of convective systems (thunderstorms, tornadoes, lightning) in your country and determine whether they are air mass storms connected with frontal lows, or mesoscale convective systems.

REFERENCES AND FURTHER READING

Books

Byers, H. R. and Braham, R. R. (1949) *The Thunderstorm*, US Weather Bureau. [Classic study of thunderstorm processes]

Cotton, W. R. and Anthes, R. A. (1989) *Storm and Cloud Dynamics,* Academic Press, San Diego, CA, 883pp. [Discusses cloud types and physical and dynamical processes, mesoscale structures, and the effects of mountains on airflow and cloud formation]

Kessler, E. (ed.) (1986) *Thunderstorm Morphology and Dynamics*, University of Oklahoma Press, Norman, OK, 411pp. [Comprehensive accounts by leading experts on convection and its modeling, all aspects of thunderstorm processes and occurrence in different environments, hail, lightning and tornadoes]

Ludlam, F. H. (1980) *Clouds and Storms. The Behavior and Effect of Water in the Atmosphere,* Pennsylvania State University, University Park and London, 405pp. [A monumental work by a renowned specialist]

Mason, B. J. (1975) *Clouds, Rain and Rainmaking* (2nd edn), Cambridge University Press, Cambridge and New York,189pp. [Valuable overview by a leading cloud physicist]

Petterssen, S. (1969) *Introduction to Meteorology,* 3rd edn, McGraw-Hill, New York, 416pp.

Strahler, A. N. (1965) *Introduction to Physical Geography,* John Wiley & Sons, New York, 455pp.

World Meteorological Organization (1956) *International Cloud Atlas*, Geneva. [Cloud classification and photographs of all sky types]

Articles

Andersson, T. (1980) Bergeron and the oreigenic (orographic) maxima of precipitation. *Pure Appl. Geophys.* 119, 558–76.

Bennetts, D. A., McCallum, E. and Grant, J. R. (1986) Cumulonimbus clouds: an introductory review. *Met. Mag.* 115, 242–56.

Bergeron, T. (1960) Problems and methods of rainfall investigation, in *The Physics of Precipitation,* Geophysical Monograph 5, Amer. Geophys. Union, Washington, DC, 5–30.

Bering, E. A. III., Few, A. A. and Benbrook, J. R. (1998 The global electric circuit. *Physics Today* 51(9), 24–30.

Braham, R. R. (1959) How does a raindrop grow? *Science* 129, 123–9.

Browning, K. A. (1980) Local weather forecasting. *Proc. Roy. Soc. Lond. Sect. A* 371, 179–211.

Browning, K. A. (1985) Conceptual models of precipitation systems. *Met. Mag.* 114, 293–319.

Browning, K. A. and Hill, F. F. (1981) Orographic rain. *Weather* 36, 326–9.

Brugge, R. (1996) Back to basics. Atmospheric stability: Part 1. Basic concepts. *Weather* 51(4), 134–40.

Chacón, R. E. and Fernandez, W. (1985) Temporal and spatial rainfall variability in the mountainous region of the Reventazón river basin,Costa Rica. *Int. J. Climatol.* 5, 176–88.

Chinn, T. J. (1979) How wet is the wettest of the wet West Coast? *New Zealand Alpine Journal* 32, 85–7.

Dudhia, J. (1996) Back to basics: thunderstorms. Part 1. *Weather* 51(11), 371–6.

Dudhia, J. (1997) Back to basics: thunderstorms. Part 2, Storm types and associated weather. *Weather* 52(1), 2–7.

Durbin, W. G. (1961) An introduction to cloud physics. *Weather* 16, 71–82 and 113–25.

East, T. W. R. and Marshall, J. S. (1954) Turbulence in clouds as a factor in precipitation. *Quart. J. Roy. Met. Soc.* 80, 26–47.

Eyre, L. A. (1992) How severe can a 'severe thunderstorm' be? *Weather* 47, 374–83.

Griffiths, D. J., Colquhoun, J. R., Batt, K. L. and Casinader, T. R. (1993) Severe thunderstorms in New South Wales: climatology and means of assessing the impact of climate change. *Climatic Change* 25, 369–88.

Henderson, R. (1993) Extreme storm rainfalls in the Southern Alps, New Zealand, in *Extreme Hydrological Events: Precipitation, Floods and Droughts (Proceedings of the Yokohama Symposium)*, IAHS Pub. No. 213, 113–20.

Hirschboeck, K. K. (1987) Catastrophic flooding and atmospheric circulation anomalies, in Mayer, L. and Nash, D. (eds) *Catastrophic Flooding*, Allen & Unwin, Boston, MA, 23–56.

Hopkins, M. M., Jr. (1967) An approach to the classification of meteorological satellite data. *J. Appl. Met.* 6, 164–78.

Houze, R. A., Jr. and Hobbs, P. V. (1982) Organization and structure of precipitating cloud systems. *Adv. Geophys.* 24, 225–315.

Jonas, P. R. (1994a) Back to basics: why do clouds form? *Weather* 49(5), 176–80.

Jonas, P. R. (1994b) Back to basics: why does it rain? *Weather* 49(7), 258–60.

Kyle, H. L. *et al.* (1993) The Nimbus Earth radiation budget (ERB) experiment: 1975 to 1992. *Bull. Amer. Met Soc.* 74, 815–30.

Latham, J. (1966) Some electrical processes in the atmosphere. *Weather* 21, 120–7.

Latham, J. *et al.* (2007) Field identification of a unique globally dominant mechanism of thunderstorm electrification. *Quart. J. Roy. Met. Soc.* 133, 1453–7

London, J., Warren, S. G. and Hahn, C. J. (1989) The global distribution of observed cloudiness – a contribution to the ISCPP. *Adv. Space Res.* 9(7), 161–5.

Mason, B. J. (1962) Charge generation in thunderstorms. *Endeavour* 21, 156–63.

Orville, R. E. *et al.* (2002) The North American lightning detection network (NALDN) – First results: 1998–2000. *Mon. Wea. Rev.* 130(8), 2098–109.

Pearce, F. (1994) Not warming, but cooling. *New Scientist* 143, 37–41.

Pike, W. S. (1993) The heavy rainfalls of 22–23 September 1992. *Met. Mag.* 122, 201–9.

Qiu, J. and Cressey, D. (2008) Taming the sky. *Nature* 453, 970–4.

Rosenfeld, D. *et al.* (2008) Flood or drought: How do aerosols affect precipitation? *Science* 321(5894): 1309–13.

Sawyer, J. S. (1956) The physical and dynamical problems of orographic rain. *Weather* 11, 375–81.

Schermerhorn, V. P. (1967) Relations between topography and annual precipitation in western Oregon and Washington. *Water Resources Research* 3, 707–11.

Schultz, D. M. *et al.* (2006) The mysteries of mamatus clouds: observations and formation mechanisms. *J. Atmos. Sci.*, 63, 2409–35.

Smith, R. B. (1989) Mechanisms of orographic precipitation. *Met. Mag.* 118, 85–8.

Stolzenburg, M., Rust, W. D. and Marshal, T. C. (1998) Electrical structure in thunderstorm convective regions. 3. Synthesis. *J. Geophys. Res.*103(D12), 14097–108.

Sumner, G. (1996) Precipitation weather. *J. Geography* 81(4), 327–45

Weston, K. J. (1977) Cellular cloud patterns. *Weather* 32, 446–50.

Wratt, D. S. *et al.* (1996) The New Zealand Southern Alps Experiment. *Bull. Amer. Met. Soc.* 77(4), 683–92.

CHAPTER SIX

Atmospheric motion: principles

LEARNING OBJECTIVES

When you have read this chapter you will:

- know the basic laws of horizontal motion in the atmosphere
- know how the Coriolis force arises and its effects
- be able to define the geostrophic wind
- know how friction modifies wind velocity in the boundary layer
- understand the principles of divergence/convergence and vorticity and their roles in atmospheric processes
- understand the thermodynamic, dynamic and topographic factors that lead to distinctive local wind regimes.

The atmosphere is in constant motion on scales ranging from short-lived, local wind gusts to storm systems spanning several thousand kilometers and lasting about a week, and to the more or less constant global-scale wind belts circling the earth. Before considering the global aspects, however, it is important to look at the immediate controls on air motion. The downward-acting gravitational field of the earth sets up the observed decrease of pressure away from the earth's surface that is represented in the vertical distribution of atmospheric mass (see Figure 2.13). This mutual balance between the force of gravity and the vertical pressure gradient is referred to as *hydrostatic equilibrium* (p. 31). This state of balance, together with the general stability of the atmosphere and its shallow depth, greatly limits vertical air motion. Average horizontal wind speeds are of the order of 100 times greater than average vertical movements, although individual exceptions occur – particularly in convective storms.

A LAWS OF HORIZONTAL MOTION

There are four controls on the horizontal movement of air near the earth's surface: the pressure-gradient force, the Coriolis force, centripetal acceleration, and frictional forces. The primary cause of air movement is the development of a horizontal pressure gradient through spatial differences in surface heating and consequent changes in air density and pressure. The fact that such a gradient can persist (rather than being destroyed by air motion towards the low pressure) results from the effect of the earth's rotation in giving rise to the Coriolis force.

1 The pressure-gradient force

The pressure-gradient force has vertical and horizontal components but, as already noted, the vertical component is more or less in balance with the force of gravity. Horizontal differences in pressure arise from thermal heating contrasts or mechanical causes such as mountain barriers and these differences control the horizontal movement of an air mass. The horizontal pressure gradient serves as the motivating force that causes air to move from areas of high pressure towards areas where it is lower, although other forces prevent air from moving directly across the isobars (lines of equal pressure). The pressure-gradient force per unit mass is expressed mathematically as

$$\frac{1}{\rho}\frac{dp}{dn}$$

where ρ = air density and dp/dn = the horizontal gradient of pressure. Hence the closer the isobar spacing the more intense is the pressure gradient and the greater the wind speed. The pressure-gradient force is also inversely proportional to air density, and this relationship is of particular importance in understanding the behavior of upper winds.

2 The earth's rotational deflective (Coriolis) force

The Coriolis force arises from the fact that the movement of masses over the earth's surface is referenced to a moving coordinate system (i.e., the latitude and longitude grid, which 'rotates' with the earth). The simplest way to visualize how this deflecting force operates is to picture a rotating disc on which moving objects are deflected. Figure 6.1 shows the effect of such a deflective force operating on a mass moving outward from the center of a spinning disc. The body follows a straight path in relation to a fixed frame of reference (for instance, a box that contains the spinning disc), but viewed relative to coordinates rotating with the disc the body swings to the right of its initial line of motion. This effect is readily demonstrated if a pencil line is drawn across a white disc on a rotating turntable. Figure 6.2 illustrates a case where the movement is not from the center of the turntable and the object possesses an initial momentum in relation to its distance from the axis of rotation. Note that the turntable model is not strictly analogous since the outwardly directed centrifugal force is involved. In the case of the rotating earth (with rotating reference coordinates of latitude and longitude), there is apparent deflection of moving objects to the right of their line of motion in the Northern Hemisphere and to the left in the Southern Hemisphere, as viewed by observers on the earth. The idea of a deflective force is credited to the work of French mathematician G.G. Coriolis in the 1830s. The 'force' (per unit mass) is expressed by:

$$-2\,\Omega\,V \sin\phi$$

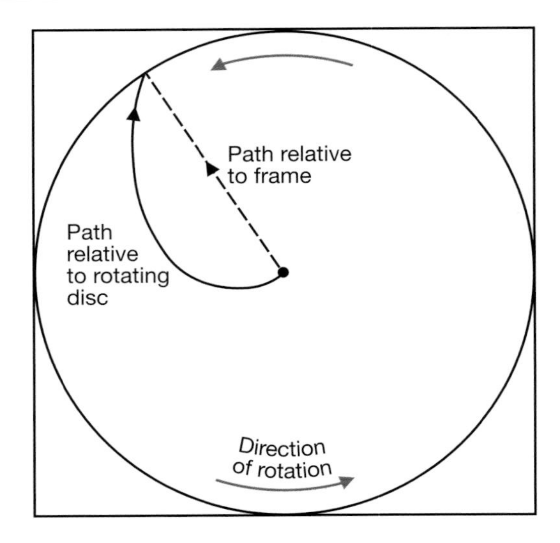

Figure 6.1 The Coriolis deflecting force operating on an object moving outward from the center of a rotating turntable.

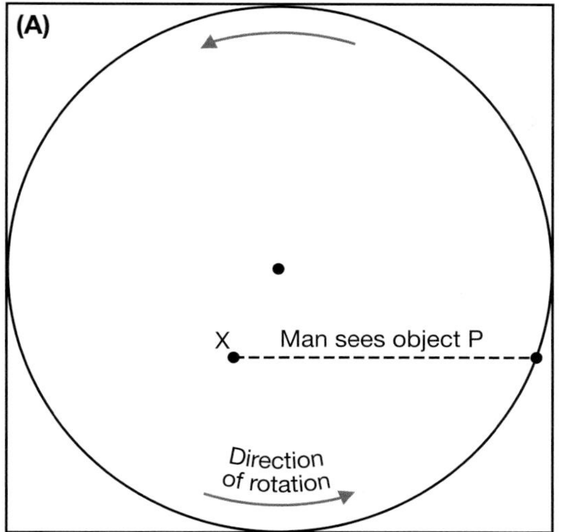

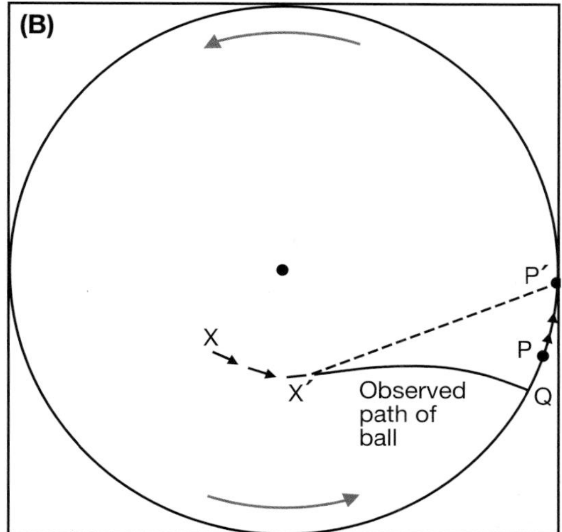

Figure 6.2 The Coriolis deflecting force on a rotating turntable. A. An observer at X sees the object P and attempts to throw a ball towards it. Both locations are rotating anticlockwise. B. the observer's position is now X′ and the object is at P′. To the observer, the ball appears to follow a curved path and lands at Q. The observer overlooked the fact that position P was moving counterclockwise and that the path of the ball would be affected by the initial impulse due to the rotation of point X.

where Ω = the angular velocity (15° hr^{-1} or $2\pi/24$ radians hr^{-1} for the earth = 7.29×10^{-5} radians s^{-1}); ϕ = the latitude and V = the velocity of the mass. $2\Omega \sin \phi$ is referred to as the Coriolis parameter (f). Angular velocity is a vector representing the rate of rotation of an object about the axis of rotation; its magnitude is the time rate of displacement of any point of the body.

The magnitude of the deflection is directly proportional to: (1) the horizontal velocity of the air (i.e., air moving at 10m s^{-1} has half the deflective force operating on it as on that moving at 20m s^{-1}); and (2) the sine of the latitude (sin 0° = 0; sine 90° = 1). The effect is thus a maximum at the poles (i.e., where the plane of the deflecting force is parallel to the earth's surface). It decreases with the sine of the latitude, becoming zero at the equator (i.e., where there is no component of the deflection in a plane parallel to the surface). The Coriolis 'force' depends on the motion itself. Hence, it affects the direction but not the speed of the air motion, which would involve doing work (i.e., changing the kinetic energy). The Coriolis force always acts at right angles to the direction of the air motion, to the right in the Northern Hemisphere (f positive) and to the left in the Southern Hemisphere (f negative). Absolute values of f vary with latitude as follows:

Latitude	0°	10°	20°	43°	90°
$f(10^{-4}\ s^{-1})$	0	0.25	0.50	1.00	1.458

The earth's rotation also produces a vertical component of rotation about a horizontal axis. This is a maximum at the equator (zero at the poles) and it causes a vertical deflection upward (downward) for horizontal west/east winds. However, this effect is of secondary importance due to the existence of hydrostatic equilibrium.

3 The geostrophic wind

Observations in the *free atmosphere* (above the level affected by surface friction at about 500 to 1000m) show that the wind blows more or less at right angles to the pressure gradient (i.e., parallel to the isobars) with, for the Northern Hemisphere, high pressure on the right and low pressure on the left when viewed downwind. This implies that for steady motion the pressure-gradient force is exactly balanced by the Coriolis deflection acting in the diametrically opposite direction (Figure 6.3A). The wind in this idealized case is called a *geostrophic wind*, the velocity (Vg) of which is given by the following formula:

$$Vg = \frac{1}{2\Omega \sin \phi} \frac{dp}{dn}$$

where dp/dn = the pressure gradient. The velocity is inversely dependent on latitude, such that the same pressure gradient associated with a geostrophic wind speed of 15m s^{-1} at latitude 43° will produce a velocity of only 10m s^{-1} at latitude 90°. Except in low latitudes, where the Coriolis parameter approaches zero, the geostrophic wind is a close approximation to the observed air motion in the free atmosphere. Since pressure systems are rarely stationary, this fact implies that air motion must continually change towards a new balance. In other words, mutual adjustments of the wind and pressure fields are constantly taking place. The common 'cause-and-effect' argument that a pressure gradient is formed and air begins to move towards low pressure before coming into geostrophic balance is an unfortunate oversimplification of reality.

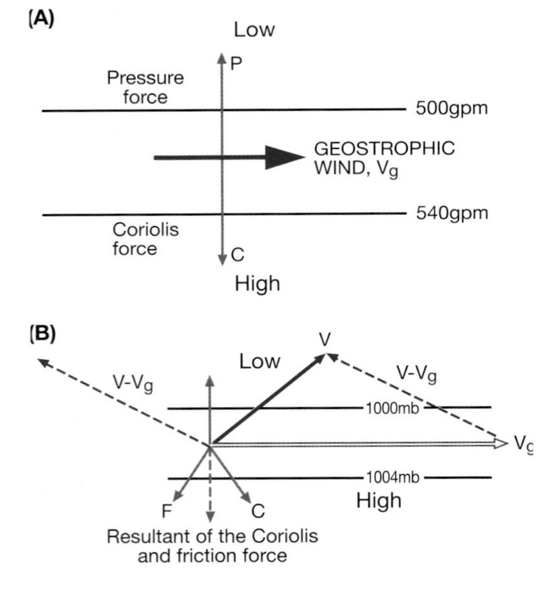

Figure 6.3 A: The geostrophic wind case of balanced motion (Northern Hemisphere) above the friction layer (contour heights are gpm) B: Surface wind **V** represents a balance between the geostrophic wind, $\mathbf{V_g}$, and the resultant of the Coriolis force (**C**) and the friction force (**F**). Note that **F** is not generally directly opposite to the surface wind.

4 The centripetal acceleration

For a body to follow a curved path there must be an inward acceleration (c) towards the center of rotation. This is expressed by:

$$c = -\frac{mV^2}{r}$$

where m = the moving mass, V = its velocity and r = the radius of curvature. This effect is sometimes regarded for convenience as a centrifugal 'force' operating radially outward (see Note 1). In the case of the earth itself, this is valid. The centrifugal effect due to rotation has in fact resulted in a slight bulging of the earth's mass in low latitudes and a flattening near the poles. The small decrease in apparent gravity towards the equator (see Note 2) reflects the effect of the centrifugal force working against the gravitational attraction directed towards the earth's center. It is therefore only necessary to consider the forces involved in the rotation of the air around a local axis of high or low pressure. Here the curved path of the air (parallel to the isobars) is maintained by an inward-acting, or centripetal, acceleration.

Figure 6.4 shows (for the Northern Hemisphere) that in a low pressure system balanced flow is maintained in a curved path (referred to as the *gradient wind*) by the Coriolis force being weaker than the pressure force. The difference between the two gives the net centripetal acceleration inward. In the high pressure case, the inward acceleration exists because the Coriolis force

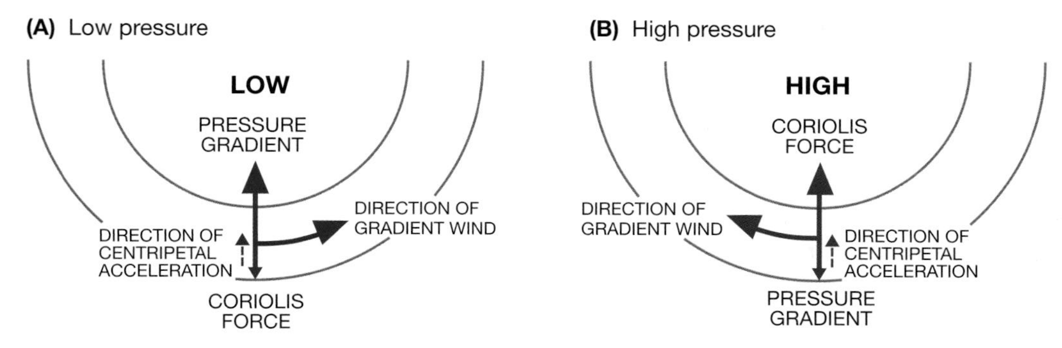

Figure 6.4 The gradient wind case of balanced motion around low pressure (A) and high pressure (B) in the Northern Hemisphere.

exceeds the pressure force. Since the pressure gradients are assumed to be equal, the different contributions of the Coriolis force in each case imply that the wind speed around the low pressure must be lower than the geostrophic value (*subgeostrophic*), whereas in the case of high pressure it is *supergeostrophic*. In reality, this effect is obscured by the fact that the pressure gradient in a high is usually much weaker than in a low. Moreover, the fact that the earth's rotation is cyclonic imposes a limit on the speed of anticyclonic flow. The maximum occurs when the angular velocity is $f/2$ (= V sin ϕ), at which value the absolute rotation of the air (viewed from space) is just cyclonic. Beyond this point anticyclonic flow breaks down ('dynamic instability'). There is no maximum speed in the case of cyclonic rotation.

The magnitude of the centripetal acceleration is generally small, but it becomes important where high-velocity winds are moving in very curved paths (i.e., around an intense low pressure vortex). Two cases are of meteorological significance: first, in intense cyclones near the equator, where the Coriolis force is negligible; and, second, in a narrow vortex such as a tornado. Under these conditions, when the large pressure-gradient force provides the necessary centripetal acceleration for balanced flow parallel to the isobars, the motion is called *cyclostrophic*.

The above arguments assume steady conditions of balanced flow. This simplification is useful, but in reality two factors prevent a continuous state of balance. Latitudinal motion changes the Coriolis parameter, and the movement or changing intensity of a pressure system leads to acceleration or deceleration of the air, causing some degree of cross-isobaric flow. Pressure change itself depends on air displacement through the breakdown of the balanced state. If air movement were purely geostrophic there would be no growth or decay of pressure systems. The acceleration of air at upper levels from a region of cyclonic isobaric curvature (subgeostrophic wind) to one of anticyclonic curvature (supergeostrophic wind) causes a fall of pressure at lower levels in the atmosphere to compensate for the removal of air aloft. The significance of this fact will be discussed in Chapter 9G. The interaction of horizontal and vertical air motions is outlined in B.2, (this chapter).

In cases where the curvature of the flow is tight, as near the eye of a tropical cyclone (see Chapter 11B.2), the centripetal acceleration may balance the pressure gradient force; the resulting wind is termed cyclostrophic.

5 Frictional forces and the planetary boundary layer

The last force that has an important effect on air movement is that due to friction from the earth's surface. Towards the surface (i.e., below about 500m for flat terrain), friction due to form drag over orography begins to reduce the wind velocity below its geostrophic value. This slowing of the wind near the surface modifies the deflective force, which is dependent on velocity, causing it also to decrease. Initially, the frictional force is opposite to the wind velocity, but in a balanced state – when the velocity and therefore the Coriolis deflection decrease (the vector sum of the Coriolis and friction components balances the pressure gradient force (Figure 6.3B). The friction force now acts to the right of the surface wind vector. Thus, at low levels, due to frictional effects, the wind blows obliquely across the isobars in the direction of the pressure-gradient. The angle of obliqueness increases with the growing effect of frictional drag due to the earth's surface averaging about 10–20° at the surface over the sea and 25–35° over land.

In summary, the surface wind (neglecting any curvature effects) represents a balance between the pressure-gradient force and the Coriolis force perpendicular to the air motion, and friction roughly parallel, but opposite, to the air motion. Where the Coriolis force is small, friction may balance the pressure gradient force and the wind (known as antitriptic) flows down the pressure gradient.

Table 6.1 Typical roughness lengths (m) associated with terrain surface characteristics

Terrain surface characteristics	Roughness length (m)
Groups of high buildings	1–10
Temperate forest	0.8
Groups of medium buildings	0.7
Suburbs	0.5
Trees and bushes	0.2
Farmland	0.05–0.1
Grass	0.008
Bare soil	0.005
Snow	0.001
Smooth sand	0.0003
Water	0.0001

Source: After Troen and Petersen (1989).

The layer of frictional influence is known as the *planetary boundary layer* (PBL). Atmospheric profilers (lidar and radar) can routinely measure the temporal variability of PBL structure. Its depth varies over land from a few hundred meters at night, when the air is stable as a result of nocturnal surface cooling, to 1–2km during afternoon convective conditions. Exceptionally, over hot, dry surfaces, convective mixing may extend to 4–5km. Over the oceans it is more consistently near 1km deep and in the tropics especially is often capped by an inversion due to sinking air. The boundary layer is typically either stable or unstable. Yet, for theoretical convenience, it is often treated as being neutrally stable (i.e., the lapse rate is that of the DALR, or the potential temperature is constant with height; see Figure 5.1). For this ideal state, the wind turns clockwise (veers) with increased height above the surface, setting up a wind spiral (Figure 6.5). This spiral profile was first demonstrated in the turning of ocean currents with depth (see Chapter 7D.1a) by V. W. Ekman; both are referred to as *Ekman spirals*. The inflow of air towards the low pressure center generates upward motion at the top of the PBL, known as *Ekman pumping*.

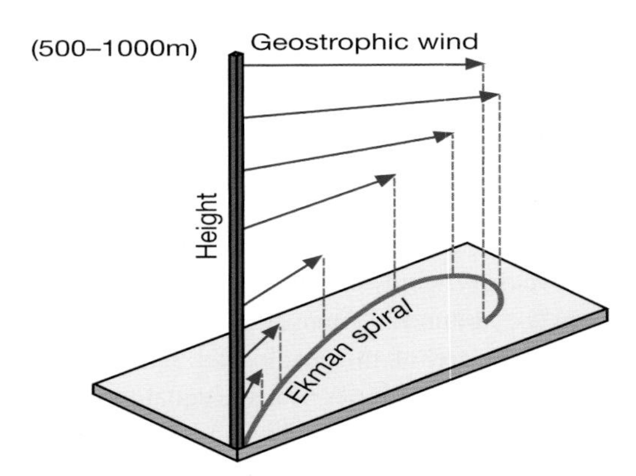

Figure 6.5 The Ekman spiral of wind with height in the Northern Hemisphere. The wind attains the geostrophic velocity at between 500 and 1000m in the middle and higher latitudes as frictional drag effects become negligible. This is a theoretical profile of wind velocity under conditions of mechanical turbulence.

Wind velocity decreases exponentially close to the earth's surface due to frictional effects. These consist of 'form drag' over obstacles (buildings, forests, hills), and the frictional stress exerted by the air at the surface interface. The mechanism of *form drag* involves the creation of locally higher pressure on the windward side of an obstacle and a lateral pressure gradient. Wind stress arises from, first, the molecular resistance of the air to the vertical wind shear (i.e., increased wind speed with height above the surface); such molecular viscosity operates in a laminar sublayer only millimeters thick. Second, turbulent eddies, a few meters to tens of meters across, brake the air motion on a larger scale (eddy viscosity). The aerodynamic roughness of terrain is described by the *roughness length* (z_0), or height at which the wind speed falls to zero based on extrapolation of the neutral wind profile. Table 6.1 lists typical roughness lengths.

Turbulence in the atmosphere is generated by the vertical change in wind velocity, (i.e., a vertical wind shear), and is suppressed by an absence of buoyancy. The dimensionless ratio of buoyant suppression of turbulence to its generation by shear, known as the Richardson number (*Ri*), provides a measure of dynamic stability. Above a critical threshold, turbulence is likely to occur.

B DIVERGENCE, VERTICAL MOTION AND VORTICITY

These three terms are the key to proper understanding of wind and pressure systems on a synoptic and global scale. Mass uplift or descent of air occurs primarily in response to dynamic factors related to horizontal airflow and is only secondarily affected by air-mass stability. Hence the significance of these factors for weather processes.

1 Divergence

Different types of horizontal flow are shown in Figure 6.6A. The first panel shows that air may accelerate (decelerate), leading to velocity divergence (convergence). When streamlines (lines of instantaneous air motion) spread out or squeeze together, this is termed diffluence or confluence, respectively. If the streamline pattern is strengthened by that of the isotachs (lines of equal wind speed), as shown in the third panel of Figure 6.6A, then there may be mass divergence or convergence at a point (Figure 6.6B). In this case, the compressibility of the air causes the density to decrease or increase, respectively. Usually, however, confluence is associated with an increase

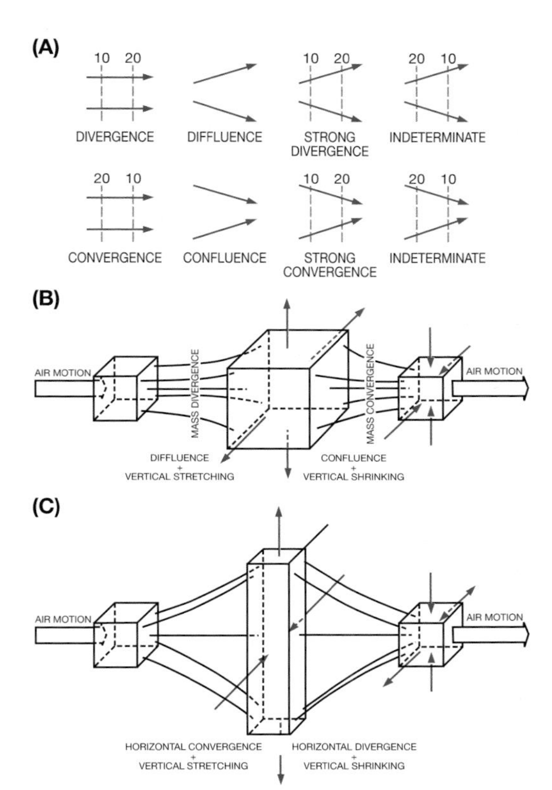

Figure 6.6 Convergence and divergence. A: Plan view of horizontal flow patterns producing divergence and convergence – the broken lines are schematic isopleths of wind speed (isotachs). B: Schematic illustration of local mass divergence and convergence, assuming density changes. C: Typical convergence-stretching and divergence-shrinking relationships in atmospheric flow.

in air velocity and diffluence with a decrease. In the intermediate case, confluence is balanced by an increase in wind velocity and diffluence by a decrease in velocity. Hence, convergence (divergence) may give rise to vertical stretching (shrinking), as illustrated in Figure 6.6C. It is important to note that if all winds were geostrophic, there could be no convergence or divergence and hence no weather!

Convergence or divergence may also occur as a result of frictional effects. Onshore winds undergo convergence at low levels when the air slows down on crossing the coastline owing to the greater friction overland, whereas offshore winds accelerate and become divergent. Frictional differences can also set up coastal convergence (or divergence) if the geostrophic wind is parallel to the coastline with, for the Northern Hemisphere, land to the right (or left) of the air current, viewed downwind.

2 Vertical motion

Horizontal inflow or outflow near the surface has to be compensated by vertical motion, as illustrated in Figure 6.7, if the low or high pressure systems are to persist and there is to be no continuous density increase or decrease. Air rises above a low pressure cell and subsides over high pressure, with compensating divergence and convergence, respectively, in the upper troposphere. In the middle troposphere, there must clearly be some level at which horizontal divergence or convergence is effectively zero; the mean 'level of non-divergence' is generally at about 600mb. Large-scale vertical motion is extremely slow compared with convective up- and downdrafts in cumulus clouds, for example. Typical rates in large depressions and anticyclones are of the order of ±5–10cm s^{-1}, whereas updrafts in cumulus may exceed 10m s^{-1}.

3 Vorticity

Vorticity implies the rotation, or angular velocity, of small (imaginary) parcels in any fluid. The air within a low pressure system may be regarded as comprising an infinite number of small air parcels, each rotating cyclonically around an axis vertical to the earth's surface (Figure 6.8). Vorticity has three elements – magnitude (defined as *twice* the angular velocity, Ω) (see Note 3), direction (the

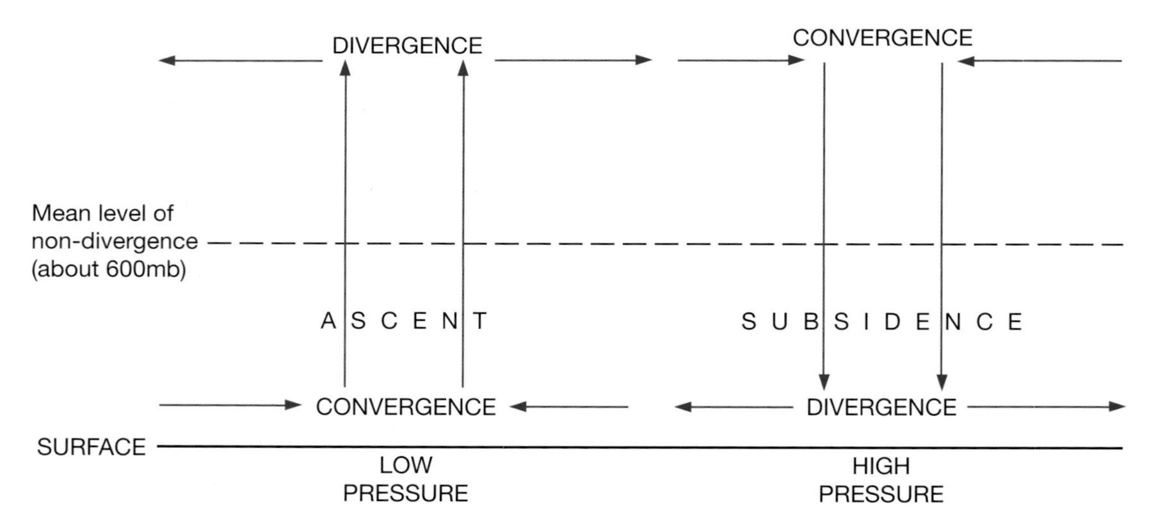

Figure 6.7 Cross-section of the patterns of vertical motion associated with (mass) divergence and convergence in the troposphere, illustrating mass continuity.

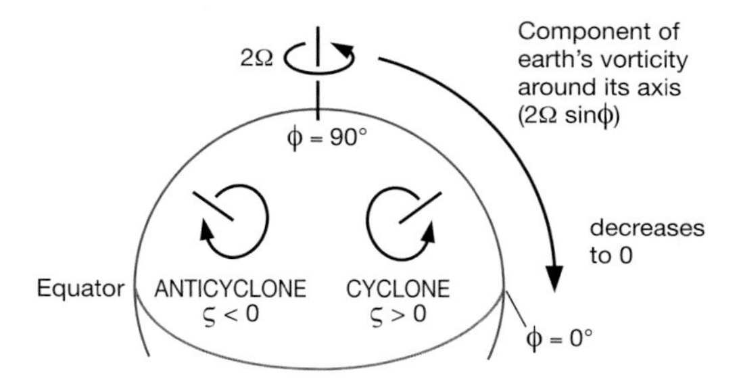

Figure 6.8 Sketch of the relative vertical vorticity (ζ) about a cyclone and an anticyclone in the Northern Hemisphere. The component of the earth's vorticity around its axis of rotation (or the Coriolis parameter, *f*), is equal to twice the angular velocity (Ω) times the sine of the latitude (f). At the pole $f = 2\Omega$, diminishing to 0 at the equator. Cyclonic vorticity is in the same sense as the earth's rotation about its own axis, viewed from above, in the Northern Hemisphere: this cyclonic vorticity is defined as positive ($\zeta > 0$).

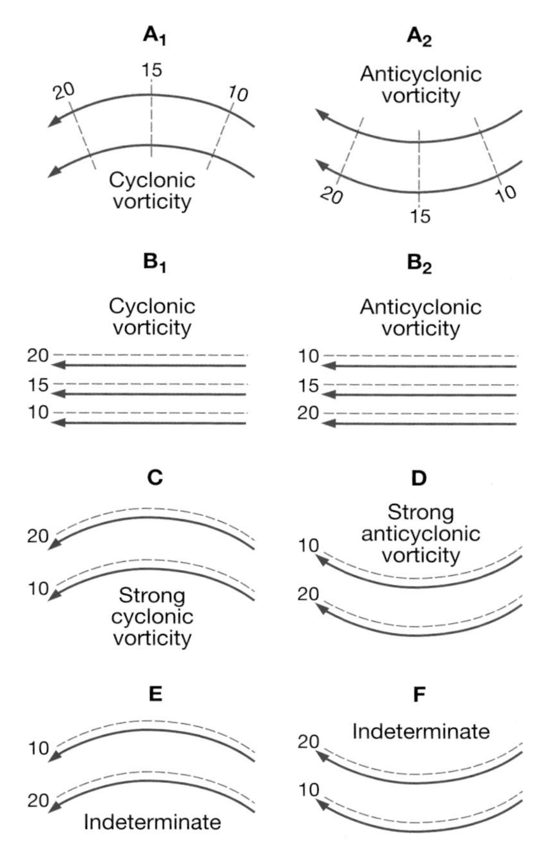

Figure 6.9 Streamline models illustrating in plan view the flow patterns with cyclonic and anti-cyclonic vorticity in the Northern Hemisphere. In C and D, the effects of curvature (a_1 and a_2) and lateral shear (b_1 and b_2) are additive, whereas in E and F they more or less cancel out. Dashed lines are schematic isopleths of wind speed.

Source: After Riehl *et al.* (1954).

horizontal or vertical axis around which the rotation occurs) and the sense of rotation. Rotation in the same sense as the earth's rotation – cyclonic in the Northern Hemisphere – is defined as positive. Cyclonic vorticity may result from cyclonic curvature of the streamlines, from cyclonic shear (stronger winds on the right side of the current, viewed downwind in the Northern Hemisphere), or a combination of the two (Figure 6.9). Lateral shear (see Figure 6.9B) results from changes in isobar spacing. Anticyclonic vorticity occurs with the corresponding anticyclonic situation. The component of vorticity around an axis vertical to the earth's surface is referred to as the vertical vorticity. This is generally the most important, but near the ground surface frictional shear causes vorticity around an axis parallel to the surface and normal to the wind direction.

Vorticity is related not only to air motion around a cyclone or anticyclone (*relative vorticity*), but also to the location of that system on the rotating earth. The vertical component of *absolute vorticity* consists of the relative vorticity (ζ) and the latitudinal value of the Coriolis parameter, $f = 2\Omega \sin \phi$ (see Chapter 6A). At the equator, the local vertical is at right angles to the earth's axis, so $f = 0$, but at the North Pole cyclonic relative vorticity and the earth's rotation act in the same sense (see Figure 6.8) and $f = 2\Omega$.

C LOCAL WINDS

For a weather observer, local controls of air movement may present more problems than the effects of the major planetary forces just discussed. Diurnal tendencies are superimposed upon both the large- and the small-scale patterns of wind velocity. These are particularly noticeable in the case of local winds. Under normal conditions, wind velocities tend to be least about dawn when there is little vertical thermal mixing and the lower air is less affected by the velocity of the air aloft (see Chapter 7A). Conversely, velocities of some local winds are greatest around 13:00–14:00 hours, when the air is most subject to terrestrial heating and vertical motion, thereby enabling coupling to the upper air movement. Air always moves more freely away from the surface, because it is not subject to the retarding effects of friction and obstruction.

Table 6.2 gives a summary classification of local winds, each of which is now discussed in detail.

1 Mountain and valley winds

Terrain features give rise to their own special meteorological conditions. On warm, sunny days, the heated air in a valley is laterally constricted compared with that over an equivalent area of lowland, and so tends to expand vertically. The volume ratio of lowland/valley air is typically about 2 or 3:1 and this difference in heating sets up a density and pressure differential, which causes air to flow from the lowland up the axis of the valley. This valley wind (Figure 6.10) is generally light and requires a weak regional pressure gradient in order to develop. This flow along the main valley develops more or less simultaneously with *anabatic* (upslope) winds, which result from greater heating of the valley sides compared with the valley floor. These slope winds rise above the ridge tops and feed an upper return current along the line of the valley to compensate for the valley wind. This feature may be obscured, however, by the regional airflow. Speeds reach a maximum around 14:00 hours.

At night, there is a reverse process as denser cold air at higher elevations drains into depressions and valleys; this is known as a *katabatic* wind. If the air drains downslope into an open valley, a 'mountain wind' develops more or less simultaneously along the axis of the valley. This flows towards the plain, where it replaces warmer, less dense air. The maximum velocity occurs just

Table 6.2 Classification of local winds

Name	Characteristics	Forcing
Anabatic	Daytime upslope warm flow	Horizontal density gradient towards the slope
Katabatic	Night-time downslope cold flow	Gravity and horizontal density gradient away from the slope
Mountain wind	Night-time cold down-valley flow	Mountains-to-plains density gradient
Valley wind	Daytime warm up-valley flow	Plains-to-mountains density gradient
Anti-mountain wind	Above the mountain wind in the opposite direction	Compensation current
Anti-valley wind	Above the valley wind in the opposite direction	Compensation current
Sea breeze	Day-time flow from the seas to the land	Density gradient from cool sea to heated land
Land breeze	Night-time flow from land to sea	Density gradient from cool land to warmer sea
Foehn (Chinook)	Down lee slope with increasing T and lower RH	Blocked flow on windward side; or flow crossing mountains with cloud/precipitation on windward slope
Bora	Down lee slope with air colder than that it replaces	Blocked flow of cold air upwind
Barrier wind	Low-level flow parallel to the mountains, directed poleward	Blocking reduces the flow speed normal to the barrier reducing the Coriolis force

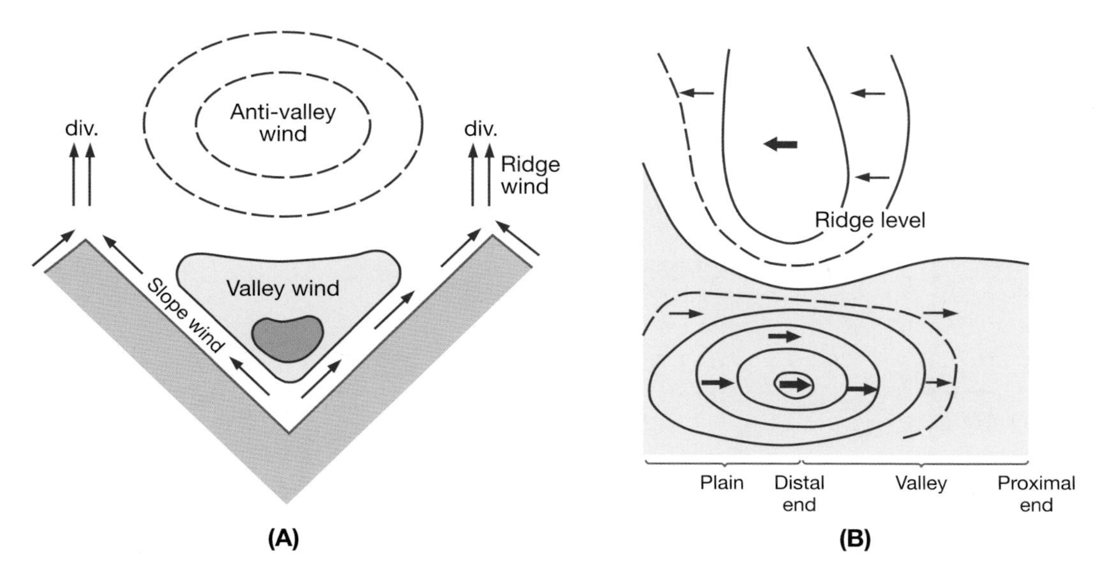

Figure 6.10 Valley winds in an ideal V-shaped valley. A: Section across the valley. The valley wind and anti-valley wind are directed at right angles to the plane of the paper. The arrows show the slope and ridge wind in the plane of the paper, the latter diverging (div.) into the anti-valley wind system. B: Section running along the center of the valley and out on to the adjacent plain, illustrating the valley wind (below) and the anti-valley wind (above).

Source: After Buettner and Thyer (1965).

before sunrise at the time of maximum diurnal cooling. As with the valley wind, an upper return current, in this case up-valley, also overlies the mountain wind.

Katabatic drainage is usually cited as the cause of frost pockets in hilly and mountainous areas. It is argued that greater radiational cooling on the slopes, especially if they are snow-covered, leads to a gravity flow of cold, dense air into the valley bottoms. Observations in California and elsewhere, however, suggest that the valley air remains colder than the slope air from the onset of nocturnal cooling, so that the air moving downslope slides over the denser air in the valley bottom. Moderate drainage winds will also act to raise the valley temperatures through turbulent mixing. Cold air pockets in valley bottoms and hollows probably result from the cessation of turbulent heat transfer to the surface in sheltered locations rather than by cold air drainage, which is often not present.

2 Land and sea breezes

Another thermally induced wind regime is the land and sea breezes (see Figure 6.11). The vertical expansion of the air column that occurs during daytime heating over the more rapidly heated land (see Chapter 3B.5) tilts the isobaric surfaces downward at the coast, causing onshore winds at the surface and a compensating offshore movement aloft. Typical land–sea pressure differences are of the order of 2mb. At night, the air over the sea is warmer and the situation is reversed, although this reversal is also the effect of downslope winds blowing off the land. Figure 6.12 shows that sea breezes can have a decisive effect on temperature and humidity on the coast of California. A basic offshore gradient flow is perturbed during the day by a westerly sea breeze. Initially, the temperature difference between the sea and the coastal mountains of central California sets up a shallow sea breeze, which by midday is

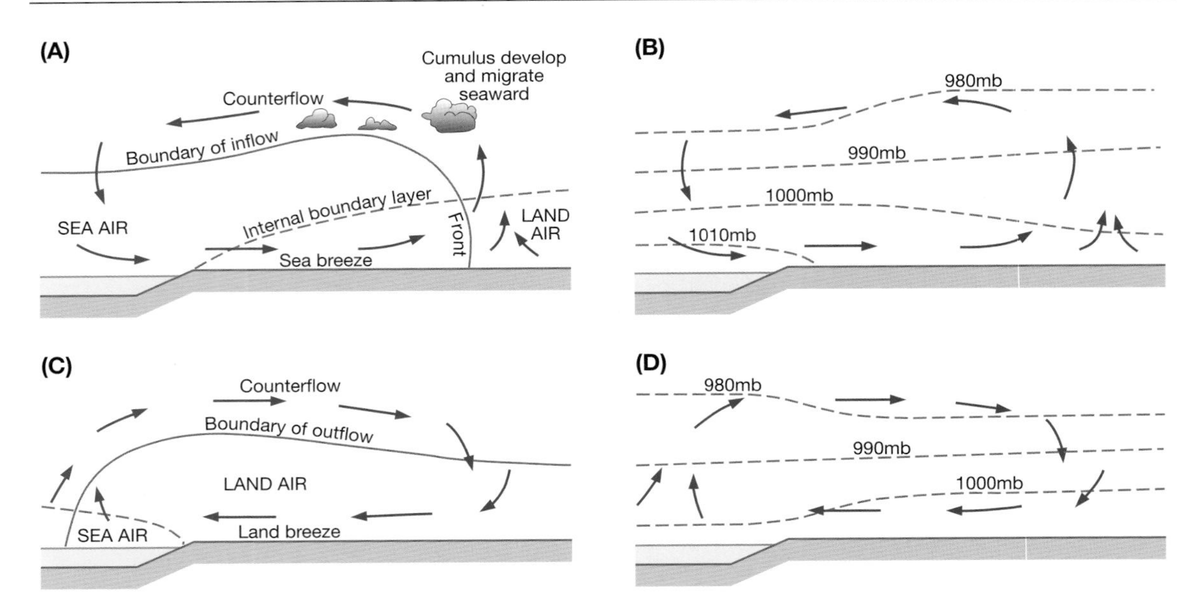

Figure 6.11 Diurnal land and sea breezes. A and B: sea breeze circulation and pressure distribution in the early afternoon during anticyclonic weather. C and D: land breeze circulation and pressure distribution at night during anticyclonic weather.
Source: A and C after Oke (1978).

300m deep. In the early afternoon, a deeper regional-scale circulation between the ocean and the hot interior valleys generates a 1km-deep onshore flow that persists until two to four hours after sunset. Both the shallow and the deeper breeze have maximum speeds of 6m s^{-1}. A shallow evening land breeze develops by 19:00 PST but is indistinguishable from the gradient offshore flow.

The advancing cool sea air may form a distinct line (or *front*; see Chapter 8D) marked by cumulus cloud development, behind which there is a distinct wind velocity maximum. This often develops in summer, for example, along the Gulf coast of Texas. On a smaller scale, such features are observed in Britain, particularly along the south and east coasts. The sea breeze has a depth of about 1km, although it thins towards the advancing edge. It may penetrate 50km or more inland by 21:00 hours. Typical wind speeds in such sea breezes are 4–7m s^{-1}, although these may be greatly increased where a well-marked low-level temperature inversion produces a 'Venturi effect' by constricting and accelerating the flow. The much shallower land breezes are usually weaker, about 2m s^{-1}. Counter-currents aloft are generally weak and may be obscured by the regional airflow, but studies on the Oregon coast suggest that under certain conditions this upper return flow may be very closely related to the lower sea breeze conditions, even to the extent of mirroring the surges in the latter. In mid-latitudes the Coriolis deflection causes turning of a well-developed onshore sea breeze (clockwise in the Northern Hemisphere) so that eventually it may blow more or less parallel to the shore. Analogous 'lake breeze' systems develop adjacent to large inland water bodies such as the Great Lakes and even the Great Salt Lake in Utah.

Small-scale circulations can be generated by local differences in albedo and thermal conductivity. Salt flats (playas) in the western deserts of the United States and in Australia, for example, cause on off-playa breeze by day and an on-playa flow at night due to differential heating. The salt

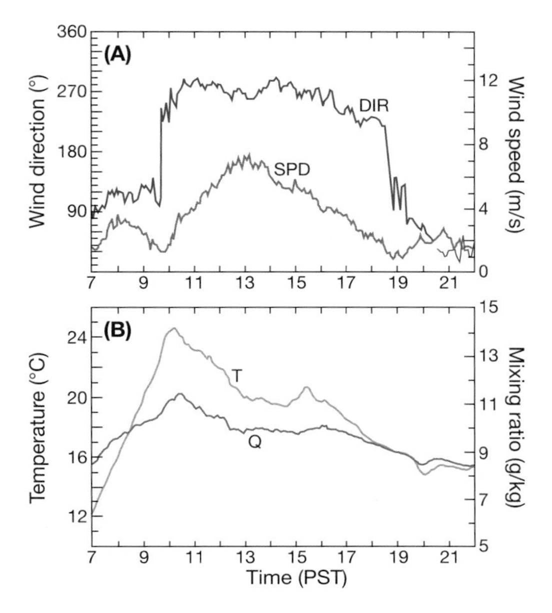

Figure 6.12 The effects of a westerly sea breeze on the California coast on 22 September 1987 on temperature and humidity. Above: Wind direction (DIR) and speed (SPD); below: air temperature (T) and humidity mixing ratio (Q) on a 27m mast near Castroville, Monterey Bay, California. The gradient flow observed in the morning and evening was easterly.

Source: Banta (1995, p. 3621, Fig. 8).

flat has a high albedo and the moist substrate results in a high thermal conductivity relative to the surrounding sandy terrain. The flows are about 100m deep at night and up to 250m by day.

3 Winds due to topographic barriers

Mountain ranges strongly influence airflow crossing them. On the upwind side of mountains perpendicular to the airflow, blocking may occur when the airflow is stable and unable to cross the barrier. As the flow approaches the barrier it slows down, thus reducing the Coriolis force. Imbalance with the pressure gradient force then causes the air to turn poleward towards the lower pressure on the left side of the flow. This sets up a low-level *barrier wind* that may feature a low-level (850mb) jet of 20m s^{-1}. Such winds are common upstream of the Sierra Nevada, California.

The displacement of air upward over an obstacle may trigger instability if the air is conditionally unstable and buoyant (see Chapter 5B), whereas stable air returns to its original level in the lee of a barrier as the gravitational effect counteracts the initial displacement. This descent often forms the first of a series of *lee waves* (or *standing waves*) downwind, as shown in Figure 6.13. The wave form remains more or less stationary relative to the barrier, with the air moving quite rapidly through it. Below the crest of the waves, there may be circular air motion in a vertical plane, which is termed a *rotor*. The formation of such features is of vital interest to pilots. The presence of lee waves is often marked by the development of lenticular clouds, and on occasion a rotor causes reversal of the surface wind direction in the lee of high mountains.

Winds on mountain summits are usually strong, at least in middle and higher latitudes. Average speeds on summits in the Colorado Rocky Mountains in winter months are around 12–15m s^{-1}, for example, and on Mount Washington, New Hampshire, an extreme value of 103m s^{-1} has been recorded. Peak speeds in excess of 40–50m s^{-1} are typical in both of these areas in winter. Airflow over a mountain range causes the air below the tropopause to be compressed and thus accelerated particularly at and near the crest line (the Venturi effect), but friction with the ground also retards the flow, compared with free air at the same level. The net result is predominantly one of retardation, but the outcome depends on the topography, wind direction and stability.

Over low hills, the boundary layer is displaced upward and acceleration occurs just above the summit. Figure 6.14 shows instantaneous airflow conditions across Askervein Hill (relief *c.* 120m) on the island of South Uist in the Scottish Hebrides, where the wind speed at a height of 10m above the ridge crest approaches 80 percent more than the undisturbed upstream velocity. In

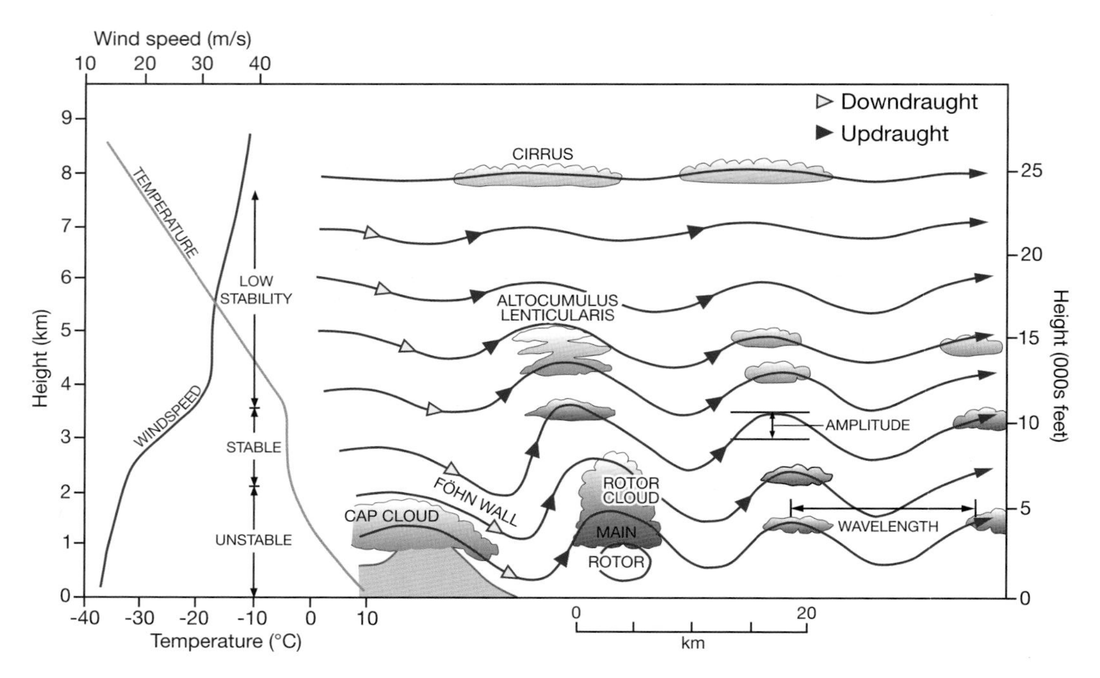

Figure 6.13 Lee waves and rotors are produced by airflow across a long mountain range. The first wave crest usually forms less than one wavelength downwind of the ridge. There is a strong surface wind down the lee slope. Wave characteristics are determined by the wind speed and temperature relationships shown schematically on the left of the diagram. The existence of an upper stable layer is particularly important.
Source: After Ernst (1976).

contrast, there was a 20 percent decrease on the initial run-up to the hill and a 40 percent decrease on the lee side, probably due to horizontal divergence. Knowledge of such local factors is critical for siting wind energy systems.

A wind of local importance near mountain areas is the *föhn*, or *chinook*. It is a strong, gusty, dry and warm wind that develops on the lee side of a mountain range when stable air is forced to flow across the barrier by the regional pressure gradient; the air descending on the lee slope warms adiabatically. Sometimes, there is a loss of moisture by precipitation on the windward side of the mountains (Figure 6.15). The air, having cooled at the saturated adiabatic lapse rate above the condensation level, subsequently warms at the greater dry adiabatic lapse rate as it descends on the lee side. This also reduces both the relative and the absolute humidity. Other investigations show that in many instances there is no loss of moisture over the mountains. In such cases, the föhn effect is the result of the blocking of air to windward of the mountains by a summit-level temperature inversion. This forces air from higher levels to descend and warm adiabatically. Southerly föhn winds are common along the northern flanks of the Alps and the mountains of the Caucasus and Central Asia in winter and spring, when the accompanying rapid temperature rise may help to trigger avalanches on the snow-covered slopes. At Tashkent in Central Asia, where the mean winter temperature is around freezing point, temperatures may rise to more than 21°C during a föhn. In the same way, the chinook is a significant feature at the eastern foot of the New Zealand Alps, the Andes in Argentina, and the Rocky

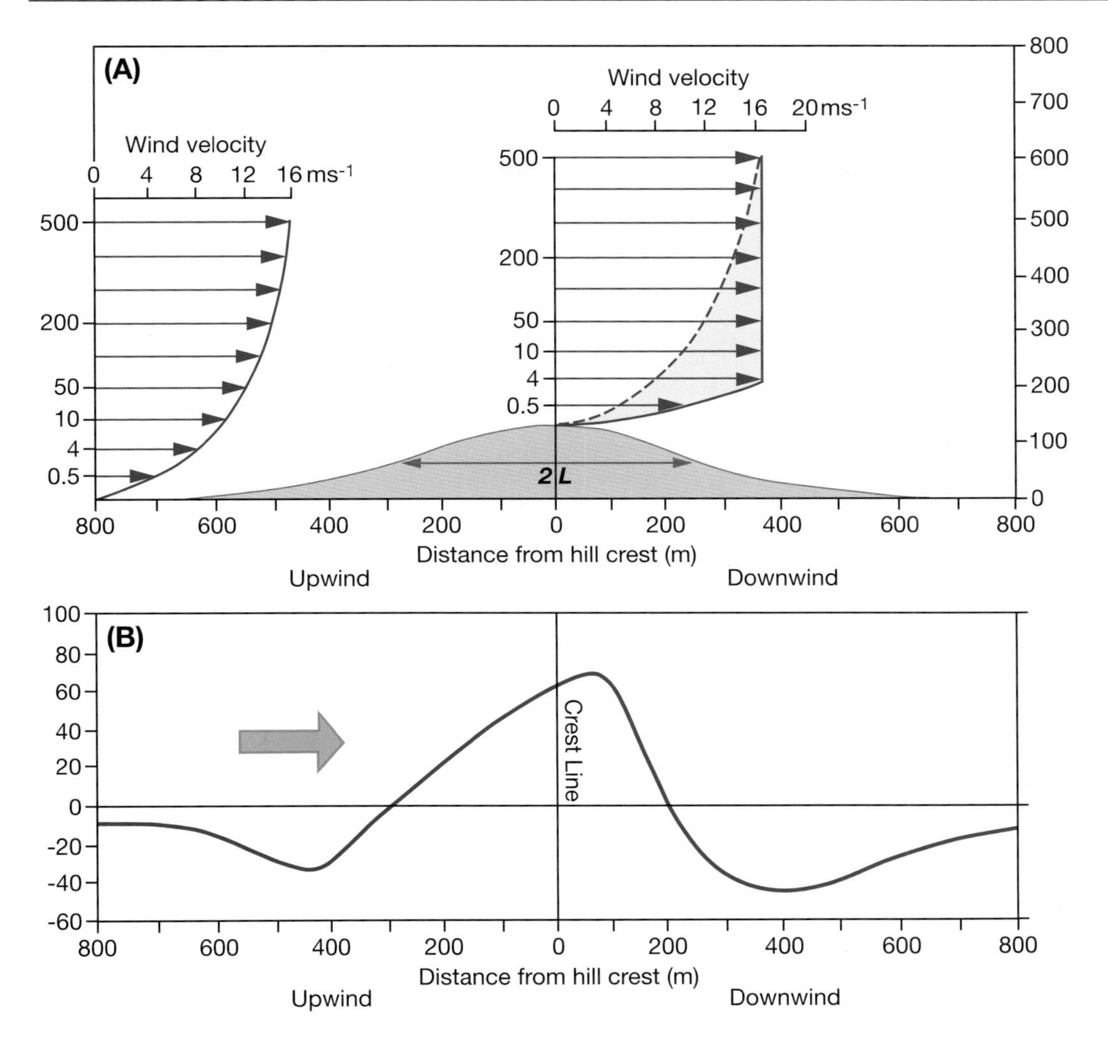

Figure 6.14 Airflow over Askervein Hill, South Uist, off the west coast of Scotland. A: Vertical airflow profiles (not true to scale) measured simultaneously 800m upwind of the crest line and at the crest line. L is the *characteristic length* of the obstruction (i.e. one-half the hill width at mid-elevation, here 500m) and is also the height above ground level to which the flow is increased by the topographic obstruction (shaded). The maximum speed-up of the airflow due to vertical convergence over the crest is to about 16.5m s^{-1} at a height of 4m. B: the relative speed-up (%) of airflow upwind and downwind of the crest line measured 14m above ground level.

Source: After Taylor, Teunissen and Salmon *et al.* From Troen and Petersen (1989).

Mountains. At Pincher Creek, Alberta, a temperature rise of 21°C occurred in four minutes with the onset of a chinook on 6 January 1966. In California, the Santa Ana is a cold season easterly wind that blows from the deserts east of the Sierra Nevada to the coast of southern California. It has an average frequency of 20 events per year and average duration of 1.5 days. It is notable for the dry air, which greatly increases the risk of chaparral fires. Less spectacular effects are also

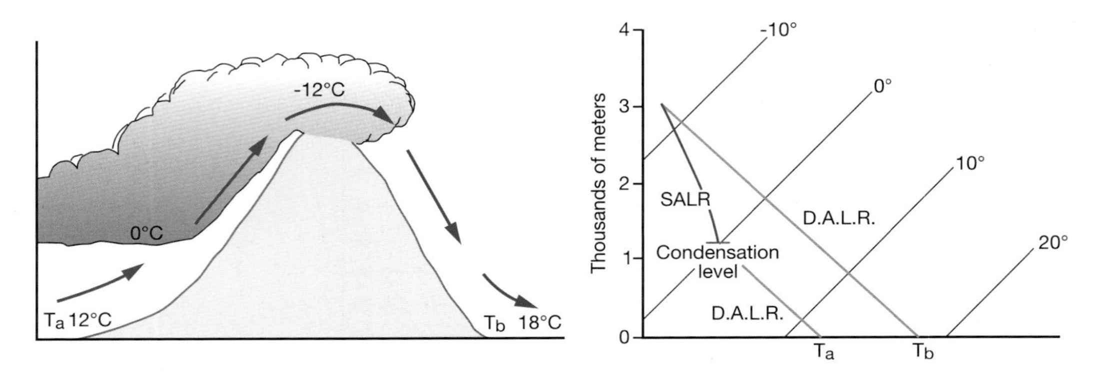

Figure 6.15 The föhn effect when an air parcel is forced to cross a mountain range 3km high. T_a refers to the temperature at the windward foot of the range and T_b to that at the leeward foot.

noticeable in the lee of the Welsh mountains, the Pennines and the Grampians in Great Britain, where the importance of föhn winds lies mainly in the dispersal of cloud by the subsiding dry air. This is an important component of so-called 'rain shadow' effects.

In some parts of the world, winds descending on the lee slope of a mountain range are colder than the air they displace (despite adiabatic warming through descent). The type example of such 'fall-winds' is the *bora* of the northern Adriatic, where cold northeasterly flows cross the Dinaric Alps, although similar winds occur on the northern Black Sea coast, in northern Scandinavia, in Novaya Zemlya and in Japan. These winds occur when cold continental air masses are forced across a mountain range by the pressure gradient and, despite adiabatic warming, displace warmer air. They are therefore primarily a winter phenomenon.

On the eastern slope of the Rocky Mountains in Colorado (and in similar continental locations), winds of either bora or chinook type can occur depending on the initial airflow characteristics. Locally, at the foot of the mountains, such winds may attain hurricane force, with gusts exceeding 45m s^{-1} (100mph). Down-slope storms of this type have caused millions of dollars of property damage in Boulder, Colorado, and the immediate vicinity. These wind storms develop when a stable layer close to the mountain-crest level prevents air to windward from crossing over the mountains. Extreme amplification of a lee wave (see Figure 6.13) drags air from above the summit level (4000m) down to the plains (1700m) over a short distance, leading to high velocities. However, the flow is not simply 'down-slope'; winds may affect the mountain slopes but not the foot of the slope, or vice versa, depending on the location of the lee wave trough. High winds are caused by the horizontal acceleration of air towards this local pressure minimum.

SUMMARY

Air motion is described by its horizontal and vertical components; the latter are much smaller than the horizontal velocities. Horizontal motions compensate for vertical imbalances between gravitational acceleration and the vertical pressure gradient.

The horizontal pressure gradient, the earth's rotational effect (Coriolis force), and the curvature of the isobars (centripetal acceleration) determine horizontal wind velocity. All three factors are accounted for in the gradient wind equation, but this can be approximated in large-scale flow by the geostrophic wind relationship. Below 1500m, the wind speed and direction are affected by surface friction.

Air ascends (descends) in association with surface convergence (divergence) of air. Air motion is also subject to relative vertical vorticity as a result of curvature of the streamlines and/or lateral shear; this, together with the earth's rotational effect, makes up the absolute vertical vorticity.

Local winds occur as a result of diurnally varying thermal differences setting up local pressure gradients (mountain–valley winds and land–sea breezes) or due to the effect of a topographic barrier on airflow crossing it (examples are the lee-side föhn and bora winds).

DISCUSSION TOPICS

- Compare the wind direction and speed reported at a station near you with the geostrophic wind velocity determined from the MSL pressure map for the same time (data sources are listed in Appendix 4).
- Why would there be no 'weather' if the winds were strictly geostrophic?
- What are the causes of mass divergence (convergence) and what roles do they play in weather processes?
- In what situations do local wind conditions differ markedly from those expected for a given large-scale pressure gradient?

REFERENCES AND FURTHER READING

Books

Barry, R. G. (2008) *Mountain Weather and Climate,* Cambridge University Press, 506 pp.) [Chapter on circulation systems related to orographic effects]

Oke, T. R. (1978) *Boundary Layer Climates,* Methuen, London, 372pp. [Prime text on surface climate processes in natural and human-modified environments]

Scorer, R. S. (1958) *Natural Aerodynamics,* Pergamon Press, Oxford 312pp.

Simpson, J. E. (1994) *Sea Breeze and Local Wind,* Cambridge University Press, Cambridge, 234pp. [A well-illustrated descriptive account of the sea breeze and its effects; on e chapter on local orographic winds]

Troen, I. and Petersen, E. L. (1989) *European Wind Atlas,* Commission of the Economic Community, Risø National Laboratory, Roskilde, Denmark, 656pp.

Wells, N. (1986) *The Atmosphere and Ocean. A Physical Introduction,* Taylor & Francis, London 345pp. [Good account of both systems and their interactions]

Articles

Banta, R.M. (1995) Sea breezes: shallow and deep on the California coast. *Mon. Wea. Rev.* 123(12). 3614–22.

Beran, W. D. (1967) Large amplitude lee waves and chinook winds. *J. Appl. Met.* 6, 865–77.

Brinkmann, W. A. R. (1971) What is a foehn? *Weather* 26, 230–9.

Brinkmann, W. A. R. (1974) Strong downslope winds at Boulder, Colorado. *Monthly Weather Review* 102, 592–602.

Buettner, K. J. and Thyer, N. (1965) Valley winds in the Mount Rainer area. *Archiv. Met. Geophys. Biokl.* B 14, 125–47.

Eddy, A. (1966) The Texas coast sea-breeze: a pilot study. *Weather* 21, 162–70.

Ernst, J. A. (1976) SMS-1 night-time infrared imagery of low-level mountain waves. *Monthly Weather Review* 104, 207–9.

Flohn, H. (1969) Local wind systems, in Flohn, H. (ed.) *General Climatology*, World Survey of Climatology 2, Elsevier, Amsterdam, 139–71.

Galvin., J.F.P. (2007)The weather and climate of the tropics. Part 2 – The subtropical jet streams. *Weather* 62, 295–99.

Geiger, R. (1969) Topoclimates, in Flohn, H. (ed.) *General Climatology*, World Survey of Climatology 2, Elsevier, Amsterdam, 105–38.

Glenn, C. L. (1961) The chinook. *Weatherwise* 14, 175–82.

Johnson, A. and O'Brien, J. J. (1973) A study of an Oregon sea breeze event. *J. Appl. Met.* 12, 1,267–83.

Lockwood, J. G. (1962) Occurrence of föhn winds in the British Isles. *Met. Mag.* 91, 57–65.

McDonald, J. E. (1952) The Coriolis effect. *Sci. American* 186, 72–8.

Persson, A. (1998) How do we understand the Coriolis force. *Weather* 79(7), 1373–85.

Persson, A. (2000) Back to basics. Coriolis: Part 1 – What is the Coriolis force? *Weather* 55(5), 165–70; Part 2 – The Coriolis force according to Coriolis. *Weather*. 55(6), 182–8; Part 3 – The Coriolis force on the physical earth. *Weather* 55(7), 234–9.

Persson, A. (2001) The Coriolis force and the geostrophic wind. *Weather* 56(8), 267–72.

Raphael, M.N. (2003) The Santa Ana winds of California. *Earth Interactions* 7, 1–13.

Riehl, H., Alaka, M. A., Jordan, C. L. and Renard, R. J. (1954) The jet stream. *Meteorol. Monogr.* 2, 23–47

Scorer, R. S. (1961) Lee waves in the atmosphere. *Sci. American* 204, 124–34.

Singleton, F. (2008) The Beaufort scale of winds – its relevance, and its use by sailors. *Weather* 63, 37–41.

Steinacker, R. (1984) Area–height distribution of a valley and its relation to the valley wind. *Contrib. Atmos. Phys.* 57, 64–74.

Thompson, B. W. (1986) Small-scale katabatics and cold hollows. *Weather* 41, 146–53.

Waco, D. E. (1968) Frost pockets in the Santa Monica Mountains of southern California. *Weather* 23, 456–61.

Wallington, C. E. (1960) An introduction to lee waves in the atmosphere, *Weather* 15, 269–76.

Wickham, P. G. (1966) Weather for gliding over Britain. *Weather* 21, 154–61.

CHAPTER SEVEN

Planetary-scale motions in the atmosphere and ocean

7

LEARNING OBJECTIVES

When you have read this chapter you will:

- learn how and why pressure patterns and wind velocity change with altitude
- become familiar with the relationships between surface and mid-tropospheric pressure patterns
- know the features of the major global wind belts
- be familiar with the basic concepts of the general circulation of the atmosphere
- understand the basic structure of the oceans, their circulation and role in climate
- know the nature and role of the thermohaline circulation.

In this chapter, we examine global-scale motions in the atmosphere and their role in redistributing energy, momentum and moisture. As noted in Chapter 3 (p. 72), there are close links between the atmosphere and oceans with the latter making a major contribution to poleward energy transport (albeit smaller than the atmospheric component). Thus, we also discuss ocean circulation and the coupling of the atmosphere–ocean system.

The atmosphere acts rather like a gigantic heat engine in which the temperature difference between the poles and the equator driven by differential solar heating drives the planetary atmospheric and ocean circulation. The conversion of heat energy into kinetic energy to produce motion must involve rising and descending air, but vertical movements are generally less obvious than horizontal movements, which may cover vast areas and persist for periods of a few days to several months. We begin by examining the relationships between winds and pressure patterns in the troposphere and those at the surface.

A VARIATION OF PRESSURE AND WIND VELOCITY WITH HEIGHT

Both pressure and wind characteristics change with height. Above the level of surface frictional effects (about 500–1000m), the wind increases in speed, and, except near the equator where the Coriolis force is very small, becomes more or less geostrophic, i.e., representing a balance between the pressure gradient and Coriolis force. In the midde and higher latitudes, meridonal temperature gradients that set up pressure gradients foster an increase of wind speed with height, in certain areas concentrated as narrow ribbons of high-velocity air termed jet streams. There are seasonal variations in wind speeds aloft, with winds being much greater in the Northern Hemisphere during winter months, when the meridional temperature gradients are at a maximum. Such seasonal variation is less pronounced in the Southern Hemisphere. In addition, the greater persistence of these gradients tends to cause the Southern Hemisphere upper winds to be more constant in direction. A history of upper air observations is given in Box 7.1.

1 The vertical variation of pressure systems

The air pressure at the surface, or at any level in the atmosphere, depends on the weight of the overlying air column. In Chapter 2B, we noted that air pressure is proportional to air density and that density varies inversely with air temperature. Accordingly, increasing the temperature of an air column between the surface and, say, 3km will reduce the air density in the column and therefore lower the air pressure at the surface without affecting the pressure at 3km altitude (the weight of the atmospheric column above 3km remains the same). Correspondingly, if we compare the heights of the 1000 and 700mb pressure surfaces, warming of the air column will lower the height of the 1000mb surface but will not affect the height of the 700mb surface (i.e., the thickness of the 1000–700mb layer increases).

The models shown in Figure 7.1 illustrate the relationships between surface and tropospheric pressure conditions. A low pressure cell at sea level with a cold core will intensify with elevation (Figure 7.1A), whereas one with a warm core will tend to weaken and may be replaced by high pressure. A warm air column of relatively low density causes the pressure surfaces to bulge upward, and conversely a cold, more dense air column leads to downward contraction of the pressure surfaces. Thus, a surface high pressure cell with a cold core (a *cold anticyclone*), such as the Siberian winter anticyclone, weakens with

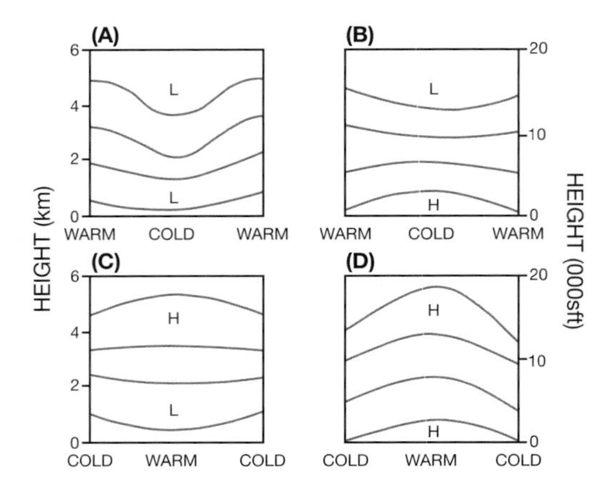

Figure 7.1 Models of the vertical pressure distribution in cold and warm air columns. A: A surface low pressure intensifies aloft in a cold air column. B: A surface high pressure weakens aloft and may become a low pressure in a cold air column. C: A surface low pressure weakens aloft and may become a high pressure in a warm air column. D: A surface high pressure intensifies aloft in a warm air column.

increasing elevation and is replaced by low pressure aloft (Figure 7.1B). Cold anticyclones are shallow and rarely extend their influence above about 2500m. By contrast, a surface high with a warm core (a *warm anticyclone*) intensifies with height (Figure 7.1D). This is characteristic of the large subtropical cells, which maintain their warmth through dynamic subsidence. The warm low (Figure 7.1C) and cold high (Figure 7.1B) are consistent with the vertical motion schemes illustrated in Figure 6.7, whereas the other two types are primarily produced by dynamic processes. The high surface pressure in a warm anticyclone is linked hydrostatically with cold, relatively dense air in the lower stratosphere. Conversely, a cold depression (Figure 7.1A) is associated with a warm lower stratosphere.

Mid-latitude low pressure cells have cold air in the rear, and hence the axis of low pressure slopes with height towards colder air to the northwest. High pressure cells slope towards the warmest air (Figure 7.2). Thus, Northern Hemisphere subtropical high pressure cells are shifted 10–15°

SIGNIFICANT 20TH-CENTURY ADVANCE

7.1 The history of upper-air measurements

Manned balloon flights during the nineteenth century attempted to measure temperatures in the upper air but the equipment was generally inadequate for the purpose. Kite measurements were common in the 1890s. During and after World War I (1914–1918), balloon, kite and aircraft measurements of temperatures and winds were collected in the lower few kilometers of the atmosphere. Forerunners of the modern radiosonde, which comprises a package of pressure, temperature and humidity sensors, suspended beneath a hydrogen-filled balloon and transmitting radio signals of the measurements during its ascent, were developed independently in France, Germany and the USSR and first used in about 1929–1930. Soundings began to be made up to about 3–4km, mainly in Europe and North America, in the 1930s and the radiosonde was widely used during and after World War II. It was improved in the late 1940s when radar tracking of the balloon enabled the calculation of upper-level wind speed and direction; the system was named the radar wind sonde or rawinsonde. There are now about 1000 upper-air sounding stations worldwide making soundings once or twice daily at 00 and 12 hours UTC, and sometimes more frequently. In addition to these systems, meteorological research programs and operational aircraft reconnaissance flights through tropical and extra-tropical cyclones commonly make use of drop-sondes that are released from the aircraft and give a profile of the atmosphere below it.

Satellites began to provide a new source of upper-air data in the early 1970s through the use of vertical atmospheric sounders. These sounders are especially valuable in providing data in areas where rawinsonde coverage is sparse, such as Antarctica, the Arctic Ocean and large areas of the global ocean. They operate in the infrared and microwave wavelengths and provide information on the temperature and moisture content of different layers in the atmosphere. They operate on the principle that the energy emitted by a given atmospheric layer is proportional to its temperature (see Figure 3.1) (and is also a function of its moisture content). The data are obtained through a complex 'inversion' technique whereby the radiative transfer relationships (p. 41) are inverted so as to calculate the temperature (moisture) from the measured radiances. Infrared sensors operate only for cloud-free conditions whereas microwave sounders record in the presence of clouds. Neither system is able to measure low-level temperatures in the presence of a low-level temperature inversion because the method assumes that temperatures are a unique function of altitude.

Ground-based remote sensing provides another means of profiling the atmosphere. Detailed information on wind velocity is available from upward-pointing high-powered radar (radio detection and ranging) systems of between 10cm (UHF) and 10m (VHF) wavelength. These wind profilers detect motion in clear air via measurements of variations in atmospheric refractivity. Such variations depend on atmospheric temperature and humidity. Radars can measure winds up to stratospheric levels, depending on their power, with a vertical resolution of a few meters. Such systems are in use in the equatorial Pacific and in North America. Information on the general structure of the boundary layer and low-level turbulence may be obtained from lidar (light detection and ranging) and sodar (sound detection and ranging) systems, but these have a vertical range of just a few kilometers.

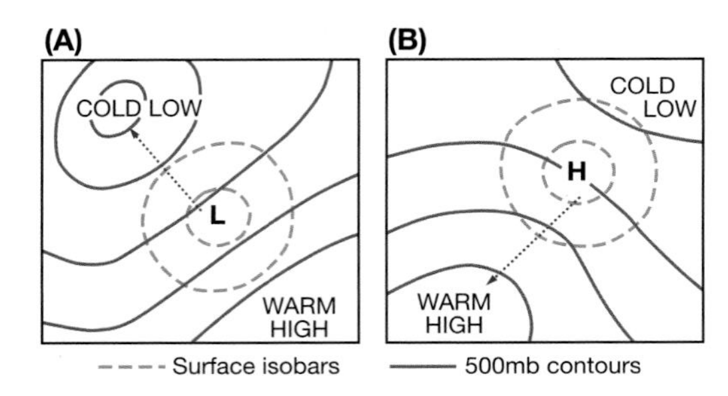

Figure 7.2 The characteristic slope of the axes of low and high pressure cells with height in the Northern Hemisphere.

latitude southward at 3km, and towards the west. Even so, this slope of the high pressure axes is not constant through time.

2 Mean upper-air patterns

The patterns of pressure and wind in the middle troposphere are less complicated in appearance than at the surface as a result of the diminished effects of the land masses. Rather than using pressure maps at a particular height, it is more convenient to depict the height of a selected pressure surface; this is termed a *contour chart* by analogy with a topographic relief map (see Note 1). Figures 7.3 and 7.4 show that in the middle troposphere of the Southern Hemisphere there is a vast circumpolar cyclonic vortex poleward of latitude 30°S in summer and winter. The vortex is more or less symmetrical around the pole, although the low center is towards the Ross Sea sector. Corresponding charts for the Northern Hemisphere also show an extensive cyclonic vortex, but one that is markedly more asymmetric, especially in winter, when centers are found over eastern Canada and eastern Siberia. The summer pattern shows a much weaker vortex, centered over the pole. The major troughs and ridges, especially well illustrated for the Northern Hemisphere winter, form what are referred to as *longwaves* (or *Rossby waves*) in the upper flow. It is worth considering why the hemispheric westerlies show such large-scale waves. The key to

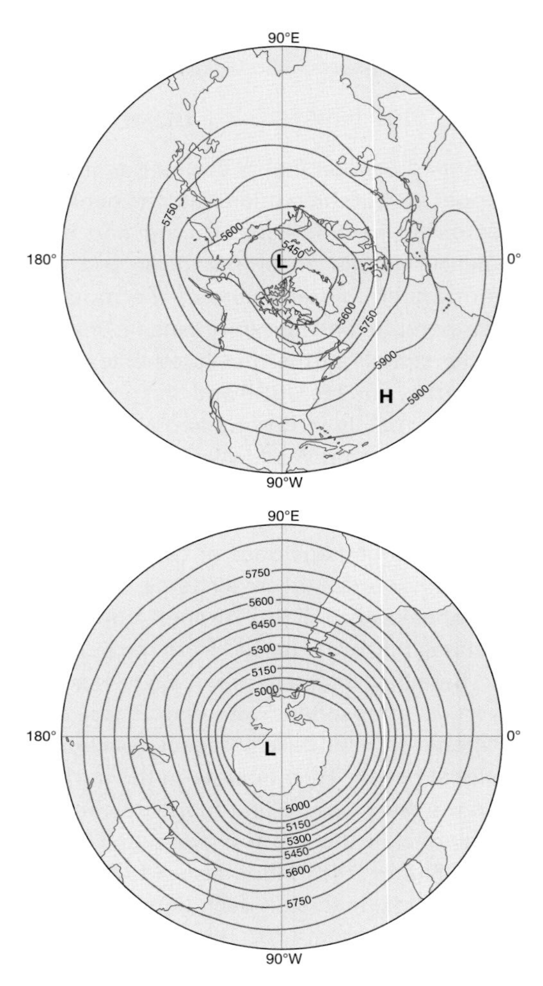

Figure 7.3 The mean contours (gpm) of the 500mb pressure surface in January and July for the Southern Hemisphere, 1970–1999. Gpm = geopotential metres.

Source: NCEP/NCAR Reanalysis Data from the NOAA-CIRES Climate Diagnostics Center.

this problem lies in the rotation of the earth and the latitudinal variation of the Coriolis parameter (Chapter 6A.2). It may be shown that for large-scale motion the absolute vorticity around a vertical axis (the sum of relative and planetary vorticity, or $f + \zeta$}) tends to be approximately conserved, i.e.,

$$d\,(f + \zeta)/dt = 0$$

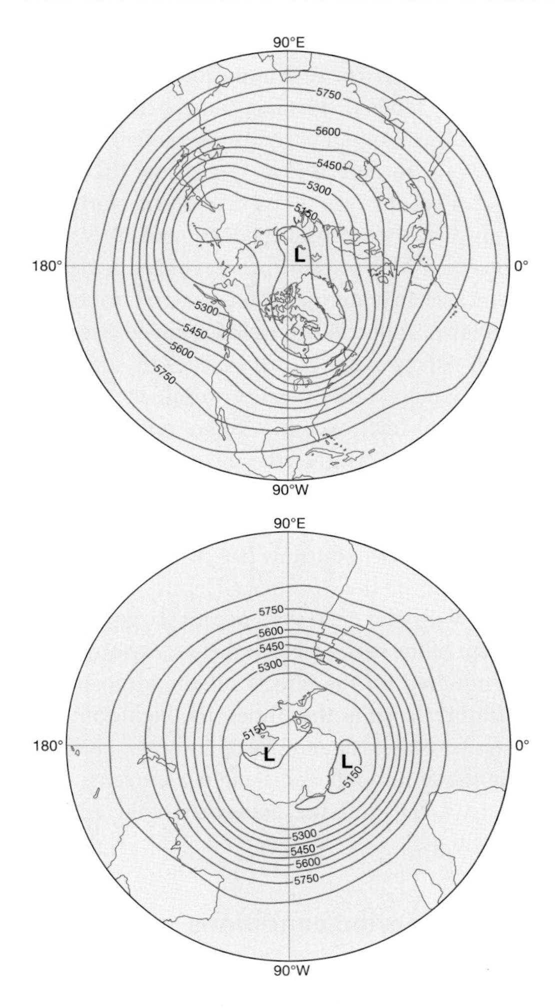

Figure 7.4 The mean contours (gpm) of the 500mb pressure surface in January and July for the Northern Hemisphere, 1970–1999.

Source: NCEP/NCAR Reanalysis from the NOAA-CIRES Climate Diagnostics Center.

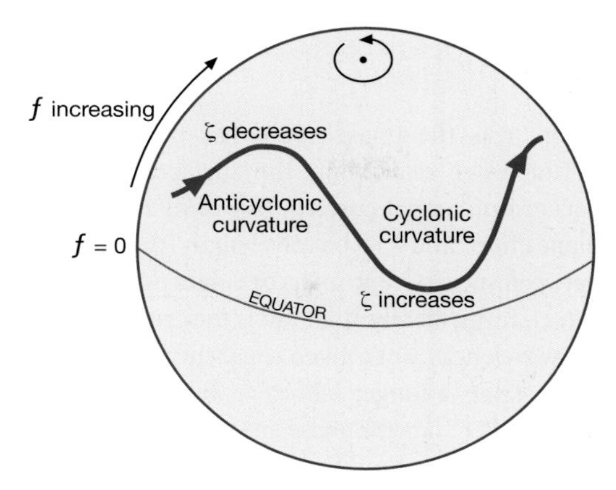

Figure 7.5 A schematic illustration of the mechanism of longwave development in the tropospheric westerlies.

The symbol d/dt denotes the time rate of change following the motion (a total differential). Consequently, if an air parcel moves poleward so that f increases, the cyclonic relative vorticity tends to decrease. The curvature thus becomes anticyclonic and the current returns towards lower latitudes. If the air moves equatorward of its original latitude, f tends to decrease (Figure 7.5), requiring ζ to increase, and the resulting cyclonic curvature again deflects the current poleward. In this manner, large-scale flow tends to oscillate in a wave pattern.

While conservation of absolute vorticity helps to explain why Rossby waves exist, an important next consideration is the motion, or propagation, of the wave form itself. Rossby waves, like their shorter wavelength cousins associated with transitory cyclones and anticyclones (see Chapter 9), can, with appropriate simplifications, be viewed as disturbances embedded in a background zonal current. The effect of the zonal current is to propagate the wave eastward with respect to the surface. In other words, the background zonal current carries the wave along with it. However, there is a competing effect. The latitudinal increase in the Coriolis parameter (known as the beta plane), associated with the concept of conservation of absolute vorticiy, acts to propagate the wave westward with respect to the surface. The relative importance of these two effects determines whether, with respect to the surface, the wave remains stationary (the two effects cancel), migrates eastward (the background zonal flow dominates) or migrates westward (the beta plane effect wins). The formal relationship, in turn built on the assumption that absolute vorticity is conserved following the motion, is:

$$c = U - \beta\left(\frac{L}{2\pi}\right)^2$$

where c is the phase speed (or propagation) of the wave relative to the surface, U is the background zonal current, $\beta = \delta f/\delta y$ is the beta plane effect, and L is the wavelength (the distance between successive troughs or ridges defining the wave). Immediately apparent is the critical role of the wavelength. For a given zonal current and beta plane value, a longer (shorter) wavelength leads to a smaller (larger) phase speed with respect to the surface. It should also be clear that if the wavelength is sufficiently long for the given zonal current, the Rossby wave may remain stationary ($c = 0$) or even move westward with respect to the surface ($c < 0$). Conversely, for two waves of the same wavelength and beta plane value, the one associated with the larger background zonal wind will propagate faster. The general observation is that long Rossby waves tend to be quasi-stationary or move slowly eastward, although westward-moving waves with respect to the surface are indeed observed. Shorter waves (often simply termed shortwaves) tend to move eastward. It is instructive to calculate the stationary wavelength, where $c = 0$ and $L = 2\pi\sqrt{(U/\beta)}$. At 45° latitude, this stationary wavelength is 3120km for a zonal velocity of 4m s^{-1}, increasing to 5400km at 12m s^{-1}. The wavelengths, at 60° latitude for zonal currents of 4 and 12m s^{-1} are, respectively, 3170 and 6430km. The pattern of waves seen in a mid-tropospheric contour chart can be quite complex, with shorter waves tending to be embedded within longwaves. An important concept is that the shorter waves (associated with transitory cyclones and anticyclones) tend to migrate along and be steered by the quasi-stationary longwaves. Knowing the pattern of the longwaves hence provides information on the path of the shorter waves.

Turning back to Figures 7.3 and 7.4, the two major Northern Hemisphere troughs at about 70°W and 150°E, again best expressed in winter, are thought to be induced by the combined influence on upper-air circulation of large orographic barriers, such as the Rocky Mountains and the Tibetan Plateau, and heat sources such as warm ocean currents (in winter) or land masses (in summer). It is noteworthy that land surfaces occupy over 50 percent of the Northern Hemisphere between latitudes 40° and 70°N. The subtropical high pressure belt has only one clearly distinct cell in January over the eastern Caribbean, whereas in July cells are well developed over the North Atlantic and North Pacific. In addition, the July map shows greater prominence of the subtropical high over the Sahara and southern North America. The Northern Hemisphere shows a marked summer to winter intensification of the mean circulation, which is explained below.

As mentioned, the flow pattern is much more symmetric in the Southern Hemisphere. This aligns with the fact that oceans comprise 81 percent of the surface. Nevertheless, asymmetries are initiated by the effects on the atmosphere of features such as the Andes, the high dome of eastern Antarctica, and ocean currents, particularly the Humboldt and Benguela currents (see Figure 7.31), and the associated cold coastal upwellings.

3 Upper wind conditions

Imagine two sets of dinner plates, one set of plates being thicker than the other. The thick and thin plates are stacked into separate piles. As we add more plates to each pile, the height of the pile of thick plates becomes increasingly greater than the height of the pile with thin plates. Similarly, as the thickness between pressure levels is greater at lower latitudes than at higher latitudes (recall from section A.1 and Figure 7.1 above that thickness is proportional to the mean temperature of the layer), the difference in height of a given pressure surface between high and low latitudes increases upward. This means that the geostrophic wind also increases with height; that is, there is a vertical wind shear. The zonal winds are strongest where and when the meridional temperature

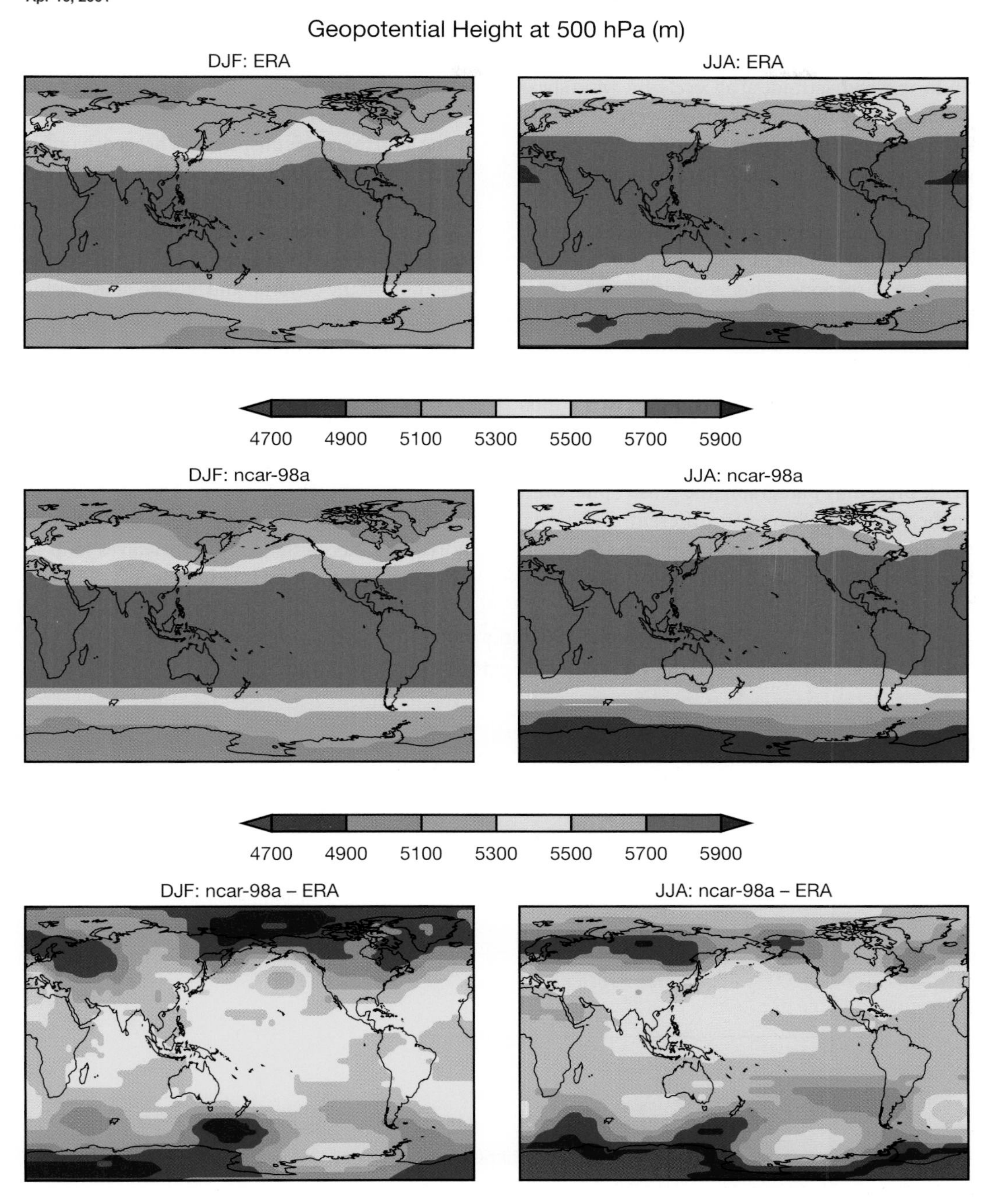

Plate 7.1 500mb geopotential heights for northern winter (DJF) and summer (JJA) in (A) ECMWF reanalysis observations. (B) NCAR CCM3 simulations and (C) the difference between CCM3 and observations.

Source: AMIP website.

gradient is at a maximum. This effect of different pressure thicknesses explains the increase in the speed of the basic mid-latitude westerlies with height. In the simple case of thicknesses decreasing uniformly with latitude at all levels, the wind shear would be only in terms of speed, with no change in direction with height. The pressure height contours would in turn parallel contours of thickness.

This is but a simplification, however. It is often observed that clouds at different levels move in different directions. This is evidence that there can be vertical wind shear not just in terms of speed, but in direction as well. This important relationship is illustrated in Figure 7.6. The diagram shows hypothetical contours of the 1000 and 500mb pressure surfaces and of the 1000 to 500mb thickness. Unlike the simple case outlined in the preceding paragraph, the geostrophic wind at the two levels blows in different directions, towards the upper right at 1000mb and from left to right at 500hPa. The height contours at 500mb also intersect the contours of 1000 to 500mb thickness. The theoretical wind vector (V_T) blowing parallel to the thickness lines, with a velocity proportional to their gradient, is termed the *thermal wind*. Facing downwind, the thermal wind blows with cold air (low thickness) to the left in the Northern Hemisphere. The geostrophic wind velocity at 500mb (G_{500}) turns out to be the vector sum of the 1000mb geostrophic wind (G_{1000}) and the thermal wind (V_T), as shown in Figure 7.6. For the simpler case in which the directions of the 1000mb and 500mb geostrophic winds are the same, the thermal wind is simply proportional to the difference in geostrophic speed between the two levels. Setting the stage for further discussion in Chapter 9, situations with directional shear can be associated with the growth of disturbances in the basic mid-latitude westerly flow, seen at the surface as traveling cyclones and anticyclones, and at upper levels as atmospheric shortwaves. Recall that these shortwaves tend to move through the long Rossby waves.

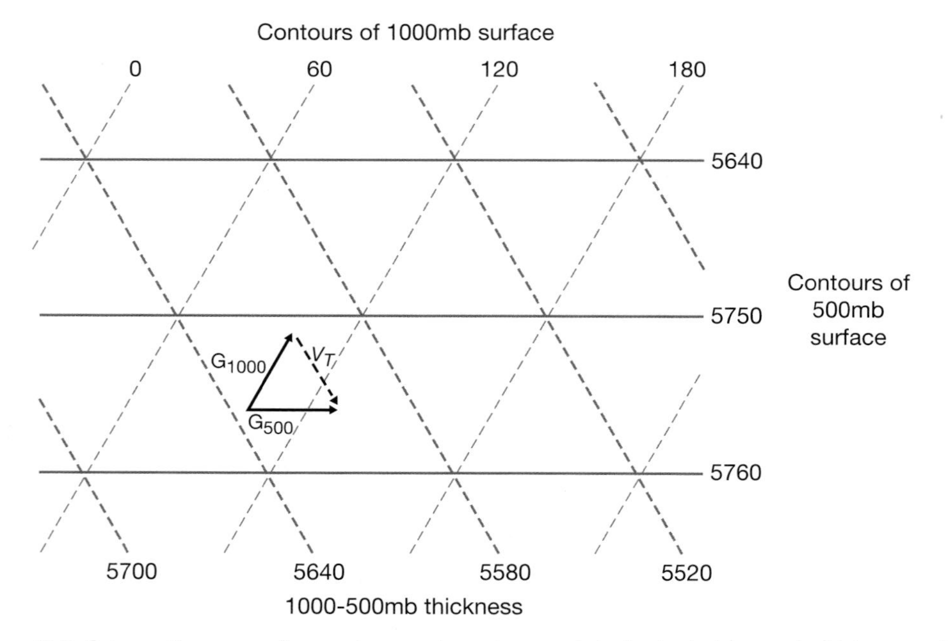

Figure 7.6 Schematic map of superimposed contours of isobaric height and thickness of the 1000–500mb layer (in meters). G_{1000} is the geostrophic velocity at 1000mb, G_{500} that at 500mb, V_T is the resultant 'thermal wind' blowing parallel to the thickness lines.

The basic westerly flow with its embedded disturbances characterizes both hemispheres in the regions poleward of the subtropical high pressure cells (centered aloft at about 15° latitude) Between the subtropical high pressure cells and the equator the winds are easterly. The dominant westerly circulation reaches maximum speeds of 45–65m s^{-1}, which even increase to 135m s^{-1} in winter. These maximum speeds are concentrated in narrow bands, typically between 9000 and 15000m, called *jet streams* (see Note 2 and Box 7.2).

A jet stream is essentially a fast-moving ribbon of air, which in turn coincides with the latitude of maximum poleward temperature gradient, or frontal zone, shown schematically in Figure 7.7. The thickness effect , as described above, is a major component of jet streams, but the basic reason for the concentration of the meridional temperature gradient in a narrow zone (or zones) is dynamical. In essence, the temperature gradient becomes accentuated when the upper wind pattern is confluent (see Chapter 6B.1). It is helpful to introduce the concept of angular momentum and its conservation. The momentum of an air parcel is the product of its mass times its velocity; the angular momentum is the product of the linear velocity of a body rotating around an axis and its perpendicular distance from the axis. Angular momentum tends to be conserved, meaning that if the radial distance of rotation of an air parcel decreases (increases), then the speed of rotation increases (decreases) Consider now a belt of westerly winds at latitude 40°. If the winds shift northward, the radial distance decreases, and so the wind velocity increases. In the atmosphere, angular momentum conservation is a major contributor to the maintenance of westerly jet streams.

Figure 7.8 shows a highly generalized north–south cross-section with three westerly jet streams in the Northern Hemisphere. The more northerly ones, termed the *Polar Front* and *Arctic Front Jet Streams* (Chapter 9E), are associated with the steep temperature gradient where polar and tropical air and polar and arctic air, respectively, interact, but the *Subtropical Jet Stream* is related to a temperature gradient confined to the upper troposphere, 12–15km (~ 200mb). Wind speeds over East Asia regularly exceed 100m s^{-1}. The Polar Front Jet Stream is very irregular in its latitudinal

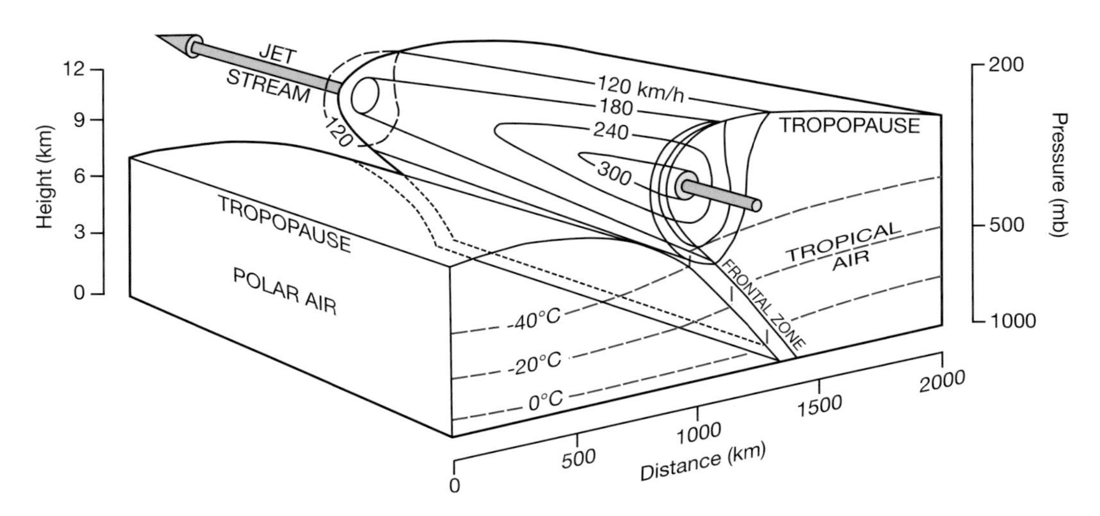

Figure 7.7 Structure of the mid-latitude frontal zone and associated jet stream showing generalized distribution of temperature, pressure and wind velocity.

Source: After Riley and Spalton (1981). Courtesy of Cambridge University Press.

SIGNIFICANT 20TH-CENTURY ADVANCE

7.2 The discovery of jet streams

Late nineteenth-century observers of high-level cloud motion noted the occasional existence of strong upper winds, but their regularity and persistence were not suspected at the time. The recognition that there are coherent bands of very strong winds in the upper troposphere was an operational discovery by Allied bomber pilots flying over Europe and the North Pacific during World War II. Flying westward, headwinds were sometimes encountered that approached the air speed of the planes. The term *jet stream*, used earlier for certain ocean current systems, was introduced in 1944 and soon became widely adopted. The corresponding German word Strahlstrome had in fact been first used in the 1930s.

Bands of strong upper winds are associated with intense horizontal temperature gradients. Locally enhanced Equator–Pole temperature gradients are associated with westerly jets and Pole–Equator gradients with easterly jets. The principal westerly jet streams are the subtropical westerly jet stream at about 150–200mb, and one associated with the main polar front at around 250–300mb. The former is located between latitudes 30–35° and the latter between 40–50° in both hemispheres. The strongest jet cores tend to occur over East Asia and eastern North America in winter. There may be additional jet stream bands associated with an Arctic frontal zone. In summer there is a persistent, albeit generally weak Arctic jet stream feature, most apparent over Eurasia, which owes its existence to differential surface heating between the cold Arctic Ocean and surrounding snow-free land. In the tropics there are strong easterly jet streams in summer over southern India and the Indian Ocean and over West Africa. These are linked to the monsoon systems.

and longitudinal locations and is commonly discontinuous, whereas the Subtropical Jet Stream is much more persistent and varies far less in latitude. For these reasons, the location of the mean jet stream in each hemisphere and season (Plate 7.2) primarily reflects the position of the Subtropical Jet Stream. The austral summer (DJF) map shows a strong zonal feature around 50°S, while the boreal summer jet is weaker and more discontinuous over Europe and North America. The winter maps (Plate 7.2 [A] and [D]) show a pronounced double structure in the Southern Hemisphere from 60°E eastward to 120°W, a more limited analogue over the eastern and central North Atlantic Ocean (0–40°W). This double structure represents the subtropical and polar jets.

The synoptic pattern of jet stream occurrence may be further complicated in some sectors by the presence of additional frontal zones (see Chapter 9E), each associated with a jet stream. This situation is common in winter over North America. Comparison of Figure 7.4 and Plate 7.2 indicates that the main jet stream cores are associated with the principal troughs of the Rossby longwaves. In summer, an *Easterly Tropical Jet Stream* forms in the upper troposphere over India and Africa due to regional reversal of the S–N temperature gradient (p. 355). The relationships between upper tropospheric wind systems and surface weather and climate will be considered later.

In the Southern Hemisphere, the mean jet stream in winter is similar in strength to its Northern Hemisphere winter counterpart and it weakens less in summer, because the meridional temperature gradient between 30° and 50°S is reinforced by heating over the southern continents (Plate 7.2).

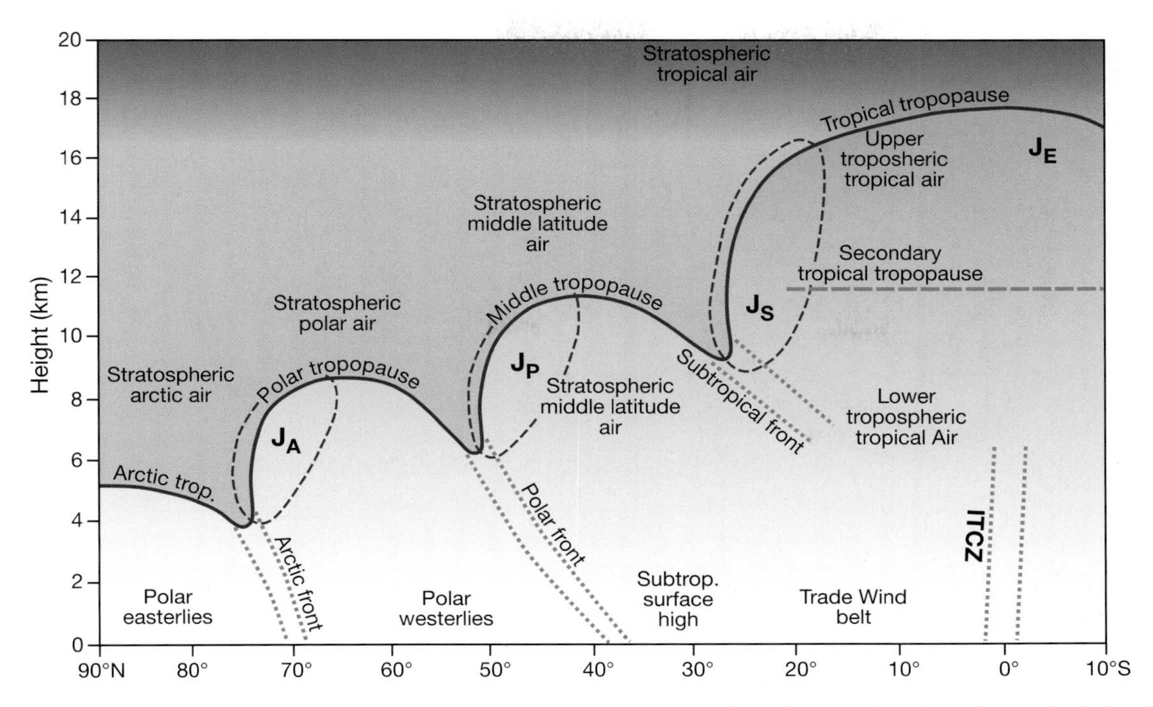

Figure 7.8 The meridional structure of the tropopause and the primary frontal zones. The 40m s^{-1} isotach (dashed) encloses the arctic (J_A), polar (J_P) and subtropical (J_S) jet streams. The tropical easterly (J_E) jet stream is also shown. Occasionally, the arctic and polar or the polar and subtropical fronts and jet streams may merge to form single systems in which about 50 percent of the pole-to-equator mid-tropospheric pressure gradient is concentrated into a singe frontal zone approximately 200km wide. The tropical easterly jet stream may be accompanied by a lower easterly jet at about 5km elevation.

Source: Shapiro *et al.* (1987). From *Monthly Weather Review* 115, p. 450 by permission of the American Meteorological Society.

4 Surface pressure conditions

The most permanent features of the mean sea-level pressure maps are the oceanic subtropical high pressure cells (Figures 7.9 and 7.10). These anticyclones are located at about 30° latitude, suggestively situated below the mean Subtropical Jet Stream. They move a few degrees equatorward in winter and poleward in summer in response to the seasonal expansion and contraction of the two circumpolar vortices. The anticyclones located in the eastern sectors of the subtropical North Atlantic and North Pacific are shallow, unlike those in the western parts. In the Northern Hemisphere, the subtropical ridges of high pressure weaken over the heated continents in summer but are thermally intensified over them in winter. The principal subtropical high pressure cells are located: (1) over the Bermuda–Azores ocean region (at 500mb the center of this cell lies over the east Caribbean); (2) over the south and southwest United States (the Great Basin or Sonoran cell) – this continental cell is seasonal, being replaced by a thermal surface low in summer; (3) over the east and north Pacific – a large and powerful cell (sometimes dividing into two, especially during the summer); and (4) over the Sahara – this, like other continental source areas, is seasonally variable in both intensity and extent, being most prominent in winter. In the Southern Hemisphere, the subtropical anticyclones are oceanic, except over southern Australia in winter.

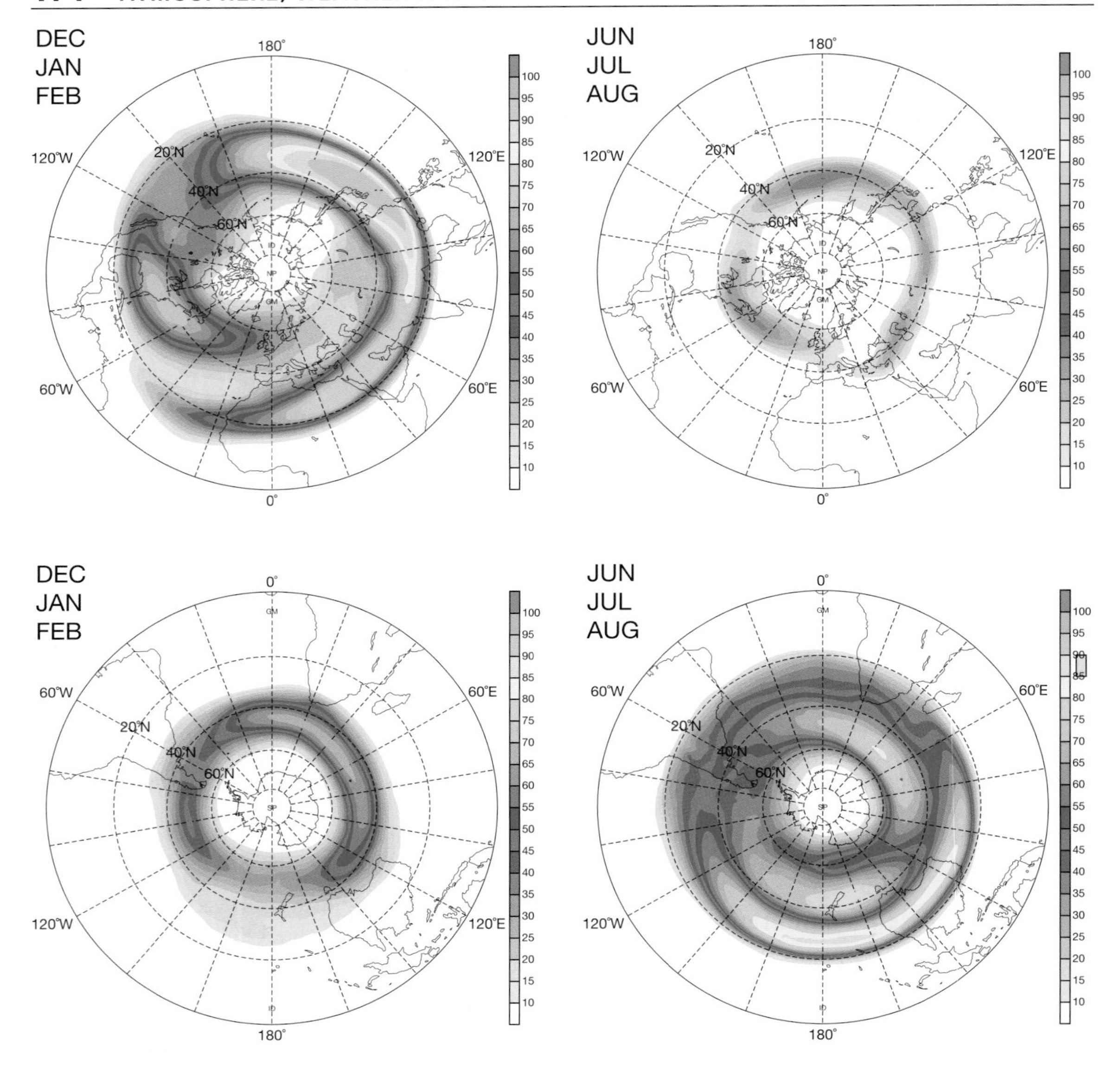

Plate 7.2 Probability of jet stream velocity (vertically integrated between 100 and 400mb) exceeding 30m s^{-1} based on ECMWF reanalyis data 1982–1992. Units are in percentages. A: Northern Hemisphere winter, DJF. B: Northern Hemisphere summer, JJA. C: Southern Hemisphere summer, DJF. D: Southern Hemisphere winter, JJA.
Courtesy of Patrick Koch and Sarah Kew, Institute for Atmospheric and Climate Science, ETH, Zurich.

The latitude of the subtropical high pressure belt depends on the meridional temperature difference between the equator and the pole and on the temperature lapse rate (i.e., vertical stability). The greater the meridional temperature difference the more equatorward is the location of the subtropical high pressure belt (Figure 7.11).

In low latitudes there is an equatorial trough of low pressure, associated broadly with the zone of maximum insolation and tending to migrate with

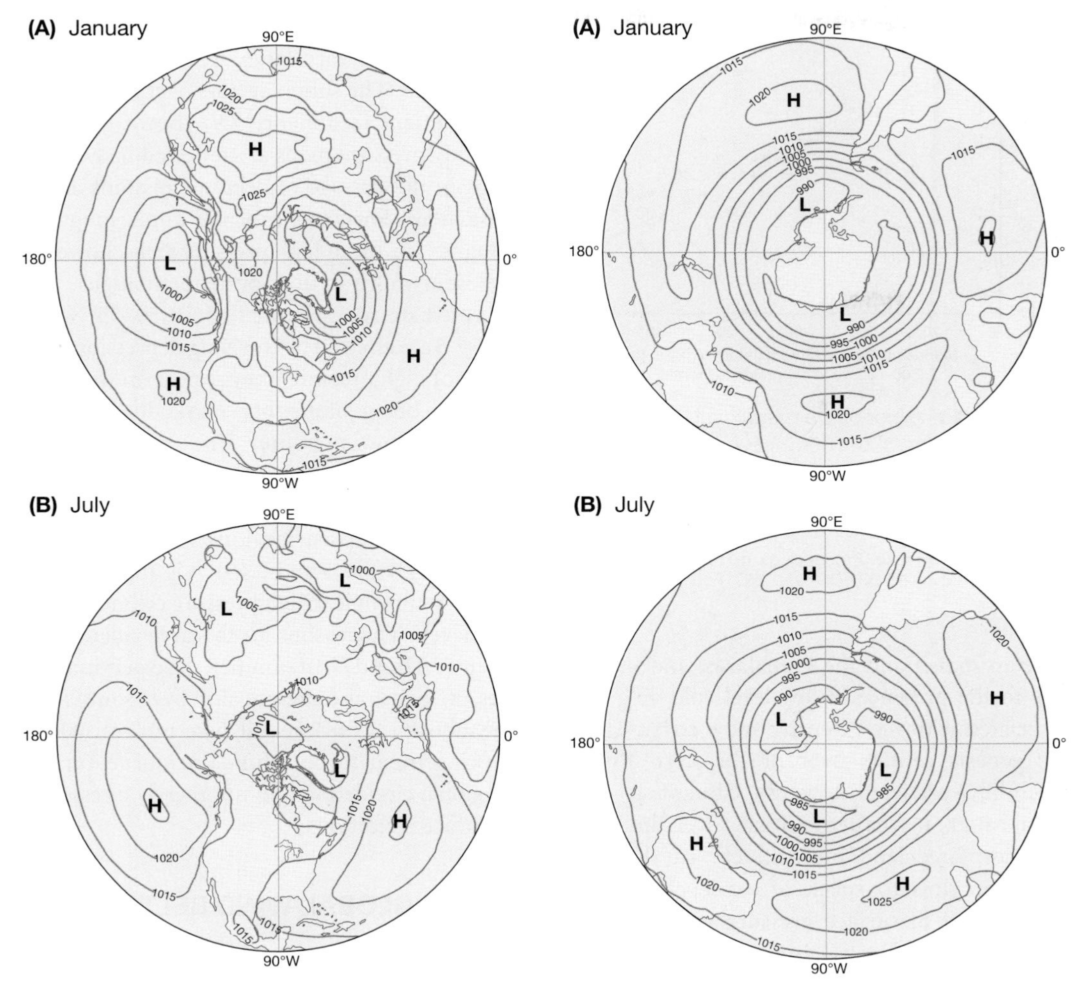

Figure 7.9 The mean sea-level pressure distribution (mb) in January and July for the Northern Hemisphere, 1970–1999.

Source: NCEP/NCAR Reanalysis Data from the NOAA-CIRES Climate Diagnostics Center.

Figure 7.10 The mean sea-level pressure distribution (mb) in January and July for the Southern Hemisphere, 1970–1999. Isobars not plotted over the Antarctic ice sheet.

Source: NCEP/NCAR Reanalysis Data from the NOAA-CIRES Climate Diagnostics Center.

it, especially towards the heated continental interiors of the summer hemisphere. Poleward of the subtropical anticyclones lies a general zone of subpolar low pressure. In the Southern Hemisphere, this sub-Antarctic Trough is virtually circumpolar (see Figure 7.10), whereas in the Northern Hemisphere the major centers are near Iceland and the Aleutians in winter and primarily over continental areas in summer. In winter, the Arctic region is affected by high and low pressure cells with semi-permanent cold air anticyclones over Siberia and, to a lesser extent, northwestern Canada. While it is still sometimes stated that anticyclonic conditions dominate the Arctic region, this is clearly not the case. The shallow Siberian high is in part a result of the exclusion of tropical air masses from the interior by the

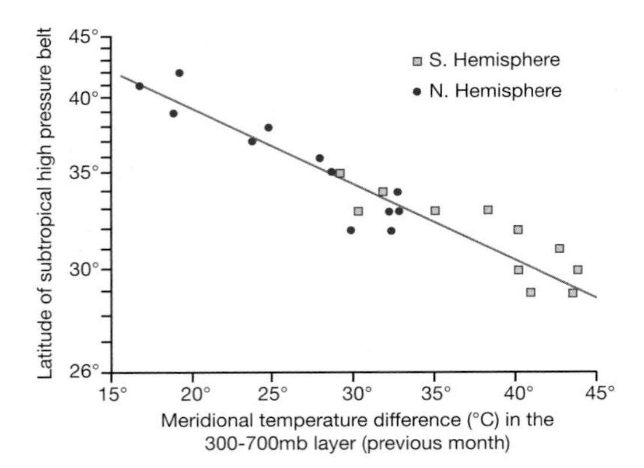

Figure 7.11 A plot of the meridional temperature difference at the 300–700mb level in the previous month against the latitude of the center of the subtropical high pressure belt, assuming a constant vertical tropospheric lapse rate.

Source: After Flohn, in Proceedings of the World Climate Conference, WMO N0.537 (1979, p. 257, Fig.2).

Tibetan massif and the Himalayas, and in part due to the presence of low-level cold air pools associated with the extensive snow cover. There are genesis centers over northeastern Russia extending eastward towards Chukotka, over Kazakhstan, and eastern China. Over the high-elevation Greenland and Antarctic ice sheets, it is meaningless to speak of sea-level pressure (adjustment of surface pressure to sea level is fraught with difficulty) but, on average, there is high pressure over the 3–4km-high eastern Antarctic plateau.

Building on earlier discussion, the mean circulation in the Southern Hemisphere is much more zonal at both 700mb and sea level than in the Northern Hemisphere, due to the more limited area and effect of land masses. There is also little difference between summer and winter circulation intensity (see Figures 7.3 and 7.10). It is important at this point to differentiate between mean pressure patterns and the highs and lows shown on daily or subdaily synoptic weather maps. In the Southern Hemisphere, the zonality of the mean circulation conceals a high degree of day-to-day variability. The *synoptic map* is a daily or subdaily 'snapshot' of the principal pressure systems over a very large area, ignoring local circulations. The subpolar lows over Iceland and the Aleutians (see Figure 7.9) shown on mean monthly pressure maps are reflected in synoptic maps in the passage of deep depressions across these areas downstream of the upper longwave troughs. The mean high pressure areas, however, represent more or less permanent highs. The intermediate zones located about 50–55°N and 40–60°S are affected by traveling depressions and ridges of high pressure; they appear on the mean maps as being of neither markedly high nor markedly low pressure. The movement of depressions is considered in Chapter 9F.

On comparing the mean surface and tropospheric pressure distributions for January (see Figures 7.4, 7.5, 7.9 and 7.10), it is apparent that only the subtropical high pressure cells extend to high levels. The reasons for this are evident from Figures 7.1B and D. In summer, the equatorial low pressure belt is also present aloft over South Asia. The subtropical cells are still discernible at 300mb, showing them to be a fundamental feature of the global circulation and not merely a response to surface conditions.

B THE GLOBAL WIND BELTS

The importance of the subtropical high pressure cells is evident from the preceding discussion. Dynamic, rather than immediately thermal in origin, and situated between 20° and 30° latitude, they seem to provide the key to the world's major wind belts, shown by the maps in Figure 7.12. In the Northern Hemisphere, the pressure gradients surrounding these cells are strongest between October and April. In terms of actual pressure, however, oceanic cells experience their highest pressure in summer, the belt being counterbalanced at low levels by thermal low pressure conditions over the continents. Their strength and persistence clearly mark them as the dominating factor controlling the position and activities of both the trades and the westerlies.

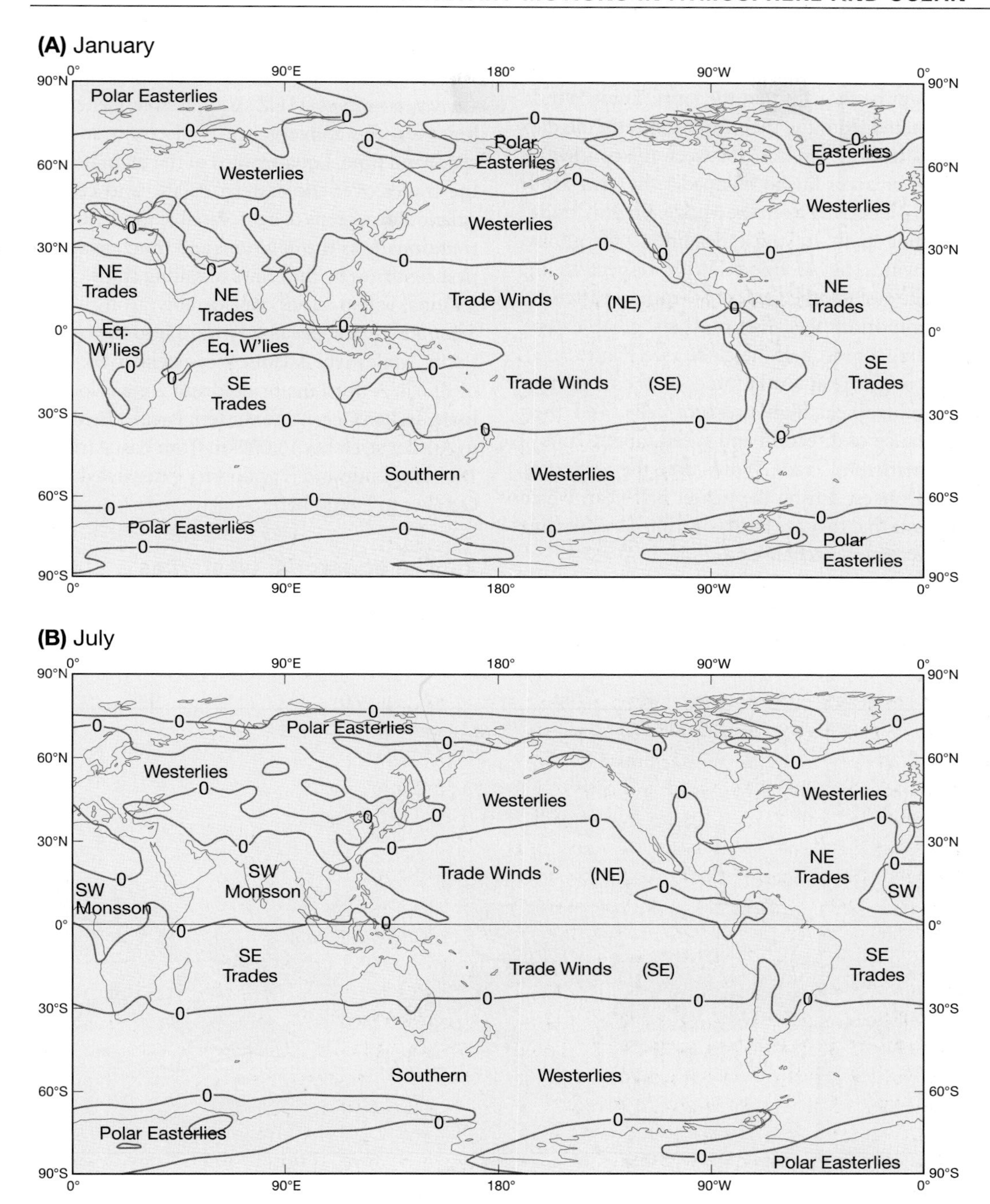

Figure 7.12 Generalized global wind zones at 1000mb in January (above) and July (below). The boundary of westerly and easterly zonal winds is the zero line. Across much of the central Pacific the Trade Winds are nearly zonal. Based on data for 1970–1999.

Source: NCEP/NCAR Reanalysis Data from the NOAA-CIRES Climate Diagnostics Center.

1 The Trade Winds

It might be thought that the term 'Trade Winds' originated from the their importance in the days of sail to fostering trade between the continents. However, according to Wikipedia, the term 'Trade Winds' is derived from the Middle English 'trade', meaning 'path' or 'track', leading to the phrase, 'the winds blows trade'. With respect to the climate system, the trades (or tropical easterlies) are important because of their great extent, affecting almost half the globe (see Figure 7.13). They originate at low latitudes on the margins of the subtropical high pressure cells, and their constancy of direction and speed (about 7m s^{-1}) is remarkable. Trade Winds, like the westerlies, are strongest during the winter half-year, which suggests that they are both controlled by the same fundamental mechanism.

The two Trade Wind systems tend to converge in the *Equatorial Trough* (of low pressure). Over the oceans, particularly the central Pacific, the convergence of these air streams is often pronounced and in this sector the term *Intertropical Convergence Zone* (ITCZ) is applicable. Generally, however, the convergence is discontinuous in space and time. Equatorward of the main belts of the trades over the eastern Pacific and eastern Atlantic are regions of light, variable winds, known traditionally as the *doldrums* and much feared in past centuries by the crews of sailing ships. Their seasonal extent varies considerably: from July to September they spread westward into the central Pacific while in the Atlantic they extend to the coast of Brazil. A third major doldrum zone is located in the Indian Ocean and western Pacific. In March to April it stretches 16,000km from East Africa to 180° longitude and is again very extensive during October to December.

2 The equatorial westerlies

In the summer hemisphere, and over continental areas especially, there is a narrow zone of generally

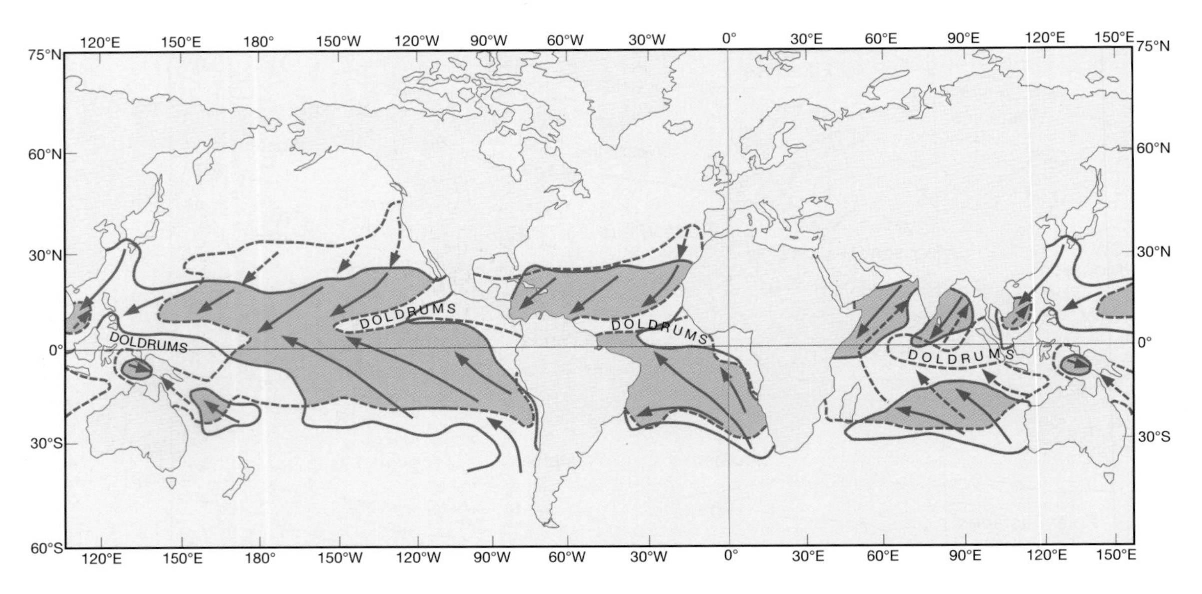

Figure 7.13 Map of the Trade Wind belts and the doldrums. The limits of the Trades – enclosing the area within which 50 percent of all winds are from the predominant quadrant – are shown by the solid (January) and the dashed (July) lines. The stippled area is affected by Trade Wind currents in both months. Schematic streamlines are indicated by the arrows – dashed (July) and solid (January, or both months).

Source: Based on Crowe (1949, 1950).

westerly winds intervening between the two Trade Wind belts (Figures 7.12 and 7.14). This westerly system is well marked over Africa and South Asia in the Northern Hemisphere summer, when thermal heating over the continents assists the northward displacement of the Equatorial Trough (see Figure 11.1). Over Africa, the westerlies reach 2–3km and over the Indian Ocean to 5–6km. In Asia, these winds are known as the 'Summer Monsoon', but this is now recognized to be a complex phenomenon, the cause of which is partly global and partly regional in origin (see Chapter 11C). The equatorial westerlies are not simply trades of the opposite hemisphere that recurve (due to the changed direction of the Coriolis deflection) on crossing the equator. There is *on average* a westerly component in the Indian Ocean at 2–3°S in June and July and at 2–3°N in December and January. Over the Pacific and Atlantic Oceans, the ITCZ does not shift sufficiently far from the equator to permit the development of this westerly wind belt.

3 The mid-latitude (Ferrel) westerlies

These are the winds of the mid-latitudes emanating from the poleward sides of the subtropical high pressure cell (see Figure 7.12). They are far more variable than the trades in both direction and intensity, since in these regions the path of air movement is frequently affected by transient cells of low and high pressure, which, although guided by the meandering long Rossby Waves discussed earlier, travel generally eastward. In addition, in the Northern Hemisphere the preponderance of land areas with their irregular relief and changing seasonal pressure patterns tend to obscure the generally westerly airflow. The Isles of Scilly, off southwest England, lying in the southwesterlies, record 46 percent of winds from between southwest and northwest, but fully 29 percent from the opposite sector, between northeast and southeast.

The westerlies of the Southern Hemisphere are stronger and more constant in direction than those of the Northern Hemisphere because the

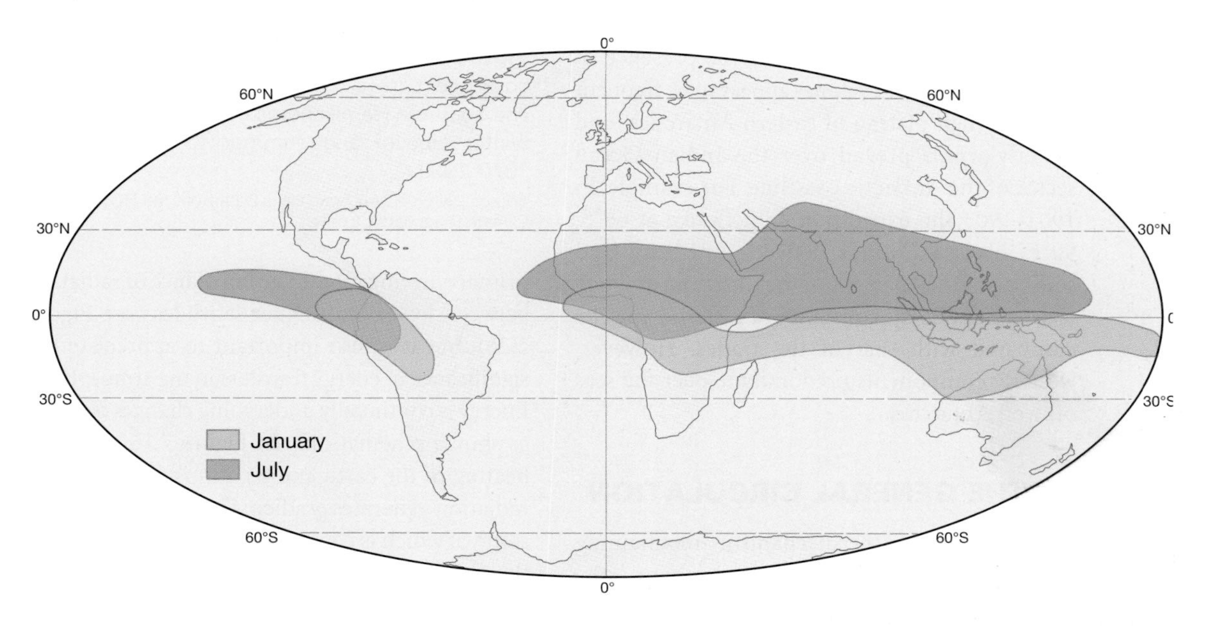

Figure 7.14 Distribution of the equatorial westerlies in any layer below 3km (about 10,000ft) for January and July.
Source: After Flohn in Indian Meteorological Department (1960).

broad expanses of ocean rule out the development of stationary pressure systems (Figure 7.15). Kerguelen Island (49°S, 70°E) has an annual frequency of 81 percent of winds from between southwest and northwest, and the comparable figure of 75 percent for Macquarie Island (54°S, 159°E) shows that this predominance is widespread over the southern oceans. However, the apparent zonality of the southern circumpolar vortex (see Figure 7.10) conceals considerable synoptic variability of wind velocity.

4 The polar easterlies

This term is applied to winds that occur between polar high pressure and subpolar low pressure. The polar high, as has already been pointed out, is by no means a quasi-permanent feature of the Arctic circulation. Easterly winds occur mainly on the poleward sides of depressions over the North Atlantic and North Pacific (Figure 7.12). If average wind directions are calculated for entire high-latitude belts there is found to be little sign of a coherent system of polar easterlies. The situation in high latitudes of the Southern Hemisphere is complicated by the presence of Antarctica, but anticyclones appear to be frequent over the high plateau of eastern Antarctica, and easterly winds prevail over the Indian Ocean sector of the Antarctic coastline. For example, in 1902–1903 the expedition ship *Gauss*, at 66°S, 90°E, observed winds between northeast and southeast for 70 percent of the time, and at many coastal stations the constancy of easterlies may be compared with that of the trades. However, westerly components predominate over the seas off west Antarctica.

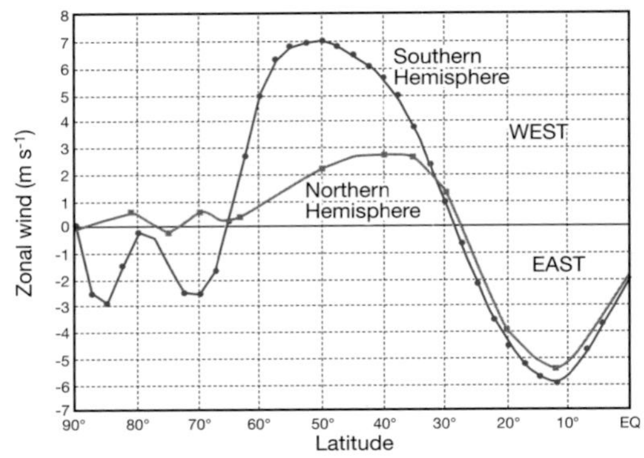

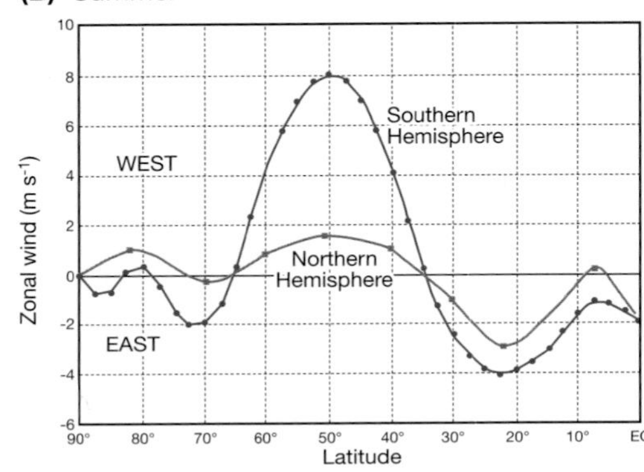

Figure 7.15 Profiles of the average west wind component (m s^{-1}) at sea level in the Northern and Southern Hemispheres during their respective winter (above) and summer (below) seasons, 1970–1999.

Source: NCEP/NCAR Reanalysis Data from the NOAA-CIRES Climate Diagnostics Center.

C THE GENERAL CIRCULATION

We next consider the mechanisms maintaining the *general circulation* of the atmosphere – the large-scale patterns of wind and pressure that persist throughout the year or recur seasonally. Reference has already been made to one of the primary driving forces, the imbalance of radiation between lower and higher latitudes (see Figure 2.26), but it is also important to appreciate the significance of energy transfers in the atmosphere. Energy is continually undergoing changes of form, as shown schematically in Figure 7.16. Unequal heating of the earth and its atmosphere by solar radiation generates gradients in potential energy, some of which is converted into kinetic energy by the rising of warm air and the sinking of cold air. Ultimately, the kinetic energy of atmospheric motion on all scales is dissipated by friction and small-scale turbulent eddies (i.e., internal

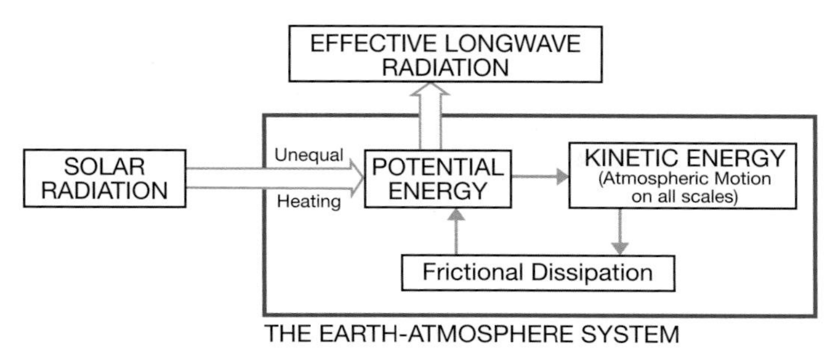

Figure 7.16 Schematic changes of energy involving the earth–atmosphere system.

viscosity). In order to maintain the general circulation, the rate of generation of kinetic energy must obviously balance its rate of dissipation. These rates are estimated to be about 2W m^{-2}, which amounts to only 1 percent of the average global solar radiation absorbed at the surface and in the atmosphere. In other words, the atmosphere is a highly inefficient heat engine (see Chapter 2E).

A second controlling factor is the angular momentum of the earth and its atmosphere. This is the tendency for the earth's atmosphere to move, with the earth, around the axis of rotation. Angular momentum is proportional to the rate of spin (that is, the angular velocity) and the square of the distance of the air parcel from the axis of rotation. With a uniformly rotating earth and atmosphere, the total angular momentum must remain constant (in other words, there is a *conservation of angular momentum*). If, therefore, a large mass of air changes its position on the earth's surface such that its distance from the axis of rotation also changes, then its angular velocity must change in a manner so as to allow the angular momentum to remain constant. Naturally, absolute angular momentum is high at the equator furthest from the axis of rotation (see Note 3) and decreases with latitude to become zero at the poles (that is, the axis of rotation), so air moving polewards tends to acquire progressively higher eastward velocities. For example, air traveling from 42° to 46° latitude and conserving its angular momentum would increase its speed relative to the earth's surface by 29m s_{-1}. This is the same principle that causes an ice skater to spin faster when the arms are progressively drawn into the body. In practice, the increase of air-mass velocity is countered or masked by the other forces affecting air movement (particularly friction), but there is no doubt that many of the important features of the general atmospheric circulation result from this poleward transfer of angular momentum.

The necessity for a poleward momentum transport is readily appreciated in terms of the maintenance of the mid-latitude westerlies (Figure 7.15). These winds continually impart westerly (eastward) relative momentum to the earth by friction, and it has been estimated that they would cease altogether due to this frictional dissipation of energy in little over a week if their momentum were not continually replenished from elsewhere. In low latitudes, the extensive tropical easterlies are gaining westerly relative momentum by friction as a result of the earth rotating in a direction opposite to their flow (see Note 4). This excess is transferred poleward with the maximum transport occurring, significantly, in the vicinity of the mean subtropical jet stream at about 250mb at 30°N and 30°S.

1 Circulations in the vertical and horizontal planes

There are two possible ways in which the atmosphere can transport heat and momentum. One is by circulation in the vertical plane as indicated in Figure 7.17, which shows three meridional cells in each hemisphere. The low-latitude *Hadley cells* were considered to be analogous to the convective circulations set up when a pan of water is heated over a flame and are referred

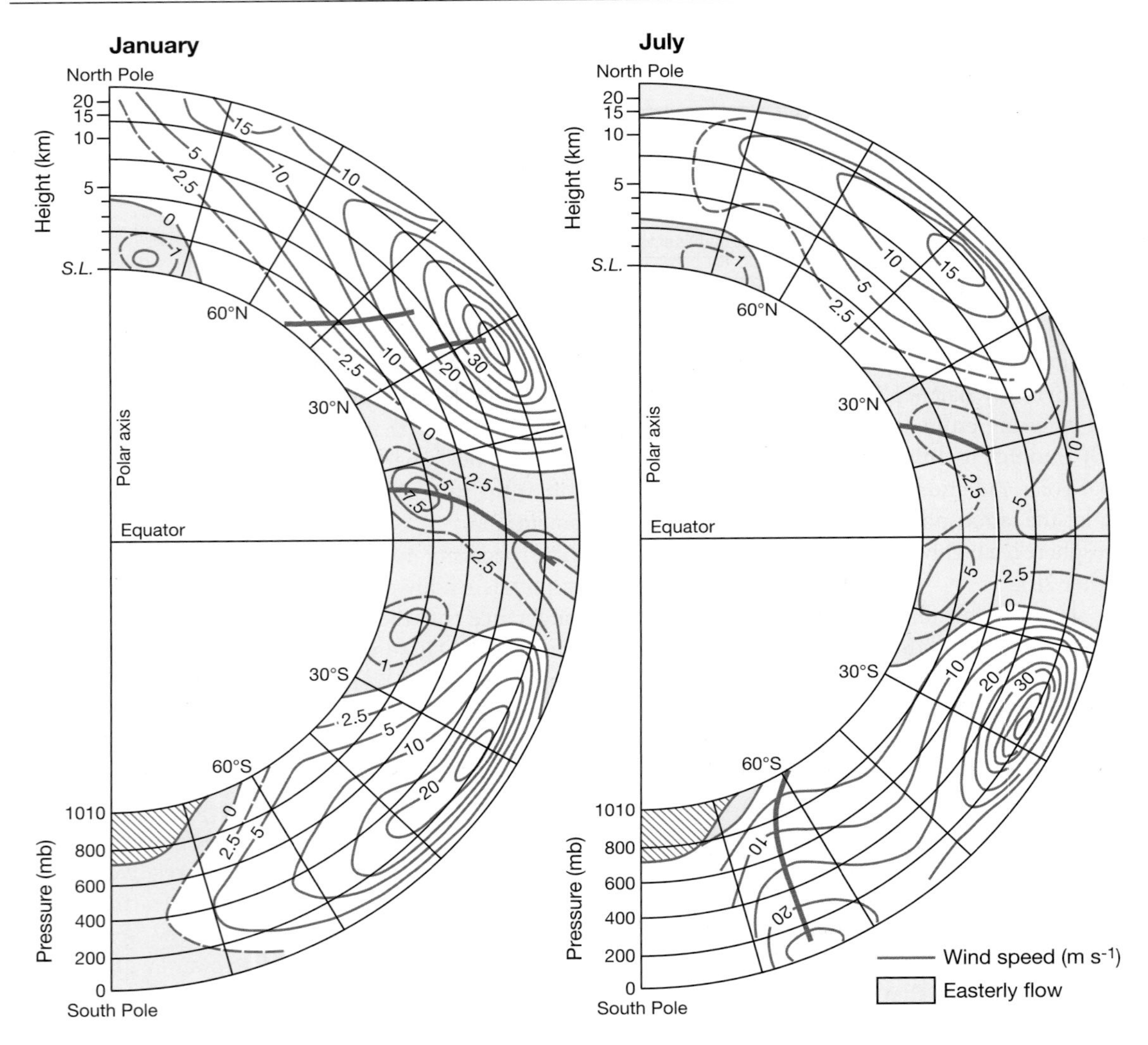

Figure 7.17 Mean zonal wind speeds (m s^{-1}) calculated for each latitude and for elevations up to more than 20km. Note the weak mean easterly flow at all levels in low latitudes dominated by the Hadley cells, and the strong upper westerly flow in mid-latitudes, localized into the subtropical jet streams.

Source: After Mintz; from Henderson-Sellers and Robinson (1986).

to as *thermally direct* cells. Warm air near the equator was thought to rise and generate a low-level flow towards the equator, the earth's rotation deflecting these currents, which thus form the northeast and southeast trades. This explanation was put forward by G. Hadley in 1735, although in 1856 W. Ferrel pointed out that the conservation of angular momentum would be a more effective factor in causing easterlies, because the Coriolis force is small in low latitudes. Poleward counter-currents aloft would complete the low-latitude cell, according to the above scheme, with the air sinking at about 30° latitude as it is cooled by radiation. However, this scheme is not entirely correct. The atmosphere does not have a simple heat source at the equator, the trades are not continuous around the globe (see Figure 7.13) and poleward upper flow occurs mainly at the western ends of the subtropical high pressure cells aloft (see Figures 7.4 and 7.10).

Figure 7.18 shows another thermally direct (polar) cell in high latitudes with cold, dense air flowing out from a polar high pressure. The reality of this is doubtful, but it is in any case of limited importance to the general circulation in view of the small mass involved. It is worth noting that a single direct cell in each hemisphere is not possible, because the easterly winds near the surface would slow down the earth's rotation. On average the atmosphere must rotate with the earth, requiring a balance between easterly and westerly winds over the globe.

The mid-latitude *Ferrel cell* in Figure 7.18 is thermally indirect and would need to be driven by the other two. Momentum considerations indicate the necessity for upper easterlies in such a scheme, yet aircraft and balloon observations during the 1930s–1940s demonstrated the existence of strong westerlies in the upper troposphere (see A.3, this chapter). Rossby modified the three-cell model to incorporate this fact, proposing that westerly momentum was transferred to mid-latitudes from the upper branches of the cells in high and low latitudes. Troughs and ridges in the upper flow could, for example, accomplish such horizontal mixing.

These views underwent radical amendment from about 1948 onward. The alternative means of transporting heat and momentum – by horizontal circulations – had been suggested in the 1920s by A. Defant and H. Jeffreys but could not be tested until adequate upper-air data became available. Calculations for the Northern Hemisphere by V. P. Starr and R. M. White at the Massachusetts Institute of Technology showed that in mid-latitudes horizontal cells transport most of the required heat and momentum poleward. This operates through (1) the mechanism of the quasi-stationary highs and (2) the traveling highs and lows near the surface, both acting in conjunction with their related wave patterns aloft. The former is referred to as the standing waves and the latter as transient eddies. The importance of such horizontal eddies for energy transport is

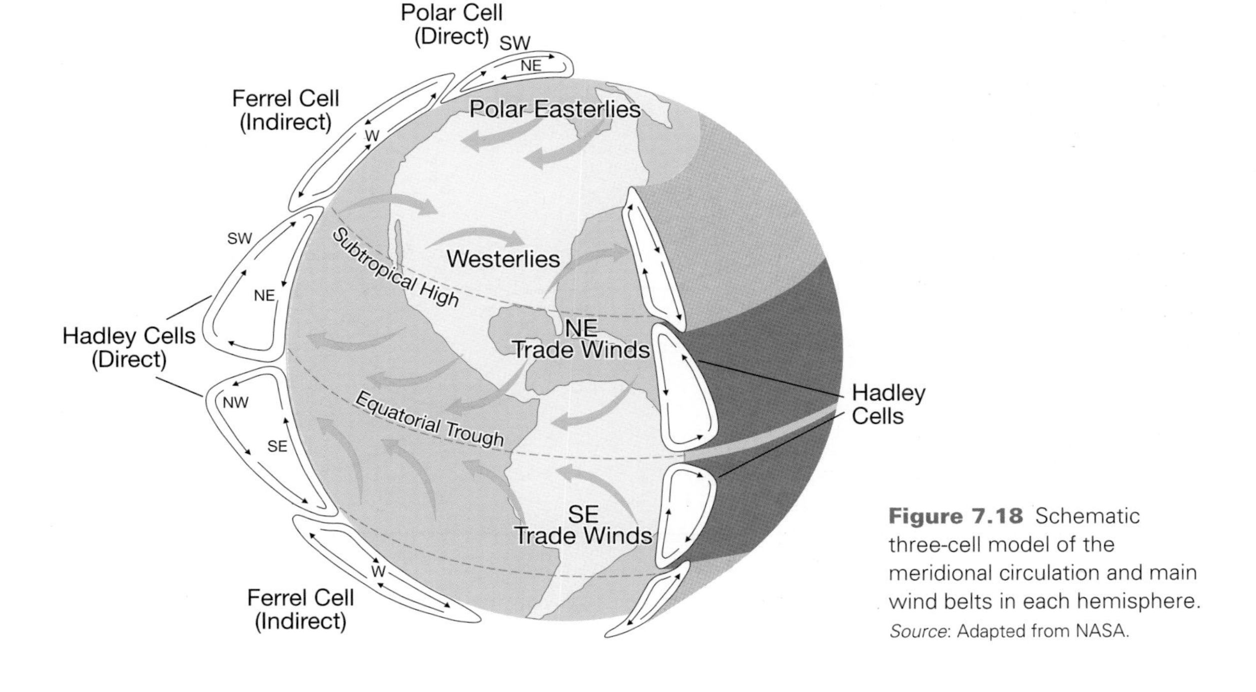

Figure 7.18 Schematic three-cell model of the meridional circulation and main wind belts in each hemisphere.
Source: Adapted from NASA.

shown in Figure 7.19 (see also Figure 3.28B). The modern concept of the general circulation therefore views the energy of the zonal winds as being derived from horizontal waves, not from meridional circulations. In lower latitudes, however, eddy transports are insufficient to account for the total energy transport required for energy balance. For this reason, the mean Hadley cell is a major feature of current representations of the general circulation, as shown in Figure 7.20. The low-latitude circulation is recognized as being complex. In particular, vertical heat transport in the Hadley cell is effected by giant cumulonimbus clouds in disturbance systems associated with the Equatorial Trough (of low pressure), which is located on average at 5°S in January and at 10°N in July (see Figure 11.1). The Hadley cell of the winter hemisphere is by far the most important, since it gives rise to low-level transequatorial flow into the summer hemisphere. The traditional model of global circulation with twin cells, symmetrical around the equator, is found only in spring/autumn.

Longitudinally, the Hadley cells are linked with the monsoon regimes of the summer hemisphere. Rising air over South Asia (and also South America and Indonesia) is associated with east–west (zonal) outflow, and these systems are known as *Walker circulations* (pp. 375–380). The poleward return transport of the meridional Hadley cells takes place in troughs that extend into low latitudes from the mid-latitude westerlies. This tends to occur at the western ends of the upper tropospheric subtropical high pressure cells. Horizontal mixing predominates in middle and high latitudes, although it is also thought that

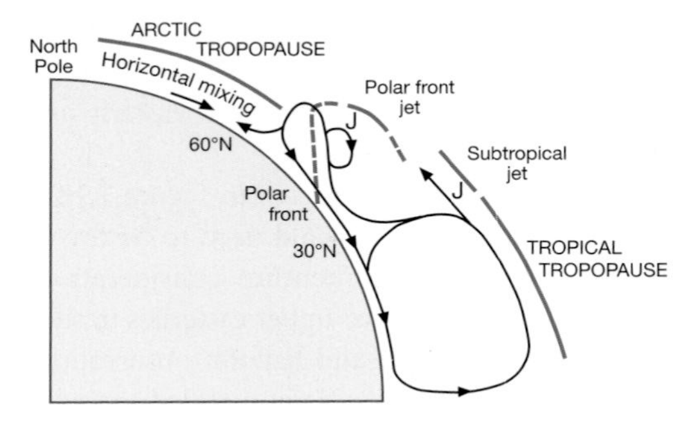

Figure 7.20 General meridional circulation model for the Northern Hemisphere in winter.
Source: After Palmén, 1951; from Barry (1967).

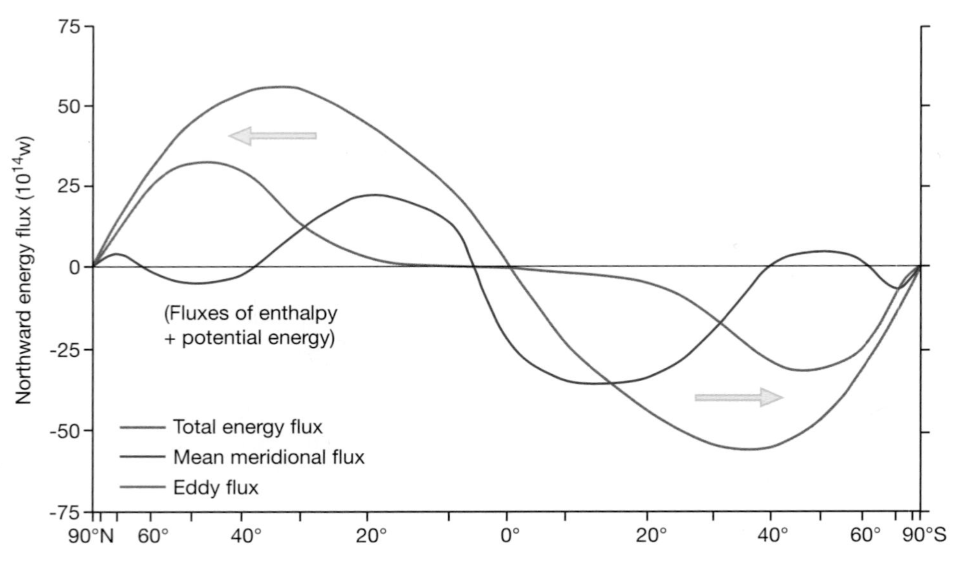

Figure 7.19 The poleward transport of energy, showing the importance of horizontal eddies in mid-latitudes.

there is a weak indirect mid-latitude cell in much reduced form (Figure 7.20). The relationship of the jet streams to regions of steep meridional temperature gradient has already been noted (see Figure 7.7). A complete explanation of the two wind maxima and their role in the general circulation is still lacking, but they undoubtedly form an essential part of the story.

In the light of these theories, the origin of the subtropical anticyclones that play such an important role in the world's climates may be re-examined. Their existence has been variously ascribed to: (1) the piling up of poleward-moving air as it is increasingly deflected eastward through the earth's rotation and the conservation of angular momentum; (2) the sinking of poleward currents aloft by radiational cooling; (3) the general necessity for high pressure near 30° latitude separating approximately equal zones of east and west winds; or to combinations of such mechanisms. An adequate theory must account not only for their permanence but also for their cellular nature and the vertical inclination of the axes. The preceding discussion shows that ideas of a simplified Hadley cell and momentum conservation are only partially correct. Moreover, recent studies rather surprisingly show no relationship, on a seasonal basis, between the intensity of the Hadley cell and that of the subtropical highs. Descent occurs near 25°N in winter, whereas North Africa and the Mediterranean are generally driest in summer, when the vertical motion is weak.

Two new ideas have recently been proposed (Figure 7.21). One suggests that the low-level subtropical highs in the North Pacific and North Atlantic in summer are remote responses to stationary planetary waves generated by heat sources over Asia. In contrast to this view of eastward downstream wave propagation, another model proposes regional effects from the heating over the summer monsoon regions of India, West Africa and southwestern North America that act upstream on the western and northern margins of these heat sources. The Indian monsoon heating leads to a vertical cell with descent over the eastern Mediterranean, eastern Sahara Desert and the Kyzylkum–Karakum desert. However, while the ascending air originates in the tropical easterlies, Rossby waves in the mid-latitude westerlies are thought to be the source of the descending air and this may provide a link with the first mechanism. Neither of these arguments addresses the winter subtropical anticyclones. Clearly, these features await a definitive and comprehensive explanation.

It is probable that the high-level anticyclonic cells that are evident on synoptic charts (these tend to merge on mean maps) are related to anticyclonic eddies that develop on the equatorward side of jet streams. Theoretical and observational studies show that, as a result of the latitudinal variation of the Coriolis parameter, cyclones in the westerlies tend to move poleward and anticyclonic cells equatorward. Hence the subtropical anticyclones are constantly regenerated. There is a statistical relationship between the latitude of the subtropical highs and the mean meridional temperature gradient (see Figure 7.11); a stronger gradient causes an equatorward shift of the high pressure, and vice versa. This shift is evident on a seasonal basis. The cellular pattern at the surface clearly reflects the influence of heat sources. The cells are stationary and elongated north–south over the Northern Hemisphere oceans in summer, when continental heating creates low pressure and also the meridional temperature gradient is weak. In winter, on the other hand, the zonal flow is stronger in response to a greater meridional temperature gradient, and continental cooling produces east–west elongation of the cells. Undoubtedly, surface and high-level factors reinforce one another in some sectors and tend to cancel out in others.

Just as Hadley circulations represent major meridional (i.e., north–south) components of the atmospheric circulation, so Walker circulations represent the large-scale zonal (i.e., east–west) components of tropical airflow. These zonal circulations are driven by major east–west pressure gradients that are set up by differences in

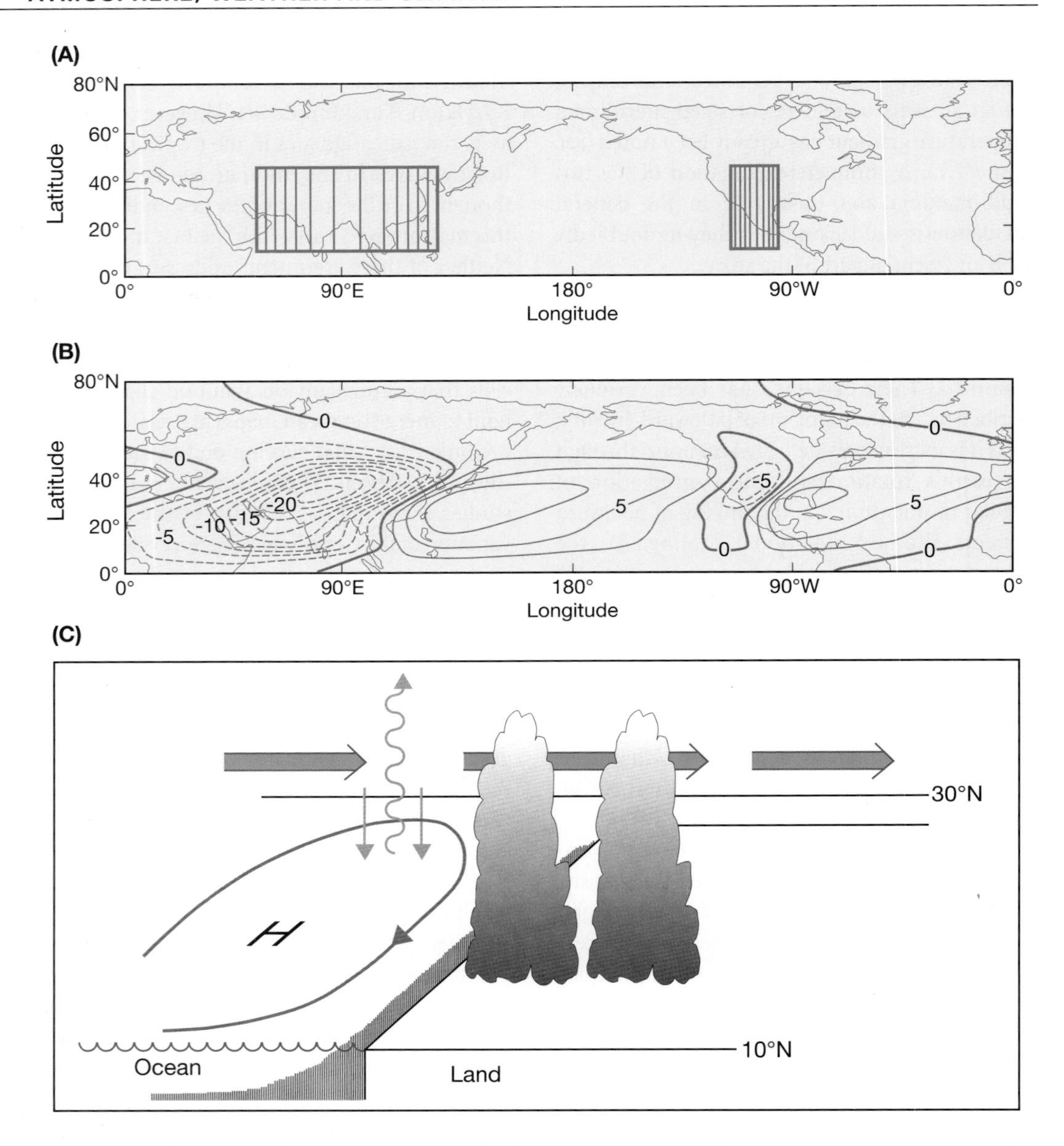

Figure 7.21 Schematic illustrations of suggested processes that form/maintain the northern subtropical anticylones in summer. A: Boxes where summer heat sources are imposed in the atmospheric model. B: Pattern of resultant stationary planetary waves (solid/dashed lines denote positive/negative height anomalies (Chen *et al.*, 2001). C: Schematic of the circulation elements proposed by Hoskins (1996); monsoon heating over the continents with descent west and poleward where there is interaction with the westerlies. The descent leads to enhanced radiative cooling acting as a positive feedback and to equatorward motion; the latter drives Ekman ocean drift and upwelling.

Sources: From Chen *et al.* (2001, *J. Atmos. Sci. 58*, p. 1832, fig. 8(a)), and from Hoskins (1996, *Bull. Amer. Met. Soc. 11*, p. 1291, fig.5). Courtesy of the American Meteorological Society.

vertical motion. On one hand, air rises over heated continents and the warmer parts of the oceans; on the other, air subsides over cooler parts of the oceans, over continental areas where deep high pressure systems have become established, and in association with subtropical high pressure cells. Sir Gilbert Walker first identified these circulations in 1922–1923 through his discovery of an inverse correlation between pressure over the eastern Pacific Ocean and Indonesia. The strength and phase of this so-called *Southern Oscillation* is commonly measured by the pressure difference between Tahiti (18°S, 150°W) and Darwin, Australia (12°S, 130°E). The Southern Oscillation Index (SOI) has two extreme phases (Figure 7.22):

- *positive* when there is a strong high pressure in the southeast Pacific and a low centered over Indonesia with ascending air and convective precipitation;
- *negative* (or low) when the area of low pressure and convection is displaced eastward towards the Date Line.

Positive (negative) SOI implies strong easterly trade winds (low-level equatorial westerlies) over the central–western Pacific. These Walker circulations are subject to fluctuations in which an oscillation (known as the El Niño–Southern Oscillation: or ENSO) between high phases (i.e., non-ENSO events) and low phases (i.e.,

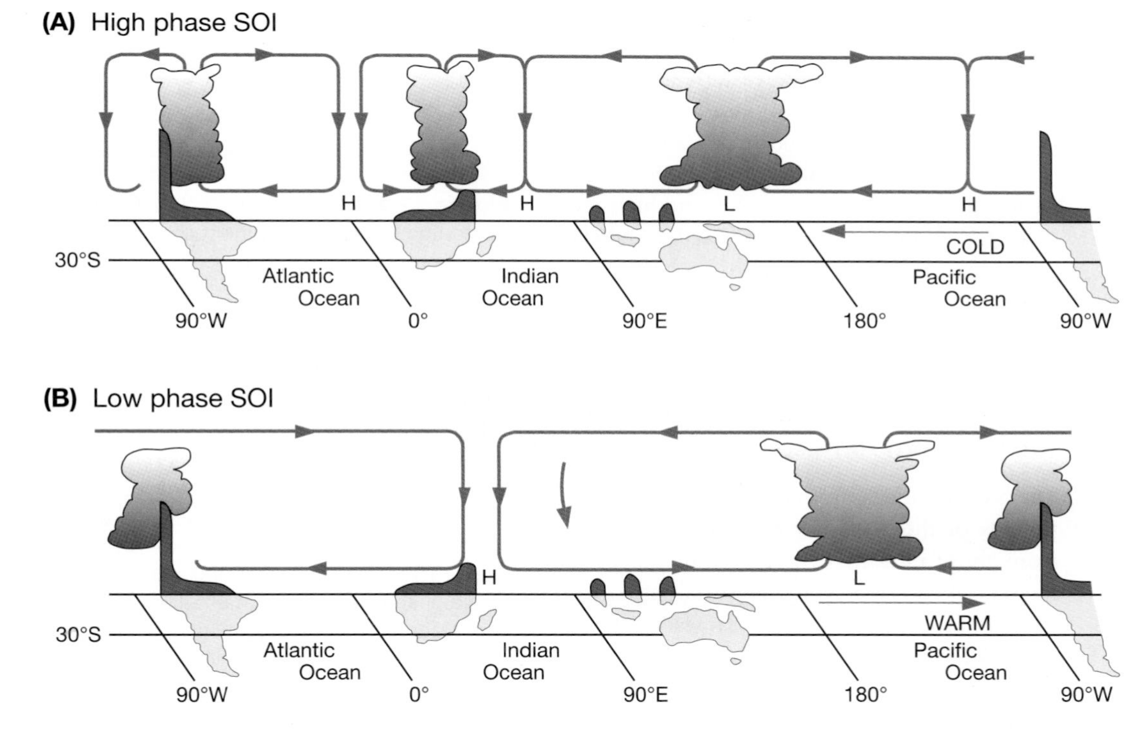

Figure 7.22 Schematic cross-sections of the Walker circulation along the equator (based on computations of Y. M. Tourre) during the high (A) and low (B) phases of the Southern Oscillation (SO). The high (low) phases correspond to non-ENSO (ENSO) patterns (see text). In the high phase there is rising air and heavy rains over the Amazon basin, central Africa and Indonesia-western Pacific. In the low phase (ENSO 1982–1983) pattern the ascending Pacific branch is shifted east of the Date Line and elesewhere convection is suppressed due to subsidence. The shading indicates the topography in exaggerated vertical scale.

Source: Based on K.Wyrtki (1985). By permission of the World Meteorological Organization.

ENSO events) is the most striking (see Chapter11G.1):

1 *High phase* (Figure 7.22A). This features four major zonal cells involving rising low pressure limbs and accentuated precipitation over Amazonia, central Africa and Indonesia/India; and subsiding high pressure limbs and decreased precipitation over the eastern Pacific, South Atlantic and western Indian Ocean. During this phase, low-level easterlies strengthen over the Pacific and subtropical westerly jet streams in both hemispheres weaken, as does the Pacific Hadley cell.
2 *Low phase* (Figure 7.22B). This phase has five major zonal cells involving rising low pressure limbs and accentuated precipitation over the South Atlantic, the western Indian Ocean, the western Pacific and the eastern Pacific; and subsiding high pressure limbs and decreased precipitation over Amazonia, central Africa, Indonesia/India and the central Pacific. During this phase, low-level westerlies and high-level easterlies dominate over the Pacific, and subtropical westerly jet streams in both hemispheres intensify, as does the Pacific Hadley cell.

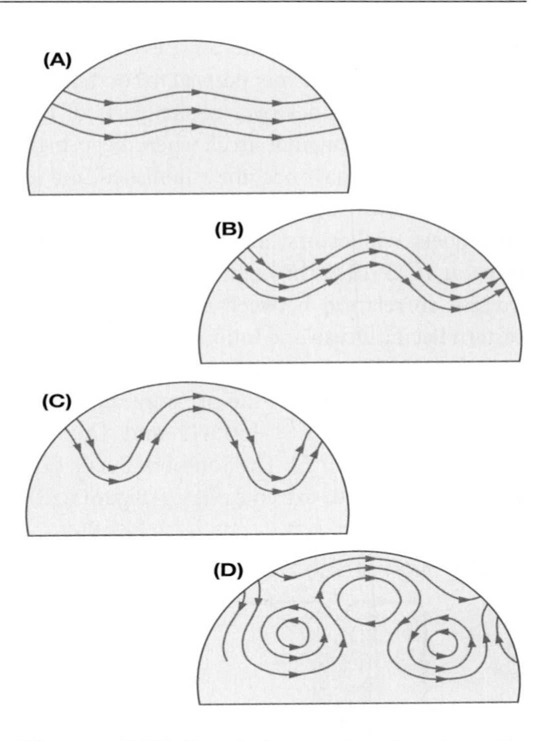

Figure 7.23 The index cycle. A schematic illustration of the development of cellular patterns in the upper westerlies, usually occupying three to eight weeks and being especially active in February and March in the Northern Hemisphere. Statistical studies indicate no regular periodicity in this sequence. A: High zonal index. The jet stream and the westerlies lie north of their mean position. The westerlies are strong, pressure systems have a dominantly east–west orientation, and there is little north–south air-mass exchange. B and C: The jet expands and increases in velocity, undulating with increasingly larger oscillations. D: low zonal index. The latter is associated with a complete breakup and cellular fragmentation of the zonal westerlies, formation of stationary deep occluding cold depressions in lower mid-latitudes and deep warm blocking anticyclones at higher latitudes. This fragmentation commonly begins in the east and extends westward at a rate of about 60° of longitude per week.

Source: After Namias; from Haltiner and Martin (1957).

2 Variations in the circulation of the Northern Hemisphere

The pressure and contour patterns during certain periods of the year may be radically different from those indicated by the mean maps (see Figure 7.4). Several kinds of variability are of special importance. On the largest scale are changes in the strength of the hemispheric zonal westerly circulation over a period of weeks. Important variations on more regional scales include oscillations in pressure over the North Atlantic and North Pacific.

Zonal index variations

Variations of three to eight weeks' duration are observed in the strength of the zonal westerlies,

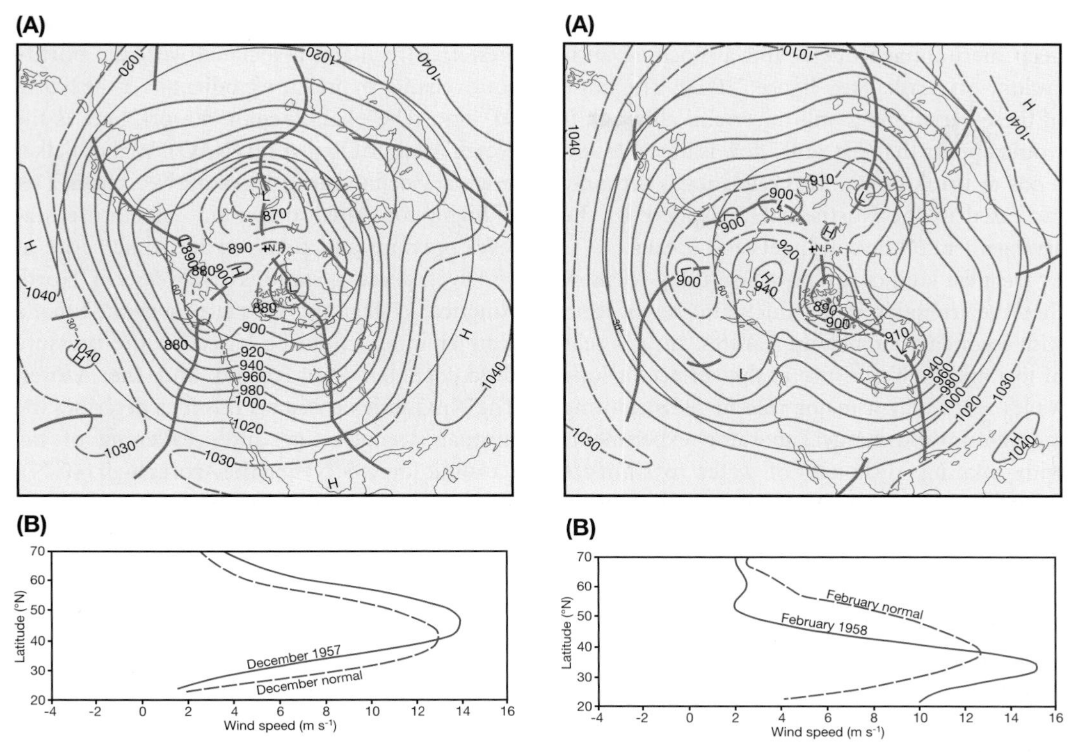

Figure 7.24 Above: Mean 700mb contours (in tens of feet) for December 1957, showing a fast, westerly, small-amplitude flow typical of a high zonal index. Below: Mean 700mb zonal wind speed profiles (m s^{-1}) in the Western Hemisphere for December 1957, compared with those of a normal December. The westerly winds were stronger than normal and displaced to the north.

Source: After Dunn (1957).

Figure 7.25 Above: Mean 700mb contours (in tens of feet) for February 1958. Below: Mean 700mb zonal wind speed profiles (m s^{-1}) in the Western Hemisphere for February 1958, compared with those of a normal February. The westerly winds were stronger than normal at low latitudes, with a peak at about 33°N.

Source: After Klein (1958).

averaged around the hemisphere. They are rather more noticeable in the winter months, when the general circulation is strongest. The nature of the changes is illustrated schematically in Figure 7.23. The mid-latitude westerlies develop waves, and the troughs and ridges become accentuated, ultimately splitting up into a cellular pattern with pronounced meridional flow at certain longitudes. The strength of the westerlies between 35° and 55°N is termed the *zonal index*; strong zonal westerlies are representative of a high index, and marked cellular patterns occur with a low index.

A relatively low index may also occur if the westerlies are well south of their usual latitudes and, paradoxically, such expansion of the zonal circulation pattern is associated with strong westerlies in lower latitudes than usual. Figures 7.24 and 7.25 illustrate the mean 700mb contour patterns and zonal wind speed profiles for two contrasting months. In December 1957 the westerlies were stronger than normal north of 40°N, and the troughs and ridges were weakly developed, whereas in February 1958 there was a low zonal index and an expanded circumpolar vortex, giving rise to strong low-latitude westerlies. The

700mb pattern shows very weak subtropical highs, deep meridional troughs and a blocking anticyclone off Alaska (see Figure 7.24D). The cause of these variations is still uncertain, although it would appear that fast zonal flow is unstable and tends to break down. This tendency is certainly increased in the Northern Hemisphere by the arrangement of the continents and oceans.

Detailed studies are now beginning to show that the irregular index fluctuations, together with secondary circulation features, such as cells of low and high pressure at the surface or long waves aloft, play a major role in redistributing momentum and energy. Laboratory experiments with rotating 'dishpans' of water to simulate the atmosphere, and computer studies using numerical models of the atmosphere's behavior, demonstrate that a Hadley circulation cannot provide an adequate mechanism for transporting heat poleward. In consequence, the meridional temperature gradient increases and eventually the flow becomes unstable in the Hadley mode, breaking down into a number of cyclonic and anticyclonic eddies. This phenomenon is referred to as *baroclinic instability*. In energy terms, the potential energy in the zonal flow is converted into potential and kinetic energy of eddies. It is also now known that the kinetic energy of the zonal flow is derived *from* the eddies, the reverse of the classical picture, which viewed the disturbances within the global wind belts as superimposed detail. The significance of atmospheric disturbances and the variations of the circulation are becoming increasingly evident. The mechanisms of the circulation are, however, greatly complicated by numerous interactions and *feedback* processes, particularly those involving the oceanic circulation discussed below.

North Atlantic Oscillation

The relative strength of the Icelandic Low and Azores High was first observed to fluctuate on annual to decadal scales by Sir Gilbert Walker in the 1920s. Fifty years later, van Loon and Rogers discussed the related west–east 'see-saw' in winter temperatures between Western Europe and western Greenland associated with the north–south change in pressure gradient over the North Atlantic. The phenomenon at work here is the North Atlantic Oscillation (NAO). While Walker originally defined an index of the NAO from a set of highly correlated surface air temperature, sea-level pressure and precipitation time series at widely separated stations over eastern North America, Walker and Bliss later suggested that a simpler index could be based on the pressure difference between Iceland and the Azores. The NAO index based on this idea describes the mutual strengthening and weakening of the Icelandic low (65°N) and the Azores high (40°N). When both are strong (weak) the NAO is taken to be in its positive (negative) mode, or phase. While the NAO can be identified in all seasons, most research has focused on the winter season when it tends to have its strongest expressions. The relationship between the positive and negative modes of the NAO noted by Walker, and the associated temperature and other anomaly patterns, are shown in Plate 7.3 for two contrasting Januarys. When the two pressure cells are well developed as in January 1984, the zonal westerlies are strong. Western Europe has a mild winter, while the intense Icelandic Low gives strong northerly flow in Baffin Bay, low temperatures in western Greenland and extensive sea ice in the Labrador Sea. In the negative phase the cells are weak, as in January 1970, and opposite anomalies are formed. In extreme cases, pressure can be higher near Iceland than to the south, giving easterlies across Western Europe and the eastern North Atlantic.

Through the late 1990s and early 2000s, following the work of D. Thompson, considerable debate arose as to whether the NAO should be considered as part of a more general pressure (mass) oscillation between the north polar region and mid-latitudes, known variously as the Arctic Oscillation (AO) or Northern Annular Mode (NAM). Part of the argument for considering the NAM as the more fundamental mode is its

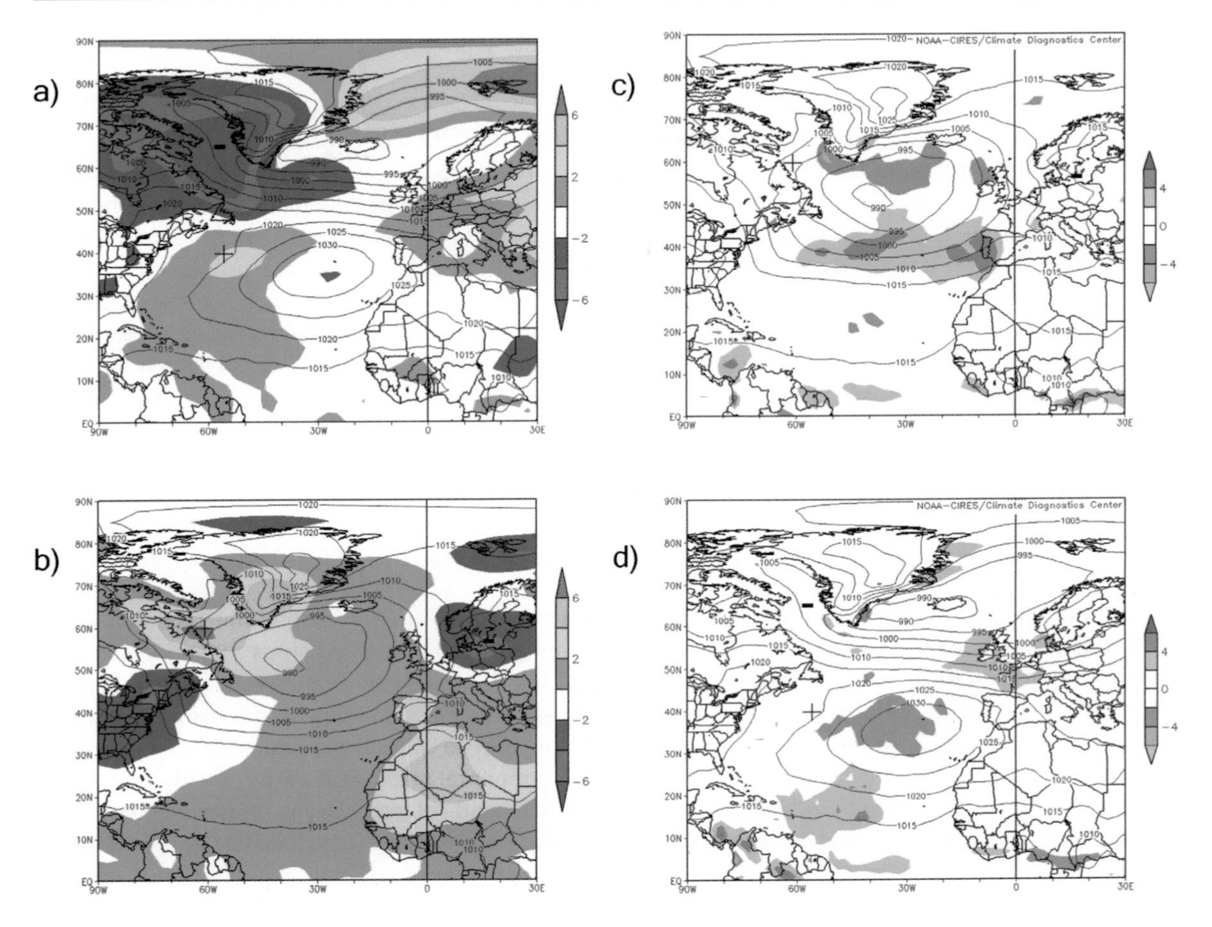

Plate 7.3 Illustration of the positive (January 1970) and negative (January 1984) phases of the North Atlantic Oscillation and their associated temperature (A and B) and precipitation (C and D) anomalies. MSL isobars at 5mb intervals; anomalies of temperature at 2°C intervals; and of daily precipitation rate at 2mm/day.

Source: Climate Diagnostics Center, NOAA, CIRES, Boulder, CO.

similarity to a corresponding oscillation of mass between the high and mid-latitudes of the Southern Hemisphere, known as the Antarctic Oscillation (AAO) or Southern Annular Mode (SAM). In comparison with the NAM, the mass oscillation associated with the SAM is much more zonally symmetric, or annular; that is, the mass oscillation is more clearly seen at all longitudes. The thinking is that if it were not for the distorting influences of orography and land–sea contrasts, the NAM would also show a fairly symmetric pattern rather than be dominated by variability in the Atlantic sector, with a much weaker center of action in the North Pacific. In other words, like the SAM, the NAM is 'inherently' a symmetric pattern, the departures from this symmetry due to the distorting effects mentioned above. Regardless, time series from the NAM and NAO. are highly correlated, and for many applications, may be viewed as different definitions of the same thing. The NAM and SAM patterns extend upward throughout the troposphere.

Based on sea-level pressure records, time series of the NAO index have been compiled back to about 1870. While the NAO has no preferred timescale of variability, a number of epochs can be defined. From about 1890–1900 it was in a most negative mode, followed by a mostly positive period from about 1900–1950. This was followed by a negative period from about 1960–1980, followed by a general rise into the mid-1990s. This recent rise fostered winters that, compared to normal, were warmer over much of northern Eurasia, and wetter (drier) conditions over Northern Europe-Scandinavia (southern Europe-Mediterranean), in association with a northward shift of storm tracks. Since the late 1990s, the NAO has regressed back to a generally more neutral phase.

The PDO, NPO and PNA

While ENSO has been previously discussed, is should be pointed out that ENSO can be linked in a variety of ways to 'ENSO-like' patterns for which multi-decadal signals are prominent. Climate signals are especially well expressed in the northwest Pacific, including Alaska. Of particular note is the Pacific Decadal Oscillation, or PDO, which has an index based on North Pacific sea surface temperatures. Its time series parallels the

SIGNIFICANT 20TH-CENTURY ADVANCE

7.3 Oceanographic observations

Meteorological and oceanographic measurements in the oceans are made on about 7000 ships of the Voluntary Observing Fleet and by moored or floating buoys. 'Selected ships' observe air and sea surface temperature, pressure and its tendency, wind, present and past weather, humidity, clouds and waves. Supplementary (and Auxillary) ships make the same observations omitting sea surface temperature, pressure tendency, waves (and clouds). The UK Meteorological Office operates seven moored deep-water buoys on the edge of the continental shelf west of the British Isles and two in the North Sea. There are similar buoys in the Pacific and Atlantic oceans off Canada and the USA, with about 65 such buoys operated by the USA. They measure pressure, air temperature, humidity, wind velocity, sea surface temperature, and wave height and period. Drifting buoys are now used worldwide. They measure barometric pressure and its tendency, and sea surface temperature while some also measure air temperature and wind velocity. The data are telemetered to the Argos satellites, which fix the buoys' position, and transmit them to Oslo, Toulouse and Søndre Strømfiord in West Greenland, as well as to Argos ground stations in the USA and France. Ocean currents are determined from the drift of ships where the difference between a ship's dead-reckoned position - determined from its previous position based on a navigational fix – and its actual location is ascribed solely to the effect of surface currents. They are also measured by current meters on surface and bottom moorings, that are installed by oceanographic research vessels and may operate for up to two years before being recovered. The temperature and salinity properties of the oceans are determined from conductivity, temperature and depth (CTD) sensors developed in the 1970s. These measure resistances of the sensors to variations in conductivity, temperature and pressure. Conductivity depends on both temperature and salinity, so with these measurements a temperature salinity profile is obtained and relayed to the research vessel. A global array of free-drifting profiling floats (Argo) that measure the temperature and salinity of the upper 2000m of the ocean began operations in 2000 and currently has 3000 floats in the world's oceans. The data are retrieved when the float periodically surfaces and are transmitted to receiving stations by satellite link.

dominant pattern of Pacific sea-level pressure variability. The basic relationship is that coooler (warmer) than average sea surface temperatures tend to occur during periods of lower (higher) than average sea-level pressure over the central North Pacific. The PDO is in turn related to the North Pacific Oscillation (NPO), which can be described from a simple index based on the area-weighted mean of sea-level pressure over the extratropical North Pacific. The PDO time series represents a good measure of the strength of the Aleutian Low. Since about 1976 the PDO has shown a general downward trend, meaning a stronger Aleutian Low, accompanied by stronger than normal westerly winds across the central North Pacific and enhanced southerly to southeasterly flow along the west coast of North America. In a larger context, variability in ENSO, the PDO and NPO link to variability in the so-called Pacific North American (PNA) teleconnection pattern. The PNA describes variations in the atmospheric longwave pattern spanning the equatorial Pacific through the northwest of North America and to the southeastern part of North America. The positive mode of the PNA is characterized by a strong Aleutian Low, a strong upper-air ridge along the west coast of Canada, and a concurrent strong trough over the southwestern United States.

D OCEAN STRUCTURE AND CIRCULATION

The oceans occupy 71 percent of the earth's surface, with over 60 percent of the global ocean area in the Southern Hemisphere. Three-quarters of the ocean area are between 3000 and 6000m deep, whereas only 11 percent of the land area exceeds 2000m altitude.

1 Above the thermocline

Vertical

The major atmosphere–ocean interactive processes (Figure 7.26) involve heat exchanges, evaporation, density changes and wind stress. The effect of these processes is to produce a vertical layering in the ocean that is of great climatic significance:

1. At the ocean surface, winds produce a *thermally mixed surface layer* averaging a few tens of meters deep poleward of latitude 60°, 400m at latitude 40° and 100–200m at the equator.
2. Below the relatively warm mixed layer is the *thermocline*, a layer in which temperature decreases and density increases (the *pycnocline*) markedly with depth. The thermocline layer, within which stable stratification tends to inhibit vertical mixing, acts as a barrier between the warmer surface water and the colder deep-layer water. In the open ocean between latitudes 60°N and 60°S the thermocline layer extends from depths of about 200m to a maximum of 1000m (at the equator from about 200 to 800m; at 40° latitude from about 400 to about 1100m). Poleward of 60° latitude, the colder deep-layer water approaches the surface. The location of the steepest temperature gradient is termed the *permanent thermocline*, which has a dynamically inhibiting effect in the ocean similar to that of a major inversion in the atmosphere. However, heat exchanges take place between the oceans and the atmosphere by turbulent mixing above the permanent thermocline, as well as by upwelling and downwelling. Below the surface mixed layer in the Arctic there is also a salinity gradient or *halocline*.

During spring and summer in the mid-latitudes, accentuated surface heating leads to the development of a *seasonal thermocline* occurring at depths of 50 to 100m. Surface cooling and wind mixing tend to destroy this layer in autumn and winter.

Below the thermocline layer is a *deep layer* of cold, dense water. Within this, water movements are mainly driven by density variations,

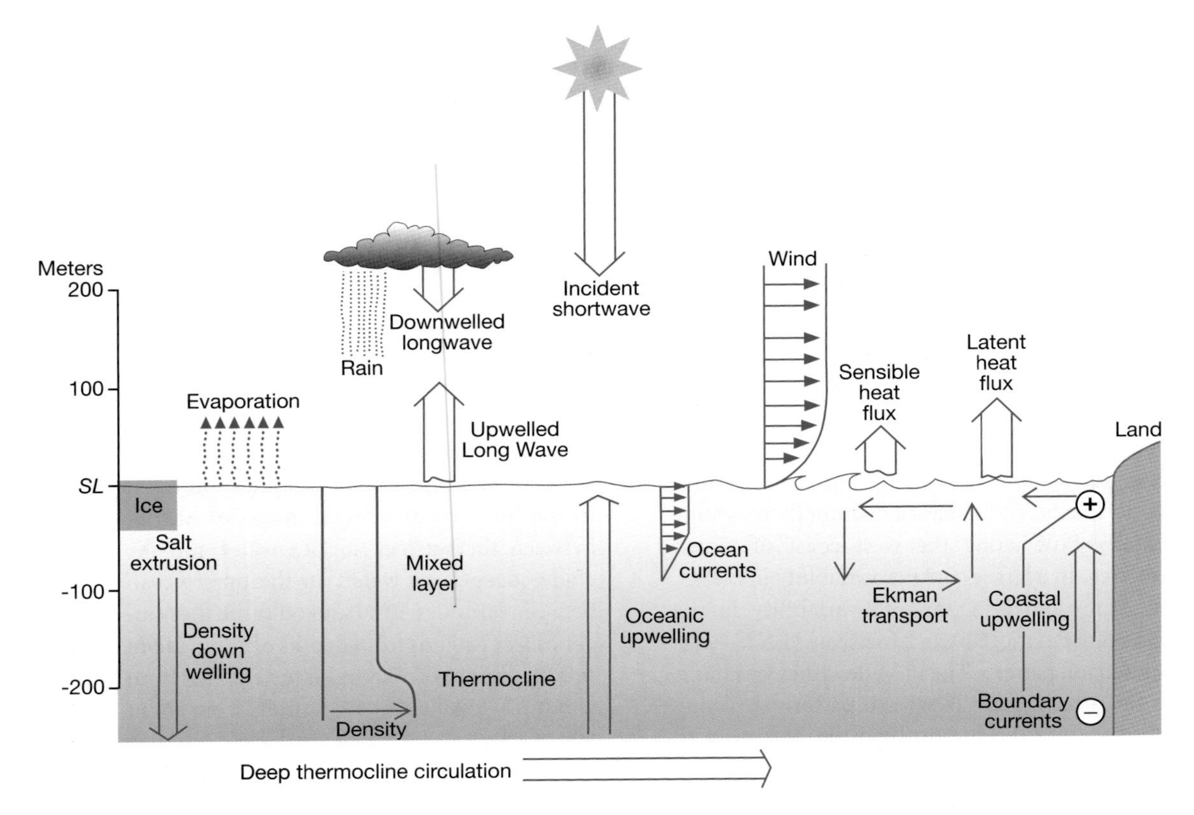

Figure 7.26 Generalized depiction of the major atmosphere–ocean interaction processes.
Source: Modified from NASA (n.d.). Courtesy NASA.

commonly due to salinity differences (i.e., a *thermohaline* mechanism).

In terms also of circulation the ocean may be viewed as consisting of a large number of layers: the topmost subject to wind stress, the next layer down to frictional drag by the layer above, and so on; all layers being acted on by the Coriolis force. The surface water tends to be deflected to the right (in the Northern Hemisphere) by an angle averaging some 45°from the surface wind direction and moving at about 3 percent of its velocity. This deflection increases with depth as the friction-driven velocity of the current decreases exponentially (Figure 7.27). On the equator where there is no Coriolis force, the surface water moves in the same direction as the surface wind. This theoretical Ekman spiral

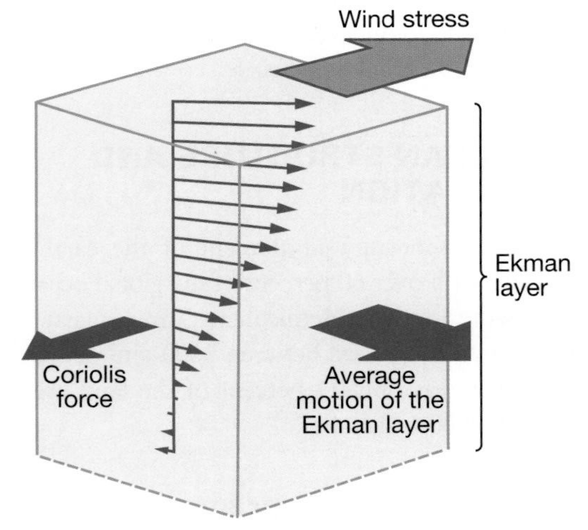

Figure 7.27 The Ekman ocean current pattern in the Northern Hemisphere.
Source: Open University 1989 from *Ocean Circulation*, Bearman (1989). Copyright © Butterworth-Heinemann, Oxford.

was developed under assumptions of idealized ocean depth, wind constancy, uniform water viscosity and constant water pressure at a given depth. This is seldom the case in reality, and under most oceanic conditions the thickness of the wind-driven Ekman layer is about 100 to 200m. North (south) of 30°N, the westerly (easterly) winds create a southward (northward) transport of water in the Ekman layer, giving rise to a convergence and sinking of water around 30°N, referred to as Ekman pumping.

Horizontal

General

Comparisons can be made between the structure and dynamics of the oceans and the atmosphere in respect of their behavior above the permanent thermocline and below the tropopause – their two most significant stabilizing boundaries. Within these two zones, fluid-like circulations are maintained by meridional thermal energy gradients, dominantly directed poleward (Figure 7. 28), and acted upon by the Coriolis force. Prior to the 1970s oceanography was studied in a coarsely averaged spatial–temporal framework similar to that applied in classical climatology. Now, however, its similarities with modern meteorology are apparent (Box 7.3). The major differences in behavior between the oceans and the atmosphere derive from the greater density and viscosity of ocean waters and the much greater frictional constraints placed on their global movement.

Many large-scale characteristics of ocean dynamics resemble features of the atmosphere. These include: the general circulation, major

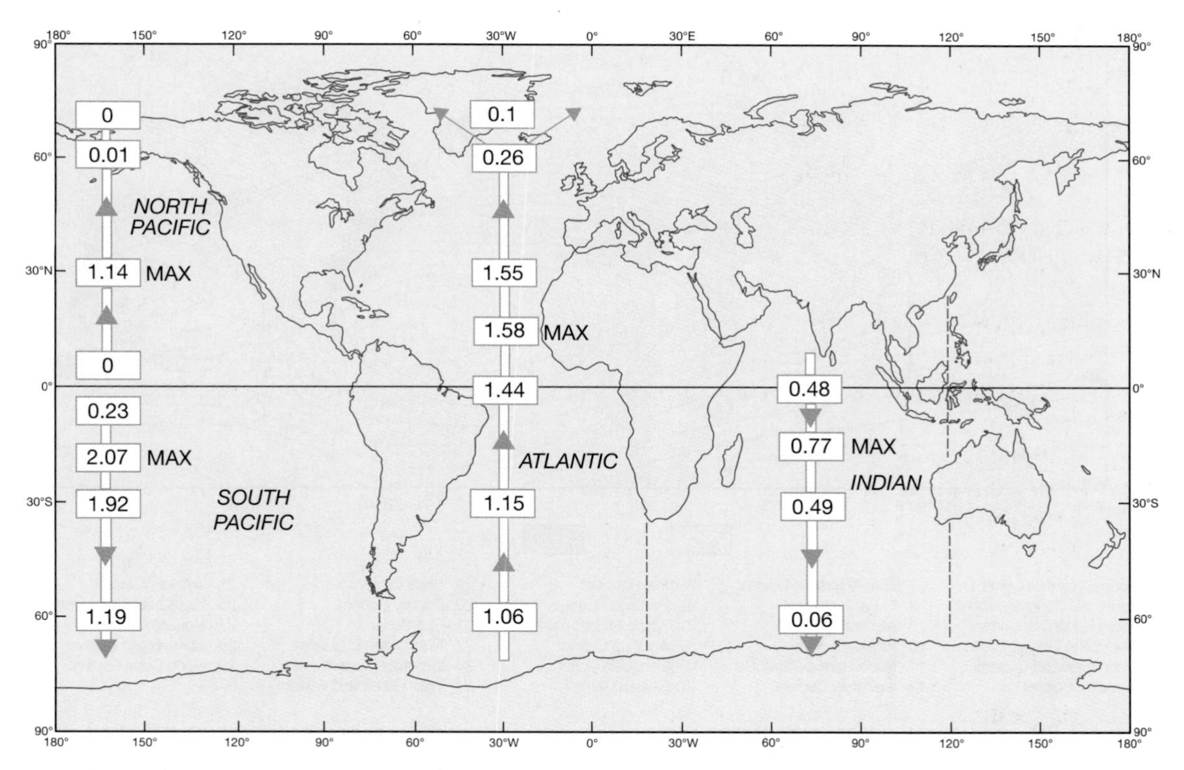

Figure 7.28 Mean annual meridional heat transport (10^{15}W) in the Pacific, Atlantic and Indian Oceans, respectively (delineated by the dashed lines). The latitudes of maximum transport are indicated.

Source: Hastenrath (1980). From *Journal of Physical Oceanography* by permission of the American Meteorological Society.

oceanic gyres (similar to atmospheric subtropical high pressure cells), major jet-like streams such as sections of the Gulf Stream (see Figure 7.29), large-scale areas of subsidence and uplift, the stabilizing layer of the permanent thermocline, boundary layer effects, frontal discontinuities created by temperature and density contrasts, and water mass ('mode water') regions.

Mesoscale characteristics that have atmospheric analogues are oceanic cyclonic and anticyclonic eddies, current meanders, cast-off ring vortices, jet filaments, and circulations produced by irregularities in the North Equatorial Current.

Macroscale

The most obvious feature of the surface oceanic circulation is the control exercised over it by the low-level planetary wind circulation, especially by the subtropical oceanic high pressure cells and the westerlies. The oceanic circulation also displays seasonal reversals of flow in the monsoonal regions of the northern Indian Ocean, off East Africa and off northern Australia (see Figure 7.29). As water moves meridionally, the conservation of angular momentum implies changes in relative vorticity (see pp. 153 and 181), with poleward-moving currents acquiring anticyclonic vorticity

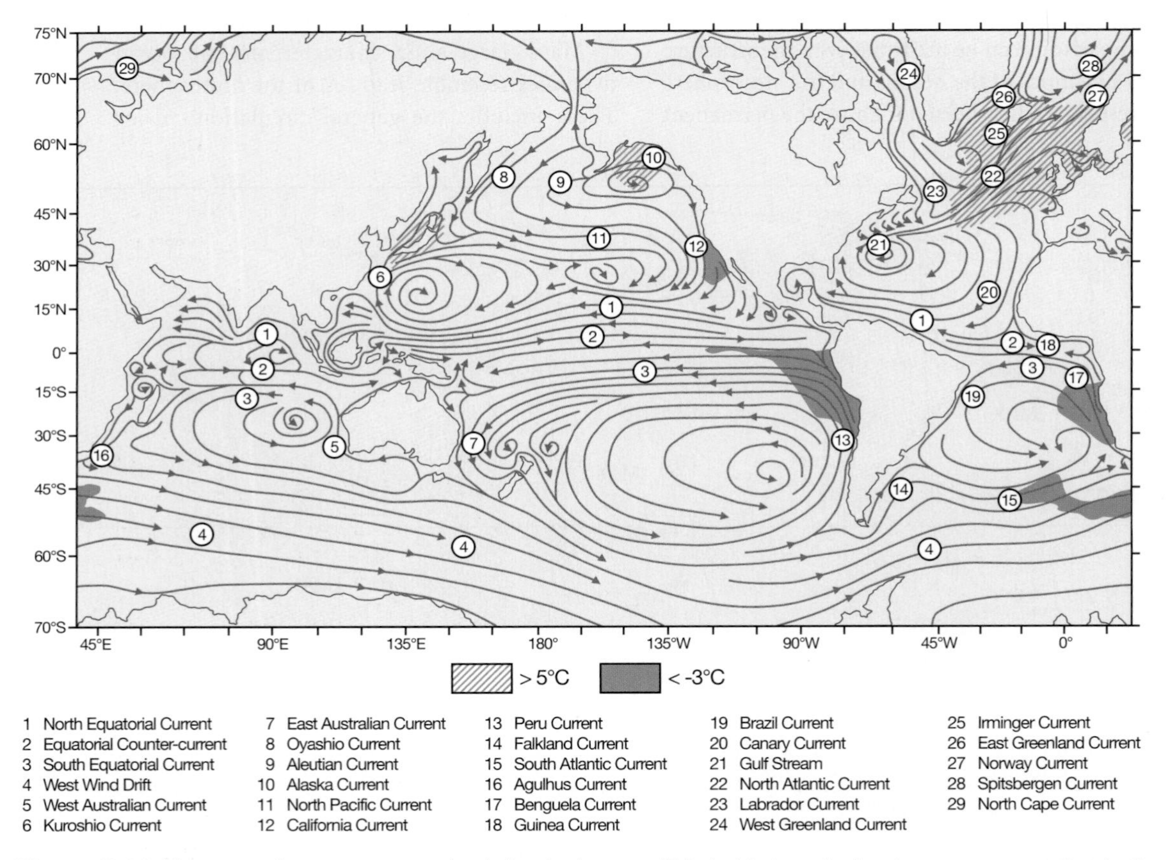

Figure 7.29 The general ocean current circulation in January. This holds broadly for the year, except that in the northern summer some of the circulation in the northern Indian Ocean is reversed by the monsoonal airflow. The shaded areas show mean annual anomalies of ocean surface temperatures (°C) of greater than +5°C and less than –3°C.

Sources: US Naval Oceanographic Office and Niiler (1992). Courtesy US Naval Oceanographic Office.

and equatorward-moving currents acquiring cyclonic vorticity.

The more or less symmetrical atmospheric subtropical high pressure cells produce oceanic gyres with centers displaced towards the west sides of the oceans in the Northern Hemisphere. The gyres in the Southern Hemisphere are more symmetrically located than those in the northern, due possibly to their connection with the powerful West Wind Drift. This results, for example, in the Brazil Current being not much stronger than the Benguela Current. The most powerful Southern Hemisphere current, the Agulhas, possesses nothing like the jet-like character of its northern counterparts.

Equatorward of the subtropical high pressure cells, the persistent Trade Winds generate the broad North and South Equatorial Currents (see Figure 7.29). On the western sides of the oceans, most of this water swings poleward with the airflow and thereafter increasingly comes under the influence of the Coriolis deflection and of the anticyclonic vorticity effect. However, some water tends to pile up near the equator on the western sides of oceans, partly because here the Ekman effect is virtually absent, with little poleward deflection and no reverse current at depth. To this is added some of the water that is displaced northward into the equatorial zone by the especially active subtropical high pressure circulations of the Southern Hemisphere. This accumulated water flows back eastward down the hydraulic gradient as compensating narrow surface equatorial counter-currents, unimpeded by the weak surface winds. Near the equator in the Pacific Ocean, upwelling raises the thermocline to only 50–100m depth, and within this layer there exist thin, jet-like Equatorial Undercurrents flowing eastward (under hydraulic gradients) at a speed of 1 to 1.5m s^{-1}.

As the circulations swing poleward around the western margins of the oceanic subtropical high pressure cells, there is the tendency for water to pile up against the continents, giving, for example, an appreciably higher sea level in the Gulf of Mexico than along the Atlantic coast of the United States. The accumulated water cannot escape by sinking because of its relatively high temperature and resulting vertical stability. Consequently, it continues poleward driven by the dominant surface airflow, augmented by the geostrophic force acting at right angles to the ocean surface slope. Through this movement, the current gains anticyclonic vorticity, reinforcing the similar tendency imparted by the winds, leading to relatively narrow currents of high velocity (for example, the Kuroshio, Brazil, Mozambique–Agulhas and, to a lesser degree, the East Australian Current). In the North Atlantic, the configuration of the Caribbean Sea and Gulf of Mexico especially favors this pile-up of water, which is released poleward through the Florida Straits as the particularly narrow and fast Gulf Stream. These poleward currents are opposed both by their friction with the nearby continental margins and by energy losses due to turbulent diffusion, such as those accompanying the formation and cutting off of meanders in the Gulf Stream. These poleward western boundary currents (e.g. the Gulf Stream and the Kuroshio Current) are approximately 100km wide and reach surface velocities greater than 2m s^{-1}. This contrasts with the slower, wider and more diffuse eastern boundary currents such as the Canary and California (approximately 1000km wide with surface velocities generally less than 0.25m s^{-1}). The northward-flowing Gulf Stream causes a heat flux of 1.2×10^{15}W, 75 percent of which is lost to the atmosphere and 25 percent in heating the Greenland–Norwegian seas area. On the poleward sides of the subtropical high pressure cells, westerly currents dominate, and where they are unimpeded by land masses in the Southern Hemisphere they form the broad and swift West Wind Drift. This strong current, driven by unimpeded winds, occurs within the zone 50 to 65°S and is associated with a southward-sloping ocean surface generating a geostrophic force, which intensifies the flow. Within the West Wind Drift, the action of the Coriolis force produces a convergence zone at about 50°S marked by westerly submarine jet streams reaching velocities of 0.5 to 1m s^{-1}. South of the West Wind Drift, the Antarctic Divergence with rising water is formed between it and the East Wind Drift closer to Antarctica. In the Northern Hemisphere, a great deal of the eastward-moving current in the Atlantic swings northwards, leading to anomalously very high sea temperatures, and is compensated for by a southward flow of cold arctic water at depth. However, more than half of the water mass comprising the North Atlantic Current, and almost all of that of the North Pacific Current, swings south around the east sides of the subtropical high pressure cells, forming the Canary and California currents. Their Southern Hemisphere equivalents are the Benguela, Humboldt (or Peru), and West Australian currents (Figure 7.29).

Ocean fronts are associated particularly with the poleward margins of the western boundary currents. Temperature gradients can be 10°C over 50km horizontally at the surface and weak gradients are distinguishable to several thousand meters depth. Fronts also form between shelf water and deeper waters where there is convergence and downwelling.

Another large-scale feature of ocean circulation, analogous to the atmosphere, is the Rossby wave. These large oscillations have horizontal wavelengths of 100s–1000s km and periods of tens of days. They develop in the open ocean of mid-latitudes in eastward-flowing currents. In equatorial, westward-flowing currents, there are faster, very long wavelength Kelvin waves (analogous to those in the lower stratosphere).

Mesoscale

Mesoscale eddies and rings in the upper ocean are generated by a number of mechanisms, sometimes by atmospheric convergence or divergence, or by the casting off of vortices by currents such as the Gulf Stream where it becomes unsteady around 65°W (Figure 7.30). Oceanographic eddies occur on the scale of 50–400km diameter and are analogous to atmospheric low and high pressure systems. Ocean mesoscale systems are much smaller than atmospheric depressions (which average about 1000km diameter), travel much slower (a few kilometers per day, compared with about 1000km per day for a depression) and persist from one to several months (compared with a depression life of about a week). Their maximum rotational velocities occur at a depth of about 150m, but the vortex circulation is observed throughout the thermocline (*ca.* 1000m depth). Some eddies move parallel to the main flow direction, but many move irregularly equatorward or poleward. In the North Atlantic, this produces a 'synoptic-like' situation in which up to 50 percent of the area may be occupied by mesoscale eddies (see Plate 7.4). Cold-core cyclonic rings (100–300km diameter) are about twice as numerous as warm-core anticyclonic eddies (100km diameter), and have a maximum rotational velocity of about 1.5m s^{-1}. About 10 cold-core rings are formed annually by the Gulf Stream and may occupy 10 percent of the Sargasso Sea.

2 Deep ocean water interactions

Upwelling

In contrast with the currents on the west sides of the oceans, equatorward-flowing eastern currents acquire cyclonic vorticity, which is in opposition to the anticyclonic wind tendency, leading to relatively broad flows of low velocity. In addition, the deflection due to the Ekman effect causes the surface water to move westward away from the coasts, leading to replacement by the upwelling of cold water from depths of 100–300m (Figure 7.31). Average rates of upwelling are low (1–2m/day), being about the same as the offshore surface current velocities, with which they are balanced. The rate of upwelling therefore varies with the surface wind stress. As the latter is proportional to the square of the wind speed, small changes in wind velocity can lead to marked variations in rates of upwelling. Although the band of upwelling is of limited width (about

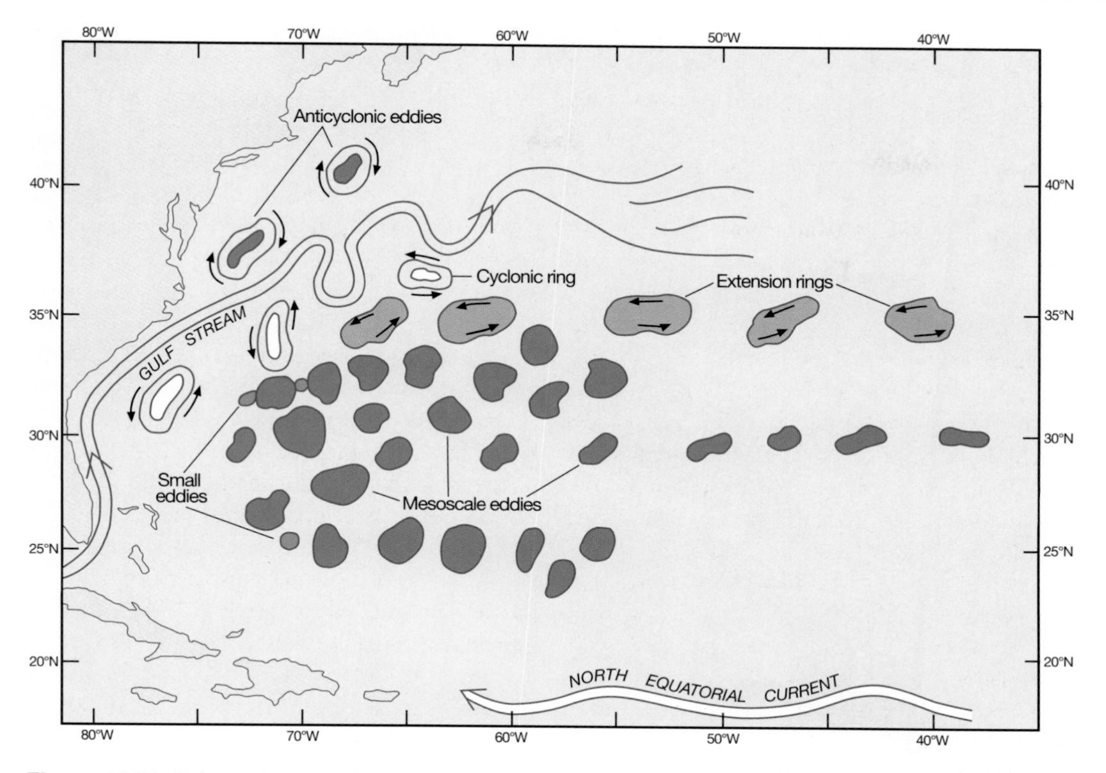

Figure 7.30 Schematic map of the western North Atlantic showing the major types of ocean surface circulation.

Source: From Tolmazin (1985). Copyright © Chapman & Hall.

200km for the Benguela Current), the Ekman effect spreads this cold water westward. On the poleward margins of these cold-water coasts, the meridional swing of the wind belts imparts a strong seasonality to the upwelling; the California Current upwelling, for example, is particularly well marked during the period March to July.

A major region of deep-water upwelling is along the West Coast of South America (Figure 11.52) where there is a narrow 20km-wide shelf and offshore easterly winds. Transport is offshore in the upper 20m but onshore at 30–80m depth. This pattern is forced by the offshore airflow normally associated with the large-scale convective Walker cell (see Chapters 7C.1 and 11G) linking Southeast Asia–Indonesia with the eastern South Pacific. Every two to ten years or so this pressure difference is reversed, producing an El Niño event with weakening Trade Winds and a pulse of warm surface water spreading eastward over the South Pacific, raising local sea surface temperatures by several degrees.

Coastal upwelling is also caused by less important mechanisms such as surface current divergence or the effect of the ocean bottom configuration (see Figure 7.31).

Deep ocean circulation

Above the permanent thermocline the ocean circulation is mainly wind driven, while in the deep ocean it is driven by density gradients due to salinity and temperature differences – a *thermohaline* circulation. These differences are mostly produced by surface processes, which feed cold, saline water to the deep ocean basins in compensation for the deep water delivered to the

Plate 7.4 False-color satellite image of the western North Atlantic indicating surface water temperatures from cold to warm (blue, green, yellow, red). Features of interest include the Gulf Stream, a Gulf Stream meander, a cyclonic cold-core ring and an anticyclonic warm-core eddy.

Source: Data of NOAA/NESDIS/NCDC/SDSD. Courtesy of Otis B. Brown, Robert Evans and M. Carle, University of Miami; Rosenstiel School of Marine and Atmospheric Science, Florida.

surface by upwelling. Although upwelling occurs chiefly in narrow coastal locations, subsidence takes place largely in two broad ocean regions – the northern North Atlantic and around parts of Antarctica (e.g., the Weddell Sea).

In the North Atlantic, particularly in winter, heating and evaporation produce warm, saline water which flows northward both in the near-surface Gulf Stream–North Atlantic Current and at intermediate depths of around 800m. In the

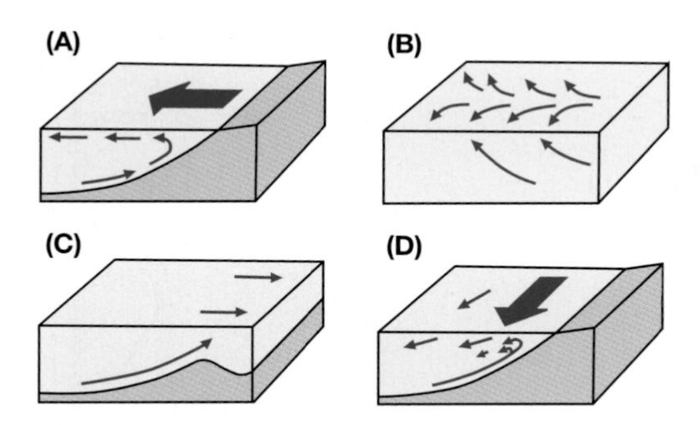

Figure 7.31 Schematic illustration of mechanisms that cause ocean upwelling. The large arrows indicate the dominant wind direction and the small arrows the currents. A: The effects of a persistent offshore wind. B: Divergent surface currents. C: Deep-current shoaling. D: Ekman motion with coastal blocking (Northern Hemisphere case).

Source: Partly modified after Stowe, *Ocean Science*, © 1983 by John Wiley & Sons, Inc. Reproduced by permission.

Norwegian, Greenland and Iceland (GIN) seas, its density is enhanced by further evaporation due to high winds, by the formation of sea ice, which expels brine during ice growth, and by cooling. Exposed to evaporation and to the chill high-latitude air masses, the surface water cools from about 10 to 2°C, releasing immense amounts of heat to the atmosphere, supplementing solar insolation there by some 25–30 percent and heating Western Europe.

The resulting dense high-latitude water, equivalent in volume to about twenty times the combined discharge of all the world's rivers, sinks to the bottom of the North Atlantic. This North Atlantic Deepwater (NADW) fuels a southward-flowing current, which forms part of a global deep-water conveyor belt (Figure 7.32). This broad, slow and diffuse flow, occurring at depths of greater than 1500m, is augmented in the South Atlantic/circum-Antarctic/Weddell Sea region by more cold, saline, dense subsiding water. The conveyor belt then flows eastward under the Coriolis influence, turning north into the Indian and, especially, the Pacific Ocean. The time taken for the conveyor belt circulation to move from the North Atlantic to the North Pacific has been estimated at 500–1000 years. In the Pacific and Indian Oceans, a decrease of salinity due to water mixing causes the conveyor belt to rise and to form a less deep return flow to the Atlantic, the whole global circulation occupying some 1500 years or so. An important aspect of this conveyor belt flow is that the western Pacific Ocean contains a deep source of warm summer water (29°C) (Figure 7.33). This heat differential with the eastern Pacific assists the high phase Walker circulation (see Figure 7.22).

The thermal significance of the conveyor belt implies that any change in it may promote climatic changes operating on timescales of several hundred or thousand years. However, it has been argued that any impediment to the rise of deep conveyor belt water may cause ocean surface temperatures to drop by 6°C within 30 years at latitudes north of 60°N. Changes to the conveyor belt circulation could be initiated by lowering the salinity of the surface water of the North Atlantic, for example, through increased precipitation, ice melting, or fresh water inflow. The role of surface freshening finds support from the paleoclimate record. There is evidence that during warmer periods of the last major glacial cycle, the thermohaline circulation was at times disrupted by massive pulses of fresh-water to the North Atlantic from the North American Laurentide ice sheet, in turn invoking periods of rapid cooling. Cause and effect, however, are still being debated. Some direct observational evidence for a role of fresh-water comes from the 'Great Salinity Anomaly' (GSA). During the late 1960s to early 1970s, the upper 100m of the waters in the Greenland, Iceland and Labrador Seas underwent reductions in salinity, apparently due to an increase in the sea ice transport (sea ice has a very low salinity) out of the Arctic and into the Greenland Sea. The GSA caused temporary cessation of oceanic convection as recorded at ocean weather station 'Bravo' (56°N, 51°W). Other lines of evidence indicate that it was associated with a reduction in the strength of the Gulf Stream system.

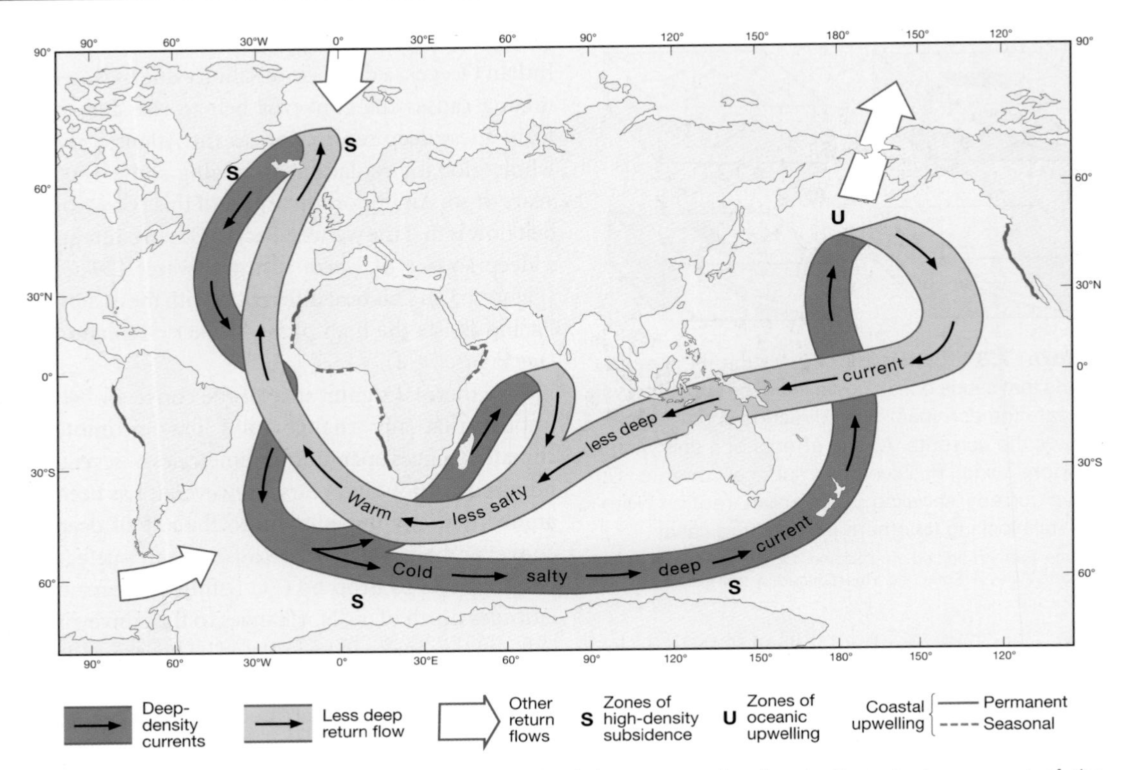

Figure 7.32 The deep ocean thermohaline circulation system leading to Broecker's concept of the oceanic conveyor belt.

Source: Kerr (1988). Reprinted with permission from *Science* 239, Fig. 259. Copyright © 1988 American Association for the Advancement of Science.

3 The oceans and atmospheric regulation

The atmosphere and the surface ocean waters are closely connected both in temperature and in CO_2 concentrations. The atmosphere contains less than 1.7 percent of the CO_2 held by the oceans, and the amount absorbed by the ocean surface rapidly regulates the concentration in the atmosphere. The absorption of CO_2 by the oceans is greatest where the water is rich in organic matter, or where it is cold. Thus the oceans can regulate atmospheric CO_2, changing the greenhouse effect and contributing to climate change. The most important aspect of the carbon cycle linking atmosphere and ocean is the difference between the partial pressure of CO_2 in the lower atmosphere and that in the upper oceanic layer. This results in atmospheric CO_2 being dissolved in the oceans, Some of this CO_2 is subsequently converted into particulate carbon, mainly through the agency of plankton, and ultimately sinks to form carbon-rich deposits in the deep ocean as part of a cycle lasting hundreds of years. Thus two of the major effects of ocean surface warming would be to increase its CO_2 equilibrium partial pressure and to decrease the abundance of plankton. Both of these effects would tend to decrease the oceanic uptake of CO_2. This would increase its atmospheric concentration, thereby producing a positive feedback (i.e., enhancing) effect on global warming. However, as will be seen in Chapter 13, the operation of the atmosphere–ocean system is complex. Thus, for example, global warming may so increase oceanic convective mixing that the resulting imports of

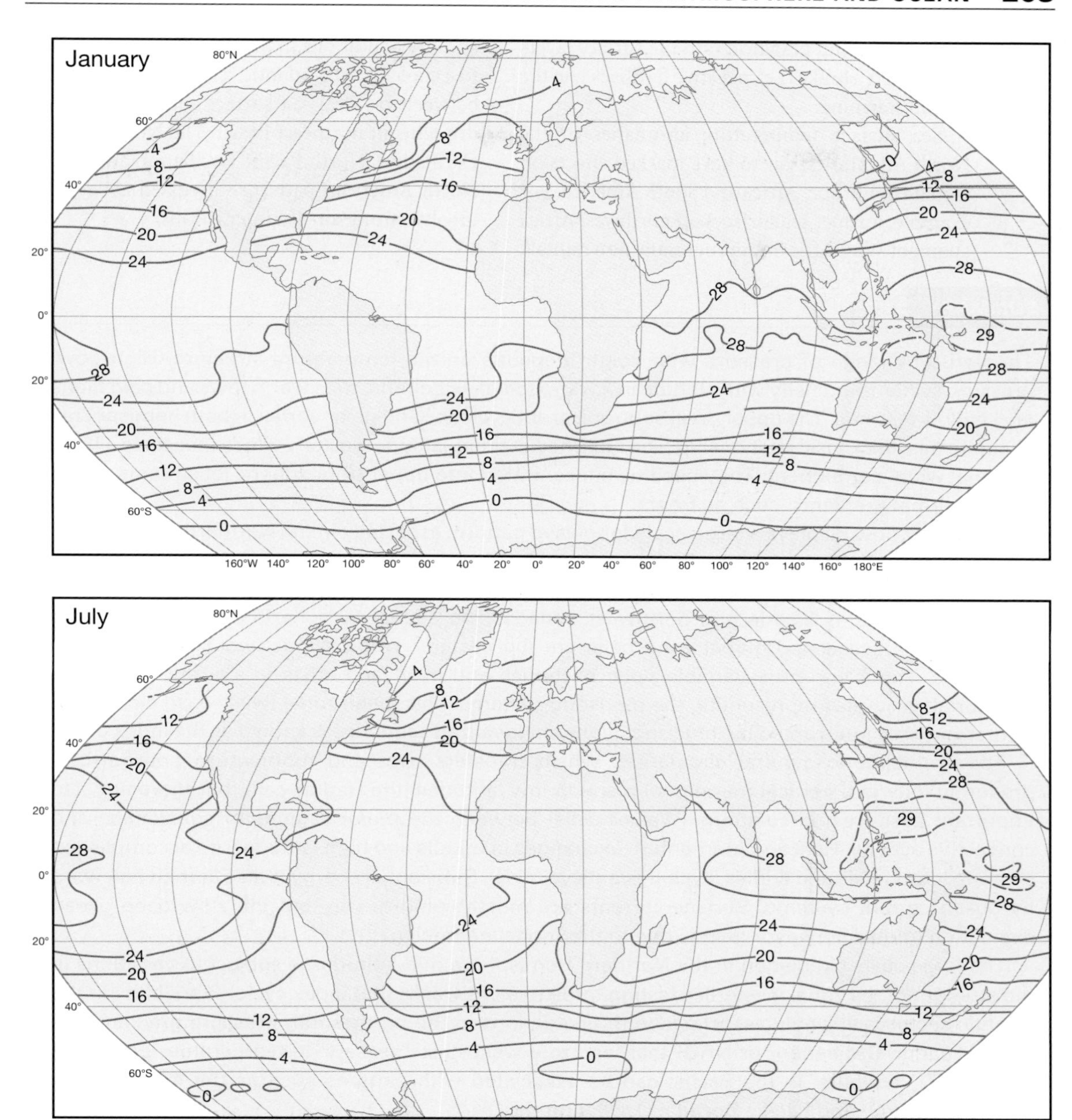

Figure 7.33 Mean ocean surface temperatures (°C) for January and July. Comparison of these maps with those of mean sea-level air temperatures (Figure 2.11) show similarities during the summer but a significant difference in the winter.

Source: Reprinted from Bottomley *et al.* (1990) *Global Ocean Surface Temperature Atlas*. By permission of the Meteorological Office. Crown copyright ©.

cooler water and plankton into the surface layers might exert a brake (i.e., negative feedback) on the system warming.

Sea surface temperature anomalies in the North Atlantic appear to have marked effects on climate in Europe, Africa and South America. For example, warmer sea surfaces off northwest Africa augment West African summer monsoon rainfall; and dry conditions in the Sahel have been linked to a cooler North Atlantic. There are similar links between tropical sea surface temperatures and droughts in northeast Brazil. The North Atlantic Oscillation, North Pacific Oscillation and Pacific North American patterns, discussed above, also involve strong air–sea interactions.

SUMMARY

The vertical change of pressure with height depends on the temperature structure. High (low) pressure systems intensify with altitude in a warm (cold) air column; thus warm lows and cold highs are shallow features. The upper-level subtropical anticyclones and polar vortex in both hemispheres illustrate this 'thickness' relationship. The intermediate mid-latitude westerly winds thus have a large 'thermal wind' component. They become concentrated into upper tropospheric jet streams above sharp thermal gradients, such as fronts.

The upper flow displays a large-scale longwave pattern, especially in the Northern Hemisphere, related to the influence of mountain barriers and land–sea differences. The surface pressure field is dominated by semi-permanent subtropical highs, subpolar lows and, in winter, shallow cold continental highs in Siberia and northwest Canada. The equatorial zone is predominantly low pressure. The associated global wind belts are the easterly Trade Winds and the mid-latitude westerlies. There are more variable polar easterlies and over land areas in summer a band of equatorial westerlies representing the monsoon systems. This mean zonal (west–east) circulation is intermittently interrupted by 'blocking' highs; an idealized sequence is known as the index cycle.

The atmospheric general circulation, which transfers heat and momentum poleward, is predominantly in a vertical meridional plane in low latitudes (the Hadley cell), but there are also important east–west circulations (Walker cells) between the major regions of subsidence and convective activity. Heat and momentum exchanges in middle and high latitudes are accomplished by horizontal waves and eddies (cyclones/anticyclones). Substantial energy is also carried poleward by ocean current systems. Surface currents are mostly wind driven, but the slow deep ocean circulation (global conveyor belt) is due to thermohaline forcing.

The large-scale circulation in the Northern Hemisphere mid-latitudes is subject to variations in the strength of the zonal westerlies lasting three to eight weeks (the index cycle). Variability in the Atlantic sector is strongly associated with fluctuations in the north–south pressure gradient (the North Atlantic Oscillation, or NAO) that lead to a west–east "seesaw' in temperature and other anomalies. Variability in the Pacific can be associated with patterns such as the North Pacific Oscillation (NPO) and Pacific Decadal Oscillation (PDO).

The ocean's vertical structure varies latitudinally and regionally. In general, the thermocline is deepest in mid-latitudes, thus permitting greater turbulent mixing and atmospheric heat exchanges. The oceans are important regulators of both atmospheric temperatures and CO_2 concentrations. Ocean dynamics and circulation features are analogous to those in the atmosphere on both the meso- and macroscale. The wind-driven Ekman layer extends to 100–200m. Ekman transport and coastal upwelling maintain normally cold sea surfaces off western South America and southwest Africa in particular.

DISCUSSION TOPICS

- What features of the global wind belts at the surface and in the upper troposphere are in accord with (differ from) those implied by the three-cell model of meridional circulation?
- What are the consequences of the westerly jet streams for transoceanic air travel?
- Examine the variation of the vertical structure of the zonal wind by creating height cross-sections for different longitudes and months using the CDC website (http://www.cdc.noaa.gov).
- Consider the effects of ocean currents on the weather and climate of coastal regions in the western and eastern sides of the Atlantic/Pacific oceans and how these effects vary with latitude.

REFERENCES AND FURTHER READING

Books

Bearman, G. (ed.) (1989) *Ocean Circulation*, The Open University, Pergamon Press, Oxford, 238pp.

Bottomley, M., Folland, C. K., Hsiung, J., Newell, R. E. and Parker, D.E. (1990) *Global Ocean Surface Temperature Atlas*, Meteorological Office, London, 20pp + 313 plates.

Corby, G. A. (ed.) (1970) *The Global Circulation of the Atmosphere*, Roy. Met. Soc., London, 257pp.

Flohn, H. and Fantechi, R. (eds) (1984) *The Climate of Europe: Past, Present and Future*, D. Reidel, Dordrecht, 356pp.

Halitner, G. J. and Martin, F. L. (1957) *Dynamical and Physical Meteorology*, McGraw-Hill, New York, 470pp. [Classic text]

Houghton, J. (ed.) (1984) *The Global Climate*, Cambridge University Press, Cambridge, 233pp.

Indian Meteorological Department (1960) *Monsoons of the World*, Delhi, 270pp.

Kuenen, Ph. H. (1955) *Realms of Water*, Cleaver-Hulme Press, London, 327pp.

Levitus, S. (1982) *Climatological Atlas of the World Ocean*, NOAA Professional Paper No. 13, Rockville, MD, 173pp.

Lorenz, E. N. (1967) *The Nature and Theory of the General Circulation of the Atmosphere*, World Meteorological Organization, Geneva, 161pp. [Classic account of the mechanisms and maintenance of the global circulation, transports of momentum, energy and moisture]

NASA (n.d.) *From Pattern to Process: The Strategy of the Earth Observing System* (Vol. II), EOS Science Steering Committee Report, NASA, Houston.

Riley, D. and Spalton, L. (1981) *World Weather and Climate* (2nd edn), Cambridge University Press, London, 128pp. [Elementary text]

Stowe, K. (1983) *Ocean Science* (2nd edn), John Wiley & Sons, New York, 673pp.

Strahler, A. N. and Strahler, A. H. (1992) *Modern Physical Geography* (4th edn), John Wiley & Sons, New York, 638pp.

Tolmazin, D. (1985) *Elements of Dynamic Oceanography*, Kluwer, Dordrecht, 182pp. [Description of ocean circulation, classical instrumental observations, satellite altimetry and acoustic tomography]

Troen, I. and Petersen, E. L. (1989) *European Wind Atlas*, Commission of the Economic Community, Risø National Laboratory, Roskilde, Denmark, 656pp.

van Loon, H. (ed.) (1984) *Climates of the Oceans*, in Landsberg, H. E. (ed.) *World Survey of Climatology* 15, Elsevier, Amsterdam, 716pp.

Wells, N. (1997) *The Atmosphere and Ocean. A Physical Introduction* (2nd edn), J. Wiley & Sons, Chichester, UK, 394pp. [Undergraduate text describing the physical properties and observed characteristics, the influence of the earth's rotation on atmospheric and ocean circulation, energy transfers and climate variability]

Articles

Barry, R. G. (1967) Models in meteorology and climatology, in Chorley, R. J. and Haggett, P. (eds) *Models in Geography*, Methuen, London, 97–144.

Borchert, J. R. (1953) Regional differences in world atmospheric circulation. *Ann. Assn Amer. Geog.* 43, 14–26.

Boville, B. A. and Randel, W. J. (1986) Observations and simulation of the variability of the stratosphere and troposphere in January. *J. Atmos. Sci.* 43, 3015–34.

Bowditch, N. (1966) American practical navigator, US Navy Hydrographic Office, Pub. No. 9, US Naval Oceanographic Office, Washington, DC.

Broecker, W. S. and Denton, G. H. (1990) What drives glacial cycles? *Sci. American* 262(1), 43–50.

Broecker, W. S., Peteet, D. M. and Rind, D. (1985) Does the ocean–atmosphere system have more than one stable mode of operation? *Nature* 315, 21–6.

Chen, P., Hoerling, M. P. and Dole, R. M. (2001) The origin of the subtropical anticyclones. *J. Atmos. Sci.* 58(13), 1827–35.

Crowe, P. R. (1949) The trade wind circulation of the world. *Trans. Inst. Brit. Geog.* 15, 38–56.

Crowe, P. R. (1950) The seasonal variation in the strength of the trades. *Trans. Inst. Brit. Geog.* 16, 23–47.

Defant, F. and Taba, H. (1957) The threefold structure of the atmosphere and the characteristics of the tropopause. *Tellus* 9, 259–74.

Dunn, C. R. (1957) The weather and circulation of December 1957: high index and abnormal warmth. *Monthly Weather Review* 85, 490–516.

Hare, F. K. (1965) Energy exchanges and the general circulation. *J. Geography* 50, 229–41.

Hastenrath, S. (1980) Heat budget of the tropical ocean and atmosphere. *J. Phys. Oceanography* 10, 159–70.

Hoskins, B. J. (1996) On the existence and strength of the summer subtropical anticyclones. *Bull. Amer. Met. Soc.* 77(6), 1287–91.

Ioannidou, L. and Yau, M.K. (2008) A climatology of the Northern Hemisphere winter anticyclones. *J. Geophys. Res.* 113: D08119, 17pp.

Kerr, R. A. (1988) Linking earth, ocean and air at the AGU. *Science* 239, 259–60.

Klein, W. H. (1958) The weather and circulation of February 1958: a month with an expanded circumpolar vortex of record intensity, *Monthly Weather Review* 86, 60–70.

Lamb, H. H. (1960) Representation of the general atmospheric circulation. *Met. Mag.* 89, 319–30.

LeMarshall, J. F., Kelly, G. A. M. and Karoly, D. J. (1985) An atmospheric climatology of the Southern Hemisphere based on 10 years of daily numerical analyses (1972– 1982): I. Overview. *Austral. Met. Mag.* 33, 65–86.

Meehl, G. A. (1987a) The annual cycle and interannual variability in the tropical Pacific and Indian Ocean regions, *Monthly Weather Review* 115, 51–74.

Meehl, G. A. (1987b) The tropics and their role in the global climate system. *Geographical Journal* 153, 21–36.

Namias, J. (1972) Large-scale and long-term fluctuations in some atmospheric and ocean variables, in Dyrssen, D. and Jagner, D. (eds) *The Changing Chemistry of the Oceans*, Nobel Symposium 20, Wiley, New York, 27–48.

Niiler, P. P. (1992) The ocean circulation, in Trenberth, K. E. (ed.) *Climate System Modelling*, Cambridge University Press, Cambridge, 117–48.

O'Connor, J. F. (1961) Mean circulation patterns based on 12 years of recent northern hemispheric data. *Monthly Weather Review* 89, 211–28.

Palmén, E. (1951) The role of atmospheric disturbances in the general circulation. *Quart. J. Roy. Met. Soc.* 77, 337–54.

Persson, A. (2002) The Coriolis force and the subtropical jet stream. *Weather* 57(7), 53–9.

Riehl, H. (1962a) General atmospheric circulation of the tropics. *Science* 135, 13–22.

Riehl, H. (1962b) *Jet streams of the atmosphere*, Tech. Paper No. 32, Colorado State University, 117pp.

Riehl, H. (1969) On the role of the tropics in the general circulation of the atmosphere. *Weather* 24, 288–308.

Riehl, H. *et al.* (1954) The jet stream. *Met. Monogr.* 2(7), American Meteorological Society, Boston, MA, 100pp.

Rodwell, M. J. and Hoskins, B. J. (1996) Monsoons and the dynamics of deserts. *Quart. J. Roy. Met. Soc.* 122, 1,385–404.

Rodwell, M. J. and Hoskins, B. J. (2001) Subtropical anticyclones and summer monsoons. *J. Clim.* 14, 3192–211.

Rossby, C-G. (1941) The scientific basis of modern meteorology. US Dept of Agriculture Yearbook *Climate and Man*, 599–655.

Rossby, C-G. (1949) On the nature of the general circulation of the lower atmosphere, in Kulper, G. P. (ed.) *The Atmosphere of the Earth and Planets*, University of Chicago Press, Chicago, IL, 16–48.

Saltzman, B. (1983) Climatic systems analysis. *Adv. Geophys.* 25, 173–233.

Sawyer, J. S. (1957) Jet stream features of the earth's atmosphere. *Weather* 12, 333–4.

Shapiro, M. A. and Keyser, D. A. (1990) Fronts, jet streams, and the tropopause, in Newton,

C. W. and Holopainen, E. D. (eds) *Extratropical Cyclones*, American Meteorological Society, Boston, MA, 167–91.

Shapiro, M. A. *et al.* (1987) The Arctic tropopause fold. *Monthly Weather Review* 115, 444–54.

Starr, V. P. (1956) The general circulation of the atmosphere. *Sci. American* 195, 40–5.

Streten, N. A. (1980) Some synoptic indices of the Southern Hemisphere mean sea level circulation 1972–77. *Monthly Weather Review* 108, 18–36.

Thompson, D. W. J. and Wallace, J. M. (1998) The Arctic oscillation signature in the wintertime geopotential height and temperature fields. *Geophys. Res. Lett.* 25(9) 1297–1300.

Tucker, G. B. (1962) The general circulation of the atmosphere. *Weather* 17, 320–40.

van Loon, H. (1964) Mid-season average zonal winds at sea level and at 500mb south of 25°S and a brief comparison with the Northern Hemisphere. *J. Appl. Met.* 3, 554–63.

van Loon, H. and Rogers, J. C. (1978) The see-saw in winter temperatures between Greenland and northern Europe Pt.1. General description: *Mon. Wea. Rev.*, 106, 296–310.

Walker, J. M. (1972) Monsoons and the global circulation. *Met. Mag.* 101, 349–55.

Wallington, C. E. (1969) Depressions as moving vortices. *Weather* 24, 42–51.

Wyrtki, K. (1985) Water displacements in the Pacific and the genesis of El Niño cycles. *J. Geophys. Res.* 90(C10), 7129–32.

Yang, S. and Webster, P.J. (1990) The effect of summer tropical heating on the location and intensity of the extratropical westerly jet streams. *J. Geophys. Res.* 95(D11), 19, 705–721.

CHAPTER EIGHT

Numerical models of the general circulation, climate and weather prediction

LEARNING OBJECTIVES

When you have read this chapter you will:

- know the basic features of atmospheric general circulation circulation models (GCMs)
- understand how simulations of the atmospheric circulation and its characteristics are performed
- be familiar with the basic approaches to weather forecasting on different timescales.

Fundamental changes in our understanding of the complex behavior of the atmosphere and climate processes have been obtained over the past four decades through the development and application of numerical climate and weather models. Numerical models simply use mathematical relationships to describe physical processes. There are many forms of climate and weather models ranging from simple point energy balance approaches to three-dimensional general circulation models (GCM) which attempt to model all the complexities of the earth climate system. We discuss in more detail the GCM in its various forms which is used to simulate both climate and weather for day-to-day forecasting.

A FUNDAMENTALS OF THE GCM

In the GCM, all dynamic and thermodynamic processes and the radiative and mass exchanges that have been treated in Chapters 2–6 are modeled using five basic sets of equations. The basic equations describing the atmosphere are:

1. The three-dimensional equations of motion (i.e., conservation of momentum; see Chapter 6A, B).
2. The equation of continuity (i.e., conservation of mass or the hydrodynamic equation, p. 152).
3. The equation of continuity for atmospheric water vapor (i.e., conservation of water vapor, Chapter 4).
4. The equation of energy conservation (i.e., the thermodynamic equation derived from the first law of thermodynamics, Chapter 7F).
5. The equation of state for the atmosphere (p. 30).
6. In addition, conservation equations for other atmospheric constituents such as sulfur aerosols may be applied in more complex models.

Model simulations of present-day and future climate conditions involve iterating the model equations for perhaps tens to hundreds of years of simulated time depending on the question at hand. In order to solve these coupled equations, additional processes such as radiative transfer through the atmosphere with diurnal and seasonal cycles, surface friction and energy transfers and cloud formation and precipitation processes must be accounted for. These are coupled in the manner shown schematically in Figure 8.1. Beginning with a set of initial atmospheric conditions usually derived from observations, the equations are integrated forward in time repeatedly using time steps of several minutes to tens of minutes at a large number of grid points over the earth and at many levels vertically in the atmosphere; typically 10–20 levels in the vertical is common. The horizontal grid is usually of the order of several degrees latitude by several degrees longitude near the equator. Another, computationally faster, approach is to represent the horizontal fields by a series of two-dimensional sine and cosine functions (a spectral model). A truncation level describes the number of two-dimensional waves that are included. The truncation procedure may be rhomboidal (*R*) or triangular (*T*); *R*15 (or *T*21) corresponds approximately to a 5° grid spacing, *R*30 (*T*42) to a 2.5° grid, and *T*102 to a 1° grid.

Realistic coastlines and mountains as well as essential elements of the surface vegetation (albedo, roughness) and soil (moisture content) are typically incorporated into the GCM. These are smoothed to be representative of the average state of an entire grid cell and therefore much regional detail is lost. Sea ice extent and sea surface temperatures have often been specified by a climatological average for each month in the past. However, in recognition that the climate system is quite interactive, the newest generation of models includes some representation of an ocean which can react to changes in the atmosphere above. Ocean models (Figure 8.2) include a so-called swamp ocean where sea surface temperatures are calculated through an energy budget and no annual cycle is possible; a slab or mixed layer ocean, where storage and release of energy can take place seasonally and the most complex

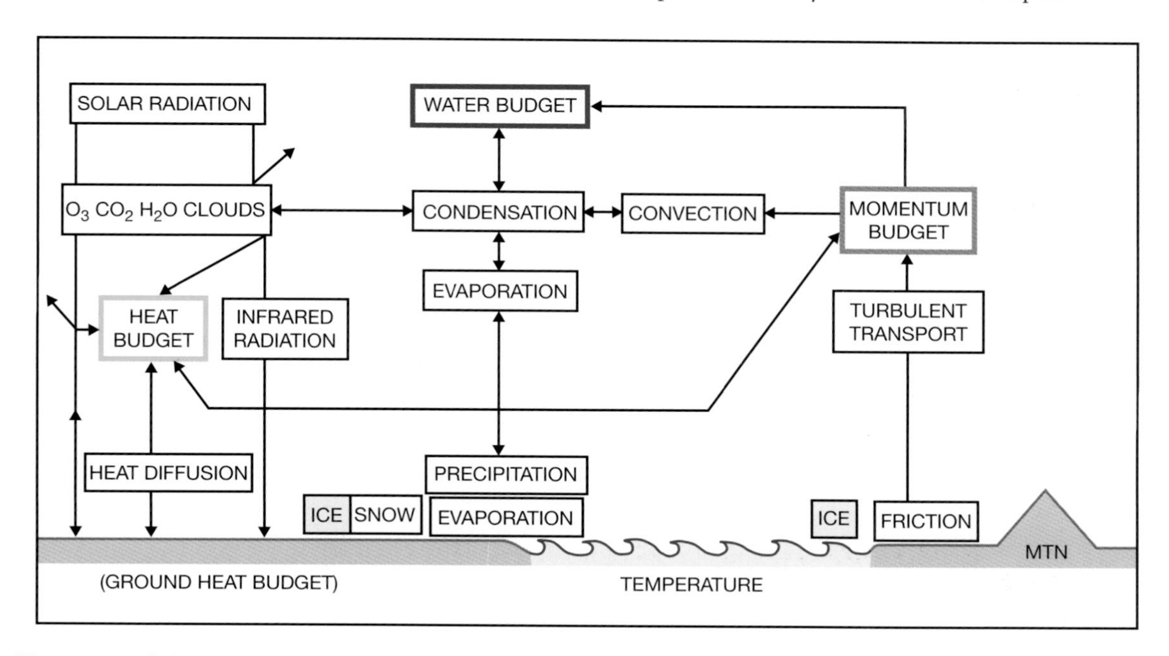

Figure 8.1 Schematic diagram of the interactions among physical processes in a general circulation model.
Source: From Druyan *et al.* (1975).

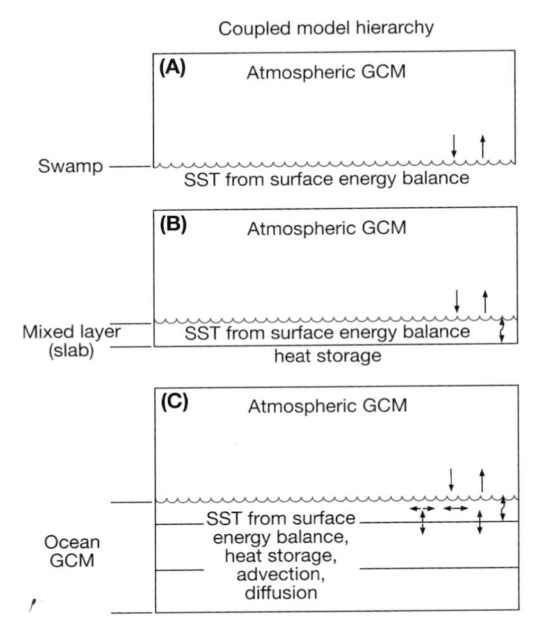

Figure 8.2 Schematic illustration of the three types of coupling of an atmospheric GCM to the ocean: (A) swamp ocean, (B) mixed layer, slab ocean, (C) ocean GCM.

Source: From Meehl in Trenberth (1992) Copyright © Cambridge University Press.

dynamic ocean models which solve appropriate equations for the ocean circulation and thermodynamic state similar to 1–5 above and which are coupled to atmospheric models. Such coupled models are referred to as Atmopshere-Ocean General Circulation Models (AOGCMs). When the global ocean is considered, seasonal freezing/melting and the effects of sea ice on energy exchanges and salinity must also be modeled. Therefore, dynamic sea ice models, which actively calculate the thickness and extent of ice, are now replacing the specification of climatological sea ice. Because of the century-long timescale of deep ocean circulations, the use of a dynamic ocean model requires large amounts of simulation time for the different model components to equilibrate which greatly increases the cost of running these models.

Because coupled AOGCMs are used in long-term (century or millennium scale) simulations, an important concern is 'model drift' (a definite tendency for the model climate to warm or cool with time) due to accumulating errors from the various component models. These tendencies are often constrained by using observed climatology at certain high-latitude or deep ocean boundaries, or by adjusting the net fluxes of heat and fresh water at each grid point on an annual basis in order to maintain a stable climate, but such arbitrary procedures are the subject of controversy, especially for climate change studies.

Many important weather and climate processes occur on a scale which is too small for the typical GCM to simulate with a grid of several degrees on a side. Examples of this would be the radiative effects or latent heating due to cloud formation or the transfer of water vapor to the atmosphere by a single tree. Both processes greatly affect our climate and must be represented for a realistic climate simulation. *Parameterizations* are methods designed to take into account the average effect of cloud or vegetation process on an entire grid cell. Parameterizations generally make use of a statistical relationship between the large-scale values calculated for the grid cell in order to determine the effect of the parameterized process.

In order to gain confidence in the performance of models in predicting future atmospheric states, it is important to evaluate how well such models perform in representing present-day climate statistics. The Atmospheric Model Intercomparison Program (AMIP) is designed to do this by comparing models from various centers around the world using common procedures and standardized data (on sea surface temperatures, for example), as well as by providing extensive documentation on the model design and the details of model parameterizations. In this way common deficencies can be detected and perhaps attributed to a single process and then addressed in future model versions. Figure 8.3 compares simulated zonally averaged surface temperature for January and July for all AMIP participants with the observed climatological mean. The general features are well represented qualitatively,

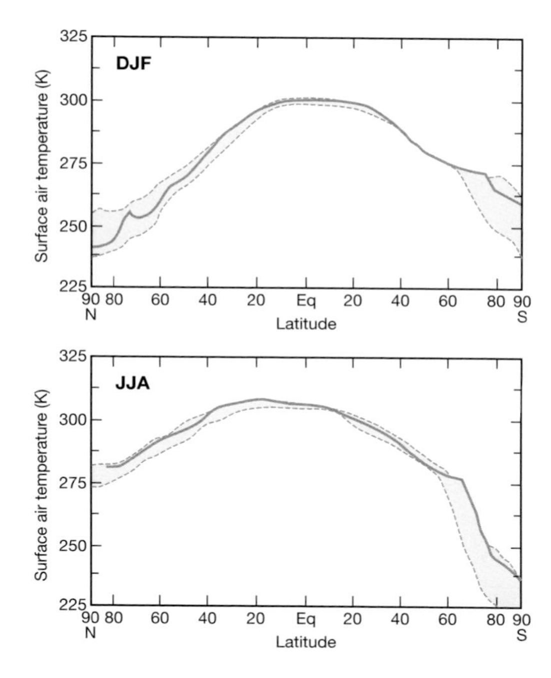

Figure 8.3 Comparison of zonally averaged surface temperatures for December to February and June to August as simulated by the AMIP models compared with observations (bold line).
Source: AMIP website.

although there can be large deviation between individual models. The evaluation of models requires analysis of their ability to reproduce interannual variability and synoptic-scale variability as well as mean conditions. A comparison project for AOGCMs similar to AMIP is now underway called the Coupled Model Intercomparison Project (CMIP).

Recent models incorporate improved spatial resolution and fuller treatment of some previously neglected physical processes. However, both changes may create additional problems as a result of the need to accurately treat complex interactions between the land surface (including soil moisture and canopy structure.) and the atmospheric boundary layer, or interactions between clouds, radiative exchanges and precipitation mechanisms. For example, fine-scale spatial resolution is necessary in the explicit treatment of cloud and rain bands associated with frontal zones in mid-latitude cyclones. Such processes require detailed and accurate representation of moisture exchanges (evaporation, condensation), cloud microphysics, radiation (and the interactions between these processes) which were all represented as averaged processes when simulated at larger spatial scales.

B MODEL SIMULATIONS

1 GCMs

Climate model simulations are used to examine possible future climates by simulating plausible scenarios (e.g., increasing atmospheric CO_2, tropical deforestation) into the future using representations of inputs (i.e., forcings), storage between components of the climate system and transfers between components (see Figure 6.38 and Chapter 11). The periods of time shown in Figure 8.4 refer to:

1. *Forcing times.* The characteristic timespans over which natural and anthropogenic changes of input occur. In the case of the former, these can be periods of solar radiation cycles or the effect of volcanism and in the case of the latter the average time interval over which significant changes of such anthropogenic effects as increased atmospheric CO_2 occur.
2. *Storage times.* For each compartment of the atmosphere and ocean subsystems these are the average times taken for an input of thermal energy to diffuse and mix within the compartment. For the earth subsystem, the average times are those required for inputs of water to move through each compartment.

These simulations can be performed in several different ways. A common procedure is to analyze the model's sensitivity to a specified change in a single variable. This may involve changes in external forcing (increased/decreased solar radiation, atmospheric CO_2 concentrations, or a

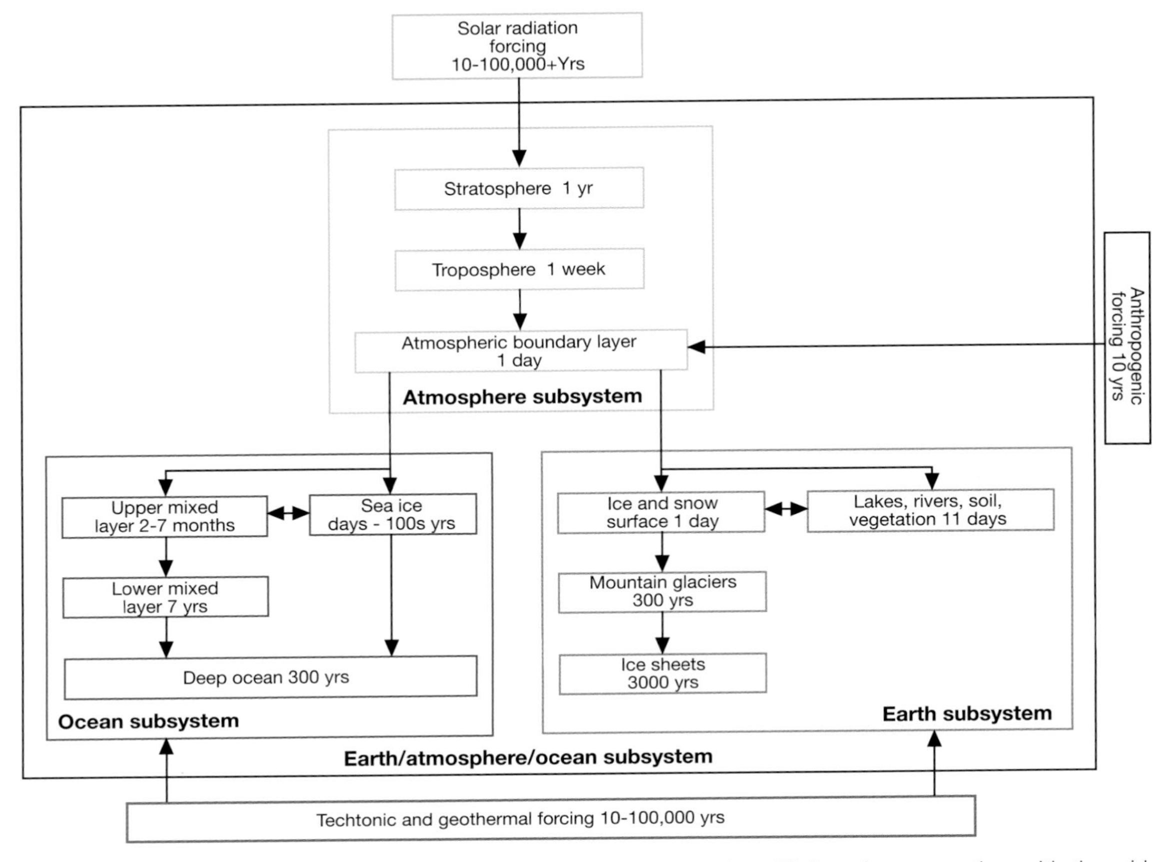

Figure 8.4 The earth–atmosphere–ocean system showing estimated equilibrium times, together with the wide time variations involving the external solar, tectonic, geothermal and anthropogenic forcing mechanisms.
Source: After Saltzman (1983).

volcanic dust layer), surface boundary conditions (orography, land surface albedo, continental ice sheets) or in the model physics (modifying the convective scheme or the treatment of biosphere exchanges). In these simulations, the model is allowed to reach a new equilibrium and the result is compared with a control experiment. A second approach is to conduct a genuine climate change experiment where, for example, the climate is allowed to evolve as atmospheric trace gas concentrations are increased at a specified annual rate (a transient experiment).

A key issue in assessments of greenhouse-gas-induced warming is the sensitivity of global climate to CO_2 doubling which is projected to occur in the mid-twenty-first century extrapolating current trends. Atmospheric GCM simulations for equilibrium condition changes, with a simple ocean treatment, indicate an increase in global mean surface air temperature of 2.5 to 5°C, comparing $1 \times CO_2$ and $2 \times CO_2$ concentrations in the models. The range is in part the result of a dependence of the temperature change on the temperature level simulated for the base state $1 \times CO_2$, and in part arises from the variations in the strength of feedback mechanisms incorporated in the models, particularly atmospheric water vapor, clouds, snow cover and sea ice. Use of coupled atmosphere–

ocean models, however, suggest only a 1–2°C surface warming for century-long transient or doubled CO_2 experiments (see Chapter 13).

2 Simpler models

Because GCMs require massive computer resources, other approaches to modeling climate have developed. A variant of the GCM is the statistical–dynamical model (SDM), in which only zonally averaged features are analyzed and north–south energy and momentum exchanges are not treated explicitly but are represented statistically through parameterization. Simpler still are the energy balance model (EBM) and the radiative convective model (RCM). The EBM assumes a global radiation balance and describes the integrated north–south transports of energy in terms of the poleward temperature gradients; EBMs can be one-dimensional (latitude variations only), two-dimensional (latitude–longitude, with simple land–ocean weightings or simplified geography) and even zero-dimensional (averaged for the globe). They are used particularly in climate change studies. The RCMs can represent a single, globally averaged vertical column. The vertical temperature structure is analyzed in terms of radiative and convective exchanges. These less complete models complement the GCMs because, for example, the RCM allows study of complex cloud–radiation interactions or the effect of atmospheric composition on lapse rates in the absence of many complicating circulation effects. Simpler models are also important for simulation paleoclimate, as these models can represent thousands or even millions of years of climate history.

3 Regional models

Because of the necessity of transferring climate and weather information representing averages over grid cells which are hundreds of kilometers on a side to point scales where information can actually be applied, a variety of downscaling techniques have been developed and applied in recent years. One methodology is to embed a regional climate model into a GCM or AOGCM in a certain region of interest and to use the global model information as a boundary condition for the regional model. The typical regional climate model has grid cells of approximately 50km on a side providing a higher resolution climate simulation over a limited area. In this way, small-scale effects such as local topography, water bodies or regionally important circulations can be represented in a climate or weather simulation. These local effects, however, are generally not transmitted back to the larger scale model at present. In addition, regional models often have a more realistic treatment of smaller scale processes (e.g., convective adjustment), which can lead to more accurate simulations.

C DATA SOURCES FOR FORECASTING

The data required for forecasting and other services are provided by worldwide standard three-hourly synoptic reports (see Appendix 3), similar observations made hourly, particularly in support of national aviation requirements. Upper-air soundings (at 00 and 12UTM), satellite data, and other specialized networks such as radar stations for severe weather provide additional data. Under the World Weather Watch program, synoptic reports are made at some 4000 land stations and by 7000 ships (Figure 8.5A). There are about 700 stations making upper-air soundings (temperature, pressure, humidity and wind) (Figure 8.5B). These data are transmitted in code via teletype and radio links to regional or national centers and into the high-speed Global Telecommunications System (GTS) connecting world weather centers in Melbourne, Moscow and Washington and eleven regional meteorological centers for redistribution. Some 184 member nations cooperate in this activity under the aegis of the World Meteorological Organization.

Meteorological information has been collected operationally by satellites of the United States and

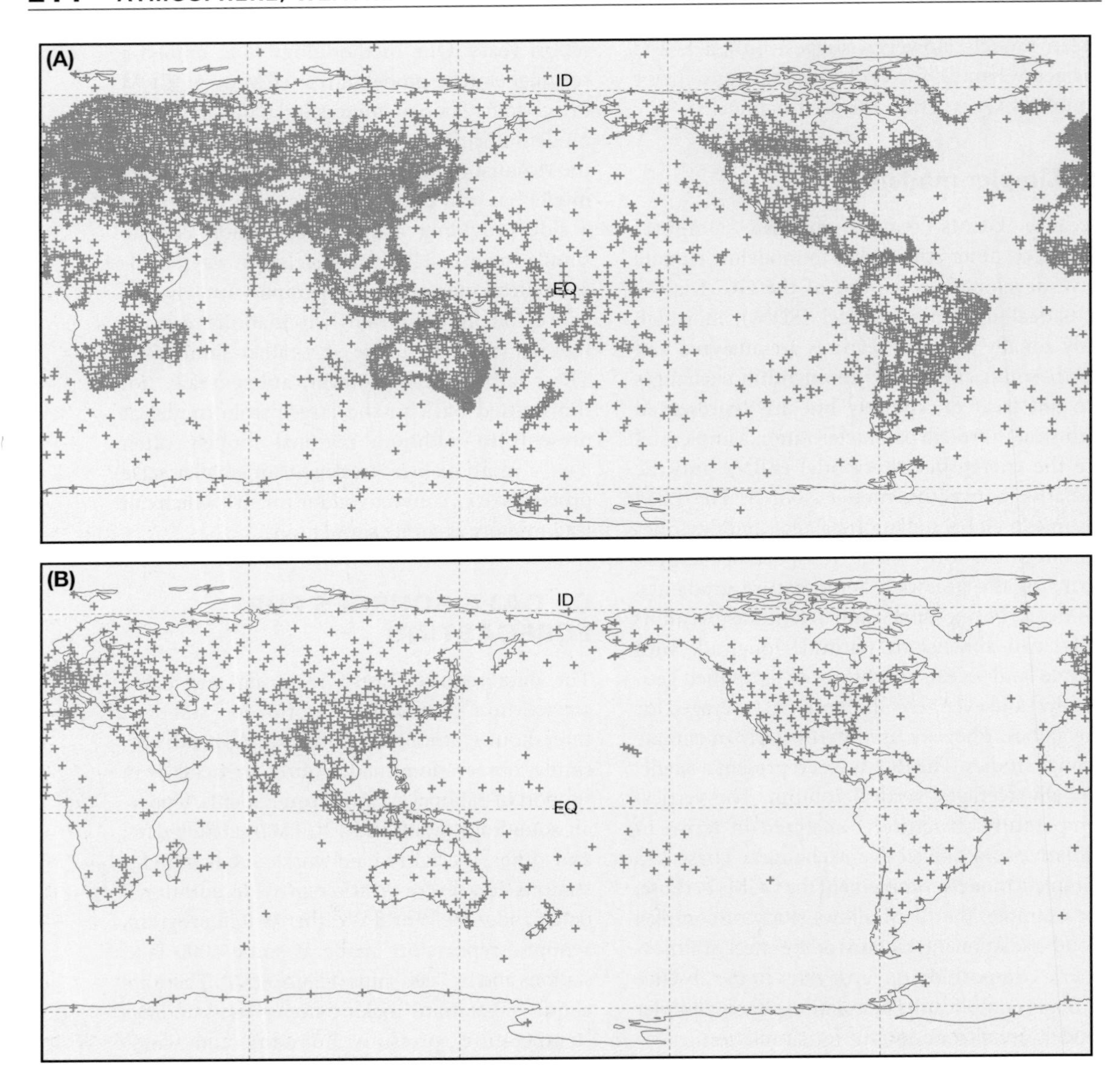

Figure 8.5 Synoptic reports from (A) surface land stations and ships, and (B) from upper-air sounding stations available over the Global Telecommunications System at the National Meteorological Center, NOAA, Washington, DC.
Source: From Barry and Carleton (2001).

Russia since 1965 and, more recently, by the European Space Agency, India and Japan (see Box 8.1). There are two general categories of weather satellite: polar orbiters providing global coverage twice per 24 hours in orbital strips over the poles (such as the United States' NOAA and TIROS series and the former USSR's Meteor); and geosynchronous satellites (such as the Geostationary Operational Environmental Satellites (GOES) and Meteosat), giving repetitive (30-minute) coverage of almost one-third of the earth's surface in low mid-latitudes (Figure 8.6). Information on

SIGNIFICANT 20TH-CENTURY ADVANCE

8.1 Satellite meteorology

The launching of meteorological satellites revoluntionized meteorology, in terms of the near-global view they provided of synoptic weather systems. The first meteorological satellite transmitted pictures on 1 April 1960. The early Television and Infrared Observing Satellites (TIROS) carried photographic camera systems and, due to their spin around an axis parallel to the earth's surface, they photographed the surface only part of the time. The types of images that were collected had been anticipated by some meteorologists, but the wealth of information exceeded expectations. New procedures for interpreting cloud features, synoptic and mesoscale weather systems were developed. Satellite pictures revealed cloud vortices, jet stream bands and other mesoscale systems that were too large to be seen by ground observers and too small to be detected by the networks of synoptic stations. Automatic Picture Transmission (APT) to ground stations began in 1963 and was soon in worldwide use for weather forecasting. In 1972 the system was upgraded to provide high-resolution (HRPT) images.

The operational polar-orbiting weather satellites in the United States were followed in 1966 by geostationary, sun-synchronous satellites positioned at fixed positions in the tropics. These give images of a wide disc of the earth at 20-minute intervals. providing valuable information on the diurnal development of cloud and weather systems. The U.S. Geostationary Operational Environmental Satellites (GOES) were positioned at 75°W and 135°W from 1974 and in 1977 the Japanese Geostationary Meteorological Satellite (GMS) and European Metosat were added at 135°E and 0° longitude, respectively. India began the Insat series in 1983, now positioned at 93.5°E.

The early photographic systems were replaced in the mid-1960s by radiometric sensors in the visible and infrared wavelengths. Initially, these were broad band sensors of moderate spatial resolution. Subsequently, narrow band sensors with improved spatial resolution replaced these; the Advanced Very High Resolution Radiometer (AVHRR) with 1.1km resolution and four channels was initiated in 1978. A further major advance took place in 1970 with the first retrieval of atmospheric temperature profiles from a Nimbus satellite. An operational system for temperature and moisture profiles (the High-resolution Infrared Radiation (HIRS) sounder became operational in 1978, followed by a system on GOES in 1980).

Satellite data are now routinely collected and exchanged between NOAA in the USA, the European Meteorological Satellite Agency (Eumetsat) and the Japanese Meteorological Agency (JMA). There are also ground-receiving stations in more than 170 countries collecting picture transmission by NOAA satellites. Satellite data collected by Russia, China and India are mostly used in those countries.

A vast suite of operational products is now available from NOAA and Department of Defense (DoD) Defense Meteorological Satellite Program (DMSP) satellites. The DMSP series are polar orbiting. They provide imagery from 1970 and digital products from 1992. NASA's Nimbus and Earth Observing System (EOS) satellites provide numerous additional research products including sea ice, vegetation indices, energy balance components, tropical rainfall amounts and surface winds.

Descriptions of available satellite data may be found at http://lwf.ncdc.naa.gov/oa/satellite/satelliteresources.html; http://eospso.gsfc.nasa.gov/; http://www.eumetsat.de/

Reference

Purdom, J. F. W. and Menzel, P. 1996. Evolution of satellite observations in the United States and their use in meteorology, in Fleming, J. R. (ed.) *Historical Essays on Meteorology 1919–1995*, Amer. Met. Soc., Boston, MA, pp. 9–155.

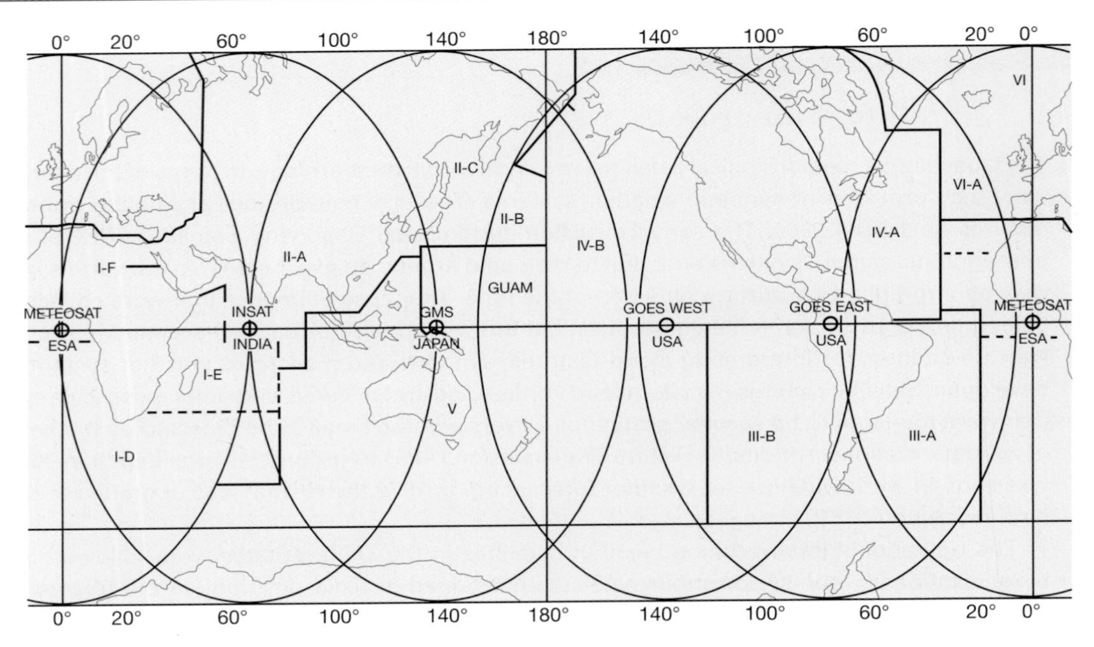

Figure 8.6 Coverage of geostationary satellites and WMO data collection areas (rectangular areas and numbers).

Source: Reproduced by courtesy of NOAA.

the atmosphere is collected as digital data or direct readout visible and infrared images of cloud cover and sea surface temperature, but it also includes global temperature and moisture profiles through the atmosphere obtained from multi-channel infrared and microwave sensors, which receive radiation emitted from particular levels in the atmosphere. In addition, satellites have a data collection system (DCS) that relays data on numerous environmental variables from ground platforms or ocean buoys to processing centers; GOES can also transmit processed satellite images in facsimile, and the NOAA polar orbiters have an automatic picture transmission (APT) system that is utilized at about 1000 stations worldwide.

D NUMERICAL WEATHER PREDICTION

General circulation models of all kinds are also applied operationally to the day-to-day prediction of weather at centers around the world. Modern weather forecasting did not become possible until weather information could be collected, assembled and processed rapidly. The first development came in the mid-nineteenth century with the invention of telegraphy, which permitted immediate analysis of weather data by the drawing of synoptic charts. These were first displayed in Britain at the Great Exhibition of 1851. Severe storm events and loss of life and property prompted the development of weather forecasting in Britain and North America in the 1860s–1870s. Sequences of weather change were correlated with barometric pressure patterns in both space and time by such workers as Fitzroy and Abercromby, but it was not until later that theoretical models of weather systems were devised, notably the Bjerknes depression model (see Figure 9.7).

Forecasts are usually referred to as short-range (up to approximately three days), medium-range (up to approximately 14 days), and long-range (monthly or seasonal) outlooks. The first two may for current purposes be considered together as their methodology is similar and owing to increasing computing power they are becoming less distinguishable as separate types of forecast.

1 Short- and medium-range forecasting

During the first half of the twentieth century, short-range forecasts were based on synoptic principles, empirical rules and extrapolation of pressure changes. The Bjerknes model of cyclone development for mid-latitudes and simple concepts of tropical weather (see Chapter 9) served as the basic tools of the forecaster. The relationship between the development of surface lows and highs and the upper-air circulation was worked out during the 1940s and 1950s by C-G. Rossby, R. C. Sutcliffe and others, providing the theoretical basis of synoptic forecasting. In this way, the position and intensities of low and high pressure cells and frontal systems were predicted.

Since 1955 in the United States – and 1965 in the United Kingdom – routine forecasts have been based on numerical models. These predict the evolution of physical processes in the atmosphere by determinations of the conservation of mass, energy and momentum. The basic principle is that the rise or fall of surface pressure is related to mass convergence or divergence, respectively, in the overlying air column. This prediction method was first proposed by L. F. Richardson, who, in 1922, made a laborious test calculation that gave very unsatisfactory results. The major reason for this lack of success was that the net convergence or divergence in an air column is a small residual term compared with the large values of convergence and divergence at different levels in the atmosphere (see Figure 6.7). Small errors arising from observational limitations may therefore have a considerable effect on the correctness of the analysis.

Numerical weather prediction (NWP) methods developed in the 1950s use a less direct approach. The first developments assumed a one-level barotropic atmosphere with geostrophic winds and hence no convergence or divergence. The movement of systems could be predicted, but not changes in intensity. Despite the great simplifications involved in the barotropic model, it has been used for forecasting 500mb contour patterns. The latest techniques employ multi-level baroclinic models and include frictional and other effects; hence the basic mechanisms of cyclogenesis are provided for. It is noteworthy that *fields* of continuous variables, such as pressure, wind and temperature, are handled and that fronts are regarded as secondary, derived features. The vast increase in the number of calculations that these models perform necessitated a new generation of supercomputers to allow the preparation of forecast maps to keep sufficiently ahead of the weather changes!

Forecast practices in the major national weather prediction centers around the globe are basically similar. As an example of the operational use of weather forcasting models we discuss the methods and procedures of the National Centers for Environmental Prediction (NCEP) in Washington, DC, established in 1995. NCEP currently runs a global spectral model operationally. The Global Forecast System (GFS) model (formerly known as the AVN/MRF for aviation/medium-range forecast) model has a spectral truncation of T170 (approximately 0.7 × 0.7-degree grid), 42 unequally spaced vertical levels out to seven days. The truncation is increased to T62 with 28 levels out to 15 days. It should be noted that typically the computer time required decreases several-fold when the grid spacing is doubled. In order to produce a forecast, an analysis of currently observed weather conditions must first be generated as an initial condition for the model. Very sophisticated data assimilation algorithms take a large amount of observational data from a variety of platforms (surface stations, rawinsondes, ship, aircraft, satellite) which are often measured at irregular intervals in both space and time and merge it into a single coherent picture of current atmospheric conditions on standard pressure levels and at regular grid intervals. The model equations are then integrated into the future from this starting point.

The GFS currently runs out 17 simulations which are identical except for very small differences in initial conditions four times a day. The repetition of numerical forecasts incorporating minor differences in the initial conditions allows the effects of uncertainties in the observations, inaccuracies in the model formulations, and 'the chaotic' nature of atmospheric behavior to be accounted for in terms of probabilities. Errors in numerical forecasts arise from several sources. One of the most serious is the limited accuracy of the initial analyses due to data deficiencies. Coverage over the oceans is sparse, and only a quarter of the possible ship reports may be received within 12 hours; even over land more than one-third of the synoptic reports may be delayed beyond six hours. However, satellite-derived information and instrumentation on commercial aircraft fill gaps in the upper-air observations. Another limitation is imposed by the horizontal and vertical resolution of the models and the need to parameterize subgrid processes such as cumulus convection. The small-scale nature of the turbulent motion of the atmosphere means that some weather phenomena are basically unpredictable, for example, the specific locations of shower cells in an unstable air mass. Greater precision than the 'showers and bright periods' or 'scattered showers' is impossible for next-day forecasts. The procedure for preparing a forecast is becoming much less subjective, although in complex weather situations the skill of the experienced forecaster still makes the technique almost as much an art as a science. Detailed regional or local predictions can only be made within the framework of the general forecast situation for the country and demand thorough knowledge of possible topographic or other local effects by the forecaster. The average of these ensembles is used for the short-term forecast. The primary analysis products issued every six hours are MSL pressure, temperature and relative humidity at 850mb and 700mb, respectively, wind velocity at 300mb, 1000–500mb thickness, and 500mb vorticity.

NCEP also computes medium-range ensemble forecasts from the 17 ensemble runs performed at each interval. For example, the probability that the 24-hour precipitation amount some days in the future will exceed a certain threshold can be computed by counting the number of model runs where the value is exceeded in a certain grid box. This is a rough estimate of the probability because 17 simulations cannot span all possible weather scenarios, given the uncertainty in initial conditions and model formulation. Current forecasts are given as a 6–10-day outlook and 8–14-day outlook of the departure of temperature and precipitation from normal.

In order to calculate forecasts with more regional detail, NCEP utilizes a limited area 'eta' model which makes up to 84-hour forecasts over North America only. Like all operational weather models the eta is in a continual cycle of improvement and redesign. At present, however, the eta model has a 12km grid spacing and 60 vertical layers. A specialized vertical coordinate is employed in order to handle the sharp changes in topography which a high-resolution model encounters. Eta has a similar suite of output variables as the GFS.

Because a typical weather forecast, even in the highest resolution regional models, is meant to depict an average over a large grid box, the actual conditions at any single point within that grid box will not generally be accurately predicted. Forecasters have always subjectively applied model information to forecasts at a single point using their own experience as to how accurate model information has been in the past under certain circumstances (i.e., a subjective assessment of model bias). An effort to make such localized use of information more objective is called model output statistics (MOS) and actual weather conditions at specific weather stations are now commonly predicted using this technique.

MOS may be applied to any model and aims to objectively interpolate gridded model output to a single station based on its climate and weather history. Multiple regression equations are developed which relate the actual weather observed at a station over the course of time with the conditions predicted by the model. With a long enough history, MOS can make a correction for local effects not simulated in the model and for certain model biases. MOS variables include daily maximum/minimum temperature, 12-hour probability of precipitation occurrences and precipitation amount, probability of frozen precipitation, thunderstorm occurrence, cloud cover and surface winds.

Various types of specialty forecasts are also regularly made. In the United States, the National Hurricane Center in Miami is responsible for issuing forecast as to hurricane intensity changes and the track the storm will follow in the Atlantic and eastern Pacific areas. Forecasts are issued for 72 hours in advance four times daily. The central Pacific Hurricane Center performs similar forecasts for storms west of 140°W and east of the dateline. The US Weather Service also uses numerical models to predict the evolution of El Niño–Southern Oscillation which is important for long-range forecasts discussed below. Special events, such as the Olympic Games, are beginning to regularly employ numerical weather forecasting in their preparations and to use regional models designed to be most accurate at the single point of interest. MOS techniques are also used to improve these very specialized forecasts.

2 Nowcasting

Severe weather is typically short-lived (<2 hours) and, due to its mesoscale character (<100km), it affects local/regional areas, necessitating site-specific forecasts. Included in this category are thunderstorms, flash floods, gust fronts, tornadoes, high winds especially along coasts, over lakes and mountains, heavy snow and freezing precipitation. Mesoscale models with grid cells which can be less than 10km on a side are regularly used to study such phenomena in detail. The development of radar networks (Plate 7.2), new instruments and high-speed communication links has provided a means of issuing warnings of severe weather within the next hour. Several countries have recently developed integrated satellite and radar systems to provide information on the horizontal and vertical extent of thunderstorms, for example. Networks of automatic weather stations (including buoys) that measure wind, temperature and humidity supplement such data. In addition, for detailed boundary layer and lower troposphere data, there is now an array of vertical sounders. These include: acoustic sounders (measuring wind speed and direction from echoes created by thermal eddies), and specialized (Doppler) radar measuring winds in clear air by returns either from insects (3.5cm wavelength radar) or from variations in the air's refractive index (10cm wavelength radar). *Nowcasting* techniques use highly automated computers and image-analysis systems to integrate data from a variety of sources rapidly. Interpretation of the data displays requires skilled personnel and/or extensive software to provide appropriate information. The prompt warning of wind shear and downburst hazards at airports is one example of the importance of nowcasting procedures.

Overall, the greatest benefits from improved forecasting can be expected in aviation and the electric power industry for forecasts less than six hours ahead, in transportation, construction and manufacturing for 12- to 24-hour forecasts and in agriculture for two- to five-day forecasts. In terms of economic losses, the last category could benefit the most from more reliable and more precise forecasts.

3 Long-range outlooks

The atmosphere-ocean system is a non-linear (chaotic) system making exact long-term prediction of individual weather events impossible. Small errors in the initial conditions used to start a model simulation invariably grow in magnitude and spatial scale, and the entire globe will generally be affected by a small observational error at a single point before long. Therefore, long-term weather prediction and climate prediction do not try to predict individual weather events, since these would certainly be in error. Instead they generally try to represent the statistics of the climate rather than the weather itself and are often associated with probabilities based on statistical relationships.

Like numerical forecasting at shorter timescales, long-range (monthly and seasonal) outlooks use a combination of dynamical and statistical approaches in order to assess the probability of certain weather situations. Long-range forecasts rely on the idea that some types of weather, despite being unpredictable in its details, may, under certain circumstances, be more likely than in others. One major recent advance in long-range forecasts is the realization that El Niño/Southern Oscillation has documented statistical effects in many parts of the globe. For any particular El Niño or La Niña it is generally not realistic to forecast increased/decreased precipitation at most points in the globe but many regions show a statistical tendency towards more or less precipitation or higher/lower temperatures depending on the phase of ENSO. Long-range forecasts make use of these statistical relationships.

ENSO has a fairly regular periodicity allowing for some skill in predicting changes in phase just from climatology. Several dynamical models also try to predict the future phase of ENSO, though these have not been dramatically more sucessful than a knowledge of the climatology. The phase of ENSO is the single most important factor going into long-range forecasts today.

The United States NCEP is again typical of the methodology used globally. It currently issues 30-day and three-month seasonal forecasts up to one year into the future. The primary information used in these outlooks is the phase of ENSO, recent and extended climate history, the pattern of soil moisture which can affect temperature and precipitation far into the future, and an ensemble of 20 GCM model runs driven with predicted SSTs from an AOGCM simulation over the period. This information is used to produce a variety of indices which predict the probability of three equally likely categories of temperature (near normal, above/below normal) and precipitation (near average, above/below the median) (see Figures 8.7 and 8.8), together with tables for many cities. Figure 8.8A illustrates the observed height field corresponding to Figure 8.7A for February 2007, showing that the pattern is well represented on the forecast chart. Figure 8.8B and C show that in this case, as is usual, the temperature forecasts are more reliable than those for precipitation.

A statistical techniqe called a canonical correlation analysis uses all the above information to produce long-range outlooks. Simulated 700mb heights, global SST patterns, US surface temperature and precipitation for the past year are all used to infer possible preferred patterns. Temperature and precipitation history give information about persistence and trends over the year. ENSO is emphasized in this analysis but other natural modes of variability such as the North Atlantic Oscillation are also accounted for.

Secondary analyses which use single predictor variables are also available and become more or less useful than the correlation analysis under differing circumstances. The composite analysis estimates ENSO effects by defining whether a La Niña, El Niño or neutral conditions are forecast for the period of interest and then taking into account whether there is confidence that this one phase of ENSO will exist. Another index predicts future tempertaure and precipitation based on persistence over the past 10–15 years. This measure emphasizes trends and long-term regimes. A third secondary index is a constructed analog forecast from soil moisture patterns.

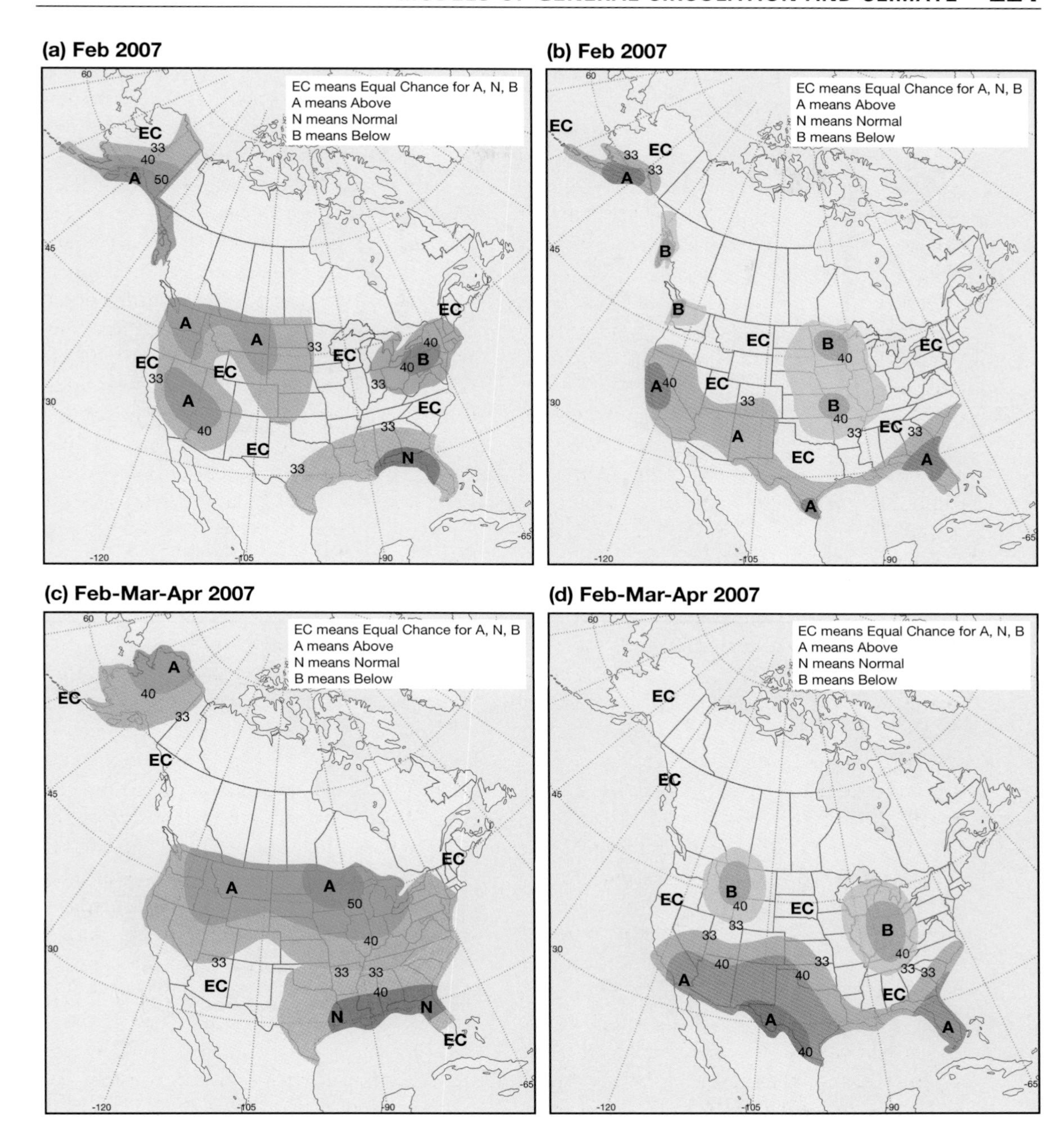

Figure 8.7 Monthly and seasonal forecasts of United States weather for: (A) February 2007 (temperature made at the end of January). (B) February 2007 (precipitation made at the end of January). (C) February, March, April (temperature made at the end of January). (D) February, March, April (precipitation made at the end of January). Predictions are made for four classes: normal (N), above normal (A), below normal (B), and equal chances for normal, above and below normal (EC).

Source: Reproduced by courtesy of NOAA.

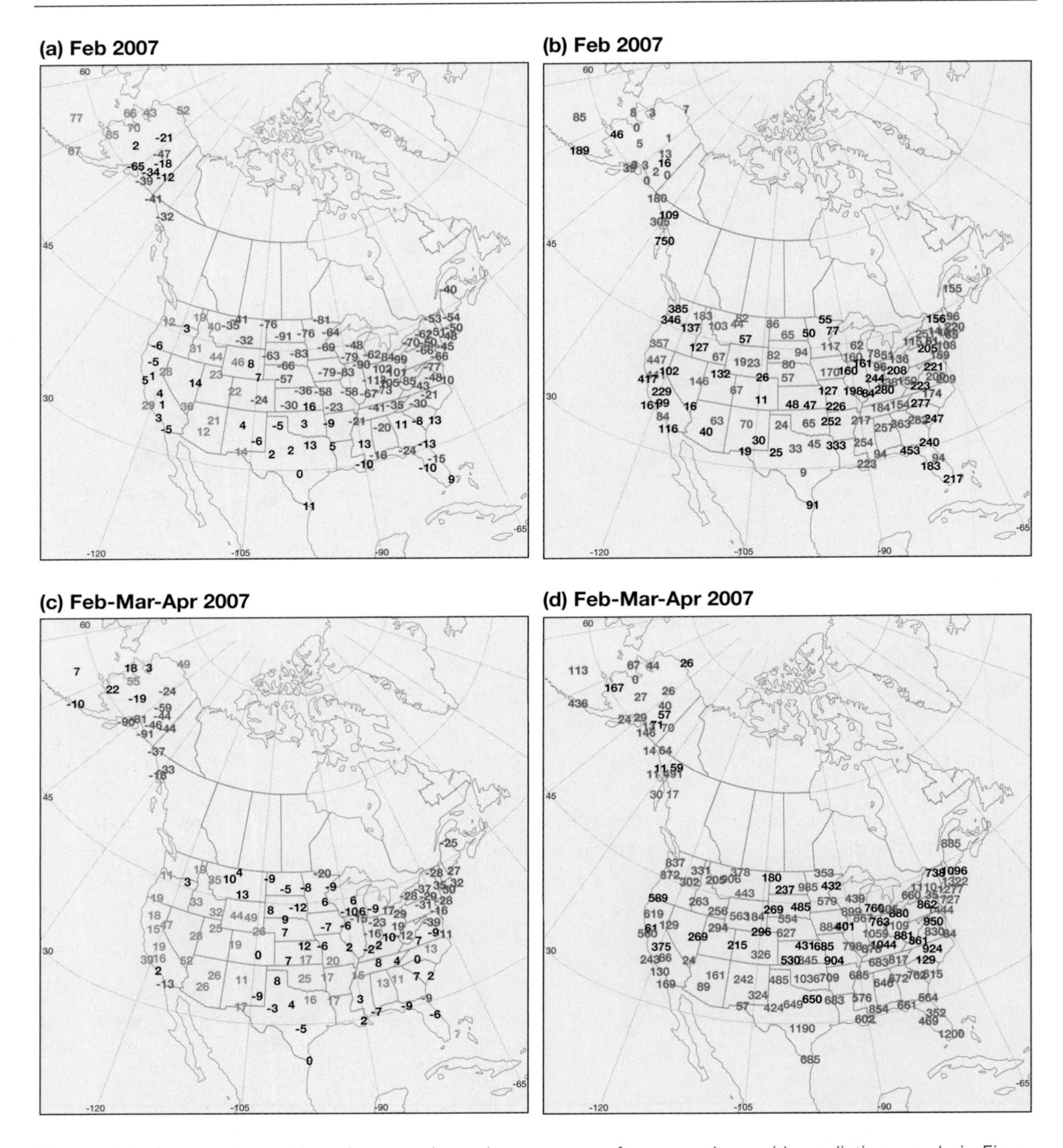

Figure 8.8 Observed monthly and seasonal weather outcomes for comparison with predictions made in Figure 8.7. (A) February 2007 temperature. (B) February 2007 precipitation. (C) February, March, April 2007 temperature. (D) February, March, April precipitation. For temperature maps red is above average, blue is below average corresponding to the colors in Figure 8.7. Similarly for the precipitation maps green is above average and brown is blow average. Note that the one-month temperature forecasts are much more accurate than the precipitation forecasts and that forecasts for both variables degrade for the three-month period.

Source: Reproduced by courtesy of NOAA.

Forecast skill for long-range outlooks is mixed. For all measures skill in temperature is higher than for precipitation. Precipitation forecasts generally show little skill unless there is a strong El Niño or La Niña. Temperature outlooks show the largest skill in late winter and late summer.

SUMMARY

Various types of numerical model are used to study the mechanisms of the atmospheric circulation, climate processes and weather forecasting. These include vertical column models of radiative and convective processes, one- and two-dimensional energy balance models and complete three-dimensional general circulation models (GCMs) which can be coupled with ocean and sea ice models or with regional climate models. While initially developed for weather forecasting such models are now widely used to study climatic anomalies and past and future changes in global climate. These uses require coupling of atmospheric and oceanic GCMs and the representation of ice and land surface processes.

Forecasts are issued for different timescales and the techniques involved differ considerably. Immediate 'nowcasts' rely heavily on current radar and satellite data. Short- and intermediate-range forecasts are now derived from numerical models with some statistical guidance while long-range forecasts use numerical models in an almost completely probabilistic manner.

DISCUSSION TOPICS

- What types of experiment can be performed with a global climate model that cannot be observed in nature?
- What are some of the problems encountered in evaluating the results of GCM experiments?
- Consider the different concepts and methodologies used in forecasting weather on time scales of a few hours, the next day and the next week.

REFERENCES AND FURTHER READING

Books

Bader, M. J. *et al.* (1995) *Images in Weather Forecasting*, Cambridge University Press, Cambridge, 499pp. [Extensive collection of imagery illustrating all types of synoptive phenomena]

Barry, R. G. and Carleton, A. M. (2001) *Synoptic and Dynamic Climatology,* Routledge, London, 620pp. [Advanced text covering climate data and analysis, the general circulation, global and regional climates, teleconnections, synoptic systems, and synoptic climatology]

Browning, K. A. (ed.) (1983) *Nowcasting*, Academic Press, New York, 256pp. [Treats the design of forecast systems, new remote sensing tools, and simple and numerical forecasts]

Conway, E. D. and the Maryland Space Grant Consortium (1997) *Introduction to Satellite Imagery Interpretation*, Johns Hopkins University Press, Baltimore, MD, 242pp [Useful, well-illustrated introduction to basics of remote sensing, satellite systems and atmospheric applications – clouds, winds, jet streams, synoptic and mesoscale systems, air quality; oceanographic applications including sea ice]

Henderson-Sellers, A. (ed.) (1995, *Future Climates of the World.: A Modelling Perspective,* Elsevier, Amsterdam, 636pp. [Provides geological perspective of past climate, observed climate variabilty and future projections, anthropogenic effects]

McGuffie, K. and Henderson-Sellers, A. (2005) *A Climate Modelling Primer* (2nd edn), John Wiley & Sons, Chichester 296pp. [Explains the basis and mechanisms of climate models; includes CD with additional resources]

Monmonier, M. (1999) *Air Apparent. How Meteorologists Learned to Map, Predict and Dramatize the Weather,* University of Chicago Press, Chicago, IL, 309pp. [A readable history of the development of the weather map and forecasting, including the tools and technologies on which they are based]

Trenberth, K. E. (ed.) (1992) *Climate System Modeling,* Cambridge University Press, Cambridge, 788pp. [Essays by specialists covering all aspects of modeling the climate system and its components – oceans, sea ice, biosphere, gas exchanges]

Washington, W. M. and Parkinson, C. L. (2005) *An Introduction to Three-dimensional Climate Modeling,* University Science Books, Mill Valley, CA, 353pp. [Comprehensive account of the basis of atmospheric general circulation models]

Websites

http://www.meto.gov.uk/research/hadleycentre/pubs/brochures/B2001/precis.pdf

Articles

Barry, R. G. (1979) Recent advances in climate theory based on simple climate models. *Prog. Phys. Geog.* 3, 259–86.

Bosart, L. (1985) Weather forecasting, in Houghton, D. D. (ed.) *Handbook of Applied Meteorology,* Wiley, New York, 205–79.

Browning, K. A. (1980) Local weather forecasting. *Proc. Roy. Soc. Lond. Sect. A* 371, 179–211.

Carson, D. J. (1999) Climate modelling: achievements and prospects. *Quart. J. Roy. Met. Soc.* 125, 1–27.

Cullen, M. J. P. (1993) The Unified Forecast/Climate model. *Met. Mag.* 122, 81–94.

Druyan, L. M., Somerville, R. C. J. and Quirk, W. J. (1975) Extended-range forcasts with the GISS model of the global atmosphere. *Mon. Wea. Rev.* 103, 779–95.

Foreman, S. J. (1992) The role of ocean models in FOAM. *Met. Mag.* 121, 113–22.

Gates, W. L. (1992) The Atmospheric Model Intercomparison Project. *Bull. Amer. Met. Soc.* 73(120), 1962–70.

Harrison, M. J. S. (1995) Long-range forecasting since 1980 – empirical and numerical prediction out to one month for the United Kingdom. *Weather* 50(12), 440–9.

Hunt, J. C. R. (1994) Developments in forecasting the atmospheric environment. *Weather* 49(9), 312–18.

Kalnay, E., Kanamitsu, M. and Baker, W. E. (1990) Global numerical weather prediction at the National Meteorological Center. *Bull. Amer. Met. Soc.* 71, 1410–28.

Kiehl, J. T. (1992) Atmospheric general circulation modeling, in Trenberth, K. E. (ed.) *Climate System Modelling,* Cambridge University Press, Cambridge, 319–69.

Klein, W. H. (1982) Statistical weather forecasting on different time scales. *Bull. Amer. Met. Soc.* 63, 170–7.

McCallum, E. and Mansfield, D. (1996) Weather forecasts in 1996. *Weather* 51(5), 181–8.

Meehl, G. A. (1984) Modelling the earth's climate. *Climatic Change* 6, 259–86.

Meehl, G. A. (1992) Global coupled models: atmosphere, ocean, sea ice, in Trenberth, K. E. (ed.) *Climate System Modelling,* Cambridge University Press, Cambridge, 555–81.

Monin, A. S. (1975) Role of oceans in climate models, in *Physical Basis of Climate: Climate Modelling,* GARP Publications Series, Report. No. 16, WMO, Geneva, 201–5.

National Weather Service (1991) *Experimental Long-lead Forecast Bulletin,* NOAA, Climate Prediction Center, Washington, DC.

Palmer, T. N. and Anderson, D. L. T. (1994) The prospects for seasonal forecasting – a review paper. *Quart. J. Roy. Met. Soc.* 120, 755–93.

Phillips, T. J. (1996) Documentation of the AMIP models on the World Wide Web. *Bull. Amer. Met. Soc.* 77(6), 1191–6.

Reed, D. N. (1995) Developments in weather forecasting for the medium range. *Weather* 50(12), 431–40.

Saltzman, B. (1983) Climatic systems and analysis. *Adv. Geophys.* 25, 175–233.

Smagorinsky, J. (1974) Global atmospheric modeling and the numerical simulation of climate, in Hess, W. N. (ed.) *Weather and Climate Modification*, Wiley, New York, 633–86.

Wagner, A. J. (1989) Medium- and long-range weather forecasting. *Weather and Forecasting* 4, 413–26.

Washington, W. M. (1992) Climate-model responses to increased CO_2 and other greenhouse gases, in Trenberth, K. E. (ed.) *Climate System Modelling*, Cambridge University Press, Cambridge, 643–68.

CHAPTER NINE

Mid-latitude synoptic and mesoscale systems

LEARNING OBJECTIVES

When you have read this chapter you will:

- understand the air mass concept, the characteristics of the major air masses and their geographical occurrence
- know the mechanisms of frontogenesis and the various frontal types
- understand the relationships between upper-air and surface processes in forming frontal cyclones
- know the major types of non-frontal cyclone and how they form
- be familiar with the role of mesoscale convective systems in severe weather

This chapter examines the classical ideas about air masses and their role in the formation of frontal boundaries and in the development of extratropical cyclones. It also discusses the limitations of those ideas and more recent models of mid-latitude weather systems. Mesoscale systems in mid-latitudes are also treated. The chapter concludes with a brief overview of severe weather phenomena.

A THE AIR-MASS CONCEPT

An air mass is defined as a large body of air whose physical properties (temperature, moisture content and lapse rate) are more or less uniform horizontally for hundreds of kilometers. The theoretical ideal is a *barotropic* atmosphere where surfaces of constant pressure are not intersected by isosteric (constant-density) surfaces, so that in any vertical cross-section, as shown in Figure 9.1, isobars and isotherms are parallel.

Three factors determine the nature and degree of uniformity of air-mass characteristics: (1) the nature of the source area where the air mass obtains its original qualities; (2) the direction of movement and changes that occur as an air mass moves over long distances; and (3) the age of the air mass. Air masses are classified on the basis of two primary factors. The first is temperature, giving arctic, polar and tropical air, and the second is the surface type in their region of origin, giving maritime and continental categories.

B NATURE OF THE SOURCE AREA

The basic idea of air-mass formation is that radiative and turbulent transfers of energy and

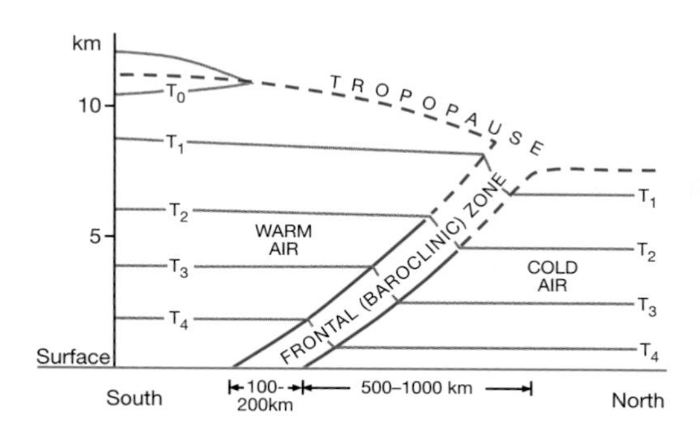

Figure 9.1 A schematic height cross-section for the Northern Hemisphere showing barotropic air masses and a baroclinic frontal zone (assuming that density decreases with height only).

moisture, between the land or ocean surface and the atmosphere, give rise to distinctive physical properties of the overlying air through vertical mixing. A degree of equilibrium between the surface conditions and the properties of the overlying air mass will be achieved if the air remains over a given geographical region for a period of about three to seven days. The chief source regions of air masses are necessarily areas of extensive, uniform surface type that are overlain by quasi-stationary pressure systems. These requirements are fulfilled where there is slow divergent flow from the major thermal and dynamic high pressure cells. In contrast, low pressure regions are zones of convergence into which air masses move (see F, this chapter).

The major cold and warm air masses will now be discussed.

1 Cold air masses

The principal sources of cold air in the Northern Hemisphere are (1) the continental anticyclones of central-eastern Siberia and northern Canada where continental Polar (cP) air masses form, and (2) the Arctic Basin, when it is dominated by high pressure in winter and spring (Figure 9.2). Sometimes Arctic Basin air is designated as continental Arctic (cA), but the differences between cP and cA air masses are limited mainly to the middle and upper troposphere, where temperatures are lower in the cA air.

The snow-covered source regions of these two air masses lead to marked cooling of the lower layers (Figure 9.3). Since the vapor content of cold air is very limited, the air masses generally have a mixing ratio of only 0.1–0.5g/kg near the surface.

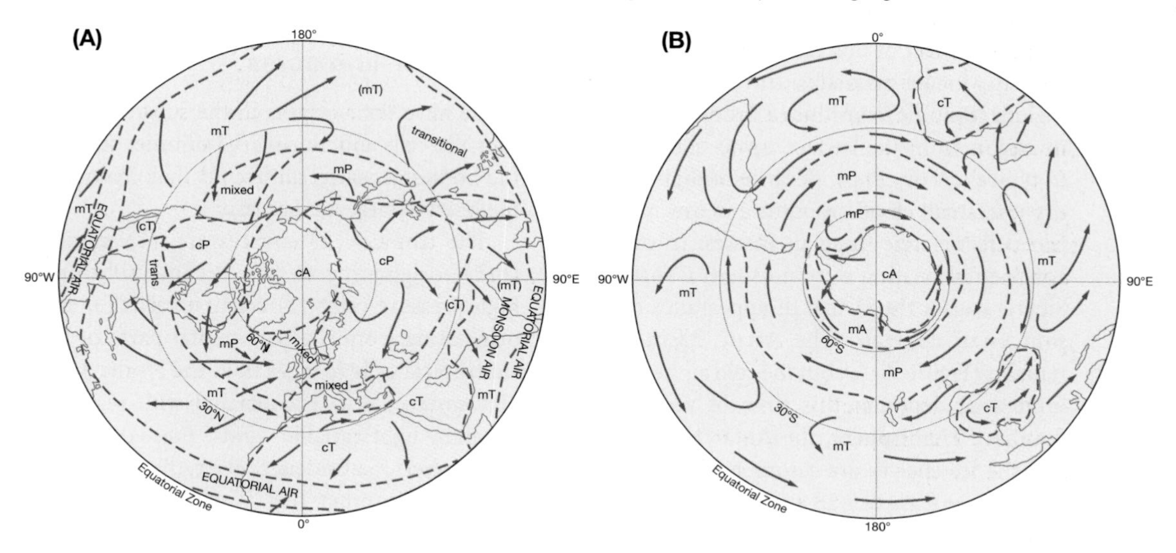

Figure 9.2 Air masses in winter. A: Northern Hemisphere; B: Southern Hemisphere.
Sources: A: After Petterssen (1950) and Crowe (1965). B: After Taljaard *et al.* (1969) and Newton (1972).

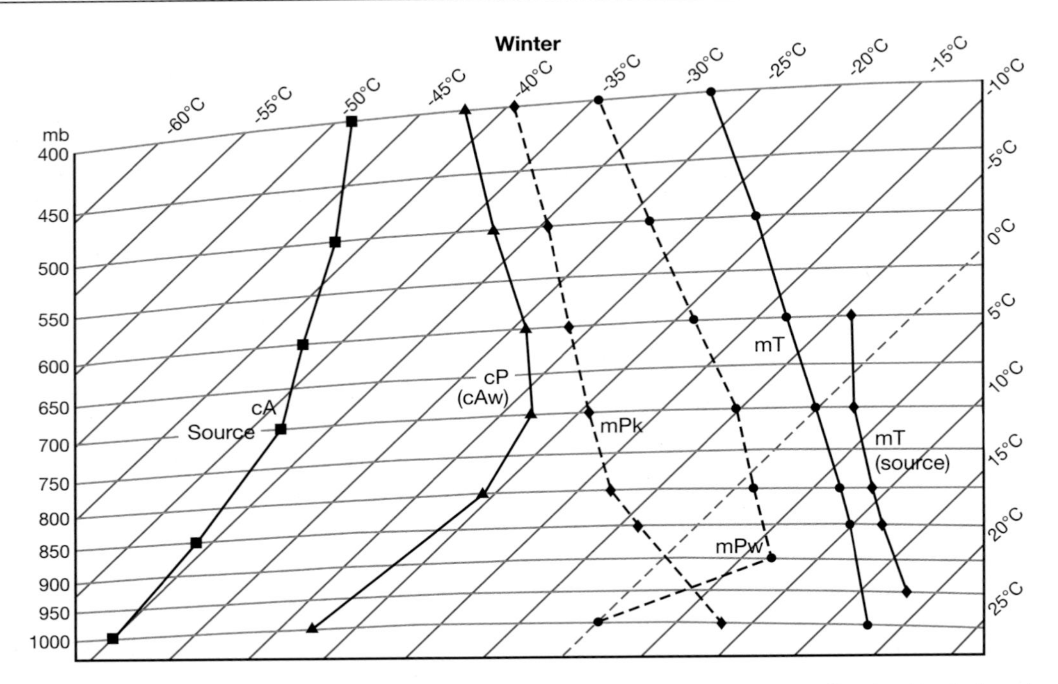

Figure 9.3 The average vertical temperature structure for selected air masses affecting North America at about 45–50°N, recorded over their source areas or over North America in winter.

Sources: After Godson (1950), Showalter (1939) and Willett.

The stability produced by the effect of surface cooling prevents vertical mixing, so further cooling occurs more slowly by radiation losses only. The effect of this radiative cooling and the tendency for air-mass subsidence in high pressure regions combine to produce a strong temperature inversion from the surface up to about 850mb in typical cA or cP air. Because of their extreme dryness, small cloud amounts and low temperatures characterize these air masses. In summer, continental heating over northern Canada and Siberia causes the virtual disappearance of their sources of cold air. The Arctic Basin source remains (Figure 9.4A), but the cold air here is very limited in depth at this time of year. In the Southern Hemisphere, the Antarctic continent and the ice shelves are a source of cA air in all seasons (see Figures 9.2B and 9.4B). There are no sources of cP air, however, due to the dominance of ocean areas in mid-latitudes. At all seasons, cA or cP air is greatly modified by a passage over the ocean. Secondary types of air mass are produced by such means and these will be considered below.

2 Warm air masses

These have their origins in the subtropical high pressure cells and, during the summer season, in the bodies of warm surface air that characterize the heart of large land areas.

The tropical (T) sources are: (1) maritime (mT), originating in the oceanic subtropical high pressure cells; (2) continental (cT), either originating from the continental parts of these subtropical cells (e.g., as does the North African *Harmattan*); or (3) associated with regions of generally light variable winds, assisted by upper tropospheric subsidence, over the major continents in summer (e.g., Central Asia). In the Southern Hemisphere, the source area of mT air covers about half of the hemisphere. There is no significant temperature gradient between the

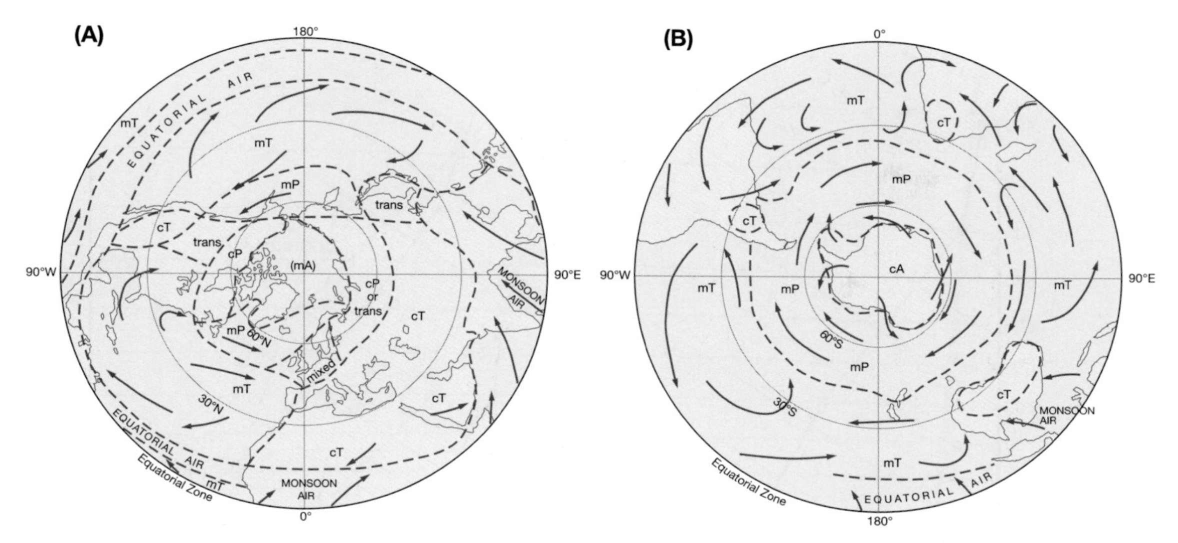

Figure 9.4 Air masses in summer. A: Northern Hemisphere; B: Southern Hemisphere.
Sources: A: After Petterssen (1950) and Crowe (1965). B: After Taljaard *et al.* (1969) and Newton (1972).

equator and the oceanic subtropical convergence at about 40°S.

The mT type is characterized by high temperatures (accentuated by the warming due to subsidence), high humidity of the lower layers over the oceans and stable stratification. Since the air is warm and moist near the surface, stratiform cloud commonly develops as the air moves poleward from its source. The continental type in winter is restricted mainly to North Africa (see Figure 9.2), where it is a warm, dry and stable air mass. In summer, warming of the lower layers by the heated land generates a steep lapse rate, but despite its instability the low moisture content prevents the development of cloud and precipitation. In the Southern Hemisphere, cT air is rather more prevalent in winter over the subtropical continents, except in South America. In summer, much of southern Africa and northern Australia is affected by mT air, while there is a small source of cT air over Argentina (see Figure 9.4B). The characteristics of the primary air masses are illustrated in Figures 9.3 and 9.5. In some cases, movement away from the source region has considerably affected their properties, and this question is discussed below (see p. 231).

Source regions can also be defined from analysis of airstreams. Streamlines of the mean resultant winds (see Note 1) in individual months can be used to analyse areas of divergence representing air-mass source regions, downstream airflow, and the confluence zones between different airstreams. Figure 9.6A shows air-mass dominance in the Northern Hemisphere in terms of annual duration. Four sources are indicated: the subtropical North Pacific and North Atlantic anticyclones, and their Southern Hemisphere counterparts. For the entire year, air from these sources covers at least 25 percent of the Northern Hemisphere; for six months of each year they affect almost three-fifths of the hemisphere. In the ocean-dominated Southern Hemisphere, the airstream climatology is much simpler (Figure 9.6B). Source areas are the oceanic subtropical anticyclones. Antarctica is the major continental source, with another mainly in winter over Australia.

C AIR-MASS MODIFICATION

As an air mass moves away from its source region it is affected by different heat and moisture

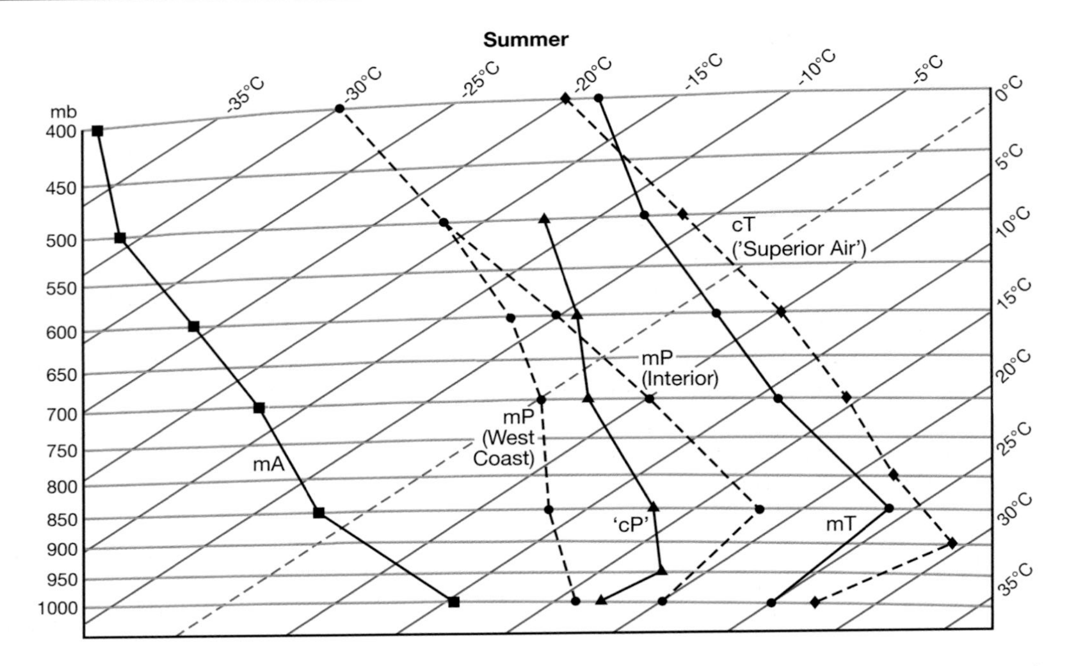

Figure 9.5 The average vertical temperature structure for selected air masses affecting North America in summer.

Sources: After Godson (1950), Showalter (1939) and Willett.

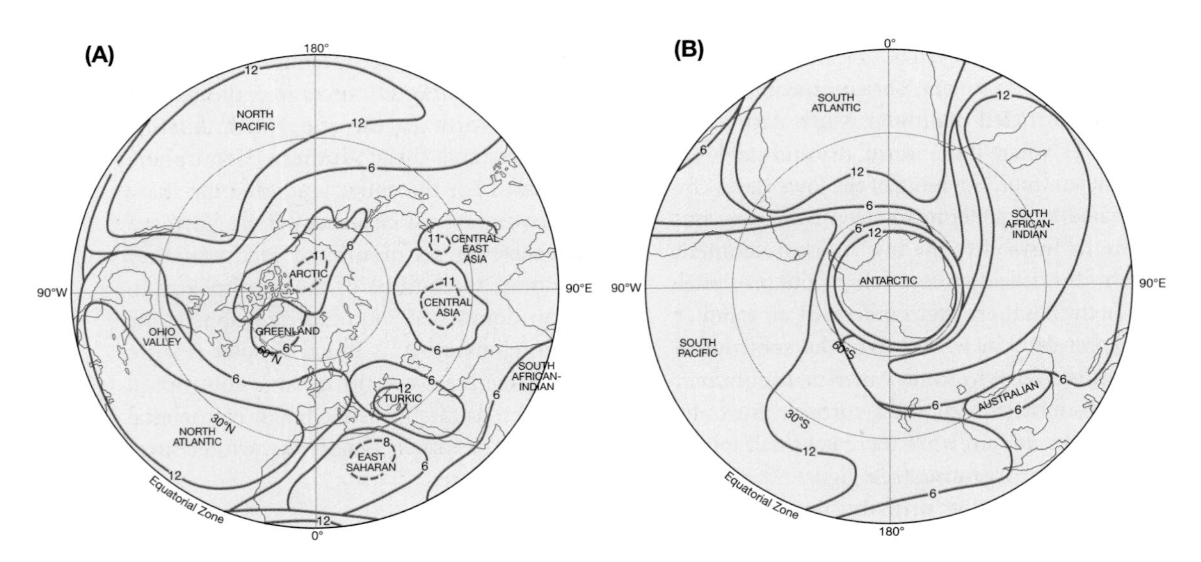

Figure 9.6 Air-mass source regions in the Northern Hemisphere (A) and the Southern Hemisphere (B). Numbers show the areas affected by each air mass in months per year.

Sources: After Wendland and Bryson (1981) and Wendland and McDonald (1986).

exchanges with the ground surface and by dynamic processes in the atmosphere. Thus, a barotropic air mass is gradually changed into a moderately *baroclinic* airstream in which isosteric and isobaric surfaces intersect one another. The presence of horizontal temperature gradients means that air cannot travel as a solid block maintaining an unchanging internal structure. The trajectory (i.e., actual path) followed by an air parcel in the middle or upper troposphere will normally be quite different from that of a parcel near the surface, due to the increase of westerly wind velocity with height in the troposphere and changes in the wind direction aloft. The structure of an airstream at a given instant is largely determined by the past history of air-mass modification processes. In spite of these qualifications, the air-mass concept retains practical value and is now used in air chemistry research.

1 Mechanisms of modification

The mechanisms by which air masses are modified are, for convenience, treated separately, although in practice they may operate together.

Thermodynamic changes

An air mass may be heated from below either by passing from a cold to a warm surface or by solar heating of the ground over which the air is located. Similarly, but in reverse, air can be cooled from below. Heating from below acts to increase air-mass instability, so the effect may be spread rapidly through a considerable thickness of air, whereas surface cooling produces a temperature inversion, which limits the vertical extent of the cooling. Thus, cooling tends to occur gradually through radiative heat loss by the air.

Changes can also occur through increased evaporation, the moisture being supplied either from the underlying surface or by precipitation from an overlying air-mass layer. In reverse, the abstraction of moisture by condensation or precipitation can also cause changes. An associated, and most important, change is the respective addition or loss of latent heat accompanying this condensation or evaporation. Annual values of latent and sensible heat transfers to the atmosphere, illustrated in Figures 3.30 and 3.31, show where these effects are important.

Dynamic changes

Dynamic (or mechanical) changes are superficially different from thermodynamic ones because they involve mixing or pressure changes associated with the actual movement of the air mass. The physical properties of air masses are considerably modified, for example, by a prolonged period of turbulent mixing (see Figure 5.7). This process is particularly important at low levels, where surface friction intensifies natural turbulence, providing a ready mechanism for the upward transfer of heat and moisture.

The radiative and advective exchanges discussed earlier are *diabatic*, but the ascent or descent of air causes adiabatic changes of temperature. Large-scale lifting may result from forced ascent by a mountain barrier or from airstream convergence. Conversely, sinking may occur when high-level convergence sets up subsidence or when stable air, that has been forced up over high ground by the pressure gradient, descends in its lee. Dynamic processes in the middle and upper troposphere are in fact a major cause of air-mass modification. The decrease in stability aloft, as air moves away from areas of subsidence, is a common example of this type of mechanism.

2 The results of modification: secondary air masses

Study of the ways in which air masses change in character tells us a great deal about many common meteorological phenomena.

Cold air

Continental polar air frequently streams out from Canada over the western North Atlantic in winter, where it undergoes rapid transformation. Heating over the Gulf Stream Drift makes the lower layers unstable, and evaporation into the air leads to sharp increases of moisture content (see Figure 3.7) and cloud formation. The turbulence associated with the convective instability is marked by gusty conditions. When the air reaches the central Atlantic, it becomes a cool, moist, maritime polar (mP) air mass. Analogous processes occur with outflow from Asia over the North Pacific (see Figure 7.2). Over mid-latitudes of the Southern Hemisphere, the circumpolar ocean gives rise to a continuous zone of mP air that, in summer, extends to the margin of Antarctica. In this season, however, a considerable gradient of ocean temperatures associated with the oceanic Antarctic Convergence makes the zone far from uniform in its physical properties.

Bright periods and squally showers, with a variable cloud cover of cumulus and cumulonimbus, typify the weather in cP airstreams. As mP air moves eastward towards Europe, the cooler sea surface may produce a neutral or even stable stratification near the surface, especially in summer, but subsequent heating over land will again regenerate unstable conditions. Similar conditions, but with lower temperatures, arise when cA air crosses high latitude oceans, producing maritime Arctic (mA) air.

When cP air moves southward in winter, over central North America, for example, it becomes more unstable, but there is little gain in moisture content. The cloud type is scattered shallow cumulus, which only rarely gives showers. Exceptions occur in early winter around the eastern and southern shores of Hudson Bay and the Great Lakes. Until these water bodies freeze over, cold airstreams that cross them are warmed rapidly and supplied with moisture, leading to locally heavy snowfalls (p. 292). Over Eurasia and North America, cP air may move southward and later recurve northward. Some schemes of air-mass classification cater for such possibilities by specifying whether the air is colder (k), or warmer (w), than the surface over which it is passing.

In some parts of the world, the surface conditions and air circulation produce air masses with intermediate characteristics. Northern Asia and northern Canada fall into this category in summer. In a general sense, the air has affinities with continental polar air masses but these land areas have extensive bog and water surfaces, so the air is moist and cloud amounts are quite high. In a similar manner, melt-water ponds and openings in the arctic pack ice make the area a source of mA air in summer (see Figure 9.4A). This designation is also applied to air over the Antarctic pack ice in winter that is much less cold in its lower levels than the air over the continent itself.

Warm air

The modification of warm air masses is usually a gradual process. Air moving poleward over progressively cooler surfaces becomes increasingly stable in the lower layers. In the case of mT air with high moisture content, surface cooling may produce advection fog, which is particularly common, for example, in the southwestern approaches to the English Channel during spring and early summer, when the sea is still cool. Similar development of advection fog in mT air occurs along the South China coast in February to April, and also off Newfoundland and over the coast of northern California in spring and summer. If the wind velocity is sufficient for vertical mixing, low stratus cloud forms in the place of fog, and drizzle may result. In addition, forced ascent of the air by high ground, or by overriding of an adjacent air mass, can produce heavy rainfall.

The cT air originating in those parts of the subtropical anticyclones situated over the arid subtropics in summer is extremely hot and dry. It is typically unstable at low levels and dust storms may occur, but the dryness and the subsidence of the upper air limit cloud development. In the case of North Africa, cT air may move out over the Mediterranean, rapidly acquiring moisture, with the consequent release of potential instability triggering off showers and thunderstorm activity.

Air masses in low latitudes present considerable problems of interpretation. The temperature contrasts found in middle and high latitudes are virtually absent, and the differences that do exist are due principally to moisture content and to the presence or absence of subsidence. *Equatorial air* is usually cooler than that subsiding in the subtropical anticyclones, for example. On the equatorward sides of the subtropical anticyclones in summer, the air moves westward from areas with cool sea surfaces (e.g., off northwest Africa and California) towards higher sea-surface temperatures. Moreover, the southwestern parts of the high pressure cells are affected only by weak subsidence due to the vertical structure of the cells. As a result, the mT air moving westward on the equatorward sides of the subtropical highs becomes much less stable than that on their northeastern margin. Eventually, such air forms the very warm, moist, unstable 'equatorial air' of the Intertropical Convergence Zone (see Figures 9.2 and 9.4). *Monsoon air* is indicated separately in these figures, although there is no basic difference between it and mT air. Modern approaches to tropical climatology are discussed in Chapter 11.

3 The age of the air mass

Eventually, the mixing and modification that accompanies the movement of an air mass away from its source causes the rate of energy exchange with its surroundings to diminish, and the various associated weather phenomena tend to dissipate. This process leads to the loss of its original identity until, finally, its features merge with those of surrounding airstreams and the air may come under the influence of a new source region.

Northwest Europe is shown as an area of 'mixed' air masses in Figures 9.2 and 9.4. This refers to the variety of sources and directions from which air may invade the region. The same is also true of the Mediterranean Sea in winter, although the area does impart its own particular characteristics to polar and other air masses that stagnate over it. Such air is termed *Mediterranean*. In winter, it is convectively unstable (see Figure 4.6) as a result of the moisture picked up over the Mediterranean Sea.

The length of time during which an air mass retains its original characteristics depends very much on the extent of the source area and the type of pressure pattern affecting the area. In general, the lower air is changed much more rapidly than that at higher levels, although dynamic modifications aloft are no less significant in terms of weather processes. Modern air-mass concepts must therefore be flexible from the point of view of both synoptic and climatological studies.

D FRONTOGENESIS

The first real advance in our understanding of mid-latitude weather variations was made with the discovery that many of the day-to-day changes are associated with the formation and movement of boundaries, or *fronts*, between different air masses. Observations of the temperature, wind direction, humidity and other physical phenomena during unsettled periods showed that discontinuities often persist between impinging air masses of differing characteristics. The term 'front' for these surfaces of air-mass conflict was a logical one, proposed during the First World War by a group of meteorologists led by Vilhelm Bjerknes working in Norway (see Box 9.1). Their ideas are still an integral part of weather analysis and forecasting in middle and high latitudes.

1 Frontal waves

The typical geometry of an air-mass interface, or front, resembles a wave form (Figure 9.7). Similar wave patterns are, in fact, found to occur on the interfaces between many different media, for example, waves on the sea surface, ripples on beach sand, eolian sand dunes and so on. Unlike these wave-forms, however, the frontal waves in the atmosphere are usually unstable; that is, they suddenly originate, increase in size, and then

SIGNIFICANT 20TH-CENTURY ADVANCE

9.1 The polar front theory of cyclones

The most significant and lasting contribution to synoptic meteorology in the twentieth century was made by the 'Bergen school of meteorologists' led by Vilhelm Bjerknes working in Norway during World War I. Isolated by the war from other sources of information, they focused on careful, systematic analysis of synoptic weather maps and time cross-sections of weather systems.

There were three components to the theory published during 1919–1922: a cyclone model (Jacob Bjerknes), the idea of a cyclone life-cycle and frontal occlusion (Tor Bergeron) and the concept of cyclone families developing along the polar front (Halvor Solberg). It was proposed that mid-latitude cyclones develop in conjunction with frontogenesis as airstream convergence leads to boundaries developing between adjacent air masses. The term front and the concept of frontal occlusion were introduced into the meteorological vocabulary. They also outlined a cross-sectional model of the distribution of clouds and precipitation in relation to frontal zones that is still widely used. In the 1930s, Bergeron distinguished between ana- and kata-types of fronts, but these ideas were not widely used until the 1960s. Although recent work has modified many aspects of the ideas of the Bergen school, several essential attributes have been clarified and reinforced. For example, in the occlusion process, the warm front may become bent back in the form of a T-bone, as noted originally by Bergeron. Theoretical and observational studies indicate that major cyclone elements are conveyor belts that transport heat and moisture within the system and lead to cellular precipitation structures.

It is well recognized that not all mid-latitude cyclones develop in frontal wave families like those forming over the oceans. Petterssen and Smeybe (1971) drew attention to the differences between waves that form in a frontal zone over the North Atlantic (type A) and those forming over North America (type B). Continental development usually involves cold air, with possibly an arctic cold front, in an upper-level trough moving eastward over a zone of low-level warm advection. Cyclogenesis can develop from a dry trough in the lee of the Rocky Mountains.

References

Friedman, R.M. (1989) *Appropriating the Weather. Vilhelm Bjerknes and the Construction of a Modern Meterology,* Cornell University Press, Ithaca, NY, 251pp.

Petterssen, S. and Smeybe, S.J. (1971) On the development of extratropical cyclones. *Quart. J. Roy. Met. Soc.* 97, 457–82.

gradually dissipate. Numerical model calculations show that in mid-latitudes waves in a baroclinic atmosphere are unstable if their wavelength exceeds a few thousand kilometers. Frontal wave cyclones are typically 1500–3000km in wavelength. The circulation of the upper troposphere plays a key role in providing appropriate conditions for their development and growth, as shown below.

2 The frontal-wave depression

A depression (also termed a low or cyclone) (see Note 2) is an area of relatively low pressure, with a more or less circular isobaric pattern. It covers an area 1500–3000km in diameter and usually has a lifespan of four to seven days. Systems with these characteristics, which are prominent on daily weather maps, are referred to as *synoptic-*

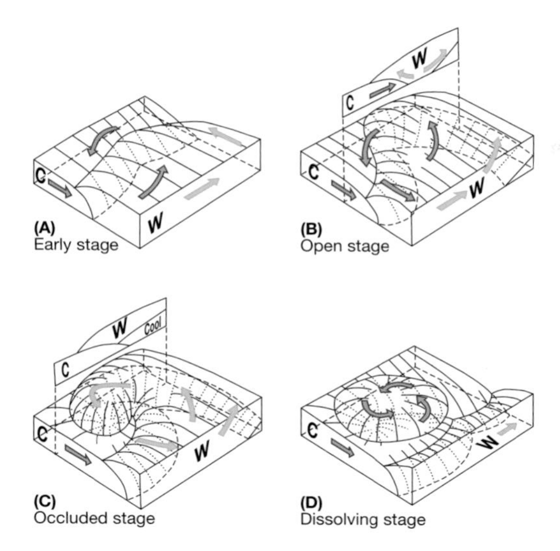

Figure 9.7 Four stages in the typical development of a mid-latitude depression. Satellite views of the cloud systems corresponding to these stages are shown in Figure 9.8.

Source: Mostly after Strahler (1965), modified after Beckinsale.

Notes: C = cold air; W = warm air.

scale features. The mid-latitude depression is usually associated with the convergence of contrasting air masses. According to the 'Norwegian cyclone model' (see Figure 9.7), the interface between these air masses develops into a wave form with its apex located at the center of the low pressure area. The wave encloses a mass of warm air between modified cold air in front and fresh, cold air in the rear. The formation of the wave also creates a distinction between the two sections of the original air-mass discontinuity since, although each section still marks the boundary between cold and warm air, the weather characteristics found within each section are very different. The two sections of the frontal surface are distinguished by the names *warm front* for the leading edge of the wave and the warm air and *cold front* for that of the cold air to the rear (see Figure 9.7B).

The boundary between two adjacent air masses is marked by a strongly baroclinic zone of large temperature gradient, 100–200km wide (see C, this chapter, and Figure 9.1). Sharp discontinuities

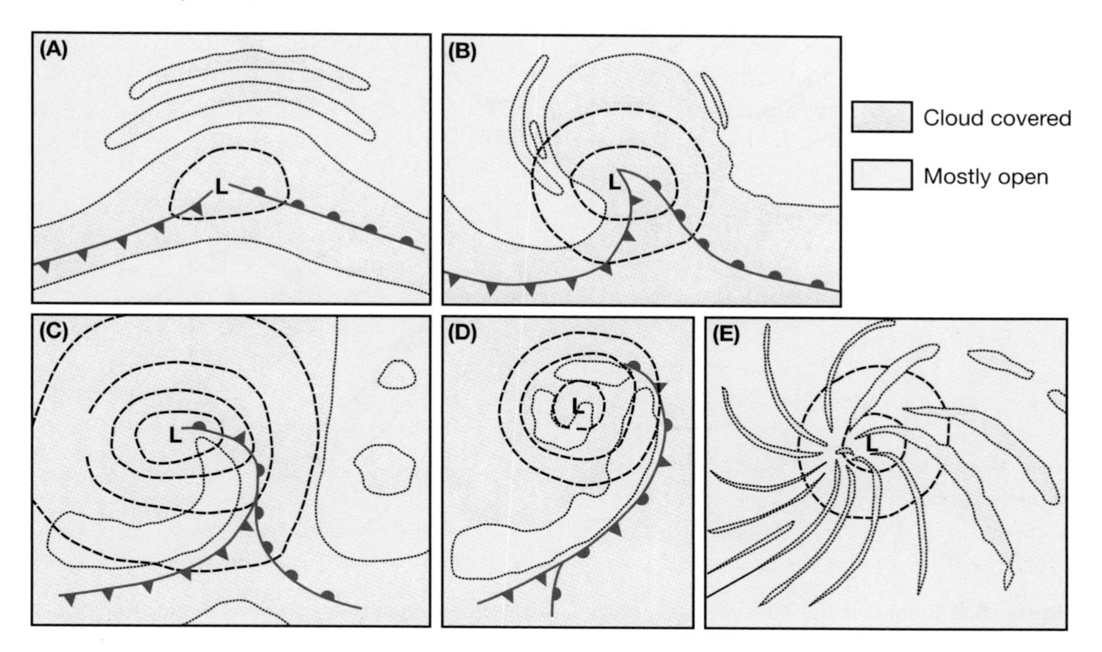

Figure 9.8 Schematic patterns of cloud cover (white) observed from satellites, in relation to surface fronts and generalized isobars. A, B, C and D correspond to the four stages in Figure 9.7.

Source: After Boucher and Newcomb (1962). Courtesy of the American Meteorological Society.

of temperature, moisture and wind properties at fronts, especially the warm front, are rather uncommon. Such discontinuities are usually the result of a pronounced surge of fresh, cold air in the rear sector of a depression, but in the middle and upper troposphere they are often caused by subsidence and may not coincide with the location of the baroclinic zone. In meteorological analysis centers, numerous criteria are used to locate frontal boundaries: 1000–500mb thickness gradients, 850mb wet-bulb potential temperature, cloud and precipitation bands, and wind shifts. However, a forecaster may have to use judgement when some of these criteria disagree.

On satellite imagery, active cold fronts in a strong baroclinic zone commonly show marked spiral cloud bands, formed as a result of the thermal advection (Figure 9.8B, C). A cirrus shield, however, typically covers warm fronts. As Figure 9.7 shows, an upper tropospheric jet stream is closely associated with the baroclinic zone, blowing roughly parallel to the line of the upper front. This relationship is examined below.

Air behind the cold front, away from the low center, commonly has an anticyclonic trajectory and hence moves at a greater than geostrophic speed (see Chapter 5A.4), impelling the cold front to acquire a supergeostrophic speed also. The

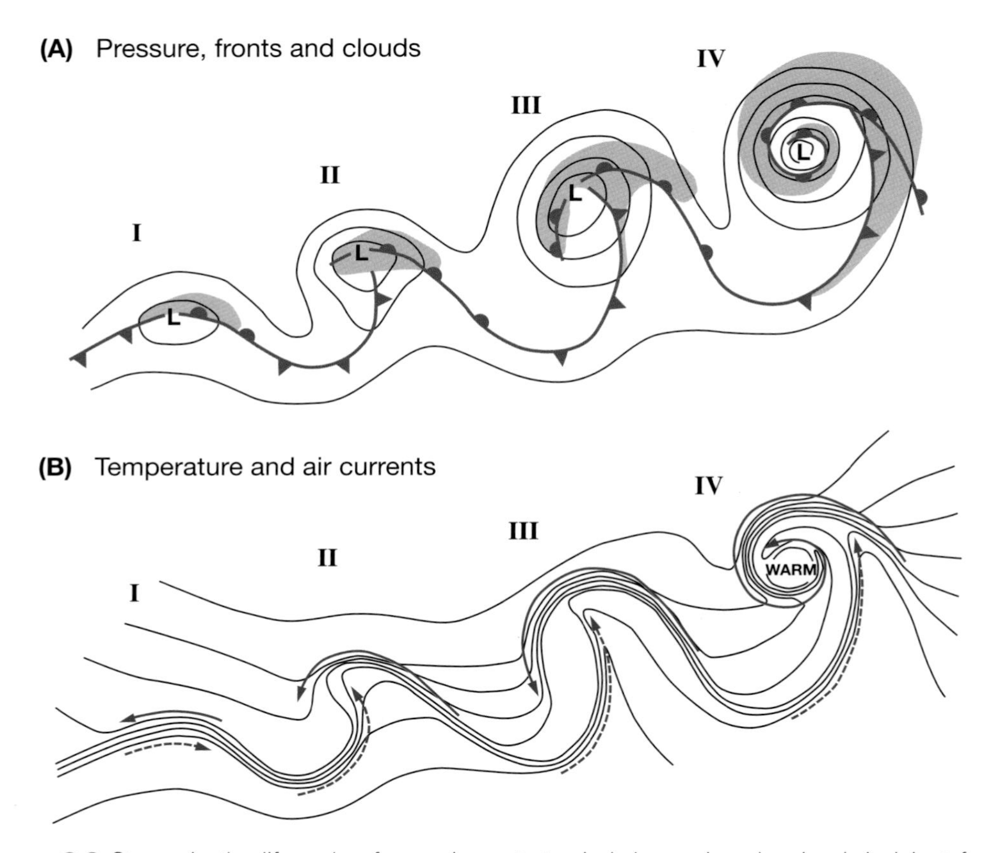

Figure 9.9 Stages in the life-cycle of a marine extratropical depression showing I: incipient frontal depression, II: frontal fracture, III: bent-back warm front and frontal T-bone, IV: warm-core seclusion. A: Schematic isobars of sea-level pressure, fronts and cloud cover (stippled). B: Isotherms and flows of cold air (solid arrows) and warm air (dashed arrows).

Source: After Shapiro and Keyser (1990). By permission of the American Meteorological Society.

wedge of warm air is pinched out at the surface and lifted bodily off the ground. This stage of *occlusion* eliminates the wave form at the surface (see Figure 9.7). The depression usually achieves its maximum intensity 12–24 hours after the beginning of occlusion. The occlusion gradually works outwards from the center of the depression along the warm front. Sometimes, the cold air wedge advances so rapidly that, in the friction layer close to the surface, cold air overruns the warm air and generates a *squall line* (see Chapter 4G).

By no means all frontal lows follow the idealized life-cycle discussed above. It is generally characteristic of oceanic cyclogenesis, although the evolution of those systems has been re-examined using aircraft observations collected during North Atlantic meteorological field programs in the 1980s. These suggest a different evolution of maritime frontal cyclones (Figure 9.9). Four stages are identified: (1) cyclone inception features a broad (400km) continuous frontal zone; (2) frontal fracture occurs near the center of the low with tighter frontal gradients; (3) a T-bone structure and bent-back warm front develop; and (4) the mature cyclone shows seclusion of the warm core within the polar airstream behind the cold front.

Over central North America, cyclones forming in winter and spring depart considerably from the Norwegian model. They often feature an outflow of cold arctic air east of the Rocky Mountains forming an arctic front, a lee trough with dry air descending from the mountains, and warm, moist southerly flow from the Gulf of Mexico (Figure 9.10). The trough superposes dry air over warm, moist air, generating instability and a rain band analogous to a warm front. The arctic air moves southward west of the low center, causing lifting of warmer, dry air but giving little precipitation. There may also be an upper cold front advancing over the trough that forms a rain band along its leading edge. Such a system is thought to have caused a record rainstorm at Holt, Missouri, on 22 June 1947, when 305mm fell in just 42 minutes!

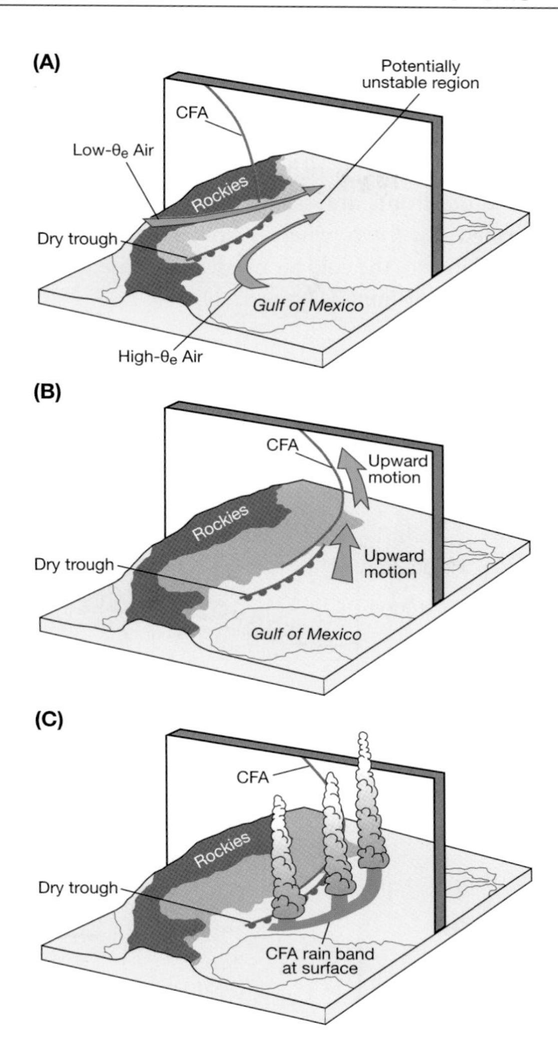

Figure 9.10 Schematic model of a dry trough and frontogenesis east of the Rocky Mountains. A: Warm, dry air with low equivalent potential temperature (u_e) from the Rockies overrides warm, moist, high u_e air from the Gulf of Mexico, forming a potentially unstable zone east of the dry trough. B: Upward motion associated with the cold front aloft (CFA). C: Location of the CFA rain band at the surface. [Equivalent potential temperature is the potential temperature of an air parcel that is expanded adiabatically until all water vapor is condensed and the latent heat released then compressed adiabatically to 1000mb pressure.]

Source: After Locatelli *et al.* (1995). By permission of the American Meteorological Society.

E FRONTAL CHARACTERISTICS

The character of frontal weather depends upon the vertical motion in the air masses. If the air in the warm sector is rising relative to the frontal zone the fronts are usually very active and are termed *ana-fronts*, whereas sinking of the warm air relative to the cold air masses gives rise to less intense *kata-fronts* (Figure 9.11).

1 The warm front

The warm front represents the leading edge of the warm sector in the wave. The frontal boundary has a very gentle slope, of the order of 0.5–1°, so the cloud systems associated with the upper portion of the front herald its approach some 12 hours or more before the arrival of the surface front. The ana-warm front, with rising warm air, has multi-layered cloud that steadily thickens and lowers towards the surface position of the front. The first clouds are thin, wispy cirrus, followed by sheets of cirrus and cirrostratus, and altostratus (Figure 9.11A). The sun is obscured as the altostratus layer thickens and drizzle or rain begins to fall. The cloud often extends through most of the troposphere and, with continuous precipita-

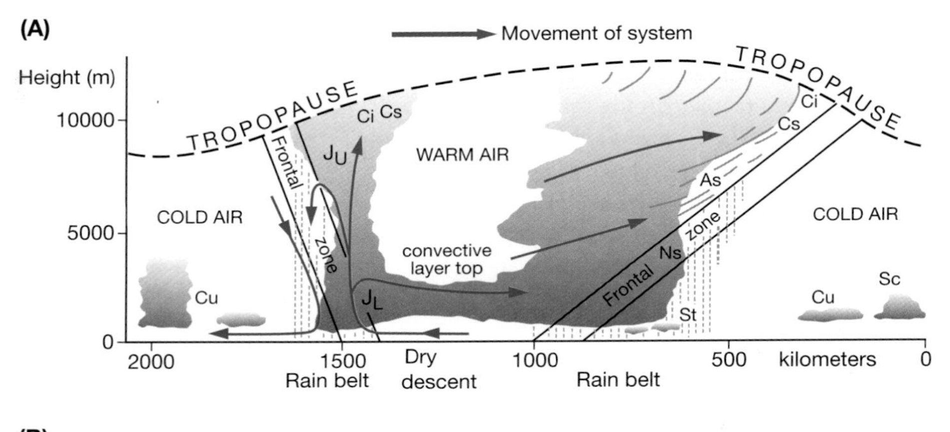

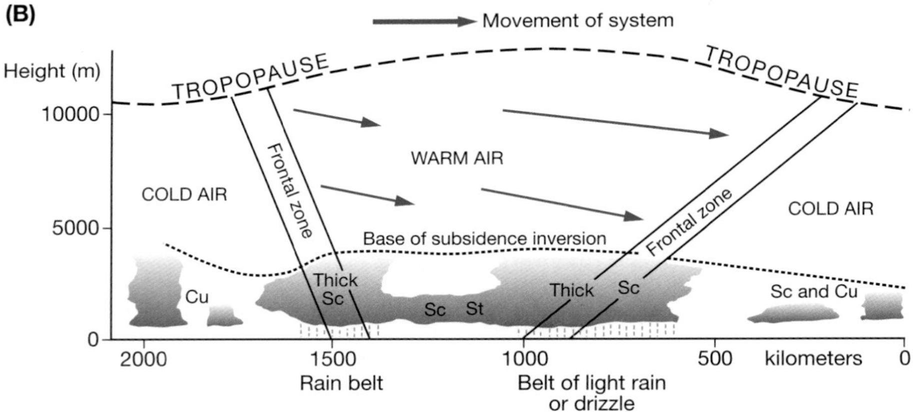

Figure 9.11 A: Cross-sectional model of a depression with ana-fronts, where the air is rising relative to each frontal surface. Note that an ana-warm front may occur with a kata-cold front and vice versa. JU and JL show the locations of the upper and lower jet streams, respectively. B: Model of a depression with kata-fronts, where the air is sinking relative to each frontal surface.

Sources: After Pedgley, *A Course in Elementary Meteorology*, and Bennetts *et al.* (1988). (Crown copyright ©), reproduced by permission of the Controller of Her Majesty's Stationery Office.

tion occurring, is generally designated as nimbostratus. Patches of fractostratus may also form in the cold air as rain falling through this air undergoes evaporation and quickly saturates it.

The descending warm air of the kata-warm front greatly restricts the development of medium- and high-level clouds. The frontal cloud is mainly stratocumulus, with a limited depth as a result of the subsidence inversions in both air masses (see Figure 9.11B). Precipitation is usually light rain or drizzle formed by coalescence.

At the passage of the warm front the wind veers, the temperature rises and the fall of pressure is checked. The rain becomes intermittent or ceases in the warm air and the thin stratocumulus cloud sheet may break up.

Forecasting the extent of rain belts associated with the warm front is complicated by the fact that most fronts are not ana- or kata-fronts throughout their length or even at all levels in the troposphere. For this reason, radar is increasingly being used to map the precise extent of rain belts and to detect differences in rainfall intensity. Such studies show that most of the production and distribution of precipitation is controlled by a broad airflow a few hundred kilometers across and several kilometers deep, which flows parallel to and ahead of the surface cold front (see Figure 9.12). Just ahead of the cold front, the flow occurs as a low-level jet with winds of up to 25–30m s^{-1} at about 1km above the surface. The air, which is warm and moist, rises over the warm front and turns southeastward ahead of the front, merging with the mid-tropospheric flow (B in Figure 9.13). This flow is termed a '*conveyor belt*' (for large-scale

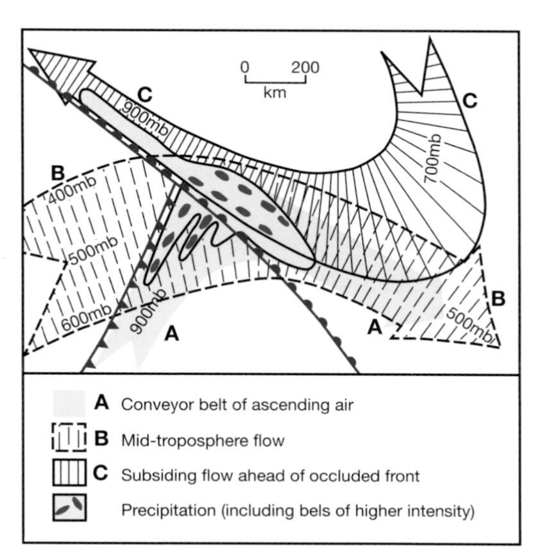

Figure 9.12 Model of the large-scale flow and mesoscale precipitation structure of a partially occluded depression typical of those affecting the British Isles. It shows the 'conveyor belt' (A) rising from 900mb ahead of the cold front over the warm front. This is overlaid by a mid-tropospheric flow (B) of potentially colder air from behind the cold front. Most of the precipitation occurs in the well-defined region shown, within which it exhibits a cellular and banded structure.

Source: After Harrold (1973). By permission of the Royal Meteorological Society.

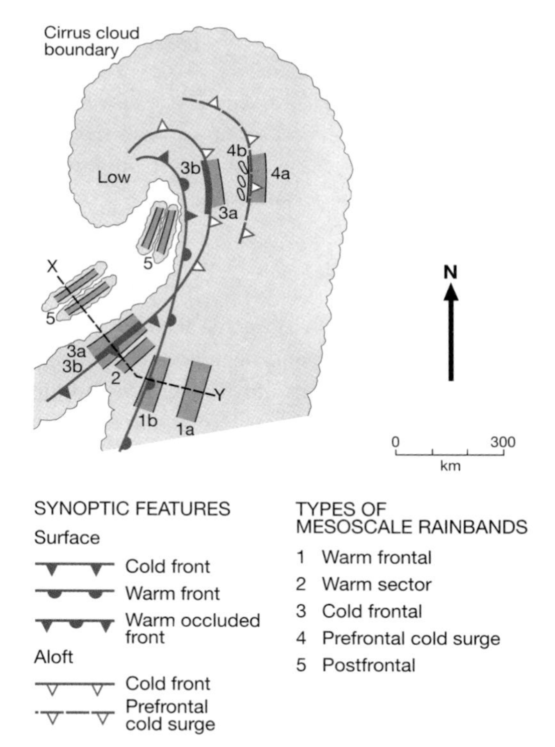

Figure 9.13 Fronts and associated rain bands typical of a mature depression. The broken line X–Y shows the location of the cross-section given in Figure 9.14.

Source: After Hobbs; from Houze and Hobbs (1982). By permission of Academic Press.

heat and momentum transfer in mid-latitudes). Broad-scale convective (potential) instability is generated by the overrunning of this low-level flow by potentially colder, drier air in the middle troposphere. Instability is released mainly in small-scale convection cells that are organized into clusters, known as mesoscale precipitation areas (MPAs). These MPAs are further arranged in bands, 50–100km wide (Figure 9.13). Ahead of the warm front, the bands are broadly parallel to the airflow in the rising section of the conveyor belt, whereas in the warm sector they parallel the cold front and the low-level jet. In some cases, cells and clusters are further arranged in bands within the warm sector and ahead of the warm front (see Figures 9.13 and 9.14). Precipitation from warm front rain bands often involves 'seeding' by ice particles falling from the upper cloud layers. It has been estimated that 20–35 percent of the precipitation originates in the 'seeder' zone and the remainder in the lower clouds (see also Figure 5.14). Orographic effects set up some of the cells and clusters, and these may travel downwind when the atmosphere is unstable.

2 The cold front

The weather conditions observed at cold fronts are equally variable, depending upon the stability of the warm-sector air and the vertical motion relative to the frontal zone. The classical cold-front model is of the ana-type, and the cloud is usually cumulonimbus. Figure 9.15 illustrates the warm conveyor belt associated with such a frontal zone and the line convection. Over the British Isles, air in the warm sector is rarely unstable, so that nimbostratus occurs more frequently at the cold front (see Figure 9.11A). With the kata-cold front the cloud is generally stratocumulus (see Figure 9.11B) and precipitation is light. With ana-cold fronts there are usually brief, heavy downpours, sometimes accompanied by thunder. The steep slope of the cold front, roughly 2°, means that the bad weather is of shorter duration than at the warm front. With the passage of the cold front the wind veers sharply, pressure begins to rise and temperature falls. The sky may clear abruptly, even before the passage of the surface cold front in some cases, although with kata-cold fronts the changes are more gradual. Forward-tilting cold fronts are sometimes observed, either due to surface friction – especially an orographic barrier – slowing the low-level motion of the front, or as a result of a cold front aloft (see Figure 9.10).

3 The occlusion

The cold front moves faster relative to the warm front, eventually catching up with it, leading to

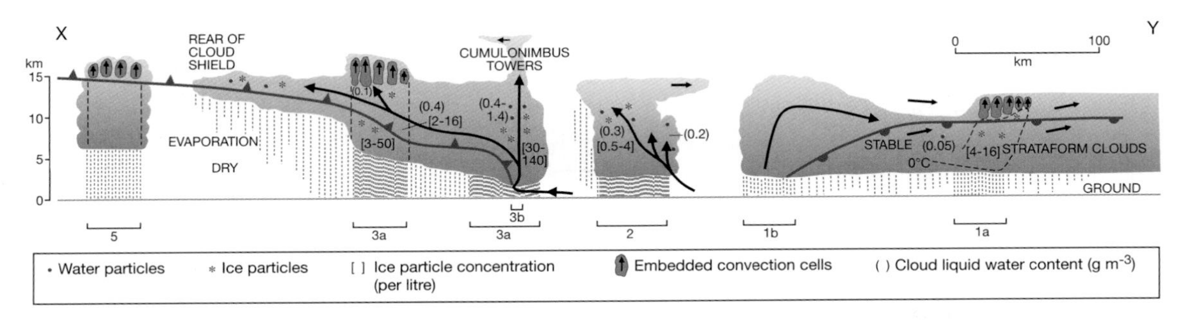

Figure 9.14 Cross-section along the line X–Y in Figure 9.13 showing cloud structures and rain bands. The vertical hatching represents rainfall location and intensity. Raindrop and ice particle regions are shown, as are ice particle concentrations and cloud liquid water content. Numbered belts refer to those shown in Figure 9.13. Scales are approximate.

Source: After Hobbs and Matejka *et al.*; from Houze and Hobbs (1982). By permission of Academic Press.

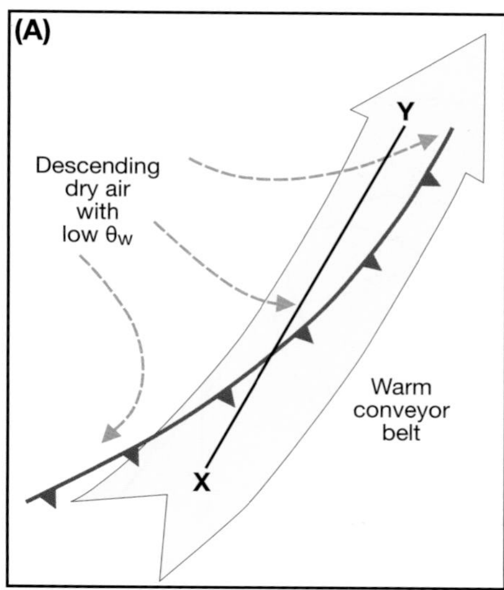

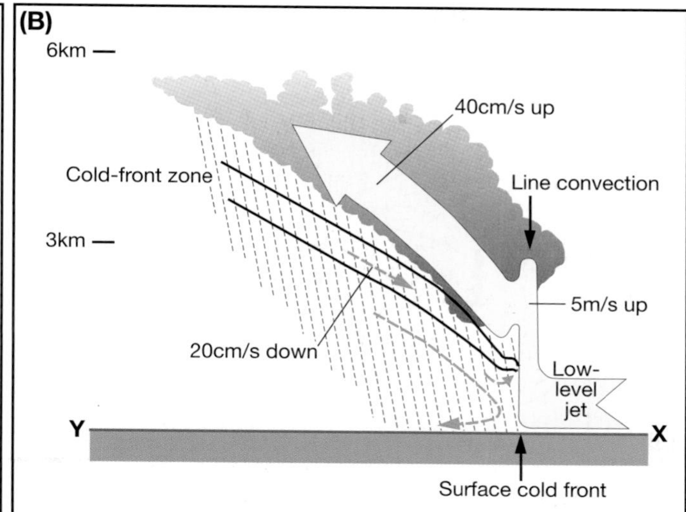

Figure 9.15 Schematic diagrams showing airflows, relative to the moving frontal system, at an ana-cold front. A warm conveyor belt (stippled) ascends above the front with cold air (dashed arrows) descending beneath it. A: Plan view. B: Vertical section along the line X–Y, showing rates of vertical motion.

Source: Browning (1990). By permission of the American Meteorological Society.

occlusion, where the warm sector air is lifted off the ground. Occlusions are classified as either *cold* or *warm*, depending on the relative states of the air masses lying in front and to the rear of the warm sector (Figure 9.16). If air mass 2 is colder than air mass 1, then the occlusion is warm, but if the reverse is so it is termed a cold occlusion. The air in advance of the depression is likely to be coldest when depressions occlude over Europe in winter and very cold cP air is affecting the continent. Recent work suggests that most occlusions are warm and that the thermal definition is often misleading. A new definition is proposed: a cold (warm) occlusion forms when the air that is more statically stable lies behind the cold front (ahead of the warm front) (Figure 9.16).

The line of the warm air wedge aloft is associated with a zone of layered cloud (similar to that found with a warm front) and often of precipitation. Hence its position is indicated separately on some weather maps and it is referred to by Canadian meteorologists as a *trowal* (trough of warm air aloft). The passage of an occluded front and trowal brings a change back to polar air-mass weather.

A different process occurs when there is interaction between the cloud bands within a polar trough and the main polar front, giving rise to an *instant occlusion*. A warm conveyor belt on the polar front ascends as an upper tropospheric jet, forming a stratiform cloud band (Figure 9.17), while a low-level polar trough conveyor belt at right angles to it produces a convective cloud band and precipitation area poleward of the main polar front on the leading edge of the cold pool.

Frontolysis (frontal decay) represents the final phase of a front's existence although it is not necessarily linked with occlusion. Decay occurs when differences no longer exist between adjacent air masses. This may arise in four ways: (1) through their mutual stagnation over a similar surface; (2) as a result of both air masses moving on parallel tracks at the same speed; (3) through their movement in succession along the same track at the same speed; or (4) by the system entraining air of the same temperature.

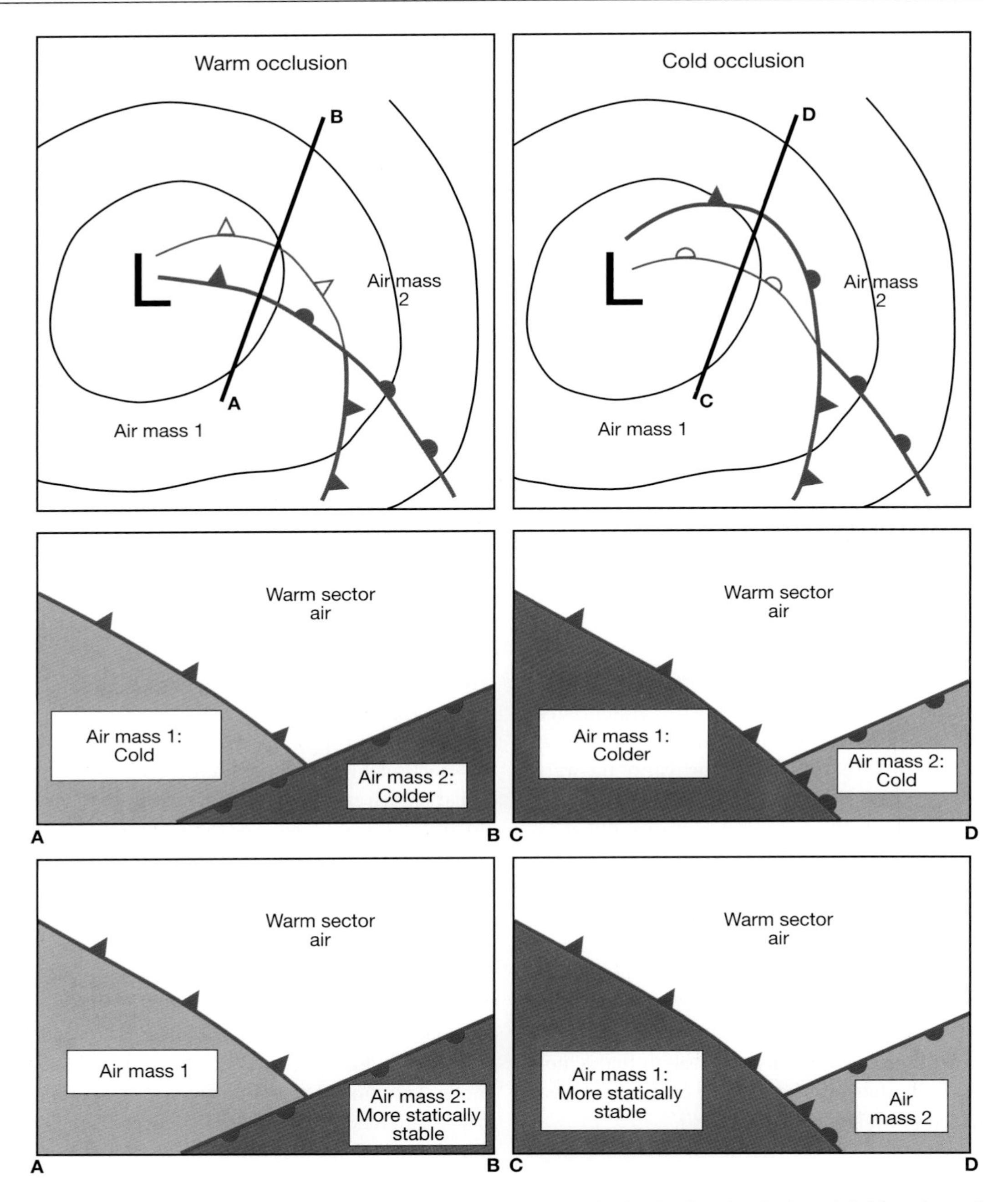

Figure 9.16 Schematic illustration of a cold and a warm occlusion in the classical model. Plan view of synoptic pattern (above) and cross-sections along line A–B (center). Colder air is shaded darker. The bottom panel illustrates proposed criteria for identifying warm and cold occlusions based on static stability.

Source: Above and center from Stoelinga *et al.*, (2002, *p. 710, Fig. 1*). Courtesy of the American Meteorological Society. The bottom panel is based on their new definition.

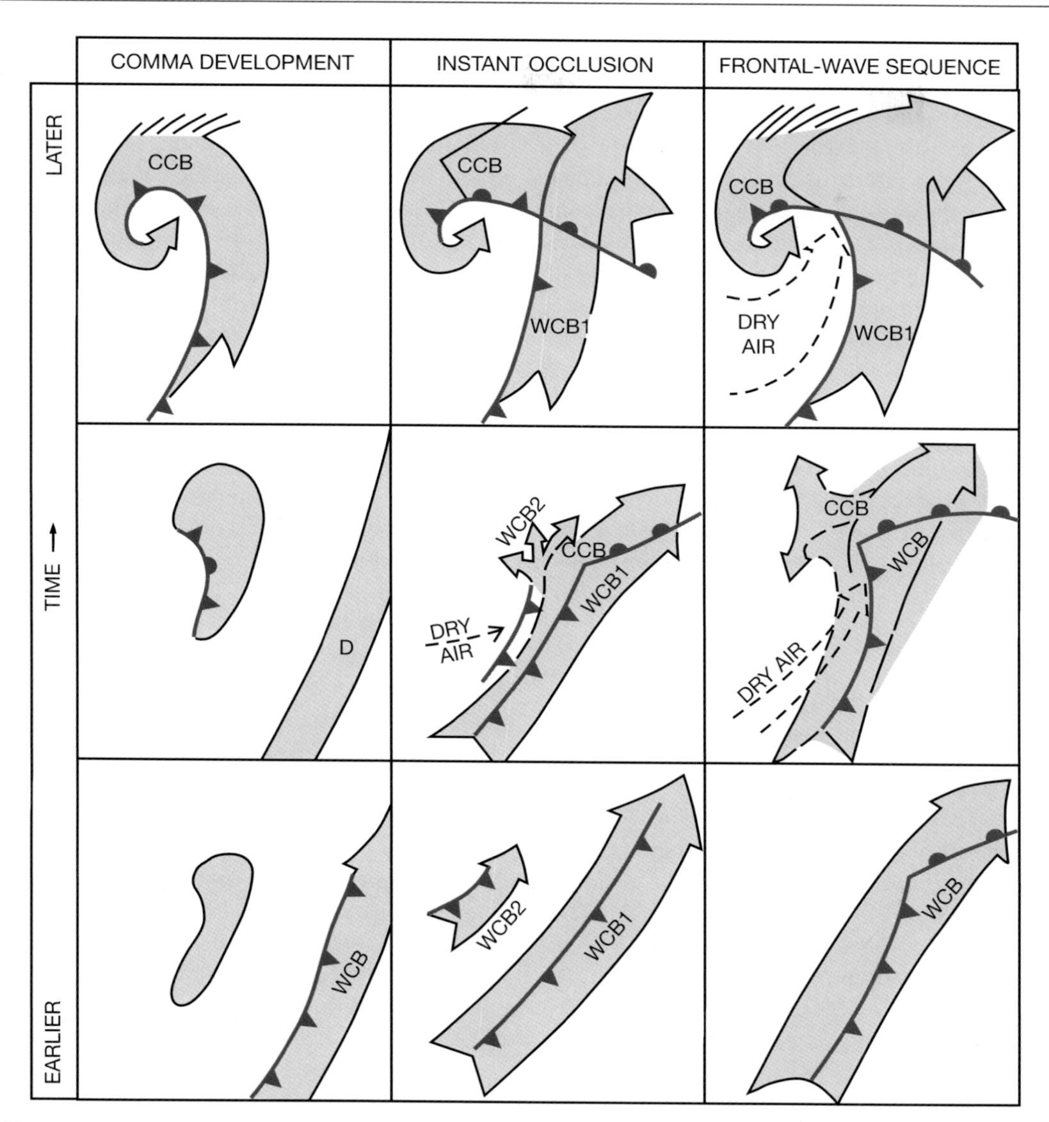

Figure 9.17 Schematic illustrations of vortex developments in satellite imagery. The sequences run from bottom to top. Left: comma cloud (C) developing in a polar airstream. Center: instant occlusion from the interaction of a polar trough with a wave on the polar front. Right: the classical frontal wave with cold and warm conveyor belts (CCB, WCB). C = enhanced convection; D = decaying cloud band; cloud cover stippled.

Source: After Browning (1990). By permission of the American Meteorological Society.

4 Frontal-wave families

Observations show that frontal waves over the oceans, at least, do not generally occur as separate units but in *families* of three or four (see Figure 9.9). The depressions that succeed the original one form as *secondary lows* along the trailing edge of an extended cold front. Each new member follows a course that is south of its progenitor as the polar air pushes further south to the rear of each depression in the series. Eventually, the front trails far to the south and the cold polar air forms an extensive meridional wedge of high pressure, terminating the sequence.

Another pattern of development may take place on the warm front, particularly at the point of occlusion, as a separate wave forms and runs ahead of the parent depression. This type of secondary is more likely with very cold (cA, mA or cP) air ahead of the warm front, and it tends to form when mountains bar the eastward movement of the occlusion. This situation often occurs when a primary depression is situated in the Davis Strait and a breakaway wave forms south of Cape Farewell (the southern tip of Greenland), moving away eastward. Analogous developments take place in the Skagerrak–Kattegat area when the occlusion is held up by the Scandinavian mountains.

F ZONES OF WAVE DEVELOPMENT AND FRONTOGENESIS

Fronts and associated depressions tend to develop in well-defined areas. The major zones of frontal wave development are areas that are most frequently baroclinic as a result of airstream confluence (Figure 9.18). This is the case off East Asia and eastern North America, especially in winter, when there is a sharp temperature gradient between the snow-covered land and warm offshore currents. These zones are referred to as the Pacific Polar and Atlantic Polar Fronts, respectively (Figure 9.19). Their position is quite variable, but they are displaced equatorward in winter, when the Atlantic Frontal Zone may extend into the Gulf of Mexico. Here there is convergence of air masses of different stability between adjacent subtropical high pressure cells. Depressions developing here commonly move northeastward, sometimes following or amalgamating with others of the northern part of the Polar Front proper or of the Canadian Arctic Front. Frontal frequency remains high across the North Atlantic, but it decreases eastward in the North Pacific, perhaps owing to a less pronounced gradient of sea surface temperature. Frontal activity is most common in the central North Pacific when the subtropical high is split into two cells with converging airflows between them.

Another section of the Polar Front, often referred to as the *Mediterranean Front*, is located over the Mediterranean–Caspian Sea areas in winter. At intervals, fresh Atlantic mP air, or cool cP air from southeast Europe, converges with warmer air masses of North African origin over the Mediterranean Basin and initiates frontogenesis. In summer, the Azores subtropical anticyclone influences the area, and the frontal zone is absent.

The summer locations of the Polar Front over the western Atlantic and Pacific are some 10° further north than in winter (see Figure 9.19), although the summer frontal zone is rather weak. There is a frontal zone over Eurasia and a corresponding one over middle North America. These reflect the general meridional temperature gradient and also the large-scale influence of orography on the general circulation (see G, this chapter).

In the Southern Hemisphere, the Polar Front is on average about 45°S in January (summer), with branches spiralling poleward towards it from about 32°S off eastern South America and from 30°S, 150°W in the South Pacific (Figure 9.20). In July (winter), there are two Polar Frontal Zones spiralling towards Antarctica from about 20°S; one starts over South America and the other at 170°W. They terminate some 4–5° latitude further poleward than in summer. It is noteworthy that the Southern Hemisphere has more cyclonic activity in summer than does the Northern Hemisphere in its summer. This appears to be related to the seasonal strengthening of the meridional temperature gradient (see p. 180).

The second major frontal zone is the Arctic Front, associated with the snow and ice margins of high latitudes (see Figure 9.19). In summer, this zone is developed at the land–sea boundary in

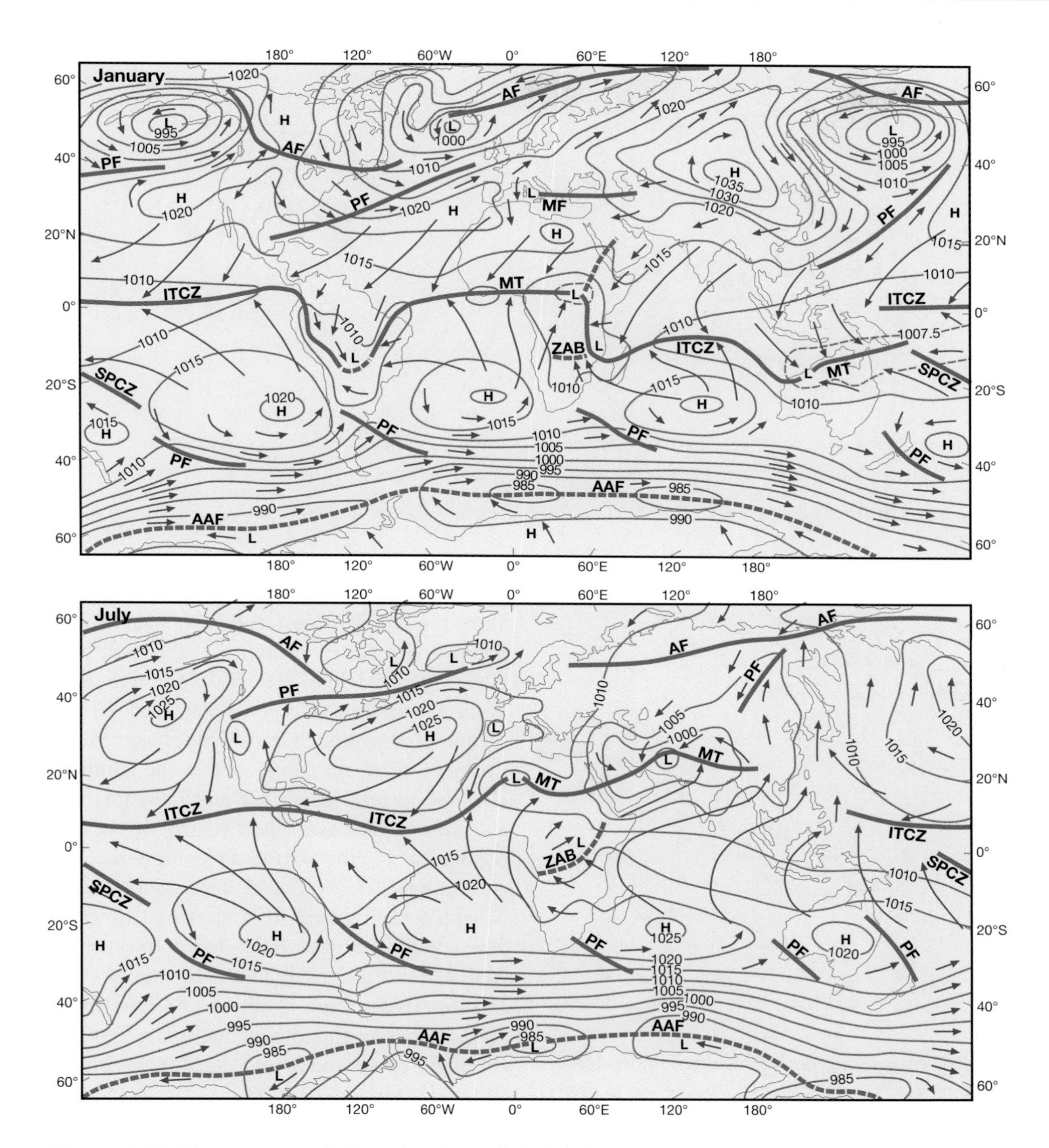

Figure 9.18 Mean pressure (mb) and surface winds for the world in January and July. The major frontal and convergence zones are shown as follows: Intertropical Convergence Zone (ITCZ), South Pacific Convergence Zone (SPCZ), Monsoon Trough (MT), Zaire Air Boundary (ZAB), Mediterranean Front (MF), Northern and Southern Hemisphere Polar Fronts (PF), Arctic Fronts (AF) and Antarctic Fronts (AAF).
Source: Partly from Liljequist (1970).

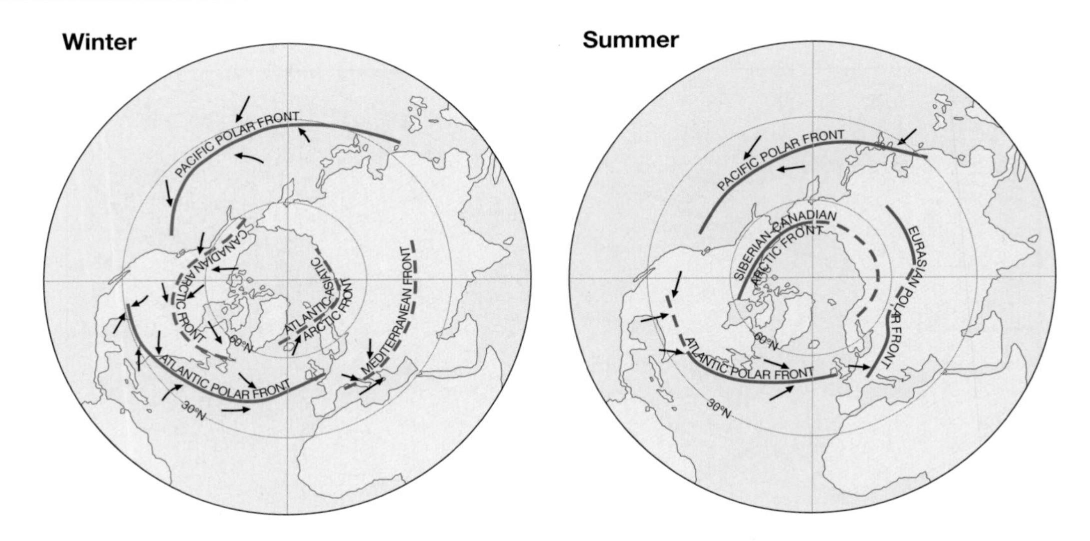

Figure 9.19 The major Northern Hemisphere frontal zones in winter and summer.

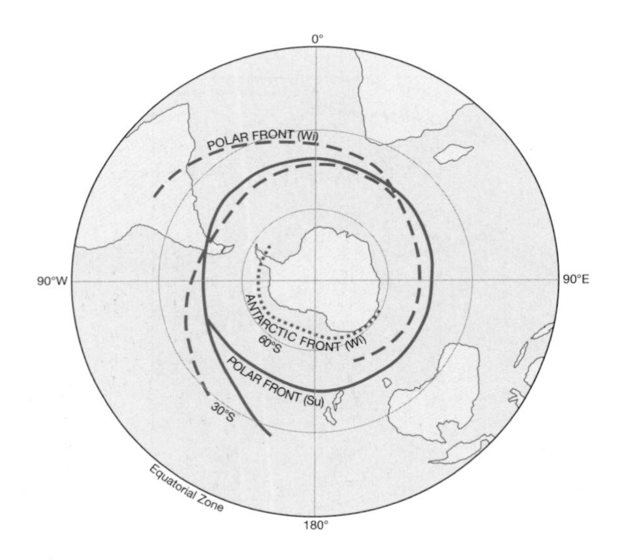

Figure 9.20 The major Southern Hemisphere frontal zones in winter (Wi) and summer (Su).

Siberia and North America where there is a strong temperature gradient between the heated snow-free land and the largely sea ice-covered, cold Arctic Ocean. In winter over North America, it is formed between cA (or cP) air and Pacific maritime air modified by crossing the Coast Ranges and the Rocky Mountains). There is also a less pronounced Arctic Frontal Zone in the North Atlantic–Norwegian Sea area, extending along the Siberian coast. A similar weak frontal zone is found in winter in the Southern Hemisphere. It is located at 65–70°S near the edge of the Antarctic pack ice in the Pacific sector (see Figure 9.20), although few cyclones form there. Zones of airstream confluence in the Southern Hemisphere (cf. Figures 9.2B and 9.4B) are fewer and more persistent, particularly in coastal regions, than in the Northern Hemisphere.

The principal tracks of depressions in the Northern Hemisphere in January are shown in Figure 9.21. The major tracks reflect the primary frontal zones discussed above. In summer, the Mediterranean route is absent and lows move across Siberia; the other tracks are similar, although more zonal and located in higher latitudes (around 60°N).

Between the two hemispherical belts of subtropical high pressure there is a further major convergence zone, the Intertropical Convergence Zone (ITCZ). Formerly this was designated as the Intertropical Front (ITF), but air-mass contrasts are not typical. The ITCZ moves seasonally away from the equator, as the subtropical high pressure cell activity alternates in opposite hemispheres.

The contrast between the converging air masses obviously increases with the distance of the ITCZ from the equator, and the degree of difference in their characteristics is associated with considerable variation in weather activity along the convergence zone. Activity is most intense in June to July over South Asia and West Africa, when the contrast between the humid maritime and dry continental air masses is at a maximum. In these sectors, the term Intertropical Front is applicable, although this does not imply that it behaves like a mid-latitude frontal zone. The nature and significance of the ITCZ are discussed in Chapter 11.

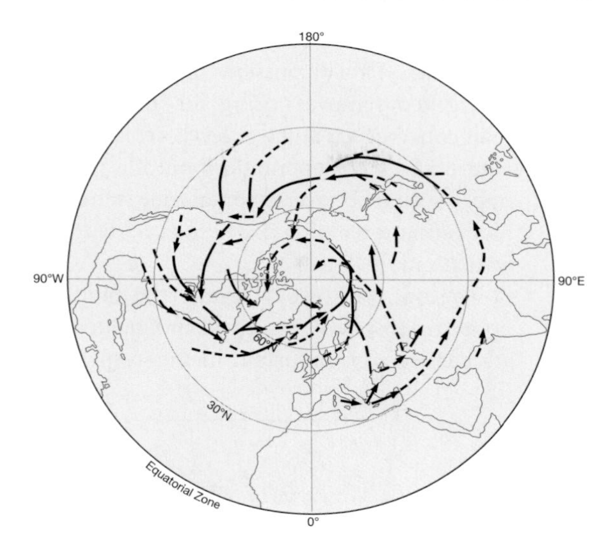

Figure 9.21 The principal Northern Hemisphere depression tracks in January. The full lines show major tracks, the dashed lines secondary tracks that are less frequent and less well defined. The frequency of lows is a local maximum where arrowheads end. An area of frequent cyclogenesis is indicated where a secondary track changes to a primary track or where two secondary tracks merge to form a primary track.
Source: After Klein (1957). Courtesy of the US Weather Bureau.

G SURFACE–UPPER-AIR RELATIONSHIPS AND THE FORMATION OF FRONTAL CYCLONES

We have noted that a wave depression is associated with air-mass convergence, yet the barometric pressure at the center of the low may

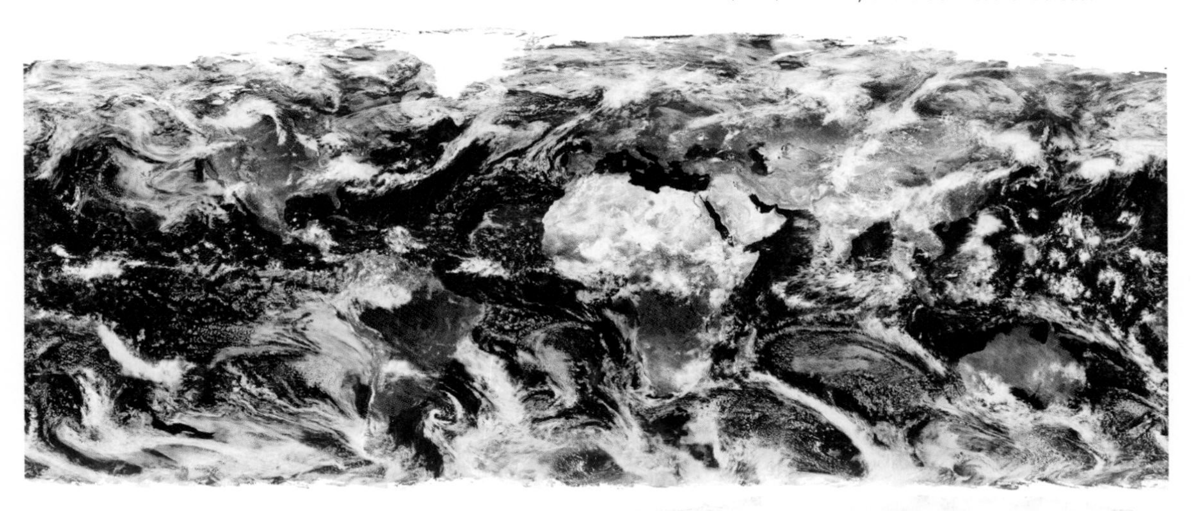

Plate 9.1 A composite image mostly from the Moderate resolution Imaging Spectrometer (MODIS) on NASA's Terra satellite orbiting 700km above the earth.

The cloud cover is a composite of visible and thermal infrared imagery for 29 July and November 2001. City lights are superimposed from Defense Meterological Satellite Program observations over a nine-month period. Topographic shading is from the US Geological Survey GTOPO 30 dataset.

Source: Blue Marble Visible Earth, NASA ftp://gloria 2-f.gsfc.nasa.gov/pub/stockli.

decrease by 10–20mb in 12–24 hours as the system intensifies. This is possible because upper-air divergence removes rising air more quickly than convergence at lower levels replaces it (see Figure 5.7). The superimposition of a region of upper divergence over a frontal zone is the prime motivating force of *cyclogenesis* (i.e., depression formation).

The long (or Rossby) waves in the middle and upper troposphere, discussed in Chapter 7A.2, are particularly important in this respect. The latitudinal circumference limits the circumpolar westerly flow to between three and six major Rossby waves, and these affect the formation and movement of surface depressions. Two primary stationary waves tend to be located about 70°W and 150°E in response to the influence on the atmospheric circulation of orographic barriers, such as the Rocky Mountains and the Tibetan plateau, and of heat sources. On the eastern limb of troughs in the upper westerlies of the Northern Hemisphere the flow is normally divergent, since the gradient wind is subgeostrophic in the trough but supergeostrophic in the ridge (see Chapter 6A.4). Thus, the sector ahead of an upper trough is a very favorable location for a surface depression to form or deepen (see Figure 9.22). It will be noted that the mean upper troughs are significantly positioned just west of the Atlantic and Pacific Polar Front Zones in winter.

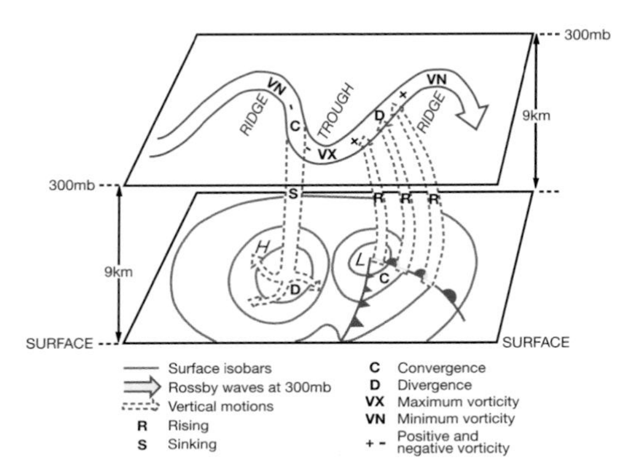

Figure 9.22 Schematic representation of the relationship between surface pressure (H and L), airflow and frontal systems, on the one hand, and the location of troughs and ridges in the Rossby waves at the 300mb level. The locations of maximum (cyclonic) and minimum (anticyclonic) relative vorticity are shown, as are those of negative (anticyclonic) and positive (cyclonic) vorticity advection.

Sources: Mostly after Musk (1988), with additions from Uccellini (1990). Courtesy of Cambridge University Press.

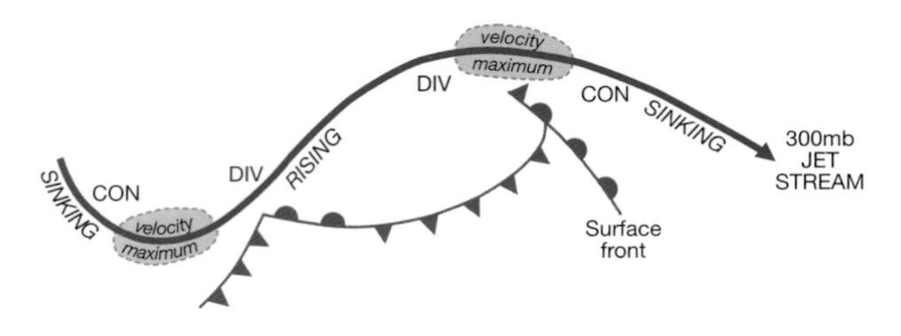

Figure 9.23 Model of the jet stream and surface fronts, showing zones of upper tropospheric divergence and convergence and the jet stream cores.

With these ideas in mind, we can examine the three-dimensional nature of depression development and the links existing between upper and lower tropospheric flow. The basic theory relates to the vorticity equation, which states that, for frictionless horizontal motion, the rate of change of the vertical component of absolute vorticity (dQ/dt or $d(f+\zeta)/dt$) is proportional to air-mass convergence ($-D$, i.e., negative divergence):

$$\frac{dQ}{dt} = DQ \text{ or } D = -\frac{1}{Q}\frac{dQ}{dt}$$

The relationship implies that a converging (diverging) air column has increasing (decreasing) absolute vorticity. The conservation of vorticity equation, which we have already discussed, is in fact a special case of this relationship.

In the sector ahead of an upper trough, the decreasing cyclonic vorticity causes divergence (i.e., D positive), since the change in ζ outweighs that in f, thereby favoring surface convergence and low-level cyclonic vorticity (see Figure 9.23). Once the surface cyclonic circulation has become established, vorticity production is increased through the effects of thermal advection. Poleward transport of warm air in the warm sector and the eastward advance of the cold upper trough act to sharpen the baroclinic zone, strengthening the upper jet stream through the thermal wind mechanism (see p. 170). The vertical relationship between jet stream and front has already been shown (see Figure 7.8); a model depression sequence is demonstrated in Figure 9.23. The actual relationship may depart from this idealized case, although the jet is commonly located in the cold air. Velocity maxima (core zones) occur along the jet stream and the distribution of vertical motion upstream and downstream of these cores is known to be quite different. In the area of the jet entrance (i.e., upstream of the core), divergence causes lower-level air to rise on the equatorward (i.e., right) side of the jet, whereas in the exit zone (downstream of the core) ascent is on the poleward side. Figure 9.24 shows how precipitation is more often related to the position of the jet stream than to that of surface fronts; maximum precipitation areas are in the right entrance sector of the jet core. This vertical motion pattern is also of basic importance in the initial deepening stage of the depression. If the upper-air pattern is unfavorable (e.g., beneath left entrance and right exit zones, where there is convergence) the depression will fill.

The development of a depression may also be considered in terms of energy transfers. A cyclone requires the conversion of potential into kinetic energy. The upward (and poleward) motion of warm air achieves this. The vertical wind shear and the superimposition of upper tropospheric divergence drive the rising warm air over a baroclinic zone. Intensification of this zone further strengthens the upper winds. The upper divergence allows surface convergence and pressure fall to occur simultaneously. Modern theory relegates the fronts to a subordinate role. They develop within depressions as narrow zones of intensified ascent, probably through the effects of cloud formation.

Recent research has identified a category of mid-latitude cyclones that develop and intensify

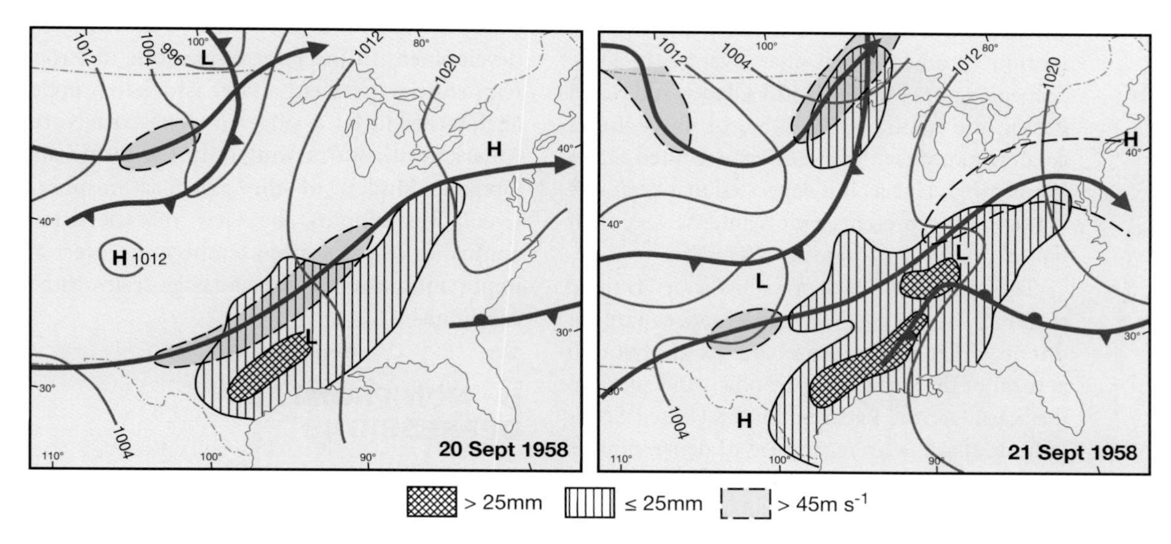

Figure 9.24 The relations between surface fronts and isobars, surface precipitation (≤25mm vertical hatching; >25mm cross-hatching), and jet streams (wind speeds in excess of about 45m s^{-1} shown by stipple) over the United States on 20 September 1958 and 21 September 1958. This illustrates how the surface precipitation area is related more to the position of the jets than to that of the surface fronts. The air over the south-central United States was close to saturation, whereas that associated with the northern jet and the maritime front was much less moist.

Source: After Richter and Dahl (1958). Courtesy of the American Meteorological Society.

rapidly, acquiring characteristics that resemble tropical hurricanes. These have been termed 'bombs' in view of their explosive rate of deepening; pressure falls of at least 24mb/24hr are observed. For example, the '*QE II* storm', which battered the ocean liner *Queen Elizabeth II* off New York on 10 September 1978, developed a central pressure below 950mb with hurricane-force winds and an eye-like storm center within 24 hours (see Chapter 11B.2). These systems are observed mainly during the cold season off the East Coast of the United States, off Japan, and over parts of the central and northeastern North Pacific, in association with major baroclinic zones and close to strong gradients of sea surface temperature. Explosive cyclogenesis is favored by an unstable lower troposphere and is often located downstream of a traveling 500mb-level trough. Bombs are characterized by strong vertical motion, associated with a sharply defined level of non-divergence near 500mb, and large-scale release of latent heat. Wind maxima in the upper troposphere, organized as jet streaks, serve to amplify the lower level instability and upward motion. Studies reveal that *average* cyclonic deepening rates over the North Atlantic and North Pacific are about 10mb/24hr, or three times greater than over the continental United States (3mb/24hr). Hence, it is suggested that explosive cyclogenesis represents a more intense version of typical maritime cyclone development.

The movement of depressions is determined essentially by the upper westerlies and, as a rule of thumb, a depression center travels at about 70 percent of the surface geostrophic wind speed in the warm sector. Records for the United States indicate that the average speed of depressions is 32km hr^{-1} in summer and 48km hr^{-1} in winter. The higher speed in winter reflects the stronger westerly circulation. Shallow depressions are mainly steered by the direction of the thermal wind in the warm sector and hence their path closely follows that of the upper jet stream (see Chapter 6A.3). Deep depressions may greatly distort the thermal pattern, however, as a result of the northward transport of warm air and the southward transport of cold air. In such cases, the depression usually becomes slow moving. The movement of a depression may be additionally guided by energy sources such as a warm sea surface that generates cyclonic vorticity, or by mountain barriers. The depression may cross obstacles, such as the Rocky Mountains or the Greenland Ice Sheet, as an upper low or trough, and subsequently redevelop, aided by the lee effects of the barrier or by fresh injections of contrasting air masses.

Ocean surface temperatures can critically influence the location and intensity of storm tracks. Figure 9.25B indicates that an extensive, relatively warm surface in the north-central Pacific in the winter of 1971–1972 caused a northward displacement of the westerly jet stream together with a compensating southward displacement over the western United States, bringing in cold air there. This pattern contrasts with that observed during the 1960s (see Figure 9.25A), when a persistent cold anomaly in the central Pacific, with warmer water to the east, led to frequent storm development in the intervening zone of strong temperature gradient. The associated upper airflow produced a ridge over western North America with warm winters in California and Oregon. Models of the global atmospheric circulation support the view that persistent anomalies of sea surface temperature exert an important control on local and large-scale weather conditions.

H NON-FRONTAL DEPRESSIONS

Not all depressions originate as frontal waves. Tropical depressions are indeed mainly non-frontal and these are considered in Chapter 11. In middle and high latitudes, four types that develop in distinctly different situations are of particular importance and interest: the lee cyclone, the thermal low, the polar low, and the cold low.

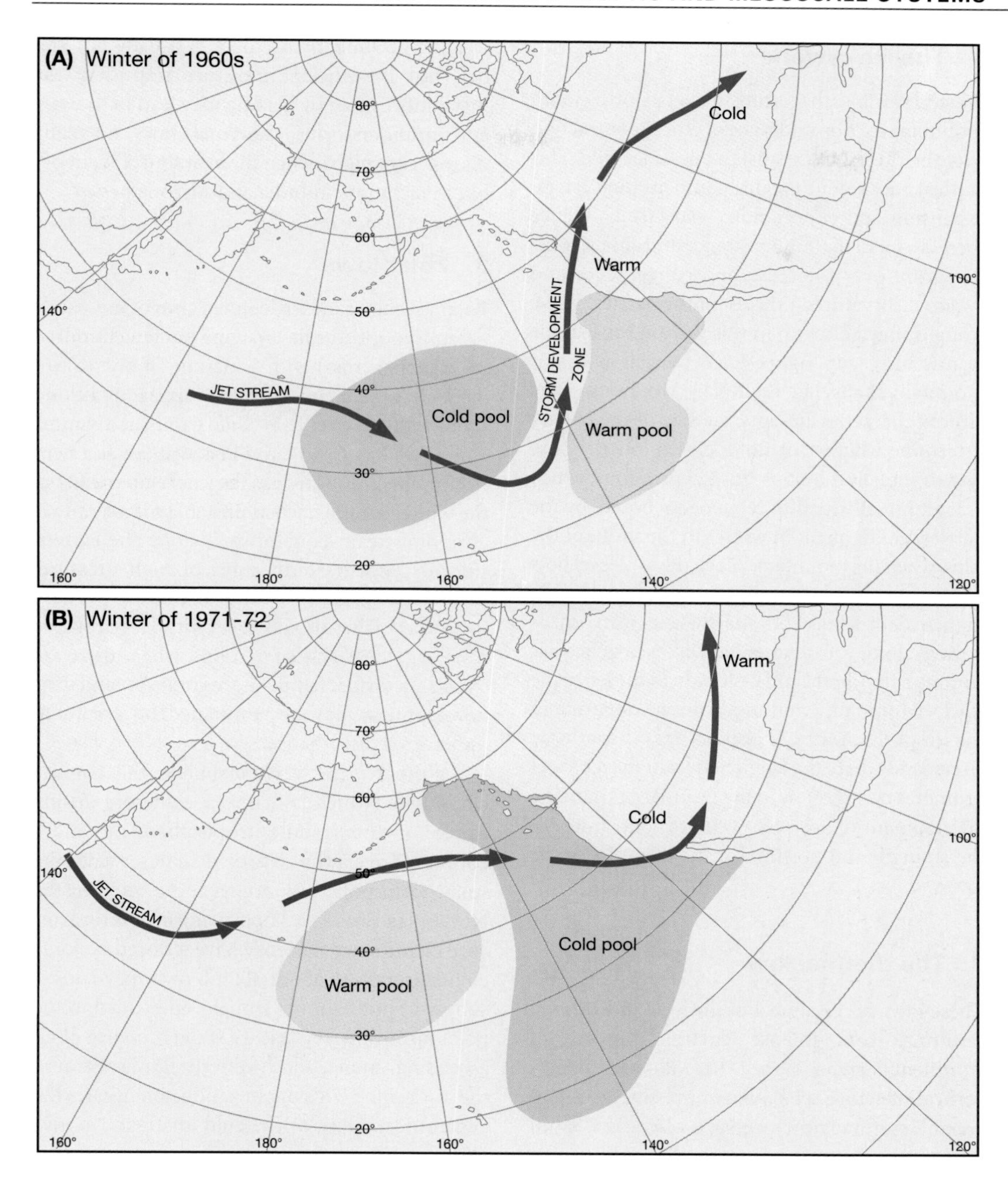

Figure 9.25 Generalized relationships between ocean surface temperatures, jet stream tracks, storm development zones and land temperatures over the North Pacific and North America during (A) average winter conditions in the 1960s, and (B) the winter of 1971–1972, as determined by J. Namias.
Source: After Wick (1973). By permission of *New Scientist*.

1 The lee cyclone

Westerly airflow that is forced over a north–south mountain barrier undergoes vertical contraction over the ridge and expansion on the lee side. This vertical movement creates compensating lateral expansion and contraction, respectively. Hence there is a tendency for divergence and anticyclonic curvature over the crest, and convergence and cyclonic curvature in the lee of the barrier. Wave troughs may be set up in this way on the lee side of low hills (see Figure 6.13) as well as major mountain chains like the Rocky Mountains. The airflow characteristics and the size of the barrier determine whether or not a closed low pressure system actually develops. Such depressions, which at least initially tend to remain 'anchored' by the barrier, are frequent in winter to the south of the Alps, when the mountains block the low-level flow of northwesterly airstreams. Fronts often develop in these depressions, but the low does not form as a wave along a frontal zone. Lee cyclogenesis is common in Alberta and Colorado in the lee of the Rocky Mountains, and in northern Argentina in the lee of the Andes. It also occurs off southeast Greenland where the barrier effect of the ice sheet promotes cyclogenesis in the Denmark Strait. The development of such lee cyclones contributes to the strength and position of the mean Icelandic low.

2 The thermal low

These lows occur almost exclusively in summer, resulting from intense daytime heating of continental areas. Figure 7.1C illustrates their vertical structure. The most impressive examples are the summer low pressure cells over Saudi Arabia, the northern part of the Indian subcontinent, and Arizona. The Iberian Peninsula is another region commonly affected by such lows. They occur over southwestern Spain on 40–60 percent of days in July and August. Typically, their intensity is only 2–4mb and they extend to about 750mb, less than in other subtropical areas. The weather accompanying them is usually hot and dry, but if sufficient moisture is present the instability caused by heating may lead to showers and thunderstorms. Thermal lows normally disappear at night, when the heat source is cut off, but in fact those in India and Arizona persist.

3 Polar lows

Polar lows are a loosely defined class of mesoscale to subsynoptic-scale systems (a few hundred kilometers across) with a lifetime of one to two days. On satellite imagery, they appear as a cloud spiral with one or several cloud bands, as a comma cloud (see Figure 9.17 and Plate 9.2), or as a swirl in cumulus cloud streets. They develop mainly in the winter months, when unstable mP or mA air currents stream equatorward along the eastern side of a north–south ridge of high pressure, commonly in the rear of an occluding primary depression. They usually form within a baroclinic zone, e.g., near sea-ice margins, where there are strong sea surface temperature gradients, and their development may be stimulated by an initial upper-level disturbance.

In the Northern Hemisphere, the comma cloud type (which is mainly a cold-core disturbance of the middle troposphere) is more common over the North Pacific, while the spiral-form polar low occurs more often in the Norwegian Sea. The latter is a low-level warm-core disturbance that may have a closed cyclonic circulation up to about 800mb or it may consist simply of one or more troughs embedded in the polar airflow. A key feature is the presence of an ascending, moist, southwesterly flow *relative* to the low center. This organization accentuates the general instability of the cold airstream to give considerable precipitation, often as snow. Latent heat release is an important mechanism for generating polar lows in the southern Norwegian Sea while stronger low-level baroclinicity and weaker convection prevail in systems in the northern Norwegian Sea and Barents Sea. Heat input to the cold air from the sea continues by

Plate 9.2 A polar low pressure system located in the Arctic Ocean on 25 February, 2008. On the right side of this image are the Queen Elizabeth Islands, Canada. Polar lows are similar to tropical cyclones if they reach sufficient strength but last only about 12 to 36 hours. They cause high surface winds and snow, and are often easily recognizable in satellite images because of their characteristic pattern, sometimes called a 'comma cloud' because of their hook-like shape.

Source: Jeff Schmaltz, Visible Earth, NASA.

night and day, so in exposed coastal districts showers may occur at any time.

In the Southern Hemisphere, polar low mesocyclones appear to be most frequent in the transition seasons, as these are the times of strongest meridional temperature and pressure gradients. In addition, over the Southern Ocean the patterns of occurrence and movement are more zonally distributed than in the Northern Hemisphere.

4 The cold low

The cold low (or *cold pool*) is usually most evident in the circulation and temperature fields of the middle troposphere. Characteristically, it displays symmetrical isotherms around the low center. Surface charts may show little or no sign of these persistent systems, which are frequent over the northeastern North America and northeast Siberia. They probably form as the result of strong vertical motion and adiabatic cooling in occluding baroclinic lows along the Arctic coastal margins. Such lows are especially important during the Arctic winter in that they bring large amounts of medium and high cloud, which offsets radiational cooling of the surface. Otherwise, they usually cause no 'weather' in the Arctic during this season. It is important to emphasize that tropospheric cold lows may be linked with either low or high pressure cells at the surface.

In mid-latitudes, cold lows may form during periods of low-index circulation pattern (see Figure 6.27) by the cutting-off of polar air from the main body of cold air to the north (these are sometimes referred to as *cut-off lows*). This gives rise to weather of polar air-mass type, although rather weak fronts may also be present. Such lows are commonly slow-moving and give persistent unsettled weather with thunder in summer. Heavy

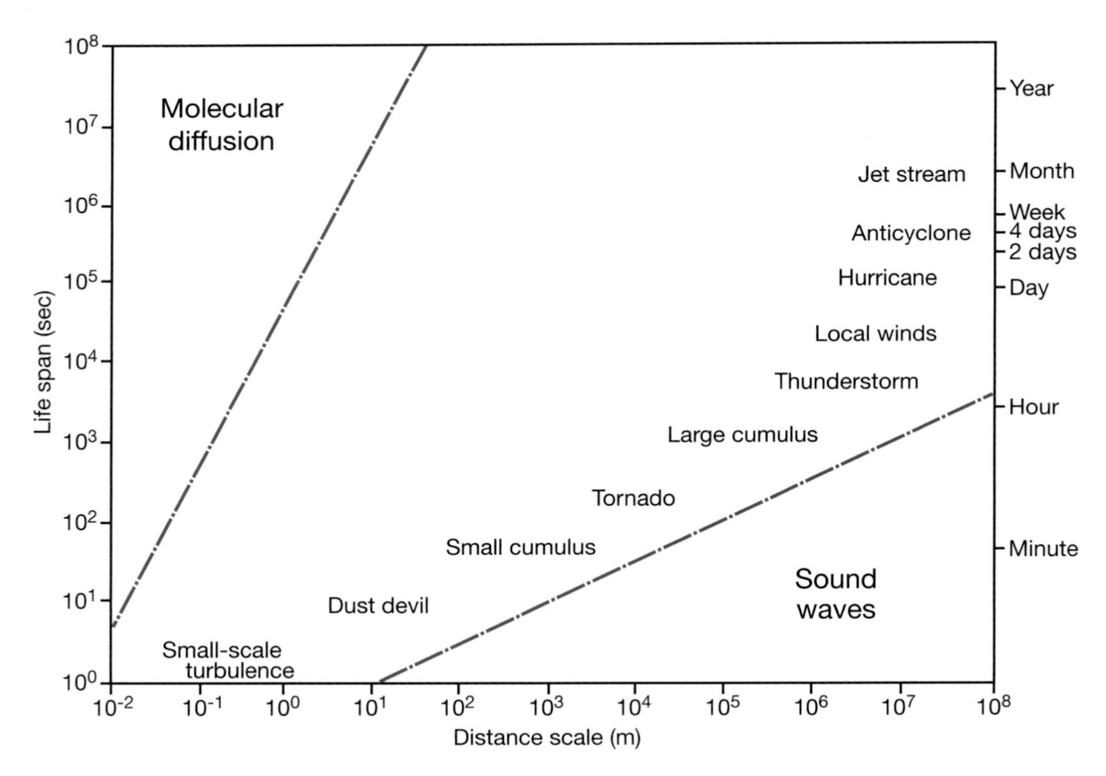

Figure 9.26 The spatial scale and lifespans of mesoscale and other meteorological systems.

precipitation over Colorado in spring and autumn is often associated with cold lows.

I MESOSCALE CONVECTIVE SYSTEMS

Mesoscale convective systems (MCSs) are intermediate in size and lifespan between synoptic disturbances and individual cumulonimbus cells (see Figure 9.26). Figure 9.27 shows the movement of clusters of convective cells, each cell about 1km in diameter, as they crossed southern Britain with a cold front. Each individual cell may be short-lived, but cell clusters may persist for hours, strengthening or weakening due to orographic and other factors.

MCSs occur seasonally in mid-latitudes (particularly the central United States, eastern China and South Africa) and the tropics (India, West and Central Africa and northern Australia) as either nearly circular clusters of convective cells or linear squall lines. The *squall line* consists of a narrow line of thunderstorm cells, which may extend for hundreds of kilometers. It is marked by a sharp veer in wind direction and very gusty conditions.

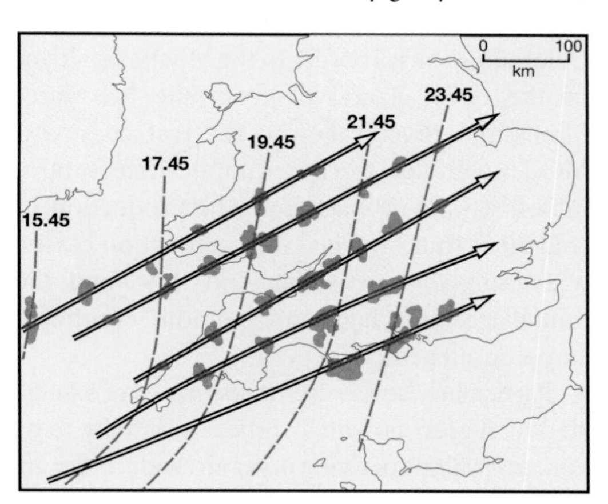

Figure 9.27 Successive positions of individual clusters of mid-tropospheric convective cells moving across southern Britain at about 50km hr^{-1} with a cold front. Cell location and intensity were determined by radar.

Source: After Browning (1990). By permission of the Royal Meteorological Society.

The squall line often occurs ahead of a cold front, maintained either as a self-propagating disturbance or by thunderstorm downdrafts. It may form a pseudo-cold front between rain-cooled air and a rainless zone within the same air mass. Mid-latitude squall lines appear to form through one of two mechanisms: (1) a pressure jump that propagates as a bore; (2) the leading edge of a cold front aloft (CFA) acting on instability present to the east of an orographic lee trough. In frontal cyclones, cold air in the rear of the depression may overrun air in the warm sector. The intrusion of this nose of cold air sets up great instability, and the subsiding cold wedge tends to act as a scoop forcing up the slower moving warm air.

Figure 9.28 demonstrates that the *relative* motion of the warm air is towards the squall line. Such conditions generate severe frontal thunderstorms such as that which struck Wokingham, England, in September 1959. This moved from the southwest at about 20m s^{-1}, steered by strong southwesterly flow aloft. The cold air subsided from high levels as a violent squall, and the updraft ahead of this produced an intense hailstorm. Hailstones grow by accretion in the upper part of the updraft, where speeds in excess of 50m s^{-1} are not uncommon, are blown ahead of the storm by strong upper winds, and begin to fall. This causes surface melting, but the stone is caught up again by the advancing squall line and reascends. The melted surface freezes, giving glazed ice as the stone is carried above the freezing level, and further growth occurs through the collection of supercooled droplets (see also Chapter 4, pp. 124 and 140).

Various types of MCS occur over the central United States in spring and summer (see Figure 9.29), bringing widespread severe weather. They may be small convective cells organized linearly, or as a large amorphous cell known as a *mesoscale convective complex* (MCC). This develops from initially isolated cumulonimbus cells. As rain falls from the thunderstorm clouds, evaporative cooling of the air beneath the cloud bases sets up cold downdrafts, and when these become

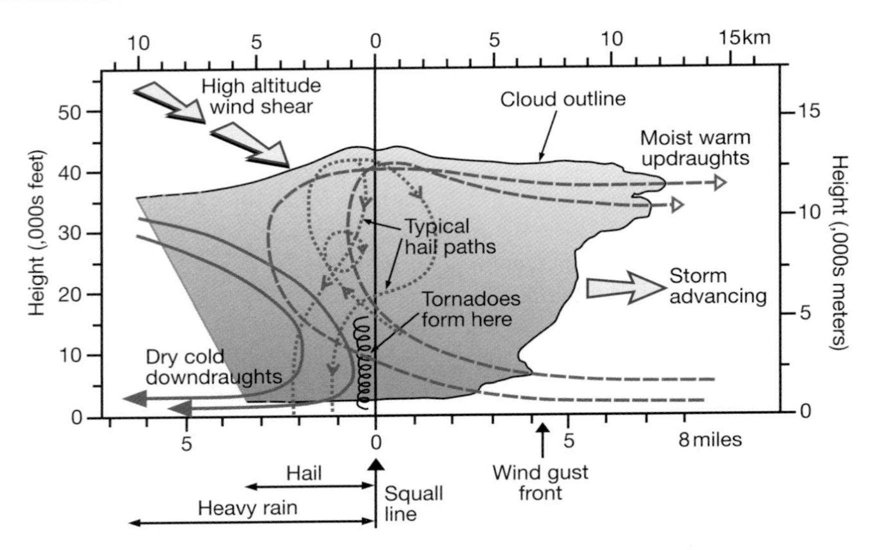

Figure 9.28 Thunder cell structure with hail and tornado formation.
Source: After Hindley (1977). By permission of *New Scientist*.

sufficiently extensive they create a local high pressure of a few millibars' intensity. The downdrafts trigger the ascent of displaced warm air, and a general warming of the middle troposphere results from latent heat release. Inflow develops towards this warm region, above the cold outflow, causing additional convergence of moist, unstable air. In some cases, a low-level jet provides this inflow. As individual cells become organized in a cluster along the leading edge of the surface high, new cells tend to form on the right flank (in the Northern Hemisphere) through interaction of cold downdrafts with the surrounding air. Through this process and the decay of older cells on the left flank, the storm system tends to move 10–20° to the right of the mid-tropospheric wind direction. As the thunderstorm high intensifies, a 'wake low', associated with clearing weather, forms to the rear of it. The system is now producing violent winds, and intense downpours of rain and hail accompanied by thunder. During the triggering of new cells, tornadoes may form as discussed below. As the MCC reaches maturity during the evening and night hours over the Great Plains, the mesoscale circulation is capped by an extensive (>100,000km^2) cold upper-cloud shield, readily identified on infrared satellite images. Statistics for 43 systems over the Great Plains in 1978 showed that the systems lasted on average 12 hours, with initial mesoscale organization occurring in the early evening (18:00–19:00 LST) and maximum extent seven hours later. During their life-cycle, systems may travel from the Colorado– Kansas border to the Mississippi River or the Great Lakes, or from the Missouri–Mississippi River valley to the east coast. An MCC usually decays when synoptic-scale features inhibit its self-propagation. The production of cold air is shut off when new convection ceases, weakening the meso-high and -low, and the rainfall becomes light and sporadic, eventually stopping altogether.

Particularly severe thunderstorms are associated with great potential vertical instability (e.g., hot, moist air underlying dryer air, with colder air aloft). This was the case with a severe storm in the vicinity of Sydney, Australia, on 21 January 1991 (Figure 9.30). The storm formed in a hot, moist, low-level airstream flowing northeast on the eastern side of the Blue Mountains escarpment. This flow was overlain by a hot, dry, northerly airstream at an elevation of 1500–6000 meters,

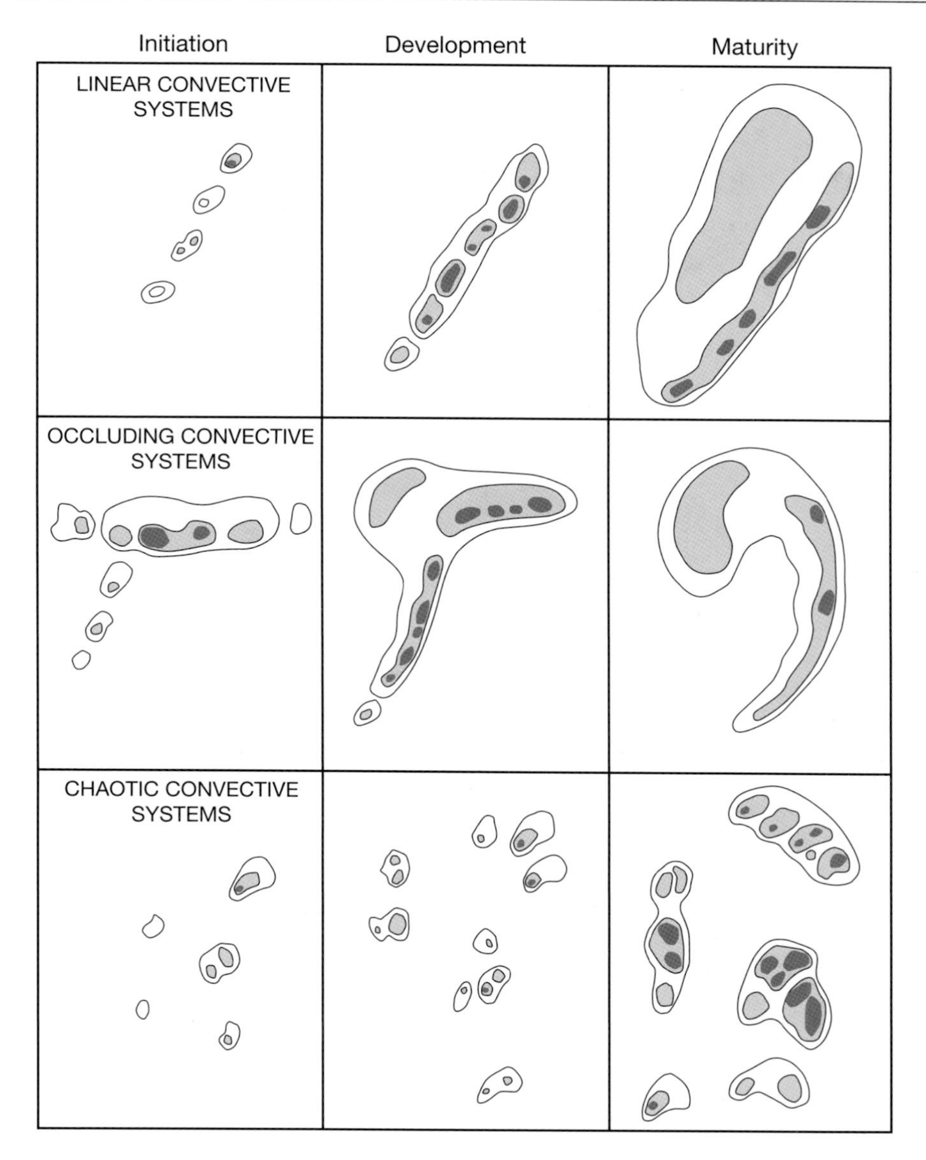

Figure 9.29 Schematic evolution of three convective modes on the US Great Plains showing several scales of cloud development (shading).
Source: Blanchard (1990, p. 996, Fig.2). Courtesy American Meteorological Society.

which, in turn, was capped by cold air associated with a nearby cold front. Five to seven such severe thunderstorms occurred annually in the vicinity of Sydney during 1950–1989.

On occasion, so-called *supercell thunderstorms* may develop as new cells, forming downstream, are swept up by the movement of an older cell. These are about the same size as thunder cell clusters but are dominated by one giant updraft and localized strong downdrafts Figure 9.31). They may give rise to large hailstones and tornadoes, although some give only moderate

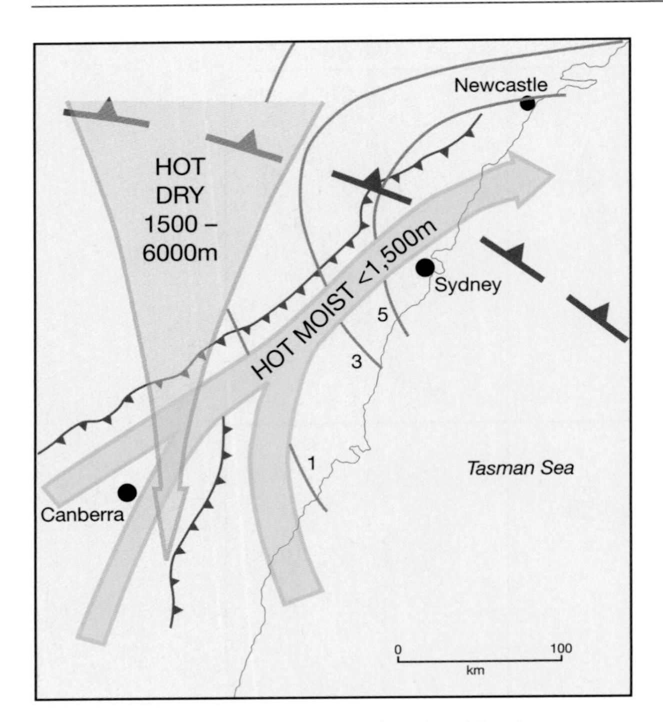

Figure 9.30 Conditions associated with the severe thunderstorm near Sydney, Australia, on 21 January 1991. The contours indicate the mean annual number of severe thunderstorms (per 25,000km^2) over eastern New South Wales for the period 1950–1989.

Source: Based on Griffiths *et al.* (1993). After Eyre (1992). Reproduced by kind permission of the NSW Bureau of Meteorology, from *Weather* by permission of the Royal Meteorological Society. Copyright ©.

rainfall amounts. A useful measure of instability in mesoscale storms is the bulk Richardson Number (Ri) which is the (dimensionless) ratio of the suppression of turbulence by buoyancy to the generation of turbulence by vertical wind shear in the lower troposphere. A high value of Ri means weak shear compared to buoyancy; Ri > 45 favors independent cell formation away from the parent updraft. For Ri < 30, strong shear supports a supercell by keeping the updraft close to its downdraft. Ri < 10 indicates weak instability and for strong vertical shear.

Tornadoes, which often develop within MCSs, are common over the Great Plains of the United States, especially in spring and early summer (see Figure 9.32 and Plate 9.3). During this period, cold, dry air from the high plateaux may override maritime tropical air (see Note 1). Subsidence beneath the upper tropospheric westerly jet (Figure 9.33) forms an inversion at about 1500–2000m, capping the low-level moist air. The moist air is extended northward by a low-level southerly jet (cf. p. 261) and, through continuing advection the air beneath the inversion becomes progressively more warm and moist. Eventually, the general convergence and ascent in the depression trigger the potential instability of the air, generating large cumulus clouds, which

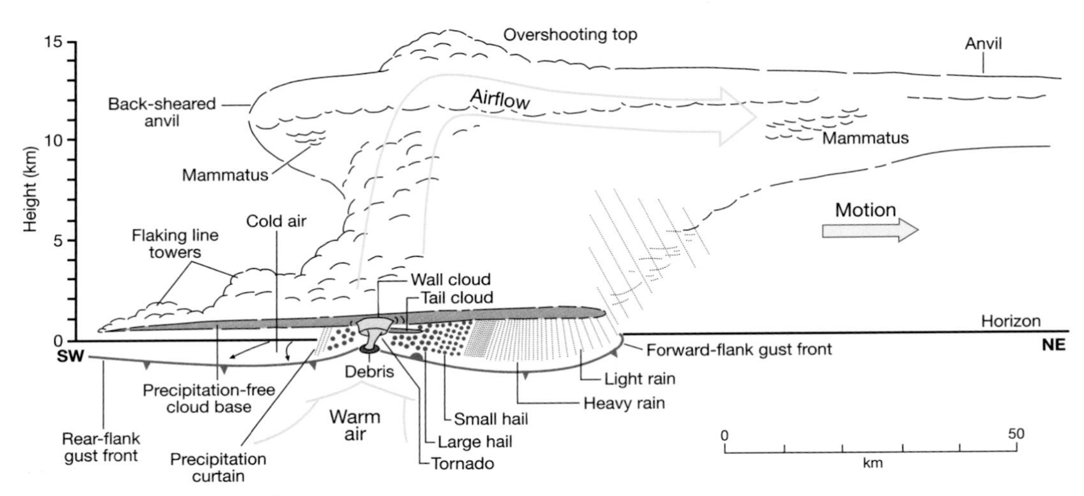

Figure 9.31 A supercell thunderstorm.

Source: After the National Severe Storms Laboratory, USA and H. Bluestein; from Houze and Hobbs (1982). Copyright © Academic Press, reproduced by permission.

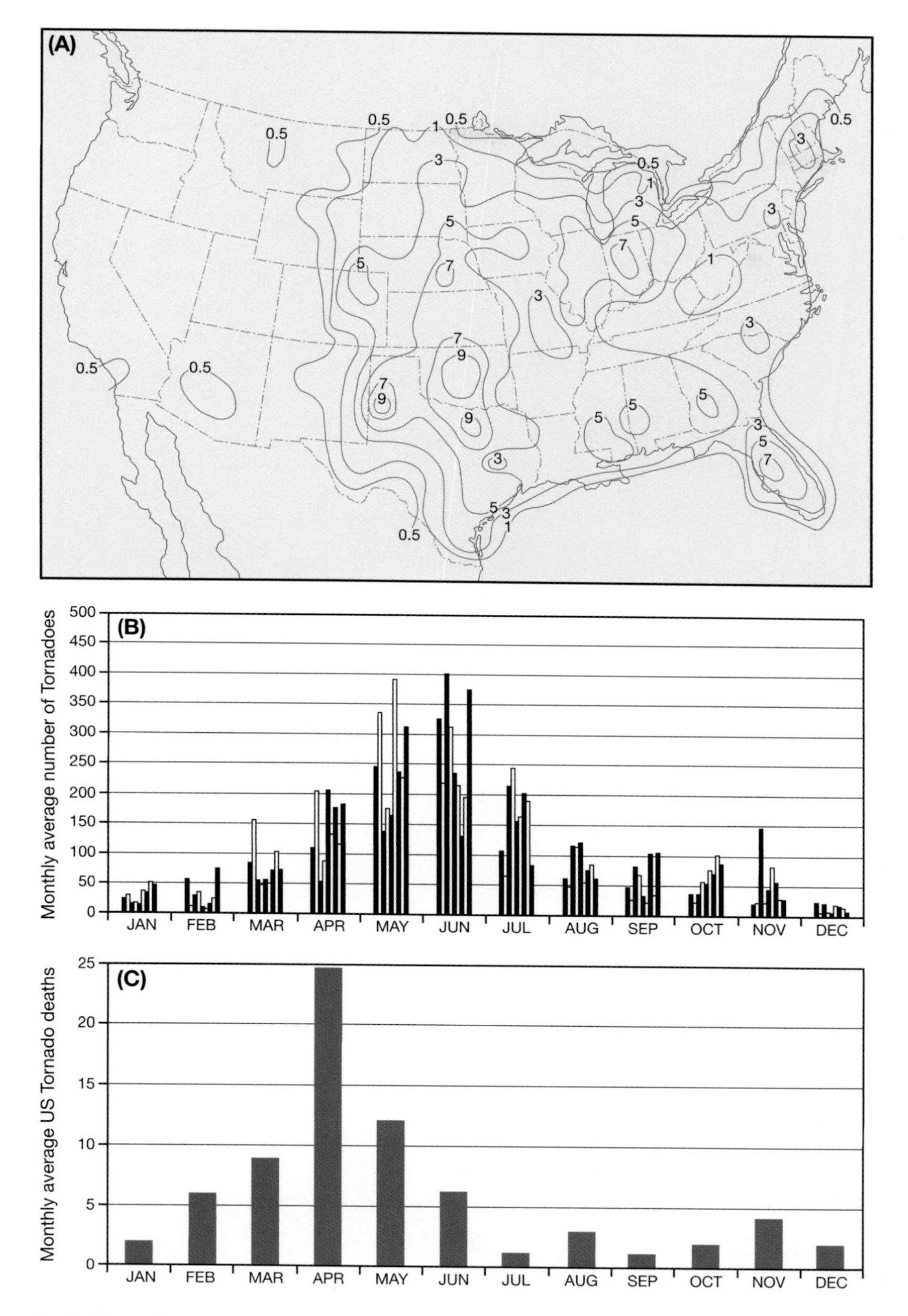

Figure 9.32 Tornado characteristics in the United States. A: Frequency of tornadoes (per 26,000km²) in the United States, 1953–1980. B: Monthly average number of tornadoes for each of the nine years shown by the alternating black and white bars (1990–1998). C: Monthly averages of resulting deaths (1966–1995).
Sources: A: From NOAA (1982). B and C: After NOAA – Storm Prediction Center.

Plate 9.3 A tornado raising surface dust in the Midwest USA.
Source: Courtesy Mark Anderson, University of Nevada

penetrate the inversion. The convective trigger is sometimes provided by the approach of a cold front towards the western edge of the moist tongue. Tornadoes may also occur in association with tropical cyclones (see p. 341) and in other synoptic situations if the necessary vertical contrast is present in the temperature, moisture and wind fields.

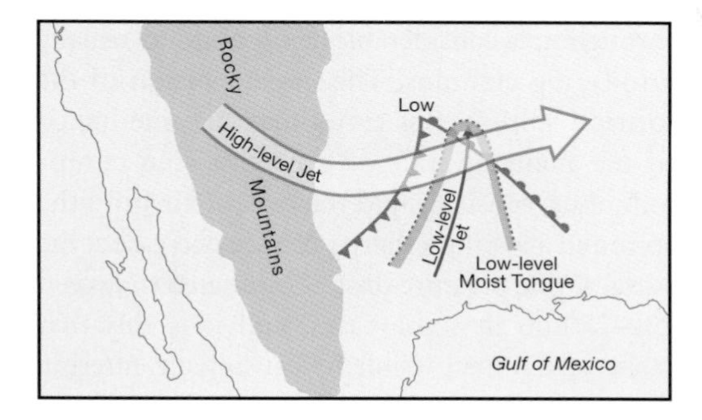

Figure 9.33 The synoptic conditions favoring severe storms and tornadoes over the Great Plains.

The exact tornado mechanism is still not fully understood owing to observational difficulties. Tornadoes tend to develop in the right-rear quadrant of a severe thunderstorm. Supercell thunderstorms are often identifiable in plan view on a radar reflectivity display by a hook echo pattern on the right-rear flank (Plate 9.4). The echo represents a (cyclonic or anticyclonic) spiral cloud band around a small central eye and its appearance may signal tornado development. The origin of the hook echo appears to involve the

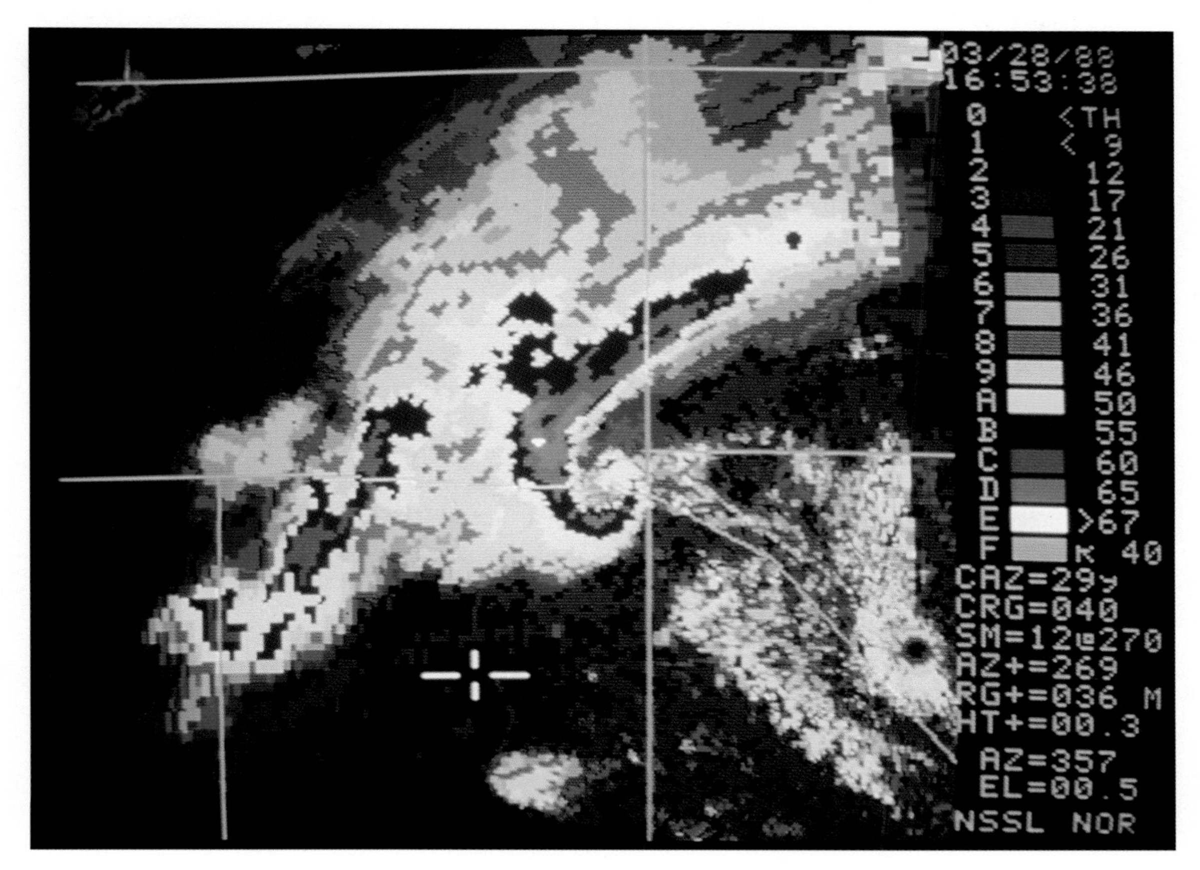

Plate 9.4 A radar image of a hook echo that is often a precursor of tornadic activity.
Source: NOAA NSS 0104.

horizontal advection of precipitation from the rear of the mesocyclone. Rotation develops where a thunderstorm updraft interacts with the horizontal airflow. Provided that the wind speed increases with height, the vertical wind shear generates vorticity (Chapter 6C) around an axis normal to the airflow, which is then tilted vertically by the updraft. Directional shear also generates vorticity that the updraft translates vertically. These two elements lead to rotation in the updraft in the lower-middle troposphere forming a meso-low 10–20km across. Pressure in the meso-low is 2–5mb lower than in the surrounding environment. At low levels, horizontal convergence increases the vorticity and rising air is replenished by moist air from progressively lower levels as the vortex descends and intensifies. The meso-low shrinks in diameter and the conservation of momentum increases the wind speed. At some point, a tornado, sometimes with secondary vortices (Figure 9.34), forms within the meso-low. The tornado funnel has been observed to originate in the cloud base and extend towards the surface (Figure 9.34). One idea is that convergence beneath the base of cumulonimbus clouds, aided by the interaction between cold precipitation downdrafts and neighboring updrafts, may initiate the funnel. Other observations suggest that the funnel forms simultaneously throughout a considerable depth of cloud, usually a towering cumulus. The upper portion of the tornado spire in this cloud may become linked to the main updraft of a neighboring cumulonimbus, causing rapid removal of air from the spire and allowing a sharp pressure decrease at the surface. The pressure drop is estimated to exceed 200–250mb in some cases, and it is this that makes the funnel visible by causing air entering the vortex to reach saturation. Over water, tornadoes are termed waterspouts; the majority rarely attain extreme intensities. The tornado vortex is usually only a few hundred meters in diameter and in an even more restricted band around the core the winds can attain speeds of 50–100m s^{-1}. Intense tornadoes may have multiple vortices rotating anticlockwise with respect to the main tornado axis, each following a cycloidal path. The whole tornado system gives a complex pattern of destruction, with maximum wind speeds on the right-side boundary (in the

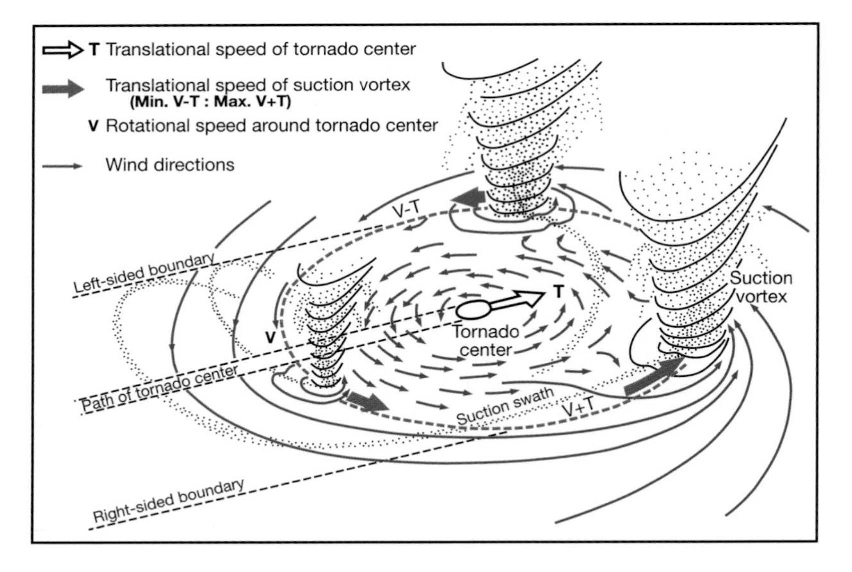

Figure 9.34 Schematic diagram of a complex tornado with multiple suction vortices.
Source: After Fujita (1981, p.1251, fig. 15). Courtesy of the American Meteorological Society.

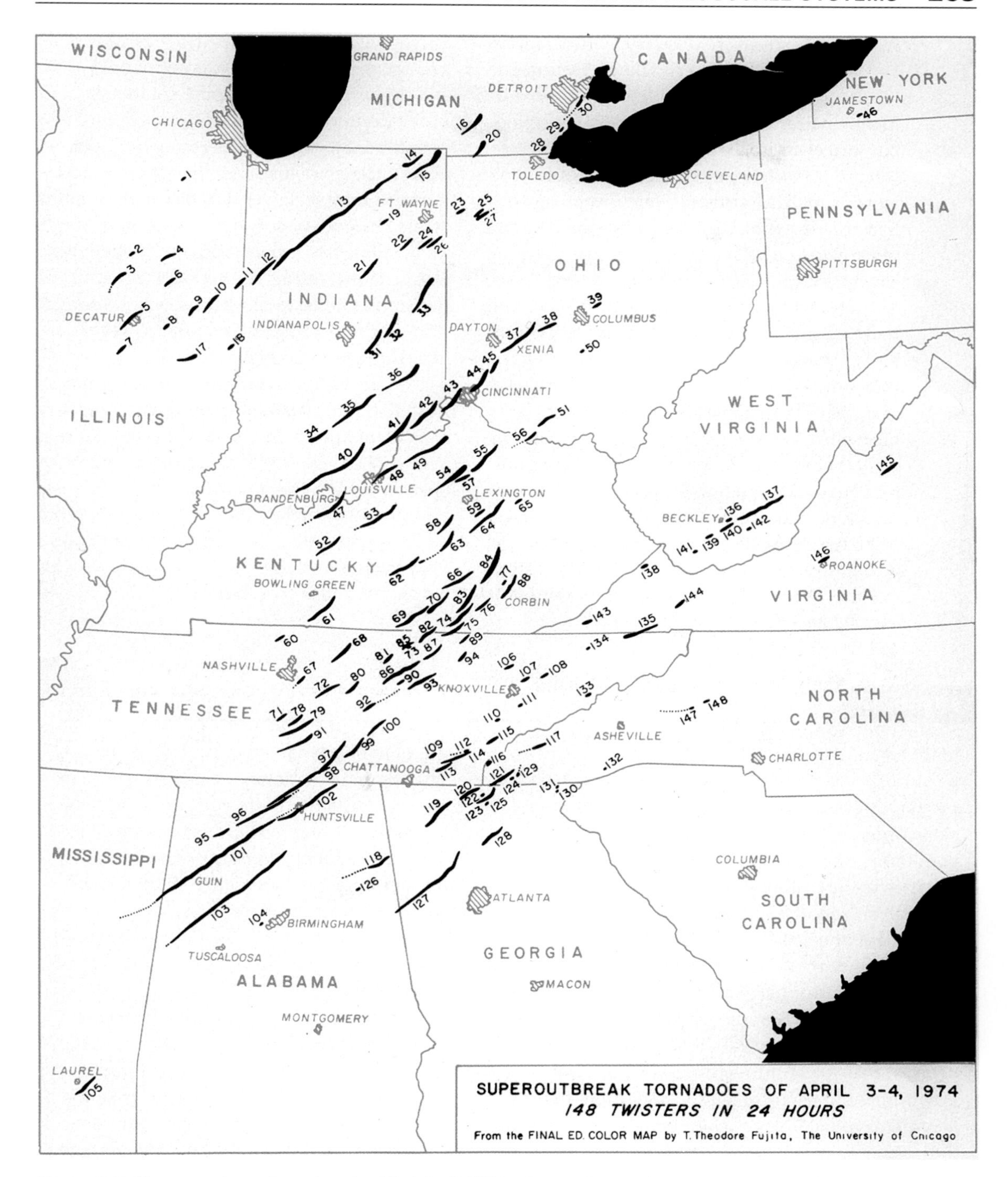

Plate 9.5 The super-tornado outbreak of 3–4 April 1974 when 148 tornadoes were reported in 24 hours. Thirty of these tornadoes were classified as F4 or F5 (winds of 92–142ms^{-1}) on the Fujita-Pearson Tornado Scale. There were 315 storm related fatalities, over 6100 storm-related injuries; and damages exceeded US$600 million (1974 dollars).

Northern Hemisphere), where the translational and rotational speeds are combined. Destruction results not only from the high winds, because buildings near the path of the vortex may explode outward owing to the pressure reduction outside. Intense tornadoes present problems as to their energy supply, and it has recently been suggested that the release of heat energy by lightning and other electrical discharges may be an additional energy source.

Tornadoes commonly occur in families and move along rather straight paths (typically between 10 and 100km long and 100m to 2km wide) at velocities dictated by the low-level jet. Thirty-year averages indicate some 750 tornadoes per year in the United States, with 60 percent of these during April to June (see Figure 9.32B). The largest outbreak in the United States occurred on 3–4 April 1974, extending from Alabama and Georgia in the south to Michigan in the north and from Illinois in the west to Virginia in the east. This 'Super Outbreak' spawned 148 tornadoes in 20 hours with a total path length of over 3200km (Plate 9.5).

Tornadoes in the United States caused an average of 59 fatalities per year during 1975–2006 and about 1800 injuries annually, although most of the deaths and destruction result from a few long-lived tornadoes, making up only 1.5 percent of the total reported. For example, the most severe recorded tornado traveled 200km in three hours across Missouri, Illinois and Indiana on 18 March 1925, killing 689 people. Tornado fatalities were almost 19,000 during the period 1880–2005 but the annual rate has declined considerably from the 1920s-1930s. The highest proportion (44 percent) of fatalities occur in mobile homes.

Tornadoes also occur in Canada, Europe, Australia, South Africa, India and East Asia. They are not unknown in the British Isles. During 1960–1982 there were 14 days per year with tornado occurrences. Most are minor outbreaks, but on 23 November 1981, 102 were reported during southwesterly flow ahead of a cold front. They are most common in autumn, when cold air moves over relatively warm seas.

SUMMARY

Ideal air masses are defined in terms of barotropic conditions, where isobars and isotherms are assumed to be parallel to each other and to the surface. The character of an air mass is determined by the nature of the source area, changes due to air-mass movement and its age. On a regional scale, energy exchanges and vertical mixing lead to a measure of equilibrium between surface conditions and those of the overlying air, particularly in quasi-stationary high pressure systems. Air masses are conventionally identified in terms of temperature characteristics (arctic, polar, tropical) and source region (maritime, continental). Primary air masses originate in regions of semi-permanent anticyclonic subsidence over extensive surfaces of similiar properties. Cold air masses originate either in winter continental anticyclones (Siberia and Canada), where snow cover promotes low temperatures and stable stratification, or over high-latitude sea ice. Some sources are seasonal, like Siberia; others are permanent, such as Antarctica. Warm air masses originate either in shallow tropical continental sources in summer or as deep, moist layers over tropical oceans. Air-mass movement causes stability changes through thermodynamic processes (heating/cooling from below and moisture exchanges) and by dynamic processes (mixing, lifting/subsidence), producing secondary air masses (e.g., mP air). The age of an air mass determines the degree to which it has lost its identity as the result of mixing with other air masses and vertical exchanges with the underlying surface.

Air-mass boundaries give rise to baroclinic frontal zones a few hundred kilometers wide. The classical (Norwegian) theory of mid-latitude cyclones considers that fronts are a key feature of their formation and life-cycle. Newer models show that instead of the frontal occlusion process, the warm front may become bent back with warm air seclusion within the polar airstream. Cyclones tend to form along major frontal zones – the polar fronts of the North Atlantic and North Pacific regions and of the southern oceans. An Arctic front lies poleward and there is a winter frontal zone over the Mediterranean. Air masses and frontal zones move poleward (equatorward) in summer (winter).

Newer cyclone theories regard fronts as rather incidental. Cloud bands and precipitation areas are associated primarily with conveyor belts of warm air. Divergence of air in the upper troposphere is essential for large-scale uplift and low-level convergence. Surface cyclogenesis is therefore favored on the eastern limb of an upper wave trough. 'Explosive' cyclogenesis appears to be associated with strong wintertime gradients of sea surface temperature. Cyclones are basically steered by the quasi-stationary long (Rossby) waves in the hemispheric westerlies, the positions of which are strongly influenced by surface features (major mountain barriers and land–sea surface temperature contrasts). Upper baroclinic zones are associated with jet streams at 300–200mb, which also follow the longwave pattern.

The idealized weather sequence in an eastward-moving frontal depression involves increasing cloudiness and precipitation with an approaching warm front; the degree of activity depends on whether the warm-sector air is rising or not (ana- or kata-fronts, respectively). The following cold front is often marked by a narrow band of convective precipitation, but rain both ahead of the warm front and in the warm sector may also be organized into locally intense mesoscale cells and bands due to the 'conveyor belt' of air in the warm sector.

Some low pressure systems form through non-frontal mechanisms. These include the lee cyclones formed in the lee of mountain ranges; thermal lows due to summer heating; polar air depressions commonly formed in an outbreak of maritime arctic air over oceans; and the upper cold low, which is often a cut-off system in upper wave development or an occluded mid-latitude cyclone in the Arctic.

Mesoscale convective systems (MCSs) have a spatial scale of tens of kilometers and a timescale of a few hours. They may give rise to severe weather, including thunderstorms and tornadoes. Thunderstorms are generated by convective uplift, which may result from daytime heating, orographic ascent or squall lines. Several cells may be organized in a mesoscale convective complex and move with the large-scale flow. Thunderstorms associated with a moving convective system provide an environment for hailstone growth and for the generation of tornadoes.

DISCUSSION TOPICS

- What are the essential differences between mesoscale and synoptic scale systems?
- Using an appropriate website with synoptic weather maps (see Appendix 4D), trace the movement of frontal and non-frontal lows/troughs and high pressure cells over a five-day period, determining rates of displacement and changes of intensity of the systems.
- In the same manner, examine the relationship of surface lows and highs to features at the 500mb level.
- Consider the geographical distribution and seasonal occurrence of different types of non-frontal low pressure systems.

REFERENCES AND FURTHER READING

Books

Church, C. R., Burgess, D., Doswell, C. and Davies-Jones, R. P. (eds) (1993) *The Tornado: Its Structure, Dynamics, Prediction, and Hazards. Geophys. Monogr.* 79, Amer. Geophys. Union, Washington, DC, 637pp. [Comprehensive accounts of vortex theory and modeling, observations of tornadic thunderstorms and tornadoes, tornado climatology , forecasting, hazards and damage surveys]

Karoly, D. I. and Vincent, D. G. (1998) *Meteorology of the* Southern Hemisphere. *Met. Monogr.* 27(49)., American Meteorological Society, Boston, MA, 410pp. [Comprehensive modern account of the circulation, meteorology of the land masses and Pacific Ocean, mesoscale processes, climate variability and change and modeling]

Kessler, E. (ed.) (1986) *Thunderstorm Morphology and Dynamics,* University of Oklahoma Press, Norman, OK, 411pp. [Comprehensive accounts by leading experts on convection and its modeling, all aspects of thunderstorm processes and occurrence in different environments, hail, lightning and tornadoes]

Newton, C. W. (ed.) (1972) *Meteorology of the* Southern Hemisphere, *Met. Monogr.* 13(35), American Meteorological Society, Boston, MA, 263pp. [Original comprehensive account now largely replaced by Karoly and Vincent, 1998]

Newton, C. W. and Holopainen, E. D. (eds) (1990) *Extratropical Cyclones: Palmén Memorial Symposium,* American Meteorological Society, Boston, MA, 262pp. [Invited and contributed conference papers and review articles by leading specialists]

Preston-Whyte, R. A. and Tyson, P. D. (1988) *The Atmosphere and Weather of Southern Africa,* Oxford University Press, Capetown, SA, 375pp. [An introductory meteorology text from a Southern Hemisphere viewpoint, with chapters on circulation and weather in Southern Africa as well as climate variability]

Riley, D. and Spolton, L. (1974) *World Weather and Climate,* Cambridge University Press, Cambridge, 120 pp.

Strahler, A. N. (1965) *Introduction to Physical Geography,* Wiley, New York, 455pp.

Taylor, J. A. and Yates, R. A. (1967) *British Weather in Maps,* 2nd edn, Macmillan, London 315pp, [Illustrates how to interpret synoptic maps and weather reports, including the lapse-rate structure]

Articles

Ashley, W. S. (2007) Spatial and temporal analysis of tornado fatalities in the United States: 1880–2005. *Weather and Forecasting* 22, 1214–28.

Belasco, J. E. (1952) Characteristics of air masses over the British Isles, Meteorological Office. *Geophysical Memoirs* 11(87) (34pp.).

Bennetts, D. A., Grant, J. R. and McCallum, E. (1988) An introductory review of fronts: Part I Theory and observations. *Met. Mag.* 117, 357–70.

Blanchard, D. O. (1990) Mesoscale convective patterns of the southern High Plains. *Bull. Amer. Met. Soc.* 71(7), 994–1005.

Boucher, R. J. and Newcomb, R. J. (1962) Synoptic interpretation of some TIROS vortex patterns: a preliminary cyclone model. *J. Appl. Met.* 1, 122–36.

Boyden, C. J. (1963) Development of the jet stream and cut-off circulations. *Met. Mag.* 92, 287–99.

Bracegirdle, T. J. and Gray, S. L. (2008) An objective climatology of the dynamical forcing of polar lows in the Nordic seas. *Int. J. Climatol.* 28, 1903–19.

Browning, K. A. (1968) The organization of severe local storms. *Weather* 23, 429–34.

Browning, K. A. (1985) Conceptual models of precipitation systems, *Met. Mag.* 114, 293–319.

Browning, K. A. (1986) Weather radar and FRONTIERS. *Weather* 41, 9–16.

Browning, K. A. (1990) Organization of clouds and precipitation in extratropical cyclones, in Newton, C. W. and Holopainen, E. D. (eds) *Extratropical Cyclones. The Erik Palmén Memorial Volume,* American Meteorological Society, Boston, MA, 129–53.

Browning, K. A. and Hill, F. F. (1981) Orographic rain. *Weather* 36, 326–9.

Browning, K. A. and Roberts, N. M. (1994) Structure of a frontal cyclone. *Quart. J. Roy. Met. Soc.* 120, 1535–57.

Browning, K. A., Bader, M. J., Waters, A. J., Young, M. V. and Monk, G. A. (1987) Application of satellite imagery to nowcasting and very short range forecasting. *Met. Mag.* 116, 161–79.

Businger, S. (1985) The synoptic climatology of polar low outbreaks. *Tellus* 37A, 419–32.

Carleton, A. M. (1985) Satellite climatological aspects of the 'polar low' and 'instant occlusion'. *Tellus* 37A, 433–50.

Carleton, A. M. (1996) Satellite climatological aspects of cold air mesocyclones in the Arctic and Antarctic. *Global Atmos. Ocean*, 5, 1–42.

Crowe, P. R. (1949) The trade wind circulation of the world. *Trans. Inst. Brit. Geog.* 15, 38–56.

Crowe, P. R. (1965) The geographer and the atmosphere. *Trans. Inst. Brit. Geog.* 36, 1–19.

Dudhia, J. (1997) Back to basics: thunderstorms. Part 2 – Storm types and associated weather. *Weather* 52, 2–7.

Eyre, J. A. (1992) How severe can a 'severe' thunderstorm be? *Weather* 47, 374–83.

Fujita, T. T. (1981) Tornadoes and downbursts in the context of generalized planetary scales. *J. Atmos. Sci.* 38, 1511–34.

Galloway, J. L. (1958a) The three-front model: its philosophy, nature, construction and use. *Weather* 13, 3–10.

Galloway, J. L. (1958b) The three-front model, the tropopause and the jet stream. *Weather* 13, 395–403.

Galloway, J. L. (1960) The three-front model, the developing depression and the occluding process. *Weather* 15, 293–309.

Godson, W. L. (1950) The structure of North American weather systems. *Cent. Proc. Roy. Met. Soc.*, London, 89–106.

Griffiths, D. J. *et al.* (1993) Severe thunderstorms in New South Wales: climatology and means of assessing the impact of climate change. *Climatic Change* 25, 369–88.

Gyakum, J. R. (1983) On the evolution of the *QE II* storm, I: Synoptic aspects. *Monthly Weather Review* 111, 1,137–55.

Hare, F. K. (1960) The westerlies. *Geog. Rev.* 50, 345–67.

Harman, J. R. (1971) *Tropical waves, jet streams, and the United States weather patterns.* Association of American Geographers, Commission on College Geography, Resource Paper No. 11, 37pp.

Harrold, T. W. (1973) Mechanisms influencing the distribution of precipitation within baroclinic disturbances. *Quart. J. Roy. Met. Soc.* 99, 232–51.

Hindley, K. (1977) Learning to live with twisters. *New Scientist* 70, 280–2.

Hobbs, P. V. (1978) Organization and structure of clouds and precipitation on the meso-scale and micro-scale of cyclonic storms. *Rev. Geophys. and Space Phys.* 16, 741–55.

Hobbs, P. V., Locatelli, J. D. and Martin, J. E. (1996) A new conceptual model for cyclones generated in the lee of the Rocky Mountains. *Bull. Amer. Met. Soc.* 77(6), 1169–78.

Houze, R. A. and Hobbs, P. V. (1982) Organization and structure of precipitating cloud systems. *Adv. Geophys.* 24, 225–315.

Hughes, P. and Gedzelman, S. D. (1995) Superstorm success. *Weatherwise* 48(3), 18–24.

Jackson, M. C. (1977) Meso-scale and small-scale motions as revealed by hourly rainfall maps of an outstanding rainfull event: 14–16 September 1968. *Weather* 32, 2–16.

Kalnay, E. *et al.* (1996) The NCEP/NCAR 40-year reanalysis project. *Bull. Amer. Met. Soc.* 77(3), 437–71.

Kelly, D.L. *et al.* (1978) An augmented tornado climatology. *Mon. Wea. Rev.,* 106, 1172–83.

Klein, W. H. (1948) Winter precipitation as related to the 700mb circulation. *Bull. Amer. Met. Soc.* 29, 439–53.

Klein, W. H. (1957) *Principal tracks and mean frequencies of cyclones and anticyclones in the* Northern Hemisphere. Research Paper No. 40, Weather Bureau, Washington, DC 60pp.

Kocin, P. J., Schumacher, P. N., Morales, R. F., Jr. and Uccellini, L. W. (1995) Overview of the 12–14 March 1993 Superstorm. *Bull. Amer. Met. Soc.* 76(2), 165–82.

Liljequist, G. H. (1970) *Klimatologi.* Generalstabens Litografiska Anstalt, Stockholm.

Locatelli, J. D., Martin, J. E., Castle, J. A. and Hobbs, P. V. (1995) Structure and evolution of winter cyclones in the central United States and their effects on the distribution of precipitation: Part III. The development of a squall line associated with weak cold frontogenesis aloft. *Monthly Weather Review* 123, 2641–62.

Ludlam, F. H. (1961) The hailstorm. *Weather* 16, 152–62.

Lyall, I. T. (1972) The polar low over Britain. *Weather* 27, 378–90.

Maddox, R. A. (1980) Mesoscale convective complexes. *Bull. Amer. Met. Soc.* 61, 1374–87.

McPherson, R. D. (1994) The National Centers for Environmental Prediction: operational climate, ocean and weather prediction for the 21st century. *Bull. Amer. Met. Soc.* 75(3), 363–73.

Miles, M. K. (1962) Wind, temperature and humidity distribution at some cold fronts over SE England. *Quart. J. Roy. Met. Soc.* 88, 286–300.

Miller, R. C. (1959) Tornado-producing synoptic patterns. *Bull. Amer. Met. Soc.* 40, 465–72.

Miller, R. C. and Starrett, L. G. (1962) Thunderstorms in Great Britain. *Met. Mag.* 91, 247–55.

Monk, G. A. (1992) Synoptic and mesoscale analysis of intense mid-latitude cyclones. *Met. Mag.* 121, 269–83.

Musk, L. F. (1988) *Weather Systems*, Cambridge University Press, Cambridge, 160pp.

Newton, C. W. (1966) Severe convective storms. *Adv. Geophys.* 12, 257–308.

NOAA (1982) *Tornado safety. Surviving nature's most violent storms.* NOAA/PA 82001 National Weather Service, NOAA, Rockville, MD, 8pp.

Parker, D. J. (2000) Frontal theory. *Weather* 55(4), 120–1.

Pedgley, D. E. (1962) A meso-synoptic analysis of the thunderstorms on 28 August 1958. *Geophys. Memo. Meteorolog. Office* 14(1) (30pp.).

Penner, C. M. (1955) A three-front model for synoptic analyses. *Quart. J. Roy. Met. Soc.* 81, 89–91.

Petterssen, S. (1950) Some aspects of the general circulation of the atmosphere. *Cent. Proc. Roy. Met. Soc.*, London, 120–55.

Portelo, A. and Castro, M. (1996) Summer thermal lows in the Iberian Peninsula: a three-dimensional simulation. *Quart. J. Roy. Met. Soc.* 122, 1–22.

Reed, R. J. (1960) Principal frontal zones of the northern hemisphere in winter and summer. *Bull. Amer. Met. Soc.* 41, 591–8.

Richter, D. A. and Dahl, R. A. (1958) Relationship of heavy precipitation to the jet maximum in the eastern United States. *Monthly Weather Review* 86, 368–76.

Roebber, P. J. (1989) On the statistical analysis of cyclone deepening rates, *Monthly Weather Review* 117, 2293–8.

Sanders, F. and Gyakum, J. R. (1980) Synoptic-dynamic climatology of the 'bomb'. *Monthly Weather Review* 108, 1,589–606.

Shapiro, M. A. and Keyser, D. A. (1990) Fronts, jet streams and the tropopause, in Newton, C. W. and Holopainen, E. O. (eds) *Extratropical Cyclones. The Erik Palmén Memorial Volume*, Amer. Met. Soc., Boston, MA., 167–91.

Showalter, A. K. (1939) Further studies of American air mass properties. *Monthly Weather Review* 67, 204–18.

Slater, P. M. and Richards, C. J. (1974) A memorable rainfall event over southern England. *Met. Mag.* 103, 255–68 and 288–300.

Smith, W. L. (1985) Satellites, in Houghton, D. D. (ed.) *Handbook of Applied Meteorology*, Wiley, New York, 380–472.

Stoelinga, M.T., Locatelli, J.D. and Hobbs, P.V. (2002) Warm occlusions, cold occlusions and forward-tilting cold fronts. *Bull. Ame. Met. Soc.* 83(5), 709–21.

Snow, J. T. (1984) The tornado. *Sci. American* 250(4), 56–66.

Snow, J. T. and Wyatt, A. L. (1997) Back to basics: the tornado, Nature's most violent wind. Part 1 – Worldwide occurrence and characterization. *Weather* 52(10), 298–304.

Sumner, G. (1996) Precipitation weather. *Geography* 81, 327–45.

Sutcliffe, R. C. and Forsdyke, A. G. (1950) The theory and use of upper air thickness patterns in forecasting. *Quart. J. Roy. Met. Soc.* 76, 189–217.

Taljaard, J. J., van Loon, H., Crutcher, H. L. and Jenne, R. L. (1969) Climate of the upper air: I. Southern hemisphere, in *Temperatures, Dew Points and Heights at Selected Pressure Levels*, vol. 1, U.S. Naval Weather Service, Washington DC NAVAIR 50–1C-55, 135pp.

Uccellini, L. W. (1990) Process contributing to the rapid development of extratropical cyclones, in Newton, C. and Holopainen, E. (eds) *Extra-tropical Cyclones: The Eric Palmen Memorial Volume. Amer. Met. Soc.* 81–107.

Vederman, J. (1954) The life cycles of jet streams and extratropical cyclones. *Bull. Amer. Met. Soc.* 35, 239–44.

Wallington, C. E. (1963) Meso-scale patterns of frontal rainfall and cloud. *Weather* 18, 171–81.

Wendland, W. M. and Bryson, R. A. (1981) Northern Hemisphere airstream regions. *Monthly Weather Review* 109, 255–70.

Wendland, W. M. and McDonald, N. S. (1986) Southern Hemisphere airstream climatology. *Monthly Weather Review* 114, 88–94.

Wick, G. (1973) Where Poseidon courts Aeolus. *New Scientist*, 18 January, 123–6.

Yoshino, M. M. (1967) Maps of the occurrence frequencies of fronts in the rainy season in early summer over east Asia. *Science Reports of the Tokyo University of Education* 89, 211–45.

Young, M.V. (1994a) Back to basics, depressions and anticyclones. Part 1 – Introduction. *Weather* 49, 306–12.

Young, M.V. (1994b) Back to basics: depressions and anticyclones. Part 2 – Life cycles and weather characteristics. *Weather* 49, 362–70.

CHAPTER TEN

Weather and climate in middle and high latitudes

10

LEARNING OBJECTIVES

When you have read this chapter you will:

- be familiar with the major factors determining climate in many regions of middle and high latitudes, and the subtropical margins
- appreciate the role of major topographic barriers in determining regional climate
- be aware of the contrasts between climatic conditions in the Arctic and Antarctic

In Chapters 7 and 8, the general structure of the atmospheric circulation has been outlined and the behavior and origin of extratropical cyclones examined. The direct contribution of pressure systems to the daily and seasonal variability of weather in the westerly wind belt is quite apparent to inhabitants of the temperate lands. Nevertheless, there are equally prominent contrasts of regional climate in mid-latitudes that reflect the interaction of geographical and meteorological factors. This chapter gives a selective synthesis of weather and climate in several extratropical regions, drawing mainly on the principles already presented. The climatic conditions of the subtropical and polar margins of the westerly wind belt, and the polar regions themselves, are examined in the final sections of the chapter. As far as possible, different themes are used to illustrate some of the more significant aspects of the climate in each area.

A EUROPE

1 Pressure and wind conditions

The dominant features of the mean pressure pattern over the North Atlantic are the Icelandic Low and the Azores High. These are present at all seasons (see Figure 7.9), although their location and relative intensity change considerably. The upper flow in this sector undergoes little seasonal change in pattern, but the westerlies decrease in strength by over half from winter to summer. The other major pressure system influencing European climates is the Siberian winter anticyclone, the occurrence of which is intensified by the extensive winter snow cover and the marked continentality of Eurasia. Atlantic depressions frequently move towards the Norwegian or Mediterranean seas in winter, but if they travel due east they occlude and fill long before they can penetrate into the heart of Siberia. Thus the Siberian high pressure is quasi-permanent at this season, and when it extends westward severe conditions affect much of Europe. In summer, pressure is low over all of Asia and depressions from the Atlantic tend to follow a more zonal path. Although the storm tracks over Europe do not shift poleward in summer, the depressions at this season are less intense and reduced air-mass contrasts produce weaker fronts.

Wind velocities over Western Europe bear a strong relationship to the occurrence and movement of depressions. The strongest winds occur on coasts exposed to the northwest airflow that follows the passage of frontal systems, or at constricted topographic locations that guide the movement of depressions or funnel airflow into them (Figure 10.1). For example, the Carcassonne Gap in southwest France provides a preferred southern route for depressions moving eastward from the Atlantic. The Rhône and Ebro valleys are funnels for strong airflow in the rear of depressions located in the western Mediterranean, generating the mistral and cierzo winds, respectively, in winter (see C.1, this chapter). Throughout Western Europe, the mean velocity of winds on hilltops is at least 100 percent greater than that in more sheltered locations. Winds in open terrain are on average 25–30 percent stronger than in sheltered locations; and coastal wind velocities are at least 10–20 percent less than those over adjacent seas (see Figure 10.1).

2 Oceanicity and continentality

Winter temperatures in northwest Europe are some 11°C or more above the latitudinal average (see Figure 3.18), a fact usually attributed to the presence of the North Atlantic Current. There is, however, a complex interaction between the ocean and the atmosphere. The current, which originates from the Gulf Stream off Florida strengthened by the Antilles Current, is primarily a wind-driven current initiated by the prevailing southwesterlies. It flows at a velocity of 16 to 32km per day and thus, from Florida, the water takes about eight or nine months to reach Ireland and about a year to reach Norway (see Chapter 7D.1). The southwesterly winds transport both sensible and latent heat acquired over the western Atlantic towards Europe, and although they continue to gain heat supplies over the northeastern North Atlantic, this local warming arises in the first place through the drag effect of the winds on the warm surface waters. Warming of air masses over the northeastern Atlantic is mainly of significance when polar or arctic air flows southeastward from Iceland. The temperature in such airstreams in winter may rise by 9°C between Iceland and northern Scotland. By contrast, maritime tropical air cools on average by about 4°C between the Azores and southwest England in winter and summer. One very evident effect of the North Atlantic Current is the absence of ice around the Norwegian coastline. However, the primary factor affecting the climate of northwestern Europe is the prevailing *onshore* winds transferring heat into the area.

The influence of maritime air masses can extend deep into Europe because there are few major topographic barriers to airflow and owing to the presence of the Mediterranean Sea. Hence the change to a more continental climatic regime is relatively gradual except in Scandinavia, where the mountain spine produces a sharp contrast between western Norway and Sweden. There are numerous indices expressing this continentality, but most are based on the annual range of temperature (see Note 1). Gorczynski's continentality index (K) is:

$$K = 1.7 \frac{A}{\sin \varphi} - 20.4$$

where A is the annual temperature range (°C) and φ is the latitude angle. (The index assumes that the annual range in solar radiation increases with latitude, but in fact the range is a maximum around 55°N.) K is scaled from 0 at extreme oceanic stations to 100 at extreme continental stations, but values occasionally fall outside of

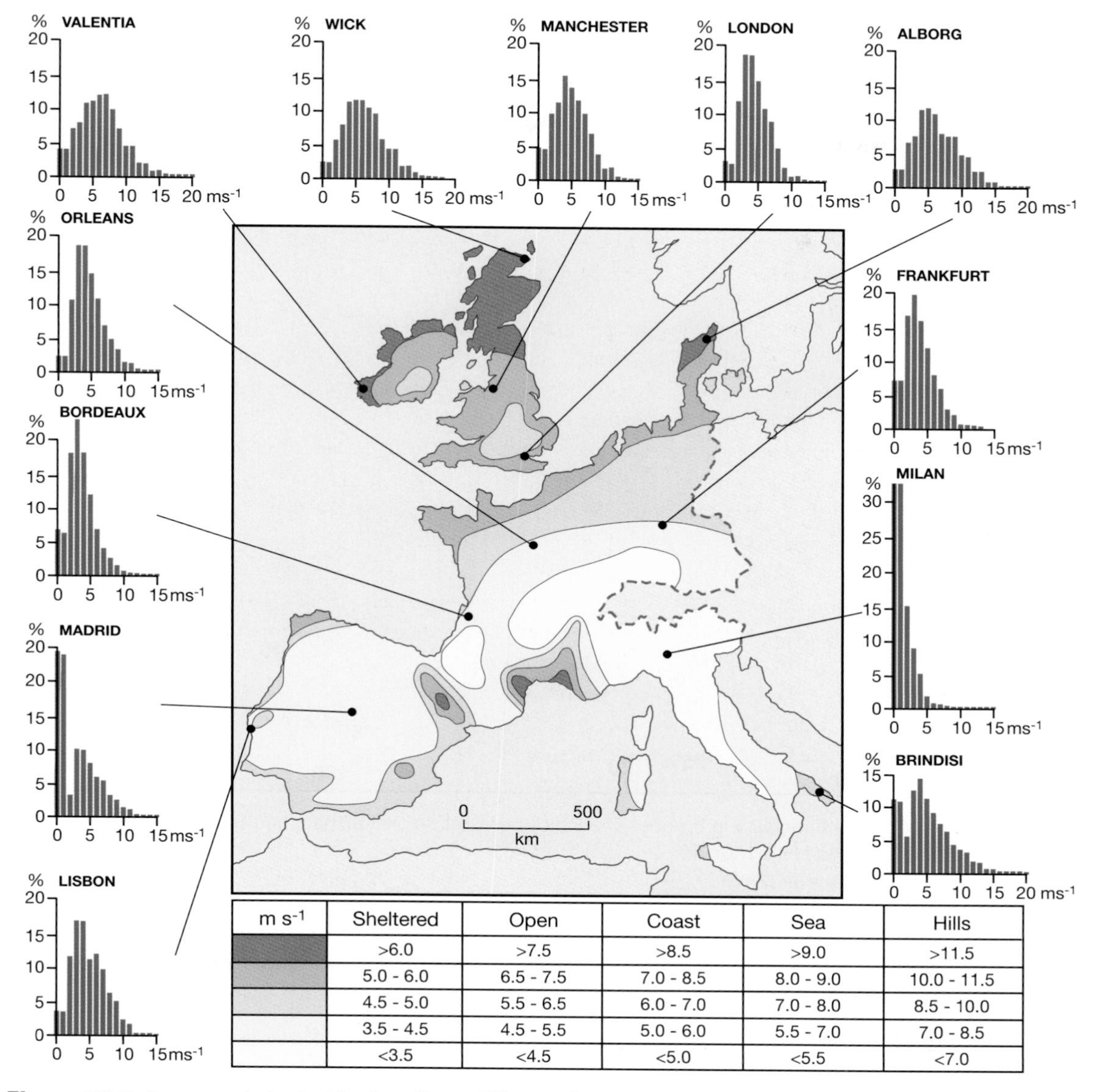

m s-1	Sheltered	Open	Coast	Sea	Hills
	>6.0	>7.5	>8.5	>9.0	>11.5
	5.0 - 6.0	6.5 - 7.5	7.0 - 8.5	8.0 - 9.0	10.0 - 11.5
	4.5 - 5.0	5.5 - 6.5	6.0 - 7.0	7.0 - 8.0	8.5 - 10.0
	3.5 - 4.5	4.5 - 5.5	5.0 - 6.0	5.5 - 7.0	7.0 - 8.5
	<3.5	<4.5	<5.0	<5.5	<7.0

Figure 10.1 Average wind velocities (m s^{-1}) over Western Europe, measured 50m above ground level for sheltered terrain, open plains, sea coast, open sea and hilltops. Frequencies (percent) of wind velocities for 12 locations are shown.

Source: From Troen and Petersen (1989). Courtesy Commission of the European Communities.

these limits. Some values in Europe are London 10, Berlin 21, and Moscow 42. Figure 10.2 shows the variation of this index over Europe.

An independent approach relates the frequency of continental air masses (C) to that of all air masses (N) as an index of continentality, i.e., $K = C / N$ (percent). Figure 10.2 shows that non-continental air occurs at least half the time over Europe west of 15°E as well as over Sweden and most of Finland.

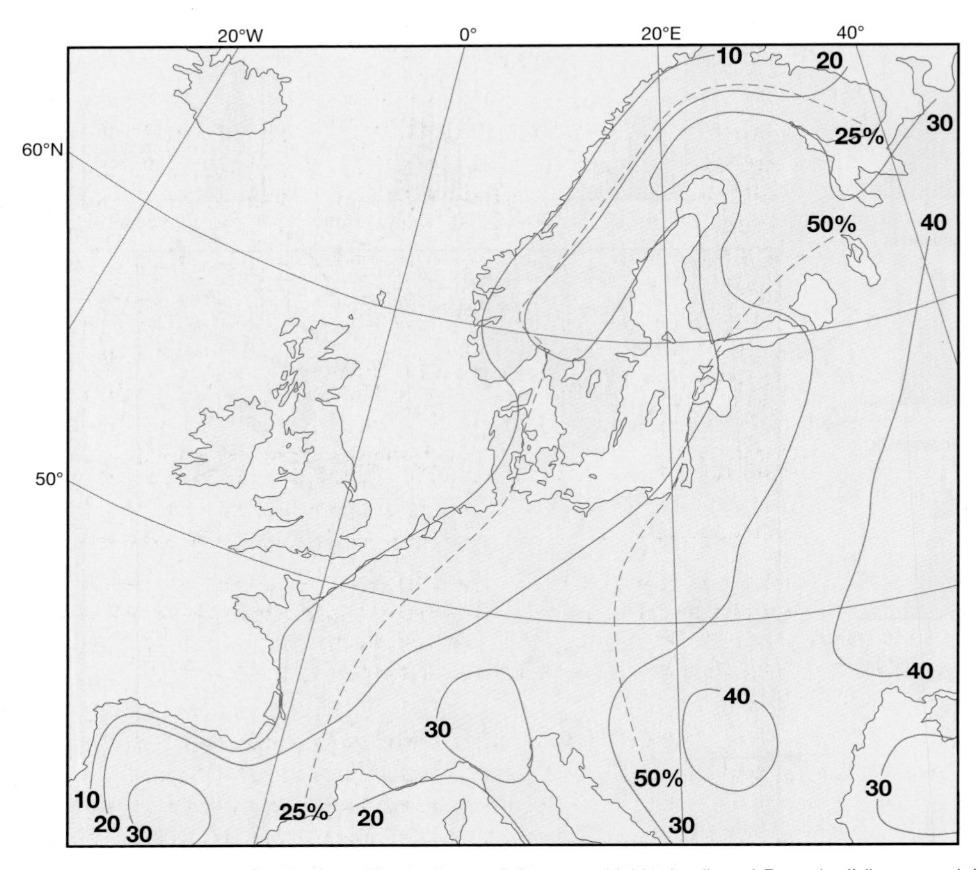

Figure 10.2 Continentality in Europe. The indices of Gorczynski (dashed) and Berg (solid) are explained in the text. See also Note 1, Ch 10.
Source: Partly after Blüthgen (1966).

A further illustration of maritime and continental regimes is provided by a comparison of Valentia (Eire), Bergen and Berlin (Figure 10.3). Valentia has a winter rainfall maximum and equable temperatures as a result of its oceanic situation (p. 60), whereas Berlin has a considerable temperature range and a summer maximum of rainfall. A theoretically ideal 'equable' climate has been defined as one with a mean temperature of 14°C in all months of the year. Bergen receives large rainfall totals due to orographic intensification and has a maximum in autumn and winter, its temperature range being intermediate between the other two. Such averages convey only a very general impression of climatic characteristics, and therefore British weather patterns are now examined in more detail.

3 British airflow patterns and their climatic characteristics

The daily weather maps for the British Isles sector (50–60°N, 2°E–10°W) from 1873 to the present day have been classified, in a scheme developed by the late Professor H. H. Lamb, according to the airflow direction or isobaric pattern. He identified seven major categories: westerly (W), northwesterly (NW), northerly (N), easterly (E) and southerly (S) types – referring to the compass directions from which the airflow and weather

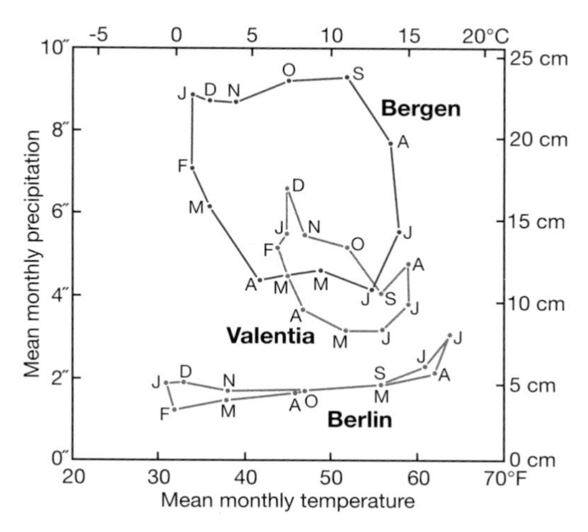

Figure 10.3 Hythergraphs for Valentia (Eire), Bergen and Berlin. Mean temperature and precipitation totals for each month are plotted.

systems are moving. Cyclonic (C) and anticyclonic (A) types denote when a low pressure or high pressure cell, respectively, dominates the weather map (Figure 10.4).

In principle, each category should produce a characteristic type of weather, depending on the season, and the term *weather type* is sometimes used to convey this idea. Statistical studies have been made of the *actual* weather conditions occurring in different localities with specific isobaric patterns – a field of study known as *synoptic climatology*. The general weather conditions and air masses that are to be associated with the airflow types identified by Lamb over the British Isles are summarized in Table 10.1.

On an annual basis, the most frequent airflow type is westerly; including cyclonic and anticyclonic subtypes, it has a 35 percent frequency in December to January and is almost as frequent in July to September (Figure 10.5). The minimum occurs in May (15 percent), when northerly and easterly types reach their maxima (about 10 percent each). Pure cyclonic patterns are most frequent (13–17 percent) in July to August and anticyclonic patterns in June and September (20 percent); cyclonic patterns have ⩾10 percent frequency in all months and anticyclonic patterns

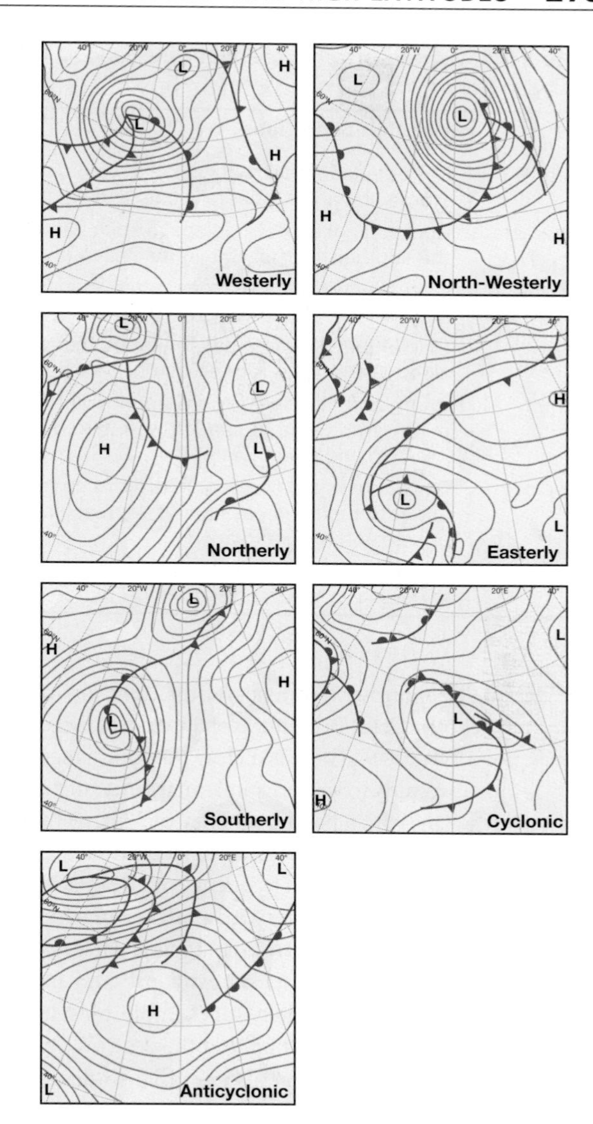

Figure 10.4 Synoptic situations over the British Isles classified according to the primary airflow types of H. H. Lamb.

Source: Lamb; O'Hare and Sweeney (1993). Copyright © The Geographical Association and G. O'Hare.

⩾13 percent. Figure 10.5 illustrates the mean daily temperature in central England and the mean daily precipitation over England and Wales for each type in the mid-season months for 1861–1979.

Table 10.1 General weather characteristics and air masses associated with Lamb's 'Airflow Types' over the British Isles

Type	Weather conditions
Westerly	Unsettled weather with variable wind directions as depressions cross the country. Mild and stormy in winter, generally cool and cloudy in summer (mP, mPw, mT).
Northwesterly	Cool, changeable conditions. Strong winds and showers affect windward coasts especially, but the southern part of Britain may have dry, bright weather (mP, mA).
Northerly	Cold weather at all seasons, often associated with polar lows. Snow and sleet showers in winter, especially in the north and east (mA).
Easterly	Cold in the winter half-year, sometimes very severe weather in the south and east with snow or sleet. Warm in summer with dry weather in the west. Occasionally thundery (cA, cP).
Southerly	Warm and thundery in summer. In winter, it may be associated with a low in the Atlantic, giving mild, damp weather especially in the southwest, or with a high over central Europe, in which case it is cold and dry in winter (mT, or cT, summer; mT or cP, winter).
Cyclonic	Rainy, unsettled conditions, often accompanied by gales and thunderstorms. This type may refer either to the rapid passage of depressions across the country or to the persistence of a deep depression (mP, mPw, mT).
Anticyclonic	Warm and dry in summer, occasional thunderstorms (mT, cT). Cold and frosty in winter with fog, especially in autumn (cP).

The monthly frequency of different air-mass types over Great Britain was analyzed by J. Belasco for 1938–1949. There is a clear predominance of northwesterly to westerly polar maritime (mP and mPw) air, which has a frequency of 30 percent or more over southeast England in all months except March. The maximum frequency of mP air at Kew (London) is 33 percent (with a further 10 percent mPw) in July. The proportion is even greater in western coastal districts, with mP and mPw occurring in the Hebrides, for example, on at least 38 percent of days throughout the year.

Air-mass types may also be used to describe typical weather conditions. Northwesterly mP airstreams produce cool, showery weather at all seasons. The air is unstable, forming cumulus clouds, although inland in winter and at night the clouds disperse, giving low night temperatures. Over the sea, heating of the lower air continues by day and by night in winter months, so showers and squalls can occur at any time, and these affect windward coastal areas. The average daily mean temperatures with mP air are within about ±1°C of the seasonal means in winter and summer, depending on the precise track of the air. More extreme conditions occur with mA air, the temperature departures at Kew being approximately –4°C in summer and winter. The visibility in mA air is usually very good. The contribution of mP and mA air masses to the mean annual rainfall over a five-year period at three stations in northern England and North Wales is given in Table 10.2, although it should be noted that both air masses may also be involved in non-frontal polar lows. Over much of southern England, and in areas to the lee of high ground, northerly and northwesterly airstreams usually give clear, sunny weather with few showers. This is illustrated in Table 10.2. At Rotherham, in the lee of the Pennines, the percentage of the rainfall occurring with mP air is much lower than on the West Coast (Squires Gate).

Maritime tropical air commonly forms the warm sector of depressions moving from between west and south towards Britain. The weather is unseasonably mild and damp with mT air in winter. There is usually a complete cover of stratus or stratocumulus cloud and drizzle or light rain may occur, especially over high ground, where low cloud produces hill fog. The clearance of

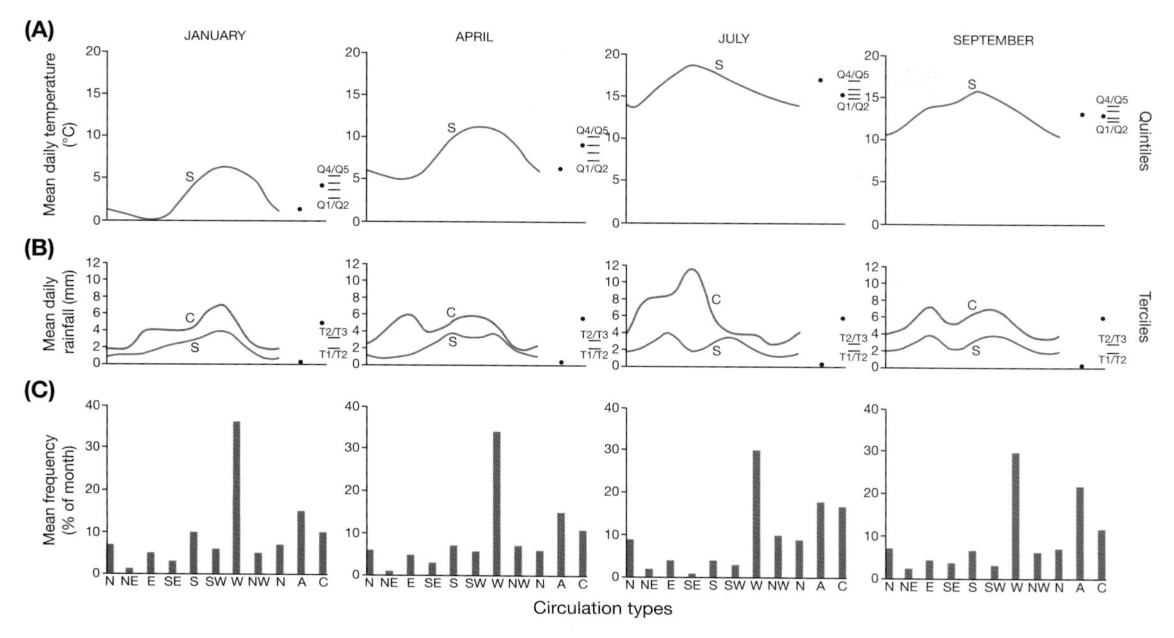

Figure 10.5 Average climatic conditions associated with Lamb's circulation types for January, April, July and September, 1861–1979. Top: Mean daily temperature (°C) in central England for the straight (S) airflow types; at the right side are the quintiles of mean monthly temperature (i.e., Q1/Q2 = 20 percent, Q4/Q5 = 80 percent). Middle: Mean daily rainfall (in millimeters) over England and Wales for the straight (S) and cyclonic (C) subdivisions of each type and terciles of the mean values (i.e., T1/T2 = 33 percent, T2/T3 = 67 percent). Bottom: Mean frequency (percent) for each circulation type, including anticyclonic (A) and cyclonic (C).

Source: After Storey (1982) By permission of the Royal Meteorological Society.

Table 10.2 Percentage of the annual rainfall (1956–1960) occurring with different synoptic situations

Station	Synoptic categories								
	Warm front	**Warm sector**	**Cold front**	**Occlusion**	**Polar low**	**mP**	**cP**	**Arctic**	**Thunder-storm**
Cwm Dyli (99m)*	18	30	13	10	5	22	0.1	0.8	0.8
Squires Gate (10m)†	23	16	14	15	7	22	0.2	0.7	3
Rotherham (21m)‡	26	9	11	20	14	15	1.5	1.1	3

Source: After Shaw (1962), and R. P. Mathews (unpublished).

Notes: *Snowdonia. †On the Lancashire coast (Blackpool). ‡In the Don Valley, Yorkshire.

cloud on nights with light winds readily cools the moist air to its dew-point, forming mist and fog. Table 10.2 shows that a large proportion of the annual rainfall is associated with warm-front and warm-sector situations and therefore is largely attributable to convergence and frontal uplift within mT air. In summer the cloud cover with this air mass keeps temperatures closer to average than in winter; night temperatures tend to be high, but daytime maxima remain rather low.

In summer, 'plumes' of warm, moist mT air may spread northward from the vicinity of Spain into Western Europe. This air is very unstable, with a significant vertical wind shear and a wet-bulb potential temperature that may exceed 18°C. Instability may be increased if cooler Atlantic air is advected under the plume from the west. Thunderstorms tend to develop along the leading northern edge of the moist plume over Britain and northwest Europe. Occasionally, depressions develop on the front and move eastward, bringing widespread storms to the region (Figure 10.6). On average, two mesoscale convective systems affect southern Britain each summer, moving northward from France.

Continental polar air occasionally affects the British Isles between December and February. Mean daily temperatures are well below average and maxima rise to only a degree or so above freezing point. The air is basically very dry and stable (see easterly type in January, Figure 10.4) but a track over the central part of the North Sea supplies sufficient heat and moisture to cause showers, often in the form of snow, over eastern England and Scotland. In total this provides only a very small contribution to the annual precipitation, as Table 10.2 shows, and on the West Coast the weather is generally clear. A transitional cP–cT type of air mass reaches Britain from southeastern Europe in all seasons, although less frequently in summer. Such airstreams are dry and stable.

Continental tropical air occurs on average about one day per month in summer, which accounts for the rarity of summer heatwaves, since

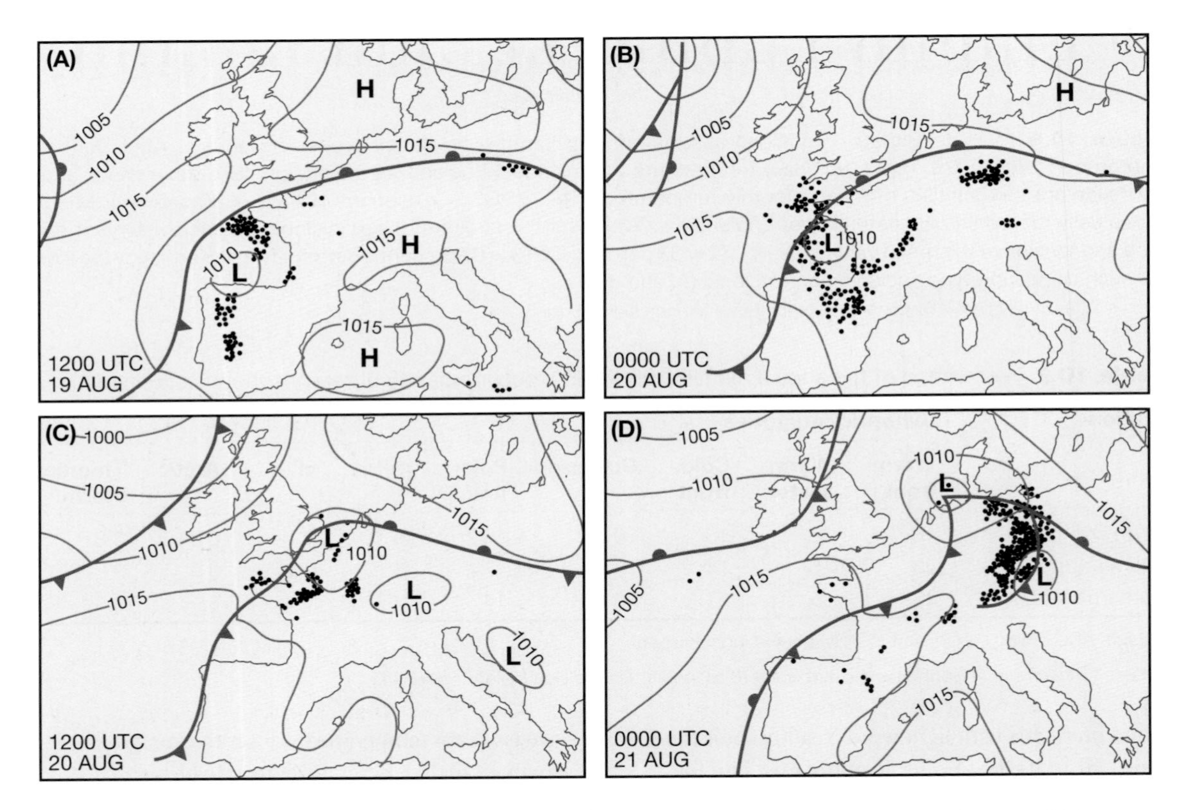

Figure 10.6 Distribution of thunderstorms over Western Europe during the period 19–21 August 1992 (storms shown for the four-hour period preceding the times given). A small depression formed over the Bay of Biscay and moved eastward along the boundary of the warm air, developing a strong squall line.

Source: Blackall and Taylor (1993). Reprinted from the *Meteorological Magazine* (Crown copyright ©) by permission of the Controller of Her Majesty's Stationery Office.

these south or southeast winds bring hot, settled weather. The lower layers are stable and the air is commonly hazy, but the upper layers tend to be unstable and surface heating may occasionally trigger off a thunderstorm (see southerly cyclonic type in July, Figure 10.4).

4 Singularities and natural seasons

Popular weather lore expresses the belief that each season has its own weather (for example, in England, 'February fill-dyke', and 'April showers'). Ancient adages suggest that even the sequence of weather may be determined by the conditions established on a given date. For example, 40 days of wet or fine weather are said to follow St Swithin's Day (15 July) in England; sunny conditions on 'Groundhog Day' (2 February) are claimed to portend six more weeks of winter in the United States. Some of these ideas are fallacious, but others contain more than a grain of truth if properly interpreted.

The tendency for a certain type of weather to recur with reasonable regularity around the same date is termed a *singularity*. Many calendars of singularities have been compiled, particularly in Europe. Early ones, which concentrated upon anomalies of temperature or rainfall, did not prove very reliable. Greater success has been achieved by studying singularities of circulation pattern; Flohn, and Hess and Brezowsky, have prepared catalogues for Central Europe and Lamb for the British Isles. Lamb's results are based on calculations of the daily frequency of the airflow categories between 1898 and 1947, some examples of which are shown in Figure 10.7. A noticeable feature is the infrequency of the westerly type in spring, the driest season of the year in the British Isles, and also in northern France, northern Germany and in the countries bordering the North Sea. The European catalogue is based on a classification of large-scale patterns of airflow in the lower troposphere (*Grosswetterlage*) over Central Europe. Some of the European

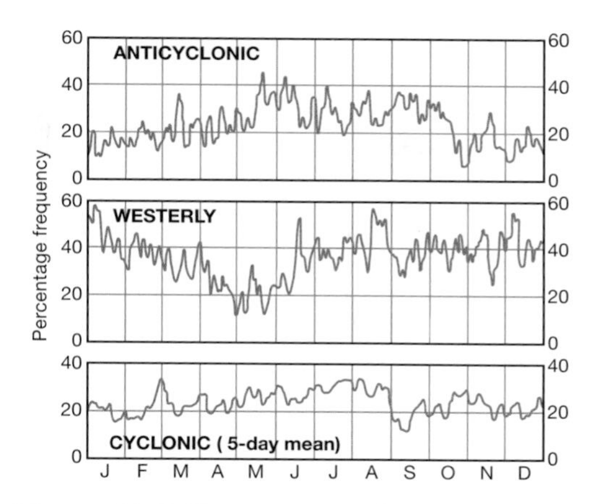

Figure 10.7 The percentage frequency of anticyclonic, westerly and cyclonic conditions over Britain, 1898–1947.
Source: After Lamb (1950).

singularities that occur most regularly are as follows:

1. A sharp increase in the frequency of westerly and northwesterly type over Britain takes place about the middle of June. These invasions of maritime air also affect Central Europe, and this period marks the beginning of the European 'summer monsoon'.
2. Around the second week in September, Europe and Britain are affected by a spell of anticyclonic weather. This may be interrupted by Atlantic depressions, giving stormy weather over Britain in late September, although anticyclonic conditions again affect Central Europe at the end of the month and Britain during early October.
3. A marked period of wet weather often affects Western Europe and also the western half of the Mediterranean at the end of October, whereas the weather in Eastern Europe generally remains fine.
4. Anticyclonic conditions return to Britain and affect much of Europe about mid-November, giving rise to fog and frost.
5. In early December, Atlantic depressions push eastward to give mild, wet weather over most of Europe.

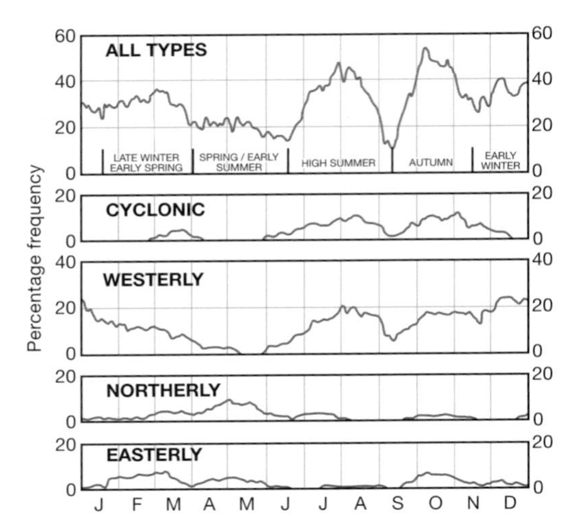

Figure 10.8 The frequency of long spells (25 days or more) of a given airflow type over Britain, 1898–1947. The diagram showing all long spells also indicates a division of the year into 'natural seasons'.

Source: After Lamb (1950). By permission of the Royal Meteorological Society.

In addition to these singularities, major seasonal trends are recognizable. For the British Isles, Lamb identified five *natural seasons* on the basis of spells of a particular type lasting for 25 days or more during the period 1898–1947 (Figure 10.8). These seasons are:

1. *Spring to early summer* (the beginning of April to mid-June). This is a period of variable weather conditions during which long spells are least likely. Northerly spells in the first half of May are the most significant feature, although there is a marked tendency for anticyclones to occur in late May to early June.
2. *High summer* (mid-June to early September). Long spells of various types may occur in different years. Westerly and northwesterly types are the most common and they may be combined with either cyclonic or anticyclonic types. Persistent sequences of cyclonic type occur more frequently than anticyclonic types.
3. *Autumn* (the second week in September to mid-November). Long spells are again present in most years. Anticyclonic spells are mainly in the first half, cyclonic and other stormy spells generally in October to November.
4. *Early winter* (from about the third week in November to mid-January). Long spells are less frequent than in summer and autumn. They are usually of westerly type, giving mild, stormy weather.
5. *Late winter and early spring* (from about the third week in January to the end of March). The long spells at this time of year can be of very different types, so that in some years it is midwinter weather, while in other years there is an early spring from about late February.

5 Synoptic anomalies

The mean climatic features of pressure, wind and seasonal airflow regime provide only a partial picture of climatic conditions. Some patterns of circulation occur irregularly and yet, owing to their tendency to persist for weeks or even months, form an essential element of the climate.

Blocking patterns are an important example. It was noted in Chapter 7 that the zonal circulation in mid-latitudes sometimes breaks down into a cellular pattern. This is commonly associated with a split of the jet stream into two branches over higher and lower mid-latitudes and the formation of a cut-off low (see Chapter 8H.4) south of a high pressure cell. The latter is referred to as a *blocking anticyclone* since it prevents the normal eastward motion of depressions in the zonal flow. Figure 10.9 illustrates the frequency of blocking for part of the Northern Hemisphere with five major blocking centers shown (H). A major area of blocking is Scandinavia, particularly in spring. Cyclones are diverted northeastward towards the Norwegian Sea or southeastward into southern Europe. This pattern, with easterly flow around the southern margins of the anticyclone, produces severe winter weather over much of northern Europe. In January to February 1947, for example, easterly flow across Britain as a result of blocking over Scandinavia led to extreme cold and frequent snowfall. Winds were almost continuously from the east between 22 January and 22 February and even daytime temperatures rose little above freezing point. Snow fell in some part of Britain every day from 22 January to 17 March 1947, and major snowstorms occurred as occluded Atlantic depressions moved slowly across the country. Other notably severe winter months – January 1881, February 1895, January 1940 and February 1986 – were the result of similar pressure anomalies with pressure well above average to the north of the British Isles and below average to the south, giving persistent easterly winds.

The effects of winter blocking situations over northwest Europe are shown in Figures 10.10 and 10.11. Precipitation amounts are above normal, mainly over Iceland and the western Mediterranean, as depressions are steered around the blocking high following the path of the upper jet streams. Over most of Europe, precipitation remains below average and this pattern is repeated with summer blocking. Winter temperatures are above average over the northeastern Atlantic and adjoining land areas, but below average over Central and Eastern Europe and the Mediterranean due to outbreaks of cP air (Figure 10.11). The negative temperature anomalies associated with cool northerly airflow in summer cover most of Europe; only northern Scandinavia has above-average values.

The exact location of the block is of the utmost importance. For instance, in the summer of 1954 a blocking anticyclone across Eastern Europe and Scandinavia allowed depressions to stagnate over the British Isles, giving a dull, wet August, whereas in 1955 the blocking was located over the North Sea and a fine, warm summer resulted. Persistent blocking over northwestern Europe caused drought in Britain and the continent during 1975–1976. Another, less common location of blocking is Iceland. A notable example was the 1962–1963 winter, when persistent high pressure southeast of Iceland led to northerly and northeasterly airflow over Britain. Temperatures in central England were the lowest since 1740, with a mean of 0°C for December 1962 to February 1963. Central Europe was affected by easterly airstreams with mean January temperatures 6°C below average.

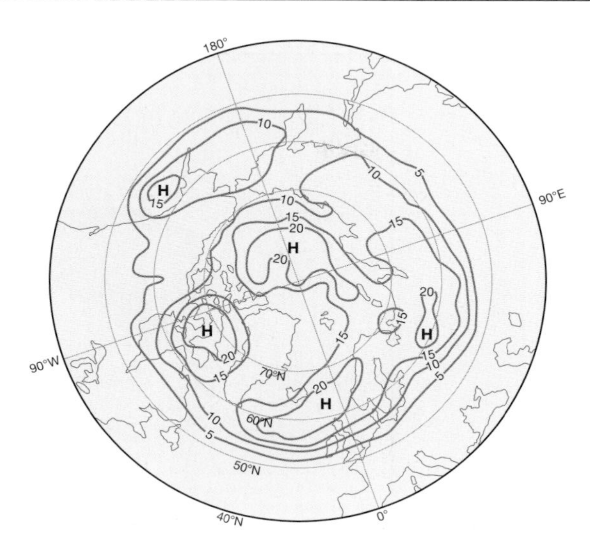

Figure 10.9 Frequency of occurrence of blocking conditions for the 500mb level for all seasons. Values were calculated as five-day means for 381 × 381km squares for the period 1946–78.

Source: From Knox and Hay (1985). By permission of the Royal Meteorological Society.

6 Topographic effects

In various parts of Europe, topography has a marked effect on the climate, not only of the uplands themselves but also of adjacent areas. Apart from the more obvious effects on temperatures, precipitation amounts and winds, the major mountain masses also affect the movement of frontal systems. Frictional drag over mountain barriers increases the slope of cold fronts and decreases the slope of warm fronts, so that the latter are slowed down and the former accelerated.

The Scandinavian mountains form one of the most significant climatic barriers in Europe as a result of their orientation with regard to westerly airflow. Maritime air masses are forced to rise over the highland zone, giving annual precipitation totals of over 2500mm on the mountains of western Norway, whereas descent in their lee

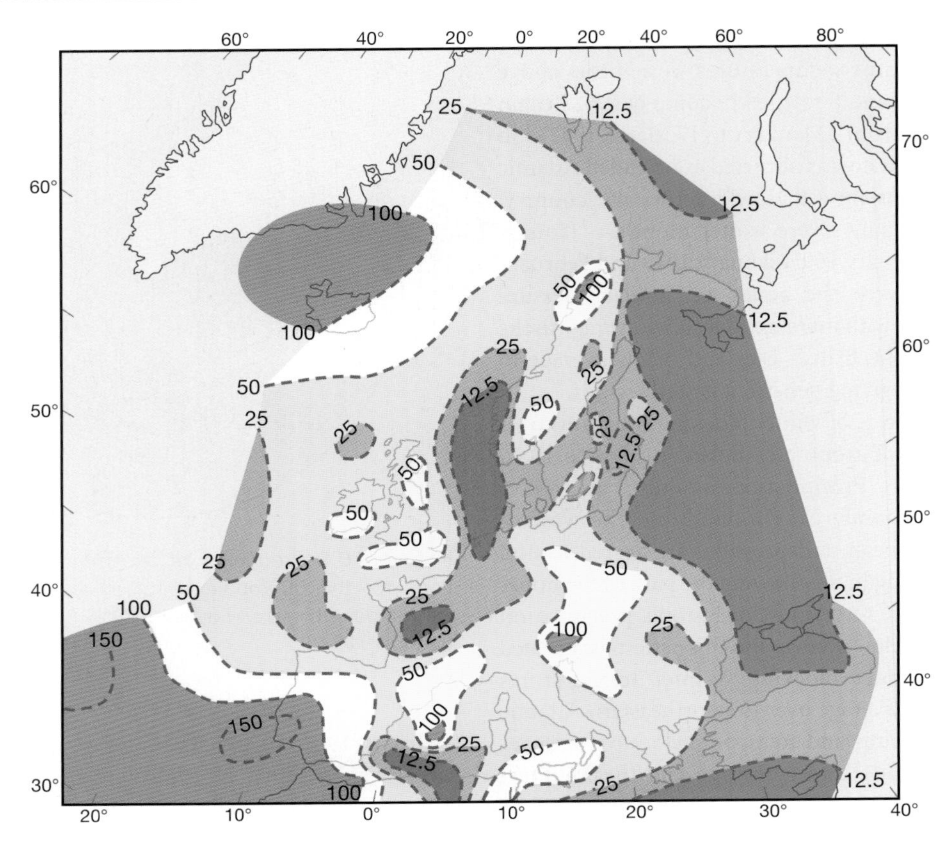

Figure 10.10 The mean precipitation anomaly, as a percentage of the average, during anticyclonic blocking in winter over Scandinavia. Areas above normal are cross-hatched, areas recording precipitation between 50 and 100 percent of normal have oblique hatching.

Source: After Rex (1950). By permission of Tellus.

produces a sharp decrease in the amounts. The upper Gudbrandsdalen and Osterdalen in the lee of the Jotunheim and Dovre Mountains receive an average of less than 500mm, and similar low values are recorded in central Sweden around Östersund.

Mountains can function equally in the opposite sense. For example, Arctic air from the Barents Sea may move southward in winter over the Gulf of Bothnia, usually when there is a depression over northern Russia, giving very low temperatures in Sweden and Finland. Western Norway is rarely affected, since the cold wave is contained to the east of the mountains. In consequence, there is a sharp climatic gradient across the Scandinavian highlands in the winter months.

The Alps illustrates other topographic effects. Together with the Pyrenees and the mountains of the Balkans, the Alps effectively separate the Mediterranean climatic region from that of Europe. The penetration of warm air masses north of these barriers is comparatively rare and short-lived. However, with certain pressure patterns, air from the Mediterranean and northern Italy is forced to cross the Alps, losing its moisture through precipitation on the southern slopes. Dry adiabatic warming on the northern side of the mountains can readily raise temperatures by 5–6°C in the upper valleys of the Aar, Rhine and Inn. At

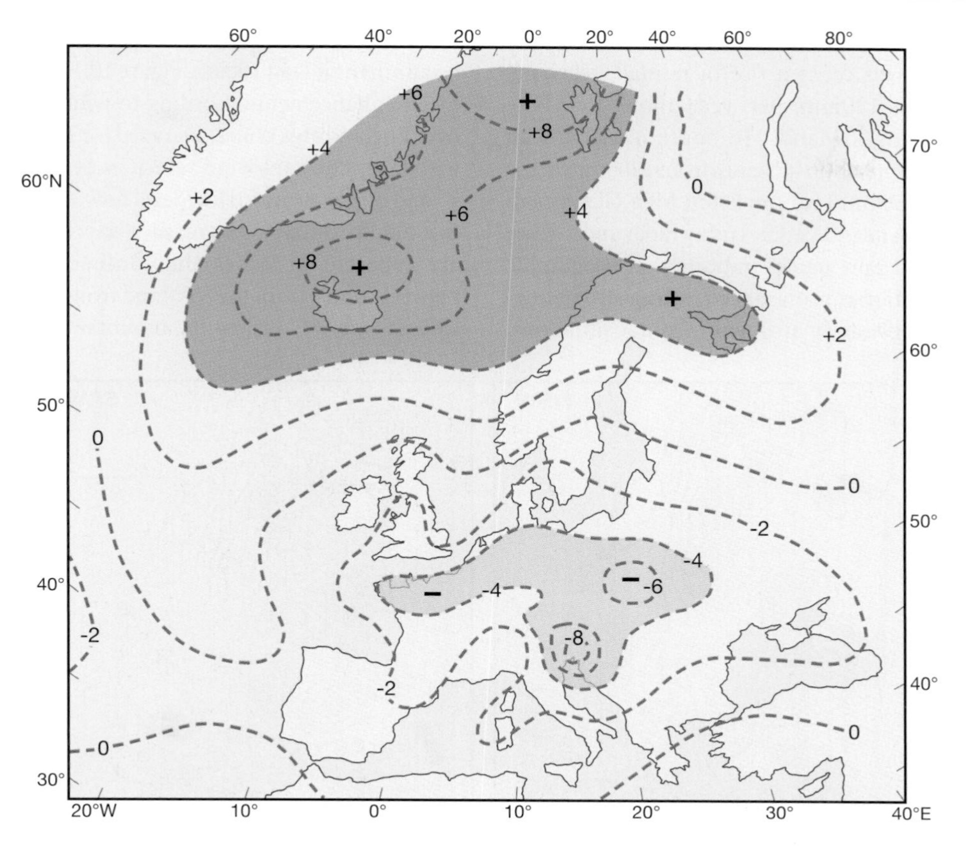

Figure 10.11 The mean surface temperature anomaly (°C) during anticyclonic blocking in winter over Scandinavia. Areas more than 4°C above normal have vertical hatching, those more than 4°C below normal have oblique hatching.

Source: After Rex (1950). By permission of Tellus.

Innsbruck, there are approximately 50 days per year with föhn winds, with a maximum in spring. Such occurrences can lead to rapid melting of the snow, creating a risk of avalanches. With northerly airflow across the Alps, föhn may occur in northern Italy, but its effects are less pronounced.

Features of upland climate in Britain illustrate some of the diverse effects of altitude. The mean annual rainfall on the west coasts near sea level is about 1140mm, but on the western mountains of Scotland, the Lake District and Wales averages exceed 3800mm per year. The annual record is 6530mm in 1954 at Sprinkling Tarn, Cumbria, and 1450mm fell in a single month (October 1909) just east of the summit of Snowdon in North Wales. The annual number of rain days (days with at least 0.25mm of precipitation) increases from about 165 in southeastern England and the south coast to over 230 days in northwest Britain. There is little additional increase in the frequency of rainfall with height on the mountains of the northwest. Hence, the mean rainfall per rain day rises sharply from 5mm near sea level in the west and northwest to over 13mm in the Western Highlands, the Lake District and Snowdonia. This demonstrates that 'orographic rainfall' here is primarily due to an intensification of the normal precipitation processes associated with frontal depressions and unstable airstreams (see Chapter 4E.3).

Even quite low hills such as the Chilterns and South Downs cause a rise in rainfall, receiving about 120–130mm per year more than the surrounding lowlands. In South Wales, mean annual precipitation increases from 1200mm at the coast to 2500mm on the 500m high Glamorgan Hills, 20km inland. Studies using radar and a dense network of rain gauges indicate that orographic intensification is pronounced during strong low-level southwesterly airflow in frontal situations. Most of the enhancement of precipitation rate occurs in the lowest 1500m. Figure 10.12 shows the mean enhancement according to wind direction over England and Wales, averaged for several days with fairly constant wind velocities of about 20m s^{-1} and nearly saturated low-level flow, attributable to a single frontal system on each day. Differences are apparent in Wales and southern England between winds from the SSW and from the WSW, whereas for SSE airflow the mountains of North

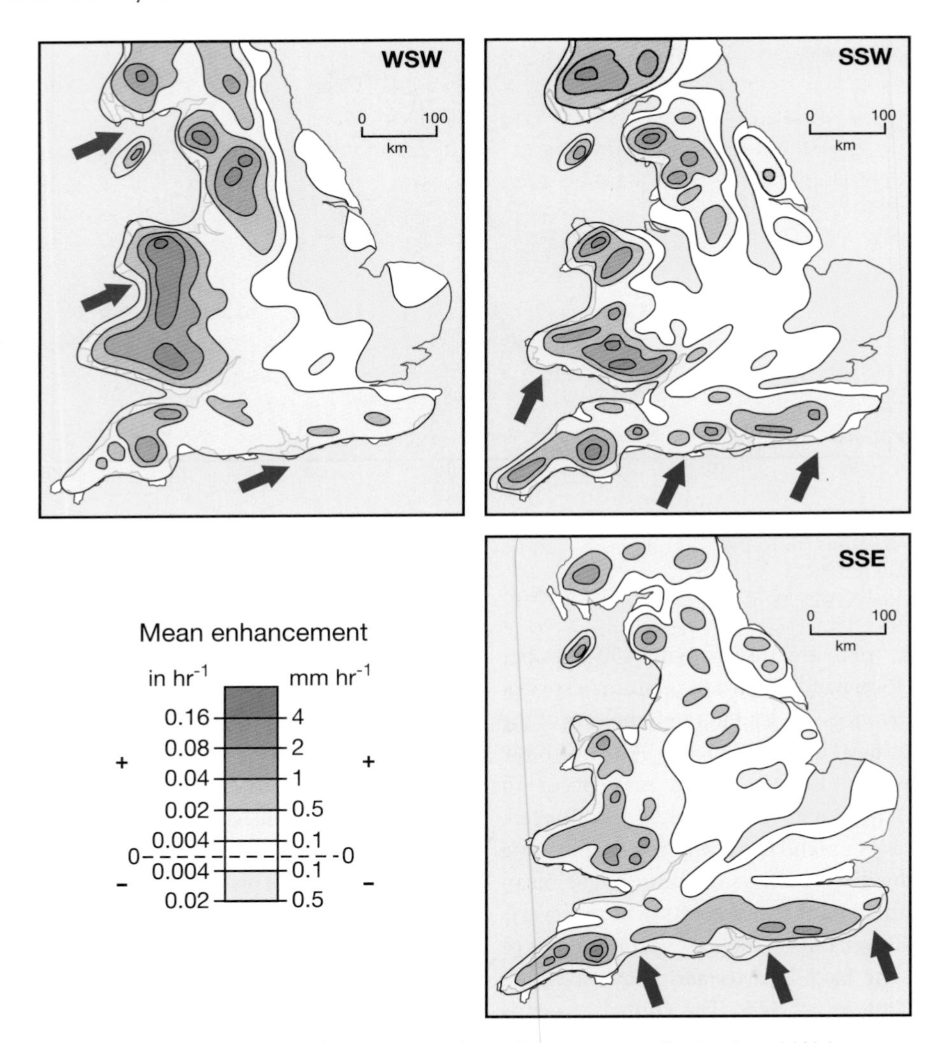

Figure 10.12 Mean orographic enhancement of precipitation over England and Wales, averaged for several days of fairly constant wind direction of about 20m s^{-1} and nearly saturated low-level airflow.

Source: After Browning and Hill (1981). By permission of the Royal Meteorological Society.

Wales and the Pennines have little effect. There are also areas of negative enhancement on the lee side of mountains. The sheltering effects of the uplands produce low annual totals on the lee side (with respect to the prevailing winds). Thus, the lower Dee valley in the lee of the mountains of North Wales receives less than 750mm, compared with over 2500mm in Snowdonia.

The complexity of the various factors affecting rainfall in Britain is shown by the fact that a close correlation exists between annual totals in northwest Scotland, the Lake District and western Norway, which are directly affected by Atlantic depressions. At the same time, there is an inverse relationship between annual amounts in the Western Highlands and lowland Aberdeenshire, less than 240km to the east. Annual precipitation in the latter area is more closely correlated with that in lowland eastern England. Essentially, the British Isles comprise two major climatic units for rainfall – first, an 'Atlantic' one with a winter season maximum, and second, those central and eastern districts with 'continental' affinities in the form of a weak summer maximum in most years. Other areas (eastern Ireland, eastern Scotland, northeast England and most of the English Midlands and the Welsh border counties) have a wet second half of the year.

The occurrence of snow is another measure of altitude effects. Near sea level, there are on average about five days per year with snow falling in southwest England, 15 days in the southeast and 35 days in northern Scotland. Between 60 and 300m, the frequency increases by about one day per 15m of elevation and even more rapidly on higher ground. Approximate figures for northern Britain are 60 days at 600m and 90 days at 900m. The number of mornings with snow lying on the ground (more than half the ground covered) is closely related to mean temperature and hence altitude. Average figures range from about five days per year or less in much of southern England and Ireland, to between 30 and 90 days on the Pennines and over 100 days on the Grampian Mountains. In the last area (on the Cairngorms) and on Ben Nevis there are several semi-permanent snow beds at about 1160m. It is estimated that the theoretical climatic snowline – above which there would be *net* snow accumulation – is at 1620m over Scotland. Since 1987 all but three years up until 2000 have seen below-average (1961–1990) snow cover duration.

Marked geographical variations in lapse rate also exist within the British Isles. One measure of these variations is the length of the 'growing season'. We can determine an index of growth opportunity by counting the number of days on which the mean daily temperature exceeds a threshold value of 6°C. Along southwestern coasts of England the 'growing season' calculated on this basis is nearly 365 days per year. Here it decreases by about nine days per 30m of elevation, but in northern England and Scotland the decrease is only about five days per 30m from between 250 to 270 days near sea level. In continental climates the altitudinal decrease may be even more gradual; in Central Europe and New England, for example, it is about two days per 30m.

B NORTH AMERICA

The North American continent spans nearly 60° of latitude and, not surprisingly, exhibits a wide range of climatic conditions. Unlike Europe, the West Coast is backed by the Pacific Coast Ranges rising to over 2750m, which lie across the path of the mid-latitude westerlies and prevent the extension of maritime influences inland. In the interior of the continent, there are no significant obstructions to air movement and the absence of any east–west barrier allows air masses from the Arctic or the Gulf of Mexico to sweep across the interior lowlands, causing wide extremes of weather and climate. Maritime influences in eastern North America are greatly limited by the fact that the prevailing winds are westerly, so that the temperature regime is continental. Nevertheless, the Gulf of Mexico is a major source of moisture supply for precipitation over the eastern half of the United States and, as a result, the precipitation regimes differ from those in East Asia.

We look first at the characteristics of the atmospheric circulation over the continent.

1 Pressure systems

The mean pressure pattern for the mid-troposphere displays a prominent trough over eastern North America in both summer and winter (see Figure 7.4). In part, this is a lee trough caused by the effect of the western mountain ranges on the upper westerlies, but at least in winter the strong baroclinic zone along the East Coast of the continent is a major contributory factor. As a result of this mean wave pattern, cyclones tend to move southeastward over the Midwest, carrying continental polar air southward, while the cyclones travel northeastward along the Atlantic coast. The planetary wave structure over the eastern North Pacific and North America is referred to as the Pacific–North America (PNA) pattern. It refers to the relative amplitude of the troughs over the central North Pacific and eastern North America, on the one hand, and the ridge over western North America on the other. In the positive (negative) mode of the PNA, there is a well-developed storm track from East Asia into the central Pacific and then into the Gulf of Alaska (cylones over East Asia move northeastward to the Bering Sea, with another area of lows off the west coast of Canada). The positive (negative) phases of PNA tend to be associated with El Niño (La Niña) events in the equatorial Pacific (see 11G).

The PNA mode has important consequences for the weather in different parts of the continent. In fact, this relationship provides the basis for the monthly forecasts of the U.S. National Weather Service. For example, if the eastern trough is more pronounced than usual, temperatures are below average in the central, southern and eastern United States, whereas if the trough is weak, the westerly flow is stronger with correspondingly less opportunity for cold outbreaks of polar air. Sometimes, the trough is displaced to the western half of the continent, causing a reversal of the usual weather pattern, since upper northwesterly airflow can bring cold, dry weather to the west while in the east there are very mild conditions associated with upper southwesterly flow. Precipitation amounts also depend on the depression tracks. If the upper trough is far to the west, depressions form ahead of it (see Chapter 7F) over the south central United States and move northeastward towards the lower St Lawrence, giving more precipitation than usual in these areas and less along the Atlantic coast.

The major features of the surface pressure map in January (see Figure 7.9A) are the extension of the subtropical high over the southwestern United States (called the Great Basin high) and the separate polar anticyclone of the Mackenzie district of Canada. Mean pressure is low off both the east and west coasts of higher mid-latitudes, where oceanic heat sources indirectly give rise to the (mean) Icelandic and Aleutian lows. It is interesting to note that, on average, in December, of any region in the Northern Hemisphere for any month of the year, the Great Basin region has the most frequent occurrence of highs, whereas the Gulf of Alaska has the maximum frequency of lows. The Pacific coast as a whole has its most frequent cyclonic activity in winter, as does the Great Lakes area, whereas over the Great Plains the maximum is in spring and early summer. Remarkably, the Great Basin in June has the most frequent cyclogenesis of any part of the Northern Hemisphere in any month of the year. Heating over this area in summer helps to maintain a shallow, quasi-permanent low pressure cell, in marked contrast with the almost continuous subtropical high pressure belt in the mid-troposphere (see Figure 7.4). Continental heating also indirectly assists in the splitting of the Icelandic low to create a secondary center over northeastern Canada. The west coast summer circulation is dominated by the Pacific anticyclone, while the southeastern United States is affected by the Atlantic subtropical anticyclone cell (see Figure 7.9B).

Broadly, there are three prominent cyclone tracks across the continent in winter (see Figure 9.21). One group, known as 'Alberta clippers', moves from the west along a more or less zonal path about 45–50°N, whereas a second loops southward over the central United States and then turns northeastward towards New England and the Gulf of St Lawrence. Some of these depressions originate over the Pacific, cross the western ranges as an upper trough and redevelop in the lee of the mountains. Alberta is a noted area for this process and also for primary cyclogenesis, since the arctic frontal zone is over northwest Canada in winter. This frontal zone involves much-modified mA air from the Gulf of Alaska and cold, dry cA (or cP) air. Cyclones of the third group form along the main polar frontal zone, which in winter is off the east coast of the United States, and move northeastward towards Newfoundland. Sometimes, this frontal zone is present over the continent at about 35°N with mT air from the Gulf and cP air from the north or modified mP air from the Pacific. Polar front depressions forming over Colorado move northeastward towards the Great Lakes; others developing over Texas follow a roughly parallel path, further to the south and east, towards New England. Anomalies in winter climate over North America are strongly influenced by the position of the jet streams and the movement of associated storm systems. Figure 10.13 illustrates their role in locating areas of heavy rain, flooding and positive/negative temperature departures in the winters of 1994–1995 and 1995–1996.

Between the Arctic and Polar Fronts, Canadian meteorologists distinguish a third frontal zone. This maritime (arctic) frontal zone is present when mA and mP air masses interact along their common boundary. The three-front (*i.e.*, four air-mass) model allows a detailed analysis to be made of the baroclinic structure of depressions over the North American continent using synoptic weather maps and cross-sections of the atmosphere. Figure 10.14 illustrates the three frontal zones and associated depressions on 29 May 1963. Along 95°W, from 60° to 40°N, the dew-point temperatures reported in the four air masses were –8°C, 1°C, 4°C and 13°C, respectively.

In summer, east coast depressions are less frequent and the tracks across the continent are displaced northward, with the main tracks moving over Hudson Bay and Labrador–Ungava, or along the line of the St Lawrence. These are associated mainly with a poorly defined maritime frontal zone. The Arctic Front is usually located along the north coast of Alaska, where there is a strong temperature gradient between the bare land and the cold Arctic Ocean and pack ice. East from here, the front is very variable in location from day to day and year to year. It occurs most often in the vicinity of northern Keewatin and Hudson Strait. One study of air-mass temperatures and airstream confluence regions suggests that an arctic frontal zone occurs further south over Keewatin in July and that its mean position (Figure 10.15) is closely related to the boreal forest–tundra boundary. This relationship reflects the importance of arctic air-mass dominance for summer temperatures and consequently for tree growth, yet energy budget differences due to land cover type appear insufficient to determine the frontal location.

Several circulation singularities have been recognized in North America, as in Europe (see A.4, this chapter). Three that have received attention in view of their prominence are: (1) the advent of spring in late March; (2) the mid-summer high-pressure jump at the end of June; and (3) the Indian summer in late September (and late October).

The arrival of spring is marked by different climatic responses in different parts of the continent. For example, there is a sharp decrease in March to April precipitation in California, due to the extension of the Pacific high. In the Midwest, precipitation intensity increases as a result of more frequent cyclogenesis in Alberta and Colorado, and northward extension of maritime tropical air from the Gulf of Mexico. These changes are part of a hemispheric readjustment of the circulation; in early April, the Aleutian

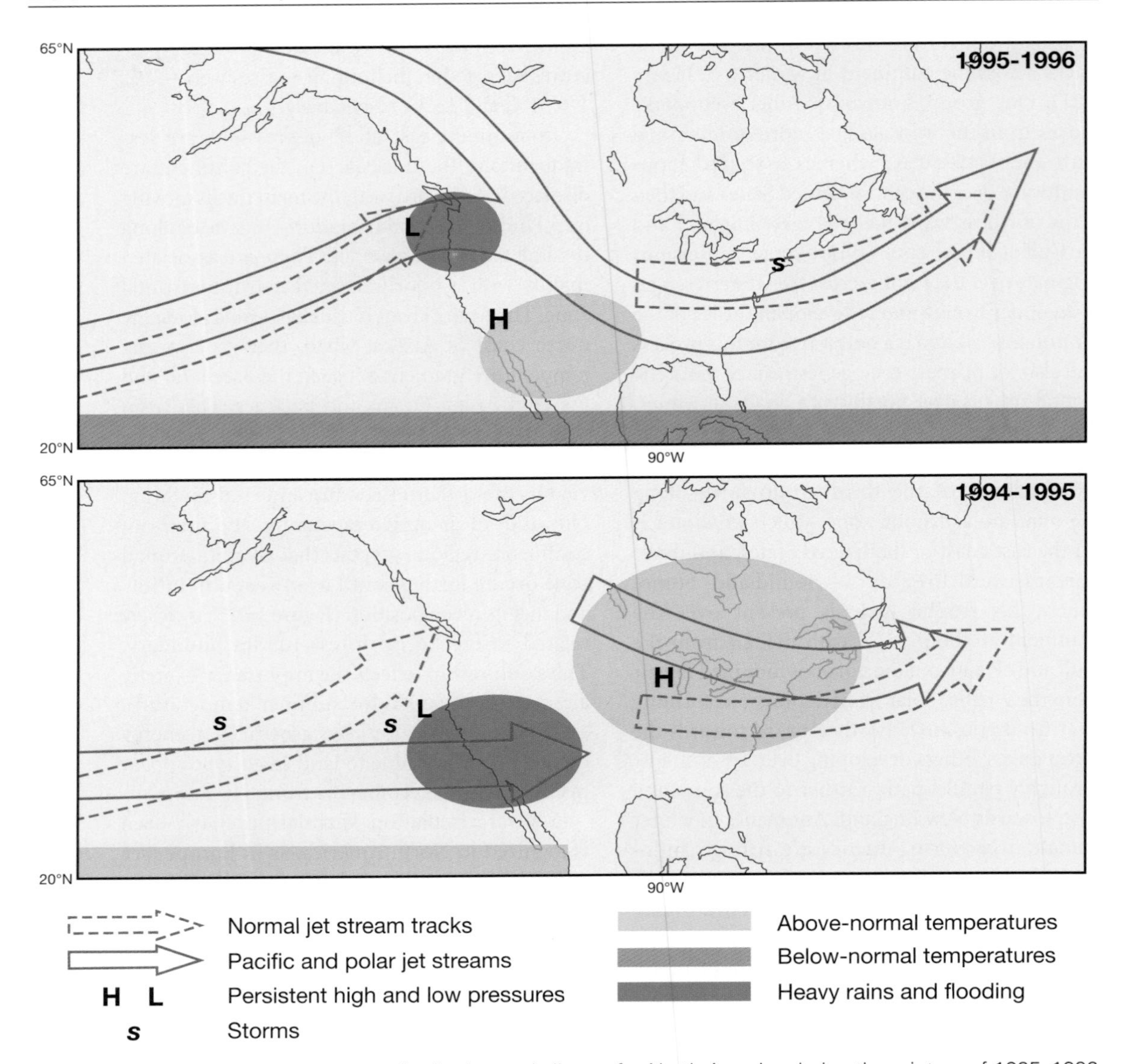

Figure 10.13 Jet streams, pressure distribution and climate for North America during the winters of 1995–1996 and 1994–1995.

Source: US Department of Commerce, Climate Prediction Center. NOAA. Courtesy US Department of Commerce.

low-pressure cell, which from September to March is located about 55°N, 165°W, splits into two, with one center in the Gulf of Alaska and the other over northern Manchuria.

In late June, there is a rapid northward displacement of the Bermuda and North Pacific subtropical high pressure cells. In North America, this also pushes the depression tracks northward with the result that precipitation decreases from June to July over the northern Great Plains, part of Idaho and eastern Oregon. Conversely, the southwesterly anticyclonic flow that affects Arizona in June is replaced by air from the Gulf of California, and this causes the onset of the

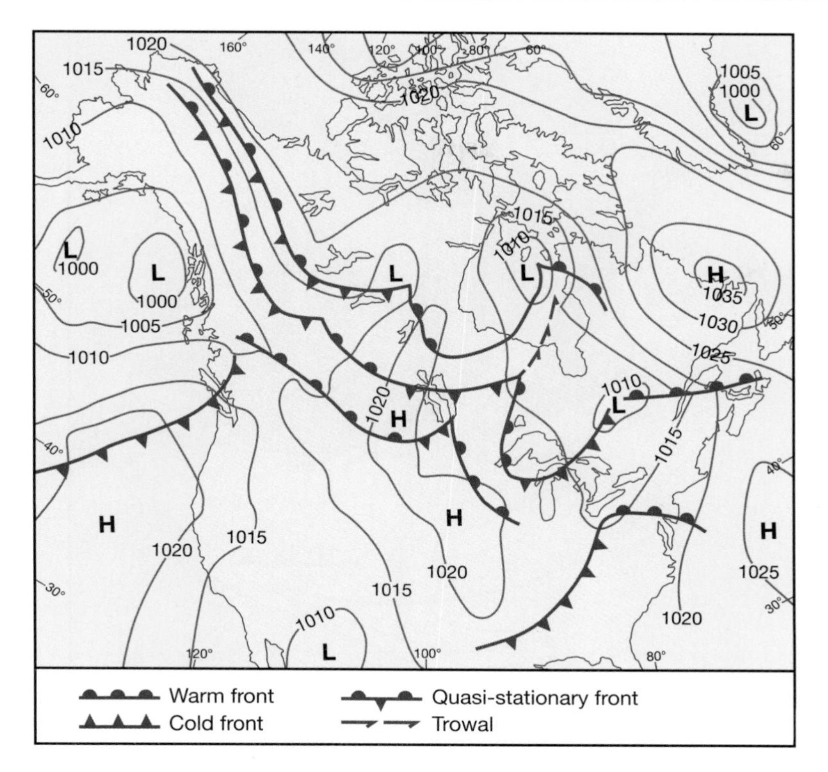

Figure 10.14 A synoptic example of depressions associated with three frontal zones on 29 May 1963 over North America.

Source: Based on charts of the Edmonton Analysis Office and the Daily Weather Report.

summer rains (see B.3, this chapter). Bryson and Lahey (1958) suggest that these circulation changes at the end of June may be connected with the disappearance of snow cover from the arctic tundra. This leads to a sudden decrease of surface albedo from about 75 to 15 percent, with consequent changes in the heat budget components and hence in the atmospheric circulation.

Frontal wave activity makes the first half of September a rainy period in the northern Midwest states of Iowa, Minnesota and Wisconsin, but after about 20 September, anticyclonic conditions return with warm airflow from the dry southwest, giving fine weather – the so-called Indian summer. Significantly, the hemispheric zonal index value rises in late September. This anticyclonic weather type has a second phase in the latter half of October, but at this time there are polar outbreaks. The weather is generally cold and dry, although if precipitation does occur there is a high probability of snowfall.

2 The temperate west coast and cordillera

The oceanic circulation of the North Pacific closely resembles that of the North Atlantic. The drift from the Kuroshio Current off Japan is propelled by the westerlies towards the west coast of North America and it acts as a warm current between 40° and 60°N. Sea surface temperatures are several degrees lower than in comparable latitudes off Western Europe, however, due to the smaller volume of warm water involved. In addition, in contrast to the Norwegian Sea, the shape of the Alaskan coastline prevents the

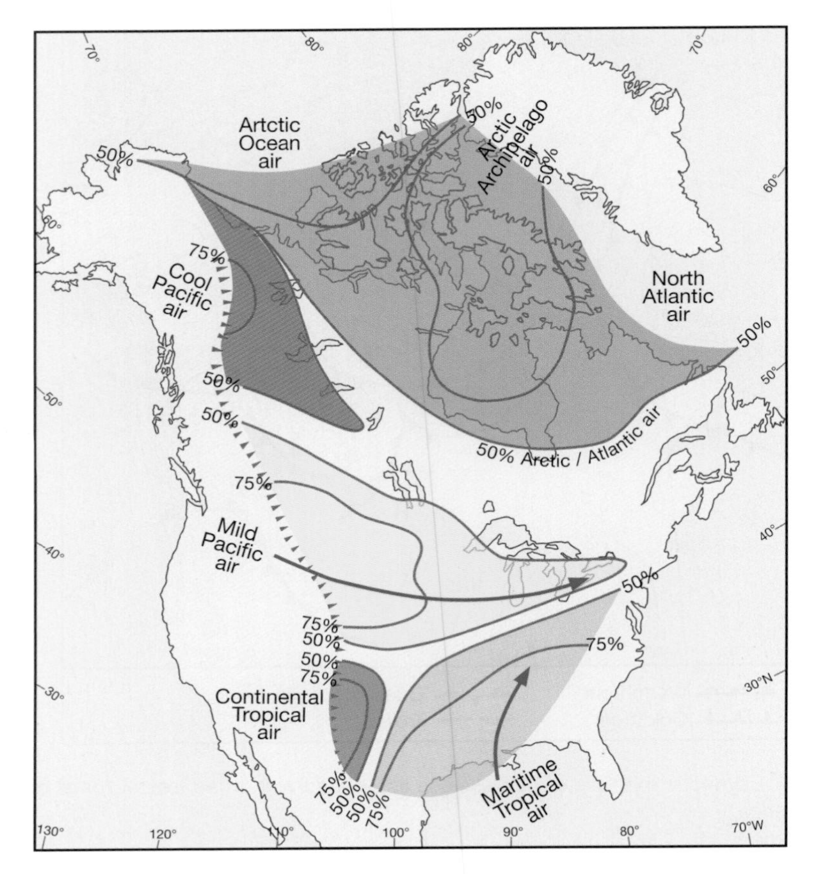

Figure 10.15 Regions in North America east of the Rocky Mountains dominated by the various air-mass types in July for more than 50 percent and 75 percent of the time. The 50 percent frequency lines correspond to mean frontal positions.

Source: After Bryson (1966).

extension of the drift to high latitudes (see Figure 7.29).

The Pacific coast ranges greatly restrict the inland extent of oceanic influences, and hence there is no extensive maritime temperate climate as in Western Europe. The major climatic features duplicate those of the coastal mountains of Norway and those of New Zealand and southern Chile in the belt of southern westerlies. Topographic factors make the weather and climate of such areas very variable over short distances, both vertically and horizontally. A few salient characteristics are selected for consideration here.

There is a regular pattern of rainy windward and drier lee slopes across the successive northwest to southeast ranges, with a more general decrease towards the interior. The Coast Range in British Columbia has mean annual totals of precipitation exceeding 2500mm, with 5000mm in the wettest places, compared with 1250mm or less on the summits of the Rockies. Yet even on the leeward side of Vancouver Island, the average figure at Victoria is only 700mm. Analogous to the 'westerlies–oceanic' regime of northwest Europe, there is a winter precipitation maximum along the littoral (Estevan Point in Figure 10.16), which also extends beyond the Cascades (in

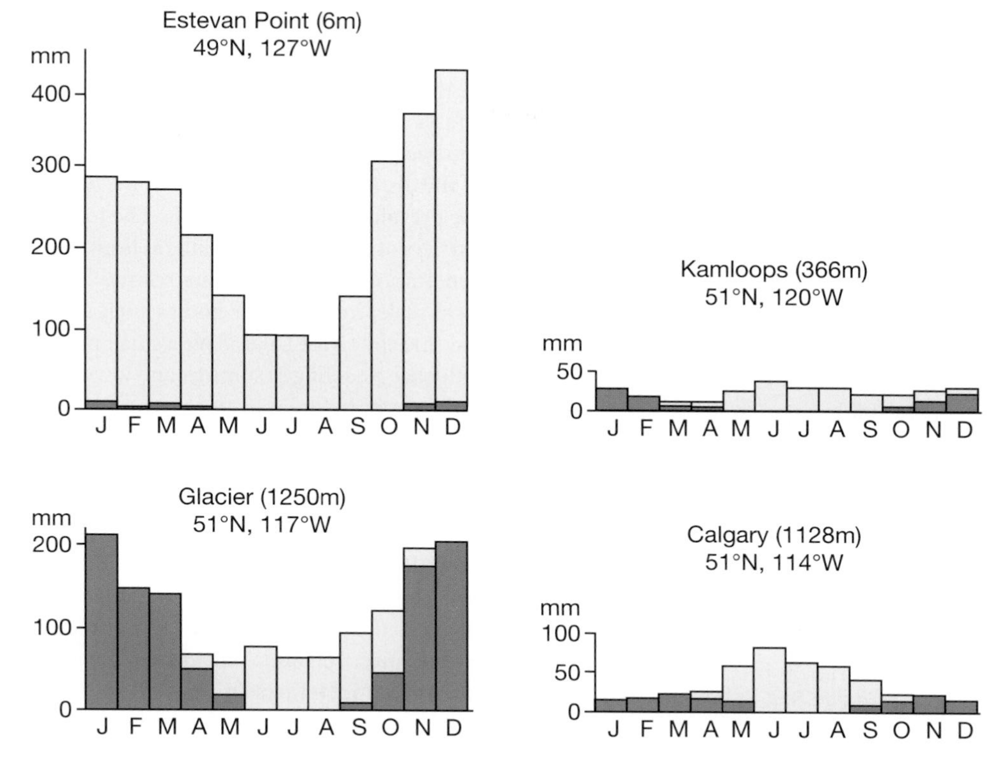

Figure 10.16 Precipitation graphs for stations in western Canada. The shaded portions represent snowfall, expressed as water equivalent.

Washington) and the Coast Range (in British Columbia), but summers are drier due to the strong North Pacific anticyclone. The regime in the interior of British Columbia is transitional between that of the coastal region and the distinct summer maximum of central North America (Calgary), although at Kamloops in the Thompson valley (annual average 250mm) there is a slight summer maximum associated with thunderstorm-type rainfall. In general, the sheltered interior valleys receive less than 500mm per year. In the driest years certain localities have recorded only 150mm. Above 1000m, much of the precipitation falls as snow (see Figure 10.16) and some of the greatest snow depths in the world are reported from British Columbia, Washington and Oregon. A US national record seasonal total of 28.96m was observed in 1998–1999 at the Mt. Baker, WA, ski area, where the annual mean amount is 16.4m. Generally, 10–15m of snow falls annually on the Cascade Range at heights of about 1500m, and even as far inland as the Selkirk Mountains snowfall totals are considerable. The mean snowfall is 9.9m at Glacier, British Columbia (elevation 1250m), and this accounts for almost 70 percent of the annual precipitation (see Figure 10.16). Near sea level on the outer coast, in contrast, very little precipitation falls as snow (for example, Estevan Point). It is estimated that the climatic snowline rises from about 1600m on the west side of Vancouver Island to 2900m in the eastern Coast Range. Inland, its elevation increases from 2300m on the west slopes of the Columbia Mountains to 3100m on the east side of the Rockies. This trend reflects the precipitation pattern referred to above.

Large diurnal variations affect the Cordilleran valleys. Strong diurnal rhythms of temperature (especially in summer) and wind direction are a feature of mountain climates and their effect is superimposed upon the general climatic characteristics of the area. Cold air drainage can produce remarkably low minima in the mountain valleys and basins. At Princeton, British Columbia (elevation 695m), where the mean daily minimum in January is –14°C, there is on record an absolute low of –45°C, for example. This leads in some cases to reversal of the normal lapse rate. Golden in the Rocky Mountain Trench has a January mean of –12°C, whereas 460m higher at Glacier (1250m) it is –10°C.

3 Interior and eastern North America

Central North America has the typical climate of a continental interior in mid-latitudes, with hot summers and cold winters (Figure 10.17), yet the weather in winter is subject to marked variability. This is determined by the steep temperature gradient between the Gulf of Mexico and the snow-covered northern plains; also by shifts of the upper wave patterns and jet stream. Cyclonic activity in winter is much more pronounced over central and eastern North America than in Asia, which is dominated by the Siberian anticyclone (see Figure 7.9A). Consequently there is no climatic type with a winter minimum of precipitation in eastern North America.

The general temperature conditions in winter and summer are illustrated in Figure 10.17, showing the frequency with which hourly temperature readings exceed or fall below certain limits. The two chief features of all four maps are: (1) the dominance of the meridional temperature gradient, away from coasts; and (2) the continentality of the interior and east compared with the 'maritime' nature of the West Coast. On the July maps, additional influences are evident and these are referred to below.

Continental and oceanic influences

The large annual temperature range in the interior of the continent shown in Figure 3.24 demonstrates the pattern of continentality of North America. The figure illustrates the key role of the distance from the ocean in the direction of the prevailing (westerly) winds. The topographic barriers of the western cordilleras limit the inland penetration of maritime airstreams. On a more local scale, inland water bodies such as Hudson Bay and the Great Lakes have a small moderating influence – cooling in summer and warming in the early winter before they freeze over.

The Labrador coast is fringed by the waters of a cold current, analogous to the Oyashio off East Asia, but in both cases the prevailing westerlies greatly limit their climatic significance. The Labrador Current maintains drift ice off Labrador and Newfoundland until June and gives very low summer temperatures along the Labrador coast (see Figure 10.18C). The lower incidence of freezing temperatures in this area in January is related to the movement of some depressions into the Davis Strait, carrying Atlantic air northward. A major role of the Labrador Current is in the formation of fog. Advection fog is very frequent between May and August off Newfoundland, where the Gulf Stream and Labrador Current meet. Warm, moist, southerly airstreams are cooled rapidly over the cold waters of the Labrador Current and with steady, light winds such fogs may persist for several days, creating hazardous conditions for shipping. Southward-facing coasts are particularly affected and at Cape Race (Newfoundland), for example, there are on average 158 days per year with fog (visibility less than 1km) at some times of day. The summer concentration is shown by the figures for Cape Race: May – 18 (days), June – 18, July – 24, August – 21, and September – 18.

Oceanic influence along the Atlantic coasts of the United States is very limited, and although there is some moderating effect of minimum temperatures at coastal stations this is scarcely evident on generalized maps such as Figure 10.17.

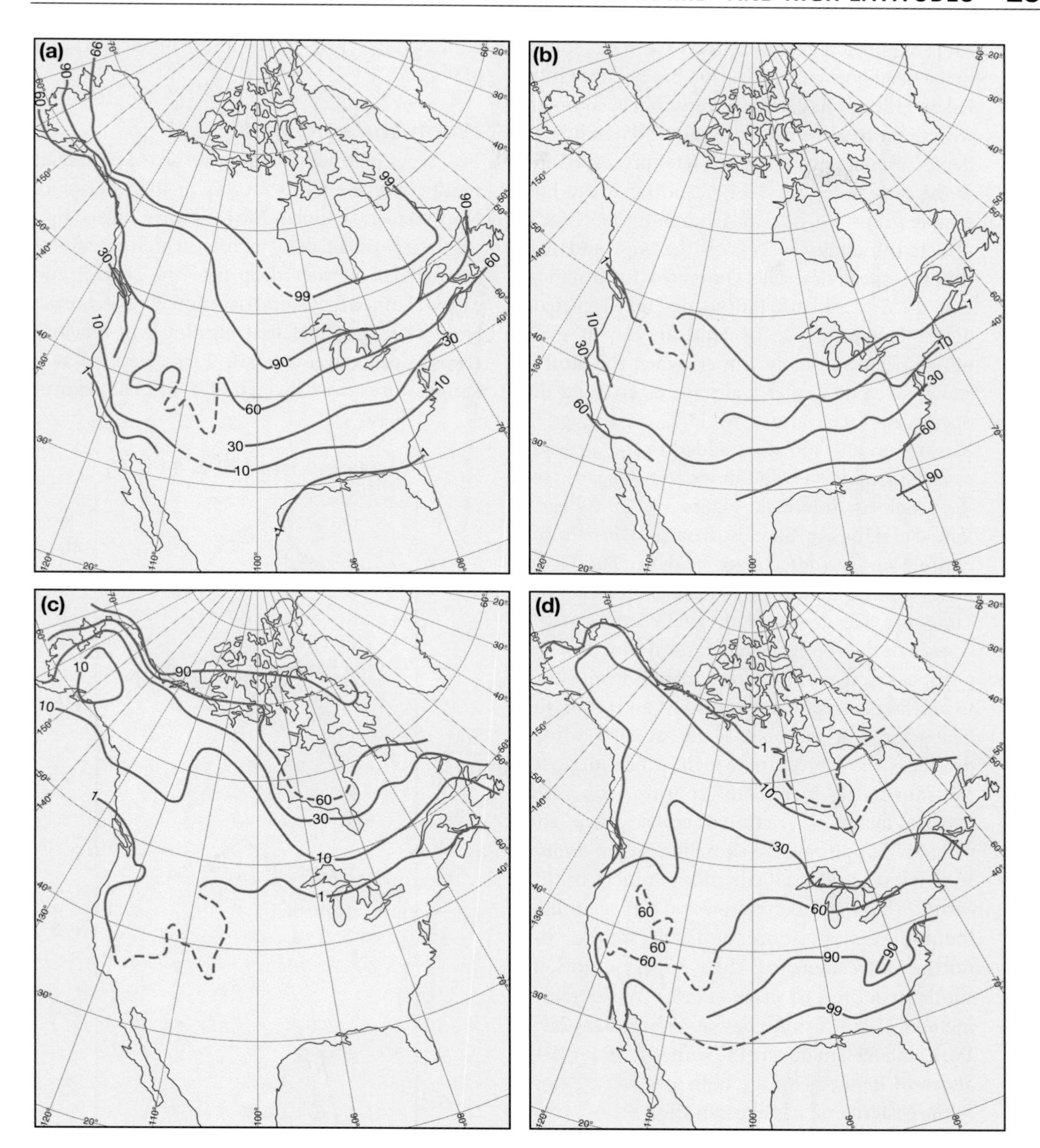

Figure 10.17 The percentage frequency of hourly temperatures above or below certain limits for North America. A: January temperatures <0°C; B: January temperatures >10°C; C: July temperatures <10°C; D: July temperatures >21°C.

Source: After Rayner (1961).

More significant climatic effects are in fact found in the neighborhood of Hudson Bay and the Great Lakes. Hudson Bay remains very cool in summer, with water temperatures of about 7–9°C, and this depresses temperatures along its shore, especially in the east (see Figure 10.17C and D). Mean July temperatures are 12°C at Churchill (59°N) and 8°C at Inukjuak (58°N), on the west and east shores respectively. This compares, for instance, with 13°C at Aklavik (68(N) on the Mackenzie delta. The influence of Hudson Bay is even more striking in early winter, when the land is snow-covered. Westerly airstreams crossing the open water are warmed by 11°C on average in November, and moisture added to the air leads to considerable snowfall in western Ungava (see the graph for Inukjuak, Figure 10.20). By early January, Hudson Bay is frozen over almost entirely and no effects are evident. The Great Lakes influence their surroundings in much the same way. Heavy winter snowfalls are a notable feature of the southern and eastern shores of the Great Lakes. In addition to contributing moisture to northwesterly streams of cold cA and cP air, the heat source of the open water in early winter produces a low pressure trough, which increases the snowfall as a result of convergence. Yet a further factor is frictional convergence and orographic uplift at the shoreline. Mean annual snowfall exceeds 250cm along much of the eastern shore of Lake Huron and Georgian Bay, the southeastern shore of Lake Ontario, the northeastern shore of Lake Superior and its southern shore east of about 90.5°W. Extremes include 114cm in one day at Watertown, New York, and 894cm during the winter of 1946–1947 at nearby Bennetts Bridge, both of which are close to the eastern end of Lake Ontario.

Transport in cities in these snow belts is quite frequently disrupted during winter snowstorms. The Great Lakes also provide an important tempering influence during winter months by raising average daily minimum temperatures at lakeshore stations by some 2–4°C above those at inland locations. In mid-December, the upper 60m of Lake Erie has a uniform temperature of 5°C.

Warm and cold spells

Two types of synoptic condition are of particular significance for temperatures in the interior of North America. One is the cold wave caused by a northerly outbreak of cP air, which in winter regularly penetrates deep into the central and eastern United States and occasionally affects even Florida and the Gulf Coast, injuring frost-sensitive crops. Cold waves are arbitrarily defined as a temperature drop of at least 11°C in 24 hours over

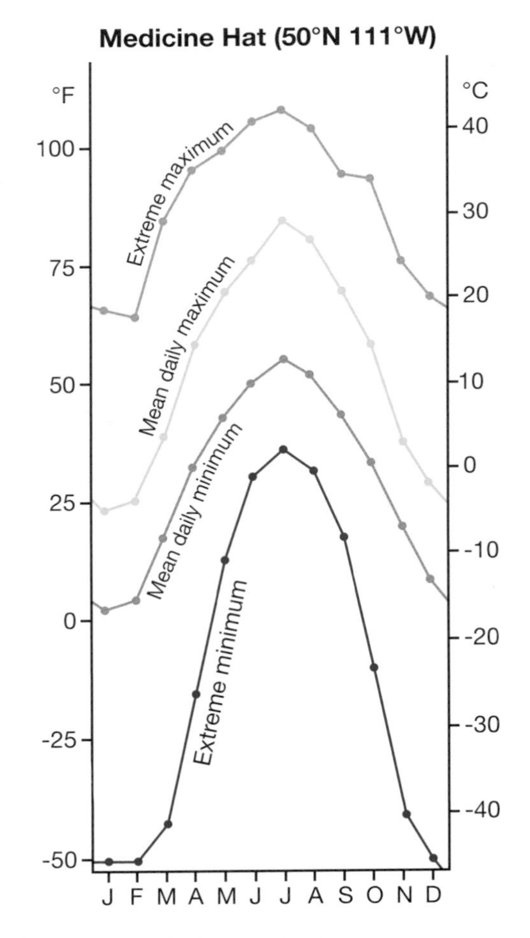

Figure 10.18 Mean and extreme temperatures at Medicine Hat, Alberta.

most of the United States, and at least 9°C in California, Florida and the Gulf Coast, to below a specified minimum depending on location and season. The winter criterion decreases from 0°C in California, Florida and the Gulf Coast to –18°C over the northern Great Plains and the northeastern states. Cold spells commonly occur with the buildup of a north–south anticyclone in the rear of a cold front. Polar air gives clear, dry weather with strong, cold winds, although if they follow snowfall, fine, powdery snow may be whipped up by the wind, creating blizzard conditions over the northern plains. These occur with winds >10m s^{-1} with falling or blowing snow reducing visibility below 400m. On average, a blizzard event affects an area of 150,000km and over 2 million people.

Another type of temperature fluctuation is associated with the *chinook* (föhn) winds in the lee of the Rockies (see Chapter 5C.2). The chinook is particularly warm and dry as air descends the eastern slopes and warms at the dry adiabatic lapse rate. The onset of the chinook produces temperatures well above the seasonal normal so that snow often thaws rapidly; in fact the Salish word 'Chinook' means snow-eater. Temperature rises of up to 22°C have been observed in five minutes. The occurrence of such warm events is reflected in the high extreme maxima in winter months at Medicine Hat (Figure 10.18). In Canada, the chinook effect may be observed at a considerable distance from the Rockies into southwestern Saskatchewan, but in Colorado its influence is rarely felt more than about 50km from the foothills. In southeastern Alberta, the belt of strong westerly chinook winds and elevated temperatures extends 150–200km east of the Rocky Mountains. Temperature anomalies average 5–9°C above winter normals, and a triangular sector southeast of Calgary, towards Medicine Hat, experiences maximum anomalies of up to 15–25°C, relative to mean daily maximum temperature values. Chinook events with westerly winds >35m s^{-1} occur on 45–50 days between November and February in this area as a result of the relatively low and narrow ridge line of the Rocky Mountains between 49 and 50°N, compared with the mountains around Banff and further north.

Chinook conditions commonly develop in a Pacific airstream that is replacing a winter high-pressure cell over the western high plains. Sometimes the descending chinook does not dislodge the cold, stagnant cP air of the anticyclone and a marked inversion is formed. On other occasions the boundary between the two air masses may reach ground level locally. Thus, for example, the western suburbs of Calgary may record temperatures above 0°C while those to the east of the city remain below –15°C.

The weather impact of very cold and very hot spells in the United States is costly, especially in terms of loss of life. In the 1990s, there were 292/282 deaths/year, respectively, attributed to extreme cold/hot conditions, more than for any other severe weather.

Precipitation and the moisture balance

Longitudinal influences are apparent in the distribution of annual precipitation, although this is in large measure a reflection of the topography. The 600mm annual isohyet in the United States approximately follows the 100°W meridian (Figure 10.19), and westward to the Rockies is an extensive dry belt in the rain shadow of the western mountain ranges. In the southeast, totals exceed 1250mm, and 1000mm or more is received along the Atlantic coast as far north as New Brunswick and Newfoundland.

The major sources of moisture for precipitation over North America are the Pacific Ocean and the Gulf of Mexico. The former need not concern us here, since comparatively little of the precipitation falling over the interior appears to derive from that source. The Gulf source is extremely important in providing moisture for precipitation over central and eastern North America, but the predominance of southwesterly airflow means that little precipitation falls over the

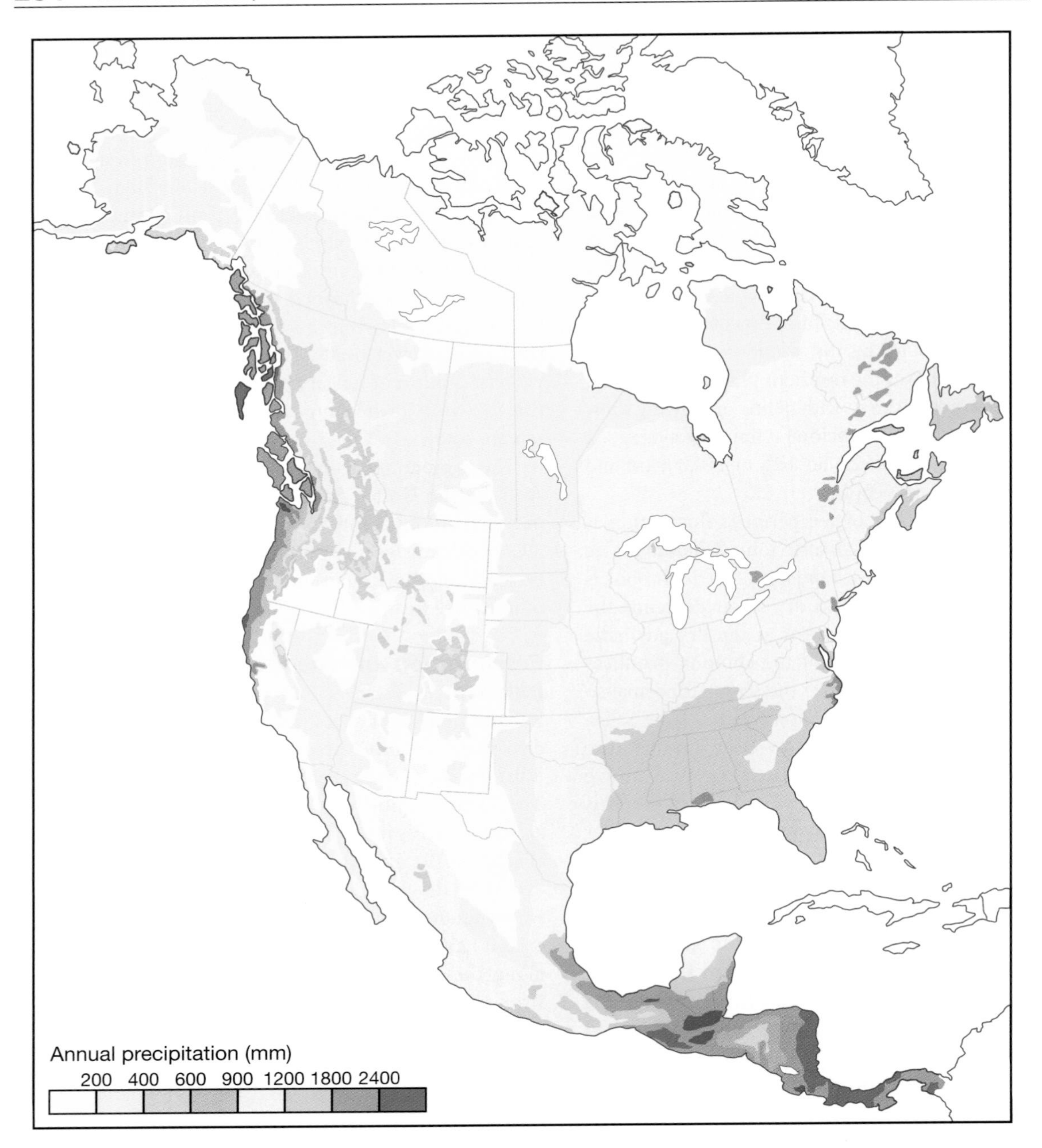

Figure 10.19 Mean annual precipitation (mm) over North America determined on a 25km grid as a function of location and elevation. Based on data from 8000 weather stations for 1951–1980. Values in the Arctic underestimate the true totals by 30–50 percent due to problems in recording snowfall accurately with precipitation gauges.

Source: Thompson *et al.* (1999). Courtesy of the US Geological Survey.

western Great Plains (see Figure 10.19). Over the southern United States, there is considerable evapotranspiration and this helps to maintain moderate annual totals northward and eastward from the Gulf by providing additional water vapor for the atmosphere. Along the east coast, the Atlantic Ocean is an additional significant source of moisture for winter precipitation.

There are at least eight major types of seasonal precipitation regime in North America (Figure 10.20); the winter maximum of the west coast and the transition type of the intermontane region in mid-latitudes have already been mentioned; the subtropical types are discussed in the next section. Four primarily mid-latitude regimes are distinguished east of the Rocky Mountains:

1 A warm season maximum is found over much of the continental interior (e.g., Rapid City). In an extensive belt from New Mexico to the Prairie provinces more than 40 percent of the annual precipitation falls in summer. In New Mexico, the rain occurs mainly with late summer thunderstorms, but May to June is the wettest time over the central and northern Great Plains due to more frequent cyclonic activity. Winters are quite dry over the plains, but the mechanism of the occasional heavy snowfalls is of interest. They occur over the northwestern plains during easterly upslope flow, usually in a ridge of high pressure. Further north, in Canada, the maximum is commonly in late summer or autumn, when

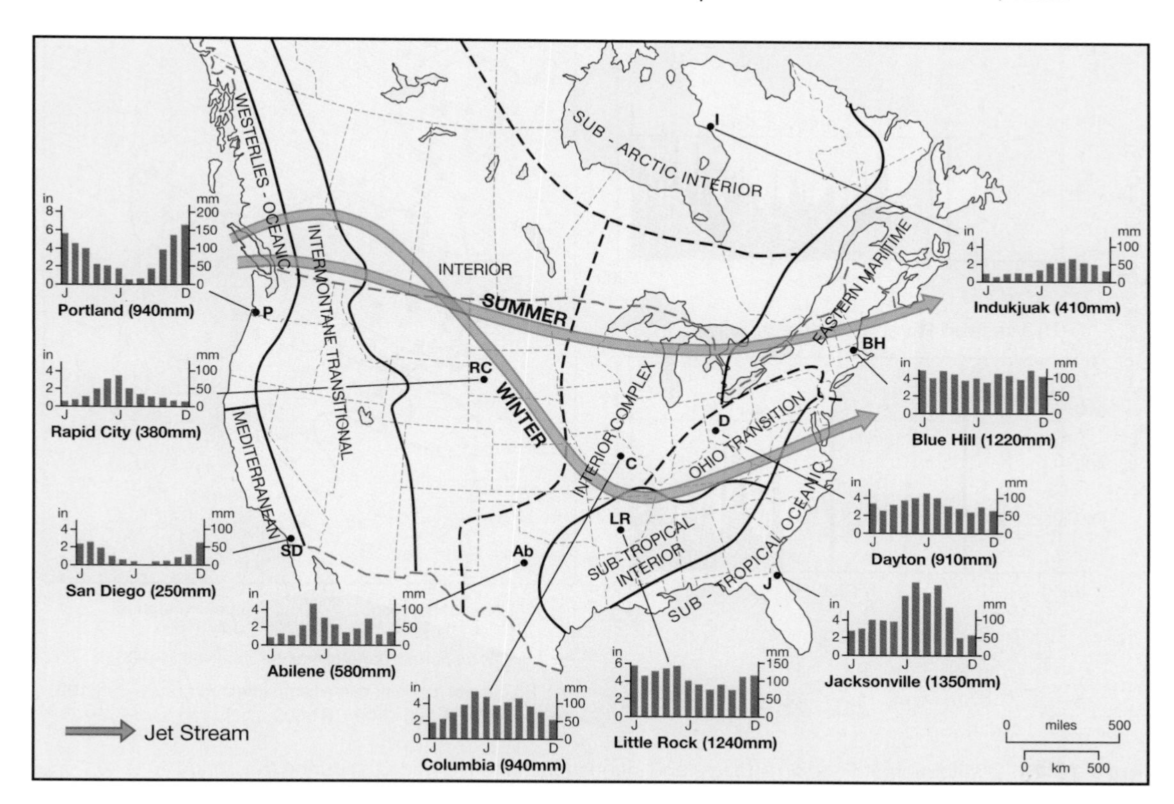

Figure 10.20 North American rainfall regime regions and histograms showing mean monthly precipitations for each region (January, June and December are indicated). Note that the jet stream is anchored by the Rockies in more or less the same position at all seasons.

Source: Mostly after Trewartha (1981); additions by Henderson-Sellers and Robinson (1986).

depression tracks are in higher mid-latitudes. There is a local maximum in autumn on the eastern shores of Hudson Bay (e.g., Inukjuak) due to the effect of open water.

2 Eastward and southward of the first zone there is a double maximum in May and September. In the upper Mississippi region (e.g., Columbia), there is a secondary minimum, paradoxically in July to August when the air is especially warm and moist, and a similar profile occurs in northern Texas (e.g., Abilene). An upper-level ridge of high pressure over the Mississippi valley seems to be responsible for reduced thunderstorm rainfall in midsummer, and a tongue of subsiding dry air extends southward from this ridge towards Texas. However, during the period June to August 1993 massive flooding occurred in the Mid-western parts of the Mississippi and Missouri Rivers as the result of up to twice the January to July average precipitation being received, with many point rainfalls exceeding amounts appropriate for recurrence intervals exceeding 100 years (Figure 10.21). The three summer months saw excesses of 500mm above the average rainfall with totals of 900mm or more. Strong, moist southwesterly airflow recurred throughout the summer with a quasi-stationary cold front oriented from southwest to northeast across the region. The flooding

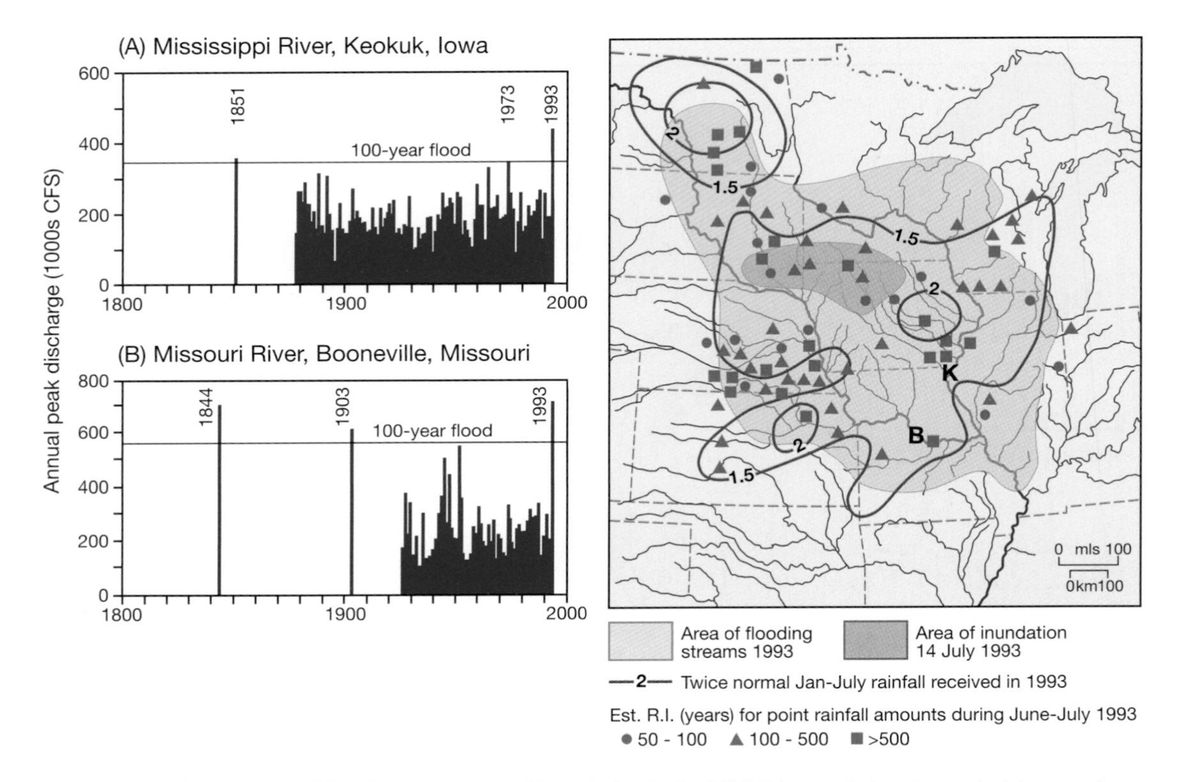

Figure 10.21 Distribution of flooding streams and inundation in the US Midwest during the period June to August 1993. Peak discharges for the Mississippi River at Keokuk, Iowa, (K) and the Missouri River at Booneville, Missouri (B) are shown, together with the historic annual peak discharge record. The isopleths indicate the multiples of the 30-year average January to July precipitation that fell in the first seven months of 1993, and the symbols the estimated recurrence intervals (RI years) for point rainfall amounts received during June to July 1993.

Sources: Parrett *et al.* (1993) and Lott (1994). Courtesy of the US Geological Survey.

resulted in 48 deaths, destroyed 50,000 homes and caused damage losses of $10 billion. In September, renewed cyclonic activity associated with the seasonal southward shift of the polar front, at a time when mT air from the Gulf is still warm and moist, typically causes a resumption of rainfall. Later in the year drier westerly airstreams affect the continental interior as the general airflow becomes more zonal.

The diurnal occurrence of precipitation in the central United States is rather unusual for a continental interior. Sixty percent or more of the summer precipitation falls during nocturnal thunderstorms (20:00–08:00 LST) in central Kansas, parts of Nebraska, Oklahoma and Texas. Hypotheses suggest that the nocturnal thunderstorm rainfall that occurs, especially with extensive mesoscale convective systems (see p. 255), may be linked to a tendency for nocturnal convergence and rising air over the plains east of the Rocky Mountains. The terrain profile appears to play a role here, as a large-scale inversion layer forms at night over the mountains, setting up a low-level jet (LLJ) east of the mountains just above the boundary layer. LLJs are most frequent in the night-time to early morning hours and occur more than 50 percent of the time in Texas in summer. This southerly flow, at 500–1000m above the surface, can supply the necessary low-level moisture influx and convergence for the storms (cf. Figure 5.27). MCSs account for 30–70 percent of the May to September rainfall over much of the area east of the Rocky Mountains to the Missouri River.

3 East of the upper Mississippi, in the Ohio valley and south of the lower Great Lakes, there is a transitional regime between that of the interior and the east coast type. Precipitation is reasonably abundant in all seasons, but the summer maximum is still in evidence (e.g. Dayton).
4 In eastern North America (New England, the Maritimes, Quebec and southeast Ontario), precipitation is fairly evenly distributed throughout the year (e.g., Blue Hill). In Nova Scotia and locally around Georgian Bay there is a winter maximum, due in the latter case to the influence of open water. In the Maritimes it is related to winter (and also autumn) storm tracks.

It is worth comparing the eastern regime with the summer maximum that is found over East Asia. There the Siberian anticyclone excludes cyclonic precipitation in winter and monsoonal influences are felt in the summer months.

The seasonal distribution of precipitation is of vital interest for agricultural purposes. Rain falling in summer, for instance, when evaporation losses are high, is less effective than an equal amount in the cool season. Figure 10.22 illustrates the effect of different regimes in terms of the moisture balance, calculated according to Thornthwaite's method (see Appendix 1B). At Halifax (Nova Scotia), sufficient moisture is stored in the soil to maintain evaporation at its maximum rate (i.e., actual evaporation = potential evaporation), whereas at Berkeley (California) there is a computed moisture deficit of nearly 50mm in August. This is a guide to the amount of irrigation water that may be required by crops, although in dry regimes the Thornthwaite method generally underestimates the real moisture deficit.

Figure 10.23 shows the ratio of actual to potential evaporation (AE/PE) for North America calculated by the methods of Thornthwaite and Mather from an equation relating PE to air temperature. It is drawn to highlight variation in the dry regions of the country. The boundary separating the moist climates of the east, where the ratio AE/PE exceeds about 8 percent or more, from the dry climates of the west (excluding the west coast), follows the 95th meridian. The major humid areas are along the Appalachians, in the northeast and along the Pacific coast, while the most extensive arid areas are in the intermontane basins, the High Plains, the southwest and parts of northern Mexico. In the west and southwest the

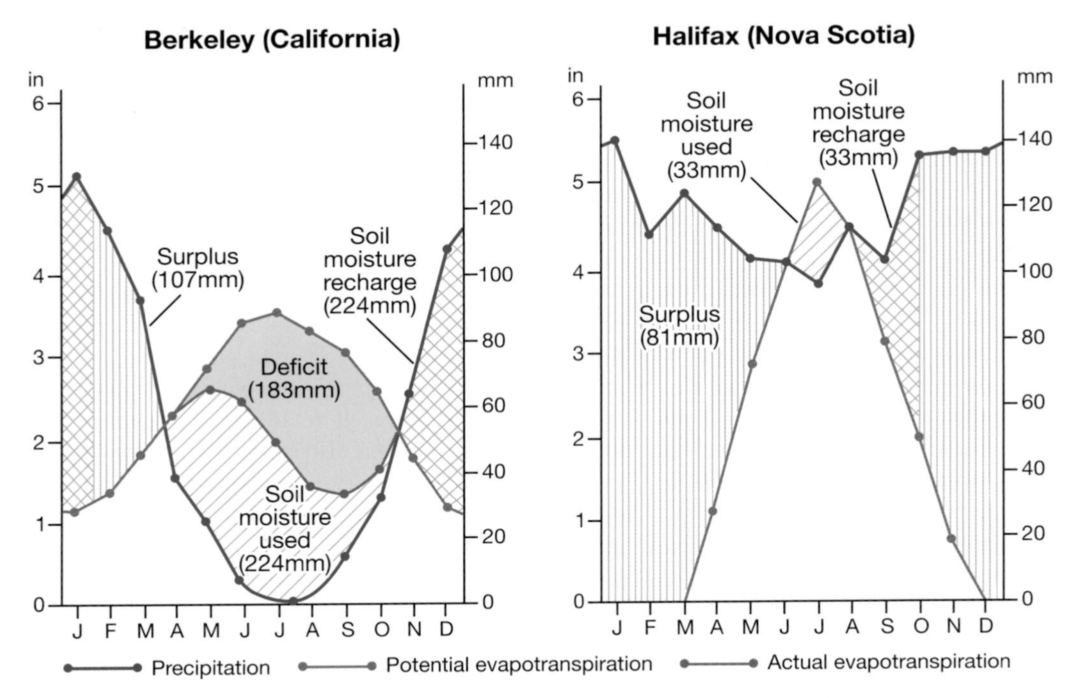

Figure 10.22 The moisture balances at Berkeley, California, and Halifax, Nova Scotia.
Source: After Thornthwaite and Mather (1955).

ratio is small due to lack of precipitation whereas in northwest Canada actual evaporation is limited by available energy.

C THE SUBTROPICAL MARGINS

1 The semi-arid southwestern United States

Both the mechanisms and patterns of the climate in areas dominated by the subtropical high pressure cells are not well documented. The inhospitable nature of these arid regions inhibits data collection, and yet the study of infrequent meteorological events requires a close network of stations maintaining continuous records over long periods. This difficulty is especially apparent in the interpretation of desert precipitation data, because much of the rain falls in local storms irregularly scattered in both space and time. The climatic conditions in the southwestern United States serve to exemplify this climatic type, based on the more reliable data for the semi-arid margins of the subtropical cells.

Observations at Tucson (730m), Arizona, between 1895 and 1957 showed a mean annual precipitation of 277mm falling on an average of about 45 days per year, with extreme annual figures of 614mm and 145mm. Two moister periods in late November to March (receiving 30 percent of the mean annual precipitation) and late June to September (50 percent) are separated by more arid seasons from April to June (8 percent) and October to November (12 percent). The winter rains are generally prolonged and of low intensity (more than half the falls have an intensity of less than 5mm per hour), falling from altostratus clouds associated with the cold fronts of depressions that are forced to take southerly routes by strong blocking to the north. This occurs during phases of equatorial displacement of the

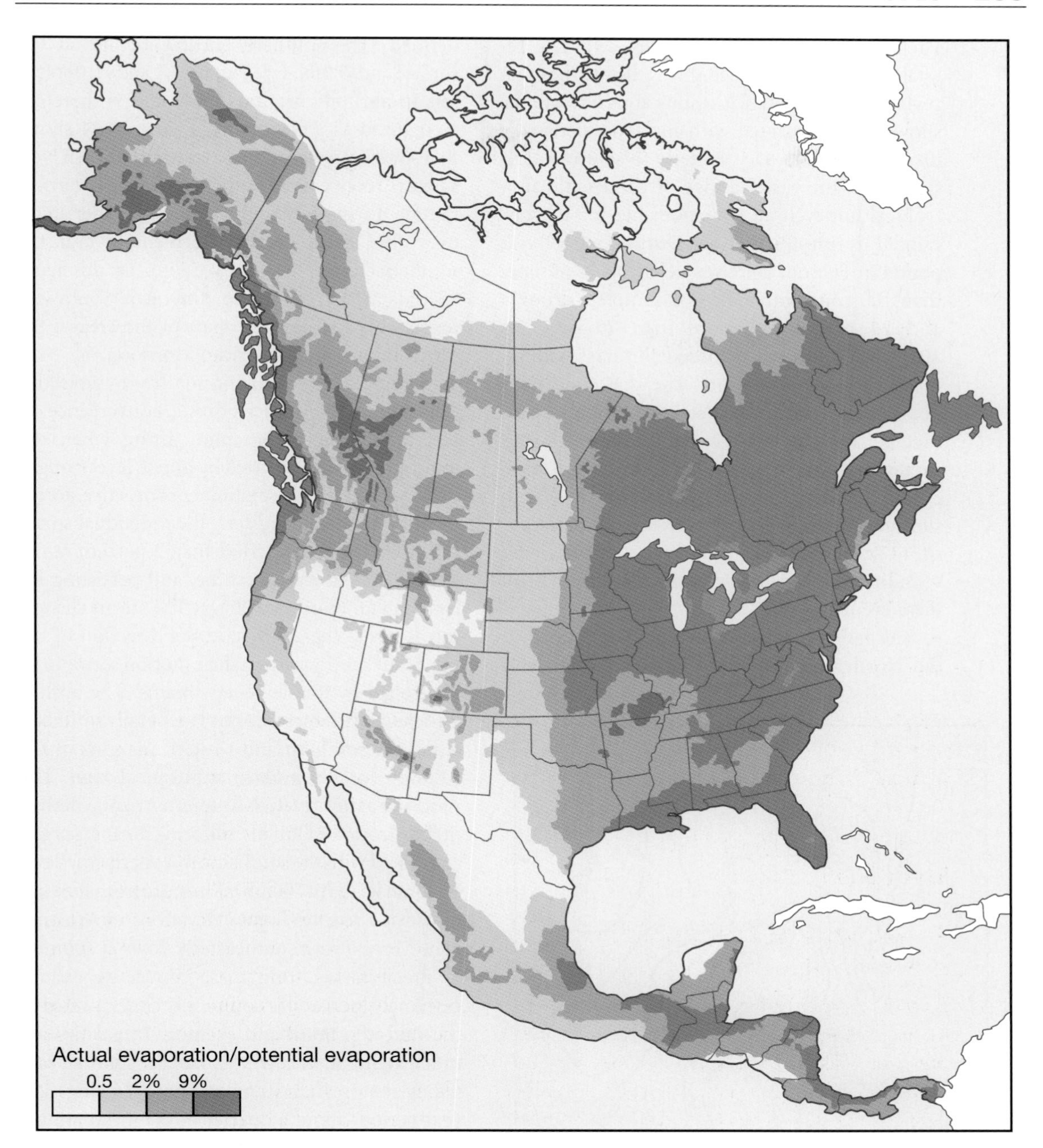

Figure 10.23 The ratio of actual/potential evaporation for North America determined using the Thornthwaite–Mather (1955) methods.

Source: Thompson *et al.* (1999). Courtesy of the US Geological Survey.

Pacific subtropical high pressure cell. The re-establishment of the cell in spring, before the main period of intense surface heating and convectional showers, is associated with the most persistent drought episodes. Dry westerly to southwesterly flow from the eastern edge of the Pacific subtropical anticyclone is responsible for the low rainfall during this season. During one 29-year period in Tucson, there were eight spells of more than 100 consecutive days of complete drought and 24 periods of more than 70 days. In 2005–2006, Phoenix recorded 143 days without measurable precipitation. The dry conditions occasionally lead to dust storms. Yuma records nine per year, on average, associated with winds averaging 10–15m s^{-1}. They occur both with cyclonic systems in the cool season and with summer convective activity. Phoenix experiences six to seven per year, mainly in summer, with visibility reduced below 1km in nearly half of these events.

The period of summer precipitation (known as the 'North American monsoon') is quite sharply defined. The southerly airflow regime at the surface and 700mb (see Figures 7.4 and 7.10) often sets in abruptly around 1 July and is therefore recognized as a singularity. Figure 10.24 shows that southeastern Arizona and southwestern New Mexico receive over 50 percent of their annual rainfall during July to September. Further south over the Sierra Madre Occidentale and the southern coast of the Gulf of California, this figure exceeds 70 percent. The American Southwest forms only the northern part of the area of the Mexican or North American Monsoon.

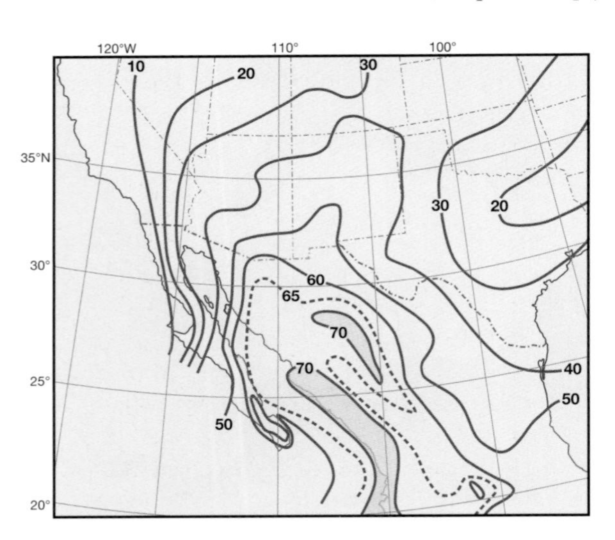

Figure 10.24 The contribution (percent) of JJA precipitation to the annual total in the southwestern United States and northern Mexico. Area greater than 50 percent stippled and greater than 70 percent hatched.

Source: After M. W. Douglas *et al.* (1993, p.1667, fig. 3). Courtesy of the American Meteorological Society.

Precipitation mainly occurs from convective cells initiated by surface heating, convergence or, less commonly, orographic lifting when the atmosphere is destabilized by upper-level troughs in the westerlies. These summer convective storms form in mesoscale clusters, the individual storm cells together covering less than 3 percent of the surface area at any one time, and persisting for less than an hour on average. The storm clusters move across the country in the direction of the upper-air motion. Often their motion seems to be controlled by low-level jet streams. The airflow associated with these storms is generally southerly along the southern and western margins of the Atlantic (or Bermudan) subtropical high. The moisture at low levels in southern Arizona derives mainly from the Gulf of California during 'surges' associated with the south-southwesterly low-level Sonoran jet (850–700mb). Moisture from the Gulf of Mexico reaches higher elevations in Arizona-New Mexico with southeasterly flows at 700mb.

Precipitation from these convective cells is extremely local and is commonly concentrated in the mid-afternoon and evening. Intensities are much higher than in winter, half the summer rain falling at more than 10mm per hour. During a 29-year period, about a quarter of the mean annual precipitation fell in storms giving 25mm or more per day. These intensities are much less than those associated with rainstorms in the humid tropics, but the sparsity of vegetation in the drier regions allows the rain to produce flash floods and considerable surface erosion.

2 The southeastern United States

The climate of the subtropical southeastern United States has no exact counterpart in Asia, which is affected by the summer and winter monsoon systems (discussed in Chapter 11). Seasonal wind changes are experienced in Florida, which is within the westerlies in winter and lies on the northern margin of the tropical easterlies in summer. The summer season rainfall maximum (see Figure 10.20 for Jacksonville) is a result of this changeover. In June, the upper flow over the Florida peninsula changes from northwesterly to southerly as a trough moves westward and becomes established in the Gulf of Mexico. This deep, moist, southerly airflow provides appropriate conditions for convection. Indeed, Florida probably ranks as the area with the highest annual number of days with thunderstorms – 90 or more, on average, in the vicinity of Tampa. These often occur in late afternoon, although two factors apart from diurnal heating are thought to be important. One is the effect of sea breezes converging from both sides of the peninsula, and the other is the northward penetration of disturbances in the easterlies (see Chapter 11). The latter may of course affect the area at any time of day. The westerlies resume control in September to October, although Florida remains under the easterlies during September, when Atlantic tropical cyclones are most frequent.

Tropical cyclones contribute around 15 percent of the average annual rainfall in the coastal Carolinas and 10–14 percent along the central Gulf Coast and in Florida. According to *Storm Data* reports for 1975–1994, hurricanes striking the southern and eastern USA account for over 40 percent of the total property damage and 20 percent of the crop damage attributed to extreme weather events in the country. Annual losses from hurricanes in the United States averaged $5.5 billion in the 1990s, with comparable national losses due to floods ($5.3 billion annually). The single most costly natural disaster up to 1989 was Hurricane Hugo ($9 billion), but this was far surpassed by the $27 billion losses caused by Hurricane Andrew over southern Florida and Louisiana in August 1992, when winds destroyed 130,000 homes, and the $81 billion losses attributed to Hurricane Katrina in August 2005. The 6m storm surge led to the failure of the levees at New Orleans and to the flooding of 80 percent of the city with widespread destruction of property and the loss of 1836 lives. In contrast, injuries (deaths) during hurricanes average only 250 (21) per year, as a result of storm warnings and the evacuation of endangered communities.

Winter precipitation along much of the eastern seaboard of the United States is dominated by an apparent oscillation between depression tracks following the Ohio valley (continental lows) and the southeast Atlantic coast (Gulf lows), only one of which is normally dominant during a single winter. The former track brings below-average winter rainfall and snowfall, but above-average temperatures, to the mid-Atlantic region, whereas the reverse conditions are associated with systems following the southeast coast track.

The region of the Mississippi lowlands and the southern Appalachians to the west and north is not simply transitional to the 'interior type', at least in terms of rainfall regime (see Figure 10.20). The profile shows a winter–spring maximum and a secondary summer maximum. The cool season peak is related to westerly depressions moving northeastward from the Gulf Coast area, and it is significant that the wettest month is commonly March, when the mean jet stream is farthest south. The summer rains are associated with convection in humid air from the Gulf, although this convection becomes less effective inland as a result of the subsidence created by the anticyclonic circulation in the middle troposphere referred to previously (see B.3, this chapter).

3 The Mediterranean

The characteristic west coast climate of the subtropics is the Mediterranean type with hot, dry summers and mild, relatively wet winters. It is interposed between the temperate maritime type and the arid subtropical desert climate. The boundary between the temperate maritime climate of Western Europe and that of the Mediterranean may be delimited on the basis of the seasonality of rainfall. However, another diagnostic feature is the relatively sharp increase in solar radiation across a zone running along northern Spain, southeast France, northern Italy and to the east of the Adriatic (Figure 10.25). The Mediterranean regime is transitional in a special way, because it is controlled by the westerlies in winter and by the subtropical anticyclone in summer. The seasonal change in position of the subtropical high and the associated subtropical westerly jet stream in the upper troposphere are evident in Figure 10.25. The type region is peculiarly distinctive, extending more than 3000km into the Eurasian continent. In addition, the configuration of seas and peninsulas produces great regional variety of weather and climate. The Californian region, with similar conditions (see Figure 10.20), is of very limited extent, and attention is therefore concentrated on the Mediterranean basin itself.

The winter season sets in quite suddenly in the Mediterranean as the summer eastward extension of the Azores high pressure cell collapses. This phenomenon may be observed on barographs throughout the region, but particularly in the western Mediterranean, where a sudden drop in pressure occurs on about 20 October and is accompanied by a marked increase in the probability of precipitation. The probability of receiving rain in any five-day period increases dramatically from 50–70 percent in early October to 90 percent in late October. This change is associated with the first invasions by cold fronts, although thundershower rain has been common since August. The pronounced winter precipitation over the Mediterranean largely results from the relatively high sea surface temperatures during that season, the sea temperature in January being about 2°C higher than the mean air temperature. Incursions of colder air into the region lead to convective instability along the cold front, producing frontal and orographic rain. Incursions of arctic air are relatively infrequent (there being, on average, six to nine invasions by cA and mA air each year), but penetration by unstable mP air is much more common. It typically gives rise to deep cumulus development and is crucial in the formation of Mediterranean depressions. The initiation and movement of these depressions (Figure 10.26) is associated with a branch of the polar front jet stream located at about 35°N. This jet develops during low index phases, when a blocking anticyclone at about 20°W distorts the westerlies over the eastern Atlantic. This leads to a deep stream of arctic air flowing southward over the British Isles and France.

Low pressure systems in the Mediterranean have three main sources. Atlantic depressions entering the western Mediterranean as surface lows make up 9 percent and 17 percent form as baroclinic waves south of the Atlas Mountains (the so-called Saharan depressions; see Figure 10.27). The latter are important sources of rainfall in late winter and spring). Fully 74 percent develop in the western Mediterranean in the lee of the Alps and Pyrenees (see Chapter 7H.1). The combination of the lee effect and that of unstable surface air over the western Mediterranean explains the frequent formation of these *Genoa-type depressions* whenever conditionally unstable mP air invades the region. These depressions are exceptional in that the instability of the air in the warm sector gives unusually intense precipitation along the warm front. The unstable mP air produces heavy showers and thunderstorm rainfall to the rear of the cold front, especially between 5 and 25°E. This warming of mP air produces air designated as *Mediterranean*. The mean boundary between this Mediterranean air mass and cT air flowing northeastward from the Sahara is referred to as the Mediterranean front (see Figure 10.26). There may be a temperature discontinuity as great as 12–16°C across it in late winter. Saharan depressions and those from the western Mediterranean move eastward, forming a belt of

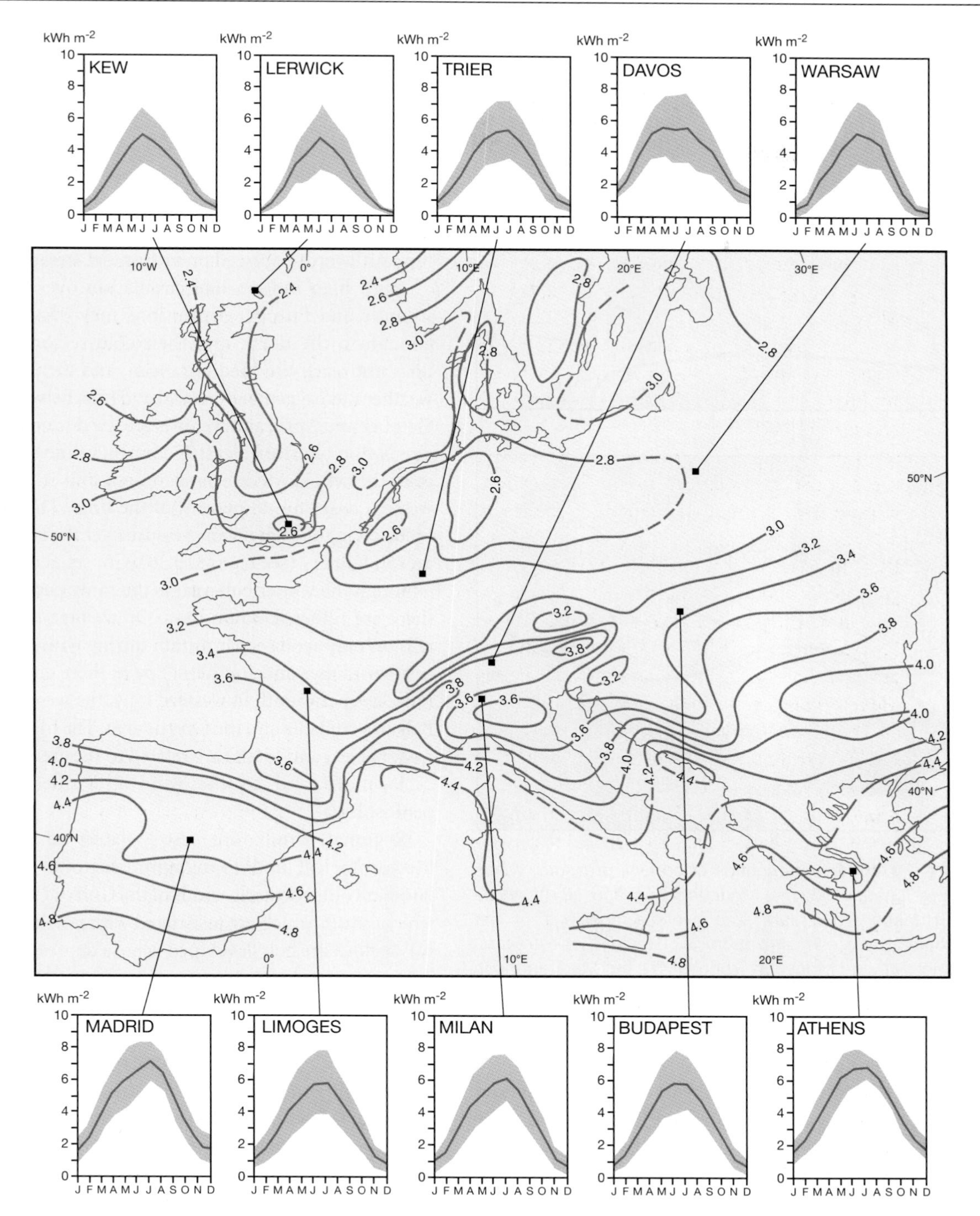

Figure 10.25 Average annual means of daily global irradiation on a horizontal surface (kWh/m^{-2}) for Western and Central Europe calculated for the period 1966–1975. Ten-year means of monthly means of daily sums, together with standard deviations (shaded band), are also shown for selected stations.

Source: Palz (1984). Reproduced by permission of the Directorate-General, Science, Research and Development, European Commission, Brussels, and W. Palz.

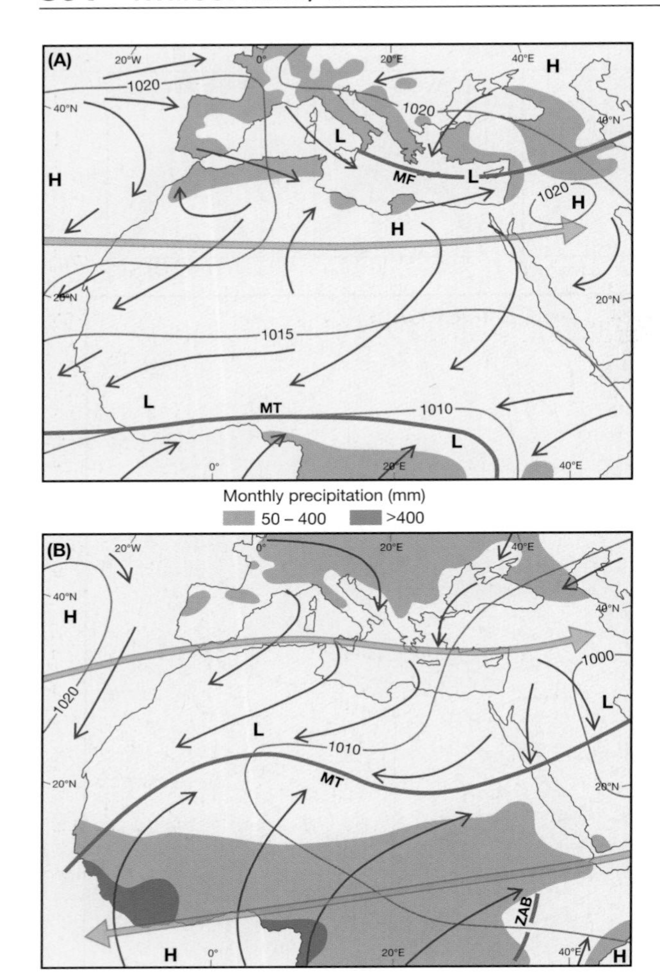

Figure 10.26 The distribution of surface pressure, winds and precipitation for the Mediterranean and North Africa during January and July. The average positions of the subtropical westerly and tropical easterly jet streams, together with the Monsoon Trough (MT), the Mediterranean Front (MF) and the Zaire Air Boundary (ZAB), are also shown.
Source: Partly after *Weather in the Mediterranean* (HMSO, 1962) (Crown Copyright Reserved).

low pressure associated with this frontal zone and frequently drawing cT air northward ahead of the cold front as the warm, dust-laden *scirocco* (especially in spring and autumn when Saharan air may spread into Europe). The movement of Mediterranean depressions is modified both by relief effects and by their regeneration in the eastern Mediterranean through fresh cP air from Russia or southeast Europe. Although many lows pass eastward into Asia, there is a strong tendency for others to move northeastward over the Black Sea and Balkans, especially as spring advances. Winter weather in the Mediterranean is quite variable, as the subtropical westerly jet stream is highly mobile and may occasionally coalesce with the southward-displaced polar front jet stream.

With high index zonal circulation over the Atlantic and Europe, depressions may pass far enough to the north that their cold-sector air does not reach the Mediterranean, and then the weather there is generally settled and fine. Between October and April, anticyclones are the dominant circulation type for at least 25 percent of the time over the whole Mediterranean area and in the western basin for 48 percent of the time. This is reflected in the high mean pressure over the latter area in January (see Figure 10.26). Consequently, although the winter half-year is the rainy period, there are rather few rain days. On average, rain falls on only six days per month during winter in northern Libya and southeast Spain; there are 12 rain days per month in western Italy, the western Balkan Peninsula and the Cyprus area. The higher frequencies (and totals) are related to the areas of cyclogenesis and to the windward sides of peninsulas.

Regional winds are also related to the meteorological and topographic factors. The familiar cold, northerly winds of the Gulf of Lions (the *mistral*), which are associated with northerly mP airflow, are best developed when a depression is forming in the Gulf of Genoa east of a high pressure ridge from the Azores anticyclone. Katabatic and funneling effects strengthen the flow in the Rhône valley and similar localities, so that violent winds are sometimes recorded. The mistral may last for several days until the outbreak of polar or continental air ceases. The frequency of these winds depends on their definition. The average frequency of strong mistrals in the south of France is shown in Table 10.3 (based on occurrence at one or more stations from

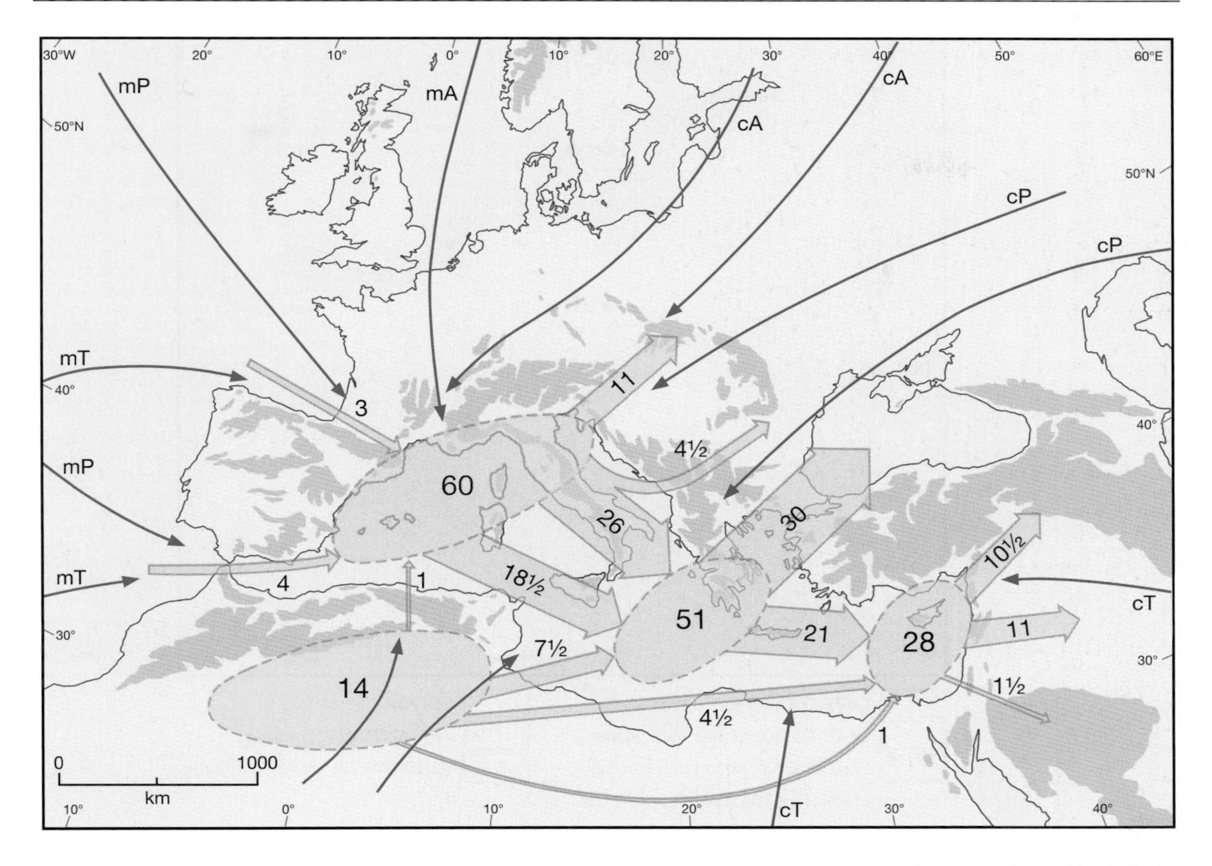

Figure 10.27 Tracks of Mediterranean depressions, showing average annual frequencies, together with air-mass sources.

Source: After *Weather in the Mediterranean* (HMSO, 1962) (Crown Copyright Reserved).

Table 10.3 Number of days with a strong mistral in the South of France

Speed	J	F	M	A	M	J	J	A	S	O	N	D	Year
$\geq$11m s^{-1} (21 kt)	10	9	13	11	8	9	9	7	5	5	7	10	103
$\geq$17m s^{-1} (33 kt)	4	4	6	5	3	2	0.6	1	0.6	0	0	4	30

Source: After *Weather in the Mediterranean* (HMSO, 1962).

Perpignan to the Rhône in 1924–1927). Similar winds may occur along the Catalan coast of Spain (the *tramontana*, see Figure 10.30) and also in the northern Adriatic (the *bora*) and northern Aegean Seas when polar air flows southward in the rear of an eastward-moving depression and is forced over the mountains (cf. Chapter 5C.2). In Spain, cold, dry, northerly winds occur in several different regions. Figure 10.28 shows the *galerna* of the north coast and the *cierzo* of the Ebro valley.

The generally wet, windy and mild winter season in the Mediterranean is succeeded by a long, indecisive spring lasting from March to May, with many false starts of summer weather. The spring period, like that of early autumn, is especially unpredictable. In March 1966, a trough

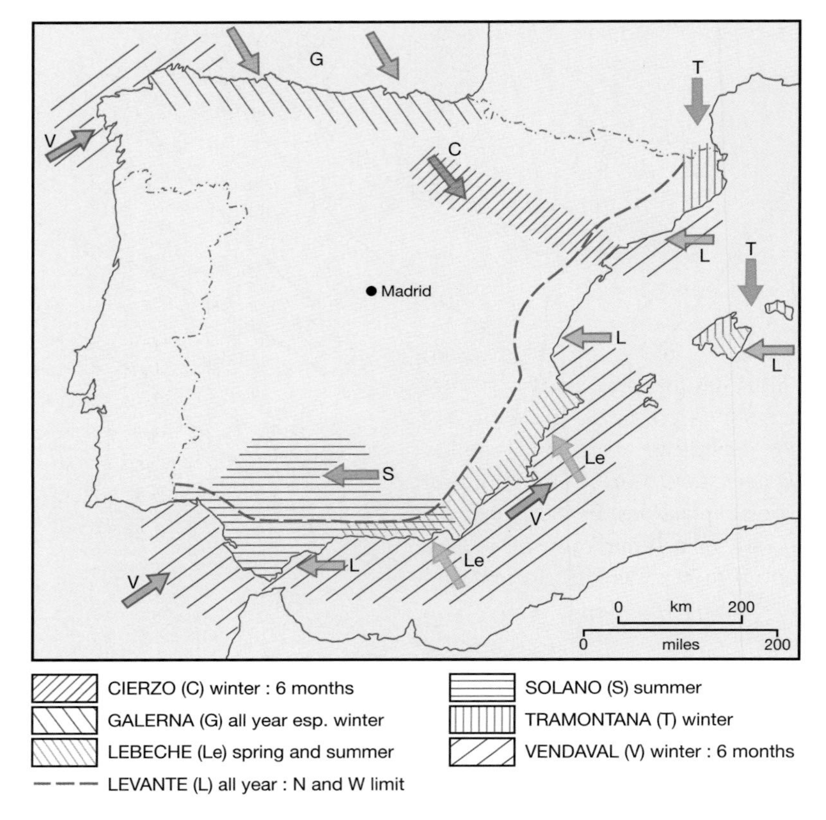

Figure 10.28 Areas affected by the major regional winds in Spain as a function of season.
Source: From Tout and Kemp (1985). By permission of the Royal Meteorological Society.

moving across the eastern Mediterranean, preceded by a warm southerly *khamsin* and followed by a northerly airstream, brought up to 70mm of rain in only four hours to an area of the southern Negev Desert. Although April is normally a dry month in the eastern Mediterranean, Cyprus having an average of only three days with 1mm of rainfall or more, high rainfalls can occur, as in April 1971 when four depressions affected the region. Two of these were Saharan depressions moving eastward beneath the zone of diffluence on the cold side of a westerly jet and the other two were intensified in the lee of Cyprus. The rather rapid collapse of the Eurasian high pressure cell in April, together with the discontinuous northward and eastward extension of the Azores anticyclone, encourages the northward displacement of depressions. Even if higher latitude air does penetrate south to the Mediterranean, the sea surface there is relatively cool and the air is more stable than during the winter.

By mid-June, the Mediterranean basin is dominated by the expanded Azores anticyclone to the west, while to the south the mean pressure field shows a low pressure trough extending across the Sahara from southern Asia (see Figure 10.26). The winds are predominantly northerly (e.g. the *etesians* of the Aegean) and represent an eastward continuation of the northeasterly trades. Locally, sea breezes reinforce these winds, but on the Levant coast they cause surface southwesterlies. Land and sea breezes, involving air up to 1500m deep, largely condition the day-to-day weather of many parts of the North African coast. Depressions are by no means absent in the summer months, but they are usually weak. The anticyclonic character of the large-scale circulation encourages subsidence, and air-mass contrasts are much reduced compared with winter. Thermal lows form from time to time over Iberia and Anatolia, although thundery outbreaks are infrequent due to the low relative humidity.

The most important regional winds in summer are of continental tropical origin. There are a variety of local names for these usually hot, dry and dusty airstreams – *scirocco* (Algeria and the Levant), *lebeche* (southeast Spain) and *khamsin* (Egypt) – which move northward ahead of eastward-moving depressions. In the Negev, the onset of an easterly *khamsin* may cause the relative humidity to drop to less than 10 percent and temperatures to rise to as much as 48°C. In southern Spain, the easterly *solano* brings hot, humid weather to Andalucia in the summer half-year, whereas the coastal *levante* – which has a long fetch over the Mediterranean – is moist and somewhat cooler (see Figure 10.28). Such regional winds occur when the Azores high extends over Western Europe with a low pressure system to the south.

Many stations in the Mediterranean receive only a few millimeters of rainfall in at least one summer month, yet the seasonal distribution does not conform to the pattern of simple winter maximum over the whole of the Mediterranean basin. Figure 10.29 shows that this is found in the eastern and central Mediterranean, whereas Spain, southern France, northern Italy and the northern Balkans have more complicated profiles with a maximum in autumn or peaks in both spring and autumn. This double maximum may be interpreted as a transition between the continental interior type with summer maximum and the Mediterranean type with winter maximum. A similar transition region occurs in the southwestern United States (see Figure 10.21), but local topography in this intermontane zone introduces irregularities into the regimes.

4 North Africa

The dominance of high pressure conditions in the Sahara is marked by the low average precipitation in this region. Over most of the central Sahara, the mean annual precipitation is less than 25mm, although the high plateaux of the Ahaggar and Tibesti receive over 100mm. Parts of western Algeria have gone at least two years without more than 0.1mm of rain in any 24-hour period, and most of southwest Egypt as much as five years. However, 24-hour storm rainfalls approaching 50mm (more than 75mm over the high plateaux) may be expected in scattered localities. During a 35-year period, excessive short-period rainfall intensities occurred in the vicinity of west-facing slopes in Algeria, such as at Tamanrasset (46mm in 63 minutes) (Figure 10.30), El Golea (8.7mm in 3 minutes) and Beni Abes (38.5mm in 25 minutes). During the summer, rainfall variability is introduced into the southern Sahara by the variable northward penetration of the Monsoon Trough (see Figure 11.2B), which on occasion allows tongues of moist southwesterly air to penetrate far north and produce short-lived low pressure centers. Study of these Saharan depressions has permitted a clearer picture to emerge of the region. In the upper troposphere at about 200mb (12km), the westerlies overlie the poleward flanks of the subtropical high pressure belt. Occasionally, the individual high pressure cells contract away from one another as meanders develop in the westerlies between them. These may extend equatorward to interact with the low-level tropical easterlies (Figure 10.31). This interaction may lead to the development of lows, which then move northeast along the meander trough associated with rain and thunder. By the time they reach the central Sahara, they are frequently 'rained out' and give rise to duststorms, but they can be reactivated further north by the entrainment of moist Mediterranean air. The interaction of westerly and easterly circulation is most likely to occur around the equinoxes or sometimes in winter if the otherwise dominant Azores high pressure cell contracts westward. The

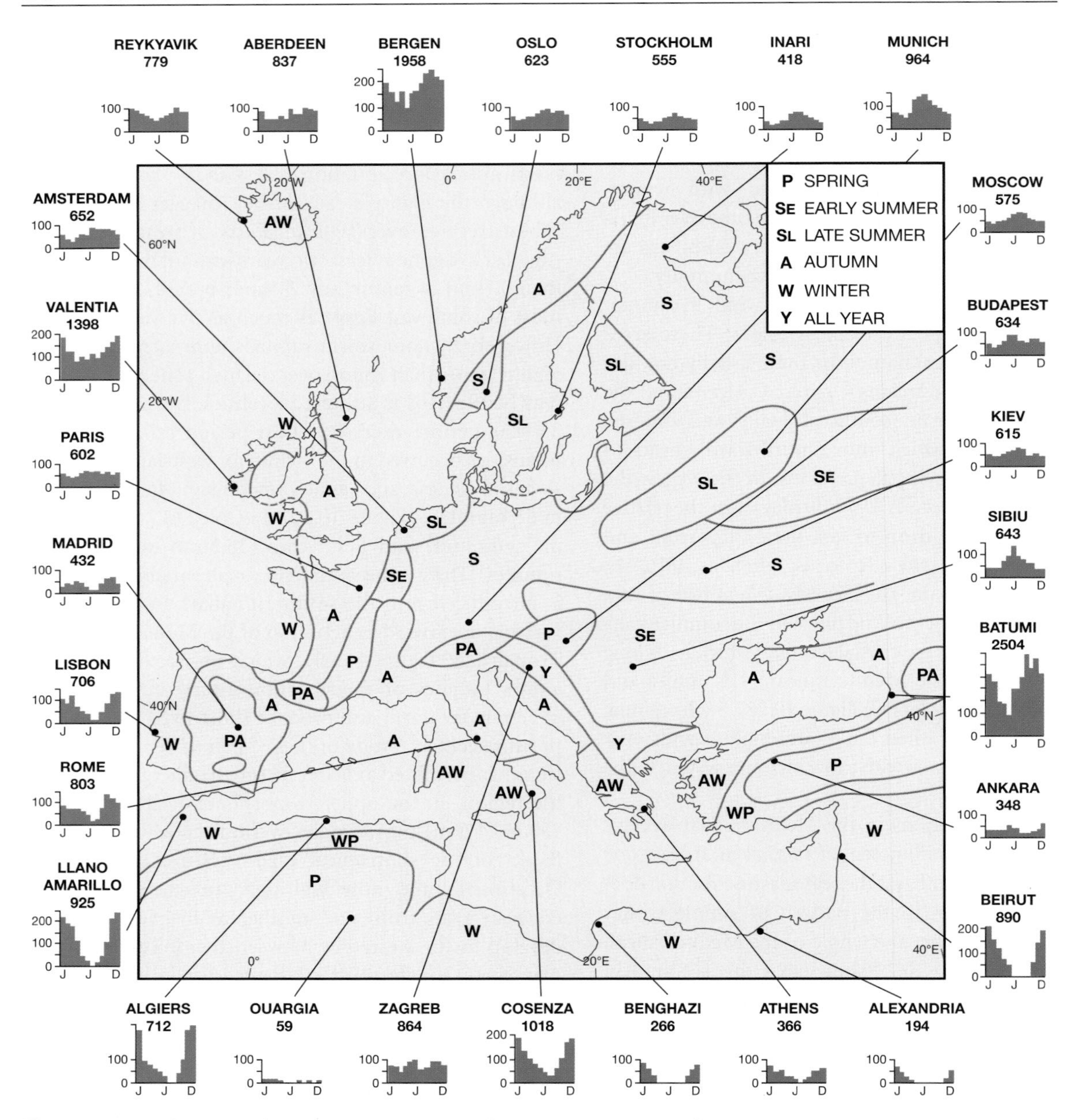

Figure 10.29 Seasons of maximum precipitation for Europe and North Africa, together with average monthly and annual figures (mm) for 28 stations.

Sources: Thorn (1965) and Huttary (1950). Reprinted from D. Martyn (1992) *Climates of the World*, with kind permission from Elsevier Science NL, Sara Burgerhartstraat 25, 1055 KV Amsterdam, the Netherlands.

westerlies may also affect the region through the penetration of cold fronts south from the Mediterranean, bringing heavy rain to localized desert areas. In December 1976, such a depression produced up to 40mm of rain during two days in southern Mauretania.

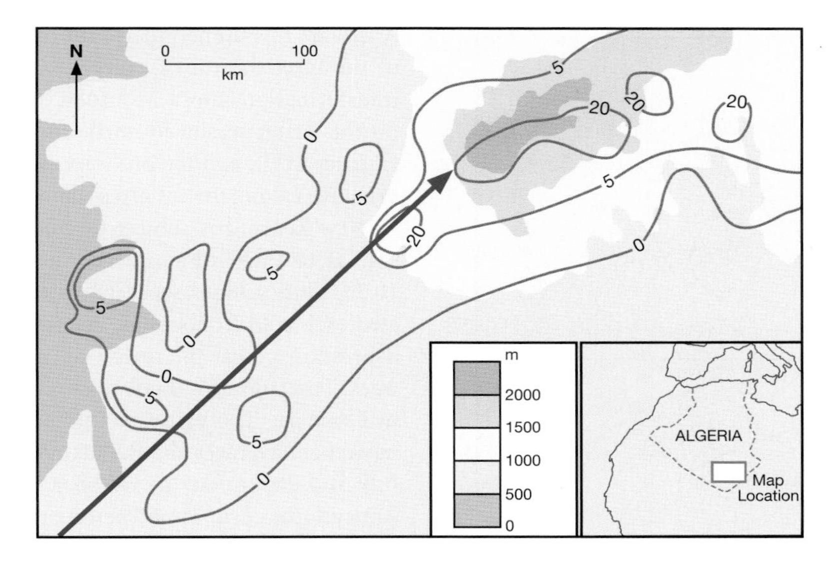

Figure 10.30 Track of a storm and the associated three-hour rainfall (mm) during September 1950 around Tamanrasset in the vicinity of the Ahaggar Mountains, southern Algeria.

Source: Partly after Goudie and Wilkinson (1977).

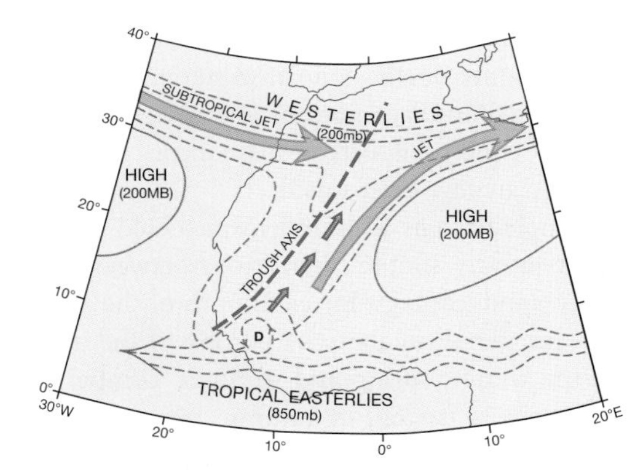

Figure 10.31 Interaction between the westerlies and the tropical easterlies leading to the production of Saharan depressions (D), which move northeastward along a trough axis.

Source: After Nicholson and Flohn (1980). Copyright © 1980/1982 by D. Reidel Publishing Company. Reprinted by permission.

5 Australasia

The subtropical anticyclones of the South Atlantic and Indian Ocean tend to generate high pressure cells which move eastward, intensifying southeast of South Africa and west of Australia. These are warm-core systems formed by descending air and extending through the troposphere. The continental intensification of the constant eastward progression of such cells causes pressure maps to give the impression of the existence of a stable anticyclone over Australia (Figure 10.32). About 40 anticyclones traverse Australia annually, being somewhat more numerous in spring and summer than in autumn and winter. Over both oceans, the frequency of anticyclonic centers is greatest in a belt around 30°S in winter and 35–40°S in summer; they rarely occur south of 45°S.

Between successive anticyclones are low pressure troughs containing inter-anticyclonic fronts (sometimes termed 'polar') (Figure 10.33). Within these troughs, the subtropical jet stream meanders equatorward, accelerates (particularly in winter, when it reaches an average velocity of 60m s^{-1} compared with a mean annual value of 39m s^{-1}) and generates upper-air depressions, which move southeastward along the front (analogous to the systems in North Africa). The variation in strength of the continental

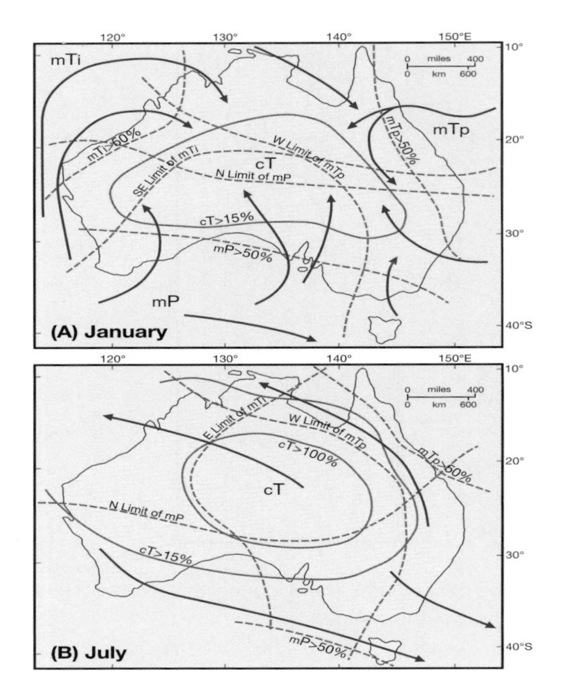

Figure 10.32 Air-mass frequencies, source areas, wind directions and dominance of the cT high pressure cell over Australia in summer (above) and winter (below).

Source: After Gentilli (1971). By permission of Elsevier Science, NL.

anticyclones and the passage of inter-anticyclonic fronts cause periodic inflows of surrounding maritime tropical air masses from the Pacific (mTp) and the Indian (mTi) Oceans. There are also incursions of maritime polar air (mP) from the south, and variations in strength of the local source of continental tropical (cT) air masses (see Figure 10.32).

The high pressure conditions over Australia promote especially high temperatures over central and western parts of the continent, towards which there is a major heat transport in summer. These pressures keep average rainfall amounts low; these normally total less than 250mm annually over 37 percent of Australia. In winter, upper-air depressions along the inter-anticyclonic fronts bring rain to southeastern regions and also, in conjunction with mTi incursions, to southwest Australia. In summer, the southward movement of the Intertropical Convergence Zone and its transformation into a Monsoon Trough brings on the wetter season in northern Australia (see Chapter 11D), and the onshore southeast trades bring rain along the eastern seaboard.

New Zealand is subject to climatic controls similar to those of southern Australia (Figure 10.33). Anticyclones, separated by troughs associated with cold fronts often deformed into wave depressions, cross the region on average once a week. Their most southerly track (38.5°S) is taken in February. The eastward rate of anticyclonic movement averages about 570km/day in May to July and 780km/day in October to December. Anticyclones occur some 7 percent of the time and are associated with settled weather, light winds, sea breezes and some fog. On the eastern (leading) edge of the high pressure cell the airflow is usually cool, maritime and southwesterly, interspersed with south or southeasterly flow producing drizzle. On the western side of the cell, the airflow is commonly north or northwesterly, bringing mild and humid conditions. In autumn, high pressure conditions increase in frequency up to 22 percent, giving a drier season.

Simple troughs with undeformed cold fronts and relatively simple interactions between the trailing and leading edge conditions of the anticyclones persist in about 44 percent of the time during winter, spring and summer, compared with only 34 percent in autumn. Wave depressions occur with about the same frequency. If a wave depression forms on the cold front to the west of New Zealand, it usually moves southeastward along the front, passing to the south of the country. In contrast, a depression forming over New Zealand may take 36–48 hours to clear the country, bringing prolonged rainy conditions (Figure 10.34). Relief, especially the Southern Alps, predominantly controls rainfall amounts. West- or northwest-facing mountains receive an average annual precipitation in excess of 2500mm, with some parts of South Island exceeding 10,000mm (see Figure 5.16). The eastern lee areas

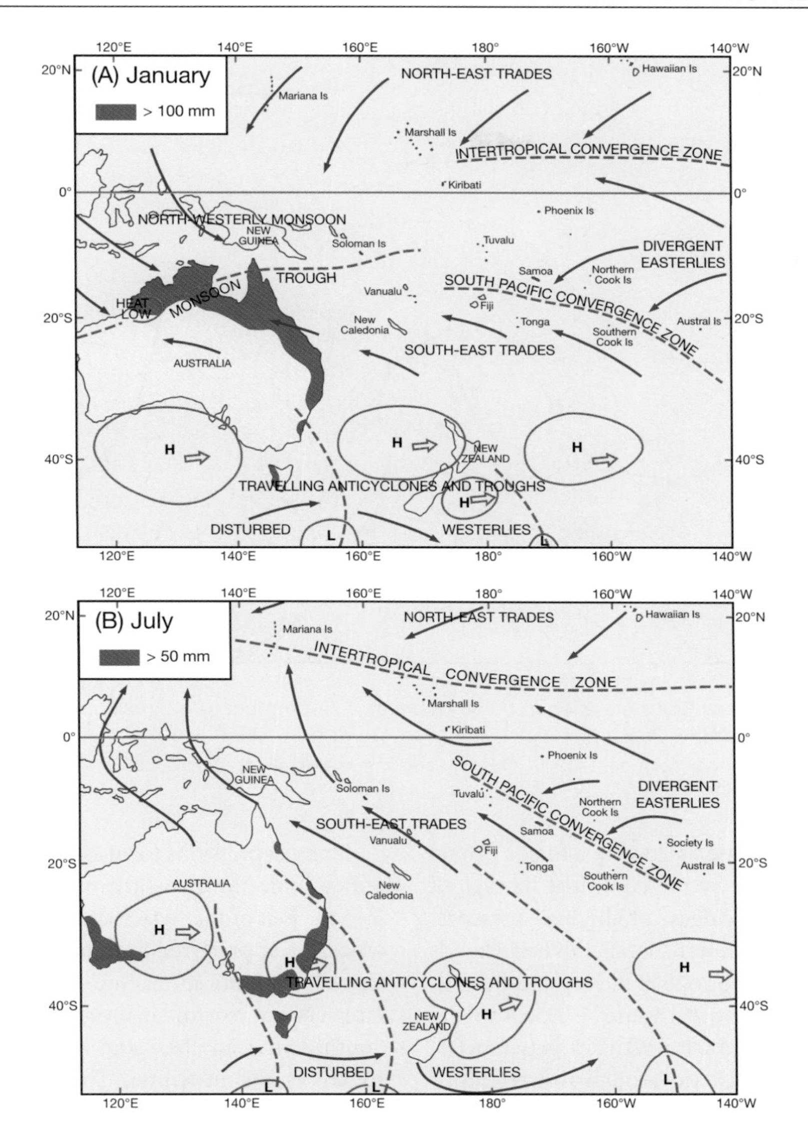

Figure 10.33 Main climatological features of Australasia and the southwest Pacific. Areas with >100mm (January) and >50mm (July) mean monthly precipitation for Australia are also shown.

Source: After Steiner, from Salinger *et al.* (1995). From *International Journal of Climatology*, copyright © John Wiley & Sons Ltd. Reproduced with permission.

have much lower amounts, with less than 500mm in some parts. North Island has a winter precipitation maximum, but South Island, under the influence of depressions in the southern westerlies, has a more variable seasonal maximum.

D HIGH LATITUDES

1 The southern westerlies

The strong zonal airflow in the belt of the southern westerlies, which is apparent only on mean

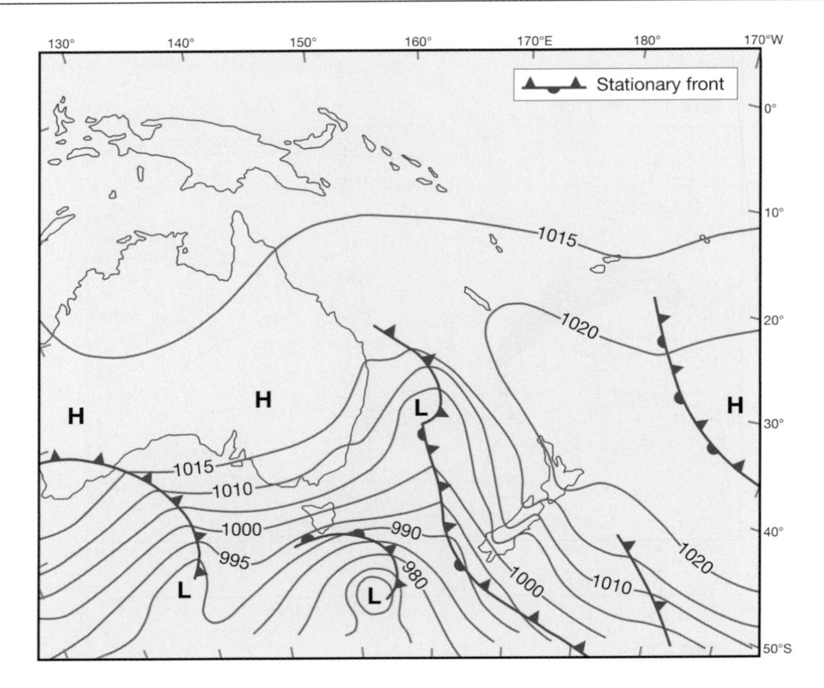

Figure 10.34 The synoptic situation at 00:00 hours on 1 September 1982, resulting in heavy rainfall in the Southern Alps of New Zealand.

Sources: After Hessell; from Wratt *et al.* (1996). From *Bulletin of the American Meteorological Society* by permission of the American Meteorological Society.

monthly maps, is associated with a major frontal zone characterized by the continual passage of depressions and ridges of higher pressure. Throughout the Southern Ocean, this belt extends southward from about 30°S in July and 40°S in January (see Figures 9.18 and 10.35B) to the Antarctic Trough which fluctuates between 60° and 72°S. The Antarctic Trough is a region of cyclonic stagnation and decay that tends to be located furthest south at the equinoxes. Around New Zealand, the westerly airflow at an elevation of 3–15km in the belt 20–50°S persists throughout the year. It becomes a jet stream at 150mb (13.5km), over 25–30°S, with a velocity of 60m s^{-1} in May–August, decreasing to 26m s^{-1} in February. In the Pacific, the strength of the westerlies depends on the meridional pressure difference between 40 and 60°S, being on average greatest all the year south of western Australia and west of southern Chile.

Many depressions form as waves on the inter-anticyclonic fronts, which move southeastward into the belt of the westerlies. Others form in the westerlies at preferred locations such as south of Cape Horn, and at around 45°S in the Indian Ocean in summer and in the South Atlantic off the South American coast and around 50°S in the Indian Ocean in winter. The Polar Front (see Figure 9.20) is most closely associated with the sea surface temperature gradient across the Antarctic convergence, whereas the sea ice boundaries further south are surrounded by equally cold surface water (Figure 10.35B).

In the South Atlantic, depressions travel at about 1300km/day near the northern edge of the belt, slowing to 450–850km/day within 5–10° latitude of the Antarctic Trough. In the Indian Ocean, eastward velocities range from 1000 to 1300km/day in the belt 40–60°S, reaching a maximum in a core at 45–50°S. Pacific

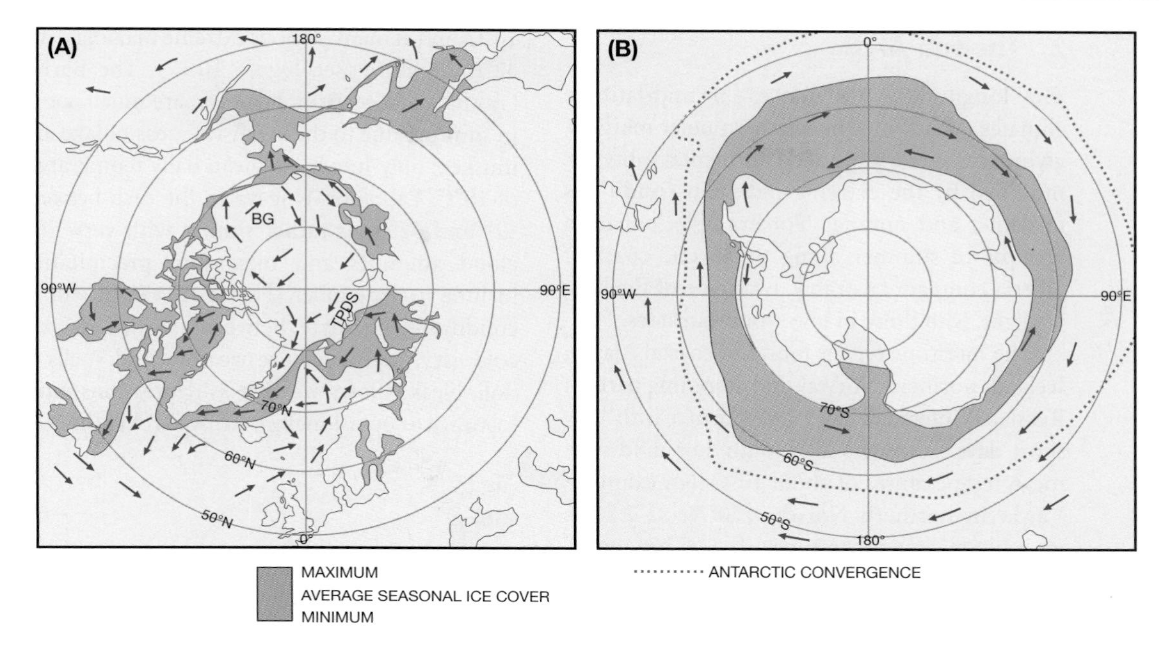

Figure 10.35 A: Surface currents in the Arctic, together with average autumn minimum and spring maximum sea ice extent for the period 1973–1990. There have been record minima in summary 2005, 2007 and 2008. B: Southern Ocean surface circulation, convergence zones and seasonal ice limits in March and September.

Sources: A: Maythum (1993) and Barry (1983). B: After Barry (1986). Copyright © Plenum Publishing Corp., New York. Published by permission.

depressions tend to be similarly located and generally form, travel and decay within a period of about a week. As in the Northern Hemisphere, high zonal index results from a strong meridional pressure gradient and is associated with wave disturbances propagated eastward at high speed with irregular and often violent winds and zonally oriented fronts. Low zonal index results in high pressure ridges extending further south and low pressure centers located further north. However, breakup of the flow, leading to blocking, is less common and less persistent in the Southern Hemisphere than in the Northern.

The southern westerlies are linked to the belt of traveling anticyclones and troughs by cold fronts, which connect the inter-anticyclonic troughs of the latter with the wave depressions of the former. Although storm tracks of the westerlies are usually well to the south of Australia (Figure 10.33), fronts may extend north into the continent, particularly from May, when the first rains occur in the southwest. On average, in midwinter (July), three depression centers skirt the southwest coast. When a deep depression moves to the south of New Zealand, the passage of the cold front causes that country to be covered first by a warm, moist westerly or northerly airflow and then by cooler southerly air. A series of such depressions may follow at intervals of 12–36 hours, each cold front being followed by progressively colder air. Further east over the South Pacific, the northern fringe of the southern westerlies is influenced by northwesterly winds, changing to west or southwest as depressions move to the south. This weather pattern is interrupted by periods of easterly winds if depression systems track along lower latitudes than usual.

2 The sub-Arctic

The longitudinal differences in mid-latitude climates persist into the northern polar margins, giving rise to maritime and continental subtypes, modified by the extreme radiation conditions in winter and summer. For example, radiation receipts in summer along the Arctic coast of Siberia compare favorably, by virtue of the long daylight, with those in lower mid-latitudes.

The maritime type is found in coastal Alaska, Iceland, northern Norway and adjoining parts of Russia. Winters are cold and stormy, with very short days. Summers are cloudy but mild with mean temperatures of about 10°C. For example, Vardø in northern Norway (70°N, 31°E) has monthly mean temperatures of –6°C in January and 9°C in July, while Anchorage, Alaska (61°N, 150°W) records –11°C and 14°C, respectively. Annual precipitation is generally between 600 and 1250mm, with a cool season maximum and about six months of snow cover.

The weather is mainly controlled by depressions, which are weakly developed in summer. In winter, the Alaskan area is north of the main depression tracks and occluded fronts and upper troughs are prominent, whereas northern Norway is affected by frontal depressions moving into the Barents Sea. Iceland is similar to Alaska, although depressions often move slowly over the area and occlude, whereas others moving northeastward along the Denmark Strait bring mild, rainy weather.

The interior, cold-continental climates have much more severe winters, although precipitation amounts are smaller. At Yellowknife (62°N, 114°W), for instance, the mean January temperature is only –28°C. In these regions, *permafrost* (permanently frozen ground) is widespread and often of great depth. In summer, only the top 1–2m of ground thaw and as the water cannot readily drain away this 'active layer' often remains waterlogged. Although frost may occur during any month, the long summer days usually give three months with mean temperatures above 10°C, and at many stations extreme maxima reach 32°C or more (see Figure 10.17). The Barren Grounds of Keewatin, however, are much cooler in summer due to the extensive areas of lake and muskeg; only July has a mean daily temperature of 10°C. Labrador–Ungava to the east, between 52° and 62°N, is rather similar with very high cloud amounts and maximum precipitation in June to September (Figure 10.36). In winter, conditions fluctuate between periods of very cold, dry, high pressure weather and spells of dull, bleak, snowy weather as depressions move eastward or occasionally northward over the area.

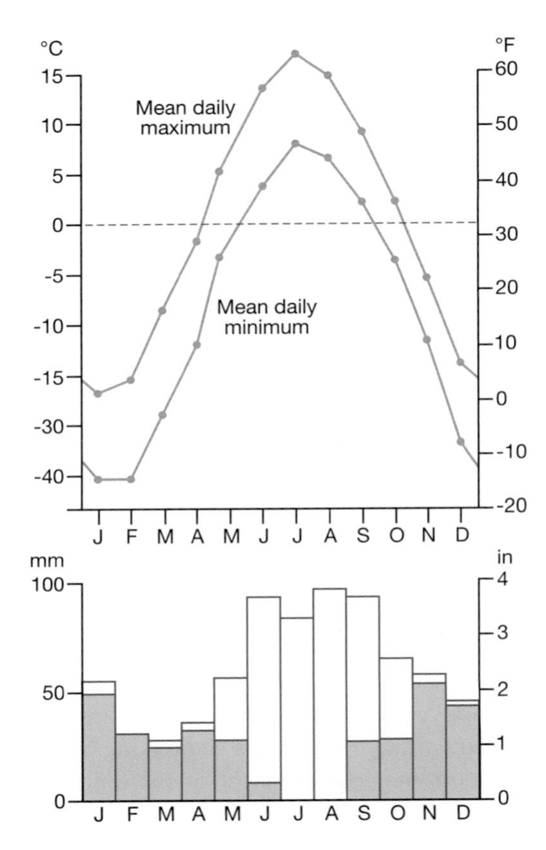

Figure 10.36 Selected climatological data for McGill Sub-Arctic Research Laboratory, Schefferville, PQ, 1955–1962. The shaded portions of the precipitation represent snowfall, expressed as water equivalent.

Source: Data from J. B. Shaw and D. G. Tout.

In spite of the very low mean temperatures in winter, there have been occasions when maxima have exceeded 4°C during incursions of maritime Atlantic air. Such variability is not found in eastern Siberia, which is intensely continental, apart from the Kamchatka Peninsula, with the Northern Hemisphere's *cold pole* located in the remote northeast (see Figure 3.11A). Verkhoyansk and Oimyakon have a January mean of –50°C, and both have recorded an absolute minimum of –67.7°C. Stations located in the valleys of northern Siberia record, on average, strong to extreme frosts 50 percent of the time during six months of the year, but very warm summers (Figure 10.37).

3 The polar regions

Common to both polar regions is the semi-annual alternation between polar night and polar day, and the prevalence of snow and ice surfaces. These factors control the surface energy budget regimes and low annual temperatures (see Chapter 10B). The polar regions are also energy sinks for the global atmospheric circulation (see Chapter 7C.1)

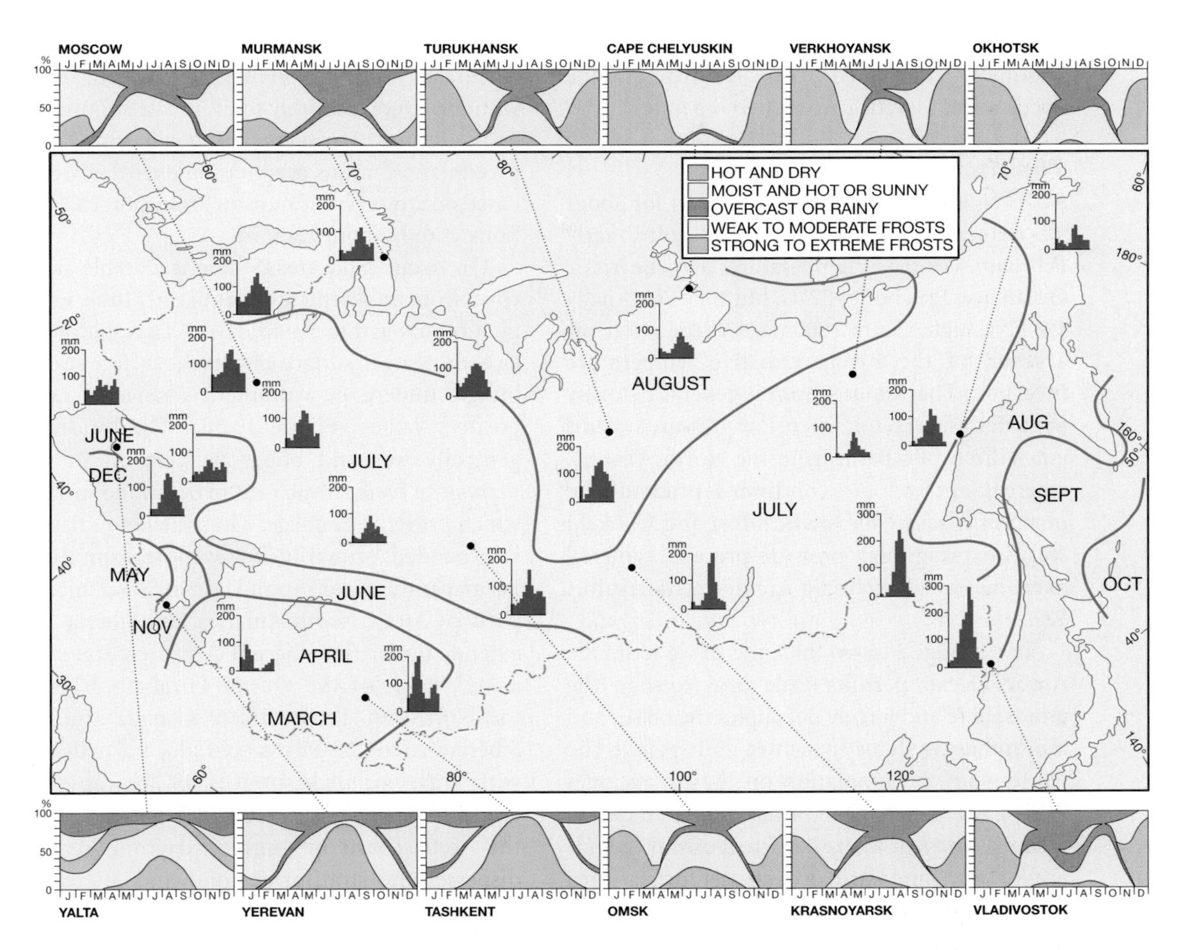

Figure 10.37 Months of maximum precipitation, annual regimes of mean monthly precipitation and annual regimes of mean monthly frequencies of five main weather types in the former USSR.

Source: Reprinted from P. E. Lydolph (1977) with kind permission from Elsevier Science NL, Sara Burgerhartstraat 25, 1055 KV Amsterdam, the Netherlands.

and in both cases they are overlain by large-scale circulation vortices in the mid-troposphere and above (see Figures 7.3 and 7.4). In many other respects, the two polar regions differ markedly owing to geographical factors. The north polar region comprises the Arctic Ocean, with its almost year-round sea ice cover (see Plate 13.5), surrounding tundra land areas, the Greenland Ice Sheet and numerous smaller ice caps in Arctic Canada, Svalbard and the Siberian Arctic Islands. In contrast, the south polar region is occupied by the Antarctic continent, with an ice plateau 3 to 4km high, floating ice shelves in the Ross Sea and Weddell Sea embayments, and surrounded by a seasonally ice-covered ocean. Accordingly, the Arctic and Antarctica are treated separately.

The Arctic

At 75°N, the sun is below the horizon for about 90 days, from early November until early February. Winter air temperatures over the Arctic Ocean average about –32°C, but they are usually 10–12°C higher some 1000m above the surface as a result of the strong radiative temperature inversion. The winter season is generally stormy in the Eurasian sector, where low pressure systems enter the Arctic Basin from the North Atlantic, whereas anticyclonic conditions predominate north of Alaska over the Beaufort and Chukchi seas. In spring, high pressure prevails, centered over the Canadian Arctic Archipelago–Beaufort Sea.

The average 2 to 4m thickness of sea ice in the Arctic Ocean permits little heat loss to the atmosphere and largely decouples the ocean and atmosphere systems in winter and spring. The winter snow accumulation on the ice averages 0.25–0.30m depth. Only when the ice fractures, forming a *lead*, or where persistent offshore winds and/or upwelling warm ocean water form an area of open water and new ice (called a *polynya*), is the insulating effect of sea ice disrupted. The ice in the western Arctic circulates clockwise in a gyre driven by the mean anticyclonic pressure field. Ice from the northern margin of this gyre, and ice from the Eurasian sector, moves across the North Pole in the Transpolar Drift Stream and exits the Arctic via Fram Strait and the East Greenland Current (see Figure 10.35A). This export largely balances the annual thermodynamic ice growth in the Arctic Basin. In late summer, the Eurasian shelf seas and the coastal section of the Beaufort Sea are mostly ice-free.

In summer, the Arctic Ocean has mostly overcast conditions with low stratus and fog. Snowmelt and extensive meltwater puddles on the ice keep air temperatures around freezing. Low pressure systems tend to predominate, entering the basin from either the North Atlantic or Eurasia. Precipitation may fall as rain or snow, with the largest monthly totals in late summer to early autumn. However, the mean annual net precipitation minus evaporation over the Arctic, based on atmospheric moisture transport calculations, is only about 180mm.

On Arctic land areas, there is a stable snow cover from mid-September until early June, when melt occurs within 10–15 days. As a result of the large decrease in surface albedo, the surface energy budget undergoes a dramatic change to large positive values (Figure 10.38). The tundra is generally wet and boggy as a result of the *permafrost table* only 0.5–1.0m below the surface, which prevents drainage. Thus the net radiation is expended primarily for evapotranspiration. Permanently frozen ground is over 500m thick in parts of Arctic North America and Siberia and extends under the adjacent Arctic coastal shelf areas. Much of the Queen Elizabeth Islands, the Northwest Territories of Canada and the Siberian Arctic Islands is cold, dry polar desert, with gravel or rock surfaces, or ice caps and glaciers. Nevertheless, 10–20km inland from the Arctic coasts in summer, daytime heating disperses the stratiform cloud and afternoon temperatures may rise to 15–20°C.

The Greenland ice sheet, 3km thick and covering an area of 1.7 million km^2, contains enough water to raise global sea level by over 7m if it were all to melt. However, there is no melting

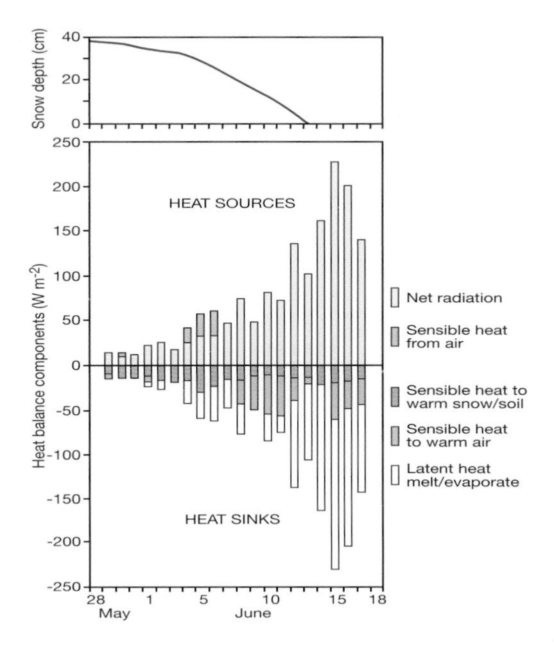

Figure 10.38 The effect of tundra snow cover on the surface energy budget at Barrow, Alaska, during the spring melt. The lower graph shows the daily net radiation and energy terms.

Source: Weller and Holmgren (1974). By permission of the American Meteorological Society.

above the equilibrium line altitude (where accumulation balances ablation), which is at about 2000m (1000m) elevation in the south (north) of Greenland. The ice sheet largely creates its own climate. It deflects cyclones moving from Newfoundland, either northward into Baffin Bay or northeastward towards Iceland. These storms give heavy snowfall in the south and on the western slope of the ice sheet. A persistent shallow inversion overlays the ice sheet with downslope katabatic winds averaging 10m s^{-1}, except when storm systems cross the area.

Antarctica

Except for protruding peaks in the Transantarctic Mountains and Antarctic Peninsula, and the Dry Valleys of Victoria Land (77°S, 160°E), over 97 percent of Antarctica is covered by a vast continental ice sheet. The ice plateau averages 1800m elevation in West Antarctica and 2600m in East Antarctica, where it rises above 4000m (82°S, 75°E). In September, sea ice averaging 0.5–1.0m in thickness covers 20 million km^2 of the Southern Ocean, but 80 percent of this melts each summer.

Over the ice sheet, temperatures are almost always well below freezing. The South Pole (2800m elevation) has a mean summer temperature of –28°C and a winter temperature of –58°C. Vostok (3500m) recorded –89°C in July 1983, a world record minimum. Mean monthly temperatures are consistently close to their winter value for the six months between equinoxes, creating a so-called 'coreless winter' pattern of annual temperature variation (Figure 10.39). Atmospheric poleward energy transfer balances the radiative loss of energy. Nevertheless, there are considerable day-to-day temperature changes associated with cloud cover increasing downward long wave radiation, or winds mixing warmer air from above the inversion down to the surface. Over the plateau, the inversion strength is about 20–25°C. Precipitation is almost impossible to measure as a result of blowing and drifting snow. Snow pit studies indicate an annual accumulation varying from less than 50mm over the high plateaux above 3000m elevation to 500–800mm in some coastal areas of the Bellingshausen Sea and parts of East Antarctica.

Lows in the southern westerlies have a tendency to spiral clockwise towards Antarctica, especially from south of Australia towards the Ross Sea, from the South Pacific towards the Weddell Sea, and from the western South Atlantic towards Kerguelen Island and East Antarctica (Figure 10.40). Over the adjacent Southern Ocean, cloudiness exceeds 80 percent year-round at 60–65°S (see Figures 3.8 and 4.13) due to the frequent cyclones, but coastal Antarctica has more synoptic variability, associated with alternating lows and highs. Over the interior, cloud cover is generally less than 40–50 percent and half of this amount in winter.

The poleward air circulation in the tropospheric polar vortex (see Figure 7.3) leads to

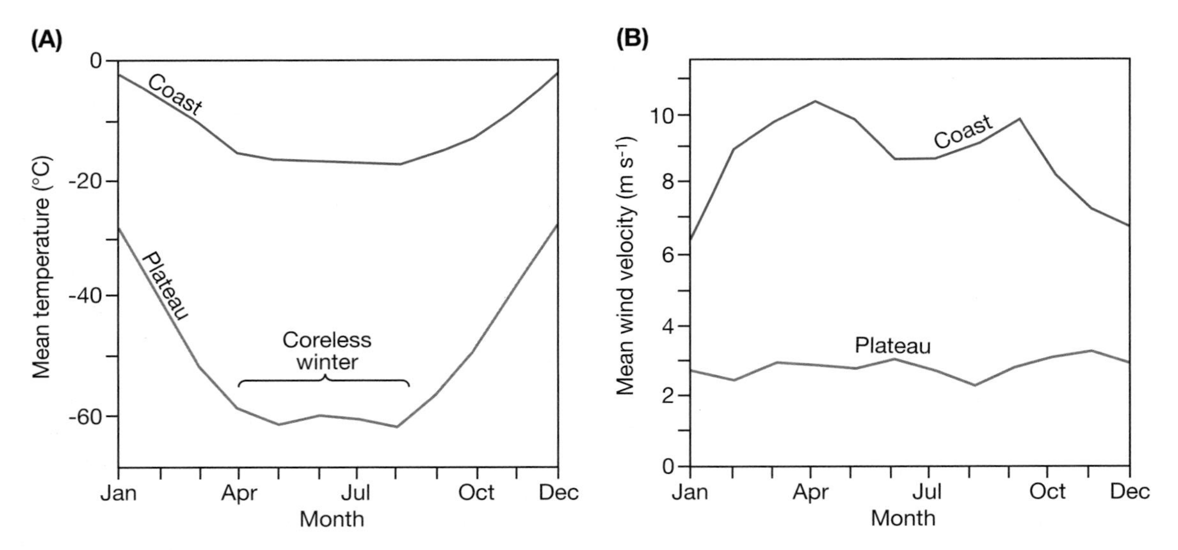

Figure 10.39 Annual course of (A) mean monthly air temperature (°C) and (B) wind speed (m s^{-1}) for 1980–1989 at Dome C (3280m), 74.5°S, 123.0°E (plateau) and D-10, an automatic weather station at 240m, 66.7°S, 139.8°E (coast).

Source: Stearns *et al.* (1993). By permission of the American Geophysical Union.

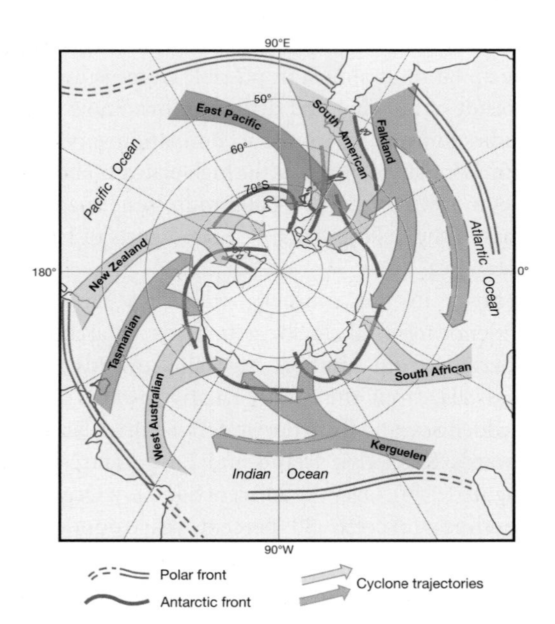

subsiding air over the Antarctic Plateau and outward flow over the ice sheet surface. The winds represent a balance between gravitational acceleration, Coriolis force (acting to the left), friction and inversion strength. On the slopes of the ice sheet, there are stronger downslope katabatic flows, and extreme speeds are observed in some coastal locations. Cape Denison (67°S, 143°E), Adelie Land, recorded average daily wind speeds of >18m s^{-1} on over 60 percent of days in 1912–1913.

Figure 10.40 Southern Hemisphere cyclone paths affecting Antarctica and major frontal zones in winter. 1. Polar Front; 2. Antarctic Front; 3. Cyclone trajectories.

Source: Carleton (1987). Copyright © Chapman & Hall, New York. Reproduced by permission.

SUMMARY

Seasonal changes in the Icelandic low and the Azores high, together with variations in cyclone activity, control the climate of Western Europe. The eastward penetration of maritime influences related to these atmospheric processes, and to the warm waters of the North Atlantic Current, is illustrated by mild winters, the seasonality of precipitation regimes and indices of continentality. Topographic effects on precipitation, snowfall, length of growing seasons and local winds are particularly marked over the Scandinavian Mountains, the Scottish Highlands and the Alps. Weather types in the British Isles may be described in terms of seven basic airflow patterns, the frequency and effects of which vary considerably with season. Recurrent weather spells about a particular date (singularities), such as the tendency for anticyclonic weather in mid-September, have been recognized in Britain and Europe and major seasonal trends in occurrence of airflow regimes may be used to define five natural seasons. Abnormal weather conditions (synoptic anomalies) are associated particularly with blocking anticyclones, which are especially prevalent over Scandinavia and may give rise to cold, dry winters and warm, dry summers.

The climate of North America is similarly affected by pressure systems that generate air masses of varying seasonal frequency. In winter, the subtropical high pressure cell extends north over the Great Basin with anticyclonic cP air to the north over Hudson Bay. Major depression belts occur at about 45–50°N, from the central USA to the St Lawrence, and along the east coast to Newfoundland. The Arctic Front is located over northwest Canada, the Polar Front lies along the northeast coast of the United States, and between the two a maritime (arctic) front may occur over Canada. In summer, the frontal zones move north, the Arctic Front lying along the north coast of Alaska, Hudson Bay and the St Lawrence being the main locations of depression tracks. Three major North American singularities concern the advent of spring in early March, the midsummer northward displacement of the subtropical high pressure cell, and the Indian summer of September to October. In western North America, the Coast Ranges inhibit the eastward spread of precipitation, which may vary greatly locally (e.g., in British Columbia), especially as regards snowfall. The strongly continental interior and east of the continent experiences some moderating effects of Hudson Bay and the Great Lakes in early winter, but with locally significant snow belts. The climate of the east coast is dominated by continental pressure influences. Cold spells are produced by winter outbreaks of high-latitude cA/cP air in the rear of cold fronts. Westerly airflow gives rise to chinook winds in the lee of the Rockies. The major moisture sources of the Gulf of Mexico and the North Pacific produce regions of differing seasonal regime: the winter maximum of the west coast is separated by a transitional intermontane region from the interior, with a general warm season maximum; the northeast has a relatively even seasonal distribution. Moisture gradients, which strongly influence vegetation and soil types, are predominantly east–west in central North America, in contrast to the north–south isotherm pattern.

The semi-arid southwestern United States comes under the complex influence of the Pacific and Bermudas high pressure cells, having extreme rainfall variations, with winter and summer maxima mainly due to depression and local thunderstorms, respectively. The interior and east coast of the United States is dominated by westerlies in winter and southerly thundery airflows in summer. Hurricanes are a major element of the summer–autumn climate of the Gulf Coast and southeast United States.

The subtropical margin of Europe consists of the Mediterranean region, lying between the belts dominated by the westerlies and the Saharan–Azores high pressure cells. The collapse of the Azores high pressure cell in October allows depressions to move and form over the relatively warm Mediterranean Sea, giving well-marked orographic winds (e.g., mistral) and stormy, rainy winters. Spring is an unpredictable season marked by the collapse of the Eurasian high pressure cell to the north and the strengthening of the Saharan–Azores anticyclone. In summer, the latter gives dry, hot conditions with strong local southerly airstreams (e.g., scirocco). The simple winter rainfall maximum is most characteristic of the eastern and southern Mediterranean, whereas in the north and west, autumn and spring rains become more important. North Africa is dominated by high pressure conditions. Infrequent rainfall may occur in the north with extratropical systems and to the south with Saharan depressions.

Australian weather is determined largely by traveling anticyclone cells from the southern Indian Ocean and intervening low pressure troughs and fronts. In the winter months, such frontal troughs give rains in the southeast. The climatic controls in New Zealand are similar to those in southern Australia, but South Island is greatly influenced by depressions in the southern westerlies. Rainfall amounts vary strongly with the relief.

The southern westerlies (30–40° to 60–70°S) dominate the weather of the Southern Ocean. The strong, mean zonal flow conceals great day-to-day synoptic variability and frequent frontal passages. The persistent low pressure systems in the Antarctic trough produce the highest year-round zonally averaged global cloudiness.

The Arctic margins have six to nine months of snow cover and extensive areas of permanently frozen ground (permafrost) in the continental interiors, whereas the maritime regions of northern Europe and northern Canada–Alaska have cold, stormy winters and cloudy, milder summers influenced by the passage of depressions. Northeast Siberia has an extreme continental climate.

The Arctic and Antarctica differ markedly owing to the types of surface – a perennially ice-covered Arctic Ocean surrounded by land areas and a high Antarctic ice plateau surrounded by the Southern Ocean and thin seasonal sea ice. The Arctic is affected by mid-latitude cyclones from the North Atlantic and in summer from northern Asia. A surface inversion dominates Arctic conditions in winter and year-round over Antarctica. In summer, stratiform cloud blankets the Arctic and temperatures are near 0°C. Sub-zero temperatures persist year-round on the Antarctic continent and katabatic winds dominate the surface climate. Precipitation amounts are low, except in a few coastal areas, in both polar regions.

DISCUSSION TOPICS

- Compare the climatic conditions in maritime and continental locations in the major continents, and in your own region of the world, using available station data from reference works or the Web.
- Consider how major topographic barriers in the Americas, Western Europe, New Zealand and so on modify the patterns of temperature and precipitation in those regions.
- Examine the seasonal distribution of precipitation in different parts of the Mediterranean Basin and consider the reasons for departures from the classical view of a wet winter/dry summer regime.
- Examine the spatial extent of 'Mediterranean-type' climates in other continents and the reasons for these conditions.
- Compare the climatic characteristics and controls of the two polar regions.
- What are the primary causes of the world's major deserts?

REFERENCES AND FURTHER READING

Books

Blüthgen, J. (1966) *Allgemeine Klimageographic*, 2nd edn, W. de Gruyter, 720pp.

Bryson, R. A. and Hare, F. K. (eds) (1974) *Climates of North America*, World Survey of Climatology 11, Elsevier, Amsterdam, 420pp. [Thorough account of the circulation systems and climatic processes; climates of Canada, the USA and Mexico are treated individually; numerous statistical data tables]

Bryson, R. A. and Lahey, J. F. (1958) *The March of the Seasons*, Meteorological Department, University of Wisconsin, 41 pp.

Chagnon, S. A. (ed) (1996) *The Great Flood of 1993*, Westview Press, Boulder, CO, 321pp. [Account of the Mississippi floods of 1993]

Chandler, T. J. and Gregory, S. (eds) (1976) *The Climate of the British Isles*, Longman, London, 390pp. [Detailed treatment by element as well as synoptic climatology, climate change, coastal, upland and urban climates; many tables and references]

Durrenberger, R. W. and Ingram, R. S. (1978) *Major Storms and Floods in Arizona 1862–1977*, State of Arizona, Office of the State Climatologist, Climatological Publications, Precipitation Series No. 4, 44 pp.

Environmental Science Services Administration (1968) *Climatic Atlas of the United States*, US Department of Commerce, Washington, DC, 80pp.

Evenari, M., Shanan, L. and Tadmor, N. (1971) *The Negev*, Harvard University Press, Cambridge, MA, 345pp. [Climate and environment of the Negev Desert]

Flohn, H. (1954) *Witterung und Klima in Mitteleuropa*, Zurich, 218pp. [Synoptic climatological approach to European climatic conditions]

Gentilli, J. (ed.) (1971) *Climates of Australia and New Zealand, World Survey of Climatology* 13, Elsevier, Amsterdam, 405pp. [Standard climatology including air masses and synoptic systems]

Goudie, A. and Wilkinson, J. (1977) *The Warm Desert Environment*, Cambridge University Press, Cambridge, 88pp.

Green, C. R. and Sellers, W. D. (1964) *Arizona Climate*, University of Arizona Press, Tucson, 503pp. [Details on the climatic variability in the State of Arizona]

Hare, F. K. and Thomas, M. K. (1979) *Climate Canada* (2nd edn), Wiley, Canada, 230pp.

Henderson-Sellers, A. and Robinson, P. J. (1999) *Contemporary Climatology*, 2nd edn., Longman, London, 317pp.

Hulme, M. and Barrow, E. (eds) (1997) *Climates of the British Isles. Present, Past and Future*, Routledge, London, 454pp. [Treats overall modern climatic conditions in terms of synoptic climatology, based on H. H. Lamb; reconstruction of historical conditions, and future projections; many useful data tables]

Keen, R. A. (2004) *Skywatch West. The Complete Weather Guide*, Fulcrum, Golden, CO, 272pp. [A popular guide to the weather of the western United States]

Linacre, W. and Hobbs, J. (1977) *The Australian Climatic Environment*, Wiley, Brisbane, 354pp. [Much broader than its title; presents weather and climate from a Southern Hemisphere perspective, including chapters on the climates of the Southern Hemipshere as well as of Australia; a chapter on climatic change and four chapters on applied climatology]

Lydolph, P. E. (1977) *Climates of the Soviet Union, World Survey of Climatology* 7, Elsevier, Amsterdam, 435pp. [The most comprehensive survey of climate for this region in English; numerous tables of climate statistics]

Manley, G. (1952) *Climate and the British Scene*, Collins, London, 314pp. [Classic description of British climate and its human context]

Meteorological Office (1952) *Climatological Atlas of the British Isles*, MO 488, HMSO, London, 139pp.

Meteorological Office (1962) *Weather in the Mediterranean I, General Meteorology* (2nd edn). MO 391, HMSO, London, 362pp.

Meteorological Office (1964a) *Weather in the Mediterranean II* (2nd edn), MO 391b, HMSO, London , 372 pp. [Classic handbook]

Meteorological Office (1964b) *Weather in Home Fleet Waters I, The Northern Seas*, Part 1, MO 732a, HMSO, London, 265pp.

Palz, W. (ed.) (1984) *European Solar Radiation Atlas*, 2 vols (2nd edn), Verlag Tüv Rheinland, Cologne, 297 and 327pp.

Rayner, J. N. (1961) *Atlas of Surface Temperature Frequencies for North America and Greenland*, Arctic Meteorological Research Group, McGill University, Montreal.

Schwerdtfeger, W. (1984) *Weather and Climate of the Antarctic*, Elsevier, Amsterdam, 261 pp. [A specialized work covering radiation balance and temperature, surface winds, circulation and disturbances, moisture budget components and ice-mass budget]

Serreze, M.C. and Barry, R.G. (2005) *The Arctic Climate System,* Cambridge University Press, Cambridge, 385pp. [An up-to-date overview of Arctic climate discussing energy and moisture balances, circulation, regional climates, sea ice, paleoclimate, recent trends, and climate projections]

Sturman, A. P. and Tapper, N. J. (1996) *The Weather and Climate of Australia and New Zealand,* Oxford University Press, Oxford, 496pp. [Undergraduate text on basic processes of weather and climate in the regional context of Australia-New Zealand; covers the global setting, synoptic and sub-synoptic systems, and climate change]

Thompson, R. S., Anderson, K. H. and Bartlein, P. J. (1999) Climate – vegetation atlas of North America. *US Geological Survey Professional Paper* 1650 A and B.

Thorn, P. (1965) *The Agro-climatic Atlas of Europe.* Elsevier, Amsterdam.

Trewartha, G. T. (1981) *The Earth's Problem Climates,* 2nd edn, University of Wisconsin Press, Madison, 371pp.

Troen, I. and Petersen, E. L. (1989) *European Wind Atlas,* Commission of the European Communities, Risø National Laboratory, Roskilde, Denmark, 656 pp.

United States Weather Bureau (1947) *Thunderstorm Rainfall,* Vicksburg, MI, 331pp.

Visher, S. S. (1954) *Climatic Atlas of the United States,* Harvard University Press, Cambridge, MA, 403pp.

Wallén, C. C. (ed.) (1970) *Climates of Northern and Western Europe, World Survey of Climatology* 5, Elsevier, Amsterdam (253 pp.). [Standard climatological handbook]

Articles

Adam, D. K. and Comrie, A. C. (1997) The North American monsoon. *Bull. Amer. Met. Soc.* 78, 2197–213.

Axelrod, D. I. (1992) What is an equable climate? *Palaeogeogr., Palaeoclim., Palaeoecol.* 91, 1–12.

Balling, R. C., Jr. (1985) Warm seasonal nocturnal precipitation in the Great Plains of the United States. *J. Climate Appl. Met.* 24, 1383–7.

Barros, A. P. and Lettenmaier, D. P. (1994). Dynamic modeling of orographically induced precpitation. *Rev. Geophysics* 32, 265–94.

Barry, R. G. (1963) Aspects of the synoptic climatology of central south England. *Met. Mag.* 92, 300–8.

Barry, R. G. (1967) The prospects for synoptic climatology: a case study, in Steel, R. W. and Lawton, R. (eds) *Liverpool Essays in Geography,* Longman, London, 85–106.

Barry, R. G. (1973) A climatological transect on the east slope of the Front Range, Colorado. *Arct. Alp. Res.* 5, 89–110.

Barry, R. G. (1983) Arctic Ocean ice and climate: perspectives on a century of polar research. *Ann. Assn Amer. Geog.* 73(4), 485–501.

Barry, R. G. (1986) Aspects of the meteorology of the seasonal sea ice zone, in Untersteiner, N. (ed.) *The Geophysics of Sea Ice,* Plenum Press, New York, 993–1020.

Barry, R. G. (1996) Arctic, in Schneider, S. H. (ed.) *Encyclopedia of Climate and Weather,* Oxford University Press, New York, 43–7.

Barry, R. G. (2002) Dynamic and synoptic climatology, in Orme, A. R. (ed.) *The Physical Geography of North America,* Oxford University Press, Oxford, 98–111.

Barry, R. G. and Hare, F. K. (1974) Arctic climate, in Ives, J. D. and Barry, R. G. (eds) *Arctic and Alpine Environments,* Methuen, London, 17–54.

Belasco, J. E. (1952) Characteristics of air masses over the British Isles, Meteorological Office. *Geophysical Memoirs* 11(87) (34pp.).

Blackall, R. M. and Taylor, P. L. (1993) The thunderstorms of 19/20 August 1992 – a view from the United Kingdom. *Met. Mag.* 122, 189.

Blake, E.S., Rappaport, E.N. and Landsea, C.W. (2007) The deadliest, costliest, and most intense United States tropical cyclones from 1851 to 2006 (and other frequently requested hurricane facts). NOAA Tech. Mem., NWS TPC-5 (43pp.).

Boast, R. and McQuingle, J. B. (1972) Extreme weather conditions over Cyprus during April 1971, *Met. Mag.* 101, 137–53.

Borchert, J. (1950) The climate of the central North American grassland. *Ann. Assn Amer. Geog.* 40, 1–39.

Browning, K. A. and Hill, F. F. (1981) Orographic rain. *Weather* 36, 326–9.

Bryson, R. A. (1966) Air masses, streamlines and the boreal forest. *Geog. Bull.* 8, 228–69.

Burbridge, F. E. (1951) The modification of continental polar air over Hudson Bay. *Quart. J. Met. Soc.* 77, 365–74.

Butzer, K. W. (1960) Dynamic climatology of large-scale circulation patterns in the Mediterranean area. *Meteorologische Rundschau* 13, 97–105.

Carleton, A. M. (1986) Synoptic-dynamic character of 'bursts' and 'breaks' in the southwest US summer precipitation singularity. *J. Climatol.* 6, 605–23.

Carleton, A. M. (1987) Antarctic climates, in Oliver, J. E. and Fairbridge, R. W. (eds) *The Encyclopedia of Climatology*, Van Nostrand Reinhold, New York, 44–64.

Chinn, T. J. (1979) How wet is the wettest of the West Coast? *New Zealand Alp. J.* 32, 84–7.

Climate Prediction Center (1996) Jet streams, pressure distribution and climate for the USA during the winters of 1995–6 and 1994–5. *The Climate Bull.* 96(3), US Department of Commerce.

Cooter, E. J. and Leduc, S. K. (1995) Recent frost date trends in the north-eastern USA. *Int. J. Climatology* 15, 65–75.

Derecki, J. A. (1976) Heat storage and advection in Lake Erie. *Water Resources Research* 12(6), 1144–50.

Douglas, M. W. *et al.* (1993) The Mexican monsoon. *J. Climate* 6(8), 1665–77.

Driscoll, D. M. and Yee Fong, J. M. (1992) Continentality: a basic climatic parameter re-examined. *Int. J. Climatol.* 12, 185–92.

Easterling, D. R. and Robinson, P. J. (1985) The diurnal variation of thunderstorm activity in the United States. *J. Climate Appl. Met.* 24, 1048–58.

Elsom, D. M. and Meaden, G. T. (1984) Spatial and temporal distribution of tornadoes in the United Kingdom 1960–1982. *Weather* 39, 317–23.

Ferguson, E. W., Ostby, F. P., Leftwich, P. W., Jr. and Hales, J. E., Jr. (1986) The tornado season of 1984. *Monthly Weather Review* 114, 624–35.

Forrest, B. and Nishenko, S. (1996) Losses due to natural hazards. *Natural Hazards Observer* 21(1), University of Colorado, Boulder, CO, 16–17.

Goodrich, G.R. and Ellis, A.W. (2008) Climatic controls and hydrologic impacts of a recent extreme seasonal precipitation reversal in Arizona. *J. Appl. Met.Climate* 47, 498–508.

Gorcynski, W. (1920) Sur le calcul du degré du continentalisme et son application dans la climatologie. *Geografiska Annaler* 2, 324–31.

Hales, J. E., Jr. (1974) South-western United States summer monsoon source – Gulf of Mexico or Pacific Ocean. *J. Appl. Met.* 13, 331–42.

Hare, F. K. (1968) The Arctic. *Quart. J. Roy. Met. Soc.* 74, 439–59.

Hawke, E. L. (1933) Extreme diurnal range of air temperature in the British Isles. *Quart. J. Roy. Met. Soc.* 59, 261–5.

Hill, F. F., Browning, K. A. and Bader, M. J. (1981) Radar and rain gauge observations of orographic rain over South Wales. *Quart. J. Roy. Met. Soc.* 107, 643–70.

Horn, L. H. and Bryson, R. A. (1960) Harmonic analysis of the annual march of precipitation over the United States. *Ann. Assn Amer. Geog.* 50, 157–71.

Hulme, M. *et al.* (1995) Construction of a 1961–1990 European climatology for climate change modelling and impact applications. *Int. J. Climatol.* 15, 1333–63.

Huttary, J. (1950) Die Verteilung der Niederschläge auf die Jahreszeiten im Mittelmeergebiet. *Meteorologische Rundschau* 3, 111–19.

Klein, W. H. (1963) Specification of precipitation from the 700mb circulation. *Monthly Weather Review* 91, 527–36.

Knappenberger, P. C. and Michaels, P. J. (1993) Cyclone tracks and wintertime climate in the mid-Atlantic region of the USA. *Int. J. Climatol.* 13, 509–31.

Knight, D. B. and Davis, R. E. (2007) Climatology of tropical cyclone rainfall in the southeastern United States. *Phys. Geogr.* 18, 126–47.

Knox, J. L. and Hay, J. E. (1985) Blocking signatures in the northern hemisphere: frequency distribution and interpretation. *J. Climatology* 5, 1–16.

Lamb, H. H. (1950) Types and spells of weather around the year in the British Isles: annual trends, seasonal structure of the year, singularities. *Quart. J. Roy. Met. Soc.* 76, 393–438.

Leffler, R. J. *et al.* (2002) Evaluation of a national seasonal snowfall record at the Mount Baker, Washington, ski area. *Nat. Wea. Digest* 25, 15–20.

Longley, R. W. (1967) The frequency of Chinooks in Alberta. *The Albertan Geographer* 3, 20–2.

Lott, J. N. (1994) The US summer of 1993: a sharp contrast in weather extremes. *Weather* 49, 370–83.

Lumb, F. E. (1961) *Seasonal variations of the sea surface temperature in coastal waters of the British Isles*, Met. Office Sci. Paper No. 6, MO 685 (21pp.).

McGinnigle, J.B. (2002) The 1952 Lynmouth floods revisited. *Weather* 57(7), 235–41.

Manley, G. (1944) Topographical features and the climate of Britain. *Geog. J.* 103, 241–58.

Manley, G. (1945) The effective rate of altitude change in temperate Atlantic climates. *Geog. Rev.* 35, 408–17.

Mather, J. R. (1985) The water budget and the distribution of climates, vegetation and soils.

Publications in Climatology 38(2), Center for Climatic Research, University of Delaware, Newark (36pp.).

Maytham, A. P. (1993) Sea ice – a view from the Ice Bench. *Met. Mag.* 122, 190–5.

Namias, J. (1964) Seasonal persistence and recurrence of European blocking during 1958–60. *Tellus* 16, 394–407.

Nicholson, S. E. and Flohn H. (1980) African environmental and climatic changes and the general atmospheric circulation in late Pleistocene and Holocene. *Climatic Change* 2, 313–48.

Nickling, W. G. and Brazel, A. J. (1984) Temporal and spatial characteristics of Arizona dust storms (1965–1980). *Climatology* 4, 645–60.

Nkemdirim, L. C. (1996) Canada's chinook belt. *Int. J. Climatol.* 16(4), 427–39.

O'Hare, G. and Sweeney, J. (1993) Lamb's circulation types and British weather: an evaluation. *Geography* 78, 43–60.

Parrett, C., Melcher, N. B. and James, R. W., Jr. (1993) Flood discharges in the upper Mississippi River basin. *U.S. Geol. Sur. Circular* 1120-A (14 pp.).

Peilke, R., Jr. and Carbone, R. E. (2002) Weather impacts, forecasts and policy: An integrated perspective. *Bull. Amer. Met. Soc.*, 83(3), 383–403.

Poltaraus, B. V. and Staviskiy, D. B. (1986) The changing continentality of climate in central Russia. *Soviet Geography* 27, 51–8.

Rex, D. F. (1950–1951) The effect of Atlantic blocking action upon European climate. *Tellus* 2, 196–211 and 275–301; 3, 100–11.

Salinger, M. J., Basher, R. E., Fitzharris, B. B., Hay, J. E., Jones, P. D., McVeigh, J. P. and Schmidely-Leleu, I. (1995) Climate trends in the southwest Pacific. *Int. J. Climatol.* 15, 285–302.

Schick, A. P. (1971) A desert flood. *Jerusalem Studies in Geography* 2, 91–155.

Schwartz, M. D. (1995) Detecting structural climate change: an air mass-based approach in the north-central United States, 1958–92. *Ann. Assn Amer. Geog.* 76, 553–68.

Schwartz, R. M. and Schmidlin, T. W. (2002) Climatology of blizzards in the conterminous United States, 1959–2000. *J. Climate* 15(13), 1765–72.

Sellers, P. *et al.* (1995) The boreal ecosystem–atmosphere study (BOREAS): an overview and early results from the 1994 field year. *Bull. Am. Met. Soc.* 76, 1549–77.

Serreze, M. C. *et al.* (1993) Characteristics of arctic synoptic activity, 1952–1989. *Met. Atmos. Phys.* 51, 147–64.

Shaw, E. M. (1962) An analysis of the origins of precipitation in Northern England, 1956–60. *Quart. J. Roy. Met. Soc.* 88, 539–47.

Sheppard, P. R. *et al.* (2002) The climate of the US Southwest. *Clim. Res.*, 21(3), 219–38.

Sivall, T. (1957) Sirocco in the Levant. *Geografiska Annaler* 39, 114–42.

Stearns, C. R. *et al.* (1993) Mean cluster data for Antarctic weather studies, in Bromwich, D. H. and Stearns, C. R. (eds) *Antarctic Meteorology and Climatology: Studies Based on Automatic Weather Stations*, Antarctic Research Series, Am. Geophys. Union 61, 1–21.

Stone, J. (1983) Circulation type and the spatial distribution of precipitation over central, eastern and southern England. *Weather* 38, 173–7, 200–5.

Storey, A. M. (1982) A study of the relationship between isobaric patterns over the UK and central England temperature and England–Wales rainfall. *Weather* 37, 2–11, 46, 88–9, 122, 151, 170, 208, 244, 260, 294, 327, 360.

Sumner, E. J. (1959) Blocking anticyclones in the Atlantic–European sector of the northern hemisphere, *Met. Mag.* 88, 300–11.

Sweeney, J. C. and O'Hare, G. P. (1992) Geographical variations in precipitation yields and circulation types in Britain and Ireland. *Trans. Inst. Brit. Geog.* (n.s.) 17, 448–63.

Thomas, M. K. (1964) *A Survey of Great Lakes Snowfall*, Great Lakes Research Division, University of Michigan, Publication No. 11, 294–310.

Thornthwaite, C. W. and Mather, J. R. (1955) The moisture balance. *Publications in Climatology* 8(1), Laboratory of Climatology, Centerton, NJ, 104pp.

Tout, D. G. and Kemp, V. (1985) The named winds of Spain. *Weather* 40, 322–9.

Trenberth, K. E. and Guillemot, C. J. (1996) Physical processes involved in the 1988 drought and 1993 floods in North America. *J. Climate* 9(6), 1288–98.

Villmow, J. R. (1956) The nature and origin of the Canadian dry belt. *Ann. Assn Amer. Geog.* 46, 221–32.

Wallace, J. M. (1975) Diurnal variations in precipitation and thunderstorm frequency over the coterminous United States. *Monthly Weather Review* 103, 406–19.

Wallén, C. C. (1960) Climate, in Somme, A. (ed.) *The Geography of Norden*, Cappelens Forlag, Oslo, 41–53.

Walters, C. K. *et al.* (2008) A long-term climatology of southerly and northerly low-level jets for the central United States. *Annals Assoc. Amer. Geogr.*, 98, 521–52.

Weller, G. and Holmgren, B. (1974) The micro-climates of the arctic tundra. *J. App. Met.* 13(8), 854–62.

Woodroffe, A. (1988) Summary of the weather pattern developments of the storm of 15/16 October 1987. *Met. Mag.* 117, 99–103.

Wratt, D. S. *et al.* (1996) The New Zealand Southern Alps Experiment. *Bull. Amer. Met. Soc.* 77(4), 683–92.

CHAPTER ELEVEN

Tropical weather and climate

11

LEARNING OBJECTIVES

When you have read this chapter you will:

- understand the characteristics and significance of the intertropical convergence zone
- be familiar with the principal weather systems that occur in low latitudes and their distribution
- know some of the diurnal and local effects that influence tropical weather
- know where and how tropical cyclones tend to occur
- understand the basic mechanisms and characteristics of El Niño and La Niña events

Tropical climates are of especial geographical interest because 50 percent of the surface of the globe lies between latitudes 30°N and 30°S, and over 75 percent of the world's population inhabit climatically tropical lands. This chapter first describes the Trade Wind systems, the intertropical convergence zone and tropical weather systems. The major monsoon regimes are then examined and the climate of Amazonia. The effects of the alternating phases of the El Niño–Southern Oscillation in the equatorial Pacific Ocean is discussed as well as other causes of climatic variation in the tropics. Finally, the problems of forecasting tropical weather are briefly considered.

The latitudinal limits of tropical climates vary greatly with longitude and season, and tropical weather conditions may reach well beyond the Tropics of Cancer and Capricorn. For example, the summer monsoon extends to 30°N in South Asia, but only to 20°N in West Africa, while in late summer and autumn tropical hurricanes may affect 'extra-tropical' areas of East Asia and eastern North America. Not only do the tropical margins extend seasonally poleward, but in the zone between the major subtropical high pressure cells there is frequent interaction between temperate and tropical disturbances. Elsewhere and on other occasions, distinct tropical and mid-latitude storms are observed. In general, however, the tropical atmosphere is far from being a discrete entity and any meteorological or climatological boundaries must be arbitrary. There are, nevertheless, a number of distinctive features of tropical weather, as discussed below.

Several basic factors help to shape tropical weather processes and also affect their analysis and interpretation. First, the Coriolis parameter approaches zero at the equator, so that winds may depart considerably from geostrophic balance.

Pressure gradients are also generally weak, except for tropical storm systems. For these reasons, tropical weather maps usually depict streamlines, not isobars or geopotential heights. Second, temperature gradients are characteristically weak. Spatial and temporal variations in moisture content are much more significant diagnostic characteristics of climate. Third, diurnal land/sea breeze regimes play a major role in coastal climates, in part as a result of the almost constant day length and strong solar heating. There are also semi-diurnal pressure oscillations of 2–3mb, with minima around 04:00 and 16:00 hours and maxima around 10:00 and 22:00 hours. Fourth, the annual regime of incoming solar radiation, with the sun overhead at the equator in March and September and over the Tropics at the respective summer solstices, is reflected in the seasonal variations of rainfall at some stations. However, dynamic factors greatly modify this conventional explanation.

A THE INTERTROPICAL CONVERGENCE

The tendency for the Trade Wind systems of the two hemispheres to converge in the Equatorial (low pressure) Trough has already been noted (see Chapter 6B). Views on the exact nature of this feature have been subject to continual revision. From the 1920s to the 1940s, the frontal concepts developed in mid-latitudes were applied in the tropics, and the streamline confluence of the northeast and southeast trades was identified as the Intertropical Front (ITF). Over continental areas such as West Africa and South Asia, where in summer hot, dry continental tropical air meets cooler, humid equatorial air, this term has some limited applicability (Figure 11.1). Sharp temperature and moisture gradients may occur, but the front is seldom a weather-producing mechanism of the mid-latitude type. Elsewhere in low latitudes, true fronts (with a marked density contrast) are rare.

Recognition of the significance of wind field convergence in tropical weather production in the 1940–1950s led to the designation of the Trade Wind convergence as the Intertropical Convergence Zone (ITCZ). This feature is apparent on a mean streamline map, but areas of convergence grow and decay, either *in situ* or within disturbances moving westward (see Plate 11.1), over periods of a few days. Moreover, convergence is infrequent even as a climatic feature in the doldrum zones (see Figure 7.13). Satellite data show that over the oceans the position and intensity of the ITCZ varies greatly from day to day.

The ITCZ is predominantly an oceanic feature where it tends to be located over the warmest surface waters. Hence, small differences of sea surface temperature may cause considerable changes in the location of the ITCZ. A sea surface temperature of at least 27.5°C seems to provide a threshold for organized convective activity; above this temperature organized convection is

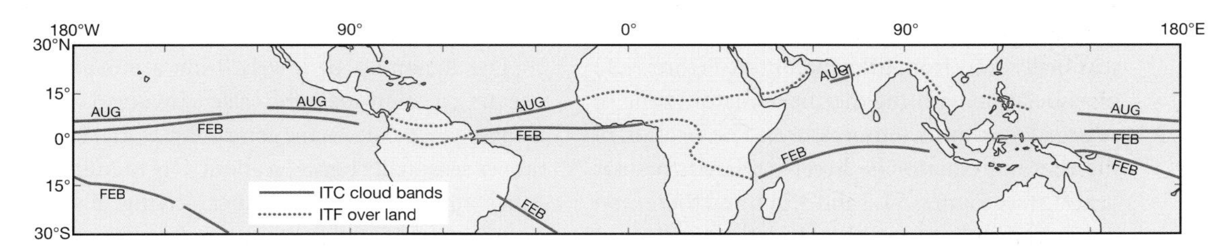

Figure 11.1 The position of the Equatorial Trough (Intertropical Convergence Zone or Intertropical Front in some sectors) in February and August. The cloud band in the southwest Pacific in February is known as the South Pacific Convergence Zone; over South Asia and West Africa the term Monsoon Trough is used.

Sources: After Saha (1973), Riehl (1954) and Yoshino (1969).

Plate 11.1 The ITCZ appears as a band of bright white clouds that cuts across the center of the image; this is a combination of cloud data from NOAA's GOES-11 and land cover classification data.
Source: GOES Project Science Office.

essentially competitive between different regions potentially available to form part of a continuous ITCZ. The convective rainfall belt of the ITCZ has very sharply defined latitudinal limits. For example, along the West African coast the following mean annual rainfalls are recorded:

12°N	1939mm
15°N	542mm
18°N	123mm

In other words, moving southwards into the ITCZ, precipitation increases by 440 percent in a meridional distance of only 330km.

As climatic features, the Equatorial Trough and the ITCZ are asymmetric around the equator, lying on average to the north. They also move seasonally away from the equator (see Figure 11.1) in association with the thermal equator (zone of seasonal maximum temperature). The location of the thermal equator is directly related to solar heating (see Figures 11.2 and 3.11), and there is an obvious link between this and the Equatorial Trough in terms of thermal lows. However, if the ITC were to coincide with the Equatorial Trough then this zone of cloudiness would decrease incoming solar radiation, reducing the surface heating needed to maintain the low pressure trough. In fact, this does not happen. Solar energy is available to heat the surface because the maximum surface wind convergence, uplift and cloud cover is commonly located several degrees equatorward of the trough. In the Atlantic (Figure 11.2B), for example, the cloudiness maximum is distinct from the Equatorial Trough in August. Figure 11.2 illustrates regional differences in the Equatorial Trough and ITCZ. Convergence of two Trade Wind systems occurs over the central North Atlantic in August and the eastern North Pacific in February. In contrast, the Equatorial Trough is defined by easterlies on its poleward side and westerlies on its equatorward side over West Africa in August and over New Guinea in February.

The dynamics of low-latitude atmosphere–ocean circulations are also involved. The Convergence Zone in the central equatorial Pacific moves seasonally between about 4°N in March to April and 8°N in September, giving a single pronounced rainfall maximum in March to April. This appears to be a response to the relative strengths of the northeast and southeast trades. The ratio of South Pacific/North Pacific Trade Wind strength exceeds 2 in September but falls to

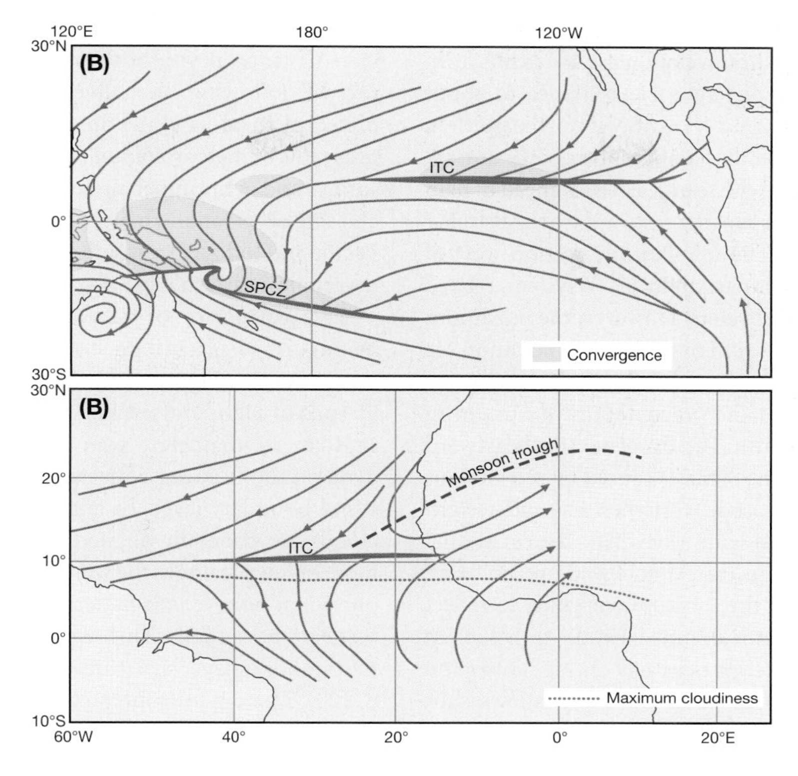

Figure 11.2 Illustrations of (A) streamline convergence forming an Intertropical Convergence (ITC) and South Pacific Convergence Zone (SPCZ) in February, and (B) the contrasting patterns of Monsoon Trough over West Africa, streamline convergence over the central tropical North Atlantic, and axis of maximum cloudiness to the south for August.

Sources: A: C. S. Ramage, personal communication (1986). B: From Sadler (1975a).

0.6 in April. Interestingly, the ratio varies in phase with the ratio of Antarctic–Arctic sea ice areas; Antarctic ice is at a maximum in September when Arctic ice is at its minimum. The convergence axis is often aligned close to the zone of maximum sea surface temperatures, but is not anchored to it. Indeed, the SST maximum located within the Equatorial Counter-current (see Figure 7.31) is a result of the interactions between the Trade Winds and horizontal and vertical motions in the ocean surface layer.

Aircraft studies show the complex structure of the central Pacific ITCZ. When moderately strong trades provide horizontal moisture convergence, convective cloud bands form, but the convergent lifting may be insufficient for rainfall in the absence of upper-level divergence. Moreover, although the southeast trades cross the equator, the mean monthly resultant winds between 115° and 180°W have, throughout the year, a more southerly component north of the equator and a more northerly one south of it, giving a zone of divergence (due to the sign change in the Coriolis parameter) along the equator.

In the southwestern sectors of the Pacific and Atlantic Oceans, satellite cloudiness studies indicate the presence of two semi-permanent confluence zones (see Figure 11.1). These do not occur in the eastern South Atlantic and South Pacific, where there are cold ocean currents. The South Pacific Convergence Zone (SPCZ) shown in the western South Pacific in February (summer) is now recognized as an important discontinuity and zone of maximum cloudiness. It extends from the eastern tip of Papua New Guinea to about 30°S, 120°W. At sea level, moist northeasterlies, west of the South Pacific subtropical anticyclone, converge with southeasterlies ahead of high pressure systems moving eastward from Australia/New Zealand. The low-latitude section west of 180° longitude is part of the ITCZ system, related to warm surface waters. However, the maximum precipitation is south of the axis of maximum sea surface temperature, and the surface convergence is south of the precipitation maximum in the central South Pacific. The southeastward orientation of the SPCZ is caused by interactions with the mid-latitude westerlies. Its southeastern end is associated with wave disturbances and jet stream clouds on the South Pacific polar front. The link across the subtropics appears to reflect upper-level tropical mid-latitude transfers of moisture and energy, especially during subtropical storm situations. Hence, the SPCZ shows substantial short-term and interannual variability in its location and development. The interannual variability is strongly associated with the phase of the Southern Oscillation (see p. 378). During the northern summer, the SPCZ is poorly developed, whereas the ITCZ is strong all across the Pacific. During the southern summer, the SPCZ is well developed, with a weak ITCZ over the western tropical Pacific. After April, the ITCZ strengthens over the western Pacific and the SPCZ weakens as it moves westward and equatorward. In the Atlantic, the ITCZ normally begins its northward movement in April to May, when South Atlantic sea surface temperatures start to fall and both the subtropical high pressure cell and the southeast trades intensify. In cold, dry years this movement may begin as early as February and in warm, wet years as late as June.

B TROPICAL DISTURBANCES

It was not until the 1940s that detailed accounts were given of types of tropical disturbances other than the long-recognized tropical cyclone. Our view of tropical weather systems was radically revised following the advent of operational meteorological satellites in the 1960s. Special programs of meteorological measurements at the surface and in the upper air, together with aircraft and ship observations, have been carried out in the Pacific and Indian Oceans, the Caribbean Sea and the tropical eastern Atlantic.

Five categories of weather system can be distinguished according to their space and timescales (see Figure 11.3). The smallest, with a lifespan of a few hours, is the individual cumulus, 1–10km in diameter, which is generated by dynamically induced convergence in the Trade Wind boundary layer. In fair weather, cumulus clouds are generally aligned in 'cloud streets', more or less parallel to the wind direction, or form polygonal honeycomb-pattern cells, rather than scattered at random. This seems to be related to the boundary-layer structure and wind speed (see p. 120). There is little interaction between the air layers above and below the cloud base under these conditions, but in disturbed weather conditions updrafts and downdrafts cause interaction between the two layers, which intensifies the convection. Individual cumulus towers, associated with violent thunderstorms, develop particularly in the Intertropical Convergence Zone, sometimes reaching above 20km in height and having updrafts of 10–14m s^{-1}. In this way, the smallest scale of system can aid the development of larger disturbances. Convection is most active over sea surfaces with temperatures exceeding 27°C, but above 32°C convection ceases to increase, due to feedbacks that are not fully understood.

The second category of system develops through cumulus clouds becoming grouped into mesoscale convective areas (MCAs) up to 100km across (see Figure 11.3). In turn, several MCAs may comprise a *cloud cluster* 100–1000km in diameter. These subsynoptic-scale systems were initially identified from satellite images as amorphous cloud areas; they have been studied primarily from satellite data over the tropical

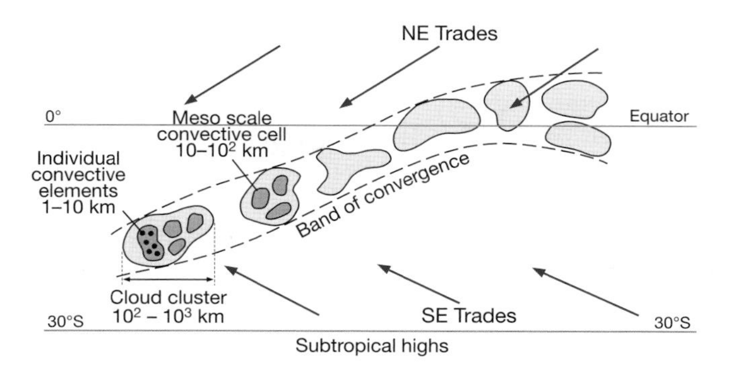

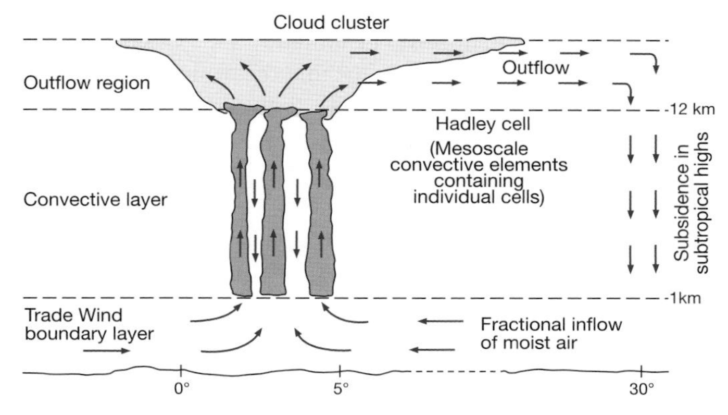

Figure 11.3 The mesoscale and synoptic structure of the equatorial trough zone (ITCZ), showing a model of the spatial distribution (above) and of the vertical structure (below) of convective elements which form the cloud clusters.

Source: From Mason (1970). By permission of the Royal Meteorological Society.

Table 11.1 Annual frequencies and usual seasonal occurrence of tropical cyclones (maximum sustained winds exceeding 25m s^{-1}), 1958–1977

Location	Annual frequency	Main occurrence
Western North Pacific	26.3	July–October
Eastern North Pacific	13.4	August–September
Western North Atlantic	8.8	August–October
Northern Indian Ocean	6.4	May–June; October–November
Northern Hemisphere total	54.6	
Southwest Indian Ocean	8.4	January–March
Southeast Indian Ocean	10.3	January–March
Western South Pacific	5.9	January–March
Southern Hemisphere total	24.5	
Global total	79.1	

Source: After Gray (1979).
Note: Area totals are rounded.

oceans. Their definition is rather arbitrary, but they may extend over an area 2° square up to 12° square. It is important to note that the peak convective activity has passed when cloud cover is most extensive through the spreading of cirrus canopies. Clusters in the Atlantic, defined as more than 50 percent cloud cover extending over an area of 3° square, show maximum frequencies of 10 to 15 clusters per month near the ITC and also at 15–20°N in the western Atlantic over zones of high sea surface temperature. They consist of a cluster of mesoscale convective cells with the system having a deep layer of convergent airflow (see Figure 9.3). Some persist for only one to two days, but others develop within synoptic-scale waves. Many aspects of their development and role remain to be determined. While convection has been stressed, studies in the western equatorial Pacific 'warm pool' region indicate that large rain areas in cloud clusters consist mainly of stratiform precipitation. This accounts for over 75 percent of the total rain area and for more than half of the rain amount. Moreover, the cloud systems are not 'warm clouds' (p. 128) but are made up of ice particles.

The fourth category of tropical weather system includes the synoptic-scale waves and cyclonic vortices (discussed more fully below) and the fifth group is represented by the planetary-scale waves. The planetary waves (with a wavelength from 10,000 to 40,000km) need not concern us in detail. Two types occur in the equatorial stratosphere and another in the equatorial upper troposphere. While they may interact with lower tropospheric systems, they do not appear to be direct weather mechanisms. The synoptic-scale systems that determine much of the 'disturbed weather' of the tropics are sufficiently important and varied to be discussed under the headings of wave disturbances and cyclonic storms.

1 Wave disturbances

Several types of wave travel westward in the equatorial and tropical tropospheric easterlies; the differences between them probably result from regional and seasonal variations in the structure of the tropical atmosphere. Their wavelength is about 2000–4000km, and they have a lifespan of one to two weeks, travelling some 6–7° longitude per day.

The first wave type to be described in the tropics was the easterly wave of the Caribbean area. This system is quite unlike a mid-latitude depression. There is a weak pressure trough, which usually slopes eastward with height (Figure 11.4). Typically the main development of cumulonimbus cloud and thundery showers is behind the trough line. This pattern is associated with horizontal and vertical motion in the

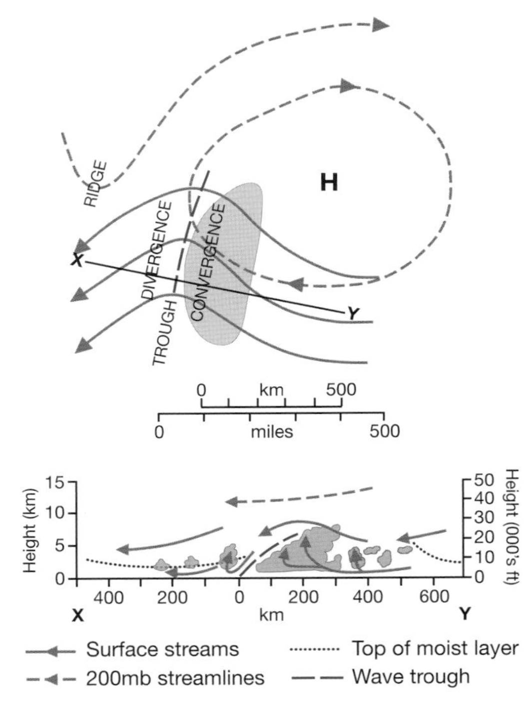

Figure 11.4 A model of the areal (above) and vertical (below) structure of an easterly wave. Cloud is stippled and the precipitation area is shown in the vertical section. The streamline symbols refer to the areal structure, and the arrows on the vertical section indicate the horizontal and vertical motions.

Source: Partly after Malkus and Riehl (1964).

easterlies. Behind the trough, low-level air undergoes convergence, while ahead of it there is divergence (see Chapter 6B.1). This follows from the equation for the conservation of potential vorticity (cf. Chapter 7G), which assumes that the air traveling at a given level does not change its potential temperature (i.e., dry adiabatic motion; see Chapter 4A):

$$\frac{f+\zeta}{\Delta p} = k$$

where f = the Coriolis parameter, ζ = relative vorticity (cyclonic positive) and Δp = the depth of the tropospheric air column. Air overtaking the trough line is moving both poleward (f increasing) and towards a zone of cyclonic curvature (ζ increasing), so that if the left-hand side of the equation is to remain constant Δp must increase. This vertical expansion of the air column necessitates horizontal contraction (convergence). Conversely, there is divergence in the air moving southward ahead of the trough and curving anticyclonically. The true divergent zone is characterized by descending, drying air with only a shallow moist layer near the surface, while in the vicinity of the trough and behind it the moist layer may be 4500m or more deep. When the easterly airflow is slower than the speed of the wave, the reverse pattern of low-level convergence ahead of the trough and divergence behind it is observed as a consequence of the potential vorticity equation. Often this is the case in the mid-troposphere, so that the pattern of vertical motion shown in Figure 11.4 is augmented.

The passage of such a transverse wave in the Trades commonly produces the following weather sequence:

1. In the ridge ahead of the trough: fine weather, scattered cumulus cloud, some haze.
2. Close to the trough line: well-developed cumulus, occasional showers, improving visibility.
3. Behind the trough: veer of wind direction, heavy cumulus and cumulonimbus, moderate or heavy, thundery showers and a decrease of temperature.

Satellite photography indicates that the classical easterly wave is less common than was supposed. Many Atlantic disturbances show an 'inverted V' waveform in the low-level wind field and associated cloud, or a 'comma' cloud related to a vortex. They are often apparently linked with a wave pattern on the ITC further south. West African disturbances that move out over the eastern tropical Atlantic usually exhibit low-level confluence and upper-level diffluence ahead of the trough, giving maximum precipitation rates in this same sector. Many disturbances in the easterlies have a closed cyclonic wind circulation at about the 600mb level.

It is difficult to trace the growth processes in wave disturbances over the oceans and in continental areas with sparse data coverage, but some generalizations can be made. At least eight out of ten disturbances develop some 2–4° latitude poleward of the Equatorial Trough. Convection is set off by convergence of moisture in the airflow, enhanced by friction, and maintained by entrainment into the thermal convective plumes (see Figure 11.3). Some 90 tropical disturbances develop during the June to November hurricane season in the tropical Atlantic, about one system every three to five days. More than half of these originate over Africa south of latitude 15°N. According to N. Frank, a high ratio of African depressions in the storm total in a given season indicates tropical characteristics, whereas a low ratio suggests storms originating from cold lows and the baroclinic zone between Saharan air and cooler, moist monsoon air. Many of these may be traced westward into the eastern North Pacific. Out of an annual total of 60 Atlantic waves, 23 percent intensify into tropical depressions and 16 percent become hurricanes.

Developments in the Atlantic are closely related to the structure of the trades. In the eastern sectors of subtropical anticyclones, active subsidence maintains a pronounced inversion at

450 to 600m (Figure 11.5). Thus, the cool eastern tropical oceans are characterized by extensive, but shallow, marine stratocumulus, which gives little rainfall. Downstream the inversion weakens and its base rises (Figure 11.6) because the subsidence decreases away from the eastern part of the anticyclone and cumulus towers penetrate through the inversion from time to time, spreading moisture into the dry air above. Easterly waves tend to develop in the Caribbean when the Trade Wind inversion is weak or even absent during summer and autumn, whereas in winter and spring subsidence aloft inhibits their growth, although disturbances may move westward above the inversion. Waves in the easterlies also originate from the penetration of cold fronts into low latitudes. In the sector between two subtropical high pressure cells, the equatorward part of the front tends to fracture generating a westward-moving wave.

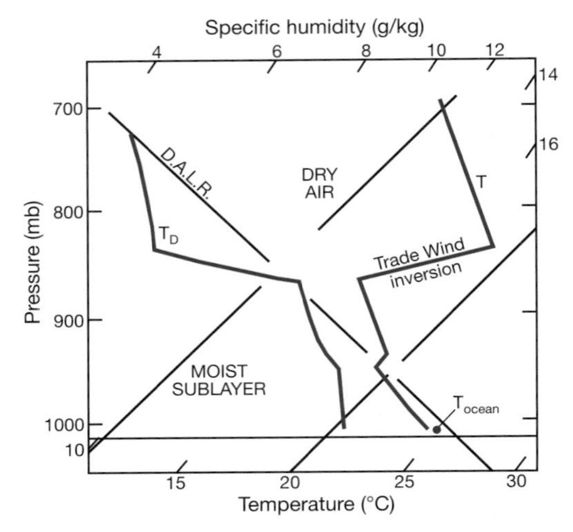

Figure 11.5 The vertical structure of Trade Wind air between the surface and 700mb in the central equatorial Atlantic, 6–12 February 1969, showing air temperature (T), dew-point temperature (T_D). The specific humidity can be read off the upper scale.

Source: After Augstein *et al.* (1973, p. 104). Courtesy of the American Meteorological Society.

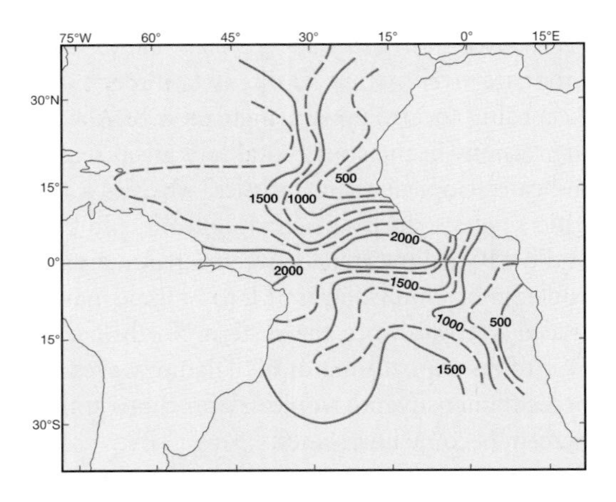

Figure 11.6 The height (in meters) of the base of the trade wind inversion over the tropical Atlantic.

Source: From Riehl (1954). By permission of McGraw Hill.

The influence of these features on regional climate is illustrated by the rainfall regime. For example, there is a late summer maximum at Martinique in the Windward Islands (15°N) when subsidence is weak, although some of the autumn rainfall is associated with tropical storms. In many Trade Wind areas, the rainfall occurs in a few rainstorms associated with some form of disturbance. Over a ten-year period, Oahu (Hawaii) had an average of 24 rainstorms per year, ten of which accounted for more than two-thirds of the annual precipitation. There is quite high variability of rainfall from year to year in such areas, since a small reduction in the frequency of disturbances can have a large effect on rainfall totals.

In the central equatorial Pacific, the Trade Wind systems of the two hemispheres converge in the Equatorial Trough. Wave disturbances may be generated if the trough is sufficiently far from the equator (usually to the north) to provide a small Coriolis force to begin cyclone motion. These disturbances quite often become unstable, forming a cyclonic vortex as they travel westward towards the Philippines, but the winds do not necessarily attain hurricane strength. The synoptic chart for part of the northwest Pacific on 17 August 1957 (Figure 11.7) shows three developmental stages of tropical low pressure systems. An incipient easterly wave formed west of Hawaii, which, however, filled and dissipated during the

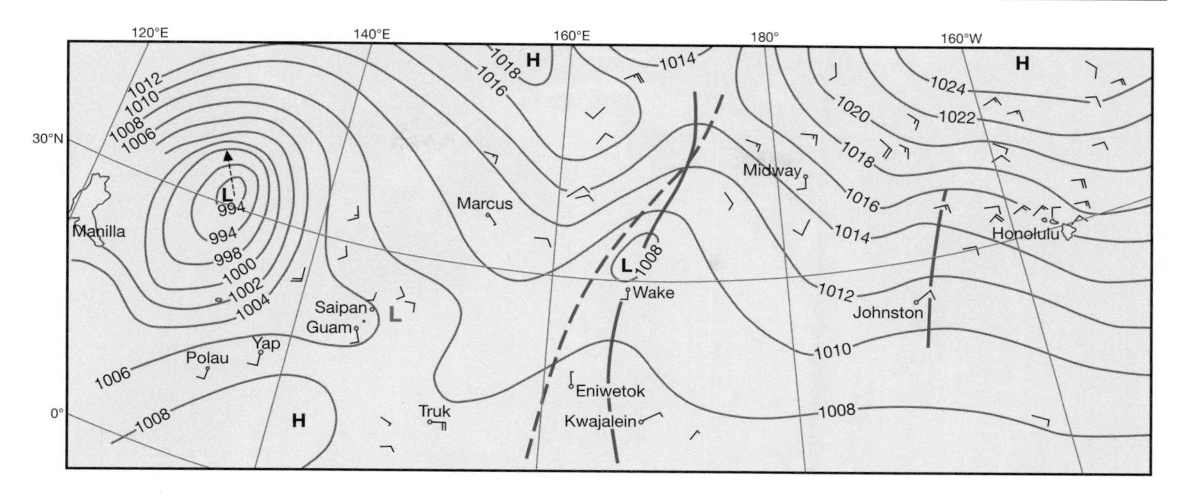

Figure 11.7 The surface synoptic chart for part of the northwest Pacific on 17 August 1957. The movements of the central wave trough and of the closed circulation during the following 24 hours are shown by the dashed line and arrow, respectively. The dashed L just east of Saipan indicates the location in which another low pressure system subsequently developed.

Source: From Malkus and Riehl (1964). By permission of the University of California Press.

next 24 hours. A well-developed wave was evident near Wake Island, having spectacular cumulus towers extending above 9km along the convergence zone some 480km east of it. This wave developed within 48 hours into a circular tropical storm with winds up to 20m s^{-1}, but not into a full hurricane. A strong, closed circulation situated east of the Philippines moved northwestward. Equatorial waves may form on both sides of the equator in an easterly current located between about 5°N and S. In such cases, divergence ahead of a trough in the Northern Hemisphere is paired with convergence behind a trough line located further to the west in the Southern Hemisphere. The reader may confirm that this should be so by applying the equation for the conservation of potential vorticity, remembering that both f and ζ operate in the reverse sense in the Southern Hemisphere.

2 Cyclones

Hurricanes and typhoons

The most notorious type of cyclone is the hurricane (or typhoon). Some 90 or so cyclones each year are responsible, on average, for 20,000 fatalities, as well as causing immense damage to property and a serious shipping hazard, due to the combined effects of high winds, high seas, flooding from the heavy rainfall and coastal storm surges. Considerable attention has been given to forecasting their development and movement, so their origin and structure are beginning to be understood. Naturally, the catastrophic force of a hurricane makes it a very difficult phenomenon to investigate, but information is obtained from aircraft reconnaissance flights sent out during the 'hurricane season', from radar observations of cloud and precipitation structure (Plate 11.1), and from satellite data.

The typical hurricane system has a diameter of about 650km, less than half that of a mid-latitude depression, although typhoons in the western Pacific are often much larger. The central pressure at sea level is commonly 950mb and exceptionally falls below 900mb. Named tropical storms are those defined as having one-minute average wind velocities of at least 18m s^{-1} at the surface. If these winds intensify to at least 33m s^{-1}, the named storm becomes a tropical cyclone. Five hurricane

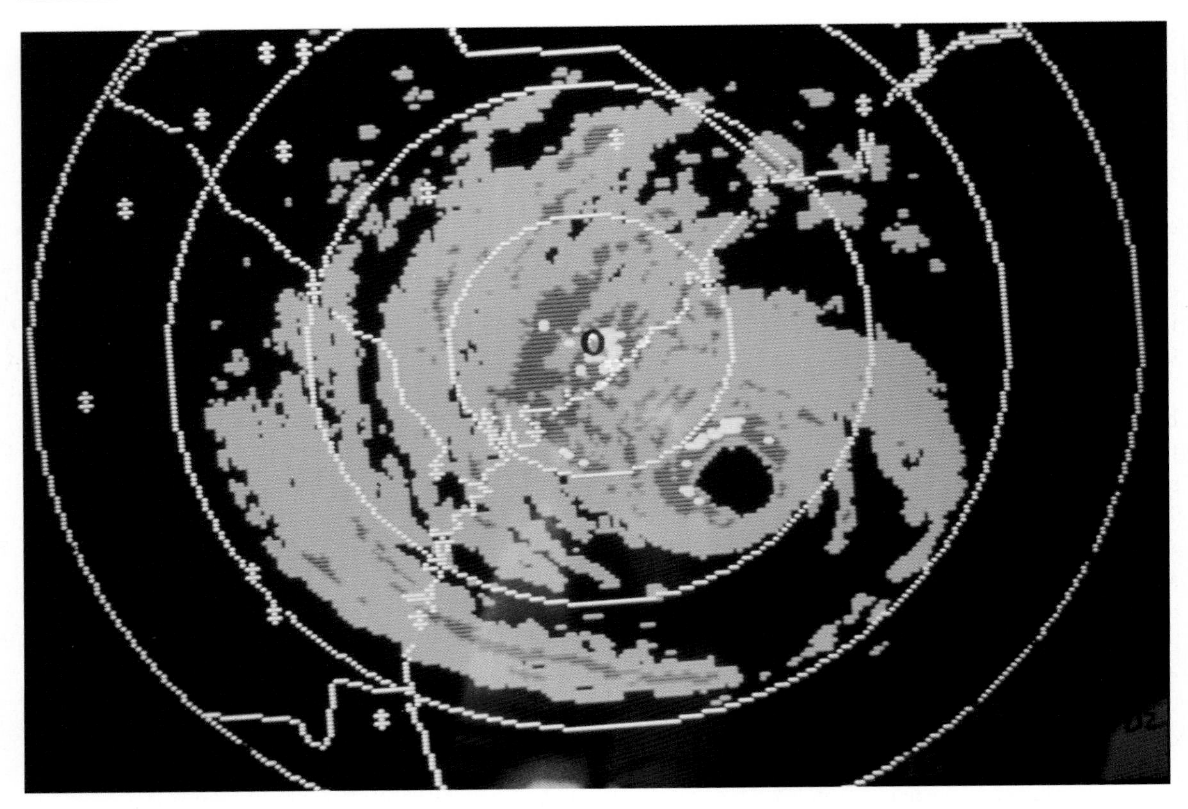

Plate 11.2 Radar image of Hurricane Hugo as observed by the South Carolina Weather Service Office, Charleston, SC on 21 September 1989. At landfall, winds were 85mph (max 160mph) in South Carolina. A storm surge coupled with high tide caused losses of $1 billion through damage to homes and timber, especially in North Carolina, with seven deaths and dozens of injuries.

Source: NOAA Central Library fly 00232.

intensity classes are distinguished: category 1, weak (winds of 33–42m s^{-1}); 2, moderate (43–49m s^{-1}); 3, strong (50–58m s^{-1}); 4, very strong (59–69m s^{-1}) and 5, devastating (70m s^{-1} or more). Hurricane Camille, which struck coastal Mississippi in August 1969, was a category 5 storm, while Hurricane Andrew, which devastated southern Florida in August 1992, has been reclassified also as a category 5 storm. Hurricane Katrina in August 2005 was the second category five storm of that season and the most destructive to date (see p. 301).

In 1997, there were 11 super-typhoons in the northwest Pacific with winds > 66m s^{-1}. The great vertical development of cumulonimbus clouds, with tops at over 12,000m, reflects the immense convective activity concentrated in such systems. Radar and satellite studies show that the convective cells are normally organized in bands that spiral inward towards the center.

Although the largest cyclones are characteristic of the Pacific, the record is held by the Caribbean hurricane 'Gilbert'. It was generated 320km east of Barbados on 9 September 1988 and moved westward at an average speed of 24–27km hr^{-1}, dissipating off the east coast of Mexico. Aided by an upper tropospheric high pressure cell north of Cuba, Hurricane Gilbert intensified very rapidly, the pressure at its center dropped to 888mb (the lowest ever recorded in the Western Hemisphere), and maximum wind speeds near the core were in excess of 55m s^{-1}. More than 500mm of rain fell on the highest parts of Jamaica in only nine hours. However, the most striking feature of this record storm was its size, being some three times that of average Caribbean hurricanes. At its maximum extent, the hurricane had a diameter of 3500km, disrupting the ITCZ along more than one-sixth of the earth's equatorial circumference and drawing in air from as far away as Florida and the Galapagos Islands.

The main tropical cyclone activity in both hemispheres is in late summer to autumn during times of maximum northward and southward shifts of the Equatorial Trough (Table 11.1, Plate 11.3). A few storms affect both the western North Atlantic and North Pacific areas as early as May and as late as December, and have occurred during every month in the latter area. In the Bay of Bengal there is also a secondary early summer maximum. Floods from a tropical cyclone that struck coastal Bangladesh on 24–30 April 1991 caused over 130,000 deaths from drowning and left over ten million people homeless. The annual frequency of cyclones shown in Table 11.1 is only approximate, since in some cases it is uncertain whether the winds actually exceeded hurricane force. In addition, storms in the more remote parts of the South Pacific and Indian Oceans frequently escaped detection prior to the use of weather satellites. A 270-year proxy record of North Atlantic hurricanes suggests a decrease in frequency from the 1760s to early 1990s with anomalously low values in the 1970s–1980s. The enhanced activity since 1995 represents a return to more normal conditions.

A number of conditions are necessary, even if not always sufficient, for cyclone formation. One requirement as shown by Figure 11.8 is an extensive ocean area with a surface temperature greater than 27°C. Cyclones rarely form near the equator, where the Coriolis parameter is close to zero, or in zones of strong vertical wind shear (*i.e.*, beneath a jet stream), as both factors inhibit the development of an organized vortex. There is also a definite connection between the seasonal position of the Equatorial Trough and zones of cyclone formation. This is borne out by the fact that only one cyclone has occurred in the South Atlantic (where the trough never lies south of 5°S) and none in the southeast Pacific (where the trough remains north of the equator). However, the northeast Pacific has an unexpected number of cyclonic vortices in summer. Many of these move westward near the trough line at about 10–15°N. About 60 percent of tropical cyclones seem to originate 5–10° latitude poleward of the Equatorial Trough in the doldrum sectors, where the trough is at least 5° latitude from the equator. The development regions of cyclones lie mainly

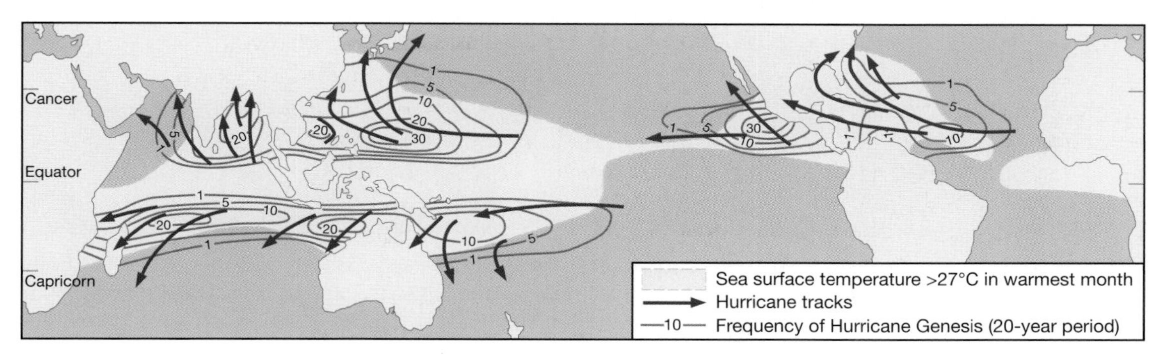

Figure 11.8 Frequency of hurricane genesis (numbered isopleths) for a 20-year period. The principal hurricane tracks and the areas of sea surface having water temperatures greater than 27°C in the warmest month are also shown.

Source: After Palmén (1948) and Gray (1979).

Plate 11.3 Three different typhoons were present over the western Pacific Ocean on 7 August, 2006, in this Moderate Resolution Imaging Spectroradiometer (MODIS) view from NASA's Aqua satellite. The strongest of the three, Typhoon Saomai (lower right), formed in the western Pacific on 4 August, 2006, as a tropical depression. Within a day, it had become organized enough to be classified as a tropical storm. While Saomai was strengthening into a storm, another tropical depression formed a few hundred kiolmeters to the north, and by 6 August, it became tropical storm Maria (upper right). Typhoon Bopha (left) formed just as Maria reached storm status and became a storm itself on 7 August. Bopha, the youngest at just a few hours old, shows only the most basic round shape of a tropical storm. Maria, a day older, shows more distinct spiral structure with arms and an apparent central eye. Despite their differences in appearance, both storms were around the same size and strength, with peak sustained winds of around 90 and 100ku/hr, respectively. A day older than Maria is the much more powerful Typhoon Saomai. At the time of this image, the typhoon had sustained winds of around 140km/hr.
Source: Jeff Schmaltz NASA Visible Earth.

over the western sections of the Atlantic, Pacific and Indian Oceans, where the subtropical high pressure cells do not cause subsidence and stability and the upper flow is divergent. About twice per season in the western equatorial Pacific, tropical cyclones form almost simultaneously in each hemisphere near 5° latitude and along the same longitude. The cloud and wind patterns in these cyclone 'twins' are roughly symmetrical with respect to the equator.

The role of convection cells in generating a massive release of latent heat to provide energy for the storm was proposed in early theories of hurricane development. However, their scale was thought to be too small for them to account for the growth of a storm hundreds of kilometers in diameter. Research indicates that energy can be transferred from the cumulus-scale to the large-scale storm circulation through the organization of the clouds into spiral bands (see Figure 11.9 and Plate 11.2), although the nature of the process is still being investigated. There is ample evidence to show that hurricanes form from pre-existing disturbances, but while many of these disturbances develop as closed low pressure cells, few attain full hurricane intensity. The key to this problem is high-level outflow (Figure 11.10). This does not require an upper tropospheric anti-cyclone but may occur on the eastern limb of an upper trough in the westerlies. This outflow in turn allows for the development of very low pressure and high wind speeds near the surface. A

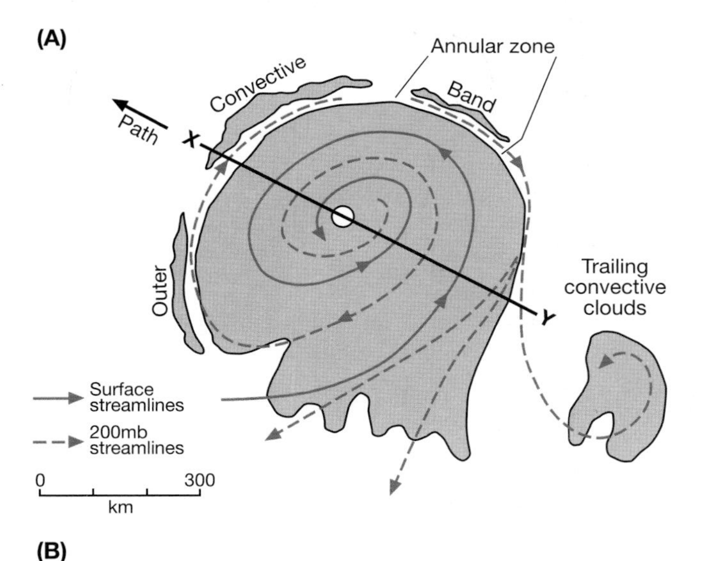

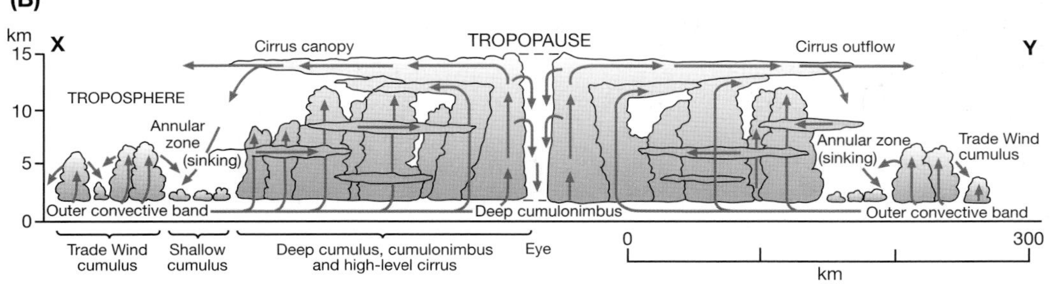

Figure 11.9 A model of the areal (A) and vertical (B) structure of a hurricane. Cloud (stippled), streamlines, convective features and path are shown.

Source: From Musk (1988). By permission of Cambridge University Press.

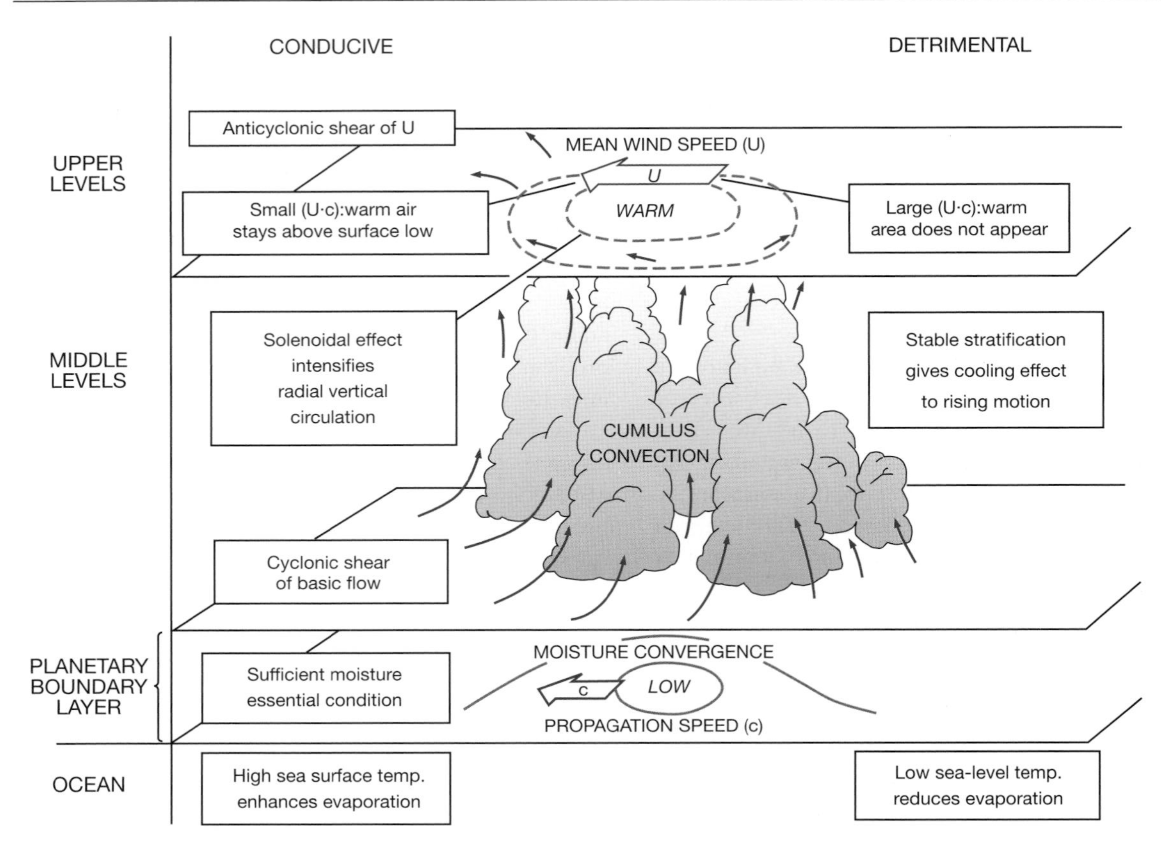

Figure 11.10 A schematic model of the conditions conducive (left) or detrimental (right) to the growth of a tropical storm in an easterly wave; *U* is the mean upper-level wind speed and *c* is the rate of propagation of the system. The warm vortex creates a thermal gradient that intensifies both the radial motion around it and the ascending air currents, termed the solenoidal effect.

Source: From Kurihara (1985). Copyright © Academic Press. Reproduced by permission.

distinctive feature of the hurricane is the warm vortex, since other tropical depressions and incipient storms have a cold core area of shower activity. The warm core develops through the action of 100–200 cumulonimbus towers releasing latent heat of condensation; about 15 percent of the area of cloud bands is giving rain at any one time. Observations show that although these 'hot towers' form less than 1 percent of the storm area within a radius of about 400km, their effect is sufficient to change the environment. The warm core is vital to hurricane growth because it intensifies the upper anticyclone, leading to a 'feedback' effect by stimulating the low-level influx of heat and moisture, which further intensifies convective activity, latent heat release and therefore the upper-level high pressure. This enhancement of a storm system by cumulus convection is termed conditional instability of the second kind, or CISK (cf. the basic parcel instability described on p. 114). The thermally direct circulation converts the heat increment into potential energy and a small fraction of this – about 3 percent – is transformed into kinetic energy. The remainder is exported by the anticyclonic circulation that exists at about the 12km (200mb) level. A major driver is the temperature difference between the ocean surface

(~300K) and the upper troposphere (~ 200K). Air spirals into the surface low, rises adiabatically in the eye wall cloud to the upper troposphere, and then descends outside the storm, completing a Carnot energy cycle (the most efficient cycle possible for converting a given amount of thermal energy into work) with an efficiency of about 33 percent. Recent work suggests that Saharan dust events may tend to influence hurricane development due to the role of the dust in suppressing cloud formation, and by the associated very dry Saharan air that is advected over the eastern tropical North Atlantic. These processes are believed to have operated during the less active 2006 Atlantic hurricane season.

In the eye, or innermost region of the storm (see Figure 11.9), adiabatic warming of descending air accentuates the high temperatures, although since high temperatures are also observed in the eye-wall cloud masses, subsiding air can only be one contributory factor. Without this sinking air in the eye, the central pressure could not fall below about 1000mb. The eye has a diameter of some 30–50km, within which the air is virtually calm and the cloud cover may be broken. The mechanics of the eye's inception are still largely unknown. If the rotating air conserved absolute angular momentum, wind speeds would become infinite at the center and clearly this is not the case. The strong winds surrounding the eye are more or less in cyclostrophic balance, with the small radial distance providing a large centripetal acceleration (see p. 149). The air rises when the pressure gradient can no longer force it further inward. It is possible that the cumulonimbus anvils play a vital role in the complex link between the horizontal and vertical circulations around the eye by redistributing angular momentum in such a way as to set up a concentration of rotation near the center.

The supply of heat and moisture combined with low frictional drag at the sea surface, the release of latent heat through condensation and the removal of the air aloft are essential conditions for the maintenance of cyclone intensity. As soon as one of these ingredients diminishes, the storm decays. This can occur quite rapidly if the track (determined by the general upper tropospheric flow) takes the vortex over a cool sea surface or over land. In the latter case, the increased friction causes greater cross-isobar air motion, temporarily increasing the convergence and ascent. At this stage, increased vertical wind shear in thunderstorm cells may generate tornadoes, especially in the northeast quadrant of the storm (in the Northern Hemisphere). However, the most important effect of a land track is that cutting off of the moisture supply removes one of the major sources of heat. Rapid decay also occurs when cold air is drawn into the circulation or when the upper-level divergence pattern moves away from the storm.

Hurricanes usually move at 16–24km hr^{-1}, controlled primarily by the rate of movement of the upper warm core. Commonly, they recurve poleward around the western margins of the subtropical high pressure cells, entering the circulation of the westerlies, where they die out or regenerate into extra-tropical disturbances.

Some of these systems retain an intense circulation and the high winds and waves can still wreak havoc. This is not uncommon along the Atlantic coast of the United States and occasionally eastern Canada. Similarly, in the western North Pacific, recurved typhoons are a major element in the climate of Japan (see D, this chapter) and may occur in any month. There is an average frequency of 12 typhoons per year over southern Japan and neighboring sea areas.

To sum up: a tropical cyclone develops from an initial disturbance, which, under favorable environmental conditions, grows first into a tropical depression and then into a tropical storm. The tropical storm stage may persist for four to five days, whereas the cyclone stage usually lasts for only two to three days (four to five days in the western Pacific). The main energy source is latent heat derived from condensed water vapor, and for this reason hurricanes are generated and continue to gather strength only within the confines of warm oceans. The cold-cored tropical storm is transformed into a warm-cored hurricane in association with the release of latent heat in cumulonimbus towers, and this establishes or intensifies an upper tropospheric anticyclonic cell. Thus high-level outflow maintains the ascent and low-level inflow in order to provide a continuous generation of potential energy (from latent heat) and the transformation of this into kinetic energy. The inner eye that forms by sinking air is an essential element in the life-cycle.

Hurricane forecasting is a complex science. Recent studies of annual North Atlantic/Caribbean hurricane frequencies suggest that three major factors are involved:

1 The west phase of the Atlantic Quasi-biennial Oscillation (QBO). The QBO involves periodic changes in the velocities of, and vertical shear between, the zonal upper tropospheric (50mb) winds and the lower stratospheric (30mb) winds. The onset of such an oscillation can be predicted with some confidence almost a year in advance. The east phase of the QBO is associated with strong easterly winds in the lower stratosphere between latitudes 10°N and 15°N, producing a large vertical wind shear. This phase usually persists for 12 to 15 months and inhibits hurricane formation. The west QBO phase exhibits weak easterly winds in the lower stratosphere and small vertical wind shear. This phase, typically lasting 13 to 16 months, is associated with 50 percent more named storms, 60 percent more hurricanes and 200 percent more major hurricanes than is the east phase.
2 West African precipitation during the previous year along the Gulf of Guinea (August to November) and in the western Sahel (August to September). The former moisture source appears to account for some 40 percent of major hurricane activity, the latter for only 5 percent. Between the late 1960s and 1980s, the Sahel drought was associated with a marked decrease in Atlantic tropical cyclones and hurricanes, mainly through strong upper-level shearing winds over the tropical North Atlantic and a decrease in the propagation of easterly waves over Africa in August and September.
3 ENSO predictions for the following year (see G, this chapter). There is an inverse correlation between the frequency of El Niños and that of Atlantic hurricanes.

Recent studies suggest that there has been an increase in the number and proportion of category 4–5 hurricanes over the past 30 years. The largest increases took place in the North Pacific, Indian and southwest Pacific oceans, and the smallest increase occurred in the North Atlantic Ocean. At the same time the number of cyclones and cyclone days has decreased in all basins except the North Atlantic. The reported increase in tropical cyclone energy, numbers and wind speeds in some regions during the past few decades has been attributed to higher sea surface temperatures. However, other studies consider that changes in observational techniques and instrumentation can account for these changes.

Other tropical depressions

Not all low pressure systems in the tropics are of the intense tropical cyclone variety. There are two other major types of cyclonic vortex. One is the monsoon depression that affects South Asia during the summer. This disturbance is somewhat unusual in that the flow is westerly at low levels and easterly in the upper troposphere (see Figure 11.27). It is more fully described in C.4, this chapter. (p. 354).

The second type of system is usually relatively weak near the surface, but well developed in the mid-troposphere. In the eastern North Pacific and northern Indian Ocean, such lows are referred to as subtropical cyclones. Some develop from the cutting off in low latitudes of a cold upper-level wave in the westerlies (cf. Chapter 9H.4). They possess a broad eye, 150km in radius with little cloud, surrounded by a belt of cloud and precipitation about 300km wide. In late winter and spring, a few such storms make a major contribution to the rainfall of the Hawaiian Islands. These cyclones are very persistent and tend eventually to be reabsorbed by a trough in the upper westerlies. Other subtropical cyclones occur over the Arabian Sea, making a major contribution to summer ('monsoon') rains in northwest India. These systems show upward motion mainly in the upper troposphere. Their development may be linked to export at upper levels of cyclonic vorticity from the persistent heat low over Arabia.

An infrequent and distinctly different weather system, known as a *temporal*, occurs along the Pacific coasts of Central America in autumn and early summer. Its main feature is an extensive layer of altostratus, fed by individual convective cells, producing sustained moderate rainfall. These systems originate in the ITCZ over the eastern tropical North Pacific Ocean and are maintained by large-scale lower tropospheric convergence, localized convection and orographic uplift.

3 Tropical cloud clusters

Mesoscale convective systems (MCSs) are widespread in tropical and subtropical latitudes. The mid-latitude mesoscale convective complexes discussed in Chapter 9I are an especially severe category of MCS. Satellite studies of cold (high) cloud-top signatures show that tropical systems typically extend over a 3000–6000km^2 area. They are common over tropical South America and the maritime continent of Indonesia–Malaysia and adjacent western equatorial Pacific Ocean warm pool. Other land areas include Australia, India and Central America, in their respective summer seasons. As a result of the diurnal regimes of convective activity, MCSs are more frequent at sunset compared with sunrise by 60 percent over the continents and 35 percent more frequent at sunrise than sunset over the oceans. Most of the intense systems (MCCs) occur over land, particularly where there is abundant moisture and usually downwind of orographic features that favor the formation of low-level jets.

Mesoscale convective systems fall into two categories: non-squall- and squall-line. The former contain one or more mesoscale precipitation areas. They occur diurnally, for example, off the north coast of Borneo in winter, where they are initiated by convergence of a nocturnal land breeze and the northeast monsoon flow (Figure 11.11). By morning (08:00 LST), cumulonimbus cells give precipitation. The cells are linked by an upper-level cloud shield, which persists when the convection dies about around noon as a sea-breeze system replaces the nocturnal convergent flow. Recent studies over the western equatorial Pacific warm pool indicate that convective cloud systems account for <50 percent of the total in large precipitation areas (boxes 240 × 240km), while stratiform precipitation is more widespread and yields over half of the total precipitation.

Tropical squall-line systems (Figure 11.12) form the leading edge of a line of cumulonimbus cells. The squall-line and gust front advance within the low-level flow and by forming new cells. These mature and eventually dissipate to the rear of the main line. The process is analogous to that of mid-latitude squall-lines (see Figure 9.28) but the tropical cells are weaker. Squall-line systems, known as sumatras, cross Malaysia from the west during the southwest monsoon season, giving heavy rain and often thunder. They appear to be initiated by the convergent effects of land breezes in the Malacca Straits.

In West Africa, systems known as disturbance lines (DLs) are an important feature of the climate in the summer half-year, when low-level southwesterly monsoon air is overrun by dry, warm Saharan air. The meridional air-mass contrast helps to set up the lower tropospheric African Easterly Jet (AEJ) (see Figure 11.38). The convective DLs are transported across West Africa by African easterly waves that are steered by the AEJ at around 600mb. The waves recur with a four- to eight-day period during the wet season (May to October). Disturbance lines tend to form when

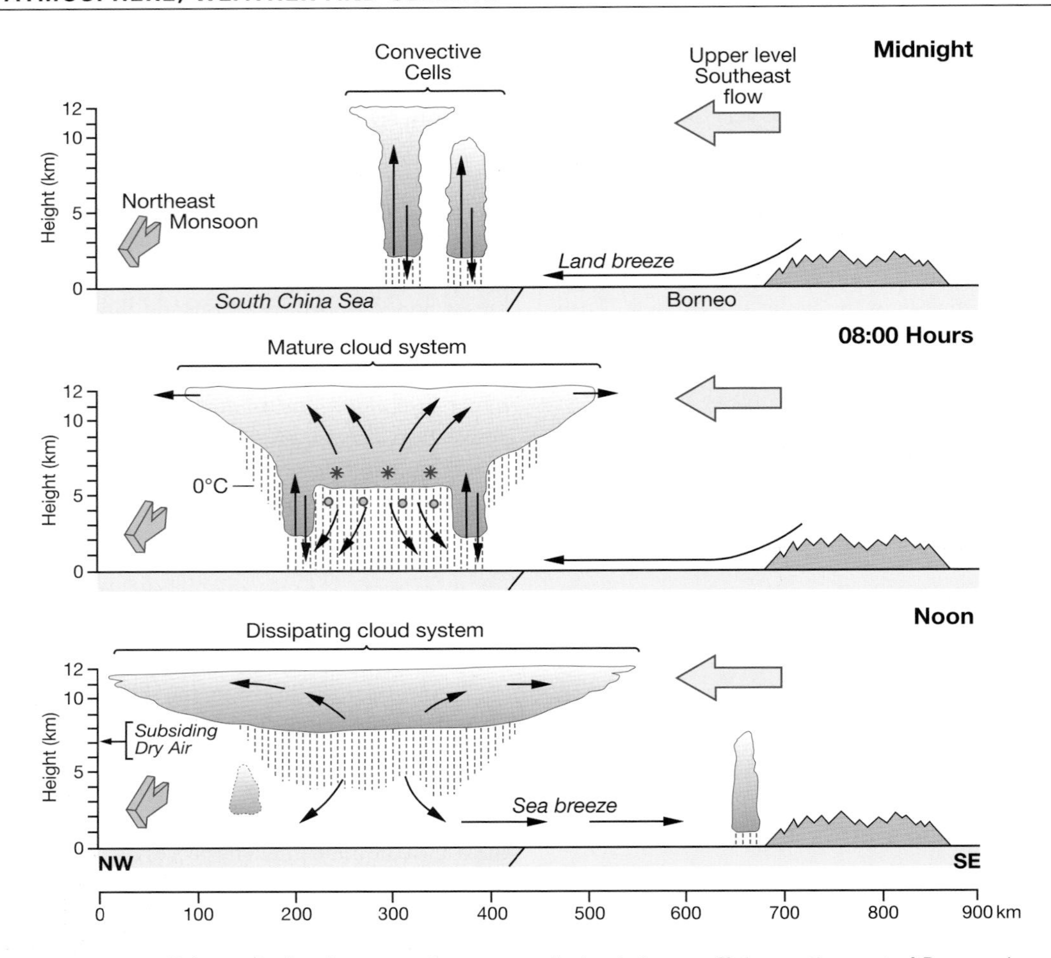

Figure 11.11 Schematic development of a non-squall cloud cluster off the north coast of Borneo: large arrows indicate the major circulation; small arrows, the local circulation; vertical shading, the zones of rain; stars, ice crystals; and circles, melted raindrops.

Source: After Houze *et al.* (1981). Courtesy of the American Meteorological Society.

there is divergence in the upper troposphere north of the Tropical Easterly Jet (see also Figure 11.40). They are several hundred kilometers long and travel westward at about 50km hr^{-1} giving squalls and thunder showers before dissipating over cold-water areas of the North Atlantic. Spring and autumn rainfall in West Africa is derived in large part from these disturbances. In wet years, when the AEJ is further north, the wave season is prolonged and the waves are stronger. Figure 11.13 for Kortright (Freetown), Sierra Leone illustrates the daily rainfall amounts in 1960–1961 associated with disturbance lines at 8°N. Here the summer monsoon rains make up the greater part of the total, but their contribution diminishes northward.

C THE SOUTH ASIAN MONSOON

The name monsoon is derived from the Arabic word *mausim*, which means season, referring to large-scale seasonal reversals of the wind regime. The Asiatic seasonal wind reversal is notable for its

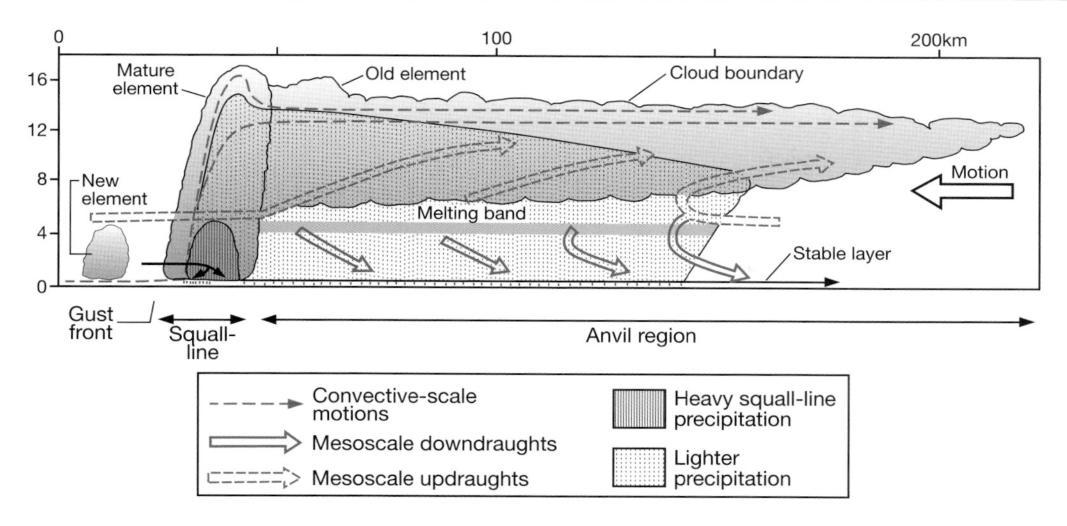

Figure 11.12 Cross-section of a tropical squall-line cloud cluster showing locations of precipitation and ice particle melting. Dashed arrows show the air motion generated by the squall-line convection and the broad arrows the mesoscale circulation.

Source: After Houze; from Houze and Hobbs (1982). By permission of Academic Press.

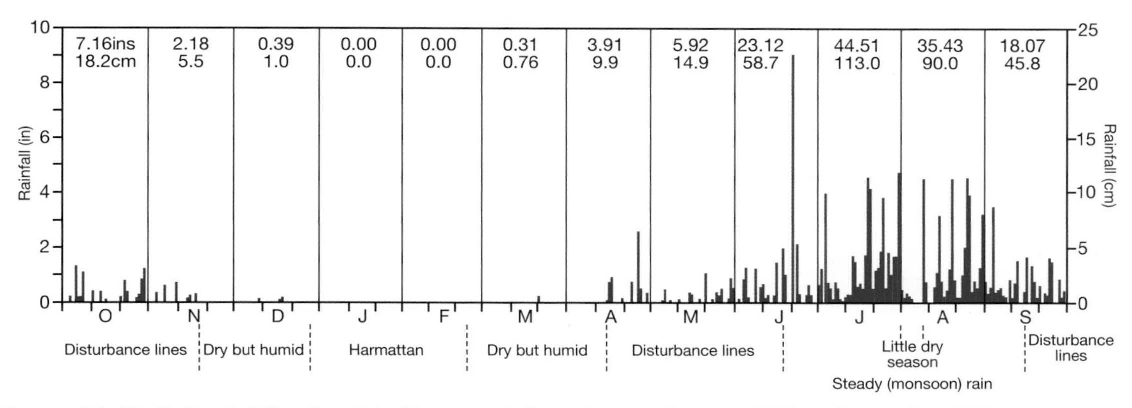

Figure 11.13 Daily rainfall at Kortright (Freetown), Sierra Leone, October 1960 to September 1961.

Source: After Gregory (1965).

vast extent and the penetration of its influence beyond tropical latitudes (Figure 11.14). However, such seasonal shifts of the surface winds occur in many regions that are not traditionally considered as monsoonal. Although there is an overlap between these traditional regions and those experiencing over 60 percent frequency of winds from the prevailing octant, it is obvious that a variety of unconnected mechanisms can lead to seasonal wind shifts. Nor is it possible to establish a simple relationship between seasonality of rainfall (Figure 11.15) and seasonal wind shift. Areas traditionally designated as 'monsoonal' include some of the tropical and near-tropical regions experiencing a summer rainfall maximum and most of those having a double rainfall maximum. It is clear that a combination of criteria is necessary for an adequate definition of monsoon areas.

In summer, the Equatorial Trough and the subtropical anticyclones are everywhere displaced

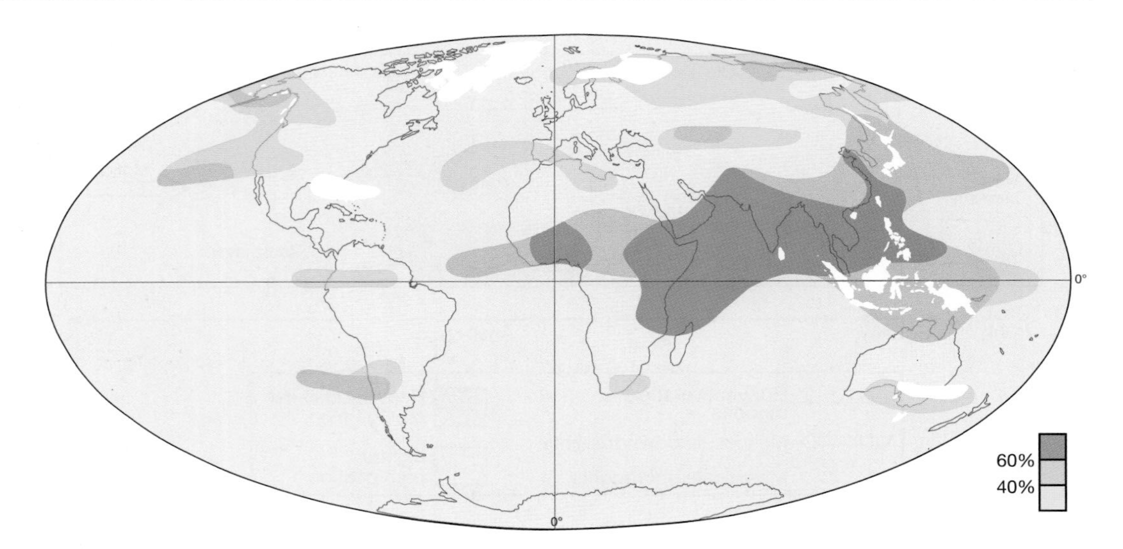

Figure 11.14 Regions experiencing a seasonal surface wind shift of at least 120°, showing the frequency of the prevailing octant.
Source: After Khromov.

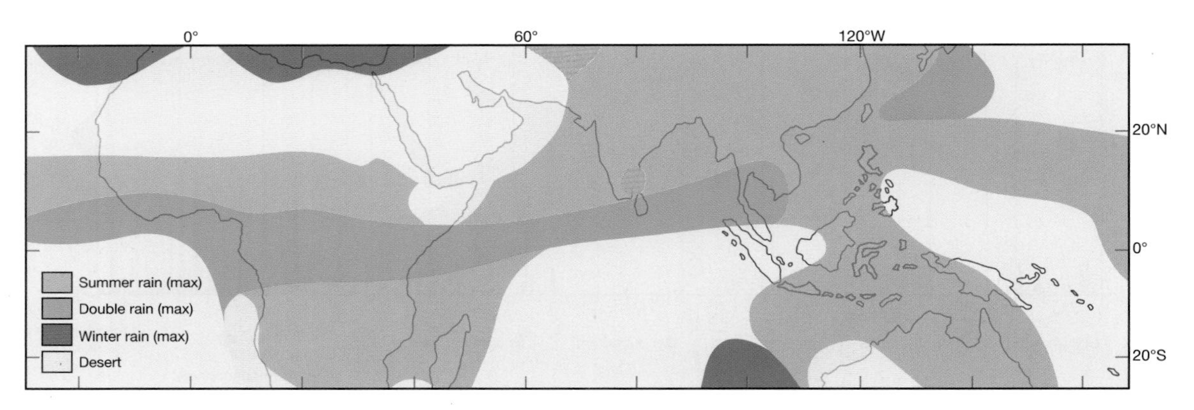

Figure 11.15 The annual distribution of tropical rainfall. The shaded areas refer to periods during which more than 75 percent of the mean annual rainfall occurs. Areas with less than 250mm y^{-1} (10in y^{-1}) are classed as deserts, and the unshaded areas are those needing at least seven months to accumulate 75 percent of the annual rainfall and are thus considered to exhibit no seasonal maximum.
Source: After Ramage (1971). By permission of Academic Press.

northward in response to the distribution of solar heating of the earth, and in South Asia this movement is magnified by the effects of the land mass. However, the attractive simplicity of the traditional explanation, which envisages a monsoonal 'sea breeze' directed towards a summer thermal low pressure over the continent, provides an inadequate basis for understanding the workings of the system. The Asiatic monsoon regime is a consequence of the interaction of planetary and regional factors, both at the surface and in the upper troposphere. It is convenient to look at each season in turn; Figure 11.16 shows the generalized meridional circulation at 90°E over India and the

Indian Ocean in winter (December to February), spring (April) and autumn (September), together with those associated with active and break periods during the June to August summer monsoon.

1 Winter

Near the surface, this is the season of the out-blowing 'winter monsoon', but aloft westerly airflow dominates. This reflects the hemispheric pressure distribution. A shallow layer of cold, high pressure air is centered over the continental interior, but this has disappeared even at 700mb (see Figure 7.4) where there is a trough over East Asia and zonal circulation over the continent. The upper westerlies split into two currents to the north and south of the high Tibetan (Qinghai–Xizang) Plateau (Figure 11.17), to reunite again off the east coast of China (Figure 11.18). The plateau, which exceeds 4000m over a vast area, is a tropospheric cold source in winter, particularly over its western part, although the strength of this source depends on the extent and duration of snow cover (snow-free ground acts as a heat source for the atmosphere in all months). Below 600mb, the tropospheric heat sink gives rise to a shallow, cold plateau anticyclone, which is best developed in December and January. The two jet-stream branches have been attributed to the disruptive effect of the topographic barrier on the airflow, but this is limited to altitudes below about 4km. In fact, the northern jet is highly mobile and may be located far from the Tibetan Plateau. Two currents are also observed farther west, where there is no obstacle to the flow. The branch over northern India corresponds to a strong latitudinal thermal gradient (from November to April) and it is probable that this factor, combined with the thermal effect of the barrier to the north, is responsible for the anchoring of the southern jet. This branch is the stronger, with an average speed of more than 40m s^{-1} at 200mb, compared with about 20–25m s^{-1} in the northern one. Where the two unite over north China and south Japan the average speed exceeds 66m s^{-1} (Figure 11.19).

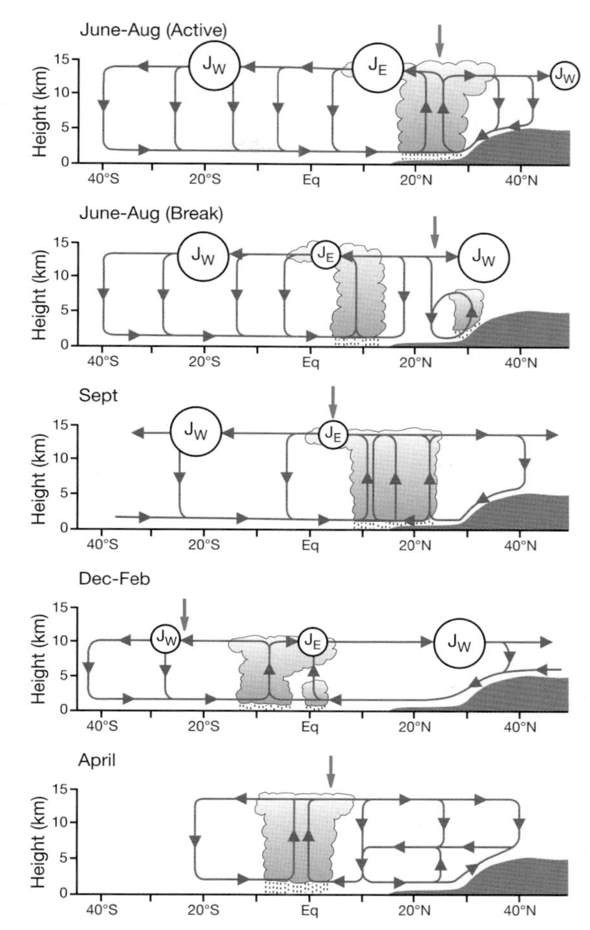

Figure 11.16 Schematic representation of the meridional circulation over India at 90°E at five characteristic periods of the year: winter monsoon (December to February); approach of the monsoon season (April); the active summer monsoon (June to August); a break in the summer monsoon (June to August); and the retreat of the summer monsoon (September). Easterly (JE) and westerly (JW) jet streams are shown at sizes depending on their strength; the arrows mark the positions of the overhead sun; and zones of maximum precipitation are indicated.

Source: After Webster (1987a). Copyright © 1987. Reproduced by kind permission of John Wiley and Sons, Inc.

Air subsiding beneath this upper westerly current gives dry out-blowing northerly winds from the subtropical anticyclone over northwest India and Pakistan. The surface wind direction is northwesterly over most of northern India,

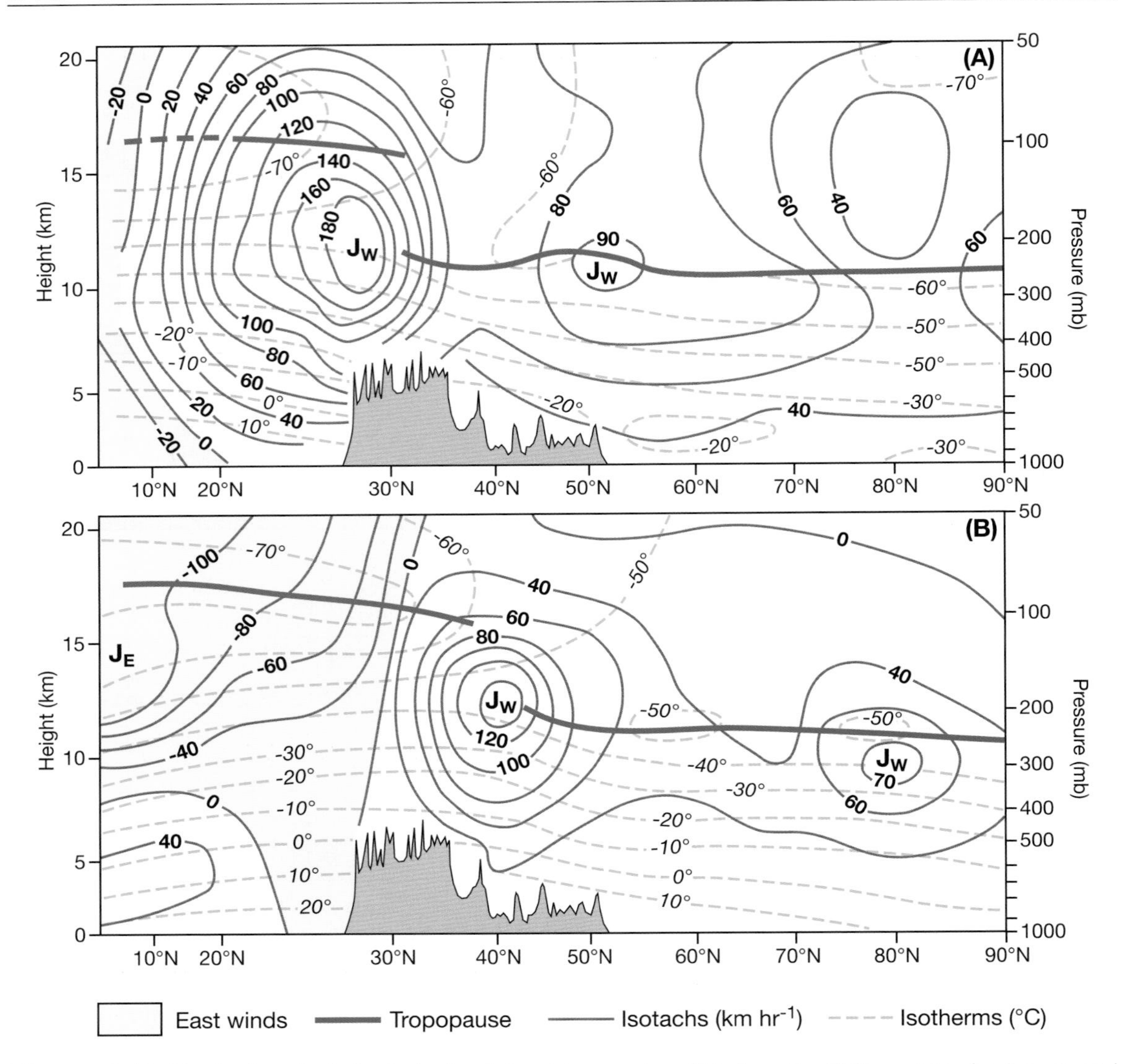

Figure 11.17 Distribution of wind velocity (km/hr) and temperature (°C) along the 90°E meridian for January and July, showing the westerly jet streams (J_W) and the tropopause. Note the variable intervals in the height and latitudinal scales.

Source: After Pogosyan and Ugarova (1959). From *Meteorologiya Gidrologiya*.

becoming northeasterly over Burma and Bangladesh and easterly over peninsular India. Equally important is the steering of winter depressions over northern India by the upper jet. The lows, which are not usually frontal, appear to penetrate across the Middle East from the Mediterranean and are important sources of rainfall for northern India and Pakistan (e.g., Kalat: Figure 11.20), especially as it falls when evaporation is at a minimum. The equatorial trough of convergence and precipitation lies between the equator and about latitude 15°S (see Figure 11.16).

Some of these westerly depressions continue eastward, redeveloping in the zone of jet stream

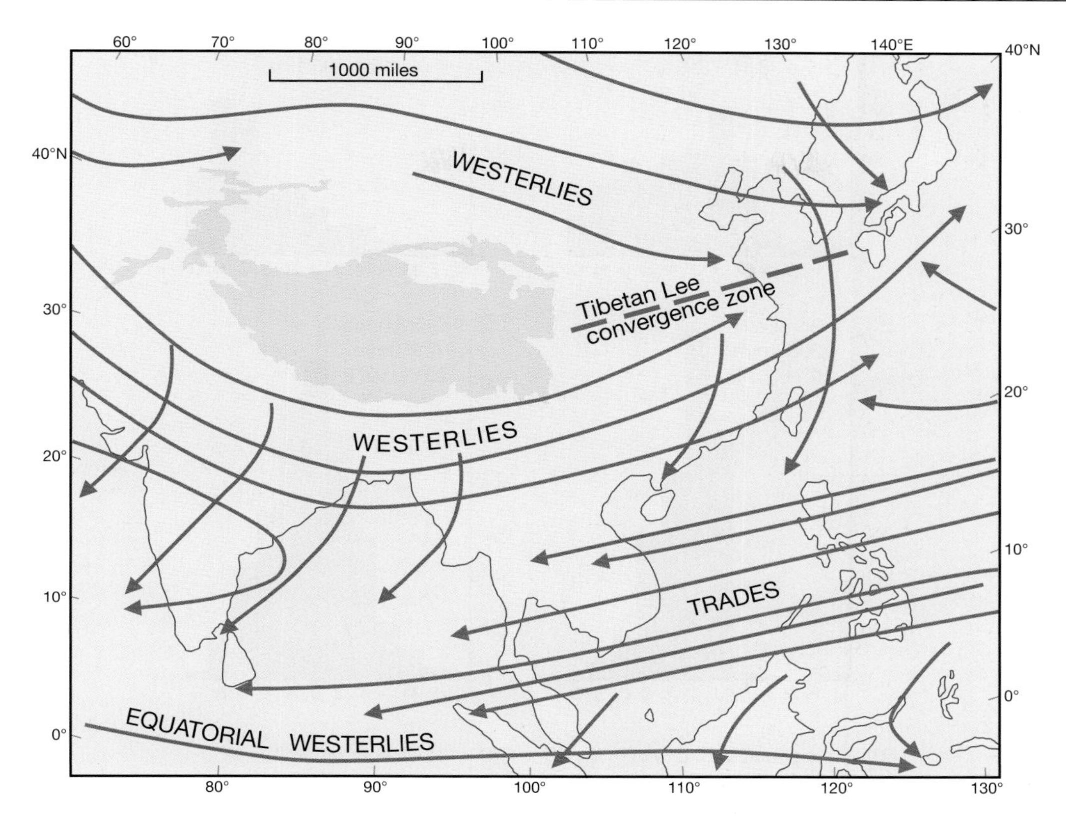

Figure 11.18 The characteristic air circulation over South and East Asia in winter. Brown lines indicate airflow at about 3000m, and green lines at about 600m. The names refer to the wind systems aloft.
Sources: After Thompson (1951), Flohn (1968), Frost and Stephenson (1965), and others.

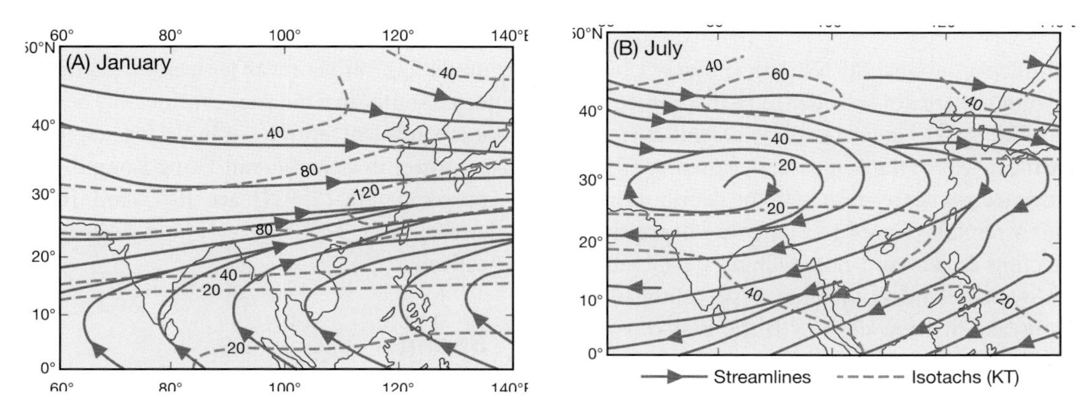

Figure 11.19 Mean 200 mb streamlines and isotachs in knots over Southeast Asia for January and July, based on aircraft reports and sounding data.
Source: From Sadler (1975b). Courtesy Dr J. C. Sadler, University of Hawaii.

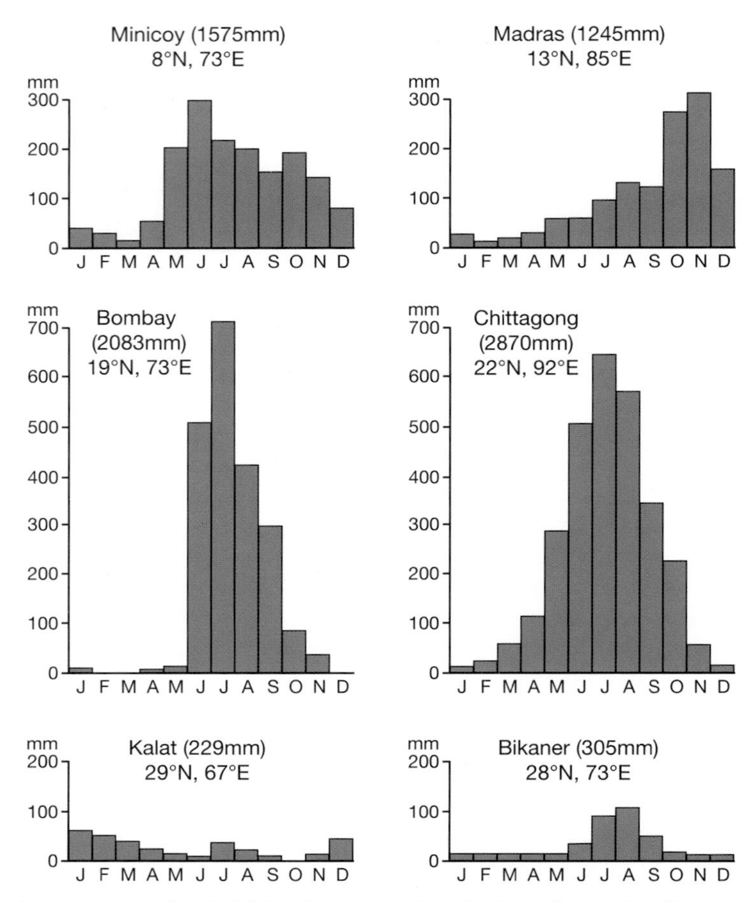

Figure 11.20 Average monthly rainfall (mm) at six stations in the Indian region. The annual total is given after the station name.

Source: Based on 'CLIMAT' normals of the World Meteorological Organization for 1931–1960.

confluence about 30°N, 105°E over China, beyond the area of subsidence in the immediate lee of Tibet (see Figure 11.18). It is significant that the mean axis of the winter jet stream over China shows a close correlation with the distribution of winter rainfall (Figure 11.21). Other depressions affecting central and north China travel within the westerlies north of Tibet or are initiated by outbreaks of fresh cP air. In the rear of these depressions are invasions of very cold air (e.g. the buran blizzards of Mongolia and Manchuria). The effect of such cold waves, comparable with the northerlies in the central and southern United States, is greatly to reduce mean temperatures (Figure 11.22). Winter mean temperatures in less-protected southern China are considerably below those at equivalent latitudes in India; for example, temperatures in Calcutta and Hong Kong (both at approximately 22.5°N) are 19°C and 16°C in January and 22°C and 15°C in February, respectively.

2 Spring

The key to change during this transition season is again found in the pattern of the upper airflow. In March the upper westerlies begin their seasonal migration northward, but whereas the northerly

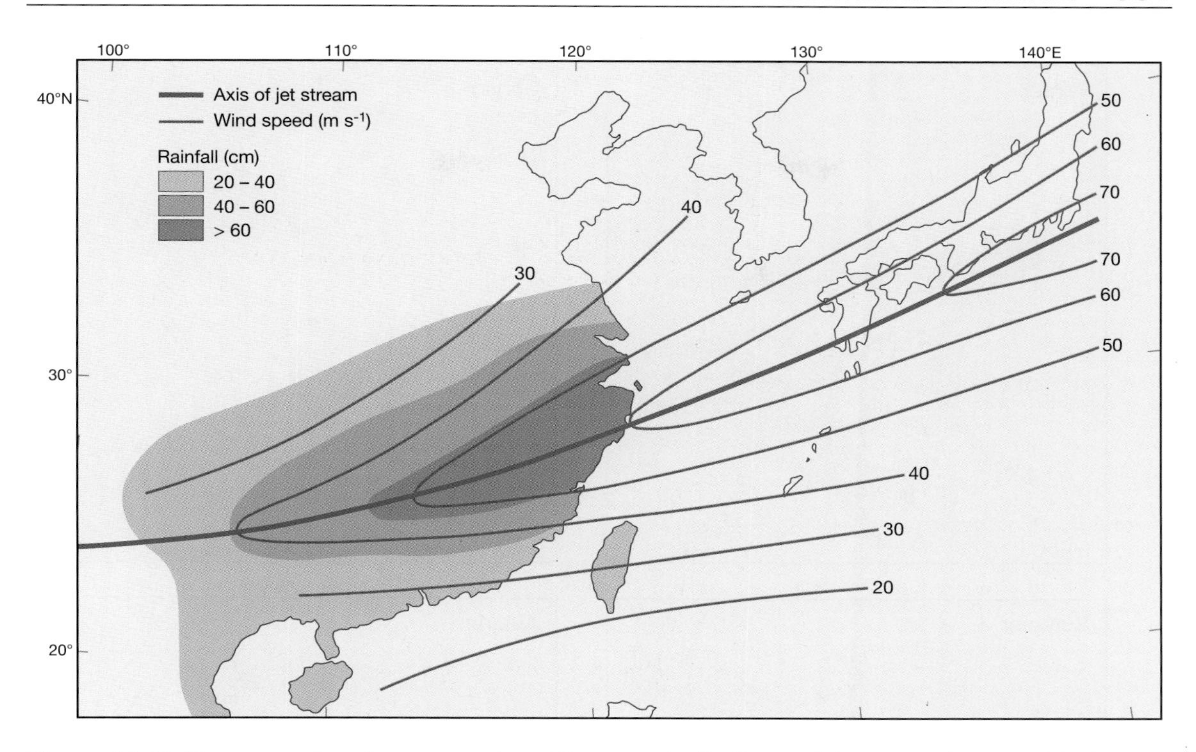

Figure 11.21 The mean winter jet stream axis at 12km over the Far East and the mean winter precipitation over China (in cm).

Source: After Mohri and Yeh; from Trewartha (1958). By permission of the University of Wisconsin Press.

jet strengthens and begins to extend across central China and into Japan, the southerly branch remains positioned south of Tibet, although weakening in intensity.

In April there is weak convection over India, where the circulation is dominated by subsiding air originating along the convective ITCZ trough centered over the equator and following the overhead sun northward over the warm Indian Ocean (see Figure 11.16). The weather over northern India becomes hot, dry and squally in response to the greater solar radiation heating. Mean temperatures in Delhi rise from 23°C in March to 33°C in May. The thermal low pressure cell (see Chapter 9H.2) now reaches its maximum intensity, but although onshore coastal winds develop, the onset of the monsoon is still a month away and other mechanisms cause only limited precipitation. Some precipitation occurs in the north with 'westerly disturbances', particularly towards the Ganges delta, where the low-level inflow of warm, humid air is overrun by dry, potentially cold air, triggering squall-lines known as nor'westers. In the northwest, where less moisture is available, the convection generates violent squalls and dust-storms termed andhis. The mechanism of these storms is not fully known, although high-level divergence in the waves of the subtropical westerly jet stream appears to be essential. The early onset of summer rains in Bengal, Bangladesh, Assam and Burma (e.g., Chittagong: Figure 11.20) is favored by an orographically produced trough in the upper westerlies, which is located at about 85–90°E in May. Low-level convergence of maritime air from the Bay of Bengal, combined with the upper-level divergence ahead of the 300mb trough, generates thunder squalls. Tropical disturbances in the Bay

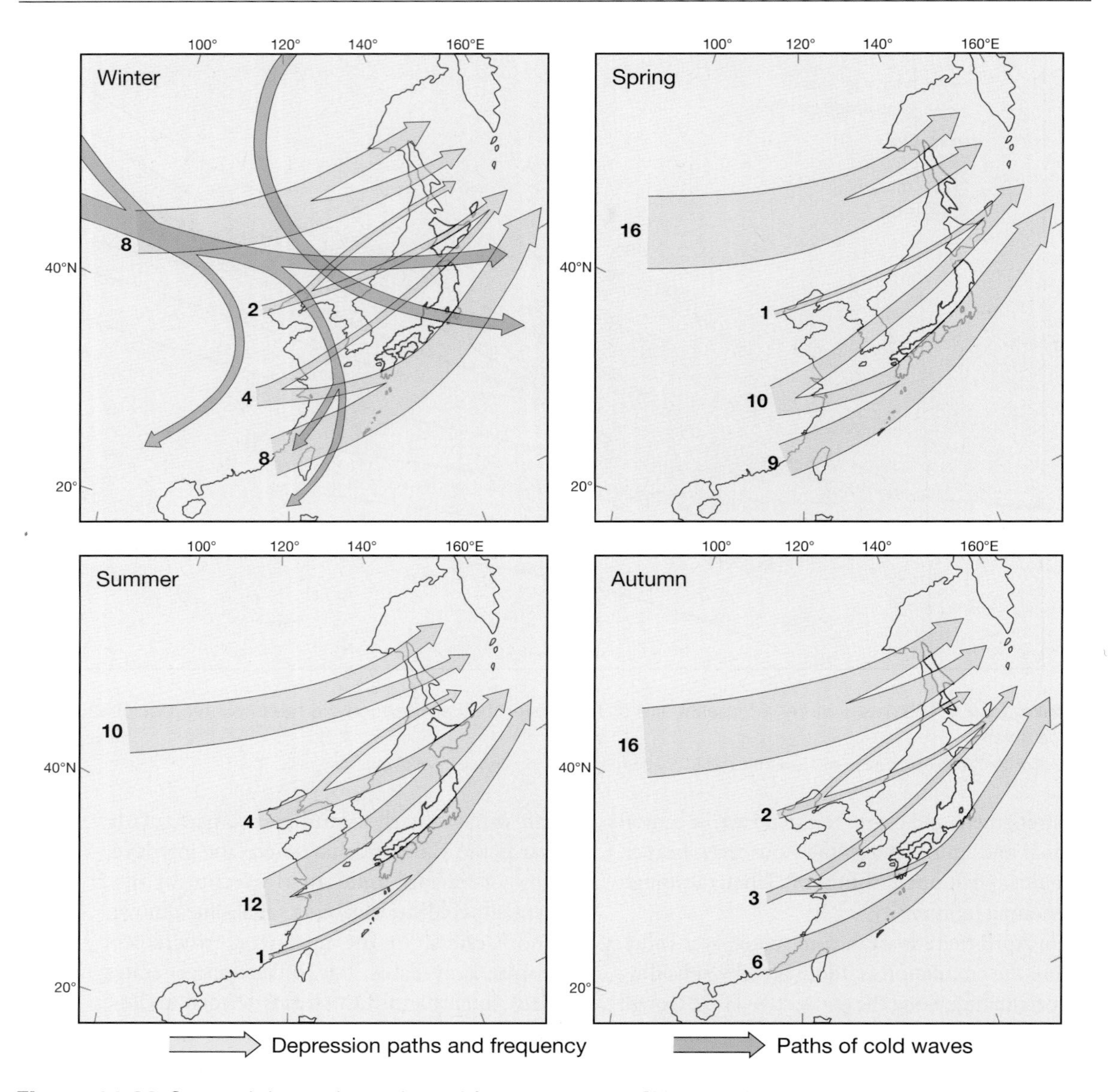

Figure 11.22 Seasonal depression paths and frequencies over China and Japan, together with typical paths of winter cold waves.

Sources: Compiled from various sources, including Tao, (1984), Zhang and Lin (1985), Sheng *et al.* (1986) and Domrös and Peng (1988). With kind permission of Springer Sciences & Business Media.

of Bengal are another source of these early rains. Rain also falls during this season over Sri Lanka and southern India (e.g., Minicoy: Figure 11.20) in response to the northward movement of the Equatorial Trough.

3 Early summer

Generally, during the last week in May the southern branch of the high-level jet begins to break down, becoming intermittent and then gradually shifting northward over the Tibetan

Plateau. At 500mb and below, however, the plateau exerts a blocking effect on the flow and the jet axis there jumps from the south to the north side of the plateau from May to June. Over India, the Equatorial Trough pushes northward with each weakening of the upper westerlies south of Tibet, but the final burst of the monsoon, with the arrival of the humid, low-level southwesterlies, is not accomplished until the upper-air circulation has switched to its summer pattern (see Figures 11.19 and 11.23). Increased continental convection overcomes the spring subsidence and the return upper-level flow to the south is deflected by the Coriolis force to produce a strengthening easterly jet located at about 10–15°N and a westerly jet to the south of the equator (see Figure 11.16). One theory suggests that this takes place in June when the col between the subtropical anticyclone cells of the west Pacific and the Arabian Sea at the 300mb level is displaced northwestward from a position about 15°N, 95°E in May towards central India. The northwestward movement of the monsoon (see Figure 11.24) is apparently related to the extension over India of the upper tropospheric easterlies.

The organization of the upper airflow has widespread effects in southern and eastern Asia. It is directly linked with the Maiyu rains of China (which reach a peak around 10–15 June), the onset of the southwest Indian monsoon and the northerly retreat of the upper westerlies over the whole of the Middle East.

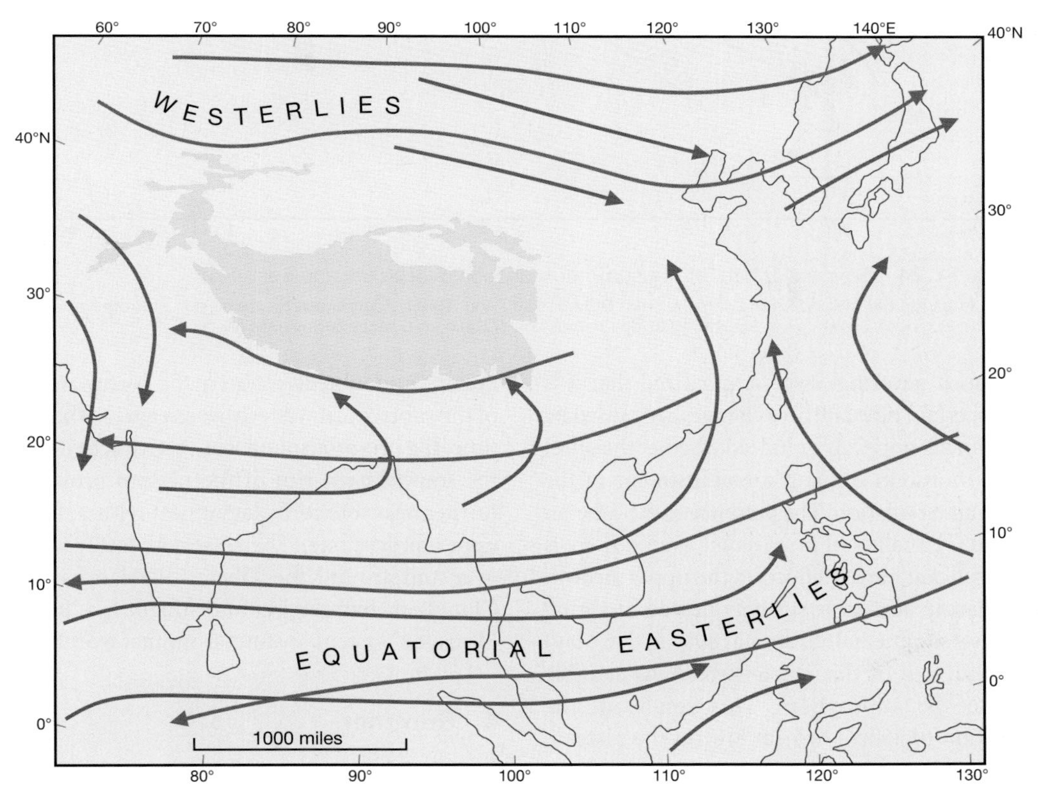

Figure 11.23 The characteristic air circulation over South and East Asia in summer. Brown lines indicate airflow at about 6000m and green lines at about 600m. Note that the low-level flow is very uniform between about 600 and 3000m.

Sources: After Thompson (1951), Flohn (1968), Frost and Stephenson (1965), and others.

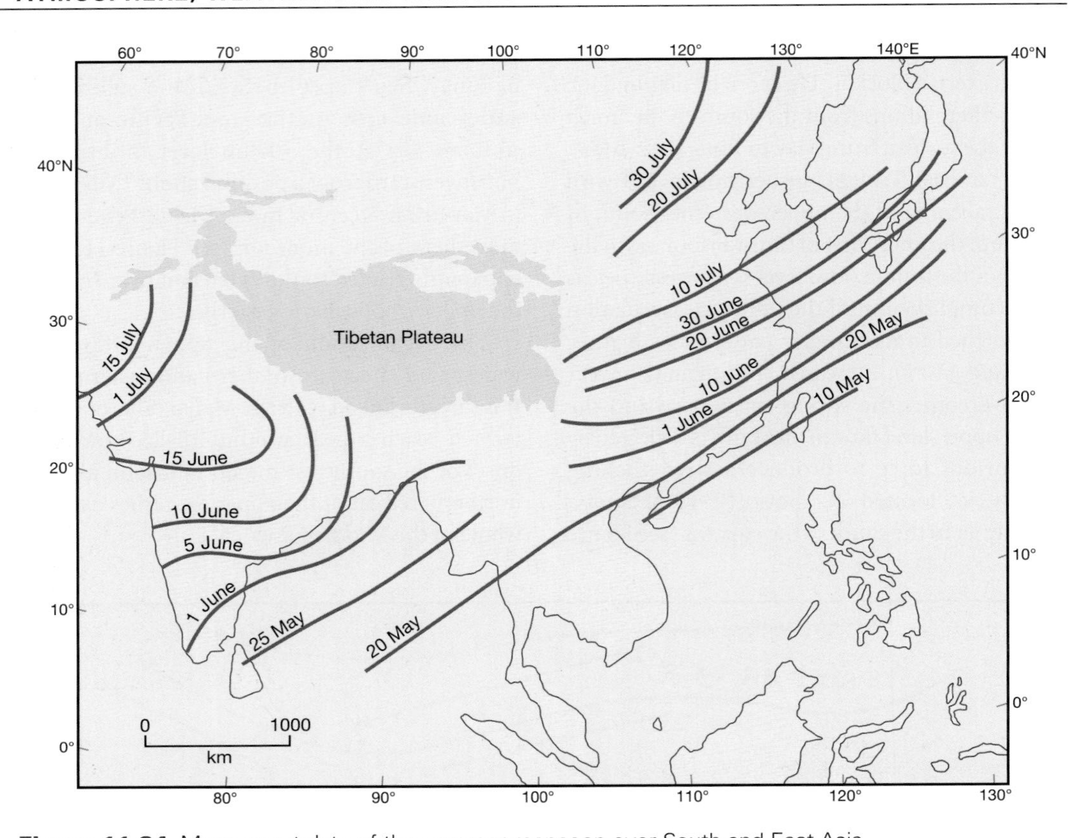

Figure 11.24 Mean onset date of the summer monsoon over South and East Asia.

Source: After Tao Shi-yan and Chen Longxun. From Domrös and Peng (1988). Reproduced by permission of Professor Tao Shi-yan and the Chinese Geographical Society. With kind permission of Springer Science & Business Media.

It must nevertheless be emphasized that it is still uncertain how far these changes are caused by events in the upper air or indeed whether the onset of the monsoon initiates a readjustment in the upper-air circulation. The presence of the Tibetan Plateau is certainly of importance even if there is no significant barrier effect on the upper airflow. The plateau surface is strongly heated in spring and early summer (Rn is about 180W m^{-2} in May) and nearly all of this is transferred via sensible heat to the atmosphere. This results in the formation of a shallow heat low on the plateau, overlain, at about 450mb, by a warm anticyclone (see Figure 7.1). The plateau atmospheric boundary layer now extends over an area about twice the size of the plateau surface itself. Easterly airflow on the southern side of the upper anticyclone undoubtedly assists in the northward shift of the subtropical westerly jet stream. At the same time, the pre-monsoonal convective activity over the southeastern rim of the plateau provides a further heat source, by latent heat release, for the upper anticyclone. The seasonal wind reversals over and around the Tibetan Plateau have led Chinese meteorologists to distinguish a 'Plateau Monsoon' system, distinct from that over India.

4 Summer

By mid-July, monsoon air covers most of South and Southeast Asia (see Figure 11.23), and in India the Equatorial Trough is located at about 25°N North of the Tibetan Plateau there is a rather weak upper westerly current with a (subtropical) high

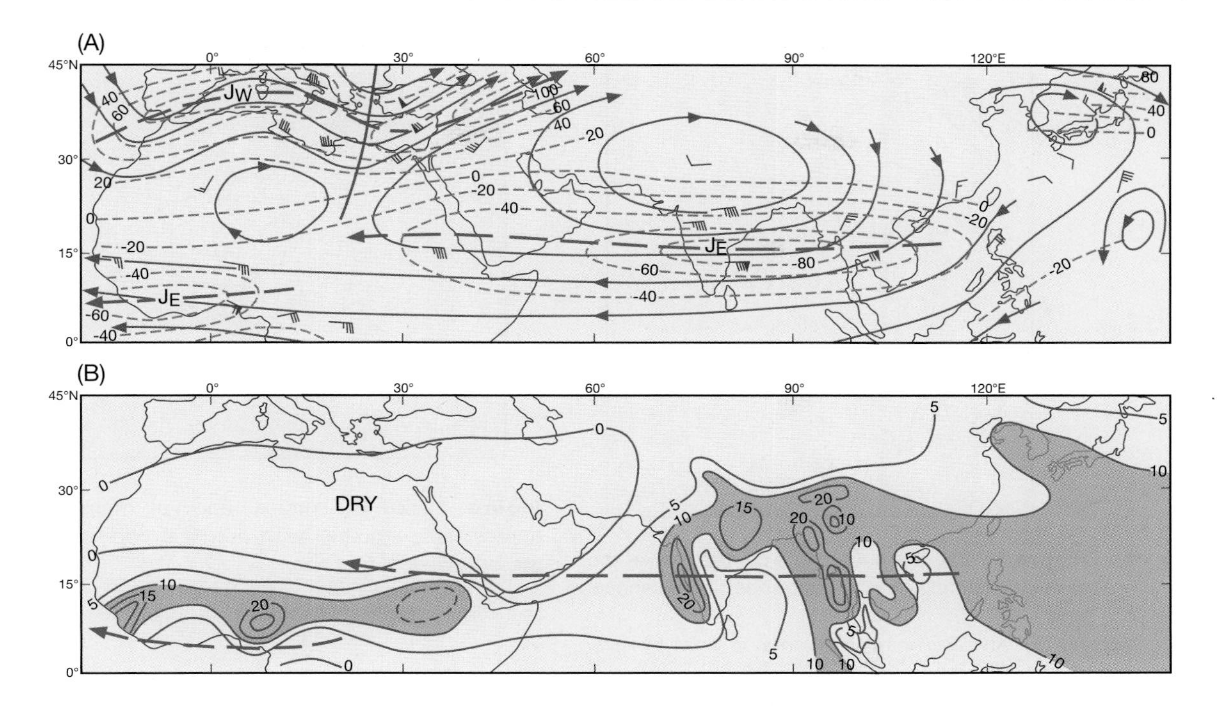

Figure 11.25 The easterly tropical jet stream. A: The location of the easterly jet streams at 200mb on 25 July 1955. Streamlines are shown in solid lines and isotachs (wind speed) dashed. Wind speeds are given in knots (westerly components positive, easterly negative). B: The average July rainfall (shaded areas receive more than 25cm) in relation to the location of the easterly jet streams.

Source: From Koteswaram (1958). By permission of Tellus.

pressure cell over the plateau. The southwest monsoon in South Asia is overlain by strong upper easterlies (see Figure 11.19) with a pronounced jet at 150mb (about 15km), which extends westward across Saudi Arabia and Africa (Figure 11.25). No easterly jets have so far been observed over the tropical Atlantic or Pacific. The jet is related to a steep lateral temperature gradient, with the upper air getting progressively colder to the south.

An important characteristic of the tropical easterly jet is the location of the main belt of summer rainfall on the right (*i.e.*, north) side of the axis upstream of the wind maximum and on the left side downstream, except for areas where the orographic effect is predominant (see Figure 11.25). The mean jet maximum is located at about 15 °N, 50–80°E.

The monsoon current does not give rise to a simple pattern of weather over India, despite the fact that much of the country receives 80 percent or more of its annual precipitation during the monsoon season (Figure 11.26). In the northwest, a thin wedge of monsoon air is overlain by subsiding continental air. The inversion prevents convection and consequently little or no rain falls in the summer months in the arid northwest of the subcontinent (e.g., Bikaner and Kalat: Figure 11.20). This is similar to the Sahel zone in West Africa, discussed below.

Around the head of the Bay of Bengal and along the Ganges valley the main weather mechanisms in summer are the 'monsoon depressions' (Figure 11.27), which usually move westward or northwestward across India, steered by the upper easterlies (Figure 11.28), mainly in

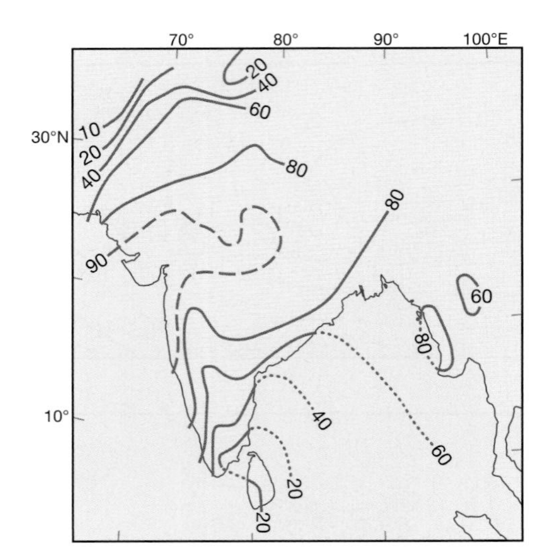

Figure 11.26 The percentage contribution of the monsoon rainfall (June to September) to the annual total.

Sources: After Rao and Ramamoorthy, in Indian Meteorological Department (1960); and Ananthakrishnan and Rajagopalachari, in Hutchings (1964).

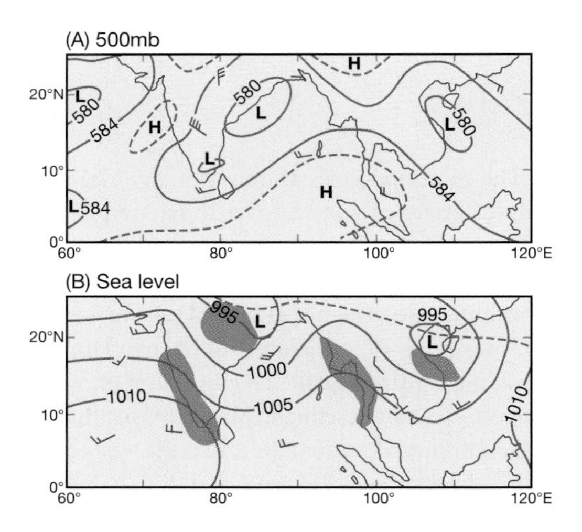

Figure 11.27 Monsoon depressions of 12:00 GMT, 4 July 1957. A: shows the height (in tens of meters) of the 500mb surface, B: the sea-level isobars. The broken line represents the Equatorial Trough, and precipitation areas are shown by the oblique shading.

Source: Based on the IGY charts of the Deutscher Wetterdienst.

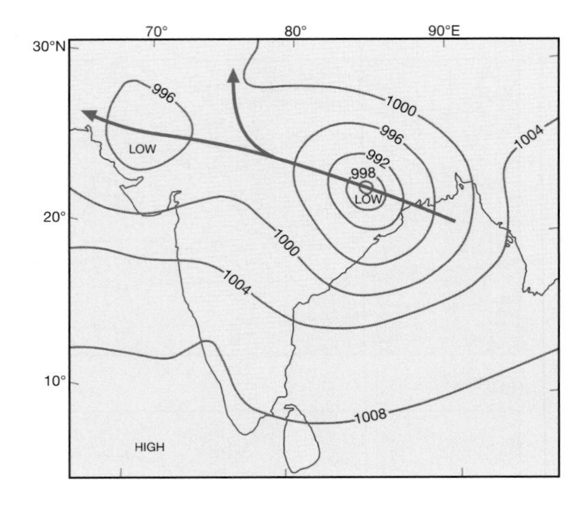

Figure 11.28 The normal track of monsoon depressions, together with a typical depression pressure distribution (mb).

Source: After Das (1987). Copyright © 1987. Reproduced by permission of John Wiley & Sons, Inc.

July and August. On average, they occur about twice a month, apparently when an upper trough becomes superimposed over a surface disturbance in the Bay of Bengal. Monsoon depressions have cold cores, are generally without fronts and are some 1000–1250km across, with a cyclonic circulation up to about 8km, and a typical lifetime of two to five days. They produce average daily rainfalls of 120–200mm, occurring mainly as convective rains in the southwest quadrant of the depression. The main rain areas typically lie south of the Equatorial or Monsoon Trough (Figure 11.29) (in the southwest quadrant of the monsoon depressions, resembling an inverted mid-latitude depression). Figure 11.30 shows the extent and magnitude of a particularly severe monsoon depression. Such storms occur mainly in two zones: (1) the Ganges valley east of 76°E; (2) a belt across central India at around 21°N, at its widest covering 6° of latitude. Monsoon depressions also tend to occur on the windward coasts and mountains of India, Burma and Malaysia. Without such disturbances, the distribution of monsoon rains would be controlled to a much larger degree by orography.

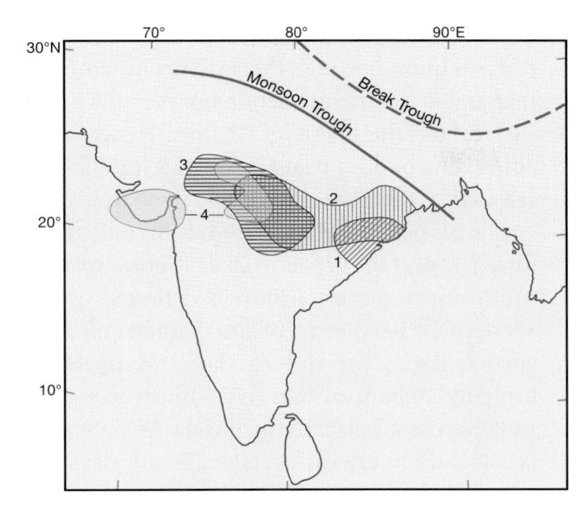

Figure 11.29 The location of the Monsoon Trough in its normal position during an active summer monsoon phase (solid) and during breaks in the monsoon (dashed).* Areas 1–4 indicate four successive daily areas of heavy rain (>50mm/day) during the period 7–10 July 1973 as a monsoon depression moved west along the Ganges valley. Areas of lighter rainfall were much more extensive.

Source: *After Das (1987). Copyright © 1987. Reproduced by permission of John Wiley & Sons, Inc.

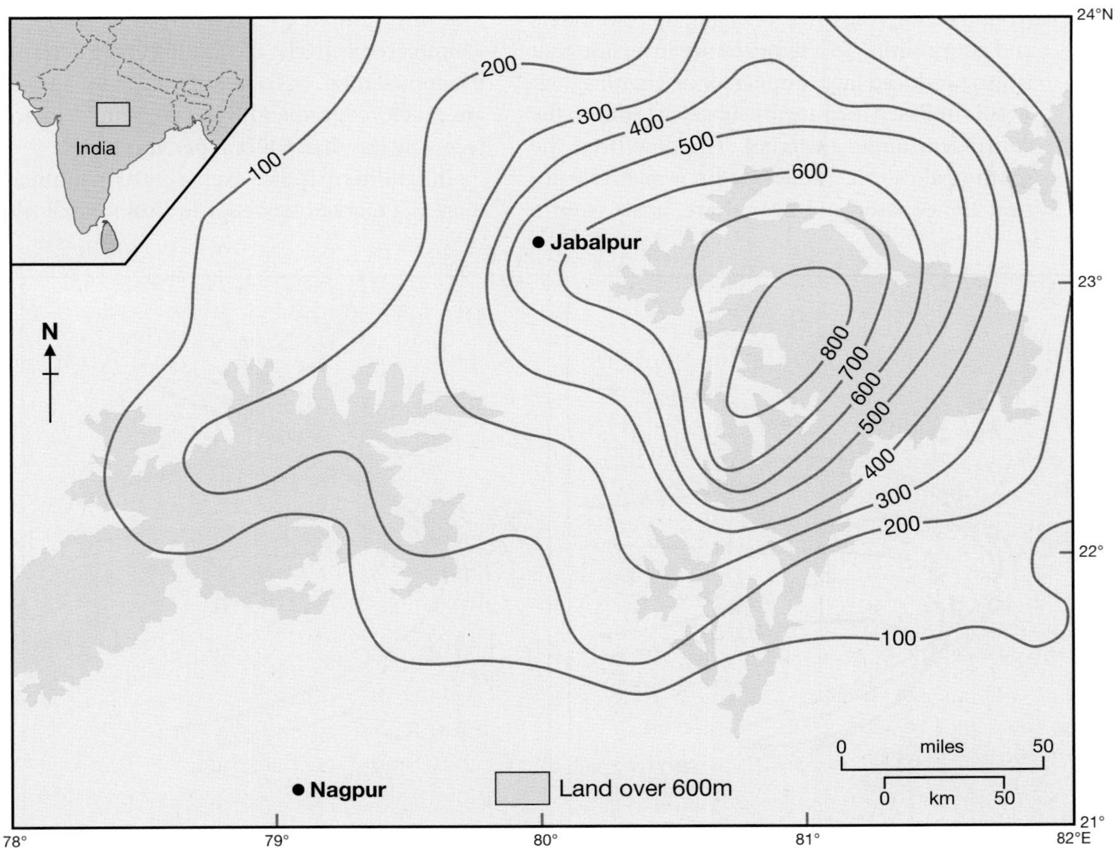

Figure 11.30 Rainfall (mm) produced in three days over a 50,000km^2 area of central India northeast of Nagpur by a severe, westward-moving monsoon depression, during September 1926.

Source: Dhar and Nandargi (1993). Copyright © John Wiley & Sons Ltd. Reproduced with permission.

A key part of the southwest monsoonal flow occurs in the form of a 15–45m s^{-1} jet stream at a level of only 1000–1500m. This jet, strongest during active periods of the Indian monsoon, flows northwestwards from Madagascar (Figure 11.31) and crosses the equator from the south over East Africa, where its core is often marked by a streak of cloud and where it may bring excessive local rainfall. The jet is displaced northward and strengthens from February to July; by May it has become constricted against the Ethiopian Highlands, it accelerates still more and is deflected eastward across the Arabian Sea towards the west coast of the Indian peninsula. This low-level jet, unique in the Trade Wind belt, flows offshore from the Horn of Africa, bringing up cool waters and contributing to a temperature inversion that is also produced by dry upper air originating over Arabia or East Africa and by subsidence due to the convergent upper easterlies. The flow from the southwest over the Indian Ocean is relatively dry near the equator and near shore, apart from a shallow, moist layer near the base. Downwind towards India, however, there is a strong temperature and moisture interaction between the ocean surface and the low-level jet flow. Hence, deep convection builds up and convective instability is released, especially as the airflow slows down and converges near the west coast of India and as it is forced up over the Western Ghats. A portion of this southwest monsoon airflow is deflected by the Western Ghats to form 100km diameter offshore vortices lasting two to three days and capable of bringing 100mm of rain in 24 hours along the western coastal belt of the peninsula. At Mangalore (13°N), there are on average 25 rain days per month in June, 28 in July and 25 in August. The monthly rainfall averages are 980, 1060 and 580mm, respectively, accounting for 75 percent of the annual total. In the lee of the Ghats, amounts are much reduced and there are semi-arid areas receiving less than 640mm per year.

In southern India, excluding the southeast, there is a marked tendency for less rainfall when

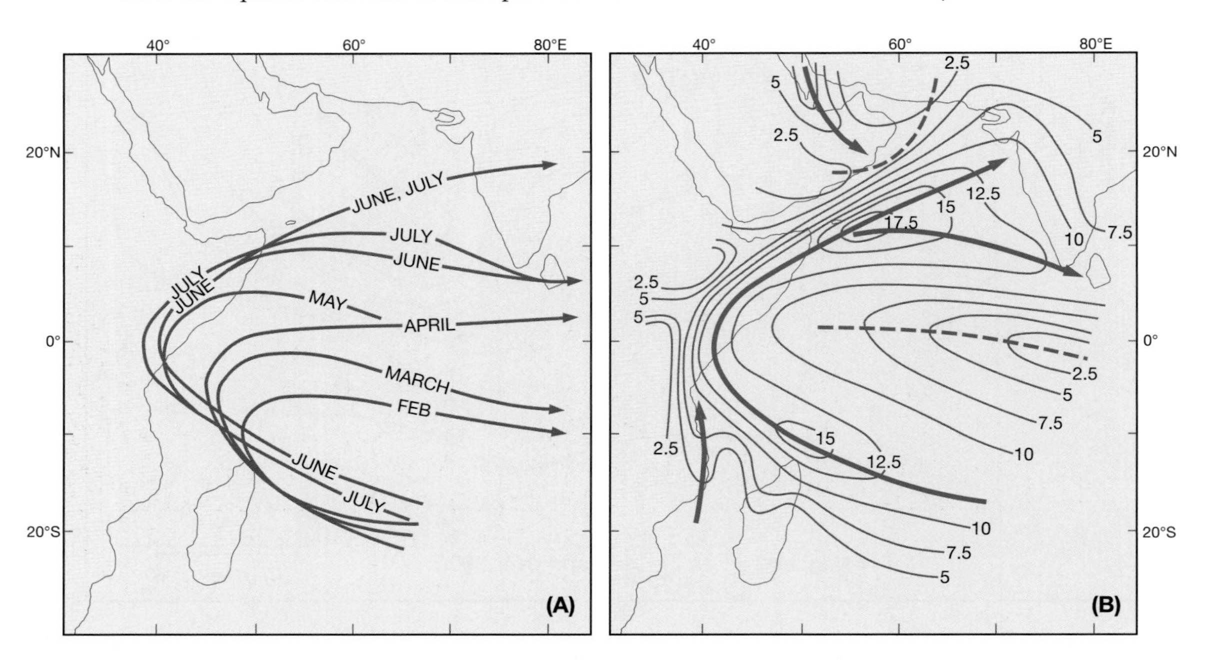

Figure 11.31 The mean monthly positions (A) and the mean July velocity (m s^{-1}) (B) of the low-level (1km) Somali jet stream over the Indian Ocean.

Source: After Findlater (1971). Reproduced by permission of the Controller of Her Majesty's Stationery Office.

the Equatorial Trough is farthest north. Figure 11.20 shows a maximum at Minicoy in June, with a secondary peak in October as the Equatorial Trough and its associated disturbances withdraw southward. This double peak occurs in much of interior peninsular India south of about 20°N and in western Sri Lanka, although autumn is the wettest period.

There is a variable pulse alternating between active and break periods in the May to September summer monsoon flow (see Figure 11.16) which, particularly at times of its strongest expression (e.g., 1971), produces periodic rainfall (Figure 11.32). During active periods, the convective Monsoon Trough is located in a southerly position, giving heavy rain over north and central India and the west coast (see Figure 11.16). Consequently, there is a strong upper-level outflow to the south, which strengthens both the easterly jet north of the equator and the westerly jet to the south over the Indian Ocean. The other upper-air outflow to the north fuels the weaker westerly jet there. Convective activity moves east from the Indian Ocean to the cooler eastern Pacific with an irregular periodicity (on average 40–50 days for strong waves; Note 1), finding maximum expression at the 850mb level and clearly being connected with the Walker circulation. After the passage of an active convective wave there is a more stable break in the summer monsoon when the ITCZ shifts to the south. The easterly jet now weakens and subsiding air is forced to rise by the Himalayas along a break trough located above the foothills (see Figure 11.16), which replaces the Monsoon Trough during break periods. This circulation brings rain to the foothills of the Himalayas and the Brahmaputra valley at a time of generally low rainfall elsewhere. The shift of the ITCZ to the south of the subcontinent is associated with a

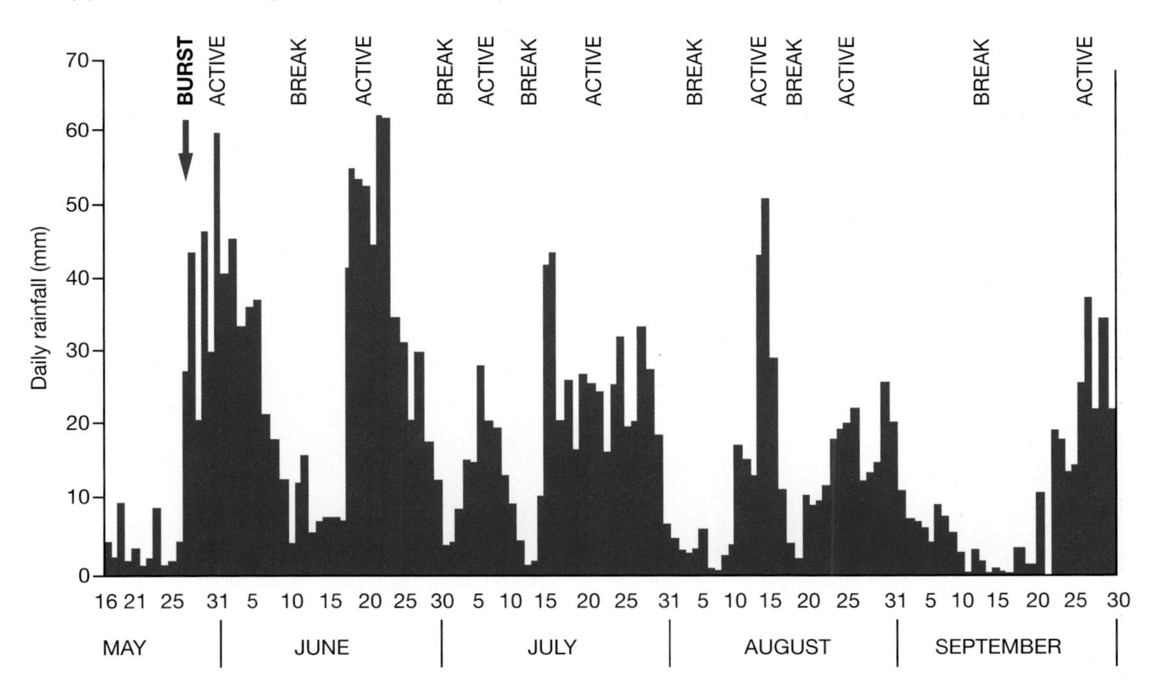

Figure 11.32 Mean daily rainfall (mm) along the west coast of India during the period 16 May to 30 September 1971, showing a pronounced burst of the monsoon followed by active periods and breaks of a periodic nature. All years do not exhibit these features as clearly.

Source: After Webster (1987b). Copyright © 1987. Reproduced by permission of John Wiley & Sons, Inc.

similar movement and strengthening of the westerly jet to the north, weakening the Tibetan anticyclone or displacing it northeastward. The lack of rain over much of the subcontinent during break periods may be due in part to the eastward extension across India of the subtropical high pressure cell centered over Arabia at this time.

It is important to realize that the monsoon rains are highly variable from year to year, emphasizing the role played by disturbances in generating rainfall within the generally moist southwesterly airflow. Droughts occur with some regularity in the Indian subcontinent: between 1890 and 1975 there were nine years of extreme drought (Figure 11.33) and at least five other years of significant drought. These droughts are brought about by a combination of a late burst of the summer monsoon and an increase in the number and length of the break periods. Breaks are most common in August to September, lasting on average five days, but they may occur at any time during the summer and may last for up to three weeks.

The strong surface heat source over the Tibetan Plateau, which is most effective during the day, gives rise to a 50–85 percent frequency of deep cumulonimbus clouds over central and eastern Tibet in July. Late afternoon rain or hail showers are generally accompanied by thunder, but half or more of the precipitation falls at night, accounting for 70–80 percent of the total in south-central and southeastern Tibet. This may be related to large-scale plateau-induced local wind systems. However, the central and eastern plateau also has a frequency maximum of shear lines and associated weak lows at 500mb during May to September. These plateau systems are shallow (2–2.5km) and only 400–1000km in diameter, but they are associated with cloud clusters on satellite imagery in summer.

5 Autumn

Autumn sees the southward swing of the Equatorial Trough and the zone of maximum convection, which lies just to the north of the weakening easterly jet (see Figure 11.16). The breakup of the summer circulation systems is associated with the withdrawal of the monsoon rains, which is much less clearly defined than their onset (Figure 11.34). By October, the easterly trades of the Pacific affect the Bay of Bengal at the 500mb level and generate disturbances at their confluence with the equatorial westerlies. This is

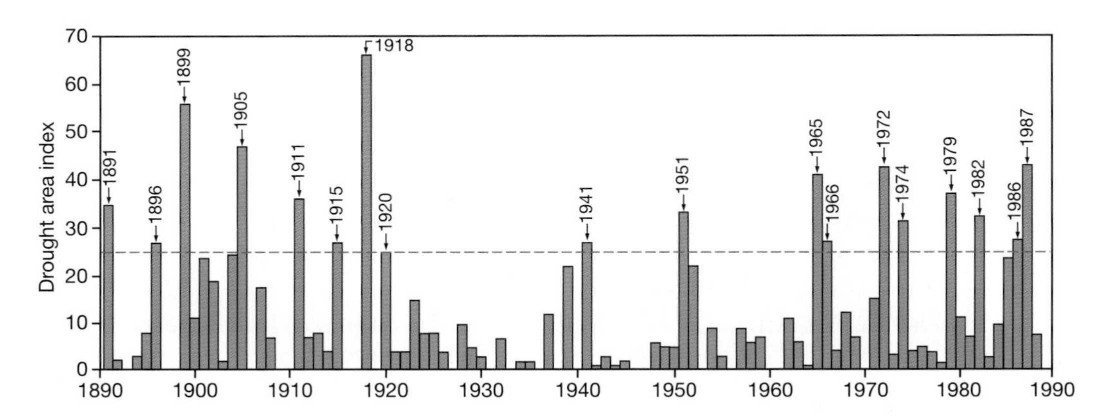

Figure 11.33 The yearly drought area index for the Indian subcontinent for the period 1891–1988, based upon the percentage of the total area experiencing moderate, extreme or severe drought. Years of extreme drought are dated. The dashed line indicates the lower limit of major droughts.

Source: After Bhalme and Mooley (1980). Updated by courtesy of H. M. Bhalme. Reproduced by permission of the American Meteorological Society.

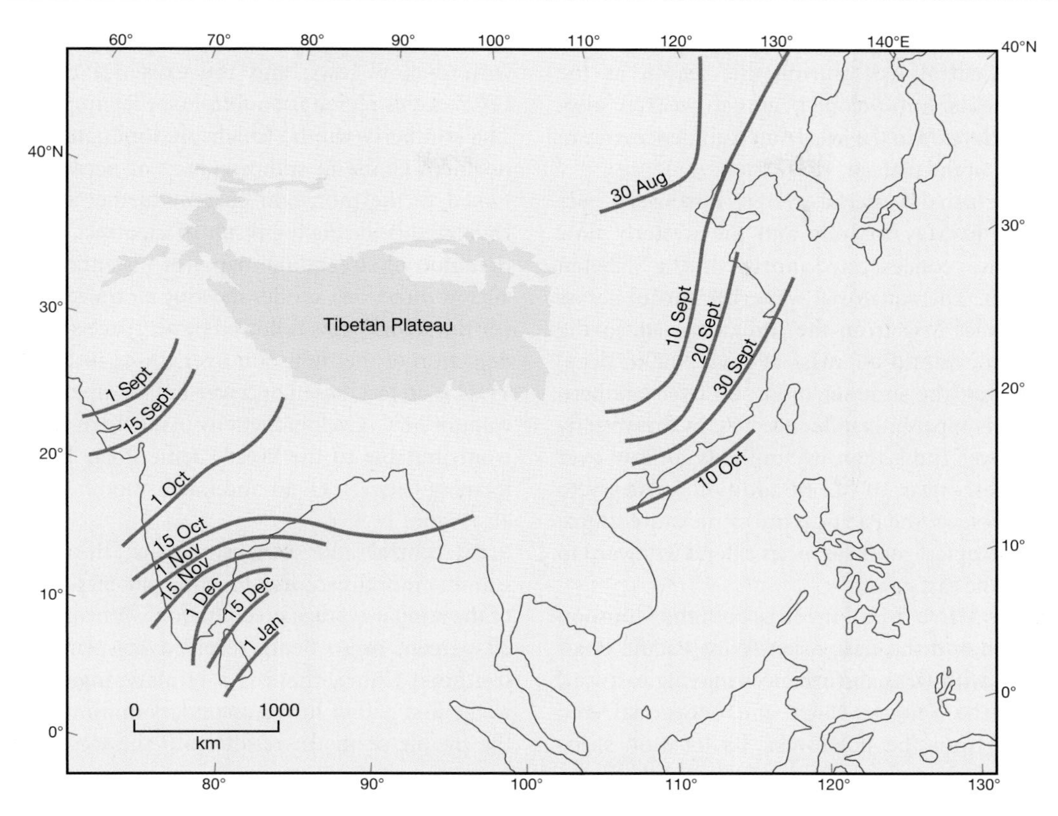

Figure 11.34 Mean onset date of the winter monsoon (i.e., retreat of the summer monsoon) over South and East Asia.

Sources: After Tao Shi-yan and Chen Longxun. Reproduced by permission of Professor Tao Shi-yan and the Chinese Geographical Society.

the major season for Bay of Bengal cyclones and it is these disturbances, rather than the onshore northeasterly monsoon, that cause the October/November maximum of rainfall in southeast India (e.g., Madras: Figure 11.20).

During October, the westerly jet re-establishes itself south of the Tibetan Plateau, often within a few days, and cool season conditions are restored over most of South and East Asia.

D EAST ASIAN AND AUSTRALIAN SUMMER MONSOONS

China has no equivalent to India's hot, pre-monsoon season. The low-level, northeasterly winter monsoon (reinforced by subsiding air from the upper westerlies) persists in north China, and even in the south it begins to be replaced by maritime tropical air only in April to May. Thus, at Guangzhou (Canton), mean temperatures rise from only 17°C in March to 27°C in May, some 6°C lower than the mean values over northern India.

The rains in western China begin earlier in the northwest in mid-May and then extend southward and eastward until mid-June. Also during this season upper-level cold lows forming east of Lake Baikal affect northeastern China, contributing 20–60 percent of the warm season rainfall and over half of the hailstorms. Westerly depressions are most frequent over China in spring (see Figure 11.22). They form more readily over Central Asia during this season as the continental anticyclone begins to weaken; also, many develop in the jet stream confluence zone in the lee of the plateau.

The zonal westerlies retreat northward over China in May to June and the westerly flow becomes concentrated north of the Tibetan Plateau. The equatorial westerlies spread across Southeast Asia from the Indian Ocean, giving a warm, humid air mass at least 3000m deep. However, the summer monsoon over southern China is apparently influenced less by the westerly flow over India than by southerly airflow over Indonesia near 100°E. In addition, contrary to earlier views, the Pacific is only a moisture source when tropical southeasterlies extend westward to affect the east coast.

The Maiyu 'front' involves both the Monsoon Trough and the East Asian–West Pacific Polar Front, with weak disturbances moving eastward along the Yangtze valley and occasional cold fronts from the northwest. Its location shifts northward in three stages, from south of the Yangtze River in early May to north of it by the end of the month and into northern China in mid-July (see Figure 11.24), where it remains until late September.

The surface airflow over China in summer is southwesterly (Table 11.2) and the upper winds are weak, with only a diffuse easterly current over southern China. According to traditional views, the monsoon current reaches northern China by July.

The annual rainfall regime shows a distinct summer maximum with, for example, 64 percent of the annual total occurring at Tianjin (Tientsin) (39°N) in July and August. Nevertheless, much of the rain falls during thunderstorms associated with shallow lows, and the existence of the ITCZ in this region is doubtful (see Figure 11.1). The southerly winds, which predominate over northern China in summer, are not necessarily linked to the monsoon current further south. Indeed, this idea is the result of incorrect interpretation of streamline maps (of instantaneous airflow direction) as ones showing air trajectories (or the actual paths followed by air parcels). The depiction of the monsoon over China in Figure 11.24 is, in fact, based on a wet-bulb temperature value of 24°C. Cyclonic activity in northern China is attributable to the West Pacific Polar Front, forming between cP air and much-modified mT air (Figure 11.35).

In central and southern China, the three summer months account for about 40–50 percent of the annual average precipitation, with another 30 percent or so being received in spring. In southeast China, there is a rainfall singularity in the first half of July; a secondary minimum in the profile seems to result from the westward extension of the Pacific subtropical anticyclone over the coast of China. A strong southeast Asia monsoon (20–30°N, 110–145°E) is related to higher SSTs in the western North Pacific, which weaken the subtropical anticyclone and allow more cyclonic circulations.

A similar pattern of rainfall maxima occurs over the Korean peninsula and over southern and central Japan (Figure 11.36), comprising two of the six natural seasons that have been recognized there. The main rains occur during the Bai-u season of the southeast monsoon resulting from waves, convergence zones and closed circulations moving mainly in the tropical airstream around the Pacific subtropical anticyclone, but partly

Table 11.2 Surface circulation over China

	January	July
North China	60% of winds from W, NW and N	57% of winds from SE, S and SW
Southeast China	88% of winds from N, NE and E	56% of winds from SE, S and SW

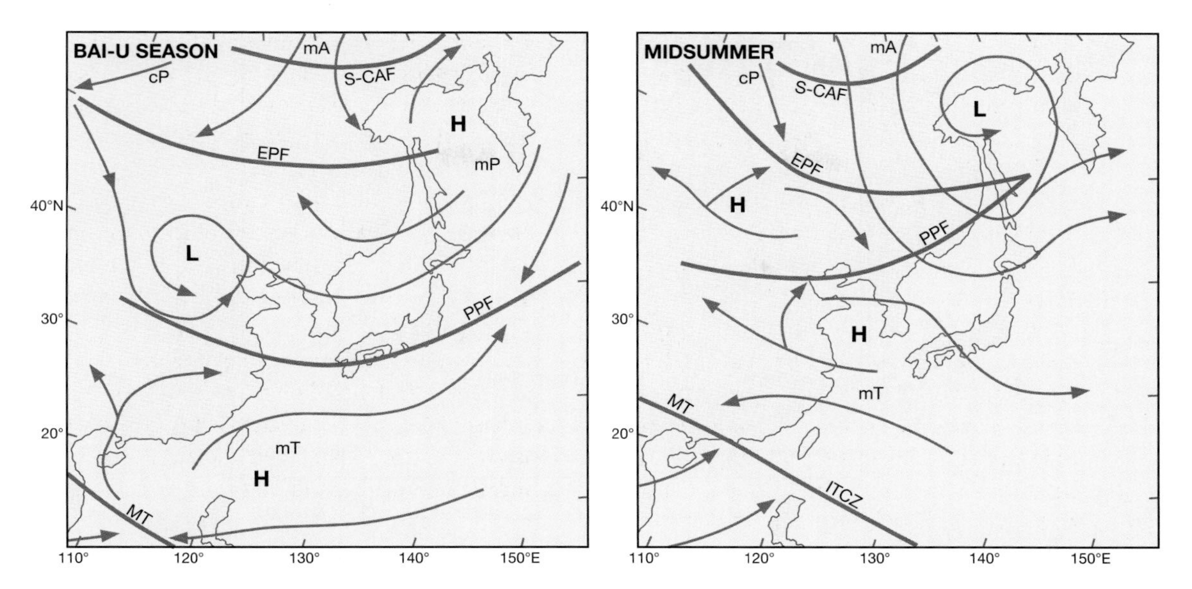

Figure 11.35 Schematic surface circulation pattern and frontal locations (Siberian–Canadian Arctic Front S–C AF, Eurasian Polar Front EPF, Pacific Polar Front PPF and Monsoon Trough MT/ Inter-tropical Convergence Zone MT/ITCZ) over East Asia during the Bai-u (i.e., July–August) season.

Source: Matsumoto (1985). Reproduced by permission, University of Tokyo.

originating in a southwesterly stream that is the extension of the monsoon circulation of Southeast Asia (Figure 11.23). The southeast circulation is displaced westward from Japan by a zonal expansion of the subtropical anticyclone during late July and August, giving a period of more settled, sunny weather. The secondary precipitation maximum of the Shurin season during September and early October coincides with an eastward contraction of the Pacific subtropical anticyclone, allowing low pressure systems and typhoons from the Pacific to swing north towards Japan. Although much of the Shurin rainfall is of typhoon origin (see Figure 11.36), some is undoubtedly associated with the southern sides of depressions moving along the southward-migrating Pacific Polar Front to the north (see Figures 11.22 and 11.35), because there is a marked tendency for the autumn rains to begin first in the north of Japan and to spread southward. The manner in which the location of the western margin of the North Pacific subtropical high pressure cell affects the climates of China and Japan is well illustrated by the changing seasonal trajectories of typhoon paths over East Asia (Figure 11.37). The northward and southward migrations of the cell zonal axis through 15° of latitude, the northwestern high pressure cell extensions over eastern China and the Sea of Japan in August, and its southeastern contraction in October are especially marked.

Northern Australia experiences a monsoon regime during the austral summer. Low-level westerlies develop in late December associated with a thermal low over northern Australia. Analogous to the vertical wind structure over Asia in July, there are easterlies in the upper troposphere. Various wind and rainfall criteria have been used to define monsoon onset. Based on the occurrence of (weighted) surface to 500mb westerly winds, overlain by 300–100mb easterlies at Darwin (12.5°S, 131°E), the main onset date is 28 December and retreat date 13 March. Despite an average duration of 75 days, monsoon

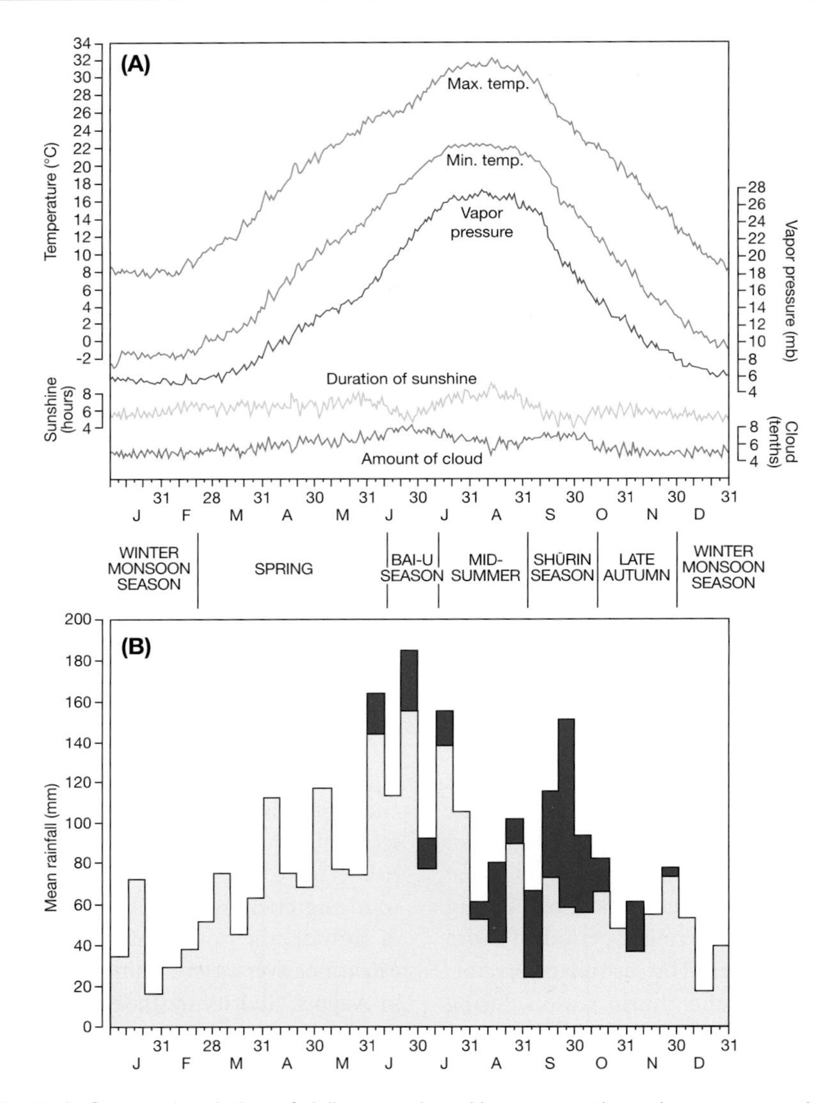

Figure 11.36 A: Seasonal variation of daily normals at Nagoya, southern Japan, suggesting six natural seasons.* B: Average 10-day precipitation amounts for a station in southern Japan, indicating in black the proportion of rainfall produced by typhoon circulations. The latter reaches a maximum during the Shurin season.†

Sources: *From Maejima (1967). †After Saito (1959), from Trewartha (1981). By permission of the University of Wisconsin Press.

conditions lasted for only 10 days in January 1961 and 1986 but for 123–125 days in 1985 and 1974. Active phases with deep westerlies and rainfall each occur on just over half of the days in a season, although there is little overlap between them. However, summer rainfall may also occur during deep easterlies associated with tropical squall-lines and tropical cyclones. Active monsoon conditions typically persist for four to 14 days, with breaks lasting for about 20 to 40 days.

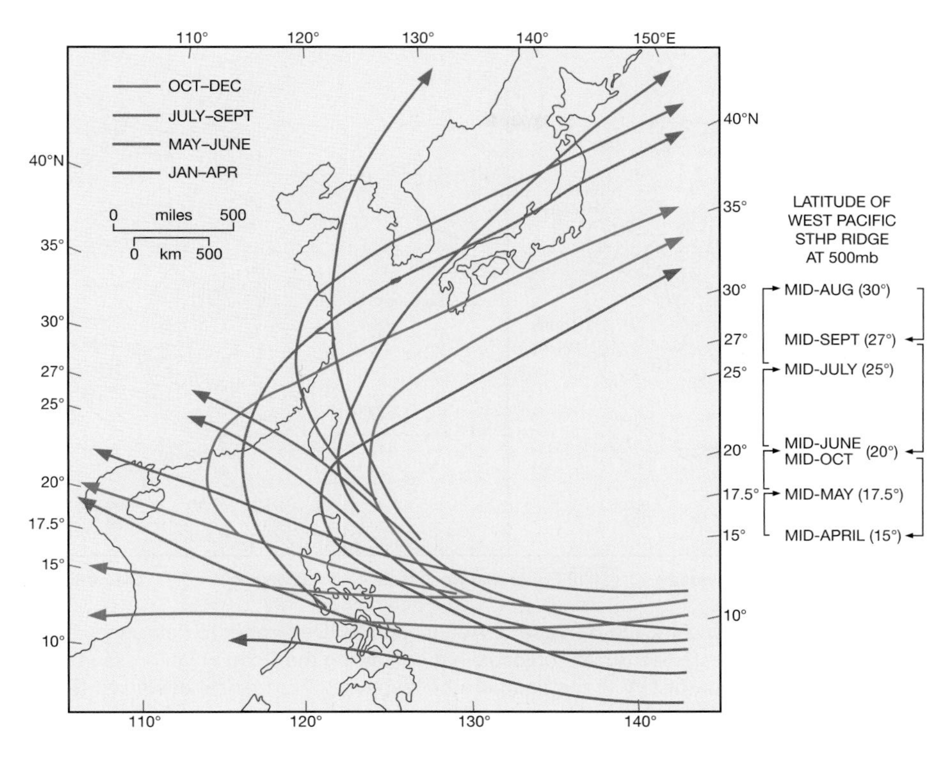

Figure 11.37 Typhoon paths over East Asia during January to April, May to June, July to September and October to December related to the mean latitude of the central ridge axis line of the subtropical high pressure cell at 500mb over the western Pacific.

Sources: Compiled from various sources, including Lin (1982) and Tao (1984). Reproduced by permission of the Chinese Geographical Society.

E CENTRAL AND SOUTHERN AFRICA

1 The African monsoon

The annual climatic regime over West Africa has many similarities to that over South Asia, the surface airflow being determined by the position of the leading edge of a Monsoon Trough (see Figure 11.2). This airflow is southwesterly to the south of the trough and easterly to northeasterly to its north (Figure 11.38). The major difference between the circulations of the two regions is due largely to the differing geography of the land–sea distribution and to the lack of a large mountain range to the north of West Africa. This allows the Monsoon Trough to migrate regularly with the seasons. In general, the West African Monsoon Trough oscillates between annual extreme locations of about 2°N and 25°N (Figure 11.39). In 1956, for example, these extreme positions were 5°N on 1 January and 23°N in August. The leading edge of the Monsoon Trough is complex in structure (see Figure 11.40B) and its position may oscillate greatly from day to day through several degrees of latitude. The classical model of a steady northward advance of the monsoon has recently been called into question. The rainy season onset in February at the coast does propagate northward to 13°N in May, but then in mid-June there is a sudden synchronous onset of rains between about 9°N and 13°N. The mechanism is not yet firmly

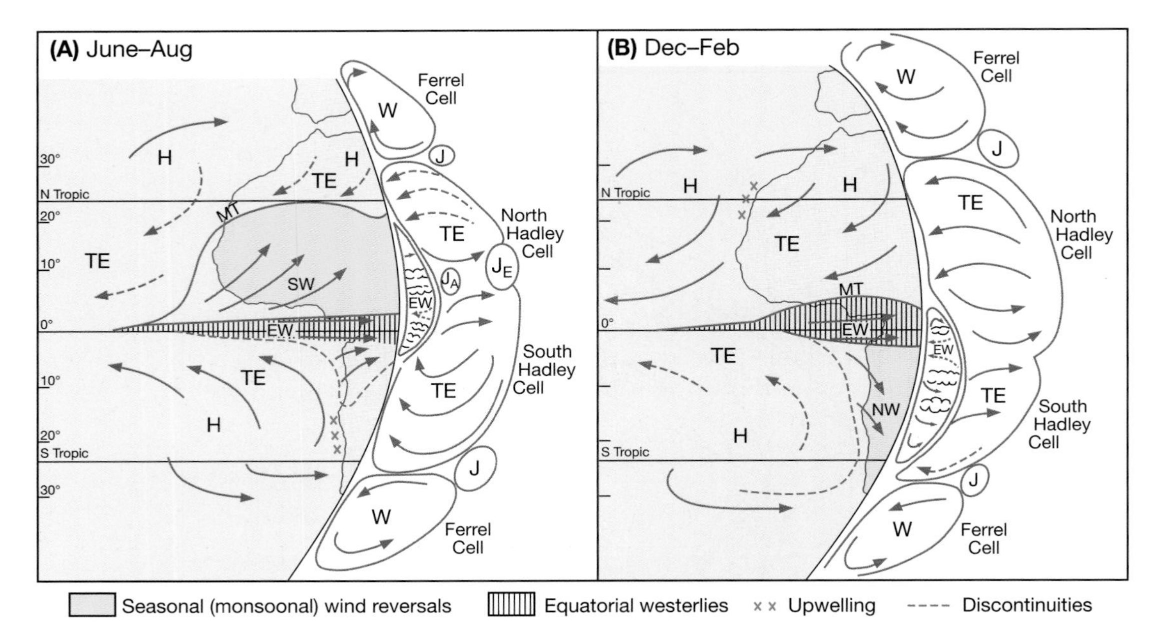

Figure 11.38 The major circulation in Africa in (A) June to August and (B) December to February. H: subtropical high pressure cells; EW: equatorial westerlies (moist, unstable but containing the Congo high pressure ridge); NW: the northwesterlies (summer extension of EW in the southern hemisphere); TE: tropical easterlies (Trades); SW: southwesterly monsoonal flow in the Northern Hemisphere; W: extratropical westerlies; J: subtropical westerly jet stream; JA and JE: the (easterly) African jet streams; and MT: Monsoon Trough.

Source: From Rossignol-Strick (1985). By permission of Elsevier Science Publishers B.V., Amsterdam.

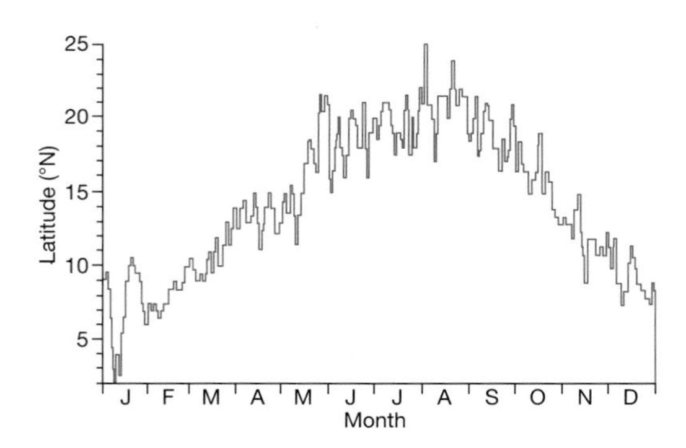

Figure 11.39 The daily position of the Monsoon Trough at longitude 3°E during 1957. This year experienced an exceptionally wide swing over West Africa, with the trough reaching 2°N in January and 25° N on 1 August. Within a few days after the latter date, the strongly oscillating trough had swung southward through 8° latitude.

Source: After Clackson (1957), from Hayward and Oguntoyinho (1987). By permission of Hutchinson.

established, but it involves a shift of the lower tropospheric African Easterly Jet (AEJ) (see Figure 11.40b).

In winter, the southwesterly monsoon airflow over the coasts of West Africa is very shallow (i.e., 1000m) with 3000m of overriding easterly winds, which are themselves overlain by strong (>20m s^{-1}) winds (see Figure 11.41). North of the Monsoon Trough, the surface northeasterlies (*i.e.*, the 2000m deep Harmattan flow) blow clockwise outward from the subtropical high pressure center. They are compensated above 5000m by an anticlockwise westerly airflow that, at about 12,000m and 20–30°N, is concentrated into a subtropical westerly jet stream of average speed 45m s^{-1}. Mean January surface temperatures decrease from about 26°C along the southern coast to 14°C in southern Algeria.

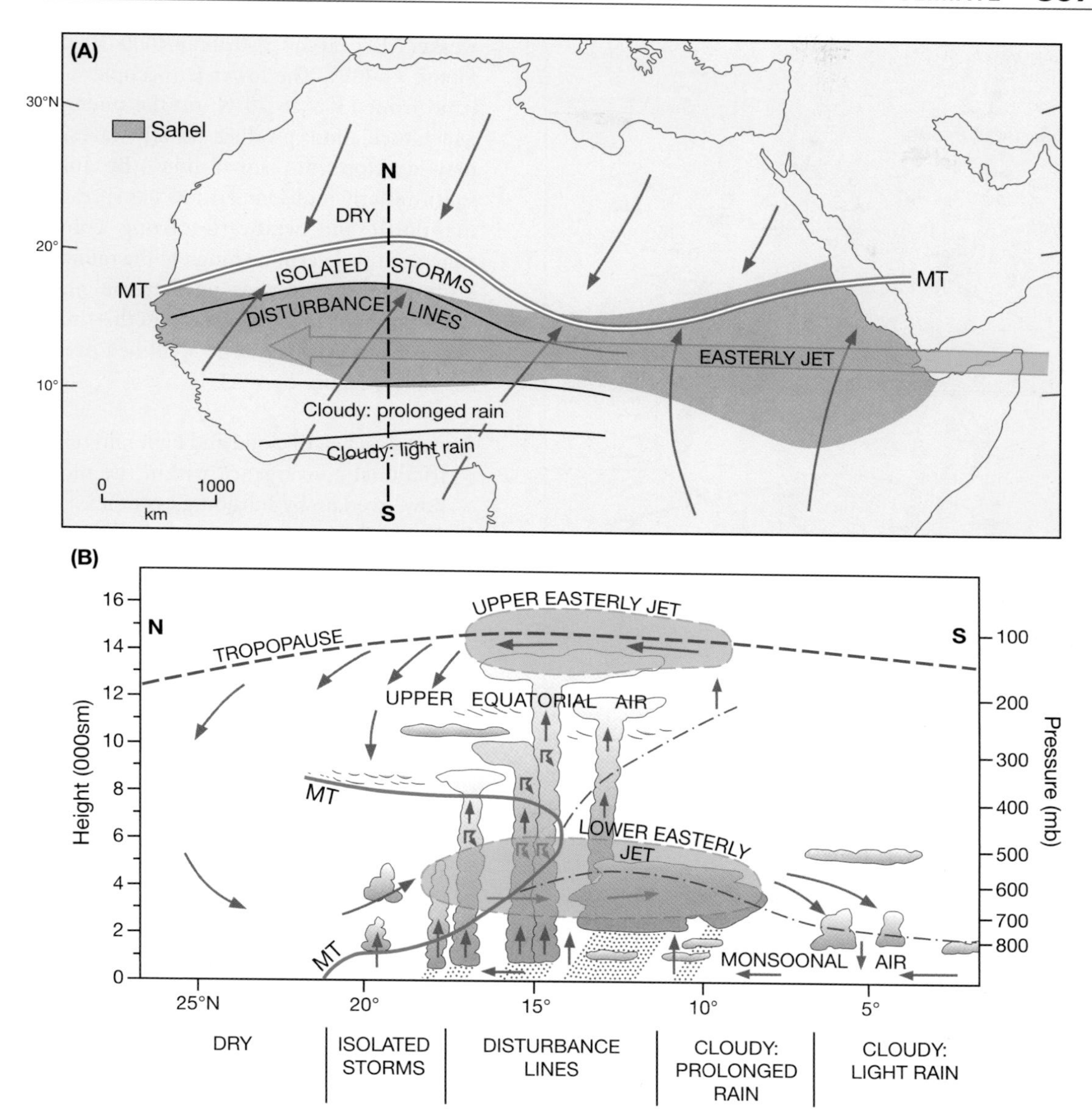

Figure 11.40 The structure of the circulation over North Africa in August. A: Surface airflow and easterly tropical jet. B: Vertical structure and resulting precipitation zones over West Africa. MT = Monsoon Trough. *Note* there is thunderstorm activity associated with the cumulonimbus towers.

Sources: A: Reproduction from the *Geographical* magazine, London. B: From Maley (1982) *Quaternary Research*. Copyright © Academic Press; reproduced by permission. Musk (1983). By permission of the *Geographical* magazine.

With the approach of the northern summer, the strengthening of the South Atlantic subtropical high pressure cell, combined with the increased continental temperatures, establishes a strong southwesterly airflow at the surface that spreads northward behind the Monsoon Trough, lagging about six weeks behind the progress of the overhead sun. The northward migration of the

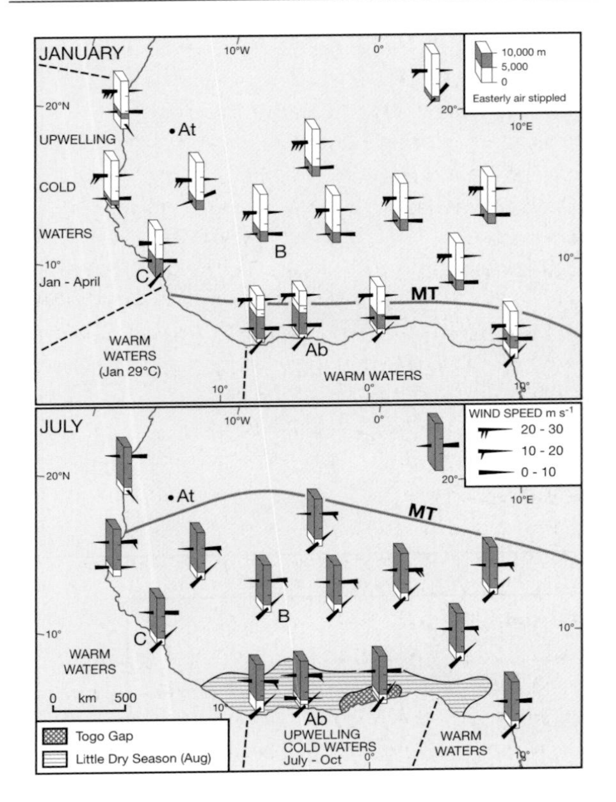

Figure 11.41 Mean wind speeds (m s^{-1}) and directions in January and July over West Africa up to about 15,000m. Ocean water temperatures and the positions of the Monsoon Trough are also shown, as are the area affected by the August Little Dry Season and the location of the anomalous Togo Gap. The locations of Abidjan (Ab), Atar (At), Bamako (B) and Conakry (C) are given (see precipitation graphs in Figure 11.42).

Source: From Hayward and Oguntoyinbo (1987). By permission of Rowman and Littlefield.

trough oscillates diurnally with a northward progress of up to 200km in the afternoons following a smaller southward retreat in the mornings. The northward spread of moist, unstable and relatively cool southwesterly airflow from the Gulf of Guinea brings rain in differing amounts to extensive areas of West Africa. Aloft, easterly winds spiral clockwise outward from the subtropical high pressure center (see Figure 11.41) and are concentrated between June and August into two tropical easterly jet streams; the stronger TEJ (>20m s^{-1}) at about 15,000–20,000m and the weaker AEJ (>10m s^{-1}) at about 4000–5000m (see Figure 11.40B). The lower jet occupies a broad band from 13°N to 20°N, on the underside of which oscillations produce easterly waves, which may develop into squall-lines. By July, the southwesterly monsoon airflow has spread far to the north and westward-moving convective systems now determine much of the rainfall. The leading trough reaches its extreme northern location, about 20°N, in August. At this time, four major climatic belts can be identified over West Africa (see Figure 11.40A):

1. A coastal belt of cloud and light rain related to frictional convergence within the monsoon flow, overlain by subsiding easterlies.
2. A quasi-stationary zone of disturbances associated with deep stratiform cloud yielding prolonged light rains. Low-level convergence south of the easterly jet axes, apparently associated with easterly wave disturbances from east central Africa, causes instability in the monsoon air.
3. A broad zone underlying the easterly jet streams, which help to activate disturbance lines and thunderstorms. North–south lines of deep cumulonimbus cells may move westward steered by the jets. The southern, wetter part of this zone is termed the Soudan, the northern part the Sahel, but popular usage assigns the name Sahel to the whole belt.
4. Just south of the Monsoon Trough, the shallow tongue of humid air is overlain by drier subsiding air. Here there are only isolated storms, scattered showers and occasional thunderstorms.

In contrast to winter conditions, August temperatures are lowest (24–25°C) along the cloudy southern coasts and increase towards the north, where they average 30°C in southern Algeria.

Both the summer airflows, the southwesterlies below and the easterlies aloft, are subject to perturbations, which contribute significantly to the rainfall during this season. Three types of perturbation are particularly prevalent:

1. Waves in the southwesterlies. These are northward surges of the humid airflow with periodicities of four to six days. They produce bands of summer monsoon rain some 160km broad and 50–80km in north–south extent, which have the most marked effect 1100–1400km south of the surface Monsoon Trough, the position of which oscillates with the surges.
2. Waves in the easterlies. These develop on the interface between the lower southwesterly and the upper easterly airflows. These waves are from 1500 to 4000km long from north to south. They move westward across West Africa between mid-June and October with a periodicity of three to five days and sometimes developing closed cyclonic circulations. Their speed is about 5–10° longitude per day (*i.e.*, 18–35km hr^{-1}). At the height of the summer monsoon, they produce most rainfall at around latitude 14°N, between 300 and 1100km south of the Monsoon Trough. On average, some 50 easterly waves per year cross Dakar. Some of these carry on in the general circulation across the Atlantic, and it has been estimated that 60 percent of West Indian hurricanes originate in West Africa as easterly waves.
3. Squall-lines. Easterly waves vary greatly in intensity. Some give rise to little cloud and rain, whereas others have embedded squall-lines when the wave extends down to the surface, producing updrafts, heavy rain and thunder. Squall-line formation is assisted where surface topographic convergence of the easterly flow occurs (e.g., the Air Mountains, the Fouta-Jallon Plateau). These disturbance lines travel at up to 60km hr^{-1} from east to west across southern West Africa for distances of up to 3000km (but averaging 600km) between June and September, yielding 40–90mm of rain per day. Some coastal locations suffer about 40 squall-lines per year, which account for more than 50 percent of the annual rainfall.

Annual rainfall decreases from 2000–3000mm in the coastal belt (e.g., Conakry, Guinea) to about 1000m at latitude 20°N (Figure 11.42). Near the coast, more than 300mm per day of rain may fall during the rainy season but further north the variability increases due to the irregular extension and movement of the Monsoon Trough. Squall-lines and other disturbances give a zone of maximum rainfall located 800–1000km south of the surface position of the Monsoon Trough (see

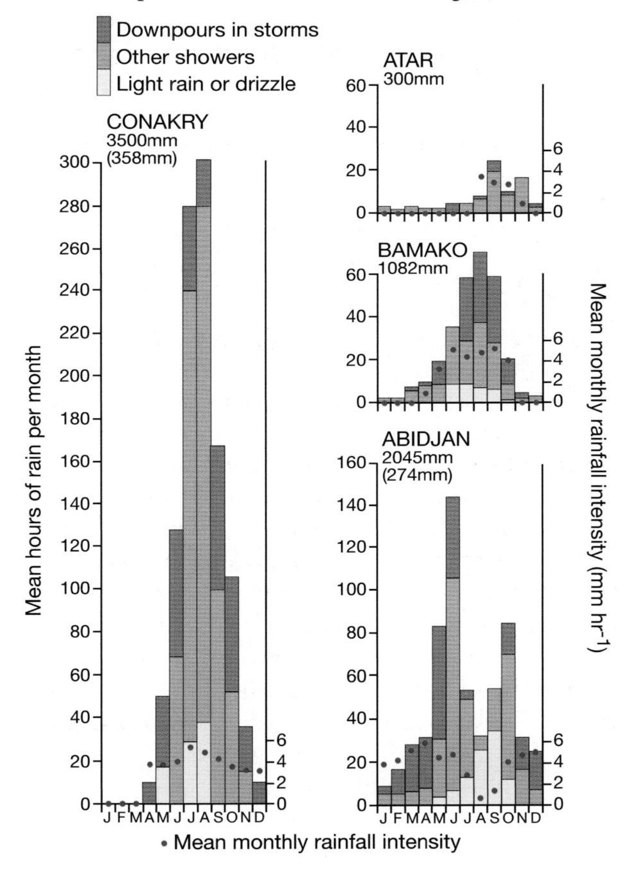

Figure 11.42 Mean number of hours of rain per month for four West African stations. Also shown are types of rainfall, mean annual totals (mm) and, in parentheses, maximum recorded daily rainfalls (m) for Conakry (August) and Abidjan (June). Dots show the mean monthly rainfall intensities (mm hr^{-1}). Note the pronounced Little Dry Season at Abidjan. Station locations are marked on Figure 11.41.

Source: From Hayward and Oguntoyinbo (1987). By permission of Rowman and Littlefield.

Figure 11.40B). Monsoon rains in the coastal zone of Nigeria (4°N) contribute 28 percent of the annual total (about 2000mm), thunderstorms 51 percent and disturbance lines 21 percent. At 10°N, 52 percent of the total (about 1000mm) is due to disturbance lines, 40 percent to thunderstorms and only 9 percent to the monsoon. Over most of the country, rainfall from disturbance lines has a double frequency maximum, thunderstorms a single one in summer (see Figure 11.43 for Minna, 9.5°N). In the northern parts of Nigeria and Ghana, rain falls in the summer months, mostly from isolated storms or disturbance lines. The high variability of these rains from year to year characterizes the drought-prone Sahel environment.

The summer rainfall in the northern Sudano–Sahelian belt is determined partly by the northward penetration of the Monsoon Trough, which may range up to 500–800km beyond its average position (Figure 11.44), and by the strength of the easterly jet streams. The latter affects the frequency of disturbance lines.

Anomalous climatic effects occur in a number of distinct West African localities at different times of the year. Although the temperatures of coastal waters always exceed 26°C and may reach 29°C in January, there are two areas of locally upwelling cold waters (see Figure 11.41). One lies north of Conakry along the coasts of Senegal and Mauretania, where dominant offshore

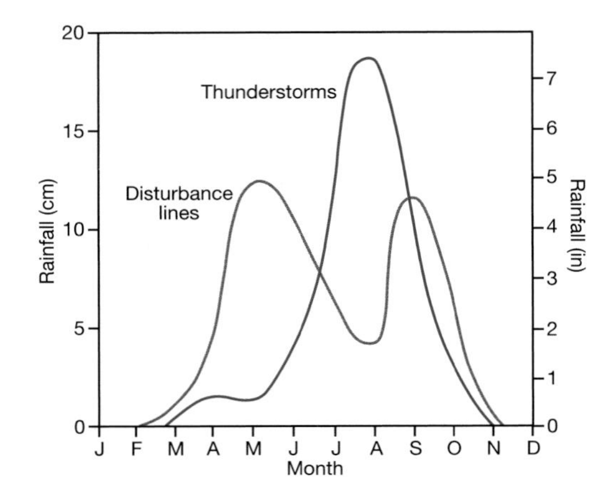

Figure 11.43 The contributions of disturbance lines and thunderstorms to the average monthly precipitation at Minna, Nigeria (9.5°N).

Source: After Omotosho (1985). By permission of the Royal Meteorological Society.

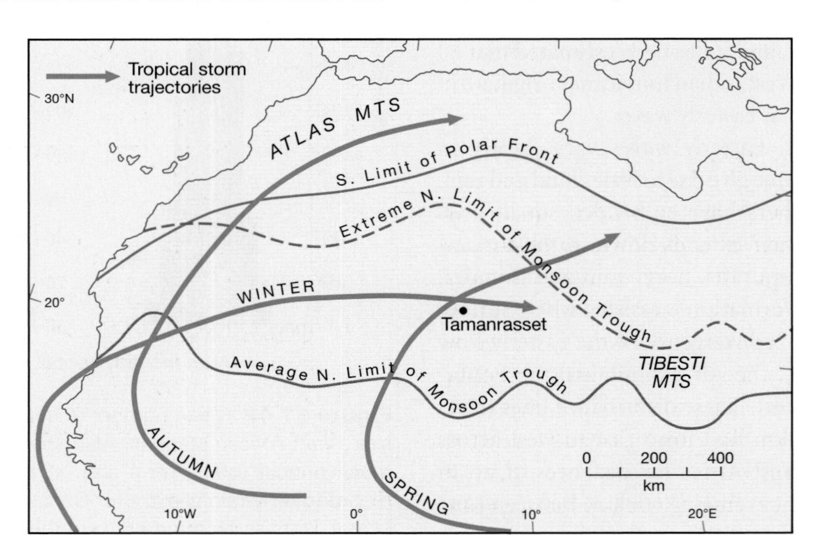

Figure 11.44 Extent of precipitation systems affecting western and central North Africa and typical tracks of Soudano–Sahelian depressions.

Source:After Dubief and Yacono; from Barry (2008).

northeasterly winds in January to April skim off the surface waters, causing cooler (20°C) water to rise, dramatically lowering the temperature of the afternoon onshore breezes. The second area of cool ocean (19–22°C) is located along the central-southern coast west of Lagos during the period July to October, for a reason that is as yet unclear. From July to September, an anomalously dry land area is located along the southern coastal belt (see Figure 11.41) during what is termed the Little Dry Season. The reason is that at this time the Monsoon Trough is in its most northerly position. The coastal zone, lying 1200–1500km to the south of it and, more important, 400–500km to the south of its major rain belt, has relatively stable air (see Figure 11.40B), a condition assisted by the relatively cool offshore coastal waters. Embedded within this relatively cloudy but dry belt is the smaller Togo Gap, between 0° and 3°E and during the summer having above-average sunshine, subdued convection, relatively low rainfall (i.e., less than 1000mm) and low thunderstorm activity. The trend of the coast here parallels the dominant low-level southwesterly winds, so limiting surface frictionally induced convergence in an area where temperatures and convection are in any case inhibited by low coastal water temperatures.

2 Southern Africa

Southern Africa lies between the South Atlantic and Indian Ocean subtropical high pressure cells in a region subject to the interaction of tropical easterly and extra-tropical westerly airflows. Both of these high pressure cells shift west and intensify (see Figure 7.10) in the southern winter. Because the South Atlantic cell always extends 3° latitude further north than the Indian Ocean cell, it brings low-level westerlies to Angola and Zaire at all seasons and high-level westerlies to central Angola in the southern summer. The seasonal longitudinal shifts of the subtropical high pressure cells are especially significant to the climate of southern Africa in respect of the Indian Ocean cell. Whereas the 7–13° longitudinal shift of the South Atlantic cell has relatively little effect, the westward movement of 24–30° during the southern winter by the Indian Ocean cell brings an easterly flow at all levels to most of southern Africa. The seasonal airflows and convergence zones are shown in Figure 11.45.

In summer (i.e., January) low-level westerlies over Angola and Zaire meet the northeast monsoon of East Africa along the Intertropical Convergence Zone (ITCZ), which extends east as the boundary between the recurved (westerly) winds from the Indian Ocean and the deep tropical easterlies further south. To the west, these easterlies impinge on the Atlantic westerlies along the Zaire Air Boundary (ZAB). The ZAB is subject to daily fluctuations and low pressure systems

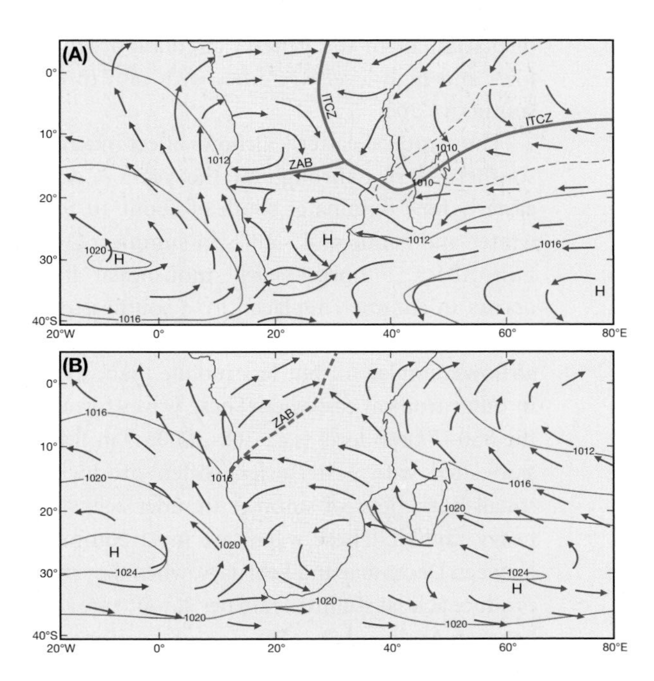

Figure 11.45 Airflow over southern Africa during January (A) and July (B), together with the locations of the Intertropical Convergence Zone (ITCZ), the Zaire Air Boundary (ZAB) and the major surface low pressure troughs.

Source: After Van Heerden and Taljaard (1988). Courtesy of the American Meteorological Society.

form along it, either being stationary or moving slowly westward. When these are deep and associated with southward-extending troughs they may produce significant rainfall. It should be noted that the complex structure of the ITCZ and ZAB means that the major surface troughs and centres of low pressure do not coincide with them but are situated some distance upwind in the low-level airflow (see Figure 11.45), particularly in the easterlies. This low-level summer circulation is dominated by a combination of these frontal lows and convectional heat lows. By March, a unified high pressure system has been established, giving a northerly flow of moist air, which produces autumn rains in western regions. In winter (*i.e.*, July), the ZAB separates the low-level westerly and easterly airflows from the Atlantic and Indian Oceans, although both are overlain by a high-level easterly flow. At this time, the northerly displacement of the general circulation brings low- and high-level westerlies with rain to the southern Cape.

Thus tropical easterly airflows affect much of southern Africa throughout the year. A deep easterly flow dominates south of about 10°S in winter and south of 15–18°S in summer. Over East Africa, a northeasterly monsoonal flow occurs in summer, replaced by a southeasterly flow in winter. Easterly waves form in these airflows, similar to, but less mobile than, those in other tropical easterlies. These waves form at the 850–700mb level (i.e., 200–3000m) in flows associated with easterly jets, often producing squall-lines, belts of summer thunder cells and heavy rainfall. These waves are most common between December and February, when they may produce at least 40mm of rain per day, but are rare between April and October. Tropical cyclones in the South Indian Ocean occur particularly around February (see Figure 11.8 and Table 11.1), when the ITCZ lies at its extreme southerly position. These storms recurve south along the east coast of Tanzania and Mozambique, but their influence is limited mainly to the coastal belt.

With few exceptions, deep westerly airflows are limited to the most southerly locations of southern Africa, especially in winter. As in northern mid-latitudes, disturbances in the westerlies involve:

1 Quasi-stationary Rossby waves.
2 Travelling waves, particularly marked at and above the 500mb level, with axes tilted westward with height, divergence ahead and convergence in the rear, moving eastwards at a speed of some 550km/day, having a periodicity of two to eight days and with associated cold fronts.
3 Cut-off low pressure centers. These are intense, cold-cored depressions, most frequent during March to May and September to November, and rare during December to February.

A feature of the climate of southern Africa is the prevalence of wet and dry spells, associated with broader features of the global circulation. Above-normal rainfall, occurring as a north–south belt over the region, is associated with a high-phase Walker circulation (see p. 376). This has an ascending limb over southern Africa; a strengthening of the ITCZ; an intensification of tropical lows and easterly waves, often in conjunction with a westerly wave aloft to the south; and a strengthening of the South Atlantic subtropical high pressure cell (see Figure 11.46). Such a wet spell may occur particularly during the spring to autumn period. Below-normal rainfall is associated with a low-phase Walker circulation having a descending limb over southern Africa; a weakening of the ITCZ; a tendency to high pressure with a diminished occurrence of tropical lows and easterly waves; and weakening of the South Atlantic subtropical high pressure cell. At the same time, there is a belt of cloud and rain lying to the east in the western Indian Ocean associated with a rising Walker limb and enhanced easterly disturbances in conjunction with a westerly wave aloft south of Madagascar (Figure 11.47).

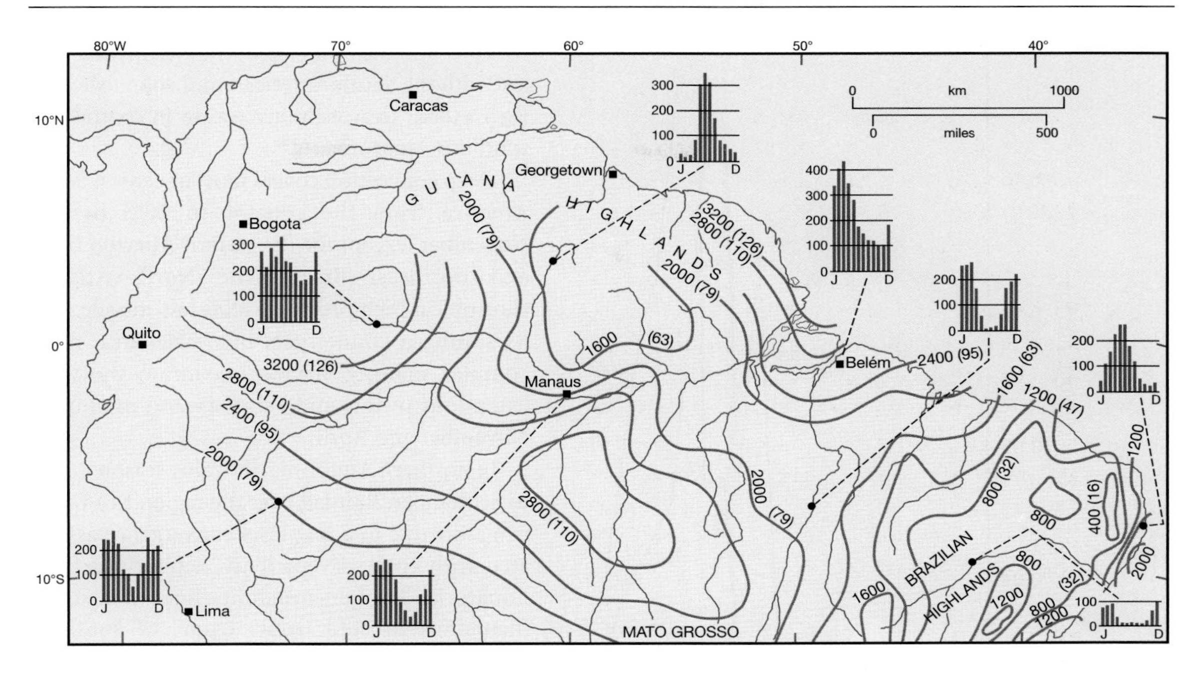

Figure 11.46 Mean annual precipitation (mm) over the Amazon basin, together with mean monthly precipitation amounts for eight stations.

Source: From Ratisbona (1976). By permission of Elsevier Science NL.

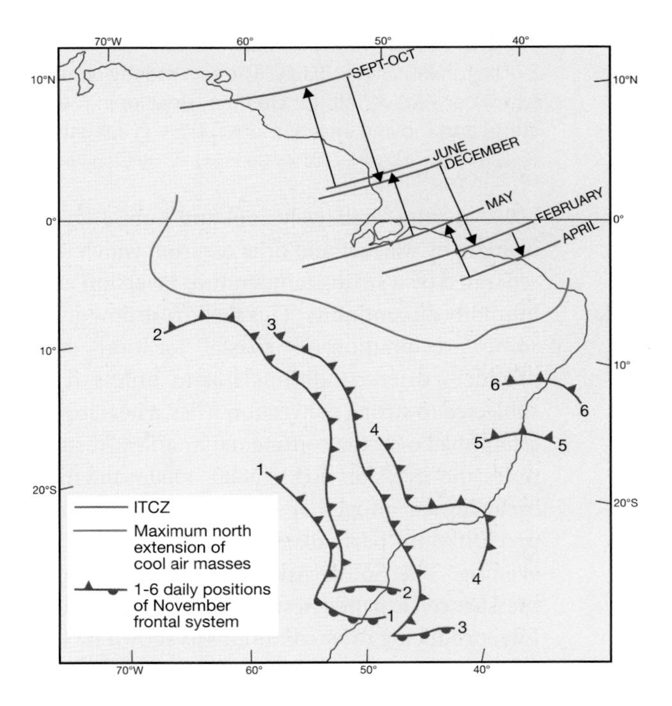

Figure 11.47 The synoptic elements of Brazil. The seasonal positions of the coastal Intertropical Convergence Zone; the maximum northerly extension of cool southerly mP air masses; and the positions of a typical frontal system during six successive days in November as the center of the low pressure moves southeastward into the South Atlantic.

Source: From Ratisbona (1976). By permission of Elsevier Science NL.

F AMAZONIA

Amazonia lies athwart the equator (Figure 11.48) and contains some 30 percent of the total global biomass. The continuously high temperatures (24–28°C) combine with the high transpiration to cause the region to behave at times as if it were a source of maritime equatorial air.

Important influences over the climate of Amazonia are the North and South Atlantic subtropical high pressure cells. From these, stable easterly mT air invades Amazonia in a shallow

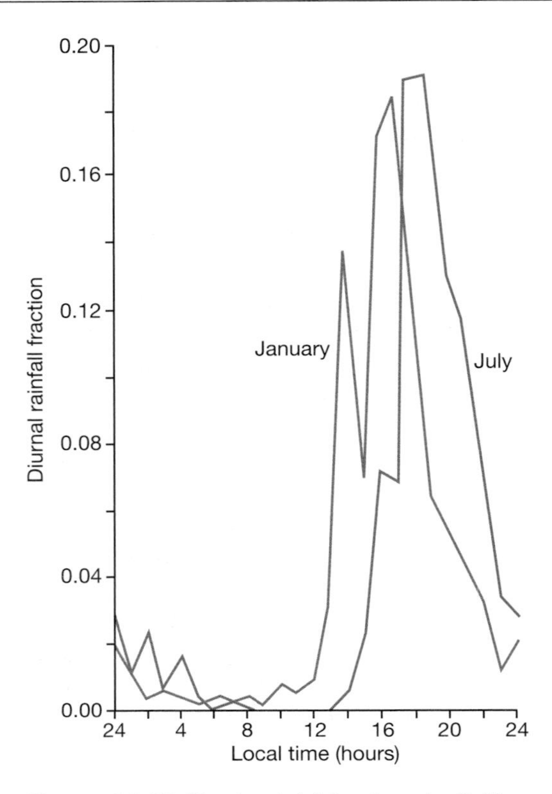

Figure 11.48 Hourly rainfall fractions for Belém, Brazil, for January and July. The rain mostly results from convective cloud clusters developing off-shore and moving inland, more rapidly in January.
Source: After Kousky (1980). Courtesy of the American Meteorological Society.

(1000–2000m), relatively cool and humid layer, overlain by warmer and drier air from which it is separated by a strong temperature inversion and humidity discontinuity. This shallow airflow gives some precipitation in coastal locations but produces drier conditions inland unless it is subjected to strong convection when a heat low is established over the continental interior. At such times, the inversion rises to 3000–4000m and may break down altogether associated with heavy precipitation, particularly in late afternoon or evening. The South Atlantic subtropical high pressure cell expands westward over Amazonia in July, producing drier conditions as shown by the rainfall at inland stations such as Manaus (see Figure 11.46), but in September it begins to contract and the build-up of the continental heat low with the South American monsoon ushers in the October to April rainy season in central and southern Amazonia.

Deep convection covers most of central South America from the equator to 20°S by late November, except for the eastern Amazon basin and northeast Brazil. The North Atlantic subtropical high pressure cell is less mobile than its southern counterpart but varies in a more complex manner, having maximum westward extensions in July and February and minima in November and April.

In northern Amazonia, the rainy season is May to September. Rainfall over the region as a whole is mainly due to a low-level convergence associated with convective activity, a poorly defined Equatorial Trough, instability lines, occasional incursions of cold fronts from the southern hemisphere, and relief effects.

Strong thermal convection over Amazonia can commonly produce more than 40mm/day of rainfall over a period of a week and much higher average intensities over shorter periods. When it is recognized that 40mm of rainfall in one day releases sufficient latent heat to warm the troposphere by 10°C, it is clear that sustained convection at this intensity is capable of fueling the Walker circulation (see Figure 11.50). During high phases of ENSO, air rises over Amazonia, whereas during the low phases the drought over northeast Brazil is intensified. In addition, convective air moving poleward may strengthen the Hadley circulation. This air tends to accelerate due to the conservation of angular momentum, and to strengthen the westerly jet streams such that correlations have been found between Amazonian convective activity and North American jet-stream intensity and location.

The Intertropical Convergence Zone (ITCZ) does not exist in its characteristic form over the interior of South America, and its passage affects rainfall only near the east coast. The intensity of this zone varies, being least when both the North and South Atlantic subtropical high pressure cells are strongest (i.e., in July), giving a pressure increase that causes the equatorial trough to fill. During October to November deep convection associated with the ITCZ is confined to the central Atlantic between 5° and 8°N. The ITCZ swings to its most northerly position during July to October, when invasions of more stable South Atlantic air are associated with drier conditions over central Amazonia, and to its most southerly in March to April (Figure 11.47). At Manaus, surface winds are predominantly southeasterly from May to August and northeasterly from September to April, whereas the upper tropospheric winds are northwesterly or westerly from May to September and southerly or southeasterly from December to April. This reflects the development in the austral summer of an upper tropospheric anticyclone that is located over the Peru–Bolivia Altiplano. This upper high is a result of sensible heating of the elevated plateau and the release of latent heat in frequent thunderstorms over the Altiplano, analogous to the situation over Tibet. Outflow from this high subsides in a broad area extending from eastern Brazil to West Africa. The drought-prone region of eastern Brazil is particularly moisture-deficient during periods when the ITCZ remains in a northerly position and relatively stable mT air from a cool South Atlantic surface is dominant (see Chapter 9B.2). Dry conditions may occur in January to May during strong ENSO events (see p. 378), when the descending branch of the Walker circulation covers most of Amazonia.

Significant Amazonian rainfall, particularly in the east, originates along mesoscale lines of instability, which form near the coast due to converging Trade Winds and afternoon sea breezes, or to the interaction of nocturnal land breezes with onshore Trade Winds. These lines of instability move westward in the general airflow at speeds of about 50km hr^{-1}, moving faster in January than in July and exhibiting a complex process of convective-cell growth, decay, migration and regeneration. Many of these instability lines reach only 100km or so inland, decaying after sunset (Figure 11.48). However, the more persistent instabilities may produce a rainfall maximum about 500km inland, and some remain active for up to 48 hours such that their precipitation effects reach as far west as the Andes. Other meso- to synoptic-scale disturbances form within Amazonia, especially between April and September. Precipitation also occurs with the penetration of cool mP air masses from the south, especially between September and November, which are heated from below and become unstable (see Figure 11.47).

Surges of cold polar air (friagens) during the winter months can cause freezing temperatures in southern Brazil, with cooling to 11°C even in Amazonia. In June to July 1994, such events caused devastation to Brazil's coffee production. Typically, an upper-level trough crosses the Andes of central Chile from the eastern South Pacific and an associated southerly airflow transports cold air northeastward over southern Brazil. Concurrently, a surface high pressure cell may move northward from Argentina, with the associated clear skies producing additional radiative cooling.

The tropical easterlies over the northern and eastern margins of Amazonia are susceptible to the formation of easterly waves and closed vortices, which move westward generating rain bands. Relief effects are naturally most noteworthy as airflow approaches the eastern slopes of the Andes, where large-scale orographic convergence in a region of high evapotranspiration contributes to the high precipitation all through the year.

G EL NIÑO–SOUTHERN OSCILLATION (ENSO) EVENTS

1 The Pacific Ocean

The Southern Oscillation is an irregular variation, see-saw or standing wave in atmospheric mass and pressure involving exchanges of air between the subtropical high pressure cell over the eastern South Pacific and a low pressure region centered on the western Pacific and Indonesia (Figure

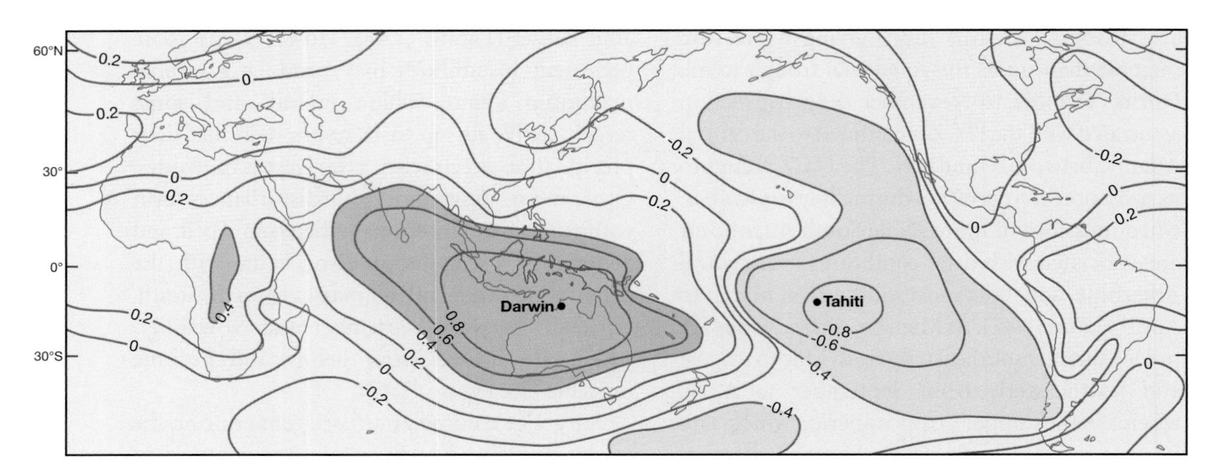

Figure 11.49 The correlation of mean annual sea-level pressures with that at Darwin, Australia, illustrating the two major cells of the Southern Oscillation.
Source: Rasmusson (1985). Copyright © *American Scientist*, (1985).

11.49). It has an irregular period of between two to ten years. Its mechanism is held by some experts to center on the control over the strength of the Pacific Trade Winds exercised by the activity of the subtropical high pressure cells, particularly the one over the South Pacific. Others, recognizing the ocean as an enormous heat energy source, believe that near-surface temperature variations in the tropical Pacific may act somewhat similar to a flywheel to drive the whole ENSO system (see Box 11.1). It is important to note that a deep (i.e., 100m+) pool of the world's warmest surface water builds up in the western equatorial Pacific between the surface and the thermocline. This is set up by the intense insolation, low heat loss from evaporation in this region of light winds, and the piling up of surface water driven westward by the easterly Trade Winds. The warm pool is dissipated periodically during El Niño by the changing ocean currents and by heat release into the atmosphere – directly and through evaporation.

The Southern Oscillation is associated with the phases of the Walker circulation that have already been introduced in Chapter 7C.1. The high phases of the Walker circulation (usually associated with non-ENSO or La Niña events), which occur on average three years out of four, alternate with low phases (*i.e.*, ENSO or El Niño events). Sometimes, however, the Southern Oscillation is not in evidence and neither phase is dominant. The level of activity of the Southern Oscillation in the Pacific is expressed by the Southern Oscillation Index (SOI), which is a complex measure involving sea surface and air temperatures, pressures at sea level and aloft, and rainfall at selected locations.

During high phase (La Niña) (Figure 11.50A), strong easterly Trade Winds in the eastern tropical Pacific produce upwelling along the west coast of South America, resulting in a north-flowing cold current (the Peru or Humboldt), locally termed La Niña – the girl – on account of its richness in plankton and fish. The low sea temperatures produce a shallow inversion, thereby further strengthening the Trade Winds (i.e., effecting positive feedback), which skim water off the surface of the Pacific, where warm surface water accumulates (Figure 11.50D). This action also causes the thermocline to lie at shallow depths (about 40m) in the east, as distinct from 100–200m in the western Pacific. The strengthening of the easterly Trades causes cold-water upwelling to spread westward and the cold tongue of surface water extends in that direction

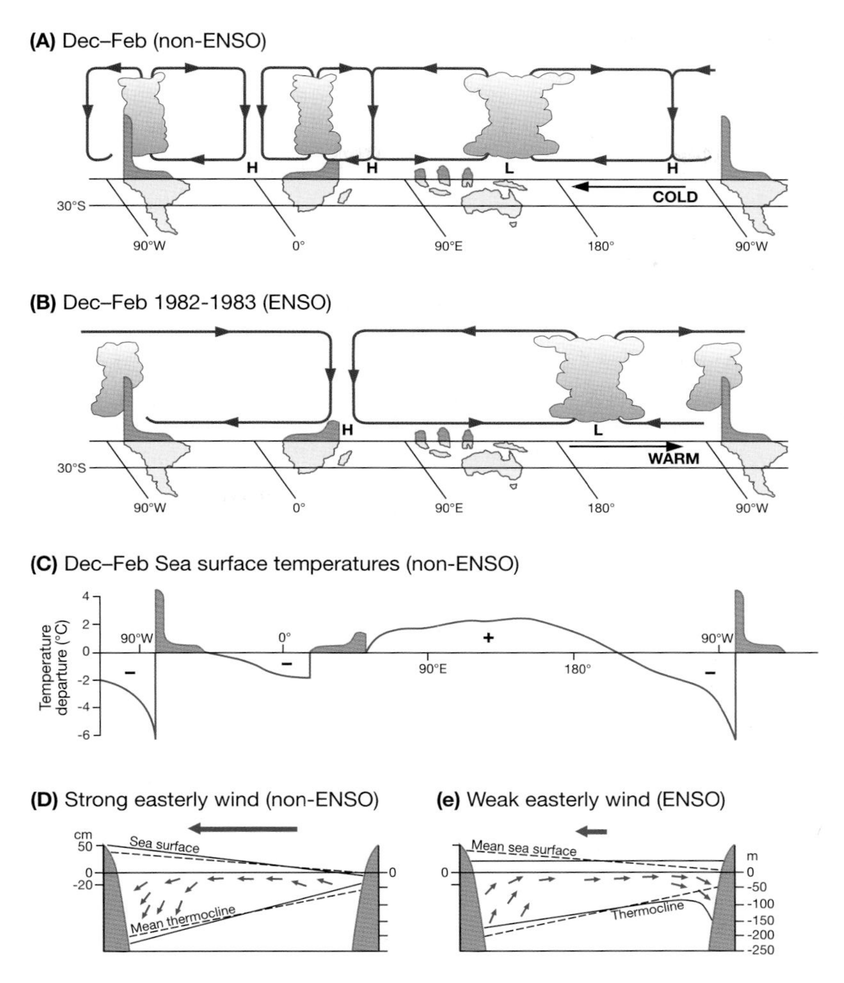

Figure 11.50 Schematic cross-sections of the Walker circulation along the equator based on computations of Y. M. Tourre. A: Mean December to February regime (non-ENSO), La Niña; rising air and heavy rains occur over the Amazon basin, central Africa and Indonesia–western Pacific. B: December to February 1982–1983 ENSO pattern; the ascending Pacific branch is shifted east of the Date Line and suppressed convection occurs elsewhere due to subsidence. C: Departure of sea surface temperature from its equatorial zonal mean, corresponding to non-ENSO case (A). D: Strong trades cause sea level to rise and the thermocline to deepen in the western Pacific for case (A). E: Winds relax, sea level rises in the eastern Pacific as water mass moves back eastward and the thermocline deepens off South America during ENSO events.

Source: Based on Wyrtki (1982). By permission World Meteorological Organization (1985).

sustained by the South Equatorial Current. This westward-flowing current is wind-driven and is compensated by a deeper surface slope. The westward contraction of warm Pacific water into the central and western tropical Pacific (Figure 11.50C) produces an area of instability and convection fed by moisture in a convergence zone under the dual influence of both the Inter-tropical Convergence Zone and the South Pacific Convergence Zone. The rising air over the western

SIGNIFICANT 20TH-CENTURY ADVANCE

11.1 El Niño and the Southern Oscillation

El Niño episodes of warm coastal currents with accompanying disastrous consequences for marine life and birds recur about every four to seven years and consequently were long known along the West Coast of South America. The related Southern Oscillation (SO) of sea-level pressure between Tahiti (normally high pressure) and Jakarta (or Darwin) (normally low pressure) was identified by Sir Gilbert Walker in 1910 and re-investigated in the mid-1950s by I. Schell and H. Berlage, and in the 1960s by A.J. Troup and J. Bjerknes. A.J. Troup linked the occurrence of El Niño conditions to an oscillation in the atmosphere over the equatorial Pacific in the 1960s. Their wider implications for air–sea interaction and global teleconnections were first proposed by Professor Jacob Bjerknes (of polar front fame) in 1966 who noted the linkages of El Niño or non-El Niño conditions with the SO. The worldwide significance of ENSO events only became fully appreciated in the 1970s–1980s with the strong El Niño events of 1972–1973 and 1982–1983. The availability of global analyses showed clear patterns of seasonal anomalies of temperature and precipitation in widely separated regions during and after the onset of warming in the eastern and central equatorial Pacific Ocean. These include droughts in northeast Brazil and in Australasia, and cool, wet winters following El Niño in the southern and southeastern United States.

The occurrence of ENSO events in the past has been studied from historical documents, inferred from tree ring data, and from coral, ice core and high-resolution sediment records. The net effect of major El Niño events on global temperature trends is estimated to be about +0.06°C between 1950 and 1998.

Reference

Diaz, H.F. and Markgraf, V. (eds) (1992) *El Niño. Historical and Paleoclimatic Aspects of the Southern Oscillation*, Cambridge University Press, Cambridge, 476pp.

Pacific feeds the return airflow in the upper troposphere (*i.e.*, at 200mb), closing and strengthening the Walker circulation. However, this airflow also strengthens the Hadley circulation, particularly its meridional component northward in the northern winter and southward in the southern winter.

Each year, usually starting in December, a weak, southward flow of warm water replaces the northward-flowing Peru Current and its associated cold upwelling southward to about 6°S along the coast of Ecuador. This phenomenon, known as El Niño (the child, after the Christ child), strengthens at irregular intervals of two to ten years (its average interval is four years) when warm surface water becomes much more extensive and the coastal upwelling ceases entirely. This has catastrophic ecological and economic consequences for fish and bird life, and for the fishing and guano industries of Ecuador, Peru and northern Chile. Figure 11.51 shows the occurrence of El Niño events between 1525 and 1987 classified according to their intensity. These offshore events, however, are part of a Pacific-wide change in sea surface temperatures. Moreover, the spatial pattern of these changes is not the same for all El Niños. Recently, K. E. Trenberth and colleagues showed that during 1950–1977, warming during an El Niño spread westward from Peru, whereas after a major shift in Pacific basin climate took place in 1976–1977, the warming spread eastward from the western equatorial Pacific. The

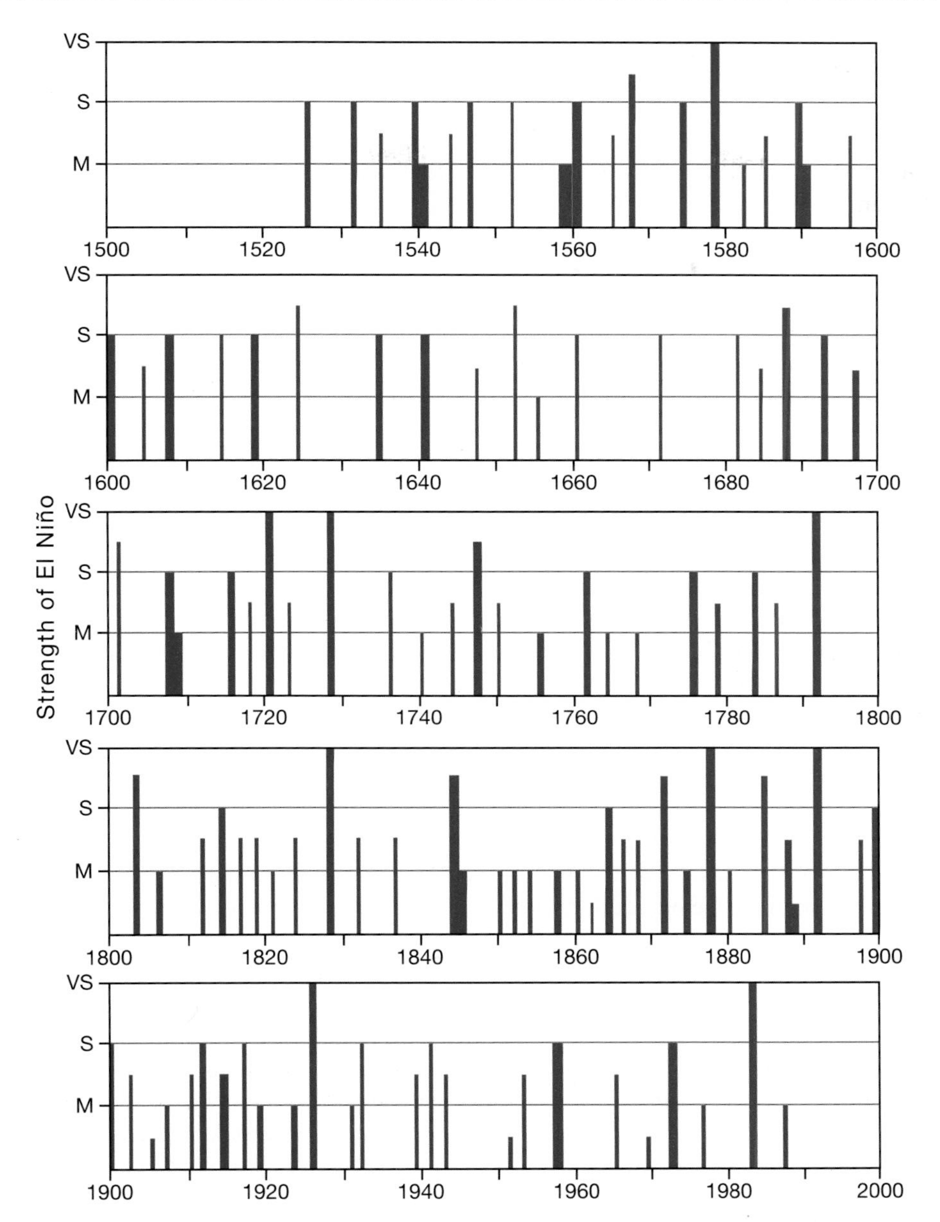

Figure 11.51 El Niño events 1525–1987 classified according to very strong, strong and medium.
Source: Quinn and Neal (1992). Copyright © Routledge, London.

atmosphere – ocean coupling during ENSO events clearly varies on multidecadal timescales.

ENSO events result from a radical reorganization of the Walker circulation in two main respects:

1 Pressure declines and the Trades weaken over the eastern tropical Pacific (Figure 11.50B), wind-driven upwelling slackens, allowing the ITCZ to extend southward to Peru. This increase in sea surface temperatures by 1–4°C

reduces the west-to-east sea surface temperature gradient across the Pacific and also tends to decrease pressure over the eastern Pacific. The latter causes a further decrease in Trade Wind activity, a decrease in upwelling of cold water, an advection of warm water and a further increase in sea surface temperatures – in other words, the onset of El Niño activates a positive feedback loop in the eastern Pacific atmosphere–ocean system.

2 Over the western tropical Pacific, the area of maximum sea temperatures and convection responds to the above weakening of the Walker circulation by moving eastward into the central Pacific (Figure 11.50B). This is partly due to an increase of pressure in the west but also to a combined movement of the ITCZ southward and the SPCZ northeastward. Under these conditions, bursts of equatorial westerly winds spread a huge tongue of warm water (i.e., warmer than 27.5°C) eastward over the central Pacific as large-scale, internal oceanic (Kelvin) waves. It has been suggested that this eastward flow may sometimes be triggered off or strengthened by the occurrence of cyclone pairs north and south of the equator. This eastward flow of warm water depresses the thermocline off South America (Figure 11.50E), preventing cold water from reaching the surface.

Thus, whether La Niña or El Niño develops, bringing westward-flowing cold surface water or eastward-flowing warm surface water, respectively, to the central Pacific, depends on the competing processes of upwelling versus advection. The most intense phase of an El Niño event commonly lasts for about one year, and the change to El Niño usually occurs about March to April, when the Trade Winds and the cold tongue are at their weakest. The changes to the Pacific atmosphere–ocean circulation during El Niño are facilitated by the fact that the time taken for ocean surface currents to adjust to major wind changes decreases markedly with decreasing latitude. This is demonstrated by the seasonal reversal of the southwest and northeast monsoon drift off the Somali coast in the Indian Ocean. Large-scale atmospheric circulation is subject to a negative feedback constraint involving a negative correlation between the strengths of the Walker and Hadley circulations. Thus the weakening of the Walker circulation during an ENSO event leads to a relative strengthening of the associated Hadley circulation.

The termination of El Niño is preceded by the return of the thermocline to shallow depths in the eastern and central equatorial Pacific, removing the positive SST anomalies. This appears to be in response to renewed easterly wind forcing or to equatorial oceanic Kelvin waves. In December to January there is a southward shift of 28°C water to south of the equator in 'normal' and El Niño years and this causes the shallowing of the thermocline.

2 Teleconnections

Teleconnections are defined as linkages over great distances of atmospheric and oceanic variables; clearly the linkages between climatic conditions in the eastern and western tropical Pacific Ocean represent a 'canonical' teleconnection. Figure 11.52 illustrates the coincidence of ENSO events with regional climates that are wetter or drier than normal.

In Chapter 7C.1 we referred to Walker's observed teleconnection between ENSO events and the lower than normal monsoon rainfall over South and Southeast Asia (Figure 11.53). This is due to the eastward movement of the zone of maximum convection over the western Pacific. However, it is important to recognize that ENSO mechanisms form only part of the South Asian monsoon phenomenon. For example, parts of India may experience droughts in the absence of El Niño and the onset of the monsoon may also depend on the control exercised by the amount of Eurasian snow cover on the persistence of the continental high pressure cell.

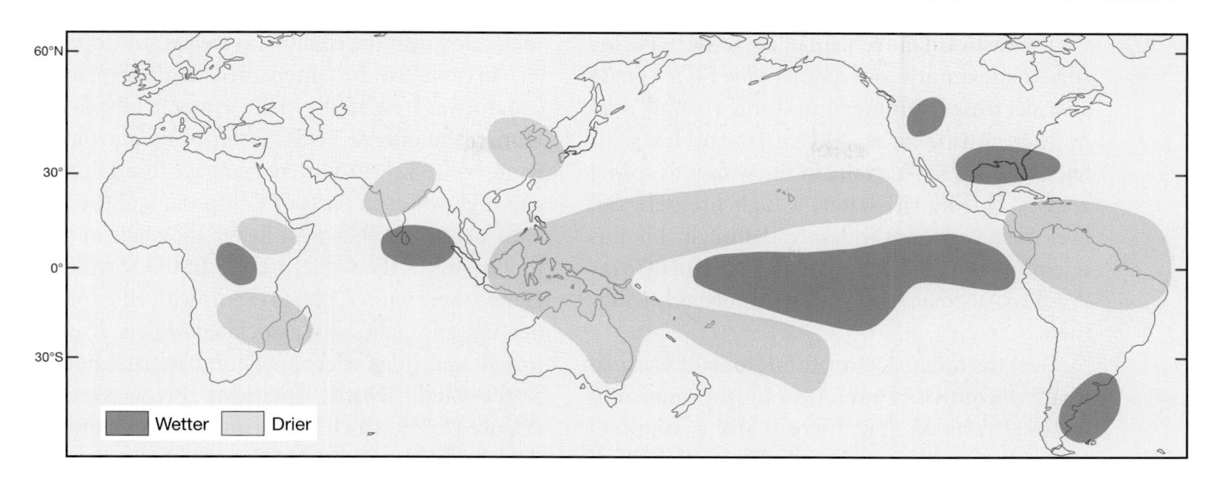

Figure 11.52 The coincidence of ENSO events with regional climates that are wetter or drier than normal.

Sources: After Rasmusson and Ropelowski, also Halpert. From Glantz *et al.* (1990). Composite reproduced by permission of Cambridge University Press.

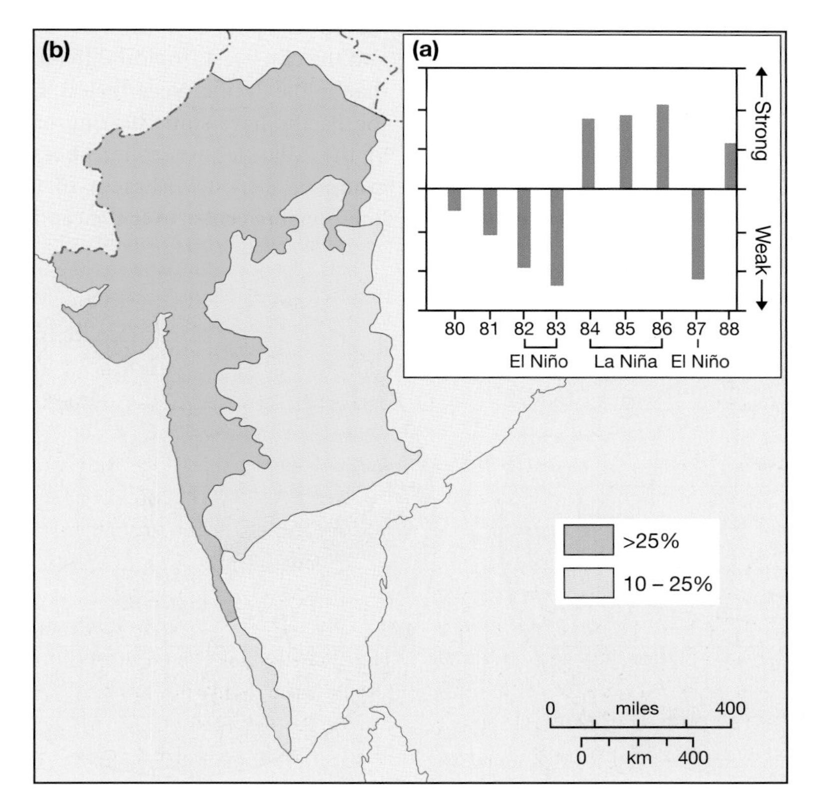

Figure 11.53 The proposed connection between the Indian summer monsoon and El Niño. A: The observed strength of the Asian summer monsoon (1980–1988) showing its weakness during the three strong El Niño years 1982, 1983 and 1987. B: Areas of India where the summer monsoonal rainfall deficits (as a percentage less than the 1901–1970 average) were significantly more frequent in the El Niño years.

Sources: A: Browning (1996). B: Gregory (1988). IGU Study Group on Recent Climate Change.

The eastward movement of the western Pacific zone of maximum convection in the ENSO phase also decreases summer monsoon rainfall over northern Australia, as well as extra-tropical rainfall over eastern Australia in the winter to spring season. During the latter, a high pressure cell over Australia brings widespread drought, but this is compensated for by enhanced rainfall over Western Australia associated with northerly winds there.

Over the Indian Ocean, the dominant seasonal weather control is exercised by the monsoon seasonal reversals, but there is still a minor El Niño-like mechanism over southeast Africa and Madagascar, which results in a decrease in rainfall during ENSO events.

It is apparent that ENSO teleconnections affect extra-tropical regions as well as tropical ones. During the most intense phase of El Niño, two high pressure cells, centered at 20°N and 20°S, develop over the Pacific in the upper troposphere, where anomalous heating of the atmosphere is at a maximum. These cells strengthen the Hadley circulation, cause upper-level tropical easterlies to develop near the equator, as well as subtropical jet streams to be intensified and displaced equatorward, especially in the winter hemisphere. During the intense ENSO event of the northern winter of 1982–1983, such changes caused floods and high winds in parts of California and the US Gulf states, together with heavy snowfalls in the mountains of the western USA. In the Northern Hemisphere winter, ENSO events with equatorial heating anomalies are associated with a strong trough and ridge teleconnection pattern, known as the Pacific–North American (PNA) pattern (Figure 11.54), which may bring cloud and rain to the southwest United States and northwest Mexico.

The Atlantic Ocean shows some tendency towards a modest effect resembling El Niño, but the western pool of warm water is much smaller, and the east–west tropical differences much less, than in the Pacific. Nevertheless, ENSO events in the Pacific have some bearing on the behavior of the Atlantic atmosphere–ocean system; for example, the establishment of the convective low-pressure center over the central and eastern

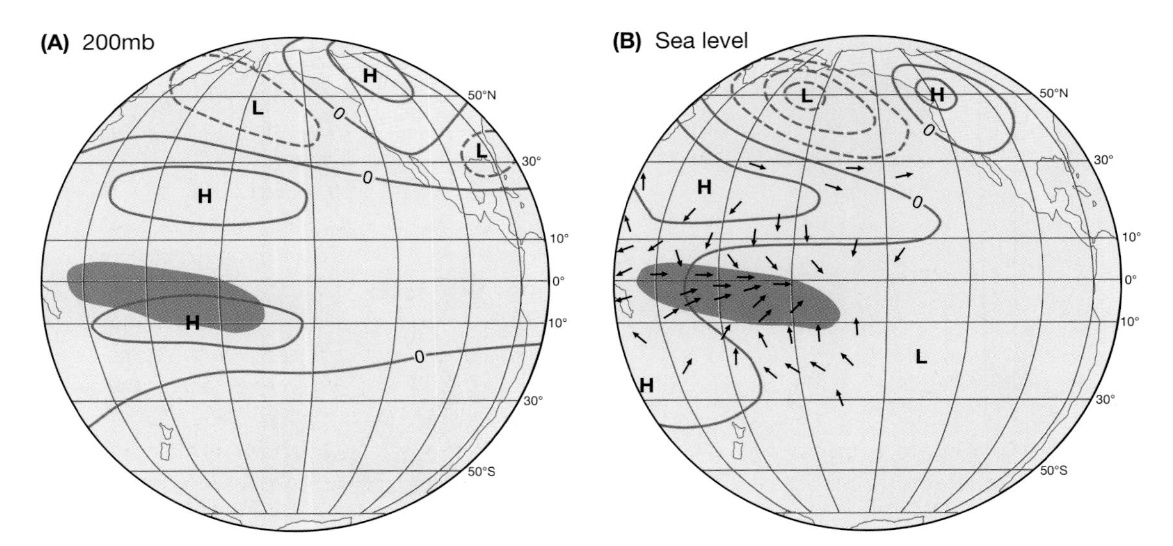

Figure 11.54 Schematic Pacific–North America (PNA) circulation pattern in the upper troposphere during an ENSO event in December to February. The shading indicates a region of enhanced rainfall associated with anomalous westerly surface wind convergence in the equatorial western Pacific.

Source: After Shukla and Wallace (1983). Courtesy of the American Meteorological Society.

Atlantic subtropical high pressure cell and of the general Trade Wind flow in the Atlantic. This results in the development of a stronger subsidence inversion layer, as well as subjecting the western tropical Atlantic to greater ocean mixing, giving lower sea surface temperatures, less evaporation and less convection. This tends to:

1 Increase drought over northeast Brazil. However, ENSO events account for only some 10 percent of precipitation variations in northeast Brazil.
2 Increase wind shear over the North Atlantic/Caribbean region such that moderate to strong ENSO events are correlated with the occurrence of some 44 percent fewer Atlantic hurricanes than occur with non-ENSO events.

A further Pacific influence involves the manner in which the ENSO strengthening of the southern subtropical jet stream may partly explain the heavy rainfall experienced over southern Brazil, Paraguay and northern Argentina during an intense El Niño. Another Atlantic teleconnection may reside in the North Atlantic Oscillation (NAO), a large-scale alternation of atmospheric mass between the Azores high pressure and the Icelandic low pressure cells (see Chapter 7C.2B). The relative strength of these two pressure systems appears to affect the rainfall of both northwest Africa and the sub-Saharan zone.

H OTHER SOURCES OF CLIMATIC VARIATIONS IN THE TROPICS

The major systems of tropical weather and climate have now been discussed, yet various other elements help to create contrasts in tropical weather in both space and time.

1 Cool ocean currents

Between the western coasts of the continents and the eastern rims of the subtropical high pressure cells the ocean surface is relatively cold (see Figure 7.33). This is the result of: the importation of water from higher latitudes by the dominant currents; the slow upwelling (sometimes at the rate of about 1m in 24 hours) of water from intermediate depths due to the Ekman effect (see Chapter 6A.5); and the coastal divergence (see Figure 7.31). This concentration of cold water gently cools the local air to dew-point. As a result, dry, warm air degenerates into a relatively cool, clammy, foggy atmosphere with a comparatively low temperature and little range along the west coast of North America off California (see Plate 11.4), off South America between latitudes 4 and 3°S, and off southwest Africa (8 and 32°S). Thus Callao, on the Peruvian coast, has a mean annual temperature of 19.4°C, whereas Bahia (at the same latitude on the Brazilian coast) has a corresponding figure of 25°C.

The cooling effect of offshore cold currents is not limited to coastal stations, since it is carried inland during the day at all times of the year by a pronounced sea-breeze effect (see Chapter 6C.2). Along the west coasts of South America and southwest Africa the sheltering effect from the dynamically stable easterly Trades aloft provided by the nearby Andes and Namib Escarpment, respectively, allows incursions of shallow tongues of cold air to roll in from the southwest. These tongues of air are capped by strong inversions at between 600 and 1500m, reinforcing the regionally low Trade Wind inversions (see Figure 11.6) and thereby precluding the development of strong convective cells, except where there is orographically forced ascent. Thus, although the cool maritime air perpetually bathes the lower western slopes of the Andes in mist and low stratus cloud and Swakopmund (southwest Africa) has an average of 150 foggy days a year, little rain falls on the coastal lowlands. Lima (Peru) has a total mean annual precipitation of only 46mm, although it suffers frequent drizzle during the winter months (June to September), and Swakopmund in Namibia has a mean annual rainfall of 16mm. Heavier rain occurs on the rare instances when

Plate 11.4 A fog bank enveloping the Golden Gate Bridge, San Francisco. NOAA wea00154

large-scale pressure changes cause a cessation of the diurnal sea breeze or when modified air from the South Atlantic or South Indian Ocean is able to cross the continents at a time when the normal dynamic stability of the Trade Winds is disturbed. In southwest Africa, the inversion is most likely to break down during either October or April, allowing convectional storms to form, and Swakopmund recorded 51mm of rain on a single day in 1934. Under normal conditions, however, the occurrence of precipitation is mainly limited to the higher seaward mountain slopes. Further north, tropical west coast locations in Angola and Gabon show that cold upwelling is a more variable phenomenon in both space and time; coastal rainfall varies strikingly with changing sea surface temperatures (Figure 11.55). In South America, from Colombia to northern Peru, the diurnal tide of cold air rolls inland for some 60km, rising up the seaward slopes of the Western Cordillera and overflowing into the longitudinal Andean valleys like water over a weir (Figure 11.56). On the west-facing slopes of the Andes of Colombia, air ascending or banked up against the mountains may under suitable conditions trigger off convectional instability in the overlying trades and produce thunderstorms. In southwest Africa, however, the 'tide' flows inland for some 130km and rises up the 1800m Namib Escarpment without producing much rain because convectional instability is not generated and the adiabatic cooling of the air is more than offset by radiational heating from the warm ground.

2 Topographic effects

Relief and surface configuration have a marked effect on rainfall amounts in tropical regions,

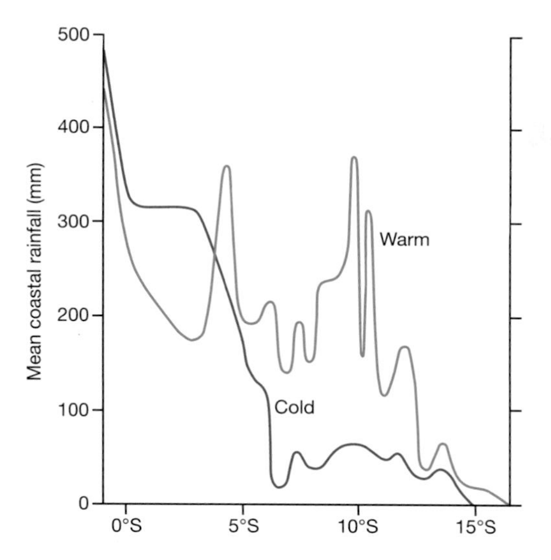

Figure 11.55 March rainfall along the southwestern coast of Africa (Gabon and Angola) associated with warm and cold sea surface conditions.

Source: After Nicholson and Entekhabi from Nicholson (1989). By permission of the Royal Meteorological Society, redrawn.

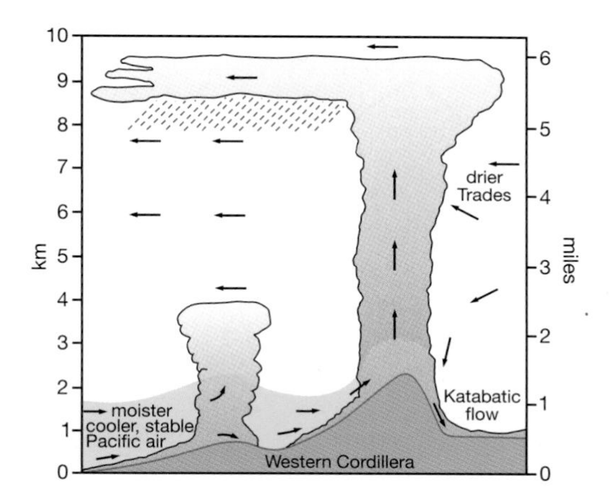

Figure 11.56 The structure of the sea breeze in western Colombia.

Source: After Lopez and Howell (1967). Courtesy of the American Meteorological Society.

where hot, humid air masses are frequent. At the southwestern foot of Mount Cameroon, Debundscha (9m elevation) receives 11,160mm yr^{-1} on average (1960–1980) from the southwesterly monsoon. In the Hawaiian Islands, the mean annual total exceeds 7600mm on the mountains, with one of the world's largest mean annual totals of 11,990 c\mm at 1569m elevation on Mount Waialeale (Kauai), but land on the lee side suffers correspondingly accentuated sheltering effects with less than 500mm over wide areas. On Hawaii itself, the maximum falls on the eastern slopes at about 900m, whereas the 4200m summits of Mauna Loa and Mauna Kea, which rise above the Trade Wind inversion, receive only 250–500mm. On the Hawaiian island of Oahu, the maximum precipitation occurs on the western slopes, just leeward of the 850m summit with respect to the easterly Trade Winds. Measurements in the Koolau Mountains, Oahu show that the orographic factor is pronounced during summer, when precipitation is associated with the easterlies, but in winter, when precipitation is from cyclonic disturbances, it is more evenly distributed (Table 11.3).

The Khasi Hills in Assam are an exceptional instance of the combined effect of relief and surface configuration. Part of the monsoon current from the head of the Bay of Bengal (see Figure 11.23) is channeled by the topography towards the high ground and the sharp ascent, which follows the convergence of the airstream in

Table 11.3 Precipitation in the Koolau Mountains, Oahu, Hawaii (mm)

Location	Elevation	Source of rainfall		
		Trade winds *28 May–3 Sep. 1957*	Cyclonic disturbances *2–29 Jan. 1957*	 *5–6 Mar. 1957*
Summit	850m	71.3cm	49.9cm	32.9cm
760m west of summit	635m	121.0cm	54.4cm	37.0cm
7,600m west of summit	350m	32.9cm	46.7cm	33.4cm

Source: After Mink (1960).

the funnel-shaped lowland to the south, results in some of the heaviest annual rainfall totals recorded anywhere. Mawsyuran (1400m elevation), 16km west of the more famous station of Cherrapunji, has a mean annual total (1941–1969) of 12,210mm and can claim to be the wettest spot in the world. Cherrapunji (1340m) during the same period averaged 11,020mm; extremes recorded there include 5690mm in July and 24,400mm in 1974 (see Figure 4.11). However, throughout the monsoon area, topography plays a secondary role in determining rainfall distribution to the synoptic activity and large-scale dynamics.

Really high relief produces major changes in the main weather characteristics and is best treated as a special climatic type. In equatorial East Africa, the three volcanic peaks of Mount Kilimanjaro (5800m), Mount Kenya (5200m) and Ruwenzori (5200m) nourish permanent glaciers above 4700–5100m. Annual precipitation on the summit of Mount Kenya is about 1140mm, similar to amounts on the plateau to the south, but on the southern slopes between 2100 and 3000m, and on the eastern slopes between about 1400 and 2400m, totals exceed 2500mm. Kabete (at an elevation of 1,800m near Nairobi) exhibits many of the features of tropical highland climates, having a small annual temperature range (mean monthly temperatures are 19°C for February and 16°C for July), a high diurnal temperature range (averaging 9.5°C in July and 13°C in February) and a large average cloud cover (mean 7–8/10ths).

3 Diurnal variations

Diurnal weather variations are particularly evident at coastal locations in the Trade Wind belt and in the Indonesia–Malaysian Archipelago. Land- and sea-breeze regimes (see Chapter 6C.1) are well developed, as the heating of tropical air over land can be up to five times that over adjacent water surfaces. The sea breeze normally sets in between 08:00 and 11:00 hours, reaching a maximum velocity of 6–15m s^{-1} at about 13:00 to 16:00 and subsiding around 20:00. It may be up to 1000–2000m in height, with a maximum velocity at an elevation of 200–400m, and it normally penetrates some 20–60km inland.

On large islands under calm conditions the sea breezes converge towards the center so that an afternoon maximum of rainfall is observed. Under steady Trade Winds, the pattern is displaced downwind so that descending air may be located over the center of the island. A typical case of an afternoon maximum is illustrated in Figure 11.57B for Nandi (Viti Levu, Fiji) in the southwest Pacific. The station has a lee exposure in both wet and dry seasons. This rainfall pattern is commonly believed to be widespread in the tropics, but over the open sea and on small islands a night-time maximum (often with a peak near dawn) seems to occur, and even large islands may display this nocturnal regime when there is little synoptic activity. Figure 11.57A illustrates this nocturnal pattern at four

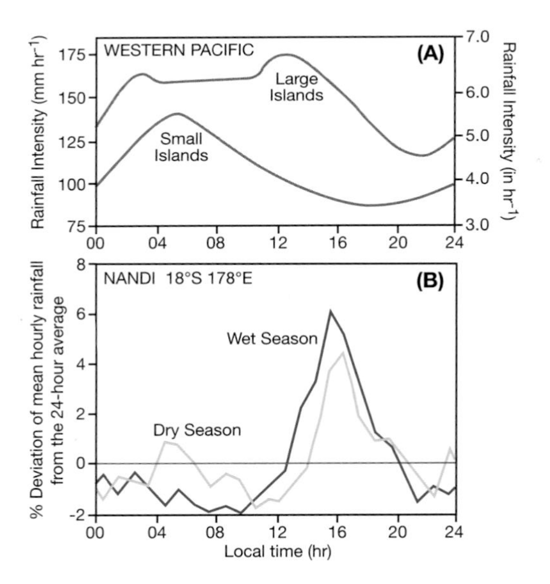

Figure 11.57 Diurnal variation of rainfall intensity for tropical islands in the Pacific. A: Large and small islands in the western Pacific. B: Wet and dry seasons for Nandi (Fiji) in the south-west Pacific (percentage deviation from the daily average).

Sources: A: After Gray and Jacobson (1977). B: After Finkelstein in Hutchings (1964).

small island locations in the western Pacific. Even large islands may show this effect, as well as the afternoon maximum associated with sea-breeze convergence and convection. There are several theories concerning the nocturnal rainfall peak. Recent studies point to a radiative effect, involving more effective nocturnal cooling of cloud-free areas around the mesoscale cloud systems. This favors subsidence, which, in turn, enhances low-level convergence into the cloud systems and strengthens the ascending air currents. Strong cooling of cloud tops, relative to their surroundings, may also produce localized destabilization and encourage droplet growth by mixing of droplets at different temperatures (see Chapter 5D). This effect would be at a maximum near dawn. Another factor is that the sea–air temperature difference, and consequently the oceanic heat supply to the atmosphere, is largest at about 03:00–06:00 hours. Yet a further hypothesis suggests that the semi-diurnal pressure oscillation encourages convergence and therefore convective activity in the early morning and evening, but divergence and suppression of convection around midday.

A large-scale survey of the Tropical Rainfall Measurement Mission (TRMM) satellite program data for 1998–2006 identifies three diurnal rainfall regimes. They are: (1) oceanic, with a peak at 06:00–09:00 LST and moderate amplitude. This is found mainly in the oceanic tropical convergence zones; (2) continental, with a peak at 15:00–18:00 LST and large amplitude, occurring in South America and Equatorial Africa; and (3) coastal, with large amplitude and phase propagation. This differs from the seaward side, where there are peaks between 21:00 and 12:00 LST with offshore propagation, and the landward side where the peaks are 12:00–21:00 LST. Pattern (3) is prominent in the Maritime continent, the Indian subcontinent, northern Australia, the west coast of equatorial Africa, northeast Brazil, and the coast from Mexico to Ecuador. Land breezes are generally weak and so may not produce much convergence. An alternative mechanism in this case may be gravity waves.

The Malaysian Peninsula displays very varied diurnal rainfall regimes in summer. The effects of land and sea breezes, anabatic and katabatic winds and topography greatly complicate the rainfall pattern by their interactions with the low-level southwesterly monsoon current. For example, there is a nocturnal maximum in the Malacca Straits region associated with the convection set off by the convergence of land breezes from Malaysia and Sumatra (cf. p. 343). However, on the east coast of Malaysia the maximum occurs in the late afternoon to early evening, when sea breezes extend about 30km inland against the monsoon southwesterlies, and convective cloud develops in the deeper sea breeze current over the coastal strip. On the interior mountains the summer rains have an afternoon maximum due to the unhindered convection process. In northern Australia, the sea-breeze phenomenon apparently extends up to 200km inland from the Gulf of Carpentaria by late evening. During the August to November dry season, this may create suitable conditions for the bore-like 'Morning Glory' – a linear cloud roll and squall-line that propagates, usually from the northeast, on the inversion created by the maritime air and nocturnal cooling. Sea breezes are usually associated with a heavy buildup of cumulus cloud and afternoon downpours.

I FORECASTING TROPICAL WEATHER

In the past two decades, significant progress has been achieved in tropical weather forecasting. This has resulted from many of the advances in observing technology and in global numerical modeling discussed in Chapter 8. Of particular importance in the tropics has been the availability of geostationary satellite data on global cloud conditions, wind vectors, sea surface temperatures, and vertical profiles of temperature and moisture. Over the past few years there have been significant improvements in ocean-observing systems. Notably, there are now over 3,300 ARGO floats operating in the world's oceans recording temperature, salinity and velocity to 200m depth and providing the data to operational worldwide in real time. Weather radar installations are also available at major centers in India, Central America and the Far East; and at some locations in Africa and the southwest Pacific; but up until now there are few in South America. A major difficulty in tropical forecasting is created by the dominant energy source of latent heat released by precipitation in convective cloud systems. This process is not easily simulated, however, due to the small-scale processes involved in cloud dynamics.

1 Short- and extended-range forecasts

The evolution and motion of tropical weather systems are primarily connected with areas of wind-speed convergence and horizontal wind shear as identified on low-level kinematic analyses depicting streamlines and isotachs and associated cloud systems, and their changes can be identified from half-hourly geostationary satellite images and weather radars; these are useful for 'now-casting' and warnings. However, cloud clusters are known to be highly irregular in their persistence beyond 24 hours. They are also subject to strong diurnal variations and orographic influences, which need to be evaluated. Analysis of diurnal variations in temperature with differing cloud states for wet and dry seasons can be a useful aid to local forecasting. Giving equal weight to persistence and climatology produces good results for low-level winds, for example. The forecasting of tropical storm movement also relies mainly on satellite imagery and radar data. For 6–12-hour forecasts, extrapolations can be made from the smoothed track over the preceding 12–24 hours. The accuracy of landfall location forecasts for the storm center is typically within about 150km. There are specialized centers for such regional forecasts and warnings in Miami, Guam, Darwin, Hong Kong, New Delhi and Tokyo. Forecasts for periods of two to five days have received limited attention. In the winter months, the tropical margins, especially of the Northern Hemisphere, may be affected by mid-latitude circulation features. Examples include cold fronts moving southward into Central America and the Caribbean, or northward from Argentina into Brazil. The motion of such systems can be anticipated from numerical model forecasts prepared at major centers such as NCEP and ECMWF.

2 Long-range forecasts

In the 15–90-day time range, it has been found that numerical models are strongly dependent on the initial conditions for several weeks. This is determined by the intra-seasonal (30–60-day) Madden–Julian Oscillation (MJO) and the slow response of the atmosphere to a change in boundary conditions. However, MJO variability is largely removed by taking monthly averages. Boundary forcing is the main determinant of forecast skill for up to one season.

For longer timescales, three areas of advance deserve attention. Predictions of the number of Atlantic tropical storms and hurricanes and of the number of days on which these occur have been developed from statistical relations with the El Niño state, mean April to May sea-level pressure over the Caribbean and the easterly or westerly phase of the stratospheric tropical winds at 30mb (see pp. 342). Cyclones in the following summer season are more numerous when during the spring season 30 and 50mb zonal winds are westerly and increasing, ENSO is in the La Niña (cold) mode and there is below-normal pressure in the Caribbean. Wet conditions in the Sahel appear to favor the development of disturbances in the eastern and central Atlantic. An initial forecast is made in November for the following season (based on stratospheric wind phase and August to November rainfall in the western Sahel) and a second forecast using information on nine predictors through July of the current year.

At least five forecast models have been developed to predict ENSO fluctuations with a lead time of up to 12 months; three involve coupled atmosphere–ocean GCMs, one is statistical and one uses analogue matching. Each of the methods shows a comparable level of moderate skill over three seasons ahead, with a noticeable decrease in skill in the northern spring. The ENSO phase strongly affects seasonal rainfall in northeast Brazil, for example, and other tropical continental areas, as well as modifying the winter climate of parts of North America through the interaction of tropical sea surface temperature anomalies and convection on mid-latitude planetary waves.

Summer monsoon rainfall in India is related to the ENSO, but the linkages are mostly simultaneous, or the monsoon events even lead the ENSO changes. El Niño (La Niña) years are associated with droughts (floods) over India. Numerous predictors of monsoon rainfall over all of India have been proposed, including spring temperatures and pressure indicative of the heat low, cross-equatorial airflow in the Indian Ocean, 500 and 200mb circulation features, ENSO phase, and Eurasian winter snow cover. A key predictor of Indian rainfall is the latitude of the 500mb ridge along 75°E in April, but the most useful operational approach seems to be a statistical combination of such parameters, with a forecast issued in May for the June to September period. The important question of the spatial pattern of monsoon onset, duration and retreat and this variability has not yet been addressed.

Rainfall over sub-Saharan West Africa is predicted by the UK Meteorological Office using statistical methods. For the Sahel, drier conditions are associated with a decreased inter-hemispheric gradient of sea surface temperatures in the tropical Atlantic and with an anomalously warm equatorial Pacific. Rainfall over the Guinea coast increases when the South Atlantic is warmer than normal.

SUMMARY

The tropical atmosphere differs significantly from that in mid-latitudes. Temperature gradients are generally weak and weather systems are mainly produced by airstream convergence triggering convection in the moist surface layer. Strong longitudinal differences in climate exist as a result of the zones of subsidence (ascent) on the eastern (western) margins of the subtropical high pressure cells. In the eastern oceans, there is typically a strong Trade Wind inversion at about 1km with dry subsiding air above, giving fine weather. Downstream, this stable lid is raised gradually by the penetration of convective clouds as the Trades flow westward. Cloud masses are frequently organized into amorphous 'clusters' on a subsynoptic scale; some of these have linear squall-lines, which are an important source of precipitation in West Africa. The Trade Wind systems of the two hemispheres converge, but not in a spatially or temporally continuous manner. This Intertropical Convergence Zone also shifts poleward over the land sectors in summer, associated with the monsoon regimes of South Asia, West Africa and northern Australia. There is a further South Pacific Convergence Zone in the southern summer.

Wave disturbances in the tropical easterlies vary regionally in character. The 'classical' easterly wave has maximum cloud buildup and precipitation behind (east of) the trough line. This distribution follows from the conservation of potential vorticity by the air. About 10 percent of wave disturbances later intensify to become tropical storms or cyclones. This development requires a warm sea surface and low-level convergence to maintain the sensible and latent heat supply and upper-level divergence to maintain ascent. Cumulonimbus 'hot towers' nevertheless account for a small fraction of the spiral cloud bands. Tropical cyclones are most numerous in the western oceans of the Northern Hemisphere in the summer to autumn seasons.

The monsoon seasonal wind reversal of South Asia is the product of global and regional influences. The orographic barrier of the Himalayas and Tibetan Plateau plays an important role. In winter, the subtropical westerly jet stream is anchored south of the mountains. Subsidence occurs over northern India, giving northeasterly surface (Trade) Winds. Occasional depressions from the Mediterranean penetrate to northwestern India–Pakistan. The circulation reversal in summer is triggered by the development of an upper-level anticyclone over the elevated Tibetan Plateau with upper-level easterly flow over India. This change is accompanied by the northward extension of low-level south-westerlies in the Indian Ocean, which appear first in southern India and along the Burma coast and then extend northwestwards. The summer 'monsoon' over East Asia also progresses from southeast to northwest, but the Mai-yu rains are mainly a result of depressions moving northeastward and thunderstorms. Rainfall is concentrated in spells associated with 'monsoon depressions', which travel westward steered by the upper easterlies. Monsoon rains fluctuate in intensity, giving rise to 'active' and 'break' periods in response to southward and northward displacements of the Monsoon Trough, respectively. There is also considerable year-to-year variability.

The West African monsoon has many similarities to that of India, but its northward advance is unhindered by a mountain barrier to the north. Four zonal climatic belts, related to the location of overlying easterly jet streams and east–west-moving disturbances, are identified. The Sahel zone is reached by the Monsoon Trough, but overlaying subsiding air greatly limits rainfall.

The climate of equatorial Africa is strongly influenced by low-level westerlies from the South Atlantic high (year-round) and easterlies in winter from the South Indian Ocean anticyclone. These flows converge along the Zaire Air Boundary (ZAB) with easterlies aloft. In summer, the ZAB is displaced southward and northeasterlies over eastern Africa meet the westerlies along the ITCZ, oriented north–south from 0° to 12°S. The characteristics of African disturbances are complex and poorly known. Deep easterly flow affects most of Africa south of 10°S (winter) or 15–18°S (summer), although the southern westerlies affect South Africa in winter.

In Amazonia, where there are broad tropical easterlies but no well-defined ITCZ, the subtropical highs of the North and South Atlantic both influence the region. Precipitation is associated with convective activity triggering low-level convergence, with meso- to synoptic-scale disturbances forming in situ, and with instability lines generated by coastal winds that move inland.

The equatorial Pacific Ocean sector plays a major role in climate anomalies throughout much of the tropics. At irregular, three- to five-year intervals, the tropical easterly winds over the eastern–central Pacific weaken, upwelling ceases off South America and the usual convection over Indonesia shifts eastward towards the central Pacific. Such warm ENSO events, which replace the normal La Niña mode, have global repercussions since teleconnection links extend to some extratropical areas, particularly East Asia and North America.

Variability in tropical climates also occurs through diurnal effects, such as land–sea breezes, local topographic and coastal effects on airflow, and the penetration of extra-tropical weather systems and airflow into lower latitudes.

Short-range tropical weather prediction is commonly limited by sparse observations and the poorly understood disturbances involved. Seasonal predictions show some success for the evolution of the ENSO regime, Atlantic hurricane activity and West African rainfall.

DISCUSSION TOPICS

- Consider the various factors that influence the damage caused by a tropical cyclone upon landfall in different parts of the world (e.g., the southeastern USA, islands in the Caribbean, Bangladesh, Northern Australia and Hong Kong).
- Use the indices of ENSO, NAO, PNA and so on available on the Web (see Appendix 4D) to compare anomalies of temperature and precipitation in a region of interest to you during positive and negative phases of the oscillations).
- Examine the similarities and differences of the major monsoon climates of the world.
- What are the similarities and differences of cyclonic systems in mid-latitudes and the tropics?
- By what mechanisms do ENSO events affect weather anomalies in the tropics and in other parts of the world?

REFERENCES AND FURTHER READING

Books

Arakawa, H. (ed.) (1969) *Climates of Northern and Eastern Asia*, World Survey of Climatology 8, Elsevier, Amsterdam, 248pp. [Comprehensive account, as of the 1960s; tables of climatic statistics]

Barry, R. G. (2008) *Mountain Weather and Climate*, 3rd edn, Cambridge University Press, Cambridge, 506pp.

Dickinson, R. E. (ed.) (1987) *The Geophysiology of Amazonia: Vegetation and Climate Interactions*, John Wiley & Sons, New York, 526pp. [Overviews of climate–vegetation–human interactions in the Amazon, forest micrometeorology and hydrology, precipitation mechanisms, general circulation modeling and the effects of land-use changes]

Domrös, M. and Peng, G-B. (1988) *The Climate of China*, Springer-Verlag, Berlin, 361pp. [Good description of climatic characteristics; climatic data tables]

Dunn, G. E. and Miller, B. I. (1960) *Atlantic Hurricanes*, Louisiana State University Press, Baton Rouge, LA, 326pp. [Classic account]

Fein, J. S. and Stephens, P. L. (eds) (1987) *Monsoons*, J.Wiley and Sons, New York, 632pp. [Theory and modeling of monsoon mechanisms considered globally and regionally; many seminal contributions by leading experts]

Gentilli, J. (ed.) (1971) *Climates of Australia and New Zealand, World Survey of Climatology* 13, Elsevier, Amsterdam, 405pp. [Detailed survey of climatic characteristics; tables of climatic statistics]

Glantz, M. H., Katz, R. W. and Nicholls, N. (eds) (1990) *Teleconnections Linking Worldwide Climate Anomalies*, Cambridge University Press, Cambridge, 535pp. [Valuable essays on ENSO characteristics, causes and worldwide effects]

Goudie, A. and Wilkinson, J. (1977) *The Warm Desert Environment*, Cambridge University Press, Cambridge, 88pp.

Griffiths, J. F. (ed.) (1972) *Climates of Africa*, World Survey of Climatology 10, Elsevier, Amsterdam, 604pp. [Detailed account of the climate of major regions of Africa; tables of climatic statistics]

Hamilton, M. G. (1979) *The South Asian Summer Monsoon*, Arnold, Australia, 72pp. [Brief account of major characteristics]

Hastenrath, S. (1985) *Climate and Circulation of the Tropics*, D. Reidel, Dordrecht, 455pp. [Comprehensive survey of weather systems, climate processes, regional phenomena and climatic change in the tropics, by a meteorologist with extensive tropical experience]

Hayward, D. F. and Oguntoyinbo, J. S. (1987) *The Climatology of West Africa*, Hutchinson, London, 271pp.

Hutchings, J. W. (ed.) (1964) *Symposium on Tropical Meteorology: Proceedings*, New Zealand Meteorological Service, Wellington.

Indian Meteorological Department (1960) *Monsoons of the World*, Delhi, 270pp. [Classic account with much valuable information]

Jackson, I. J. (1977) *Climate, Water and Agriculture in the Tropics*, Longman, London, 248pp.

[Material on precipitation and the hydrological cycle in the tropics]

Lighthill, J. and Pearce, R. P. (eds) (1981) *Monsoon Dynamics*, Cambridge University Press, Cambridge, 735pp. [Conference proceedings; specialist papers on observations and modeling of the Asian monsoon]

Logan, R. F. (1960) *The Central Namib Desert, South-west Africa*, National Academy of Sciences, National Research Council, Publication 758, Washington, DC (162 pp.).

Lopez, M. E. and Howell, W. E. (1967) Katabatic winds in the equatorial Andes. *J. Atmos.* Sci. 24, 29–35.

Malkus, J. S. and Riehl, H. (1964) *Cloud Structure and Distributions over the Tropical Pacific Ocean*, University of California Press, Berkeley and Los Angeles (229 pp.).

Musk, L. F. (1988) *Weather Systems,* Cambridge University Press, Cambridge, 160pp. [Basic introduction to weather systems]

NOAA (1992) *Experimental Long-lead Forecast Bulletin*, NOAA, Washington, DC.

Philander, S. G. (1990) *El Niño, La Niña, and the Southern Oscillation*, Academic Press, New York, 289pp. [Comprehensive survey]

Pielke, R. A. (1990) *The Hurricane,* Routledge, London and New York, 228pp. [Brief descriptive presentation of hurricane formation, distribution and movement; annual track maps for all Atlantic hurricanes, 1871–1989]

Ramage, C. S. (1971) *Monsoon Meteorology*, Academic Press, New York and London, 296pp. [Excellent overview of the Asian monsoon and its component weather systems by a tropical specialist

Ramage, C. S. (1995) *Forecaster's Guide to Tropical Meteorology*, AWS/TR–95/001, Air Weather Service, Scott Air Force Base, IL, 392pp. [Useful overview of tropical weather processes and valuable local information]

Riehl, H. (1954) *Tropical Meteorology*, McGraw-Hill, New York, 392pp. [Classic account of weather systems in the tropics by the discoverer of the easterly wave]

Riehl, H. (1979) *Climate and Weather in the Tropics*, Academic Press, New York, 611pp. [Extends his earlier work with more a climatological view; extensive material on synoptic scale weather systems]

Sadler, J. C. (1975b) *The Upper Tropospheric Circulation over the Global Tropics*, UHMET–75–05, Department of Meteorology, University of Hawaii, 35 pp.

Schwerdtfeger, W. (ed.) (1976) *Climates of Central and South America*, World Survey of Climatology 12, Elsevier, Amsterdam, 532pp. [Chapters on the climate of six regions/countries and one on Atlantic tropical storms provide the most comprehensive view of the climates of this continent; many useful diagrams and data tables]

Shaw, D. B. (ed.) (1978) *Meteorology over the Tropical Oceans*, Royal Meteorological Society, Bracknell, 278pp. [Symposium papers covering a range of important topics]

Shukla, J. and Wallace, J. M. (1983) Numerical simulation of the atmospheric response to equatorial Pacific sea surface temperature anomalies. *J. Atmos.* Sci. 40, 1613–30.

Tao, S. Y. (ed. for the Chinese Geographical Society) (1984) *Physical Geography of China*, Science Press, Beijing, 161pp. (in Chinese).

Trewartha, G. T. (1981) *The Earth's Problem Climates*, 2nd edn, University of Wisconsin Press, Madison, WI, 371pp. [Worldwide examples]

Tyson, P. D. (1986) *Climatic Change and Variability in Southern Africa*, Oxford University Press, Cape Town, 220pp. [Includes subtropical and tropical circulation systems affecting Africa south of the Equator]

World Meteorological Organization (n.d.) *The Global Climate System. A Critical Review of the Climate System during 1982–1984*, World Climate Data Programme, WMO, Geneva, 52pp.

Yoshino, M. M. (ed.) (1971) *Water Balance of Monsoon Asia*, University of Tokyo Press, Tokyo, 308pp. [Essays by Japanese climatologists focusing on moisture transport and precipitation]

Young, J. A. (coordinator) (1972) *Dynamics of the Tropical Atmosphere* (Notes from a Colloquium), National Center for Atmospheric Research, Boulder, CO, 587pp. [Summer school proceedings with presentations and discussion by leading specialists]

Zhang, J. and Lin, Z. (1985) *Climate of China*, Science and Technology Press, Shanghai, 603pp. (in Chinese). [Source of some useful diagrams]

Articles

Academica Sinica (1957–1958) On the general circulation over eastern Asia. *Tellus* 9, 432–46; 10, 58–75 and 299–312.

Anthes, R. A. (1982) Tropical cyclones: their evolution, structure, and effects. *Met. Monogr.* 19(41), Amer. Met. Soc., Boston, MA (208pp.).

Augstein, A. *et al.* (1973) Mass and energy transports in an undisturbed Atlantic trade wind flow. *Mon. Wea. Rev.* 101, 101–11.

Avila, L. A. (1990) Atlantic tropical systems of 1989. *Monthly Weather Review* 118, 1178–85.

Barnston, A. G. (1995) Our improving capability in ENSO forecasting. *Weather* 50(12), 419–30.

Barry, R. G. (1978) Aspects of the precipitation characteristics of the New Guinea mountains. *J. Trop. Geog.* 47, 13–30.

Beckinsale, R. P. (1957) The nature of tropical rainfall. *Tropical Agriculture* 34, 76–98.

Bhalme, H. N. and Mooley, D. A. (1980) Large-scale droughts/floods and monsoon circulation. *Monthly Weather Review* 108, 1,197–211.

Blumenstock, D. I. (1958) Distribution and characteristics of tropical climates. *Proc. 9th Pacific Sci. Congr.* 20, 3–23.

Breed, C. S. *et al.* (1979) Regional studies of sand seas, using Landsat (ERTS) imagery. US Geological Survey Professional Paper No. 1052, 305–97.

Browning, K. A. (1996) Current research in atmospheric science. *Weather* 51(5), 167–72.

Chang, J-H. (1962) Comparative climatology of the tropical western margins of the northern oceans. *Ann. Assn Amer. Geog.* 52, 221–7.

Chang, J-H. (1967) The Indian summer monsoon. *Geog. Rev.* 57, 373–96.

Chang, J-H. (1971) The Chinese monsoon. *Geog. Rev.* 61, 370–95.

Chopra, K. P. (1973) Atmospheric and oceanic flow problems introduced by islands. *Adv. Geophys.* 16, 297–421.

Clackson, J. R. (1957) The seasonal movement of the boundary of northern air. Nigerian Meteorological Service, Technical Note 5 (see Addendum 1958).

Crowe, P. R. (1949) The trade wind circulation of the world. *Trans. Inst. Brit. Geog.* 15, 37–56.

Crowe, P. R. (1951) Wind and weather in the equatorial zone. *Trans. Inst. Brit. Geog.* 17, 23–76.

Cry, G. W. (1965) Tropical cyclones of the North Atlantic Ocean. Tech. Paper No. 55, Weather Bureau, Washington, DC (148pp.).

Curry, L. and Armstrong, R. W. (1959) Atmospheric circulation of the tropical Pacific Ocean. *Geografiska Annaler* 41, 245–55.

Das, P. K. (1987) Short- and long-range monsoon prediction in India, in Fein, J. S. and Stephens, P. L. (eds) *Monsoons*, John Wiley & Sons, New York, 549–78.

Dhar, O. N. and Nandargi, S. (1993) Zones of extreme rainstorm activity over India. *Int. J. Climatology* 13, 301–11.

Dubief, J. (1963) Le climat du Sahara. Memoire de l'Institut de Recherches Sahariennes, Université d'Alger, Algiers (275pp.).

Eldridge, R. H. (1957) A synoptic study of West African disturbance lines. *Quart. J. Roy. Met. Soc.* 83, 303–14.

Emmanuel, K. (2005) Increasing destructiveness of tropical cyclones over the past 30 years. *Nature* 436, 686–8.

Fett, R. W. (1964) Aspects of hurricane structure: new model considerations suggested by TIROS and Project Mercury observations. *Monthly Weather Review* 92, 43–59.

Findlater, J. (1971) Mean monthly airflow at low levels over the Western Indian Ocean. *Geophysical Memoirs* 115, Meteorological Office (53pp.).

Findlater, J. (1974) An extreme wind speed in the low-level jet-stream system of the western Indian Ocean. *Met. Mag.* 103, 201–5.

Flohn, H. (1968) *Contributions to a Meteorology of the Tibetan Highlands.* Atmos. Sci. Paper No. 130, Colorado State University, Fort Collins (120 pp.).

Flohn, H. (1971) Tropical circulation patterns. *Bonn. Geogr. Abhandl.* 15 (55pp.).

Fosberg, F. R., Garnier, B. J. and Küchler, A. W. (1961) Delimitation of the humid tropics, *Geog. Rev.* 51, 333–47.

Fraedrich, K. (1990) European Grosswetter during the warm and cold extremes of the El Niño/Southern Oscillation. *Int. J. Climatology* 10, 21–31.

Frank, N. L. and Hubert, P. J. (1974) Atlantic tropical systems of 1973. *Monthly Weather Review* 102, 290–5.

Frost, R. and Stephenson, P. H. (1965) Mean streamlines and isotachs at standard pressure levels over the Indian and west Pacific oceans and adjacent land areas. *Geophys. Mem.* 14(109), HMSO, London (24pp.).

Galvin, J. F. P. (2007) The weather and climate of the tropics. Part 1 – Setting the scene. *Weather* 62(9), 245–51.

Galvin, J. F. P. (2008a) The weather and climate of the tropics. Part 3 – Synoptic scale weather systems. *Weather* 63, 16–21.

Galvin, J. F. P. (2008b) The weather and climate of the tropics. Part 6 – Monsoons. *Weather* 63(5), 129–37.

Gao, Y-X. and Li, C. (1981) Influence of Qinghai-Xizang plateau on seasonal variation of general atmospheric circulation, in *Geoecological and Ecological Studies of Qinghai-Xizang Plateau*, Vol. 2, Science Press, Beijing, 1477–84.

Garnier, B. J. (1967) *Weather Conditions in Nigeria*, Climatological Research Series No. 2, McGill University Press, Montreal (163pp.).

Gray, W. M. (1968) Global view of the origin of tropical disturbances and hurricanes. *Monthly Weather Review* 96, 669–700.

Gray, W. M. (1979) Hurricanes: their formation, structure and likely role in the tropical circulation, in Shaw, D. B. (ed.) *Meteorology over the Tropical Oceans*, Royal Meteorological Society, Bracknell, 155–218.

Gray, W. M. (1984) Atlantic seasonal hurricane frequency. *Monthly Weather Review* 112, 1649–68; 1669–83.

Gray, W. M. and Jacobson, R. W. (1977) Diurnal variation of deep cumulus convection, *Monthly Weather Review* 105, 1171–88.

Gray, W. M., Mielke, P. W. and Berry, K. J. (1992) Predicting Atlantic season hurricane activity 6–11 months in advance. *Weather Forecasting* 7, 440–55.

Gregory, S. (1965) *Rainfall over Sierra Leone*. Geography Department, University of Liverpool, Research Paper No. 2 (58pp.).

Gregory, S. (1988) El Niño years and the spatial pattern of drought over India, 1901–70, in Gregory, S. (ed.) *Recent Climatic Change*, Belhaven Press, London, 226–36.

Halpert, M. S. and Ropelewski, C. F. (1992) Surface temperature patterns associated with the Southern Oscillation. *J. Climate* 5, 577–93.

Hastenrath, S. (1995) Recent advances in tropical climate prediction. *J. Climate* 8(6), 1519–32.

Houze, R. A. and Hobbs, P. V. (1982) Organization and structure of precipitating cloud systems. *Adv. Geophys.* 24, 225–315.

Houze, R. A., Goetis, S. G., Marks, F. D. and West, A. K. (1981) Winter monsoon convection in the vicinity of North Borneo. *Mon. Wea. Rev.* 109, 591–614.

Jalu, R. (1960) Etude de la situation météorologique au Sahara en Janvier 1958. *Ann. de Géog.* 69(371), 288–96.

Jordan, C. L. (1955) Some features of the rainfall at Guam. *Bull. Amer. Met. Soc.* 36, 446–55.

Kamara, S. I. (1986) The origins and types of rainfall in West Africa. *Weather* 41, 48–56.

Kikuchi, K, and Wang, B. (2008) Diurnal precipitation regimes in the global tropics. *J. Climate* 21(11), 2680–96.

Kiladis, G. N. and Diaz, H. F. (1989) Global climatic anomalies associated with extremes of the Southern Oscillation. *J. Climate* 2, 1069–90.

Knox, R. A. (1987) The Indian Ocean: interaction with the monsoon, in Fein, J. S. and Stephens, P. L. (eds) *Monsoons*, John Wiley & Sons, New York, 365–97.

Koteswaram, P. (1958) The easterly jet stream in the tropics. *Tellus* 10, 45–57.

Kousky, V. E. (1980) Diurnal rainfall variation in northeast Brazil, *Monthly Weather Review* 108, 488–98.

Kreuels, R., Fraedrich, K. and Ruprecht, E. (1975) An aerological climatology of South America. *Met. Rundsch.* 28, 17–24.

Krishna Kumar, K., Soman, M. K. and Rupa Kumar, K. (1996) Seasonal forecasting of Indian summer monsoon rainfall: a review. *Weather* 50(12), 449–67.

Krishnamurti, T. N. (ed.) (1977) Monsoon meteorology. *Pure Appl. Geophys.* 115, 1087–1529.

Kurashima, A. (1968) Studies on the winter and summer monsoons in east Asia based on dynamic concept. *Geophys. Mag.* (Tokyo) 34, 145–236.

Kurihara, Y. (1985) Numerical modeling of tropical cyclones, in Manabe, S. (ed.) *Issues in Atmospheric and Oceanic Modeling. Part B, Weather Dynamics, Advances in Geophysics*, Academic Press, New York, 255–87.

Lander, M. A. (1990) Evolution of the cloud pattern during the formation of tropical cyclone twins symmetrical with respect to the Equator. *Monthly Weather Review* 118, 1194–202.

Landsea, C. W., Gray, W. M., Mielke, P. W., Jr. and Berry, K. J. (1994) Seasonal forecasting of Atlantic hurricane activity. *Weather* 49, 273–84.

Lau, K-M. and Li, M-T. (1984) The monsoon of East Asia and its global associations – a survey. *Bull. Amer. Met. Soc.* 65, 114–25.

Le Barbe, L., Lebel, T. and Tapsoba, D. (2002) Rainfall variability in West Africa during the years 1950–90. *J. Climate* 15(2), 187–202.

Le Borgue, J. (1979) Polar invasion into Mauretania and Senegal. *Ann. de Géog.* 88(485), 521–48.

Lin, C. (1982) The establishment of the summer monsoon over the middle and lower reaches of the Yangtze River and the seasonal transition of circulation over East Asia in early summer. *Proc. Symp. Summer Monsoon South East Asia*, People's Press of Yunnan Province, Kunming, 21–8 (in Chinese).

Lockwood, J. G. (1965) The Indian monsoon – a review. *Weather* 20, 2–8.

Lowell, W. E. (1954) Local weather of the Chicama Valley, Peru. *Archiv. Met. Geophys. Biokl.* B 5, 41–51.

Lydolph, P. E. (1957) A comparative analysis of the dry western littorals. *Ann. Assn Amer. Geog.* 47, 213–30.

Maejima, I. (1967) Natural seasons and weather singularities in Japan. Geography Report No. 2, Tokyo Metropolitan University, 77–103.

Maley, J. (1982) Dust, clouds, rain types, and climatic variations in tropical North Africa. *Quaternary Res.* 18, 1–16.

Malkus, J. S. (1955–1956) The effects of a large island upon the trade-wind air stream. *Quart. J. Roy. Met. Soc.* 81, 538–50; 82, 235–8.

Malkus, J. S. (1958) Tropical weather disturbances: why do so few become hurricanes? *Weather* 13, 75–89.

Mason, B. J. (1970) Future developments in meteorology: an outlook to the year 2000. *Quart. J. Roy. Met. Soc.* 96, 349–68.

Matsumoto, J. (1985) Precipitation distribution and frontal zones over East Asia in the summer of 1979. *Bull. Dept Geog., Univ. Tokyo* 17, 45–61.

Meehl, G. A. (1987) The tropics and their role in the global climate system. *Geog. J.* 153, 21–36.

Memberry, D.A. (2001) Monsoon tropical cyclones. Part I. *Weather* 56, 431–8.

Mink, J. F. (1960) Distribution pattern of rainfall in the leeward Koolau Mountains, Oahu, Hawaii. *J. Geophys. Res.* 65, 2869–76.

Mohr, K. and Zipser, E. (1996) Mesoscale convective systems defined by their 85–GHz ice scattering signature: size and intensity comparison over tropical oceans and continents. *Monthly Weather Review* 124, 2417–37.

Molion, L. C. B. (1987) On the dynamic climatology of the Amazon Basin and associated rain-producing mechanisms, in Dickinson, R. E. (ed.) *The Geophysiology of Amazonia*, John Wiley & Sons, New York, 391–405.

Musk, L. (1983) Outlook – changeable. *Geog. Mag.* 55, 532–3.

Neal, A. B., Butterworth, L. J. and Murphy, K. M. (1977) The morning glory. *Weather* 32, 176–83.

Nicholson, S. E. (1989) Long-term changes in African rainfall. *Weather* 44, 46–56.

Nicholson, S. E. and Flohn, H. (1980) African environmental and climatic changes and the general atmospheric circulation in late Pleistocene and Holocene. *Climatic Change* 2, 313–48.

Nyberg, J. *et al.* (2007) Low Atlantic hurricane activity in the 1970s and 1980s compared to the past 270 years. *Nature* 447, 698–701.

Omotosho, J. B. (1985) The separate contributions of line squalls, thunderstorms and the monsoon to the total rainfall in Nigeria. *J. Climatology* 5, 543–52.

Palmén, E. (1948) On the formation and structure of tropical hurricanes. *Geophysica* 3, 26–38.

Palmer, C. E. (1951) Tropical meteorology, in Malone, T. F. (ed.) *Compendium of Meteorology*, American Meteorological Society, Boston, MA, 859–80.

Physik, W. L. and Smith, R. K. (1985) Observations and dynamics of sea breezes in northern Australia. *Austral. Met. Mag.* 33, 51–63.

Pogosyan, K. P. and Ugarova, K. F. (1959) The influence of the Central Asian mountain massif on jet streams. *Meteorol. Gidrol.* 11, 16–25 (in Russian).

Quinn, W. H. and Neal, V. T. (1992) The historical record of El Niño, in Bradley, R. S. and Jones, P. D. (eds) *Climate Since A.D. 1500*, Routledge, London, 623–48.

Raghavan, K. (1967) Influence of tropical storms on monsoon rainfall in India. *Weather* 22, 250–5.

Ramage, C. S. (1952) Relationships of general circulation to normal weather over southern Asia and the western Pacific during the cool season. *J. Met.* 9, 403–8.

Ramage, C. S. (1964) Diurnal variation of summer rainfall in Malaya. *J. Trop. Geog.* 19, 62–8.

Ramage, C. S. (1968) Problems of a monsoon ocean. *Weather* 23, 28–36.

Ramage, C. S. (1986) El Niño. *Sci. American* 254, 76–83.

Ramage, C. S., Khalsa, S. J. S. and Meisner, B. N. (1980) The central Pacific near-equatorial convergence zone. *J. Geophys. Res.* 86(7), 6,580–98.

Ramaswamy, C. (1956) On the sub-tropical jet stream and its role in the department of large-scale convection. *Tellus* 8, 26–60.

Ramaswamy, C. (1962) Breaks in the Indian summer monsoon as a phenomenon of interaction between the easterly and the sub-tropical westerly jet streams. *Tellus* 14, 337–49.

Rasmusson, E. M. (1985) El Niño and variations in climate. *Amer. Sci.* 73, 168–77.

Ratisbona, L. R. (1976) The climate of Brazil, in Schwerdtfeger, W. (ed.) *Climates of Central and South America*, World Survey of Climatology 12, Elsevier, Amsterdam, 219–93.

Reynolds, R. (1985) Tropical meteorology. *Prog. Phys. Geog.* 9, 157–86.

Riehl, H. (1963) On the origin and possible modification of hurricanes. *Science* 141, 1001–10.

Rodwell, M. J. and Hoskins, B. J. (1996) Monsoons and the dynamics of deserts. *Quart. J. Roy. Met. Soc.* 122, 1385–404.

Ropelewski, C. F. and Halpert, M. S. (1987) Global and regional scale precipitation patterns associated with the El Niño/Southern Oscillation. *Monthly Weather Review* 115, 1606–25.

Rossignol-Strick, M. (1985) Mediterranean Quaternary sapropels, an immediate response of the African monsoon to variation in isolation. *Palaeogeog., Palaeoclim., Palaeoecol.* 49, 237–63.

Sadler, J. C. (1975) The monsoon circulation and cloudiness over the GATE area. *Monthly Weather Review* 103, 369–87.

Saha, R. R. (1973) Global distribution of double cloud bands over the tropical oceans. *Quart. J. Roy. Met. Soc.* 99, 551–5.

Saito, R. (1959) The climate of Japan and her meteorological disasters. *Proceedings of the International Geophysical Union*, Regional Conference in Tokyo, Japan, 173–83.

Sawyer, J. S. (1970) Large-scale disturbance of the equatorial atmosphere. *Met. Mag.* 99, 1–9.

Sikka, D. R. (1977) Some aspects of the life history, structure and movement of monsoon depressions. *Pure and Applied Geophysics* 15, 1501–29.

Suppiah, R. (1992) The Australian summer monsoon: a review. *Prog. Phys. Geog.* 16(3), 283–318.

Thompson, B. W. (1951) An essay on the general circulation over South-East Asia and the West Pacific. *Quart. J. Roy. Met. Soc.* 569–97.

Trenberth, K. E. (1976) Spatial and temporal oscillations in the Southern Oscillation. *Quart. J. Roy. Met. Soc.* 102, 639–53.

Trenberth, K. E. (1990) General characteristics of El Niño–Southern Oscillation, in Glantz, M. H., Katz, R. W. and Nicholls, N. (eds) *Teleconnections Linking Worldwide Climate Anomalies*, Cambridge University Press, Cambridge, 13–42.

Trewartha, G. T. (1958) Climate as related to the jet stream in the Orient. *Erdkunde* 12, 205–14.

Vera, C. *et al.* (2006) Toward a unified view of the American monsoon systems. *J. Clim.*, 19(20), 4977–5000.

Vincent, D. G. (1994) The South Pacific Convergence Zone (SPCZ): a review. *Monthly Weather Review* 122(9), 1949–70.

Webster, P. J. (1987a) The elementary monsoon, in Fein, J. S. and Stephens, P. L. (eds) *Monsoons*, John Wiley & Sons, New York, 3–32.

Webster, P. J. (1987b) The variable and interactive monsoon, in Fein, J. S. and Stephens, P. L. (eds) *Monsoons*, John Wiley & Sons, New York, 269–330.

Webster, P. J. *et al.* (2005) Changes in tropical cyclone number, duration, and intensity in a warming environment. *Science* 309, 1844–6.

World Meteorological Organization (1972) Synoptic analysis and forecasting in the tropics of Asia and the south-west Pacific. WMO No. 321, Geneva (524pp.).

Wyrtki, K. (1982) The Southern Oscillation, ocean–atmosphere interaction and El Niño. *Marine Tech. Soc. J.* 16, 3–10.

Yarnal, B. (1985) Extratropical teleconnections with El Niño/Southern Oscillation (ENSO) events. *Prog. Phys. Geog.* 9, 315–52.

Ye, D. (1981) Some characteristics of the summer circulation over the Qinghai–Xizang (Tibet) plateau and its neighbourhood. *Bull. Amer. Met. Soc.* 62, 14–19.

Ye, D. and Gao, Y-X. (1981) The seasonal variation of the heat source and sink over Qinghai–Xizang plateau and its role in the general circulation, in *Geoecological and Ecological Studies of Qinghai–Xizang Plateau*, Vol. 2, Science Press, Beijing, 1453–61.

Yihui, D. and Zunya, W. (2008) A study of rainy seasons in China. *Met. Atmos. Phys.*, 100, 121–38.

Yoshino, M. M. (1969) Climatological studies on the polar frontal zones and the intertropical convergence zones over South, South-east and East Asia. *Climatol. Notes* 1, Hosei University (71pp.).

Yuter, S.E. and Houze, R.A., Jr. (1998) The natural variability of precipitating clouds over the western Pacific warm pool. *Quart. J. R. Met. Soc.* 124, 53–99.

Zhang, C. *et al.* (2008) Climatology of warm season cold vortices in East Asia, 1979–2005. *Met. Atmos. Phys.*,100, 291–301.

CHAPTER TWELVE

Boundary layer climates

12

LEARNING OBJECTIVES

When you have read this chapter you will:

- understand the significance of surface characteristics for energy and moisture exchanges and thus small-scale climates
- appreciate how forest and urban environments modify atmospheric conditions and the local climate
- know the characteristics of an urban heat island

Meteorological phenomena encompass a wide range of space and timescales, from gusts of wind that swirl up leaves and litter to the global-scale wind systems that shape the planetary climate. Their time and length scales, and their kinetic energy, are illustrated in Figure 12.1 in comparison with those for a range of human activities. Small-scale turbulence, with wind eddies of a few meters dimension and lasting only for a few seconds, represents the domain of *micrometeorology*, or boundary layer climates. Small-scale climates occur within the planetary boundary layer (see Chapter 5) and have vertical scales in the order of 10^3m, horizontal scales of some 10^4m, and timescales of about 10^5 seconds (i.e., one day). The boundary layer is typically 1km thick, but varies between 20m and several kilometers in different locations and at different times in the same location. Within this layer mechanical and convective diffusion processes transport mass, momentum and energy, as well as exchanging aerosols and chemicals between the lower atmosphere and the earth's surface. The boundary layer is especially prone to nocturnal cooling and diurnal heating, and within it the wind velocity decreases through friction from the free air velocity aloft to lower values near the surface, and ultimately to the zero-velocity *roughness length* height (see Chapter 5).

Diffusion processes within the boundary layer are of two types:

1. *Eddy diffusion.* Eddies involve parcels of air that transport energy, momentum and moisture from one location to another. Usually, they can be resolved into upward-spiraling vortices leading to transfers from the earth's surface to the atmosphere or from one vertical

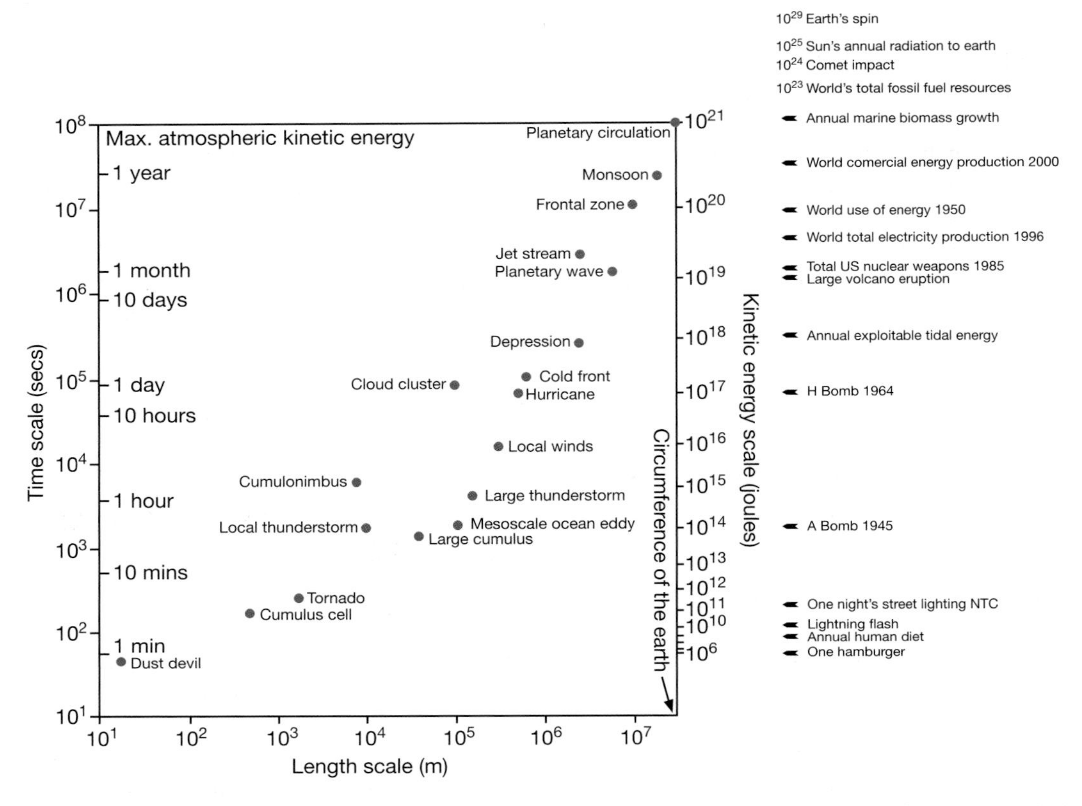

Figure 12.1 The relationship between the time and length scales of a range of meteorological phenomena together with their equivalent kinetic energy (KE) (joules). The equivalent KE values are shown for some other human and natural phenomena. 'Comet impact' refers to the KT (Cretaceous/Tertiary event). The Big Bang had an estimated energy equivalent to 10^{62} hamburgers!

layer of air to another. These eddies can be defined by generalized streamlines (i.e., resolved fluctuations). They range in size from a few centimeters (10^{-2}m) in diameter above a heated surface to 1–2m (10^0m) resulting from small-scale convection and surface roughness, and grade into dust devils (10^1m, lasting 10^1–10^2s) and tornadoes (10^3m, lasting 10^2–10^3s).

2 *Turbulent diffusion.* These are apparently random (i.e., unresolved) fluctuations of instantaneous velocities having variations of a second or less.

A SURFACE ENERGY BUDGETS

We first review the process of energy exchange between the atmosphere and an unvegetated surface. The surface energy budget equation, discussed in Chapter 3D, is usually written as:

$$Rn = H + LE + G$$

where Rn, the net all-wavelength radiation = $[S(1 - \alpha)] + Ln$

S = incoming shortwave radiation,

α = fractional albedo of the surface, and
Ln = the net (outgoing) longwave radiation.

Rn is usually positive by day, since the absorbed solar radiation exceeds the net outgoing longwave radiation; at night, when $S = 0$, Rn is determined by the negative magnitude of Ln. since the outgoing longwave radiation from the surface invariably exceeds the downward component from the atmosphere.

The surface energy flux terms are defined as positive away from the surface interface:

G = heat flux into the ground,
H = turbulent sensible heat flux to the atmosphere,
LE = turbulent latent heat flux to the atmosphere (E = evaporation; L = latent heat of vaporization).

By day, the available net radiation is balanced by the outgoing turbulent fluxes of sensible heat (H) and latent heat (LE) into the atmosphere and by conductive heat flux into the ground (G). At night, the negative Rn caused by net outgoing longwave radiation is offset by the supply of conductive heat from the soil (G) and turbulent heat from the air (H) (Figure 12.2A). Occasionally, condensation may contribute heat to the surface.

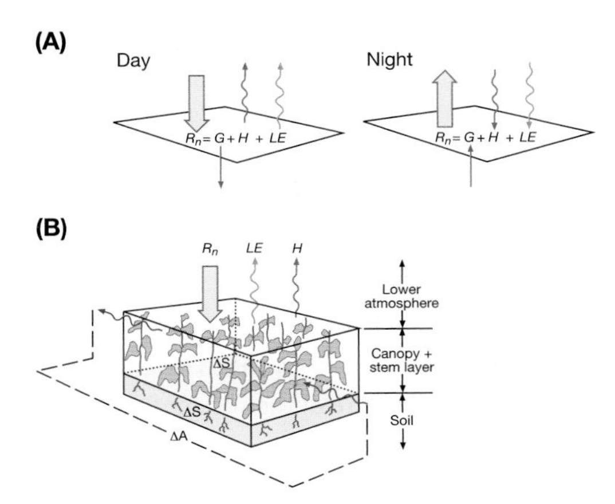

Figure 12.2 Energy flows involved in the energy balance of a simple surface during day and night (A) and a vegetated surface (B).
Source: After Oke (1978). Courtesy of Routledge and Methuen.

Commonly, there is a small residual heat storage (ΔS) in the soil in spring/summer and a return of heat to the surface in autumn/winter. Where a vegetation canopy is present there may be a small additional biochemical heat storage, due to photosynthesis, as well as physical heat storage by leaves and stems. An additional energy component to be considered in areas of mixed canopy cover (forest/grassland, desert/oasis), and in water bodies, is the horizontal transfer (*advection*) of heat by wind and currents (ΔA; see Figure 12.2B). The atmosphere transports sensible and latent heat both vertically and horizontally.

B NON-VEGETATED NATURAL SURFACES

1 Rock and sand

The energy exchanges of dry desert surfaces are relatively simple. A representative diurnal pattern of energy exchange over desert surfaces is shown in Figure 12.3. The 2m air temperature varies between 17 and 29°C, although the surface of the dry lake-bed reaches 57°C at midday. Rn reaches a maximum at about 13:00 hours when most of the heat is transferred to the air by turbulent convection; in the early morning the heat goes into the ground. At night this soil heat is returned to the surface, offsetting radiative cooling. Over a 24-hour period, about 90 percent of the net radiation goes into sensible heat, 10 percent into ground flux. Extreme surface temperatures exceeding 88°C (190°F) have been measured in Death Valley, California, and it seems that an upper limit is about 93°C (200°F). Record maximum air temperatures are 56.7°C at Greenland Ranch, Death Valley, California, and 57.8°C (136°F) at El Azizia, Libya.

Surface properties modify the heat penetration, as shown by mid-August measurements in the Sahara (Figure 12.4). Maximum surface

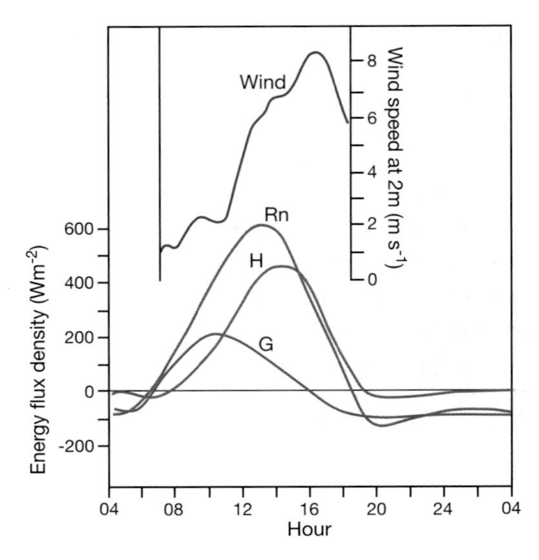

Figure 12.3 Energy flows involved at a dry-lake surface at El Mirage, California (35°N), on 10–11 June 1950. Wind speed due to surface turbulence was measured at a height of 2m.

Source: After Vehrencamp (1953) and Oke (1978).

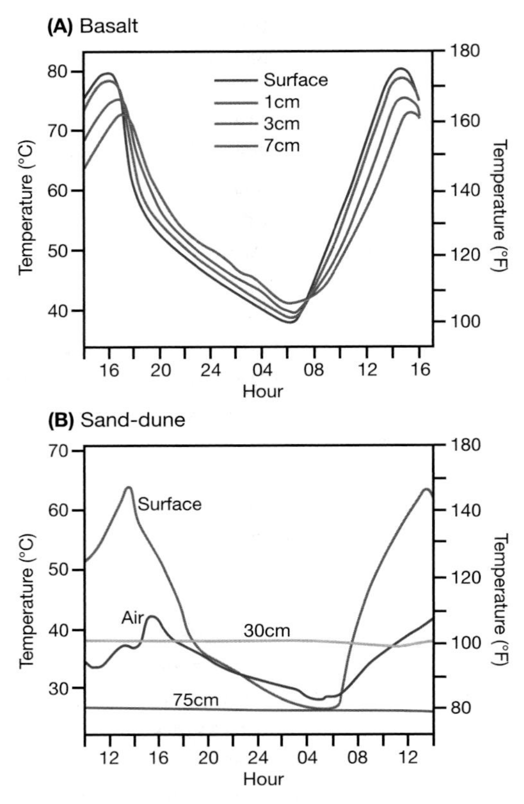

Figure 12.4 Diurnal temperatures near, at and below the surface in the Tibesti region, central Sahara, in mid-August 1961. A: At the surface and at 1cm, 3cm and 7cm below the surface of a basalt. B. In the surface air layer, at the surface and at 30cm and 75cm below the surface of a sand-dune.

Source: After Peel (1974). Courtesy of Zeitschrift für Geomorphologie.

temperatures reached on dark-colored basalt and light-colored sandstone are almost identical, but the greater thermal conductivity of basalt (3.1W m^{-1} K^{-1}) versus sandstone (2.4W m^{-1} K^{-1}) gives a larger diurnal range and deeper penetration of the diurnal temperature wave, to about 1m in the basalt. In sand, the temperature wave is negligible at 30cm due to the low conductivity of intergranular air. Note that the surface range of temperature is several times that in the air. Sand also has an albedo of 0.35, compared with about 0.2 for a rock surface.

2 Water

For a water body, the energy fluxes are very differently apportioned. Figure 12.5 illustrates the diurnal regime for the tropical Atlantic Ocean averaged for 20 June to 2 July 1969. The simple energy balance is based on the assumption that the horizontal advective term due to heat transfer by currents is zero and that the total energy input is absorbed in the upper 27m of the ocean. Thus, between 06:00 and 16:00 hours, almost all of the net radiation is absorbed by the water layer (i.e., ΔW is positive) and at all other times the ocean water is heating the air through the transfer of sensible and latent heat of evaporation. The afternoon maximum is determined by the time of maximum temperature of the surface water.

3 Snow and ice

Surfaces that have snow or ice cover for much of the year present more complex energy budgets. The surface types include: ice-covered ocean;

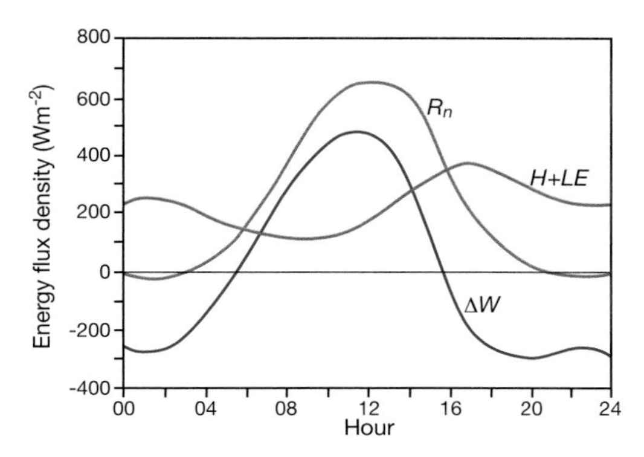

Figure 12.5 Average diurnal variation of the energy balance components in and above the tropical Atlantic Ocean during the period 20 June to 2 July 1969.

Source: After Holland. From Oke (1987). By permission of Routledge and Methuen, London, and T. R. Oke.

glaciers, tundra; boreal forests and steppe, all of which are snow-covered during the long winter. Similar energy balances characterize the winter months (Figure 12.6). An exception is the local areas of ocean covered by thin sea ice and open leads in the ice have 300W m^{-2}available – more than the net radiation for boreal forests in summer. The spring transition on land is very rapid (see Figure 10.38). During the summer, when albedo becomes a critical surface parameter, there are important spatial contrasts. In summer, the radiation budget of sea ice more than 3 meters thick is quite low and for ablating glaciers is lower still. Melting snow involves the additional energy balance component (ΔM), which is the net latent heat storage change (positive) due to melting

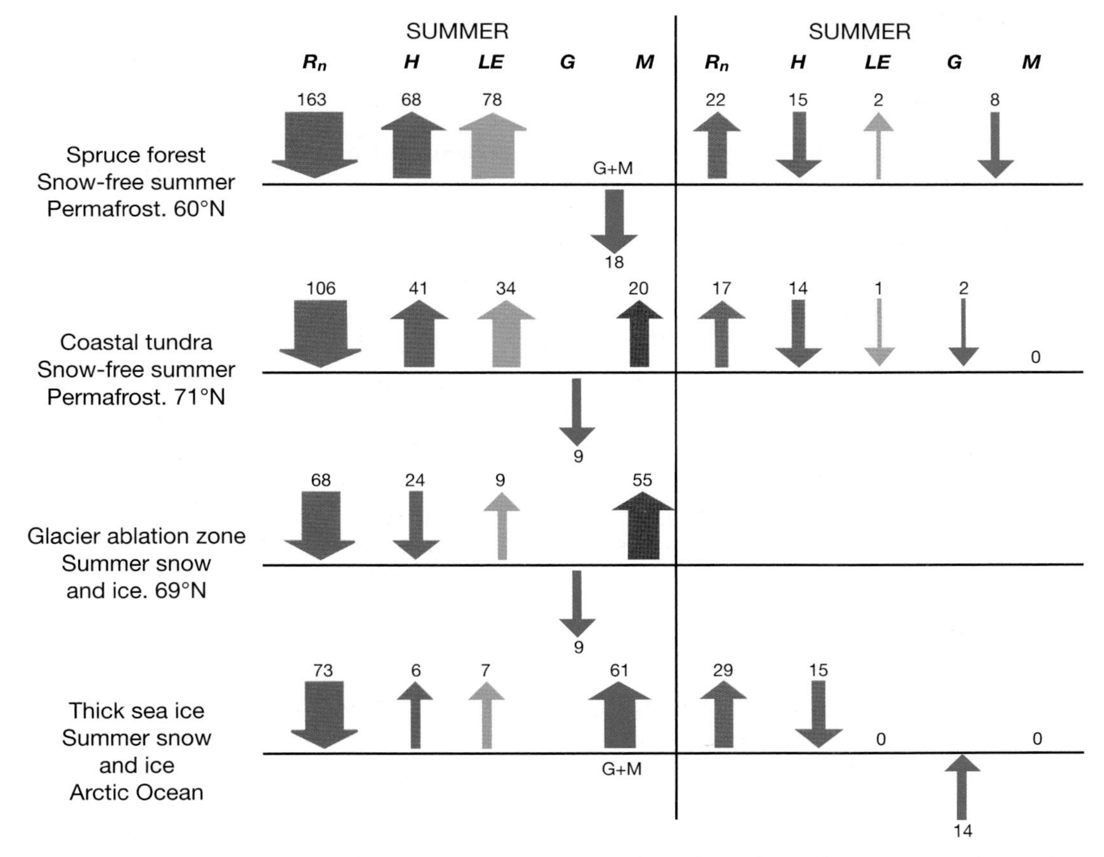

Figure 12.6 Energy balances (Wm^{-2}) over four terrain types in the polar regions. M = energy used to melt snow.

Source: Weller and Wendler (1990). Reprinted from *Annals of Glaciology* with the permission of the International Glaciological Society.

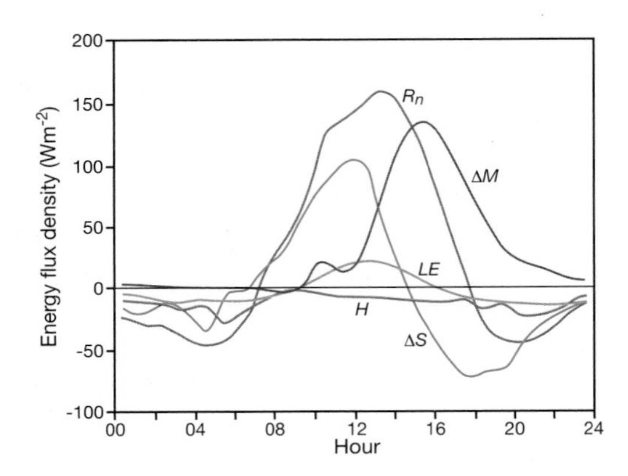

Figure 12.7 Energy balance components for a melting snow cover at Bad Lake, Saskatchewan (51°N) on 10 April 1974.

Source: Granger and Male. Modified by Oke (1987). By permission of Routledge and Methuen, London, and T. R. Oke.

(Figure 12.7). In this example of snow-melt at Bad Lake, Saskatchewan on 10 April 1974, the value of *Rn* was kept low by the high albedo of the snow (0.65). As the air was always warmer than the melting snow, there was a flow of sensible heat from the air at all times (i.e., *H* negative). Prior to noon, almost all the net radiation went into snow heat storage, causing melting, which peaked in the afternoon (ΔM maximum). Net radiation accounted for about 68 percent of the snow-melt and convection ($H + LE$) for 31 percent. Snow melts earlier in the boreal forests than on the tundra, and as the albedo of the uncovered spruce forest tends to be lower than that of the tundra, the net radiation of the forest can be significantly greater than for the tundra. Thus, south of the arctic tree-line the boreal forest acts as a major heat source.

Table 12.1 Rates of energy dispersal (W m^{-2}) at noon in a 20cm stand of grass (in higher mid-latitudes on a June day)

Net radiation at the top of the crop	550
Physical heat storage in leaves	6
Biochemical heat storage (i.e. growth processes)	22
Received at soil surface	200

C VEGETATED SURFACES

From the viewpoint of energy regime and plant canopy microclimate, it is useful to consider short crops and forests separately.

1 Short green crops

Short green crops, up to a meter or so high, supplied with sufficient water and exposed to similar solar radiation conditions, all have a similar net radiation (*Rn*) balance. This is largely due to the small range of albedos, 20–30 percent for short green crops compared with 9–18 percent for forests. Canopy structure appears to be the primary reason for this albedo difference.

General figures for rates of energy dispersal at noon on a June day in a 20cm high stand of grass in the higher mid-latitudes are shown in Table 12.1.

Figure 12.8 shows the diurnal and annual energy balances of a field of short grass near Copenhagen (56°N). For an average 24-hour period in June, about 58 percent of the incoming radiation is used in evapotranspiration. In December the small net outgoing radiation (i.e., *Rn* negative) is composed of 55 percent heat supplied by the soil and 45 percent sensible heat transfer from the air to the grass.

We can generalize the microclimate of short growing crops according to T. R. Oke (see Figure 12.9):

1 *Temperature.* In early afternoon, there is a temperature maximum just below the vegetation crown, where the maximum energy absorption is occurring. The temperature is lower near the soil surface, where heat flows into the soil. At night, the crop cools mainly by longwave emission and by some continued transpiration, producing a temperature minimum at about two-thirds the height of the crop. Under calm conditions, a temperature inversion may form just above the crop.

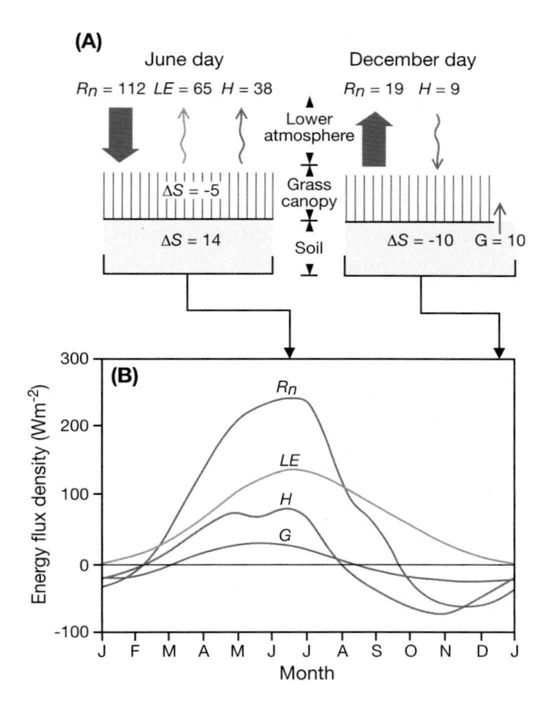

Figure 12.8 Energy fluxes over short grass near Copenhagen (56°N). A: Totals for a day in June (17 hours' daylight; maximum solar altitude 58°) and December (7 hours' daylight; maximum solar altitude 11°). Units are Wm^{-2}. B: Seasonal curves of net radiation (*Rn*), latent heat (*LE*), sensible heat (*H*) and ground-heat flux (*G*).

Source: Data from Miller (1965); and after Sellers (1965).

2. *Wind speed.* This is at a minimum in the upper crop canopy where the foliage is most dense. There is a slight increase below and a marked increase above.
3. *Water vapor.* The maximum diurnal evapotranspiration rate and supply of water vapor occurs at about two-thirds the crop height where the canopy is most dense.
4. *Carbon dioxide.* By day, CO_2 is absorbed through photosynthesis of growing plants and at night is emitted by respiration. The maximum sink and source of CO_2 is at about two-thirds the crop height.

Finally, we examine the conditions accompanying the growth of irrigated crops. Figure 12.10 illustrates the energy relationships in a 1m-high stand of irrigated Sudan grass at Tempe, Arizona, on 20 July 1962. The air temperature varied between 25 and 45°C. By day, evapotranspiration in the dry air was near its potential and *LE* (anomalously high due to a local temperature inversion) exceeded *Rn*, the deficiency being made up by a transfer of sensible heat from the air (*H* negative). Evaporation continued during the night due to a moderate wind (7m s^{-1}) sustained by the continued heat flow from the air. Thus evapotranspiration leads to comparatively low diurnal temperatures within irrigated desert crops. Where the surface is inundated with water,

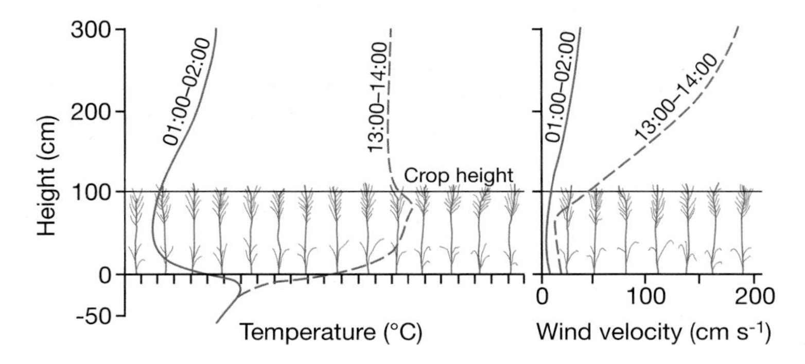

Figure 12.9 Temperature and wind velocity profiles within and above a meter-high stand of barley at Rothamsted, southern England, on 23 July 1963 at 01:00–02:00 hours and 13:00–14:00 hours.

Source: After Long *et al.* (1964). By permission of Meteorologische Rundschau.

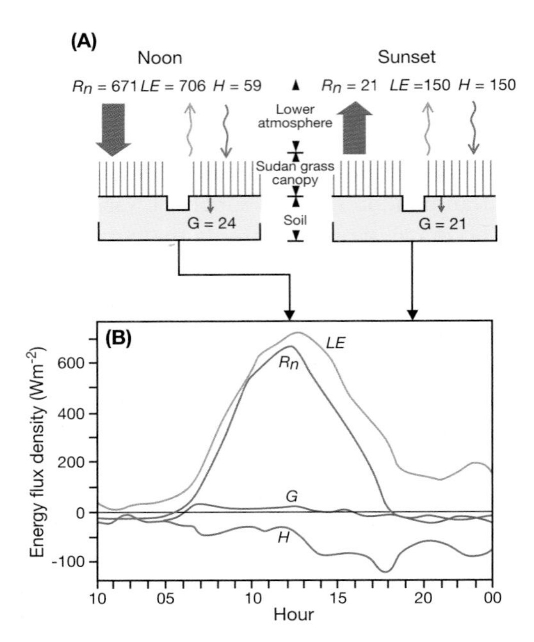

Figure 12.10 Energy flows involved in the diurnal energy balance of irrigated Sudan grass at Tempe, Arizona, on 20 July 1962.

Source: After Sellers (1965). By permission of the University of Chicago Press.

as in a rice paddy-field, the energy balance components and thus the local climate take on something of the character of water bodies (see B, this chapter). In the afternoon and at night the water becomes the most important heat source and turbulent losses to the atmosphere are mainly in the form of the latent heat.

2 Forests

The vertical structure of a forest, which depends on the species composition, the ecological associations and the age of the stand, largely determines the forest microclimate. The climatic influence of a forest can be explained in terms of the geometry of the forest, including morphological characteristics, size, cover, age and stratification. Morphological characteristics include amount of branching (bifurcation), the periodicity of growth (i.e., evergreen or deciduous), together with the size, density and texture of the leaves. Tree size is obviously important. In temperate forests the sizes may be closely similar, whereas in tropical forests there may be great local variety. Crown coverage determines the physical obstruction presented by the canopy to radiation exchange and airflow.

Different vertical structures in tropical rainforests and temperate forests can have important microclimatic effects. In tropical forests the average height of the taller trees is around 46–55m, with individuals rising to over 60m. The dominant height of temperate forest trees is generally up to 30m. Tropical forests possess a great variety of species, seldom fewer than 40 per hectare (100 hectares = 1km^2) and sometimes over 100, compared with fewer than 25 (occasionally only one) tree species with a trunk diameter greater than 10cm in Europe and North America. Some British woodlands have almost continuous canopy stratification, from low shrubs to the tops of 36m beeches, whereas tropical forests are strongly stratified with dense undergrowth, simple trunks, and commonly two upper strata of foliage. This stratification results in more complex microclimates in tropical forests than in temperate ones.

It is convenient to describe the climatic effects of forest stands in terms of their modification of energy transfers, the airflow, humidity environment and thermal environment.

Modification of energy transfers

Forest canopies significantly change the pattern of incoming and outgoing radiation. The shortwave reflectivity of forests depends partly on the characteristics of the trees and their density. Coniferous forests have albedos of about 8–14 percent, while values for deciduous woods range between 12 and 18 percent, increasing as the canopy becomes more open. Values for semi-arid savanna and scrub woodland are much higher.

Besides reflecting energy, the forest canopy traps energy. Measurements made in summer in a 30-year-old oak stand in the Voronezh district of Russia indicate that 5.5 percent of the net radiation at the top of the canopy is stored in the soil and the trees. Dense red beeches (*Fagus sylvatica*) intercept 80 percent of the incoming radiation at the treetops and less than 5 percent reaches the forest floor. The greatest trapping occurs in sunny conditions, because when the sky is overcast the diffuse incoming radiation has greater possibility of penetration laterally to the trunk space (Figure 12.11). Visible light, however, does not give an altogether accurate picture of total energy penetration, because more ultraviolet than infrared radiation is absorbed in the crowns. As far as light penetration is concerned, there are great variations depending on type of tree, tree spacing, time of year, age, crown density and height. About 50–75 percent of the outside light intensity may penetrate to the floor of a birch–beech forest, 20–40 percent for pine and 10–25 percent for spruce and fir, but for tropical forests in the Congo the figure may be as low as 0.1 percent, and 0.01 percent has been recorded for a dense elm stand in Germany. One of the most important effects of this is to reduce the length of daylight. For deciduous trees, more than 70 percent of the light may penetrate when they are leafless. Tree age is also important in that this controls both crown cover and height. Figure 12.11 shows this rather complicated effect for spruce in the Thuringian Forest, Germany.

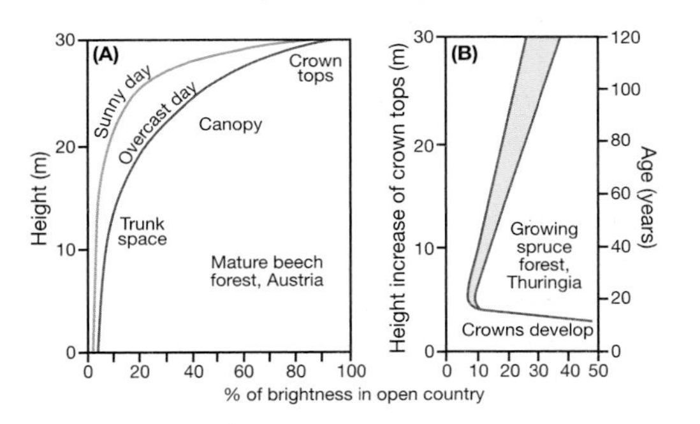

Figure 12.11 The amount of light beneath the forest canopy as a function of cloud cover and crown height: (A) For a thick stand of 120–150-year-old red beeches (*Fagus sylvatica*) at an elevation of 1000m on a 20° southeast-facing slope near Lunz, Austria; (B) For a Thuringian spruce forest in Germany over more than 100 years of growth, during which the crown height increased to almost 30m.

Source: After Geiger (1965). By permission of Rowman and Littlefield.

Modification of airflow

Forests impede both the lateral and the vertical movement of air. In general, air movement within forests is slight compared with that in the open, and quite large variations of outside wind velocity have little effect inside woods. Measurements in European forests show that 30m of penetration reduces wind velocities to 60–80 percent, 60m to 50 percent and 120m to only 7 percent. A wind of 2.2m s^{-1} outside a Brazilian evergreen forest was reduced to 0.5m s^{-1} at 100m within it, and was negligible at 1000m. In the same location, external storm winds of 28m s^{-1} were reduced to 2m s^{-1} some 11km deep in the forest. Where there is a complex vertical structuring of the forest, wind velocities become more complex. Thus, in the crowns (23m) of a Panama rainforest the wind velocity was 75 percent of that outside, while it was only 20 percent in the undergrowth (2m). Other influences include the density of the stand and the season. The effect of season on wind velocities in deciduous forests is shown in Figure 12.12. In a Tennessee mixed oak forest, forest wind velocities were 12 percent of those in the open in January, but only 2 percent in August.

Knowledge of the effect of forest barriers on winds has been utilized in the construction of windbreaks to protect crops and soil. Cypress breaks of the southern Rhône valley and Lombardy poplars (*Populus nigra*) of the Netherlands form distinctive features of the landscape. It has been found that the denser the obstruction the greater the shelter immediately behind it, although the downwind extent of its effect is reduced by lee turbulence set up by the barrier. A windbreak of about 40 percent penetrability (Figure 12.13) gives the maximum protection. An obstruction begins to have an effect about 18 times its own height upwind, and the downwind effect can be increased by the *back coupling* of more than one belt (see Figure 12.13).

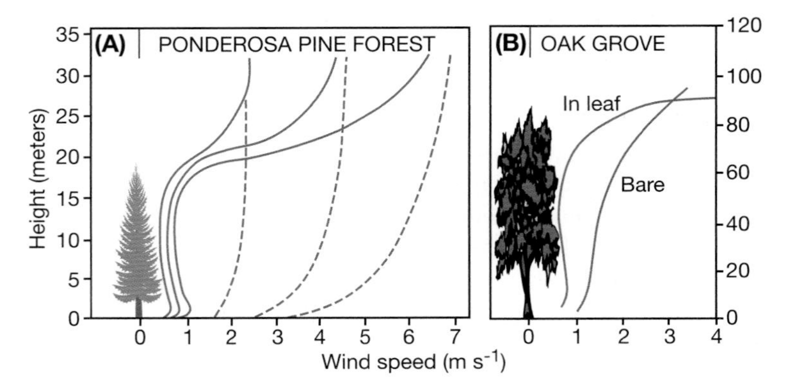

Figure 12.12 Influence on wind velocity profiles exercised by: (A) a dense stand of 20m high ponderosa pines (*Pinus ponderosa*) in the Shasta Experimental Forest, California. The dashed lines indicate the corresponding wind profiles over open country for general wind speeds of about 2.3, 4.6 and 7.0m s^{-1}, respectively; (B) a grove of 25m high oak trees, both bare and in leaf.

Sources: A: After Fons; and Kittredge (1948). B: After R. Geiger and H. Amann, and Geiger (1965). By permission of Rowman and Littlefield.

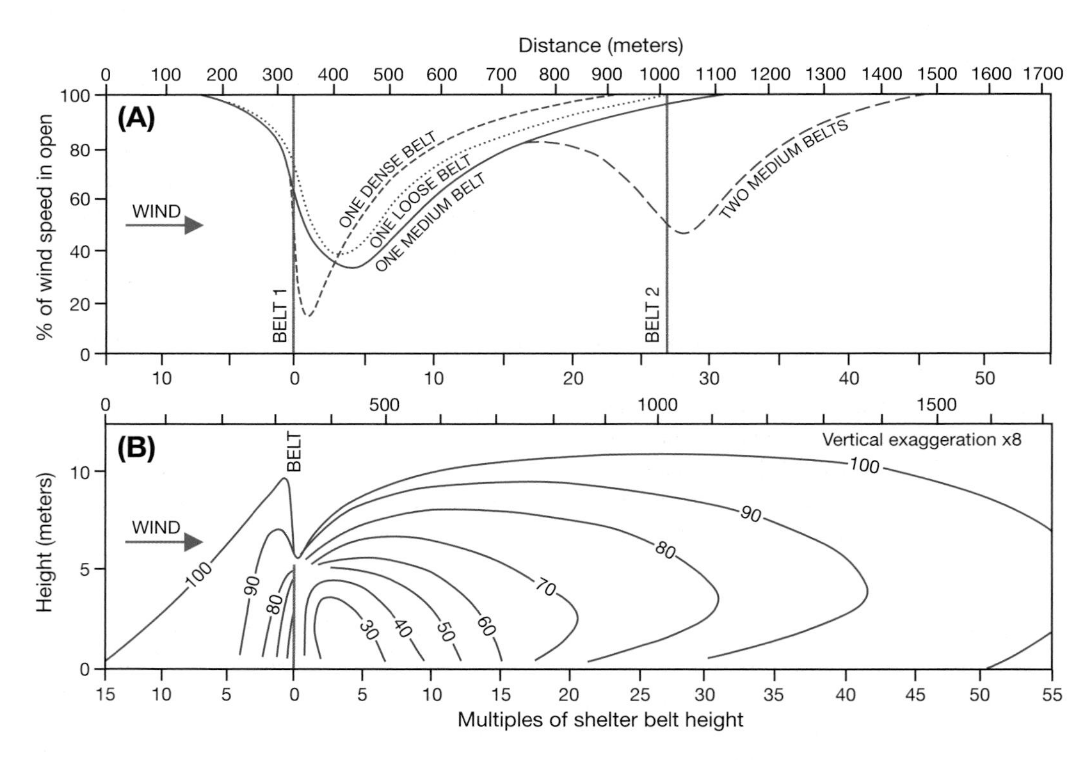

Figure 12.13 The influence of shelter belts on wind velocity distributions (expressed as percentages of the velocity in the open). A: The effects of one shelter belt of three different densities, and of two back-coupled medium-dense shelter belts. B: The detailed effects of one half-solid shelter belt.

Sources: A: After W. Nägeli; and Geiger (1965). B: After Bates and Stoeckeler; and Kittredge (1948).

There are some less obvious microclimatic effects of forest barriers. One of the most important is that the reduction of wind speed in forest clearings increases the frost risk on winter nights. Another is the removal of dust and fog droplets from the air by the filtering action of forests. Measurements 1.5km upwind, on the lee side and 1.5km downwind of a kilometer-wide German forest gave dust counts (particles per liter) of 9000, less than 2000 and more than 4000, respectively. Fog droplets can be filtered from laterally moving air, resulting in a higher precipitation catch within a forest than outside. The winter rainfall catch outside a eucalyptus forest near Melbourne, Australia was 500mm, whereas inside the forest it was 600mm.

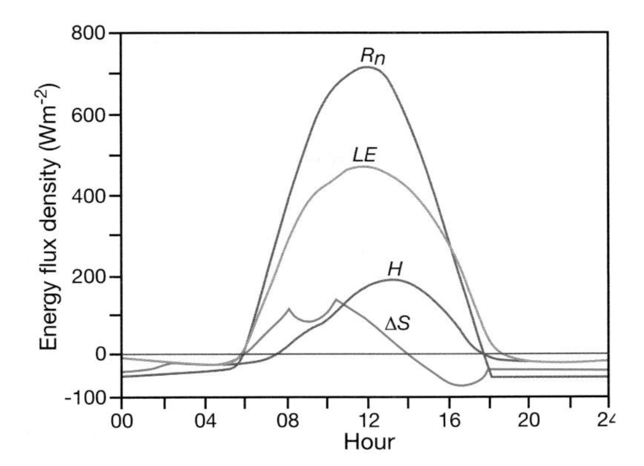

Figure 12.14 A computer simulation of energy flows involved in the diurnal energy balance of a primary tropical broad-leaved forest in the Amazon during a high-sun period on the second dry day following 22mm daily rainfall.

Source: After a Biosphere Atmosphere Transfer Scheme (BATS) model from Dickinson and Henderson-Sellers (1988). By permission of the Royal Meteorological Society, redrawn.

Modification of the humidity environment

The humidity conditions within forest stands contrast strikingly with those in the open. Evaporation from the forest floor is usually much less owing to the decreased direct sunlight, lower wind velocity, lower maximum temperature, and generally higher forest air humidity. Evaporation from the bare floor of pine forests is 70 percent of that in the open for Arizona in summer and only 42 percent for the Mediterranean region.

Unlike many cultivated crops, forest trees exhibit a wide range of physiological resistance to transpiration processes and, hence, the proportions of forest energy flows involved in evapotranspiration (*LE*) and sensible heat exchange (*H*) vary. In the Amazonian tropical broad-leaved forest, estimates suggest that after rain up to 80 percent of the net solar radiation (*Rn*) is involved in evapotranspiration (*LE*) (Figure 12.14). Figure 12.15 compares diurnal energy flows during July for a pine forest in eastern England and a fir forest in British Columbia. In the former case, only 0.33*Rn* is used for *LE* due to the high resistance of the pines to transpiration, whereas 0.66*Rn* is similarly employed in the British Columbia fir forest, especially during the afternoon. Like short green crops, only a very small proportion of *Rn* is ultimately used for tree growth, an average figure being about 1.3W m^{-2}, some 60 percent of which produces wood tissue and 40 percent forest litter.

During daylight, leaves transpire water through open pores or *stomata*. This loss is controlled by the length of day, the leaf temperature (modified by evaporational cooling), surface area, the tree species and its age, as well as by the meteorological factors of available radiant energy, atmospheric vapor pressure and wind speed. Total evaporation figures are therefore extremely varied. The evaporation of water intercepted by the vegetation surfaces also enters into the totals, in addition to direct transpiration. Calculations made for a catchment covered with Norway spruce (*Picea abies*) in the Harz Mountains of Germany showed an annual evapotranspiration of 340mm and additional interception losses of 240mm.

The humidity of forest stands is closely linked to the amount of evapotranspiration and increases with the density of vegetation present The increase in forest relative humidity over that outside averages 3–10 percent and is especially marked in summer. Vapor pressures were higher within an oak stand in Tennessee than outside for every

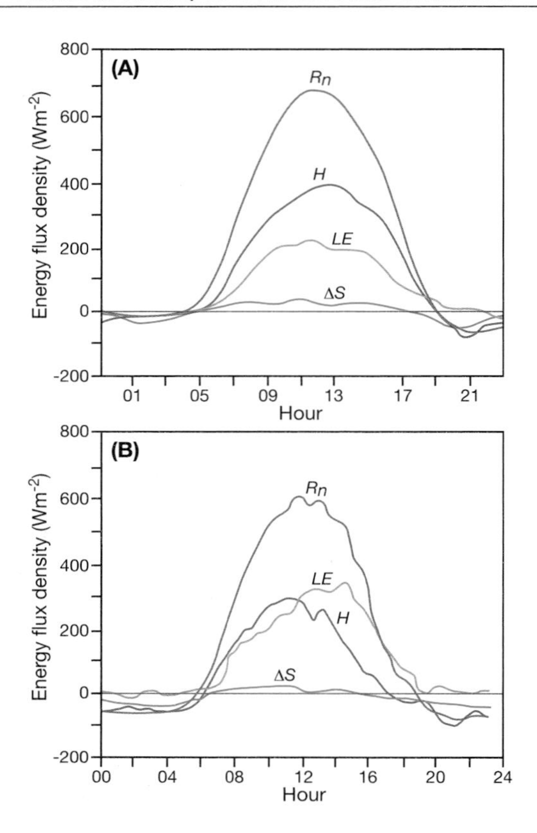

Figure 12.15 Energy components on a July day in two forest stands. A: Scots and Corsican pine at Thetford, England (52°N), on 7 July 1971. Cloud cover was present during the period 00:00–05:00 hours. B: Douglas fir stand at Haney, British Columbia (49°N), on 10 July 1970. Cloud cover was present during the period 11:00–20:00 hours.

Sources: A: Data from Gay and Stewart (1974); after Oke (1978); B: data from McNaughton and Black (1973); after Oke (1978). By permission of Routledge and Methuen.

month except December. Tropical forests exhibit almost complete night saturation irrespective of elevation in the trunk space, whereas by day humidity is inversely related to elevation. Measurements in Amazonia show that in dry conditions daytime specific humidity in the lower trunk space (1.5m) is near 20g kg^{-1}, compared with 18g kg^{-1} at the top of the canopy (36m).

Recent research in boreal forests shows that they have low photosynthetic and carbon drawdown rates, and consequently low transpiration rates. Over the year, the uptake of CO_2 by photosynthesis is balanced by its loss through respiration. During the growing season, the evapotranspiration rate of boreal (mainly spruce) forests is surprisingly low (less than 2 mm per day). The low albedo, coupled with low energy use for evapotranspiration, leads to high available energy, high sensible heat fluxes and the development of a deep convective planetary boundary layer. This is particularly marked during spring and early summer due to intense mechanical and convective turbulence. In autumn, in contrast, soil freezing increases its heat capacity, leading to a lag in the climate system. There is less available energy and the boundary layer is shallow.

The influence of forests on precipitation is still unresolved. This is partly due to the difficulties of comparing rain-gauge catches in the open with those in forests, within clearings or beneath trees. In small clearings, low wind speeds cause little turbulence around the opening of the gauge and catches are generally greater than outside the forest. In larger clearings, downdrafts are more prevalent and consequently the precipitation catch increases. In a 25m-high pine and beech forest in Germany, catches in 12m diameter clearings were 87 percent of those upwind of the forest, but the catch rose to 105 percent in clearings of 38m. An analysis of precipitation records for Letzlinger Heath (Germany) before and after afforestation suggested a mean annual increase of 6 percent, with the greatest excesses occurring during drier years. It seems that forests have little effect on cyclonic rain, but they may slightly increase orographic precipitation, by lifting and turbulence, in the order of 1–3 percent in temperate regions.

A more important influence of forests on precipitation is through the direct interception of rainfall by the canopy. This varies with crown coverage, season and rainfall intensity. Measurements in German beech forests indicate that, on average, they intercept 43 percent of precipitation in summer and 23 percent in winter. Pine forests may intercept up to 94 percent of low-intensity precipitation but as little as 15 percent of high intensities, the average for temperate pines being about 30 percent. In tropical rainforest, about 13 percent of annual rainfall is intercepted. The intercepted precipitation either evaporates on the canopy, runs down the trunk or drips to the ground. Assessment of the total precipitation reaching the ground (the through-fall) requires careful measurements of the stem flow and the contribution of drips from the canopy. Canopy interception contributes 15–25 percent of total evaporation in tropical rainforests. It is not a total loss of moisture from the forest, since the solar energy used in the evaporating process is not available to remove soil moisture or transpiration water. However, the vegetation does not derive the benefit of water cycling through it via the soil. Canopy evaporation depends on net radiation receipts and the type of species. Some Mediterranean oak forests intercept 35 percent of rainfall and almost all evaporates from the canopy. Water balance studies indicate that evergreen forests allow 10–50 percent more evapotranspiration than grass in the same climatic conditions. Grass normally reflects 10–15 percent more solar radiation than coniferous tree species and hence less energy is available for evaporation. In addition, trees have a greater surface roughness, which increases turbulent air motion and, therefore, the evaporation efficiency. Evergreens allow transpiration to occur year-round. Nevertheless, research to verify these results and test various hypotheses is needed.

Modification of the thermal environment

Forest vegetation has an important effect on microscale temperature conditions. Shelter from the sun, blanketing at night, heat loss by evapotranspiration, reduction of wind speed and the impeding of vertical air movement all influence the temperature environment. The most obvious effect of canopy cover is that, inside the forest, daily maximum temperatures are lower and minima are higher (Figure 12.16). This is particularly apparent during periods of high summer evapotranspiration, which depress daily maximum temperatures and cause mean monthly temperatures in tropical and temperate forests to fall well below that outside. In temperate forests at sea-level, the mean annual temperature may be about 0.6°C lower than that in surrounding open country, the mean monthly differences may reach 2.2°C in summer but not exceed 0.1°C in winter. On hot summer days the difference can be more than 2.8°C. Mean monthly temperatures and diurnal ranges for temperate beech, spruce and pine forests are given in Figure 12.17. This also shows that when trees transpire little in the summer (e.g., the *forteto* oak maquis of the Mediterranean), the high daytime temperatures reached in the sheltered woods may cause the pattern of mean monthly values to be the reverse of temperate forests. Even within individual climatic regions it is difficult to generalize, however. At elevations of 1000m the lowering of temperate forest mean temperatures below those in the open may be double that at sea-level.

The vertical structure of forest stands gives rise to a complex temperature structure, even in relatively simple stands. For example, in a ponderosa pine forest (*Pinus ponderosa*) in Arizona the recorded mean June to July

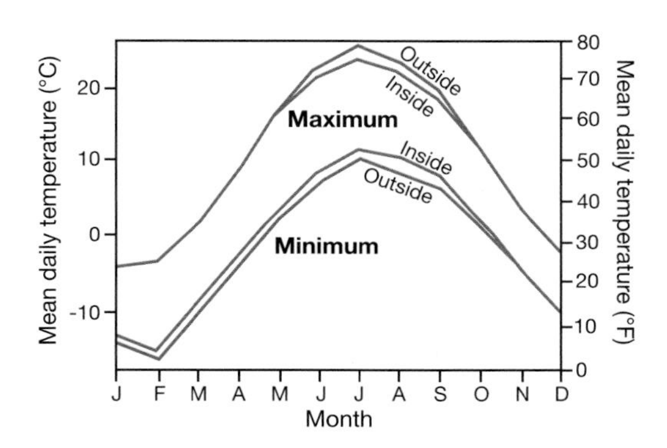

Figure 12.16 Seasonal regimes of mean daily maximum and minimum temperatures inside and outside a birch–beech–maple forest in Michigan.
Source: After US Department of Agriculture Yearbook (1941).

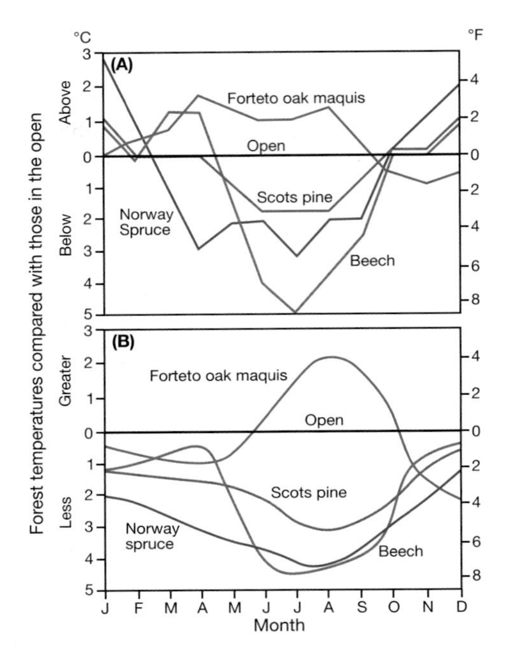

Figure 12.17 Seasonal regimes of (A) mean monthly temperatures and (B) mean monthly temperature ranges, compared with those in the open, for four types of Italian forest. Note the anomalous conditions associated with the *forteto* oak scrub (*maquis*), which transpires little.

Source: Food and Agriculture Organization of the United Nations (1962).

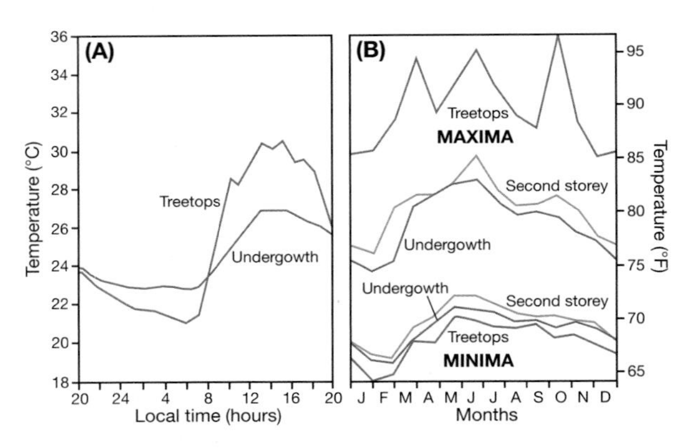

Figure 12.18 The effect of tropical rainforest stratification on temperature.* A: Daily march of temperature (10–11 May 1936) in the treetops (24m) and in the undergrowth (0.7m) during the wet season in primary rainforest at Shasha Reserve, Nigeria. B: Average weekly maximum and minimum temperatures in three layers of primary (Dipterocarp) forest, Mount Maquiling, Philippine Islands.

Sources: *After Richards (1952); A: After Evans; B: After Brown.

maximum was increased by 0.8°C simply by raising the thermometer from 1.5 to 2.4m above the forest floor. In stratified tropical forests the thermal picture is more complex. The dense canopy heats up considerably during the day and quickly loses its heat at night, showing a much greater diurnal temperature range than the undergrowth (Figure 12.18A). Whereas daily maximum temperatures of the second storey are intermediate between those of the treetops and the undergrowth, the nocturnal minima are higher than either treetops or undergrowth because the second storey is insulated by trapped air both above and below (Figure 12.18B). During dry conditions in the Amazonian rainforest, there is a similar decoupling of the air in the lower storey from the upper two-thirds of the canopy, as reflected by the reduced amplitude of diurnal temperature range. At night the pattern is reversed: temperatures respond to radiative cooling in the lowest two-thirds of the vegetation canopy. Temperature variations within a layer up to 25m in height are now decoupled from those in the treetops and above.

D URBAN SURFACES

From a total of 6.6 billion in 2007, world population is projected to increase to 8.2 billion in 2025, with the proportion of urban dwellers rising from 45 to 60 percent during the same period. Thus, in this century the majority of the human race will live and work in association with urban climatic influences (see Box 12.1).The construction of every house, road or factory destroys existing microclimates and creates new ones of great complexity that depend on the design, density and function of the building. Despite the internal variation of urban climatic influences, it is possible to treat the effects of urban structures in terms of:

1 modification of atmospheric composition;
2 modification of the heat budget;
3 modification of surface characteristics.

SIGNIFICANT 20TH-CENTURY ADVANCE

12.1 Urban climates

The first recognition of the role of cities in modifying local climate was made by Luke Howard in a book *The Climate of London* published in 1818. Howard made observations in the city region between 1806 and 1830 and drew attention to the heat island effect. In his classic book *Climate Near the Ground* Rudolf Geiger reported many such findings. However, dedicated urban climate studies began in the 1950s. To supplement data from the few existing city weather stations, T. J. Chandler examined urban–rural temperature differences around London, England at different times of the day and year by making traverses in an instrumented vehicle. By repeating the journey in the opposite direction, and averaging the results, the effect of time changes was essentially eliminated. Chandler (1965) wrote a classic book on the climate of London. Similar methods were adopted elsewhere and the vertical structure of the urban atmosphere was also investigated by mounting instruments on tall buildings and towers. Helmut Landsberg in the United States focused on European and North American cities with long historical records while Tim Oke in Canada conducted observational and modeling studies of urban energy budgets and radiative and turbulent transfers in urban 'canyons'.

The number of modern cities with populations in excess of 10 million inhabitants was at least 25 in 2007, with many of these in the tropics and subtropics, but our current knowledge of urban effects in these climatic zones is more limited.

1 Modification of atmospheric composition

Urban pollution modifies the thermal properties of the atmosphere, cuts down the passage of sunlight, and provides abundant condensation nuclei. The modern urban atmosphere comprises a complex mixture of gases including ozone, sulphur dioxide, nitrogen oxides and particulates such as mineral dust, carbon and complex hydrocarbons. First, we examine its sources under two main headings:

1. *Aerosols.* Suspended particulate matter (measured in mg m^{-3} or μg m^{-3}) consists chiefly of carbon, lead and aluminium compounds, and silica. Studies of the effects of chronic exposure to air pollution have identified fine particulate matter as the major factor in the life-shortening effects of polluted air.
2. *Gases.* The production of gases (expressed in parts per million by volume – ppmv) may be viewed in terms of industrial and domestic coal burning releasing such gases as sulfur dioxide (SO_2), or from the standpoint of gasoline and oil combustion producing carbon monoxide (CO), hydrocarbons (Hc), nitrogen oxides (NO_x), ozone (O_3) and the like. A three-year survey of 39 urban areas in the United States identified 48 hydrocarbon compounds: 25 paraffins (60 percent of the total with a median concentration of 266ppb carbon), 15 aromatics (26 percent of the total, 116ppb C) and seven biogenic olefins (11 percent, 47ppb C). Biogenic hydrocarbons (olefins) emitted by vegetation are highly reactive. They destroy ozone and form aerosols in rural conditions, but cause ozone to form under urban conditions. Pine forests emit monoterpenes, $C_{10}H_{16}$ and deciduous woodlands isoprene, C_3H_8; rural concentrations of these hydrocarbons are in the range 0.1–1.5ppb and 0.6–2.3ppb, respectively.

In dealing with atmospheric pollution it must be remembered, first, that the diffusion or concentration of pollutants is a function both of atmospheric stability (especially the presence of inversions) and of the horizontal air motion. It is also generally greater on weekdays than at the weekends or on holidays. Second, aerosols are removed from the atmosphere by settling out and by washing out. Third, certain gases are susceptible to complex chains of photochemical changes, which may destroy some gases but produce others.

Aerosols

As discussed in Chapter 2A.2 and A.4, the global energy budget is affected significantly by the natural production of aerosols that are deflated from deserts, erupted from volcanoes, produced by fires and so on (see Chapter 13D.3). Over the past century the average dust concentration has increased, particularly in Eurasia, due only in part to volcanic eruptions. The proportion of atmospheric dust directly or indirectly attributable to human activity has been estimated at 30 percent (see Chapter 2A.4). As an example of the latter, the North African tank battles of World War II disturbed the desert surface to such an extent that the material subsequently deflated was visible in clouds over the Caribbean. Soot aerosols generated by the Indonesian forest fires of September 1997 and March 2000 were transported across Southeast Asia.

The background concentration of fine particles (PM_{10}, radius <10μm) currently averages 20–30μg m^{-3} in the British countryside but daily average values in industrial cities of Eastern Europe and in many developing nations regularly exceed 50–100μg m^{-3} near ground level. The greatest concentrations of smoke generally occur with low wind speed, low vertical turbulence, temperature inversions, high relative humidity and air moving from the pollution sources of factory districts or areas of high-density housing. The character of domestic heating and power demands causes city smoke pollution to take on striking seasonal and diurnal cycles, with the greatest concentrations occurring at about 08:00 hours in early winter (Figure 12.19). The sudden morning increase is also partly due to natural processes. Pollution trapped during the night beneath a stable layer a few hundred meters above the surface may be brought back to ground level (a process termed *fumigation*) when thermal convection sets off vertical mixing.

The most direct effect of particulate pollution is to reduce visibility, incoming radiation and sunshine. In Los Angeles, carbon aerosol accounts for 40 percent of the total fine particle mass and is the major cause of severe visibility decreases, yet it is not routinely monitored. Half of this total is from vehicle exhausts and the remainder from industrial and other stationary fuel burning. Pollution, and the associated fogs (termed *smog*) used to cause some British cities to lose 25–55 percent of incoming solar radiation during the period November to March. In 1945, it was estimated that the city of Leicester lost 30 percent of incoming radiation in winter, as against 6 percent in summer. These losses are naturally greatest when the sun's rays strike the smog layer at a low angle. Compared with the radiation received in the surrounding countryside, Vienna loses 15–21 percent of radiation when the sun's altitude is 30°, but the loss rises to 29–36 percent with an altitude of 10°. The effect of smoke pollution is dramatically illustrated by Figure 12.20, which compares conditions in London before and after the enforcement of the UK Clean Air Act of 1956 and the move to burning cleaner fuels and the decline of heavy industry. Before 1950 there was a striking difference of sunshine between the surrounding rural areas and the city center (see Figure 12.19A), which could mean a loss of mean daily sunshine of 16 minutes in the outer suburbs, 25 minutes in the inner suburbs and 44 minutes in the city center. It must be remembered, however, that smog layers also impeded the re-radiation of surface heat at night and that this blanketing effect contributed to higher night-time city temperatures. Occasionally,

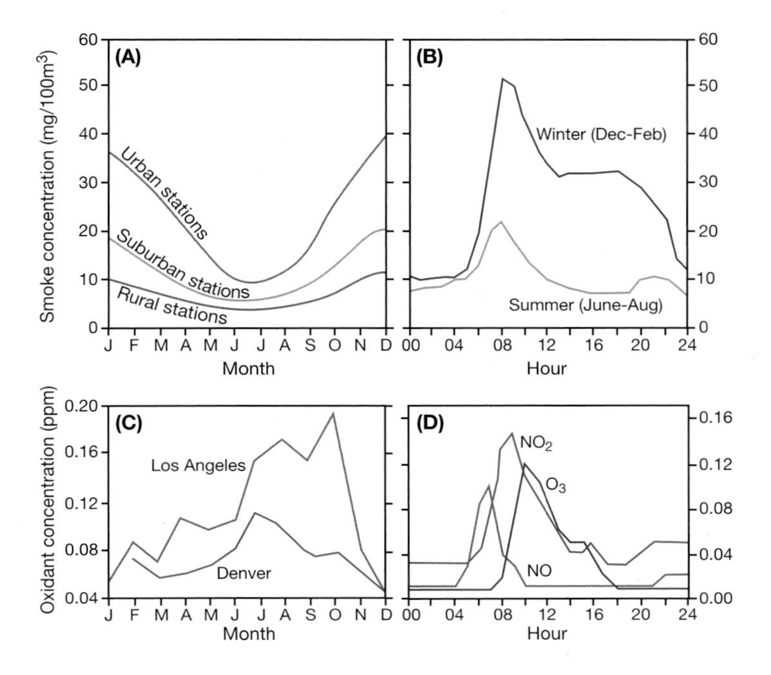

Figure 12.19 Annual and daily pollution cycles. A: Annual cycle of smoke pollution in and around Leicester, England, during the period 1937–1939, before smoke abatement legislation was introduced. B: Diurnal cycle of smoke pollution in Leicester during summer and winter, 1937–1939. C: Annual cycle of mean daily maximum one-hour average oxidant concentrations for Los Angeles (1964–1965) and Denver (1965) (dashed). D: Diurnal cycles of nitric oxide (NO), nitrogen dioxide (NO_2) and ozone (O_3) concentrations in Los Angeles on 19 July 1965.

Sources: A, B: After Meetham (1952) [*et al.* 1980]. C, D: After US DHEW (1970) and Oke (1978).

very stable atmospheric conditions combine with excessive pollution production to give dense smog of a lethal character. During the period 5–9 December 1952, a temperature inversion over London caused a dense fog with visibility of less than 10 m for 48 consecutive hours. There were 12,000 more deaths (mainly from chest complaints) during the period December 1952 to February 1953 compared with the same period the previous year. The close association of the incidence of fog with increasing industrialization and urbanization was well shown by the city of Prague, where the mean annual number of days with fog rose from 79 during the period 1860–1880 to 217 during 1900–1920.

The use of smokeless fuels and other pollution controls cut London's total smoke emission from 1.4×10^8kg (141,000 tons) in 1952 to 0.9×10^8kg (89,000 tons) in 1960. Figure 12.20B shows the increase in average monthly sunshine figures for 1958–1967 compared with those of 1931–1960. Since the early 1960s annual average concentrations of smoke and sulfur dioxide in the UK have fallen from 160ppm and 60ppm, respectively, to below 20ppm and 10ppm in the 1990s.

Visibility in the UK improved at many measuring sites during the second half of the twentieth century. In the 1950s–1960s, days with visibility at midday in the lowest 10th percentile were generally in the 4–5km range, whereas in the 1990s this had improved to 6–9km. Annual average 12 UTC visibility at Manchester airport was 10km in 1950, but near 30km in 1997. The improvements are attributed to improved fuel

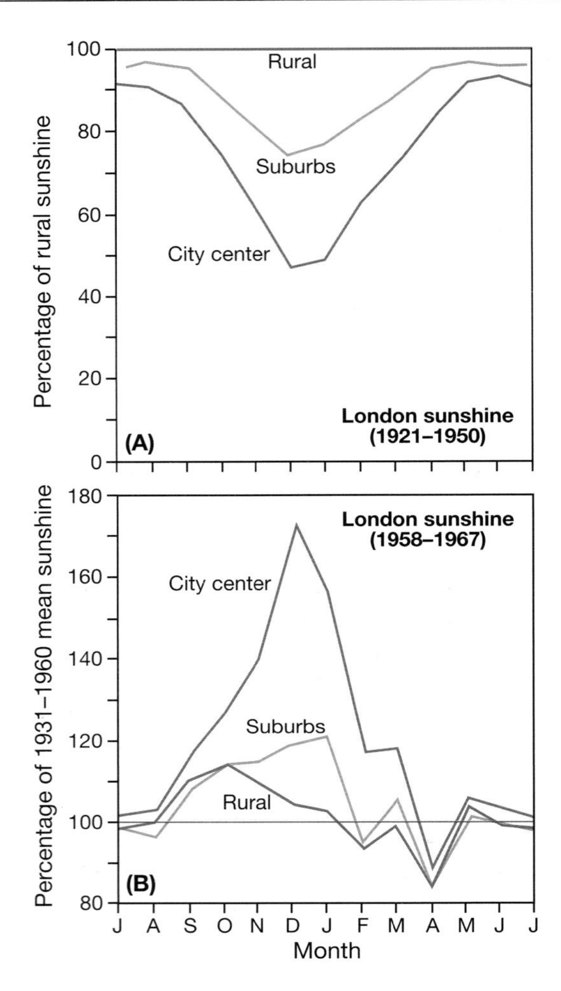

Figure 12.20 Sunshine in and around London. A: Mean monthly bright sunshine recorded in the city and suburbs for the years 1921–1950, expressed as a percentage of that in adjacent rural areas. This shows clearly the effects of winter atmospheric pollution in the city. B: Mean monthly bright sunshine recorded in the city, suburbs and surrounding rural areas during the period 1958–1967, expressed as a percentage of the averages for the period 1931–1960. This shows the effect of the 1956 Clean Air Act in increasing the receipt of winter sunshine, in particular in central London.

Sources: A: After Chandler (1965); B: after Jenkins (1969). By permission of Hutchinson and of the Royal Meteorological Society.

efficiency by vehicles and catalytic converter installation in the 1970s.

Gases

As well as particulate pollution produced by urban and industrial activities involving coal and coke combustion, there is the associated generation of pollutant gases. Before the Clean Air Act in the UK, it was estimated that domestic fires produced 80–90 percent of London's smoke. However, these were responsible for only 30 percent of the sulfur dioxide released into the atmosphere – the remainder being contributed by electricity power stations (41 percent) and factories (29 percent). After the early 1960s, improved technology, the phasing out of coal fires and anti-pollution regulations brought about a striking decline in sulfur dioxide pollution in many European and North American cities (Figure 12.21). Nevertheless, the effect of the regulations was not always clear. The decrease in London's atmospheric pollution was not apparent until eight years after the introduction of the 1956 Clean Air Act,

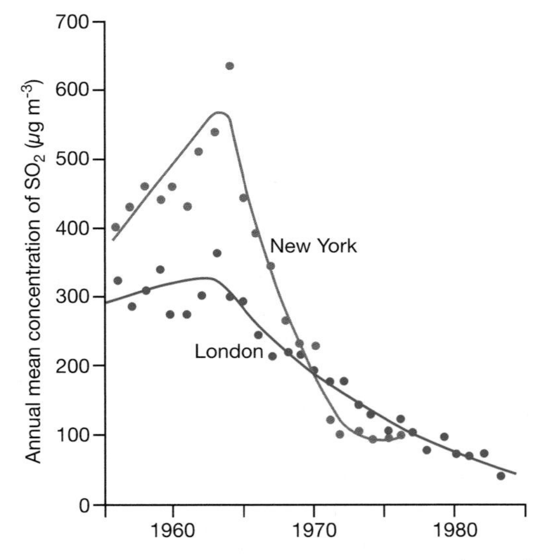

Figure 12.21 Annual mean concentration of sulfur dioxide (mg m^{-3}) measured in New York and London during a 25–30-year period. These show dramatic decreases of urban pollution by SO_2.

Source: From Brimblecombe (1986). By permission of Cambridge University.

whereas in New York City the observed decrease began in the same year (1964) – *prior* to the air pollution control regulations there. Daily mean concentrations in most of Western Europe and North America now seldom exceed about 0.04ppm (125μg m^{-3}), but where coal is still widely used for domestic heating and industry and there is heavy diesel traffic, as in Eastern Europe, Asia and South America, levels may be 5–10 times higher.

Urban complexes in many parts of the world are significantly affected by pollution resulting from the combustion of gasoline and diesel fuel by vehicles and aircraft, as well as from petrochemical industries. Los Angeles, lying in a topographically constricted basin and often subject to temperature inversions, is the prime example of such pollution, although this affects all modern cities. Even with controls, 7 percent of the gasoline from private cars is emitted in an unburned or poorly oxidized form, another 3.5 percent as photochemical smog and 33–40 percent as carbon monoxide. Smog involves at least four main components: carbon soot, particulate organic matter (POM), sulfate (SO_4^{2-}) and peracyl nitrates (PANs). Half of the aerosol mass is typically POM and sulfate. However, there are important regional differences. For example, the sulfur content of fuels used in California and Australia is lower than in the eastern United States and Europe, and NO_2 emissions greatly exceed those of SO_2 in California. The production of the Los Angeles smog, which, unlike traditional city smogs, occurs characteristically during the daytime in summer and autumn, is the result of a very complex chain of chemical reactions termed the disrupted photolytic cycle (Figure 12.22). Ultraviolet radiation dissociates natural NO_2 into NO and O. Monatomic oxygen (O) may then combine with natural oxygen (O_2) to produce ozone (O_3). The ozone in turn reacts with the artificial NO to produce NO_2 (which goes back into the photochemical cycle forming a dangerous positive feedback loop) and oxygen. The hydrocarbons produced by the combustion of petrol combine

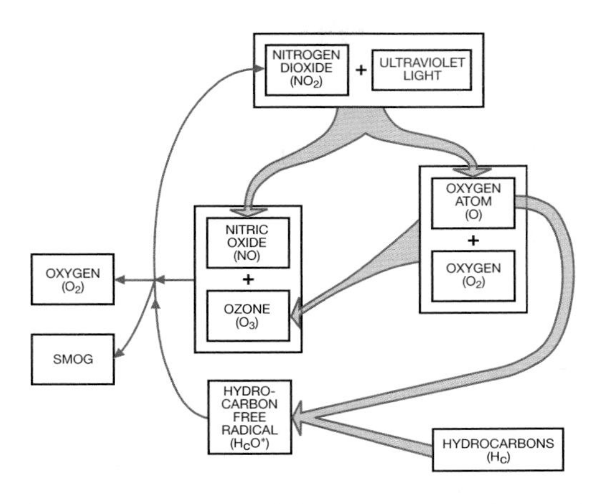

Figure 12.22 The NO_2 photolytic cycle disrupted by hydrocarbons to produce photochemical smog.
Sources: US DHEW (1970) and Oke (1978).

with oxygen atoms to produce the hydrocarbon free radical HcO*, and these react with the products of the O_3–NO reaction to generate oxygen and photochemical smog. This smog exhibits well-developed annual and diurnal cycles in the Los Angeles basin (see Figures 12.19C and D). Annual levels of photochemical smog pollution in Los Angeles (from averages of the daily highest hourly figures) are greatest in late summer and autumn, when clear skies, light winds and temperature inversions combine with high amounts of solar radiation. The diurnal variations in individual components of the disrupted photolytic cycle indicate complex reactions. For example, an early morning concentration of NO_2 occurs due to the buildup of traffic and there is a peak of O_3 when incoming radiation receipts are high. The effect of smog is not only to modify the radiation budget of cities but also to produce a human health hazard.

Evolving state and city regulations in the United States have given rise to considerable differences in the type and intensity of urban pollution. For example, Denver, Colorado, situated in a basin at 1500m altitude, regularly had a winter 'brown cloud' of smog and high summer ozone levels in the 1970–1980s. By the beginning of this century, substantial improvements had been achieved through the mandatory use of gasoline additives in winter, restrictions on wood burning, and scrubbers installed on power plants.

Pollution distribution and impacts

Polluted atmospheres often display well-marked physical features around urban areas that are very dependent upon environmental lapse rates, particularly the presence of temperature inversions, and on wind speed. A pollution dome develops as pollution accumulates under an inversion that forms the urban boundary layer (Figure 12.23A). A wind speed as low as 2m s^{-1} is sufficient to displace the Cincinnati pollution dome downwind, and a wind speed of 3.5m s^{-1} will disperse it into a plume. Figure 12.23B shows a section of an urban plume with the volume

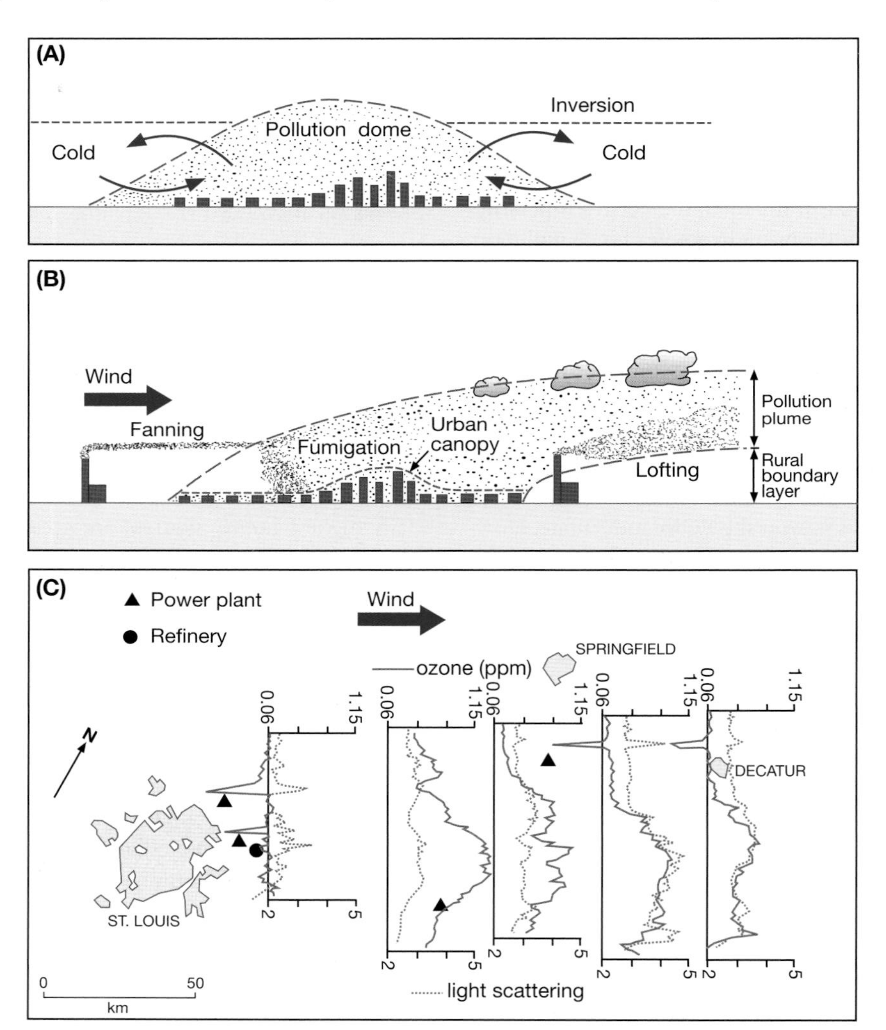

Figure 12.23 Configurations of urban pollution. A: Urban pollution dome. B: Urban pollution plume in a stable situation (i.e. early morning following a clear night). Fanning is indicative of vertical atmospheric stability. C: Pollution plume northeast of St Louis, Missouri, on 18 July 1975.

Sources: B: After Oke (1978); C: after White *et al.* (1976) and Oke (1978). By permission of Routledge and Methuen.

above the urban canopy of the building tops filled by buoyant mixing circulations. When an inversion lid prevents upward dispersion, but lapse conditions due to morning heating of the surface air allow convective plumes and associated downdrafts to bring pollution back to the surface, this process is termed *fumigation*. Downwind, *lofting* occurs above the temperature inversion at the top of the rural boundary layer, dispersing the pollution upward. Figure 12.23C illustrates some features of a pollution plume up to 160km downwind of St Louis on 18 July 1975. In view of the complexity of photochemical reactions, it is of note that ozone increases downwind due to photochemical reactions within the plume, but decreases over power plants as a result of other reactions with the emissions. This plume was observed to stretch for a total distance of 240km, but under conditions of an intense pollution source, steady large-scale surface airflow and vertical atmospheric stability, pollution plumes may extend downwind for hundreds of kilometers. Plumes originating in the Chicago–Gary conurbation have been observed from high-flying aircraft to extend almost to Washington, DC, 950km away.

The impacts of air pollution include: direct meteorological effects (on radiative transfer, sunshine, visibility, fog and cloud development), greenhouse gas production (by release of CO_2, CH_4, NO_x CFCs and HFCs), photochemical effects (tropospheric ozone formation), acidification (processes involving SO_2 , NO_x , and NH_3), and societal nuisance (dust, odor, smog) affecting health and the quality of life, especially in urban areas.

2 Modification of the heat budget

The energy balance of the built surface is similar to soil surfaces described earlier, except for the heat production resulting from energy consumption by combustion, which in some cities may even exceed *Rn* during the winter. Values in Toulouse, France are reported to be around 70W m^{-2} during winter and 15W m^{-2} during summer. Although *Rn* may not be greatly different from that in nearby rural areas (except during times of significant pollution) heat storage by surfaces is greater (20–30 percent of *Rn* by day), leading to greater nocturnal values of *H*; *LE* is much less in city centers. After long, dry periods, evapotranspiration may be zero in city centers, except for certain industrial operations, and in the case of irrigated parks and gardens, where *LE* may exceed *Rn*. This lack of *LE* means that by day 70–80 percent of *Rn* may be transferred to the atmosphere as sensible heat (*H*). Beneath the urban canopy, the effects of elevation and aspect on the energy balance, which may vary strikingly even within one street, determine the microclimates of the streets and 'urban canyons'.

The complex nature of the urban modification of the heat budget is demonstrated by observations made in and around the city of Vancouver. Figure 12.24 compares the summer diurnal energy

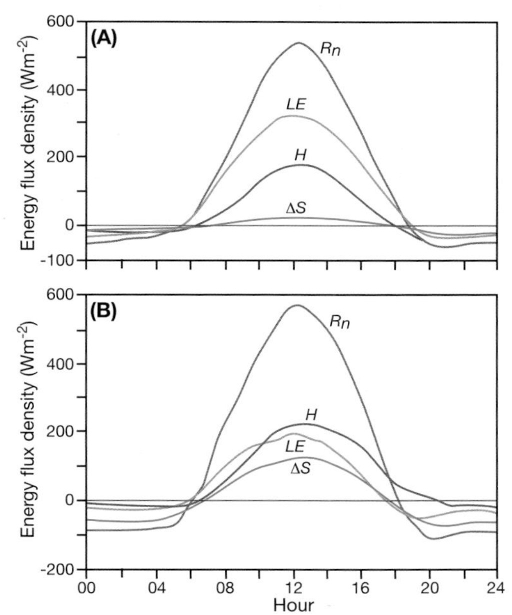

Figure 12.24 Average diurnal energy balances for (A) rural and (B) suburban locations in Greater Vancouver for 30 summer days.

Source: After Clough and Oke, from Oke (1988). By permission of T. R. Oke.

balances for rural and suburban locations. Rural areas show considerable consumption of net radiation (*Rn*) by evapotranspiration (*LE*) during the day, giving lower temperatures than in the suburbs. While the suburban gain of net radiation is greater by day, the loss is greater during the evening and night due to release of turbulent sensible heat from the suburban fabric (i.e., ΔS negative). The diurnal energy balance for the dry top of an urban canyon is symmetrical about midday (Figure 12.25C) and two-thirds of the net radiation is transferred into atmospheric sensible heat and one-third into heat storage in the building material (ΔS). Figure 12.25A–C explains this energy balance symmetry in terms of the behavior of its components (i.e., canyon floor, east-facing wall and west-facing wall); these make up a white, windowless urban canyon in early September aligned north–south and with a canyon height equal to its width. The east-facing wall receives the first radiation in the early morning, reaching a maximum at 10:00 hours, but being totally in shadow after 12:00 hours. Total *Rn* is low because the east-facing wall is often in shadow. The street level (i.e., canyon floor) is sunlit only in the middle of the day and *Rn* and *H* dispositions are symmetrical. The third component of the urban canyon total energy balance is the west-facing wall, which is a mirror image (centered on noon) of that of the east-facing wall. Consequently, the symmetry of the street level energy balance and the mirror images of the east- and west-facing walls produce the symmetrical diurnal energy balance of *Rn*, *H* and (*S* observed at the canyon top.

The thermal characteristics of urban areas contrast strongly with those of the surrounding countryside; the generally higher urban temperatures are a result of the interaction of the following factors:

1. Changes in the radiation balance due to atmospheric composition.
2. Changes in the radiation balance due to the albedo and thermal capacity of urban surface materials, and to canyon geometry.

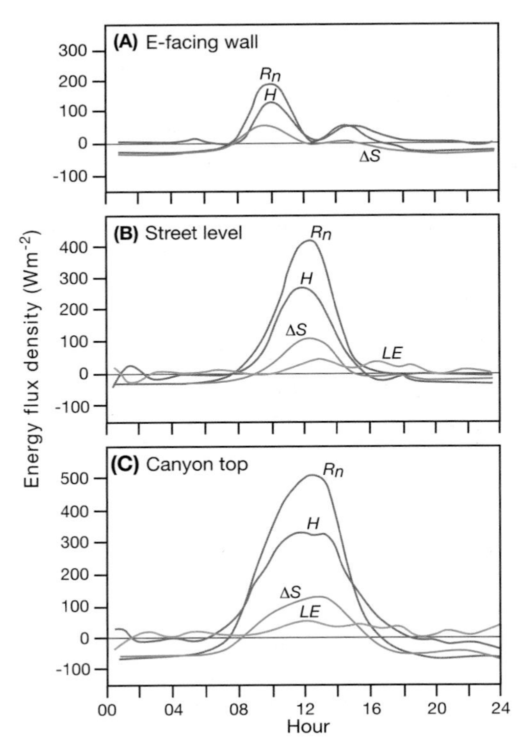

Figure 12.25 Diurnal variation of energy balance components for a N–S-oriented urban canyon in Vancouver, British Columbia, having white concrete walls, no windows, and a width/height ratio of 1:1, during the period 9–11 September 1973. A: The average for the E-facing wall. B: The average for the floor. C: Averages of fluxes through the canyon top.
Source: After Nunez and Oke, from Oke (1978). By permission of Routledge and Methuen.

3. The production of heat by buildings, traffic and industry.
4. The reduction of heat diffusion due to changes in airflow patterns caused by urban surface roughness.
5. The reduction in thermal energy required for evaporation and transpiration due to the surface character, rapid drainage and generally lower wind speeds of urban areas.

Consideration of the last two factors will be left to D.3, this chapter.

The demand for energy to heat or cool houses and businesses is measured by heating and cooling degree days. This quantitative index is defined in the United States with respect to daily mean temperatures above or below 18°C (65°F). Heating (cooling) degree days are summations of the negative (positive) differences between the mean daily temperature and the 18°C base. Heating degree days are accumulated from 1 July to 30 June and cooling degree days from 1 January to 31 December.

Atmospheric composition

Air pollution makes the *transmissivity* of urban atmospheres significantly lower than that of nearby rural areas. During the period 1960–1969, the atmospheric transmissivity over Detroit averaged 9 percent less than that for nearby areas, and reached 25 percent less under calm conditions. The increased absorption of solar radiation by aerosols plays a role in daytime heating of the boundary layer pollution dome (see Figure 12.23A) but is less important within the urban canopy layer, which extends to mean rooftop height (see Figure 12.23B). Table 12.2 compares urban and rural energy budgets for the Cincinnati region during summer 1968 under anticyclonic conditions with <3/10 cloud and a wind speed of <2m s^{-1}. The data show that pollution reduces the incoming shortwave radiation, but a lower albedo and the greater surface area within urban canyons counterbalance this. The increased urban *Ln* at 12:00 and 20:00 LST is largely offset by anthropogenic heating (see below).

Urban surfaces

Primary controls over a city's thermal climate are the character and density of urban surfaces, that is, the total *surface* area of buildings and roads, as well as the building geometry. Table 12.2 shows the relatively high heat absorption of the city surface. A problem of measurement is that the stronger the urban thermal influence, the weaker the heat absorption *at street level*, and, consequently, observations made only in streets may lead to erroneous results. The geometry of urban canyons is particularly important. It involves an increase in effective surface area and the trapping by multiple reflection of shortwave radiation, as well as a reduced 'sky view' (proportional to the areas of the hemisphere open to the sky), which decreases the loss of infrared radiation. From analyses by T. R. Oke, there appears to be an inverse linear relationship on calm, clear summer nights between the sky view factor (0–1.0) and the maximum urban–rural temperature difference. The difference is 10–12°C for a sky view factor of 0.3, but only 3°C for a sky view factor of 0.8–0.9.

Human heat production

Numerous studies show that urban conurbations now produce energy through combustion at rates comparable with incoming solar radiation in winter. Solar radiation in winter averages around 25W m^{-2} in Europe, compared with similar heat production from large cities. Figure 12.26

Table 12.2 Energy budget figures (W m^{-2}) for the Cincinnati region during the summer of 1968

Area	Central business district			Surrounding country		
Time	08:00	13:00	20:00	08:00	13:00	20:00
Shortwave, incoming ($Q + q$)	288*	763	–	306	813	–
Shortwave, reflected [$(Q + q)a$]	42†	120†	–	80	159	–
Net longwave radiation (L_n)	–61	–100	–98	–61	–67	–67
Net radiation (*Rn*)	184	543	–98	165	587	–67
Heat produced by human activity	36	29	26‡	0	0	0

Source: From Bach and Patterson (1966).

Notes: *Pollution peak. †An urban surface reflects less than agricultural land, and a rough skyscraper complex can absorb up to six times more incoming radiation. ‡Replaces more than 25 percent of the longwave radiation loss in the evenings.

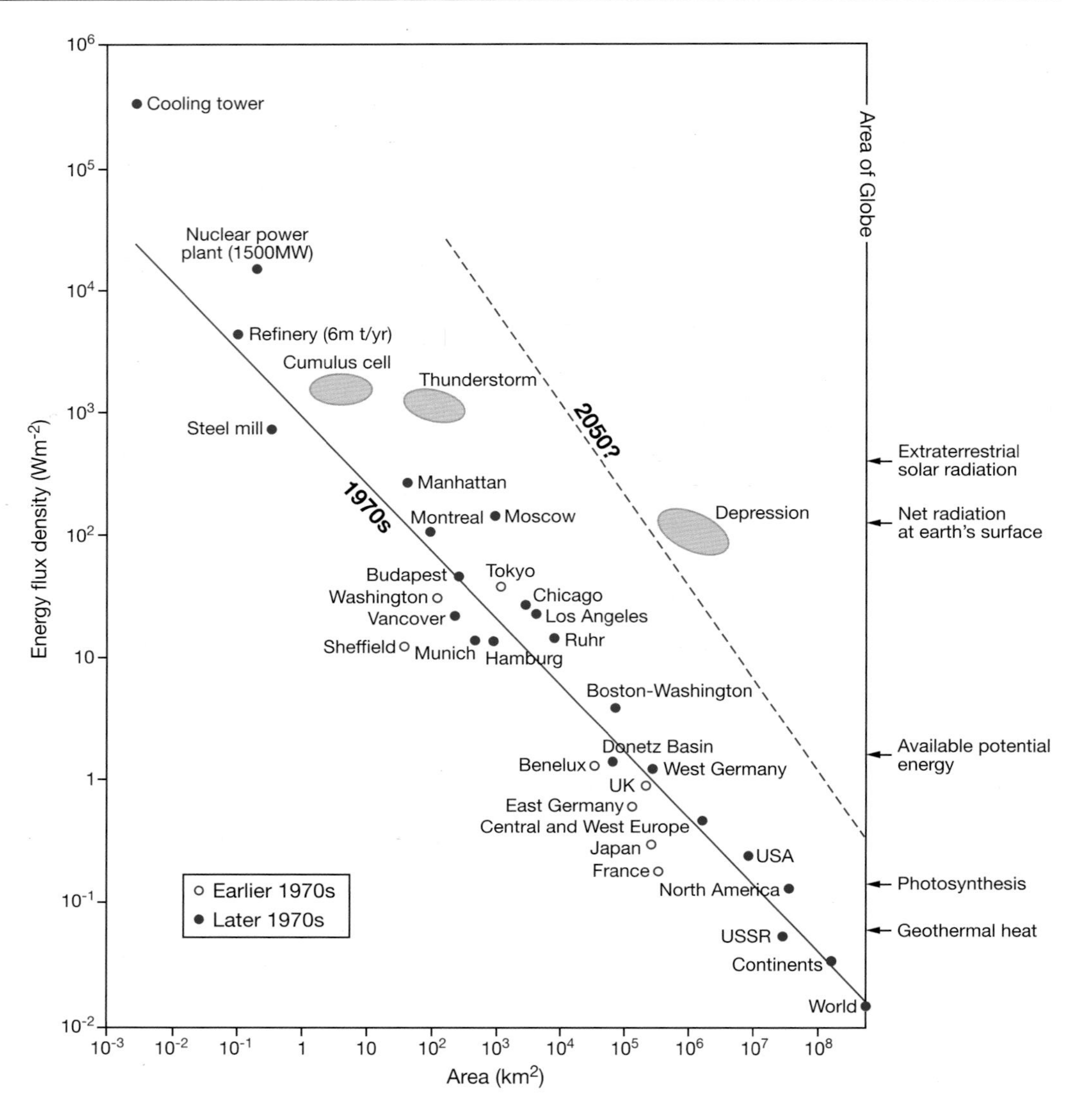

Figure 12.26 A comparison of natural and artificial heat sources in the global climate system on small, meso- and synoptic scales. Generalized regressions are given for artificial heat releases in the 1970s (early 1970s: circles, late 1970s: dots), together with predictions for 2050.

Sources: Modified after Pankrath (1980) and Bach (1979).

illustrates the magnitude and spatial scale of artificial and natural energy fluxes and projected increases. In Cincinnati, a significant proportion of the energy budget is generated by human activity, even in summer (see Table 12.2). This averages 26W m^{-2}or more, two-thirds of which was produced by industrial, commercial and domestic sources and one-third by cars. In the extreme situation of Arctic settlements during polar darkness, the energy balance during calm conditions depends only on net longwave radiation and heat production by anthropogenic activities. In Reykjavik, Iceland (population 100.000) the anthropogenic heat release is 35W m^{-2}

mainly as a result of geothermal pavement heating and hot water pipelines.

Heat islands

The net effect of urban thermal processes is to make city temperatures in mid-latitudes generally higher than in the surrounding rural areas. This is most pronounced after sunset during calm, clear weather, when cooling rates in the rural areas greatly exceed those in the urban areas. The energy balance differences that cause this effect depend on the radiation geometry and thermal properties of the surface. It is believed that the canyon geometry effect dominates in the urban canopy layer, whereas the sensible heat input from urban surfaces determines the boundary layer heating. The urban heat island intensity is mainly a function of the ratio of canyon height to width (H/W), though it increases also with an increasing difference in thermal inertia of the urban and rural surfaces, and downward infrared radiation from pollution layers. By day the urban boundary layer is heated by increased absorption of shortwave radiation due to the pollution, as well as by sensible heat transferred from below and entrained by turbulence from above.

The *heat island* effect may result in minimum urban temperatures being 5–6°C greater than those of the surrounding countryside. These differences may reach 6–8°C in the early hours of calm, clear nights in large cities, when the heat stored by urban surfaces during the day (augmented by combustion heating) is released. Because this is a *relative* phenomenon, the heat island effect also depends on the rate of rural cooling, which is influenced by the magnitude of the regional environmental lapse rate.

For the period 1931–1960, the center of London had a mean annual temperature of 11.0°C, compared with 10.3°C for the suburbs, and 9.6°C for the surrounding countryside. Calculations for London in the 1950s indicated that domestic fuel consumption gave rise to a 0.6°C warming in the city in winter and that accounted for one-third to one-half of the average city heat excess compared with adjacent rural areas. Differences are most evident during still air conditions, especially at night under a regional inversion (Figure 12.27). For the heat island effect to operate effectively there must be wind speeds of less than 5–6m s^{-1}. It is especially apparent on calm nights during summer and early autumn, when it has steep cliff-like margins on the upwind edge of the city and the highest temperatures are associated with the highest density of urban dwellings. In the absence of regional winds, a well-developed heat island may generate its own inward local wind circulation at the surface. Thus the thermal contrasts of a city, like many of its climatic features, depend on its topographic situation and are greatest for sheltered sites with light winds. The fact that urban–rural temperature differences are greatest for London in summer, when direct heat combustion and atmospheric pollution are at a minimum, indicates that heat loss from buildings by radiation is the most important single factor contributing to the heat island effect. Seasonal differences are not necessarily the same, however, in other macroclimatic zones.

The effects on minimum temperatures are especially marked. For central Moscow, winter extremes below –28°C occurred only 11 times during 1950–1989 compared with 23 cases at Nemchinovka west of the city. Cologne, Germany has an average of 34 percent fewer days with minima below 0°C than its surrounding area. In London, Kew has an average of 72 more days with frost-free screen temperatures than rural Wisley. Precipitation characteristics are also affected; incidences of rural snowfall are often associated with either sleet or rain in the city center.

Although it is difficult to isolate changes in temperatures that are due to urban effects from those due to other climatic factors (see Chapter 13), it has been suggested that city growth is often accompanied by an increase in mean annual temperature. At Osaka, Japan, temperatures have risen by 2.6°C in the past 100 years. Under calm conditions, the maximum difference in

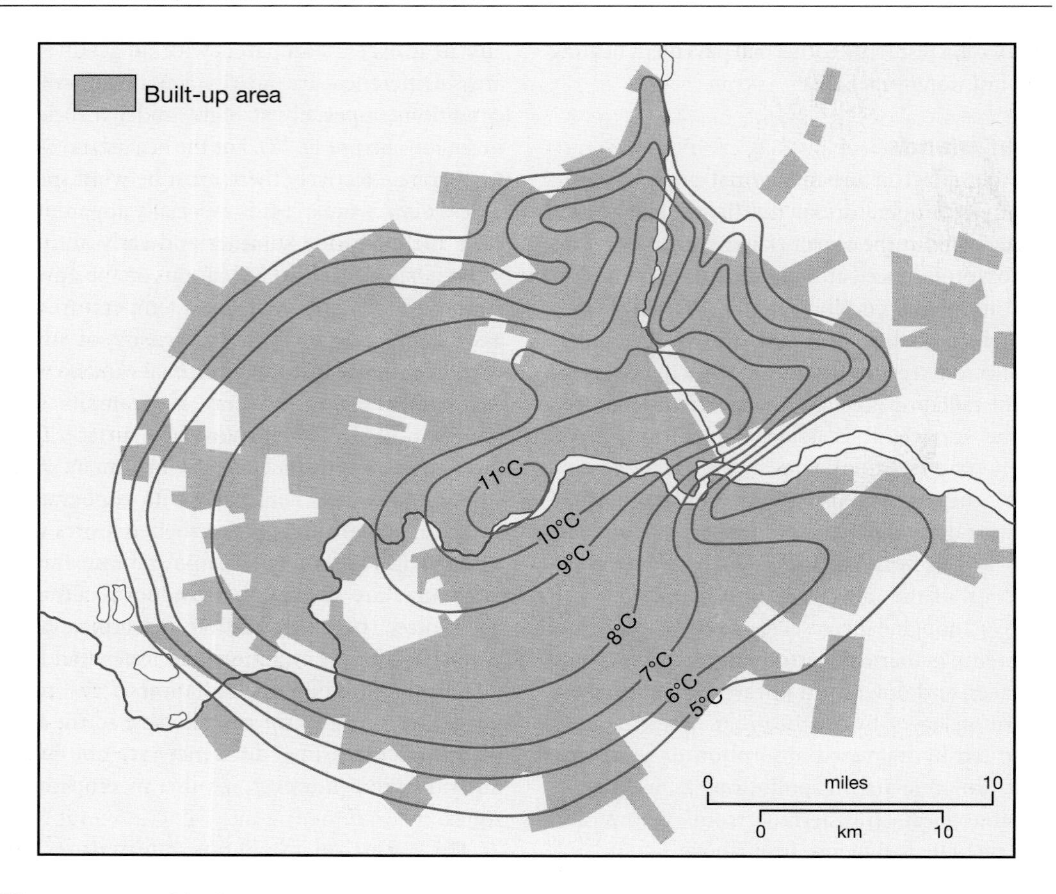

Figure 12.27 Distribution of minimum temperatures (°C) in London on 14 May 1959, showing the relationship between the 'urban heat island' and the built-up area.
Source: After Chandler (1965). By permission of Hutchinson.

urban–rural temperatures is related statistically to population size, being nearly linear with the logarithm of the population. For New York, about one-third of the warming since 1900 is attributed to the heat island effect and the rest to regional climate change. At Central Park, New York the heat island strength in 2007 was ~ 2.5°C. In North America, the maximum urban–rural temperature difference reaches 2.5°C for towns of 1000, 8°C for cities of 100,000 and 12°C for cities of one million people. European cities show a smaller temperature difference for equivalent populations, perhaps as a result of the generally lower building height.

A convincing example of the relationship between urban growth and climate is for Tokyo, which expanded greatly after 1880 and particularly after 1946 (Figure 12.28A). The population increased to 10.4 million in 1953 and to 11.7 million in 1975. During the period 1880–1975, there was a significant increase in mean January minimum temperatures and a decrease in the number of days with minimum temperatures below 0°C (Figures 12.28B and C). Although the graphs suggest a reversal of these trends during War II (1942–1945), when evacuation almost halved Tokyo's population, it is clear that the basis of correlations of urban climate with population is complex. Urban density, industrial activity and the production of anthropogenic heat are all involved. Leicester, England, for example, when it had a population of 270,000, exhibited warming comparable in intensity with that of central London over smaller sectors. This suggests that the thermal influence of city size is not as important as that of urban density. The vertical extent of the heat island is little known, but is thought to exceed 100–300m, especially early in the night. In the case of cities with skyscrapers, the vertical and horizontal patterns of wind and temperature are very complex (see Figure 12.29).

In some high-latitude cities there is a reverse 'cold island' effect of 1–3°C in summer. In the United States this effect has been reported in Boston, MA, Dallas, TX, Detroit, MI and Seattle, WA, when corrections are made to temperature for latitude and elevation differences. At the microscale, low solar elevation angle causes shading of urban streets, in contrast to locations outside the built-up area. A similar cool island is observed in cities in tropical and subtropical deserts where it is attributed to the high thermal inertia of the built-up area, and the sharp diurnal temperature fluctuations. Its intensity depends on the orientation of the street canyons - increasing as street-axis orientation approaches north–south and decreasing to near zero in the east–west direction (see below). One recent study, where inhomogeneities and biases in the temperature data were carefully examined, finds no statistically significant impact of urbanization in the contiguous United States; as many cities showed a cold island as a heat island. A suggested reason for this is that micro- and local-scale impacts

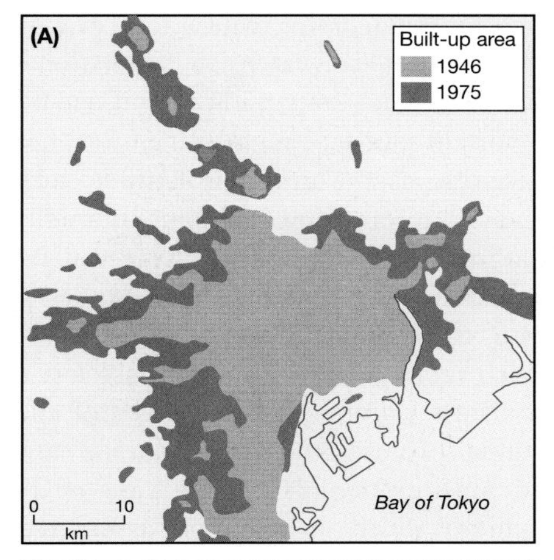

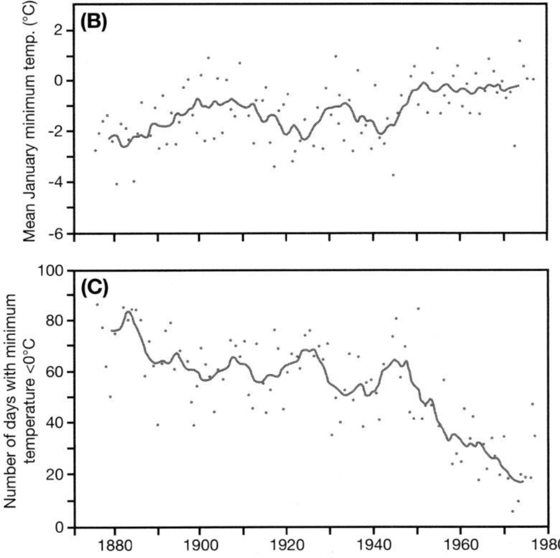

Figure 12.28 A: The built-up area of Tokyo in 1946. B: The mean January minimum temperature. C: Number of days with sub-zero temperatures between 1880 and 1975. During World War II, the population of the city fell from 10.36 million to 3.49 million and then increased to 10.4 million in 1953 and 11.7 million in 1975.

Source: "After", Maejima *et al.* (1980). Courtesy of Professor J. Matsumoto. Courtesy of Professor J. Matsumoto.

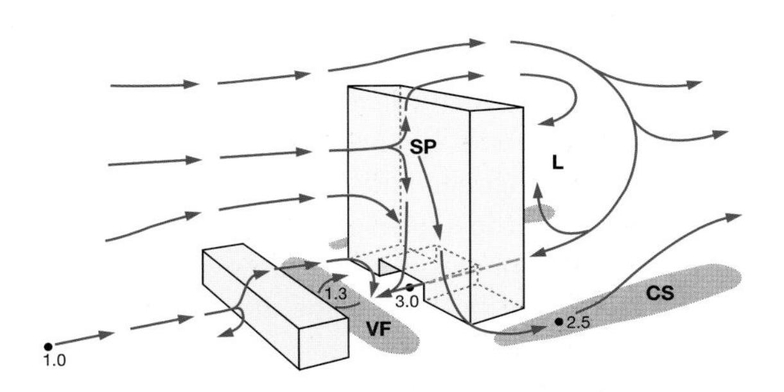

Figure 12.29 Details of urban airflow around two buildings of differing size and shape. Numbers give relative wind speeds; stippled areas are those of high wind velocity and turbulence at street level.

Sources: After Plate (1972) and Oke (1978).

Notes: SP = stagnation point; CS = corner stream; VF = vortex flow; L = lee eddy.

dominate over the mesoscale urban heat island. More work is needed on these important scale effects.

3 Modification of surface characteristics

Airflow

On average, city wind speeds are lower than those recorded in the surrounding open country owing to the sheltering effect of the buildings. Average city-center wind speeds are usually at least 5 percent less than those of the suburbs. However, the urban effect on air motion varies greatly depending on the time of day and the season. During the day, city wind speeds are considerably lower than those of surrounding rural areas, but during the night the greater mechanical turbulence over the city means that the higher wind speeds aloft are transferred to the air at lower levels by turbulent mixing. During the day (13:00 hours), the mean annual wind speed for the period 1961–1962 at Heathrow Airport (open country within the suburbs) was 2.9m s^{-1}, compared with 2.1m s^{-1} in central London. The comparable figures at night (01:00 hours) were 2.2m s^{-1} and 2.5m s^{-1}. Rural–urban wind speed differences are most marked with strong winds, and the effects are therefore more evident during winter when a higher proportion of strong winds is recorded in mid-latitudes.

Urban structures affect the movement of air both by producing turbulence as a result of their surface roughness and by the channeling effects of the urban canyons. Figure 12.29 gives some idea of the complexity of airflow around urban structures, illustrating the great differences in ground-level wind velocity and direction, the development of vortices and lee eddies, and the reverse flows that may occur. Structures play a major role in the diffusion of pollution within the urban canopy; for example, narrow streets often cannot be flushed by vortices. The formation of high-velocity streams and eddies in the usually dry and dusty urban atmosphere, where there is an ample debris supply, leads to general urban airflows of only 5m s^{-1} being annoying, and those of more than 20m s^{-1} being dangerous.

Moisture

The absence of large bodies of standing water in urban areas and the rapid removal of surface runoff through drains reduces local evaporation. The lack of an extensive vegetation cover eliminates much evapotranspiration, and this is an important source of augmenting urban heat. For these reasons, the air of mid-latitude cities has a tendency towards lower absolute humidity than that of their surroundings, especially under conditions of light winds and cloudy skies. During calm, clear weather, the streets trap warm air, which retains its moisture because less dew is deposited on the warm surfaces of the city. Humidity contrasts between urban and rural areas are most noticeable in the case of relative humidity, which can be as much as 30 percent less in the city by night as a result of the higher temperatures.

Urban influences on precipitation (excluding fog) are much more difficult to quantify, partly because there are few rain gauges in cities and partly because turbulent flow makes their 'catch' unreliable. Ground-based weather radar has been used in a study of Atlanta, Georgia. It is fairly certain that urban areas in Europe and North America are responsible for local conditions that, in summer especially, can trigger excesses of precipitation under marginal conditions. Such triggering involves both thermal effects and the increased frictional convergence of built-up areas. European and North American cities tend to record 6–7 percent more days with rain per year than their surrounding regions, giving a 5–10 percent increase in urban precipitation. Over southeast England during 1951–1960, summer thunderstorm rain (which comprised 5–15 percent of the total precipitation) was especially concentrated in west, central and south London, and contrasted strikingly with the distribution of mean annual total rainfall. During this period, London's thunderstorm rain was 200–250mm greater than that in rural southeast England. Urban areas in the Midwest and the southeastern USA significantly increase summer convective activity. Areas in eastern metropolitan Atlanta received 30 percent more rainfall during days of mT air in June to August 2002–2006 than areas west of the city. Both precipitation amount and frequency were enhanced up to 80km to the east of the urban core of Atlanta. The enhanced rainfall was most evident between 19:00 and midnight LST. More frequent thunderstorms and hail occur for 30–40km downwind of industrial areas of St Louis compared with rural areas (Figure 12.30). The anomalies illustrated here are among the best-documented urban effects. Many of the urban effects here are based on case studies. Table 12.3 gives a summary of average climatic differences between cities and their surroundings.

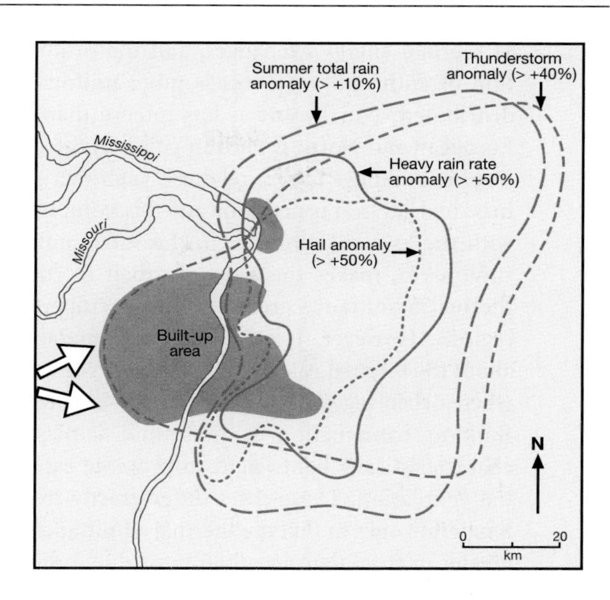

Figure 12.30 Anomalies of summer rainfall, rate of heavy rains, hail frequency and thunderstorm frequency downwind of the St Louis, MO, metropolitan area. Large arrows indicate the prevailing direction of motion of summer rain systems.
Source: After Changnon (1979). Courtesy of the American Meteorological Society.

Table 12.3 Average mid-latitude urban climatic conditions compared with those of surrounding rural areas

Atmospheric composition	carbon dioxide	×2
	sulfur dioxide	×50-200
	nitrogen oxides	×10
	carbon monoxide	×200(+)
	total hydrocarbons	×20
	particulate matter	×3 to 7
Radiation	global solar	–15 to 20%
	ultraviolet (winter)	–30%
	sunshine duration	–5 to 15%
Temperature	winter minimum (average)	+1 to 2°C
	heating degree days	–10%
Wind speed	annual mean	–20 to 30%
	number of calms	+5 to 20%
Fog	winter	+100%
	summer	+30%
Cloud		+5 to 10%
Precipitation	total	+5 to 10%
	days with <5mm	+10%

Source: Partly after the World Meteorological Organization (1970).

4 Tropical urban climates

A striking feature of recent and projected world population growth is the relative increase in the tropics and subtropics. Today there are 45 world megacities with more than five million people inhabitants. By AD 2025 it is predicted that, of the 13 cities that will have populations in the 20–30 million range, eleven will be in less-developed countries (Mexico City, São Paulo, Lagos, Cairo, Karachi, Delhi, Bombay, Calcutta, Dhaka, Shanghai and Jakarta).

Despite the difficulties in extrapolating knowledge of urban climates from one region to another, the ubiquitous high-technology architecture of most modern city centers and multi-storey residential areas will tend to impose similar influences on their differing background climates. Nevertheless, most tropical urban built land differs from that in higher latitudes; it is commonly composed of high-density, single-storey buildings with few open spaces and poor drainage. In such a setting, the composition of roofs is more important than that of walls in terms of thermal energy exchanges, and the production of anthropogenic heat is more uniformly distributed spatially and is less intense than in European and North American cities. In the dry tropics, buildings have a relatively high thermal mass to delay heat penetration and this, combined with the low soil moisture in the surrounding rural areas, makes the ratio of urban to rural thermal admittance greater than in temperate regions. However, it is difficult to generalize about the thermal role of cities in the dry tropics where urban vegetation can lead to 'oasis' effects. Building construction in the humid tropics is characteristically lightweight to promote essential ventilation. These cities differ greatly from temperate ones in that the thermal admittance is greater in rural than in urban areas due to high rural soil moisture levels and high urban albedos.

Tropical heat island characteristics are similar to those of temperate cities but are usually weaker, typically 4°C for the nocturnal maximum – compared with 6°C in mid-latitudes, and they are best developed in the dry season. There are also different timings for temperature maxima, and with complications introduced by the effects of afternoon and evening convective rainstorms and by diurnal breezes.

The thermal characteristics of tropical cities differ from those in mid-latitudes because of dissimilar urban morphology (e.g., building density, materials, geometry, green areas) and because they have fewer sources of anthropogenic heat. Urban areas in the tropics tend to have slower rates of cooling and warming than do the surrounding rural areas, and this causes the major nocturnal heat island effect to develop later than in mid-latitudes – i.e., around sunrise (Figure 12.31A). Urban climates in the subtropics are well illustrated by four cities in Mexico (Table 12.4). The heat island effect is, as expected, greater for larger cities and best exemplified at night during the dry season (November to April), when anticyclonic conditions, clear skies and inversions are most common (Figure 12.31B). It is of note that in some tropical coastal cities (e.g., Veracruz; Figure 12.31A), afternoon urban heating may produce instability that reinforces the sea-breeze effect to the point where there is a 'cool island' urban effect (Figure 12.31A). Elevation may play a significant thermal role (Table 12.4), as in Mexico City, where the urban heat island may be accentuated by rapid nocturnal cooling of the surrounding countryside. Quito, Ecuador (2851m) shows a maximum heat island effect by

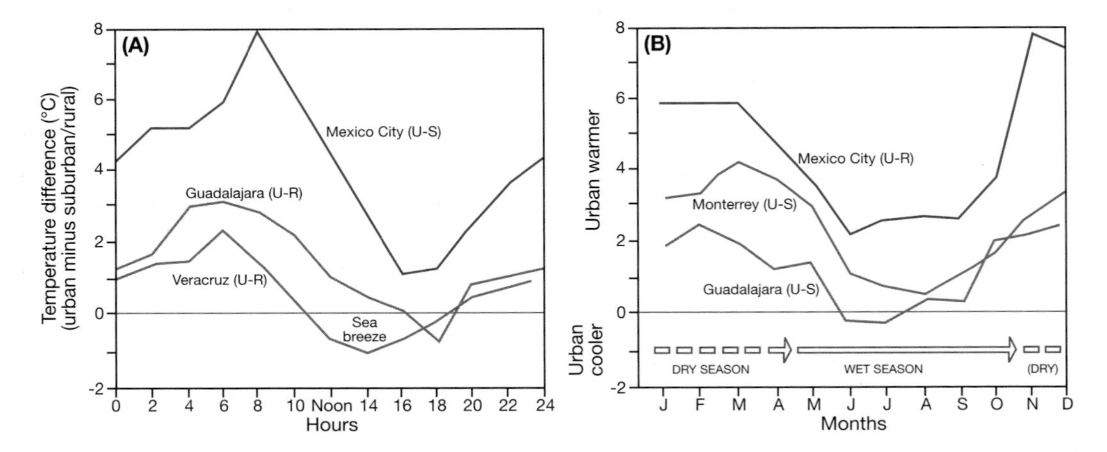

Figure 12.31 Diurnal (A) and seasonal (B) heat island intensity variations (i.e. urban minus rural or suburban temperature differences) for four Mexican cities.

Source: Jauregui (1987). Copyright © *Erdkunde*. Published by permission.

Table 12.4 Population (1990) and elevation for four Mexican cities

	Population (millions)	**Elevation (meters)**
Mexico City (19°25′N)	15.05	2,380
Guadalajara (20°40′N)	1.65	1,525
Monterey (25°49′N)	1.07	538
Veracruz (19°11′N)	0.33	Sea level

Source: Jauregui (1987).

day (as much as 4°C) and weaker night-time effects, probably due to the nocturnal drainage of cold air from the nearby volcano Pichincha.

Ibadan, Nigeria (population over one million; elevation 210 m), at 7°N, records higher rural than urban temperatures in the morning and higher urban temperatatures in the afternoon, especially in the dry season (November to mid-March). In December, the harmattan dust haze tends to reduce city maximum temperatures. During this season, mean monthly minimum temperatures are significantly greater in the urban heat island than in rural areas (March +12°C, but December only +2°C due to the atmospheric dust effect). In general, urban–rural minimum temperature differences vary between –2° and +15°C. Two other tropical cities exhibiting urban heat islands are Nairobi, Kenya (+3.5°C for minimum temperatures and +1.6°C for maximum temperatures) and Delhi, India (+3 to 5°C for minimum temperatures and +2 to 4°C for maximum temperatures).

Despite insufficient data, there seems to be some urban precipitation enhancement in the tropics, which is maintained for more of the year than that associated with summer convection in mid-latitudes.

SUMMARY

Small-scale climates are determined largely by the relative importance of the surface energy budget components, which vary in amount and sign depending on time of day and season. Bare land surfaces may have wide temperature variations controlled by *H* and *G*, whereas those of surface water bodies are strongly conditioned by *LE* and advective flows. Snow and ice surfaces have small energy transfers in winter with net outgoing radiation offset by transfers of *H* and *G* towards the surface. After snow-melt, the net radiation is large and positive, balanced by turbulent energy losses. Vegetated surfaces have more complex exchanges usually dominated by *LE*; this may account for >50 percent of the incoming radiation, especially where there is an ample water supply (including irrigation). Forests have a lower albedo (<0.10 for conifers) than most other vegetated surfaces (0.20–0.25). Their vertical structure produces a number of distinct microclimatic layers, particularly in tropical rainforests. Wind speeds are characteristically low in forests and trees form important shelter-belts. Unlike short vegetation, various types of tree exhibit a variety of rates of evapotranspiration and thereby differentially affect local temperatures and forest humidity. Forests may have a marginal topographic effect on precipitation under convective conditions in temperate regions, but fog drip is more significant in foggy/cloudy areas. The disposition of forest moisture is very much affected by canopy interception and evaporation, but forested catchments appear to have greater evapotranspiration losses than ones with a grass cover. Forest microclimates have lower temperatures and smaller diurnal ranges than their surroundings.

Urban climates are dominated by the geometry and composition of built-up surfaces and by the effects of human urban activities. The composition of the urban atmosphere is modified by the addition of aerosols, producing smoke pollution and fogs, by industrial gases such as sulfur dioxide, and by a chain of chemical reactions initiated by automobile exhaust fumes, which causes smog and inhibits both incoming and outgoing radiation. Pollution domes and plumes are produced around cities under appropriate conditions of vertical temperature structure and wind velocity. *H* and *G* dominate the urban heat budget, except in city parks, and as much as 70–80 percent of incoming radiation may become sensible heat, which is very variably distributed between the complex urban built forms. Urban influences combine to give generally higher temperatures than in the surrounding countryside, not least due to the growing importance of heat production by human activities. These factors lead to the urban heat island, which may be 6–8°C warmer than surrounding areas in the early hours of calm, clear nights, when heat stored by urban surfaces is being released. The urban–rural temperature difference under calm conditions is statistically related to the city population size; the urban canyon geometry and sky view are major controlling factors. The heat island may be a few hundred meters deep, depending on the building configuration. In some cases, summer daytime cold islands are observed. Urban wind speeds are generally lower than in rural areas by day, but the wind flow is complex, depending on the geometry of city structures. Cities tend to be less humid than rural areas, but their topography, roughness and thermal qualities can intensify summer convective activity over and downwind of the urban area, giving more thunderstorms and heavier storm rainfall. Tropical cities have heat islands, but the diurnal phase tends to be delayed relative to mid-latitude ones. The temperature amplitude is largest during dry season conditions.

DISCUSSION TOPICS

- In what ways do vegetated surfaces modify the surface climate compared with unvegetated ones and what processes are involved?
- What are the major effects of urban environments on atmospheric composition? (Data from air sampling sites in North America and Europe are available on the Web.)
- Look for evidence from local weather station reports and/or vegetation types for topoclimatic differences in locations where you live/travel and consider whether these arise from differences in solar radiation, day/night temperatures, moisture balance, wind speed or combinations of these factors.
- Look for evidence of urban–rural climatic differences in cities near you using weather reports of day/night temperatures, visibility, snowfall events and so on.

REFERENCES AND FURTHER READING

Books

Bailey, W. G., Oke, T. R. and Rouse, W. R. (eds) (1997) *The Surface Climates of Canada*, McGill-Queen's University Press, Montreal and Kingston, 369pp. [Sections on surface climate concepts and processes, the climatic regimes of six different natural surfaces, as well as agricultural and urban surfaces]

Brimblecombe, P. (1986) *Air: Composition and Chemistry*, Cambridge University Press, Cambridge, 224pp. [Suitable introduction to atmospheric composition, gas phase chemistry, aerosols, air pollution sources and effects, and stratospheric ozone for environmental science students]

Chandler, T. J. (1965) *The Climate of London*, Hutchinson, London, 292pp. [Classic account of the urban effects of the city of London in the 1950–1960s]

Cotton, W. R. and Pielke, R. A. (1995) *Human Impacts on Weather and Climate*, Cambridge University Press, Cambridge, 288pp. [Treats intentional and accidental weather and climate modification on regional and global scales]

Garratt, J. R. (1992) *The Atmospheric Boundary Layer*, Cambridge University Press, Cambridge 316pp. [Advanced level text on the atmospheric boundary layer and its modeling]

Geiger, R., Aron, R. H., and Todhunter, P. (2003) *The Climate Near the Ground*, 6th edn, Rowman & Littlefield, Lanham, Md, 584pp. [Classic descriptive text on local, topo- and microclimates; extensive references to European research]

Kittredge, J. (1948) *Forest Influences*, McGraw-Hill, New York, 394pp. [Classic account of the effects of forests on climate and other aspects of the environment]

Landsberg, H. E. (ed.) (1981) *General Climatology 3*, World Survey of Climatology 3, Elsevier, Amsterdam 408pp. [Focuses on applied climatology with chapters on human bioclimatology, agricultural climatology and city climates]

Lowry, W. P. (1969) *Weather and Life: An Introduction to Biometeorology*, Academic Press, New York, 305pp. [A readable introduction to energy and moisture in the environment, energy budgets of systems, the biological environment and the urban environment]

Meetham, A. R. *et al.* (1980) *Atmospheric Pollution*, 4th edn, Pergamon Press, Oxford and London.

Monteith, J. L. (1973) *Principles of Environmental Physics*, Arnold, London 241pp. [Discusses radiative fluxes and radiation balance at the surface and in canopies, exchanges of heat, mass and momentum, and the micrometeorology of crops]

Monteith, J. L. (ed.) (1975) *Vegetation and the Atmosphere*, Vol. 1: *Principles*, Academic Press, London, 278pp. [Chapters by specialists on micrometerology and plants – radiation, exchanges of momentum, heat, moisture and particles, micrometerological models and instruments]

Munn, R. E. (1966) *Descriptive Micrometeorology*, Academic Press, New York, 245pp. [Readable introduction to processes in micrometerology including urban pollution]

Oke, T. R. (1987) *Boundary Layer Climates*, 2nd edn, Methuen, London, 435pp. [Prime text on surface climate processes in natural and human-modified environments by a renowned urban climatologist]

Richards, P. W. (1952) *The Tropical Rain Forest, An Ecological Study*, Cambridge University Press, Cambridge, 450pp. [A classic text on tropical forest biology and ecology that includes chapters on climatic and microclimatic conditions]

Sellers, W. D. (1965) *Physical Climatology*, University of Chicago Press, Chicago, 272pp. [Classic and still valuable treatment of physical processes in meteorology and climatology]

Sopper, W. E. and Lull, H. W. (eds) (1967) *International Symposium on Forest Hydrology*, Pergamon, Oxford and London, 813pp. [Includes contributions on moisture budget components in forests]

Sukachev, V. and Dylis, N. (1968) *Fundamentals of Forest Biogeocoenology*, Oliver and Boyd, Edinburgh, 672pp.

Articles

Adebayo, Y. R. (1991) 'Heat island' in a humid tropical city and its relationship with potential evaporation. *Theoret. and App. Climatology* 43, 137–47.

Anderson, G. E. (1971) Mesoscale influences on wind fields. *J. App. Met.* 10, 377–86.

Atkinson, B. W. (1968) A preliminary examination of the possible effect of London's urban area on the distribution of thunder rainfall 1951–60. *Trans. Inst. Brit. Geog.* 44, 97–118.

Atkinson, B. W. (1977) *Urban Effects on Precipitation: An Investigation of London's Influence on the Severe Storm of August 1975*. Department of Geography, Queen Mary College, London, Occasional Paper 8 (31pp.).

Atkinson, B. W. (1987) Precipitation, in Gregory, K. J. and Walling, D. E. (eds) *Human Activity and Environmental Processes*, John Wiley & Sons, Chichester, 31–50.

Bach, W. (1971) Atmospheric turbidity and air pollution in Greater Cincinnati. *Geog. Rev.* 61, 573–94.

Bach, W. (1979) Short-term climatic alterations caused by human activities. *Prog. Phys. Geog.* 3(1), 55–83.

Bach, W. and Patterson, W. (1966) Heat budget studies in Greater Cincinnati. *Proc. Assn. Amer. Geog.* 1, 7–16.

Betts, A. K., Ball, J. H. and McCaughey, H. (2001). Near-surface climate in the boreal forest. *J. Geophys. Res.* 106(D24), 33,529–541.

Brimblecombe, P. (2006) The Clean Air Act after 50 years. *Weather* 61(11), 311–14.

Brimblecombe, P. and Bentham, G. (1997) The air that we breathe: smogs, smoke and health. in Hulme M. and Barrow E. (eds) *The Climates of the British Isles. Present, Past and Future*, Routledge, London, 243–61.

Caborn, J. M. (1955) The influence of shelter-belts on microclimate. *Quart. J. Roy. Met. Soc.* 81, 112–15.94:1–11.

Chandler, T. J. (1967) Absolute and relative humidities in towns. *Bull. Amer. Met. Soc.* 48, 394–9.

Changnon, S. A. (1969) Recent studies of urban effects on precipitation in the United States. *Bull. Amer. Met. Soc.* 50, 411–21.

Changnon, S. A. (1979) What to do about urban-generated weather and climate changes. *J. Amer. Plan. Assn.* 45(1), 36–48.

Coutts, J. R. H. (1955) Soil temperatures in an afforested area in Aberdeenshire. *Quart. J. Roy. Met. Soc.* 81, 72–9.

Dickinson, R. E. and Henderson-Sellers, A. (1988) Modelling tropical deforestation: a study of GCM land-surface parameterizations. *Quart. J. Roy. Met. Soc.* 114, 439–62.

Duckworth, F. S. and Sandberg, J. S. (1954) The effect of cities upon horizontal and vertical temperature gradients. *Bull. Amer. Met. Soc.* 35, 198–207.

Food and Agriculture Organization of the United Nations (1962) *Forest Influences*, Forestry and Forest Products Studies No. 15, Rome (307pp.).

Gaffin, S. R. *et al.* (2008) Variability in New York city's urban heat island strength over time and space. *Theor. Appl. Climatol.* 94, 1–11

Garnett, A. (1967) Some climatological problems in urban geography with special reference to air pollution. *Trans. Inst. Brit. Geog.* 42, 21–43.

Garratt, J. R. (1994) Review: the atmospheric boundary layer. *Earth-Science Reviews*, 37, 89–134.

Gay, L. W. and Stewart, J. B. (1974) *Energy balance studies in coniferous forests*, Report No. 23, Inst. Hydrol., Nat. Env. Res. Coun., Wallingford.

Goldreich, Y. (1984) Urban topo-climatology. *Prog. Phys. Geog.* 8, 336–64.

Harriss, R. C. *et al.* (1990) The Amazon boundary layer experiment: wet season 1987. *J. Geophys. Res.* 95(D10), 16,721–736.

Heintzenberg, J. (1989) Arctic haze: air pollution in polar regions. *Ambio* 18, 50–5.

Hewson, E. W. (1951) Atmospheric pollution, in Malone, T. F. (ed.) *Compendium of Meteorology*, American Meteorological Society, Boston, MA, 1139–57.

Jauregui, E. (1987) Urban heat island development in medium and large urban areas in Mexico. *Erdkunde* 41, 48–51.

Jenkins, I. (1969) Increases in averages of sunshine in Greater London. *Weather* 24, 52–4.

Kessler, A. (1985) Heat balance climatology, in Essenwanger, B. M. (ed.) *General Climatology*, World Survey of Climatology 1A, Elsevier, Amsterdam (224pp.).

Koppány, Gy. (1975) Estimation of the life span of atmospheric motion systems by means of atmospheric energetics. *Met. Mag.* 104, 302–6.

Kubecka, P. (2001) A possible world record maximum natural ground surface temperature. *Weather* 56 (7), 218–21.

Landsberg, H. E. (1981) City climate, in Landsberg, H. E. (ed.) *General Climatology 3, World Survey of Climatology* 3, Elsevier, Amsterdam, 299–334.

Long, I. F., Monteith, J. L., Penman, H. L. and Szeicz, G. (1964) The plant and its environment. *Meteorol. Rundschau* 17(4), 97–101.

McNaughton, K. and Black, T. A. (1973) A study of evapotranspiration from a Douglas fir forest using the energy balance approach. *Water Resources Research* 9, 1579–90.

Maejima, I. *et al.* (1980) Recent climatic change and urban growth in Tokyo and its environs. *Geog. Reports,Tokyo Metropolitan University*, 14/15: 27–48.

Marshall, W. A. L. (1952) *A Century of London Weather*, Met. Office, Air Ministry Report

Miess, M. (1979) The climate of cities, in Laurie, I. C. (ed.) *Nature in Cities*, Wiley, Chichester, 91–104.

Miller, D. H. (1965) The heat and water budget of the earth's surface. *Adv. Geophys.* 11, 175–302.

Mills, G. (2008) Luke Howard and the climate of London. *Weather* 63(6), 153–57.

Montavez, J. P., Gonzalez-Rouce, J. F. and Valero, F. (2008) A simple model for estimating the maximum intensity of nocturnal heat island. *Int. J. Climatol.*, 28, 235–42.

Mote, T. L., Lacke, M. C. and Shepherd, J. M. (2007) Radar signatures of the urban effect on precipitation distribution: a case study for Atlanta, Georgia. *Geophys. Res. Lett.* 34, L20710.

Nicholas, F. W. and Lewis, J. E. (1980) Relationships between aerodynamic roughness and land use and land cover in Baltimore, Maryland. US Geol. Surv. Prof. Paper 1099–C (36pp.).

Nunez, M. and Oke, T. R. (1977) The energy balance of an urban canyon. J. Appl. Met. 16,11–19.

Oke, T. R. (1979) *Review of Urban Climatology 1973–76*, WMO Technical Note No. 169, World Meteorological Organization, Geneva (100pp.).

Oke, T. R. (1980) Climatic impacts of urbanization, in Bach, W., Pankrath, J. and Williams, J. (eds) *Interactions of Energy and Climate*, D. Reidel, Dordrecht, 339–56.

Oke, T. R. (1982) The energetic basis of the heat island. *Quart. J. Roy. Met. Soc.* 108, 1–24.

Oke, T. R. (1986) *Urban Climatology and its Applications with Special Regard to Tropical Areas*, World Meteorological Organization Publication No. 652, Geneva (534pp.).

Oke, T. R. (1988) The urban energy balance. *Prog. Phys. Geog.* 12(4), 471–508.

Oke, T. R. and East, C. (1971) The urban boundary layer in Montreal. *Boundary-Layer Met.* 1, 411–37.

Pankrath, J. (1980) Impact of heat emissions in the Upper-Rhine region, in Bach, W., Pankrath, J. and Williams, J. (eds) *Interactions of Energy and Climate*, D. Reidel, Dordrecht, 363–81.

Pearlmutter, D., Berliner, P. and Shaviv, E. (2007) Urban climatology in arid regions: current research in the Negev desert. *Int. J. Climatol.* 27, 1875–85.

Pease, R. W., Jenner, C. B. and Lewis, J. E. (1980) The influences of land use and land cover on climate analysis: an analysis of the Washington–Baltimore Area, US Geol. Surv. Prof. Paper 1099–A (39pp.).

Peel, R. F. (1974) Insolation and weathering: some measures of diurnal temperature changes in exposed rocks in the Tibesti region, central Sahara. *Zeit. für Geomorph. Supp.* 21, 19–28.

Peterson, J. T. (1971) Climate of the city, in Detwyler, T. R. (ed.) *Man's Impact on Environment*, McGraw-Hill, New York, 131–54.

Peterson, T. C. (2003) Assessment of urban versus rural in situ temperatures in the contiguous United States ; no difference found. *J. Clim.* 16, 2941–59.

Pigeon, G. *et al.* (2007) Anthropogenic heat release in an old European agglomeration (Toulouse, France). *Int. J. Climatol.* 27, 1969–81.

Plate, E. (1972) Berücksichtigung von Windströmungen in der Bauleitplanung, in *Seminarberichte Rahmenthema Unweltschutz*, Institut für Stätebau und Landesplanung, Selbstverlag, Karlsruhe, 201–29.

Reynolds, E. R. C. and Leyton, L. (1963) Measurement and significance of throughfall in forest stands, in Whitehead, F. M. and Rutter, A. J. (eds) *The Water Relations of Plants*, Blackwell Scientific Publications, Oxford, 127–41.

Roth, M. (2007) Review of urban climate research in (sub)tropical regions. *Int. J. Climatol.* 27, 1859–73.

Rutter, A. J. (1967) Evaporation in forests. *Endeavour* 97, 39–43.

Seinfeld, J. H. (1989) Urban air pollution: state of the science. *Science* 243, 745–52.

Shuttleworth, W. J. (1989) Micrometeorology of temperate and tropical forest. *Phil. Trans. Roy. Soc. London* B324, 299–334.

Shuttleworth, W. J. *et al.* (1985) Daily variation of temperature and humidity within and above the Amazonian forest. *Weather* 40, 102–8.

Steinecke, K. (1999) Urban climatological studies in the Reykjavik subarctic environment, Iceland. *Atmos. Environ.* 33 (24–5), 4157–62.

Terjung, W. H. and Louis, S. S-F. (1973) Solar radiation and urban heat islands. *Ann. Assn Amer. Geog.* 63, 181–207.

Terjung, W. H. and O'Rourke, P. A. (1980) Simulating the causal elements of urban heat islands. *Boundary-Layer Met.* 19, 93–118.

Terjung, W. H. and O'Rourke, P. A. (1981) Energy input and resultant surface temperatures for individual urban interfaces. *Archiv. Met. Geophys. Biokl.* B, 29, 1–22.

Tyson, P. D., Garstang, M. and Emmitt, G. D. (1973) *The Structure of Heat Islands*, Occasional Paper No. 12, Department of Geography and Environmental Studies, University of the Witwatersrand, Johannesburg (71pp.).

US Department of Health, Education and Welfare (1970) *Air Quality Criteria for Photochemical Oxidants*, National Air Pollution Control Administration, US Public Health Service, Publication No. AP–63, Washington, DC.

Vehrencamp, J. E. (1953) Experimental investigation of heat transfer at an air–earth interface. *Trans. Amer. Geophys. Union* 34, 22–30.

Weller, G. and Wendler, G. (1990) Energy budgets over various types of terrain in polar regions. *Ann. Glac.* 14, 311–14.

White, W. H., Anderson, J. A., Blumenthal, D. L., Husar, R. B., Gillani, N. V., Husar, J. D. and Wilson, W. E. (1976) Formation and transport of secondary air pollutants: ozone and aerosols in the St Louis urban plume. *Science* 194, 187–9.

World Meteorological Organization (1970) *Urban Climates*, WMO Technical Note No. 108 (390pp.).

Zon, R. (1941) Climate and the nation's forests, in *Climate and Man*, US Department of Agriculture Yearbook, 477–98.

CHAPTER THIRTEEN

Climate change

Mark C. Serreze and Roger G. Barry

LEARNING OBJECTIVES

When you have read this chapter you will:

- understand the difference between climate variability and climate change and know the characteristic features that may constitute a change of climate
- be aware of the different timescales on which past climate conditions are studied and the sources of evidence that may be used
- recognize the major climate forcing factors and feedback mechanisms and the timescales over which they operate
- understand the anthropogenic contributions to climate change
- appreciate the possible impacts of climate change on environmental systems.

A GENERAL CONSIDERATIONS

In this final chapter we examine climate variability and change, climate forcing factors, feedbacks and projected future states of the climate system. In many parts of the world, the climate has varied sufficiently within the past few thousand years to affect patterns of agriculture and settlement. As will become clear, the evidence is now overwhelming that human activities have begun to influence climate.

Realization that climate is far from being constant came only during the 1840s, when indisputable evidence of former Ice Ages was obtained. Studies of past climate began with a few individuals in the 1920s and gained momentum in the 1950s (see Box 13.1). Instrumental records for most parts of the world span only the past 100 to 150 years, and are typically assembled at monthly, seasonal or annual time resolution. However, proxy indicators from tree rings, pollen in bog and lake sediments, ice core records of physical and chemical parameters, and ocean foraminifera in sediments provide a wealth of paleoclimatic data. Tree rings and ice cores can give seasonal or annual records. Peat bog and ocean sediments may provide records with 100 to 1000-year time resolution.

In any study of climate variability and change, one must pay careful attention to possible artifacts in the records. For instrumental records, these include changes in instrumentation (e.g., rain gauge types), observational practices, station

SIGNIFICANT 20TH-CENTURY ADVANCE

13.1 Pioneers of climatic change research

In the late nineteenth century, it was widely accepted that climate conditions were described by long-term averages (sometimes termed normals). The longer the record, the better would be the approximation of the long-term mean. The standard interval for computing climate means from instrumental records adopted by the World Meteorological Organization is 30 years: 1971–2000, for example. Geologists and a few meteorologists were aware that climates of the past had been very different from the present and sought explanations of Ice Ages in astronomical and solar variations. Two classic works by C. E. P. Brooks – *The Evolution of Climate* (1922) and *Climates of the Past* (1926) – provided a remarkably comprehensive picture of variations through geologic time and set out the possible forcing factors, external and internal to the earth's climate system. It was not until the 1950–1960s, however, that awareness grew of substantial decadal-century-scale climatic fluctuations. Historical weather records and proxy climatic data began to be assembled. Pioneers in historical climatology included Gordon Manley and Hubert Lamb in England, Herman Flohn in Germany, Emmanuel LeRoy Ladurie in France, and J. Murray Mitchell and Reid Bryson in the United States.

In the 1970s attention turned initially to the possibility of a renewed Ice Age and then to concern over the effect of increasing carbon dioxide concentrations in the atmosphere. The possibility of global cooling arose from two main sources; the first was paleoclimatological evidence that previous interglacial conditions lasted for only about 10,000 years and that the post-glacial Holocene period was already of that length. A conference titled 'The Present Interglacial – how and when will it end?' took place at Brown University, Providence, RI, in 1972 (G. Kukla, R. Matthews and J. M. Mitchell). The second source was concern over the role of aerosols in reducing incoming solar radiation. Adding to concern over potential cooling, the early 1970s saw an increase in the extent of Northern Hemisphere snow cover (Kukla and Kukla, 1974). Almost simultaneously, however, the first conferences on carbon dioxide and greenhouse warming were taking place! The occurrence of abrupt climatic shifts during the last Late Pleistocene and Holocene began to be identified in the 1970s–1980s. Most notable is the 1000-year-long severe cooling known as the Younger Dryas that occurred around 12,000 years ago.

Interest in past climates was driven by the concept that 'the past is the key to the future'. Hence, efforts were made to document and understand conditions during historical times and the remote geological past, when global climate varied over a much wider range of extremes. As a final note, recent calculations of orbital forcings indicate that the present interglacial will last for another 30,000 years.

location, or the surroundings of the instrumental site, or even errors in transcribed data. Proxy records may suffer from errors in dating or interpretation. Even when climate signals are real, it may be difficult to ascribe them to unique causes owing to the complexity of the climate system, a system which is characterized by myriad interactions between its various components on a suite of spatial and temporal scales (Figure 13.1).

What is the distinction between climate variability and change? Climate variability, as defined by the Intergovernmental Panel on

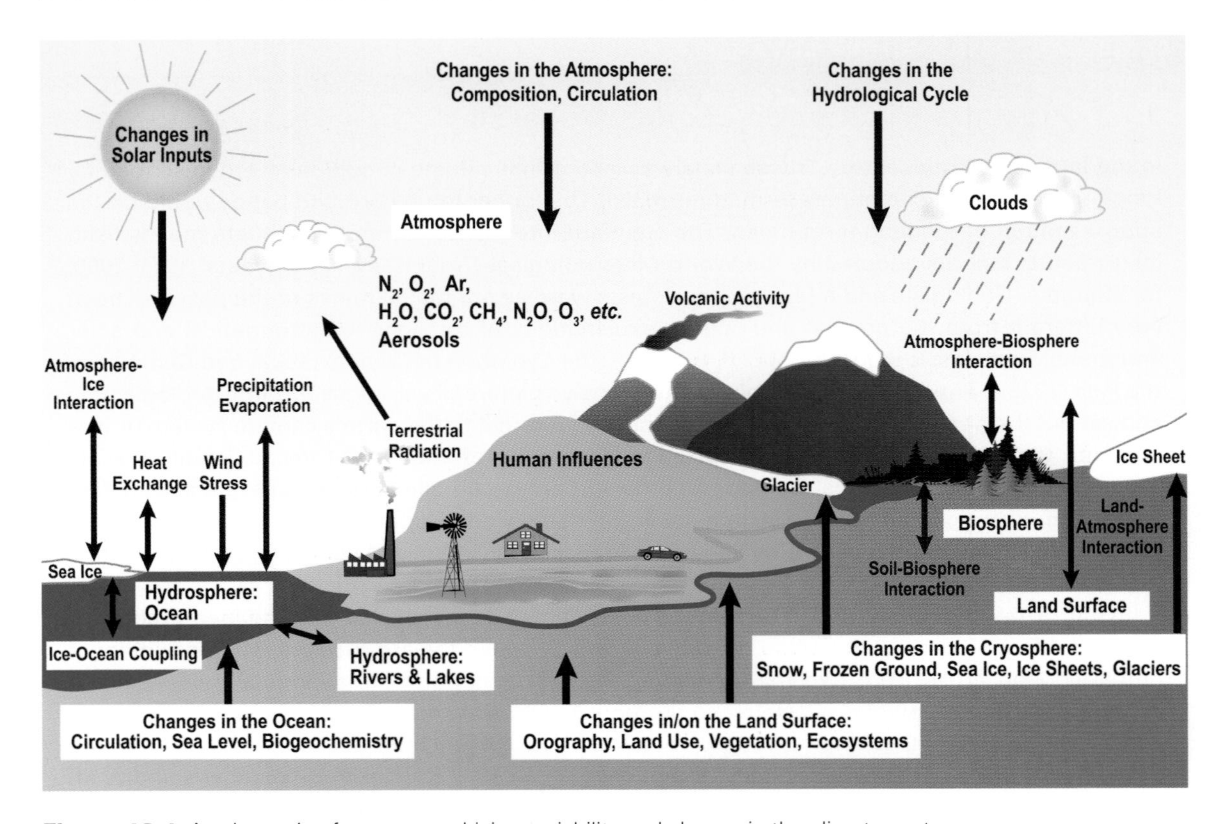

Figure 13.1 A schematic of processes driving variability and change in the climate system.
Source: IPCC (2007). Reproduced by permission of the IPCC (ch. 1, Historical overview of climate change science, Report of WG1 1, IPCC, P. 104, FAQ 1.2, fig. 1).

Climate Change (IPCC), refers to fluctuations in the mean state and other statistics (such as the standard deviation, extremes, or shape of frequency distribution, see Note 1) of climate elements on all spatial and temporal scales beyond those of individual weather events. Variability can be associated with either natural internal processes within the climate system, or with variations in natural or anthropogenic climate forcing. Climate change, by contrast, is viewed by the IPCC as a statistically significant variation in the mean state of the climate or in its variability persisting over an extended period, typically decades or longer. Climate change may be due to natural internal processes, natural external forcings, or persistent anthropogenic-induced changes in atmospheric composition or land use.

The student may be excused if the distinction seems fuzzy. Consider Figure 13.2. A given climate record, whether from instrumental or from proxy sources, may exhibit a suite of behaviors. It may document a rapid shift from one mean state to another (B), a gradual trend, followed by a new mean state (C) or a change in the variance with no change in the mean over the period of record (D). Even within a fairly stable mean state, there can be fluctuations about that state that are quasi-periodic (B) or non-periodic (C). In turn, a record might be characterized only by long periodic oscillation (A). Given that climate variability as viewed by the IPCC includes fluctuations on all spatial and temporal scales beyond synoptic weather events, one could legitimately view all of the behaviors in the figure as expressions of

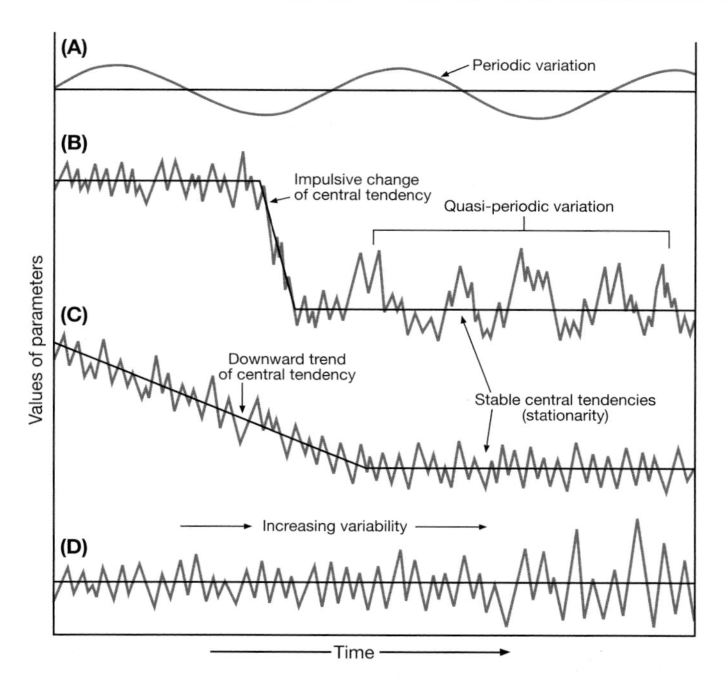

Figure 13.2 Different types of climatic variation. The scales are arbitrary.
Source: Hare (1979). Courtesy World Meteorological Organization.

variability. On the other hand, while one can correctly say (for example) that the major glacial and interglacial cycles of the Pleistocene are expressions of climate variability within the past two million years, it is also appropriate to consider the evolution from full glacial to interglacial conditions as an expression of climate change. Similarly, while we usually view the global temperature rise over the past 100 years as climate change, reserving the term variability for embedded shorter timescale features, the century-long warming could also be viewed as an aspect of climate variability over the past 1000 years. The distinction between variability and change is hence dependent on the time frame over which one considers the climate statistics.

The United Nations Framework Convention on Climate Change (UNFCCC) offers a different definition that can help to resolve some of these problems. They define climate change as 'a change of climate which is attributed directly or indirectly to human activity that alters the composition of the atmosphere and which is in addition to natural climate variability observed over comparable timescales'. This definition is useful in that it makes a clear distinction between natural processes and anthropogenic influences. The remainder of this chapter will view climate change in this context. Variability, in turn, will be viewed as associated with natural processes.

B CLIMATE FORCING, FEEDBACK AND RESPONSE

The most fundamental measure of the earth's climate state is the global mean, annually averaged surface air temperature. Year-to-year and even decadal-scale variations in this value can occur due to processes purely internal to the climate system. The warm phase of ENSO, for example, may be viewed as an internal process in which heat in the ocean reservoir (i.e., heat already within the climate system) is transferred to the atmosphere, expressed as a rise in global mean surface temperature. When considering timescales of decades or longer, thinking must turn to climate *forcings* and attendant *feedbacks*. Forcing factors represent imposed perturbations to the global system, and are defined as positive when they induce an increase in global mean surface temperature, and negative when they induce a decrease. Forcing factors may in turn be of natural or anthropogenic origin. The magnitude of the global temperature response to forcing depends on the feedbacks. Positive feedbacks amplify the temperature change while negative feedbacks dampen the change.

1 Climate forcing

Many different types of climate forcing can be identified. Key forcings are associated with the following processes:

- *Plate Tectonics.* On geological timescales, plate tectonics have resulted in great changes in continental positions and sizes, the configuration of ocean basins and (through associated phases in volcanic activity) atmospheric composition. While there is little doubt that such changes altered the globally averaged surface albedo and greenhouse gas concentrations, plate movements have also altered the size and location of mountain ranges and plateaus. As a result, the global circulation of the atmosphere and the pattern of ocean circulation were modified. In 1912, Alfred Wegener proposed continental drift as a major determinant of climates and biota, but this idea remained controversial until the motion of crustal plates was identified in the 1960s. Alterations in continental location have contributed substantially to major Ice Ages of the distant past (such as the Permo-Carboniferous glaciation of Gondwanaland) as well as to intervals with extensive arid (Permo-Triassic) or humid (coal deposits) environments during other geological periods. Over the past few million years, the uplift of the Tibetan Plateau and the Himalayan ranges has caused the onset, or intensification, of desert conditions in western China and Central Asia.
- *Astronomical periodicities.* As noted in Chapter 3A.2, the earth's orbit around the sun is subject to long-term variations, leading to changes in the seasonal and spatial distribution of solar radiation incident to the surface. These are known as Milankovich forcings after the astronomer Milutan Milankovich, whose careful calculations of their effects built upon the work of nineteenth-century astronomers and geologists. There are three principal effects: the eccentricity (or stretch) of the orbit influencing the strength of the contrast in solar radiation received at perihelion (closest to sun) and aphelion (furthest from sun), with periods of approximately 95,000 years and 410,000 years; the tilt of the earth's axis (approximately 41,000 years) influencing the strength of the seasons; and a wobble in the earth's axis of rotation, which causes seasonal changes in the timing of perihelion and aphelion (Figure 13.3). This precession effect, with a period of about 21,000 years, is further illustrated in Figure 3.3. The range of variation of these three components and their consequences are summarized in Table 13.1. Astronomical periodicities are associated with global temperature fluctuations of ±2–5°C per 10,000 years. The timing of orbital forcing is clearly represented in glacial–interglacial fluctuations with the last four major glacial cycles spanning roughly 100,000 years (or 100ka). The astronomical theory of glacial cycles became widely accepted in the 1970s after Hays, Imbrie and Shackleton provided convincing evidence from ocean core records.
- *Solar variability.* The sun is a variable star. The approximately 11-year solar (sunspot) cycle (and 22-year magnetic field cycle) are well known. As discussed in Chapter 2, the 11-year sunspot cycle is associated with ±1W m^{-2}

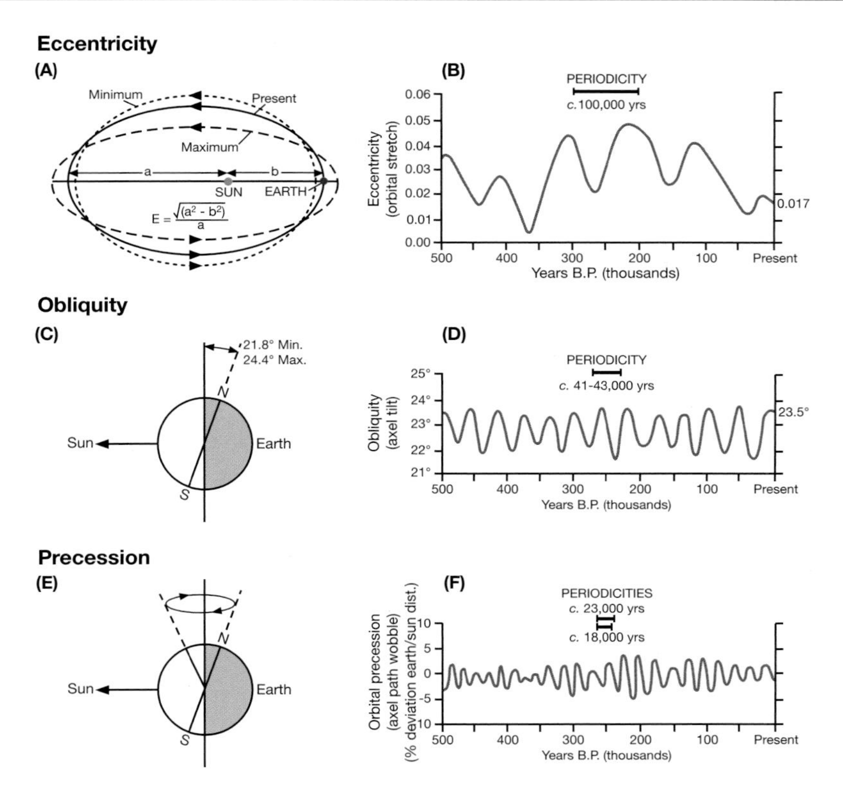

Figure 13.3 Summary of astronomical (orbital) effects on solar irradiance and their relevant timescales over the past 500,000 years. A and B: Eccentricity or orbital stretch; C and D: Obliquity or axial tilt; E and F: Precession or axial path wobble.

Sources: Partly after Broecker and Van Donk 1970, and Henderson-Sellers and McGuffie 1984. B, D and F: from *Review of Geophysics and Space Physics* 8 (1970). Reproduced by kind permission of the American Geophysical Union.

fluctuations in solar irradiance (i.e., a departure from the solar constant; in terms of radiation receipts globally averaged over the top of the atmosphere, the effective value is only 0.25W m^{-2}) (see below and Chapter 3A.1). Effects on ultraviolet radiation are proportionally larger in terms of percent change. There is also evidence for longer-term variations. Intervals when sunspot and solar flare activity were much reduced (especially the Maunder Minimum of AD 1645–1715) may have been associated with global temperature decreases of about 0.5°C. Solar variability also seems to have played a role in decadal-scale variations of global temperature until the latter part of the twentieth century, when anthropogenic effects became dominant. Turning to the distant past, it is known that solar irradiance three billion years ago (during the Archean) was about 80 percent of the modern value. Interestingly, the effect of this faint early sun was offset, most likely, by a concentration of carbon dioxide that was perhaps 100 times higher than now, and perhaps also by the effects of a largely water-covered earth (meaning lots of water vapor in the atmosphere).

- *Volcanic eruptions.* Major individual explosive eruptions inject dust and sulfur gases (especially sulfur dioxide) into the stratosphere, the latter forming sulfuric acid droplets.

Table 13.1 Orbital forcings and characteristics

Element		Index range	Present value	Average periodicity
Obliquity of Ecliptic (′) (Tilt of axis of rotation) Effects equal in both hemispheres, effect intensifies poleward (for caloric seasons)		22–24.5°	23.4°	41 ka
Low ′ Weak seasonality, steep poleward radiation gradient	High ′ Strong seasonality, more summer radiation at poles, weaker radiation gradient			
Precession of Equinox (v) (Wobble of axis of rotation) Changing earth–sun distance alters seasonal cycle structure; complex effect, modulated by eccentricity of orbit		0.05 to –0.05	0.0164	19, 23 ka
Eccentricity of Orbit (e) Gives 0.02% variation in annual incoming radiation; modifies amplitude of precession cycle changing seasonal duration and intensity; effects opposite in each hemisphere; greatest in low latitudes		0.005 to 0.0607	0.0167	410, 95 ka

Equatorial eruption plumes spread into both hemispheres, whereas plumes from eruptions in mid-to high latitudes are confined to that hemisphere. Observational evidence from the past 100 years demonstrates that major eruptions can be associated with global averaged cooling of several tenths of a degree C in the year following the event and much larger changes on a regional to hemispheric basis. The cooling is primarily from the sulfuric acid droplets which reflect solar radiation. Dust also causes surface cooling by absorbing solar radiation in the stratosphere, but compared to the sulfuric acid these effects are short-lived (weeks to months) Stratospheric aerosols may also cause brilliant sunsets (see Figure 2.12). The most recent major volcanic eruption with significant climate impacts was Mt. Pinatubo in 1991.

- *Human-induced changes in atmospheric composition and land cover.* The effect of greenhouse gases such as carbon dioxide and methane on the radiation budget has already been introduced (see also Chapter 2). The observed buildup of these gases since the dawn of the industrial age represents a positive forcing. Human activities have also led to a buildup of tropospheric aerosols, which induce a partly compensating cooling. Changes in land use and land cover have also led to a small increase in surface albedo that promotes cooling.

While the common feature of all of these forcings is that they influence aspects of the earth's radiation budget, they are obviously distinguished in large part by the timescales at which they operate. In terms of inducing global temperature change over the past 100 years, as well as changes projected through the twenty-first century, the effects of plate tectonics (operating on timescales of millions of years) and Milankovich forcings (operating on timescales of tens of thousands of years) are irrelevant. Note also that while Milankovich forcings are associated with very significant impacts on the seasonal and spatial distribution of solar radiation incident on the surface, impacts on incident radiation when globally averaged through the annual cycle are quite small. For example, while a decrease in obliquity means less summer radiation in the Northern Hemisphere summer, it means more in the Southern Hemisphere winter, with these seasonal effects largely canceling out.

Milankovich forcings hence contrast fundamentally with the effects of changing solar irradiance, volcanic eruptions, or human-induced changes in atmospheric greenhouse gas concentrations and surface albedo, all of which, considered in terms of their immediate effect, have a globally and annually averaged impact on the radiation balance at the top of the atmosphere. Because of this property, they are termed *radiative forcings*. For example, an increase in solar output will lead to more radiation incident to the top of the earth's atmosphere, irrespective of latitude or season. The immediate effect will be a globally averaged radiation imbalance at the top of the atmosphere (more energy coming in than going out), leading to a rise in temperature that would eventually bring the earth/atmosphere system into a new radiative equilibrium. Similarly, the immediate response to increasing the concentration of greenhouse gases will be a globally averaged decrease in longwave emission to space, a radiation imbalance promoting warming, also eventually leading to a new radiative balance (provided that the forcing remains constant).

Global climate change (change due to human influences by our adopted conventions) is best viewed in the context of global radiative forcing. In the IPCC framework adopted here, radiative forcing specifically refers to the amount by which a factor alters the globally and annually averaged radiation balance at the top of the atmosphere, expressed in units of W m^{-2}, evaluated as forcing relative to the year 1750, the start of the Industrial Revolution. In 2005, there was an estimated radiative forcing from human activities of 1.6W m^{-2}.

2 Climate feedbacks

Building on the framework of radiative forcing, consider further the change in global average surface temperature resulting from increasing the atmospheric concentration of carbon dioxide. As just discussed, because of the imposed perturbation, more of the longwave radiation emitted upward from the surface is absorbed by the atmosphere, and directed back towards the surface. The result is a radiation imbalance at the top of the atmosphere – net solar radiation entering the top of the atmosphere exceeds the longwave loss to space. The climate forcing from adding carbon dioxide is hence positive. Now consider the feedbacks. The most important of these is the water vapor feedback. Warming results in more evaporation, and a warmer atmosphere can carry more water vapor. However, water vapor is also a greenhouse gas, so it causes further warming. Some of the earth's snow cover and sea ice will melt, reducing the earth's surface albedo, also causing further warming. These are examples of positive feedbacks, as they amplify the global surface temperature change induced by the climate forcing. If the carbon dioxide concentration in the atmosphere were lowered, thereby imposing a negative climate forcing, the positive feedbacks would foster further cooling.

A fascinating aspect of the global climate system is that positive feedbacks dominate. For example, one of the responses to increasing greenhouse gases could be an increase in cloud cover, which through increasing the planetary albedo would represent a negative feedback. However, this and other potential negative feedbacks would only appear to be capable of slowing the rate of warming, not reversing it.

While climate feedbacks can be either positive or negative, they can also be broadly differentiated regarding how quickly they operate. In the framework of global radiative forcing appropriate to understanding human-induced global climate change, it is the fast feedbacks which are relevant. The most important are changes in water vapor and albedo (mentioned above). Both can operate over timescales of days and even less. Cloud cover can also change very quickly (hours). Examples of slow feedbacks are changes in the extent of continental ice sheets (influencing planetary albedo) and greenhouse gases during the Pleistocene in response to Milankovich periodicities. Records from ice cores show that these glacial-interglacial cycles were nearly coincident with fluctuations in both atmospheric carbon dioxide (±50ppm) and methane (±150ppb). The nature of these trace gas feedbacks remains incompletely resolved. Potential mechanisms include changes in ocean chemistry, increased plankton growth acting to sequester carbon dioxide, suppression of air–sea gas exchange by sea ice, changes in ocean temperature that affect the solubility of carbon dioxide, and altered ocean circulation. Most likely a suite of processes worked in concert. Negative (positive) excursions in greenhouse gas concentrations are associated with cold (warm) intervals, as illustrated in Figure 2.6.

3 Climate response

How much does the global mean surface temperature change in response to a radiative forcing of a given magnitude? How long does it take for the change to occur? These are among the most important, pressing questions in climate change science.

The first question deals with the issue of equilibrium climate sensitivity. In the IPCC framework, equilibrium climate sensitivity is the equilibrium change in annual mean global averaged surface air temperature following a doubling of the atmospheric equivalent carbon dioxide. Doubling the carbon dioxide concentration equates to a radiative forcing (top of atmosphere radiation imbalance) of about 4W m^{-2}. In response to this doubling the surface and atmosphere would warm up. Eventually, radiative balance would be restored again with a new and higher surface temperature. Estimates of equilibrium climate sensitivity obtained from the current generation of global climate models range from 2–4.5°C, with a best estimate of 3.0°C. The uncertainly lies largely in the spread of model estimates of the climate feedbacks, particularly in the cloud feedbacks. Cloud feedbacks are complex and hard to model. Negative feedbacks may operate when increased global heating leads to greater evaporation and greater amounts of high-altitude cloud cover, which reflect more incoming solar radiation. However, other types of clouds, and clouds in the polar regions, can induce surface warming

Expressed in a more convenient fashion, the best estimate of 3°C for carbon dioxide doubling equates to 0.75°C global mean surface temperature increase per W m^{-2} of forcing. It is stressed that the climate simulations used to obtain these sensitivity numbers only deal with the fast feedbacks. If there were no feedbacks present in the climate system, the climate sensitivity would be only about 0.30°C per W m^{-2}. While equilibrium climate sensitivity in the IPCC framework is based on a doubling of atmospheric equivalent carbon dioxide, it appears that the equilibrium temperature response to any radiative forcing is roughly the same. This is an important concept, since it means that to a first approximation, one can linearly add different forcings to obtain a net value from which an equilibrium temperature change can be estimated.

It also appears that most of the equilibrium temperature response to a radiative forcing with the fast feedbacks at work occurs over a time span of 30 to 50 years. Most of the time lag is due to the large thermal inertia of the oceans. The basic issue is that the oceans can absorb and store a great deal of heat without a large rise in the surface (radiating) temperature. Consider what is happening in response to the current radiative forcing from human activities of 1.6Wm^{-2}. Using the equilibrium climate sensitivity of 0.75 implies that this radiative forcing, if maintained, will eventually yield about 1.2°C of warming. Over the instrumental record, the global mean temperature has risen by about 0.7°C, implying another 0.5°C remaining after the ocean sufficiently heats up. How much has the heat content of the ocean already increased? Based on available hydrographic data from 1955–1998, the world ocean between the surface and 3000m depth gained ~1.6 × 10^{22}J. Compared with atmospheric kinetic energy (p. 70), this is a very large number.

An obvious shortcoming of the concept of equilibrium climate sensitivity is that radiative forcing is always changing. Consider explosive volcanic eruptions. While the global radiative forcing from a single eruption can be very significant (2–3W m^{-2} at peak), the forcing is short-lived (several years) such that the system can never come into equilibrium with it (while the global surface temperature can be temporarily reduced by several tenths of a degree, this is much smaller than the calculated temperature change in equilibrium with the peak forcing). Similarly, the system could never be in equilibrium with solar variability associated with the 11-year sunspot cycle. If we were to somehow freeze the current radiative forcing from human activities at its present value, the climate system would eventually approach a new temperature in equilibrium with it (assuming no complications like multiple volcanic eruptions). However, radiative forcing from human activities has grown over the past century and will continue to grow in the future, meaning that the equilibrium temperature value has changed and will continue to change. Put differently, the picture over the past 100 years and into the future is a climate system constantly trying to catch up with a growing radiative forcing but always lagging behind it.

4 The importance of framework

While the distinction between climate forcing and feedback is fairly straightforward when considering changes in globally averaged temperature, it must be stressed that this distinction may change if adopting a different framework, such as the evaluation of regional climate variability and change. For example, due to loss of its sea ice cover, rises in surface air temperature are expected to be especially pronounced over the Arctic Ocean. In the framework of human-induced global climate change (see Table 13.2), this may be viewed as part of the feedback process that amplifies the global average temperature response to increased greenhouse gas concentrations. However, if one were to conduct a regional study of the Arctic, one could legitimately view the sea ice loss as a forcing on Arctic temperature change. Similarly, global climate change may be attended by shifts in patterns of atmospheric circulation, precipitation and cloud cover. While on the global scale these would be viewed as feedbacks, investigations of regional impacts could view them as forcings.

Another framework issue regards how one views transitions between glacial and interglacial conditions. While changes in ice sheet area and greenhouse gas concentrations during these transitions are appropriately viewed as slow feedbacks, if one considers full glacial and interglacial conditions as two equilibrium states, these slow feedbacks may instead be thought of as climate forcings. With estimates of the global temperature change between the equilibrium states and the forcings, one then has another way to estimate equilibrium climate sensitivity. Numbers obtained from this approach turn out to agree fairly well with those coming from global climate models. Hence, in summary, depending on the chosen framework, one person's feedback may be another person's forcing.

C THE CLIMATIC RECORD

1 The geological record

Understanding the significance of climatic trends over the past 100 years requires that they be viewed against the backdrop of earlier conditions.

On geological timescales, global climate has undergone major shifts between generally warm, ice-free states and Ice Ages with continental ice sheets. There have been at least seven major Ice Ages through geological time. The first occurred 2500 million years ago (Ma) in the Archean period, followed by three more between 900 and 600Ma, in the Proterozoic. There were two Ice Ages in the Paleozoic era (the Ordovician, 500–430Ma; and the Permo-Carboniferous, 345–225Ma). The most recent Ice Age began about 34Ma in Antarctica at the Eocene/Oligocene boundary and about three million years ago in northern high latitudes. At present, we are considered to be still within this most recent Ice Age, albeit in the warm part of it known as the Holocene, which began about 11.5ka. While the total volume of land ice today (mostly comprising the Antarctic and Greenland ice sheets) is certainly much smaller than it was at 20ka, it is still substantial compared to other times of the earth's past.

Major Ice Ages and ice-free periods can be linked to a combination of external and internal climate forcing (plate tectonics, greenhouse gas concentrations, solar irradiance). The ice sheets of the Ordovician and Permo-Carboniferous periods formed in high southern latitudes on the former mega-continent of Gondwanaland. Uplift of the western cordilleras of North America and the Tibetan Plateau by plate movements during the Tertiary period (50–2Ma) caused regional aridity to develop in the respective continental interiors. However, geographical factors are only part of the explanation of climate variations. For example, warm high-latitude conditions during the mid-Cretaceous period, about 100Ma, may be attributable to atmospheric concentrations of carbon dioxide three to seven times higher than at present, augmented by the effects of alterations in land–sea distribution and ocean heat transport.

Much more is known about ice conditions and climate forcings through the Quaternary, which began about 2.6 million years ago, comprising the Pleistocene (2.6Ma–11.5 ka) and the Holocene (11.5ka-present) epochs. It is abundantly clear that this most recent Ice Age we live in was far from being uniformly cold. Instead it was characterized by oscillations between glacial and interglacial conditions (see Box 13.2). Eight cycles of global ice volume are recorded in land and ocean sediments during the last 0.8–0.9Ma, each averaging roughly 100ka, with only 10 percent of each cycle as warm as the twentieth century (Figure 13.4D and E). Each glacial period was in turn characterized by abrupt terminations. Because of reworking of sediments, only four or five of these glaciations are identified from terrestrial records. Nevertheless, it is likely that all were characterized by large ice sheets covering northern North America and northern Europe. Sea-levels were also lowered by about 130m due to the large volume of water locked up in the ice. Records from tropical lake basins show that these regions were generally arid at those times. Prior to 0.9Ma the timing of glaciations is more complex. Ice volume records show a dominant 41ka periodicity, while ocean records of calcium carbonate indicate fluctuations of 400ka.

These periodicities are linked to the Milankovich forcings discussed earlier (see also Chapter 3A). The precession signature (19 and 23ka) is most apparent in low-latitude records, whereas that of obliquity (41ka) is represented in high latitudes. However, the 100ka orbital eccentricity signal is generally dominant overall. The basic idea is that onset of glacial conditions is initiated by Milankovich forcings that yield summer cooling over the northern land masses. This favors survival of snow cover through summer, a feedback promoting further cooling and ice sheet growth, leading to even further cooling through slow feedbacks in the carbon cycle discussed earlier. Onset of an interglacial works the other way, with Milankovich forcings promoting initial warming over the northern land masses, setting feedbacks into motion to give further warming and ice melt.

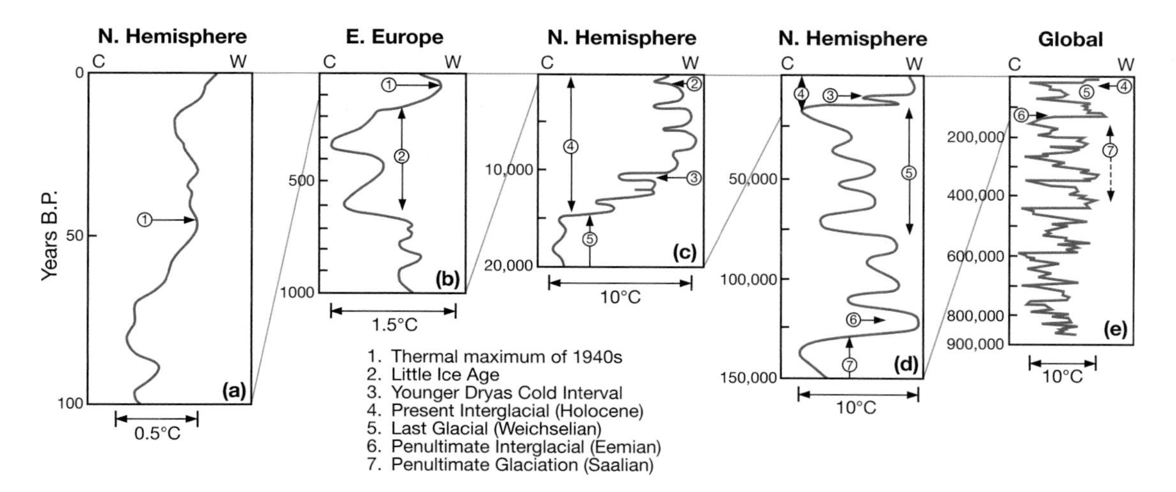

Figure 13.4 Main trends in global climate during the past million years or so. A: Northern Hemisphere, average land–air temperatures. B: Eastern Europe, winter temperatures. C: Northern Hemisphere, average land–air temperatures. D: Northern Hemisphere, average air temperatures based partly on sea surface temperatures. E: Global average temperatures derived from deep-sea cores.

Source: From *Understanding Climatic Change: A Program for Action* (1975). By permission of the National Academy Press, Washington, DC.

Table 13.2 The four categories of climatic variable subject to anthropogenically induced change.

Variable changed	Scale of effect	Sources of change
Atmospheric composition temperature	Local–global	Release of aerosols and trace gases
Surface properties; energy budgets	Regional	Deforestation; desertification; urbanization
Wind regime	Local–regional	Deforestation; urbanization
Hydrological cycle components urbanization	Local–regional	Deforestation; desertification, irrigation;

2 The last glacial cycle and post-glacial conditions

The last interglacial, known as the Eemian, peaked about 125ka. The last glacial cycle following the Eemian was itself characterized by periods of extensive ice (known as stades) and less extensive ice (known as interstades), Maximum global ice volume (the Last Glacial Maximum, or LGM) occurred around 25–18ka. The LGM ended with abrupt warming between about 15 and 13ka, depending on latitude and area, interrupted by a cold regression called the Younger Dryas (13–11.7ka). This was then followed by a renewed sharp warming trend (Figure 13.4). The Holocene (our present interglacial) is considered to begin at 11.5ka, after the close of the Younger Dryas event. Based on assessments of Milankovich forcings, the present interglacial should last for at least another 30,000 years. A particularly striking feature of the last glacial cycle is rapid millennial-scale changes between warm and cold conditions, known as Dansgaard-Oeschger (D-O) cycles. The Younger Dryas event is considered to be the last of these D-O cycles. As is evident in a number of proxy records (see Figure 13.5), the onset and termination of the Younger Dryas cold event, with a switch from near glacial to interglacial

SIGNIFICANT 20TH-CENTURY ADVANCE

13.2 Documenting Paleoclimates

Geologists documenting the last Ice Age made the first studies of paleoclimate in the late nineteenth century. Early progress was hampered by uncertainty as to the age of the earth and the length of the geological record. However, by 1902 it was accepted that there had been at least four or five glacial episodes in the Alps and in North America during the Pleistocene epoch. Explanations were sought in variations of the astronomical periods affecting the earth's orbit, notably by J. Croll (1875) and M. Milankovitch (1920, 1945), and in variations of the solar constant (G. C. Simpson, 1934, 1957). Confirmation that astronomical periodicities act as a 'pacemaker' of the Ice Ages was not forthcoming until the timing of major changes in planktonic foraminiferae in ocean sediment records could be accurately deciphered and dated in the 1970s (J. Hays, J. Imbrie and N. Shackleton, 1976)

The use of proxy evidence to investigate past climate began almost a century ago. In 1910 the Swedish scientist Baron G. de Geer used the annual deposits of sediments (varves) in glacial lakes to date changes in vegetation inferred from the pollen record. Pollen cores spanning the post-glacial interval, extracted from peat bogs and lake sediments, began to be widely studied in Europe and North America in the 1950s–1960s following the development of radiocarbon dating of organic materials by W. Libby in 1951. At the same time, the ocean sedimentary record of changes in marine microfauna – both surface-dwelling (planktonic) and bottom-dwelling (benthic) foraminifera – began to be investigated. Assemblages of fauna associated with different water masses (polar, subpolar, midlatitude, tropical) enabled wide latitudinal shifts in ocean temperatures during the Quaternary epoch to be traced. The use of oxygen isotopic ratios (O^{18}/O^{16}) by C. Emiliani and S. Epstein provided independent estimates of ocean temperature and particularly changes in global ice volume. These records showed that there had been eight glacial/interglacial cycles during the past 800,000 years.

In the southwestern United States, counts of annual tree rings had been used by archeologists early in the twentieth century to date timbers in paleo-Indian structures. In the 1950s–1960s, ring width was investigated as a signal of summer drought in the desert margins and summer temperature at high elevations. The field of dendroclimatology, employing statistical methods, was developed under the leadership of H. C. Fritts. Subsequently, F. Schweingruber introduced the use of ring density variations analyzed by X-ray techniques as a seasonal indicator. The 1970s–1980s saw numerous sophisticated biological indicators in use. These included insects, particularly beetles, diatoms, ostracods, pack rat middens containing plant macrofossils, and corals.

The most comprehensive information on the paleo-atmosphere over the past 1000 to 100,000 years has been retrieved from deep ice cores in Greenland, Antarctica, and plateau ice caps in low latitudes. The principal types of proxy data are: atmospheric temperatures from δO^{18} (developed for glacier ice by W. Dansgaard), accumulation from the annual layer thickness, carbon dioxide and methane concentrations from air bubbles trapped in the ice, volcanic activity from electrical conductivity variations caused by the sulfates, aerosol load and sources (continental, marine and volcanic). The earliest deep cores were collected at Camp Century in northwest Greenland and Byrd station in West Antarctica in the 1960s. Subsequently many cores have been recovered and analyzed. Those of particular note are the GISP II and GRIP cores from Summit, Greenland, spanning about 140,000 years, and the Vostok and EPICA cores from Antarctica, spanning about 450,00 years and 720,000 years, respectively.

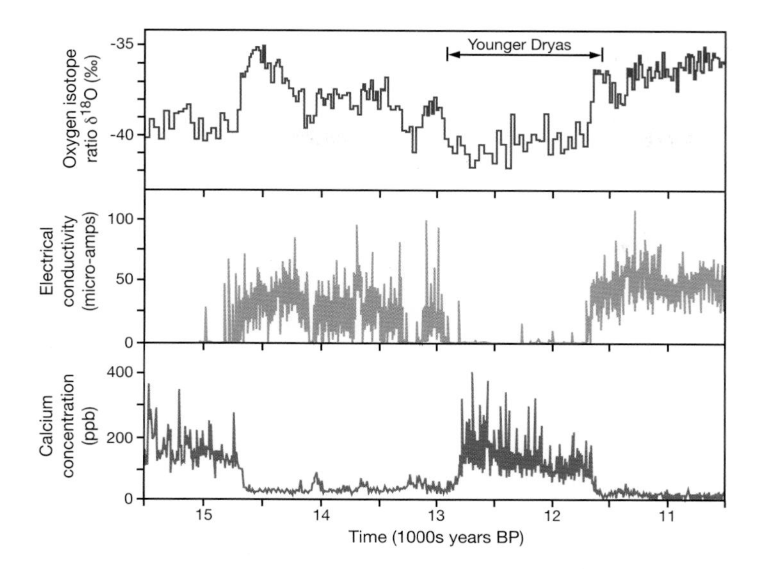

Figure 13.5 The late glacial to interglacial transition (14.7 to 11.6ka) as indicated by d^{18}O (ppm), electrical conductivity (microamps) and calcium concentrations (ppb) in the Greenland Ice Sheet Program (GISP) ice core from central Greenland.

Source: Reprinted by permission of Grootes (1995). Copyright © National Academy of Sciences. Courtesy of the National Academy of Sciences, Washington, DC.

climate conditions and back again, apparently occurred within a five-year time interval for both transitions! The processes driving D-O events like the Younger Dryas are still incompletely understood, but likely in some way involve massive discharges of fresh water from melting ice sheets to the North Atlantic that disrupted the Atlantic thermohaline circulation (see Figure 7.32).

Early Holocene warmth around 10ka is attributed to July solar radiation being 30–40W m^{-2} greater than now in northern mid-latitudes, again due to Milankovich effects. Following the final retreat of the continental ice sheets from Europe and North America between 10,000 and 7000 years ago, the climate rapidly ameliorated in middle and higher latitudes. In the subtropics this interval was also generally wetter, with high lake levels in Africa and the Middle East. A Holocene Thermal Maximum (HTM) was reached in the mid-latitudes about 5000 years ago, when summer temperatures were 1–2°C higher than today (see Figure 13.5B) and the Arctic tree line was several hundred kilometers further north in Eurasia and North America. By this time, subtropical desert regions were again very dry and were largely abandoned by primitive peoples.

A temperature decline set in around 2000 years ago with colder, wetter conditions in Europe and North America. Although temperatures have not since equaled those of the HTM (we are getting close), a relatively warmer interval (or intervals) occurred between the ninth and mid-fifteenth centuries AD. Summer temperatures in Scandinavia, China, the Sierra Nevada (California), Canadian Rocky Mountains and Tasmania exceeded those that prevailed until the late twentieth century.

3 The past 1000 years

Temperature reconstructions for the Northern Hemisphere over the past millennium are based on several types of proxy data, but especially dendrochronology, ice cores and historical

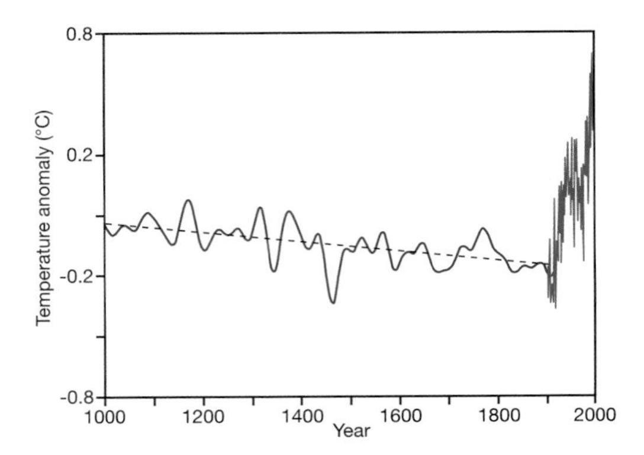

Figure 13.6 Variation in surface air temperature for the Northern Hemisphere over the past millennium. The reconstructed 40-year smoothed values are plotted for 1000–1880 together with the linear trend 1000–1850, and observed temperatures for 1902–1998. The reconstruction is based on estimates from ice cores, tree rings and historical records, and has two standard error limits of about ±0.5°C during 1000–1600. The values are plotted as anomalies relative to 1961–1990.

Source: Adapted from Mann *et al.* (1999). Courtesy of M. E. Mann, Pennsylvania State University.

records. Figure 13.6 shows a reconstruction based on such proxies for the past millennium. Until about AD 1600 there is still considerable disparity in different estimates of decadal mean values and their range of variation. Conditions appear to have been slightly warmer between AD 1050 and 1330 than between 1400 and 1900. There is evidence in Western and Central Europe for a warm phase around AD 1300. Icelandic records indicate mild conditions up until the late twelfth century, and this phase was marked by the Viking colonization of Greenland and the occupation of Ellesmere Island in the Canadian Arctic by the Inuit.

Deteriorating conditions followed. This cool period, known as the 'Little Ice Age', was associated with extensive Arctic sea ice and glacier advances in some areas to maximum positions since the end of the last glacial cycle. These advances occurred at dates ranging from the mid-seventeenth to the late nineteenth century in Europe, as a result of the lag in glacier response and regional variability. The coldest interval of the Little Ice Age in the Northern Hemisphere was AD 1570–1730. What caused the Little Ice Age is not entirely clear. Reduced solar output associated with the Maunder Minimum in sunspot activity (1645–1715) likely played a role, as did increased volcanic activity.

Long instrumental records for stations in Europe and the eastern United States indicate that the warming trend that ended the Little Ice Age began at least by the mid-nineteenth century. The time series of global annual averaged surface air temperature from instrumental records shows a significant temperature rise of about 0.7°C from 1880 through 2007. Both hemispheres have participated in this warming, but it is most pronounced in the Northern Hemisphere (Figure 13.7). Warming in turn encompasses both land and ocean regions, being stronger over land (Figure 13.8). Warming has been smallest in the tropics and largest in northern high latitudes. Warming is in turn strongest during winter. The general temperature rise has not been continuous, however, and four basic phases may be identified in the global record:

1 1880–1920, during which there was an oscillation within extreme limits of about 0.3°C but no trend.
2 1920–mid-1940s, during which there was considerable warming of approximately 0.4°C; this warming was most strongly expressed in northern high latitudes.
3 Mid-1940s–early 1970s, during which there were oscillations within extreme limits of about 0.4°C, with the Northern Hemisphere cooling slightly on average and the Southern Hemisphere remaining fairly constant in temperature. Regionally, northern Siberia, the eastern Canadian Arctic and Alaska experienced a mean lowering of winter temperatures by 2–3°C between 1940 and 1949 and 1950 and 1959; this was partly compensated by a slight warming in the western United States, Eastern Europe and Japan.

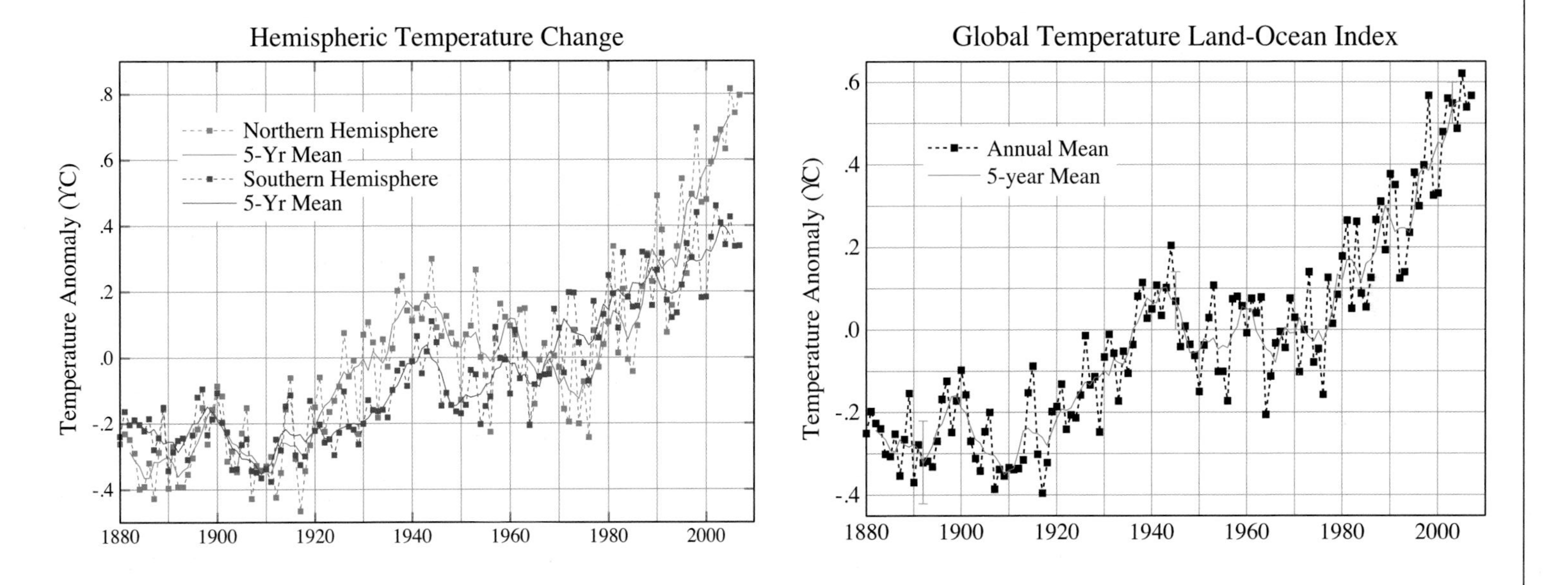

Figure 13.7 Long-term instrumental records of annual average surface air temperature, expressed as anomalies with respect to the base period 1951–1980. A: Global average. B: Averages for the Northern and Southern Hemispheres. The red lines depict the time series smoothed with five-year means.

Source: NASA, Goddard Institute for Space Sciences (http://data.giss.nasa.gov/gistemp/).

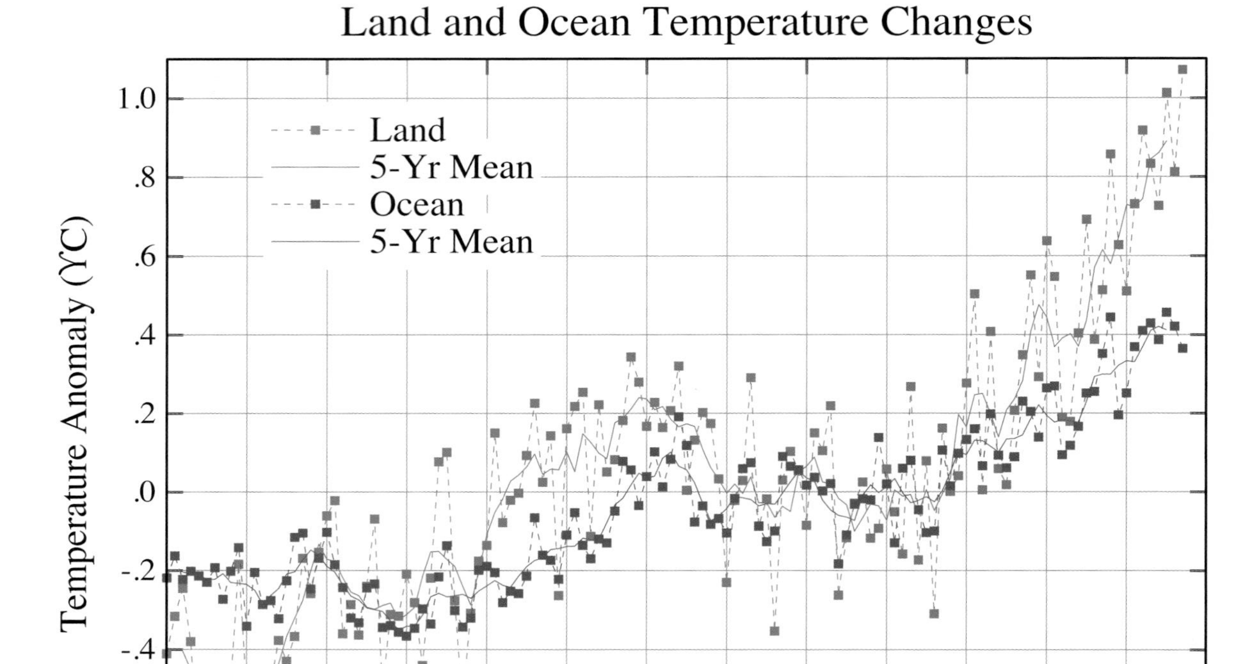

Figure 13.8 Long-term instrumental records of annual average surface air temperature, expressed as anomalies with respect to the base period 1951–1980 for the global land and ocean areas. The solid lines depict the time series smoothed with five-year means.

Source: NASA, Goddard Institute for Space Sciences (http://data.giss.nasa.gov/gistemp/).

4 Mid-1970s–2008, during which there was a marked overall warming of about 0.5°C, but with strong regional variability (see Plate 13.1).

Based on balloon soundings and assessments from satellite sounders, lower tropospheric temperatures over the period 1958 to the present have increased at slightly higher rates than at the surface. This interpretation, however, must acknowledge discontinuities and biases in the time series introduced by changing satellites, orbit decay, drift and other factors. There is evidence that balloon soundings may have a cooling bias.

Global mean surface temperatures during the past decade reached their highest levels on record and probably for the last millennium. In the NASA GISS analysis used to compile Figure 13.7, the warmest year on record was 2005, with 2007 and 1998 tied for second warmest. Rankings based on other global temperature analyses (e.g., from the Climatic Research Unit of the UK) differ somewhat, but tell the same basic story of very warm conditions for the past decade. The key spatial feature of change over the past decade (Figure 13.9) is very strong warming over northern high latitudes. This is especially apparent

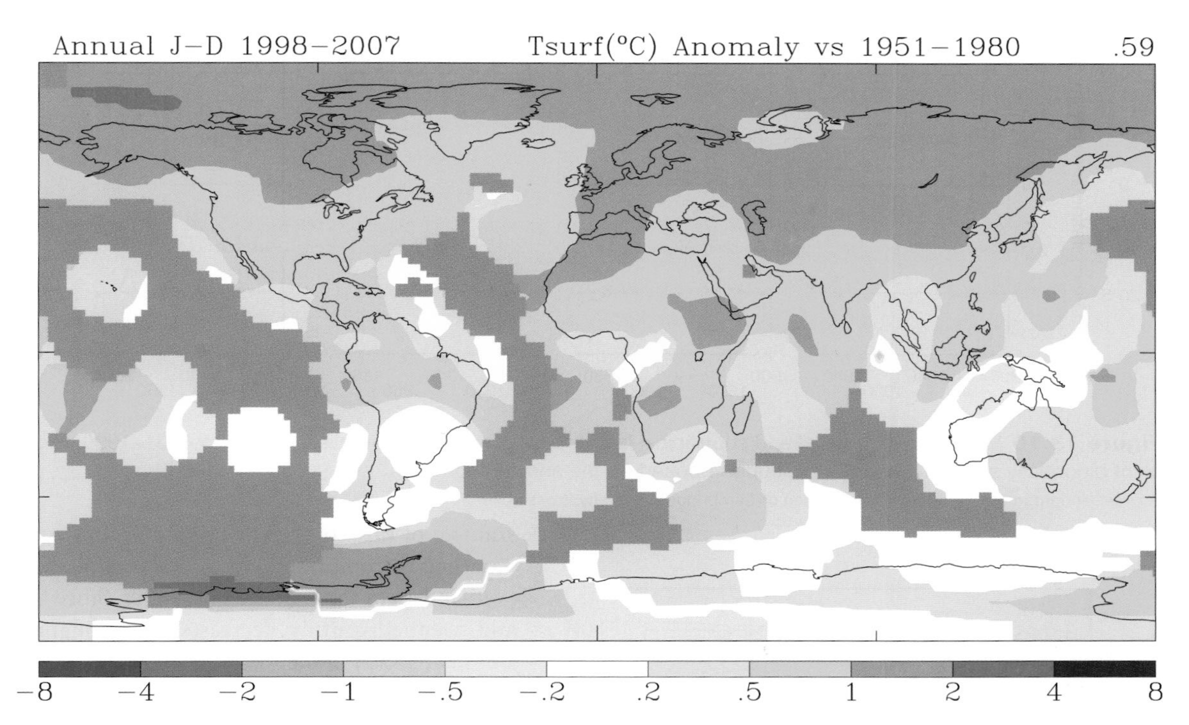

Figure 13.9 Annual mean surface air temperatures for the decade 1998–2007, expressed as anomalies with respect to the base period 1951–1980. Areas in grey have insufficient data to compute anomalies. *Source:* NASA, Goddard Institute for Space Sciences (http://data.giss.nasa.gov/gistemp/).

in autumn and winter and over the Arctic Ocean. It is linked to both changes in atmospheric circulation and declining sea ice extent. Regarding the latter, anomalous areas of open water in autumn and winter allow for large heat fluxes from the ocean surface to the lower atmosphere. Note also the strong warming over the Antarctic Peninsula.

One of the manifestations of recent warming is a longer growing season. For example, in central England, the growing season (defined as daily mean temperature >5°C for five days in succession) lengthened by 28 days over the twentieth century and was about 270 days in the 1990s compared with around 230–250 days in the eighteenth to nineteenth centuries. In the Arctic, there is strong evidence of links between recent warming and regional transitions from tundra to shrub vegetation. A further tendency of the past 50 years or so is a decrease in the diurnal temperature range; night-time minimum temperatures increased by 0.8°C during 1951–1990 over at least half of the northern land areas compared with only 0.3°C for daytime maximum temperatures. This appears to be mainly a result of increased cloudiness, which, in turn, *may* be a response to increased greenhouse gases and tropospheric aerosols. However, the linkages are not yet adequately determined.

Precipitation changes are much more difficult to characterize. The period since 1900 has seen an overall increase in precipitation north of about 30ºN. By contrast, since the 1970s, there have been decreases over much of the tropics and subtropics. However, these general features mask strong seasonal, regional and temporal variations. As an

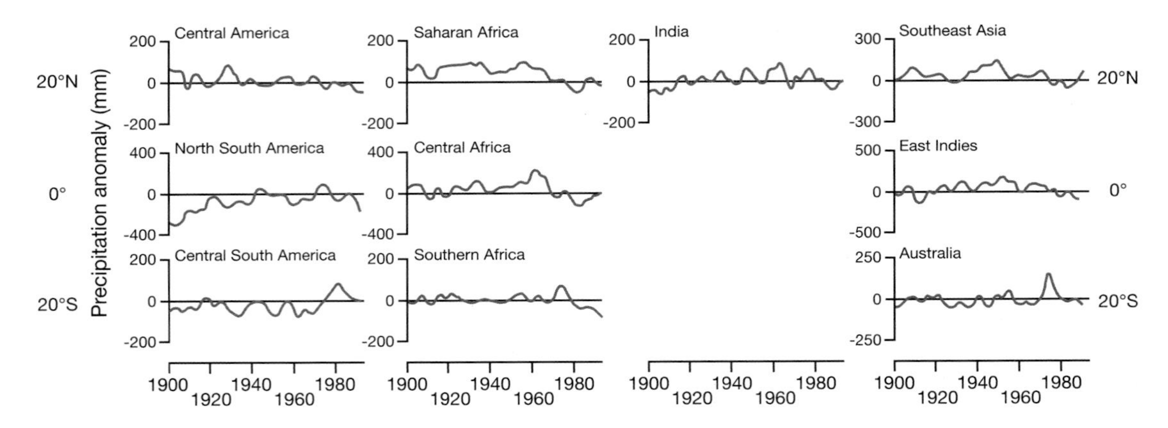

Figure 13.10 Variations of tropical and subtropical land area precipitation anomalies relative to 1961–1990. Nine-point binomially smoothed curves are superimposed on the annual anomalies.

Source: Houghton *et al.* (1996). By permission of Cambridge University Press.

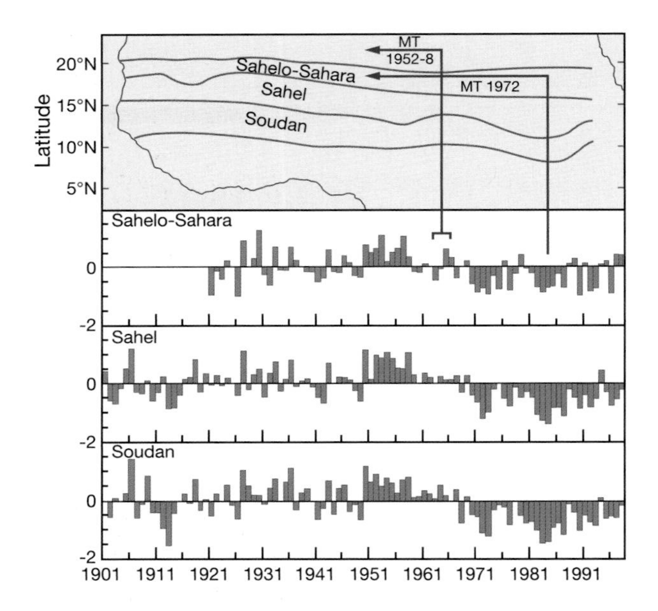

Figure 13.11 Rainfall variations (percentage of standard departures) during 1901–1998 for the Sahelo–Sahara, Sahel and Soudan zones of West Africa. The average positions of the Monsoon Trough (MT) in northern Nigeria during 1952–1958 and in 1972 (El Niño year) are shown.

Source: From Nicholson (2000, p. 2630, fig. 3). Courtesy of American Meteorological Society.

example of this complexity, Figure 13.10 shows variations in tropical and subtropical precipitation over land areas through the mid-1990s. Since the mid-twentieth century, decreases in precipitation dominate much of the region from North Africa eastward to Southeast Asia and Indonesia. Many of the dry episodes are associated with El Niño events. Equatorial South America and Australasia also show ENSO influences. The Indian monsoon area shows wetter and drier intervals; the drier periods are evident in the early twentieth century and during 1961–1990.

As a further example of complexity, West African records for the twentieth century (Figure 13.11) show a tendency for both wet and dry years to occur in runs of up to 10 to 18 years. Precipitation minima were experienced in the 1910s, 1940s and post-1968, with intervening wet years, in all of sub-Saharan West Africa. Throughout the two northern zones outlined in Figure 13.11, means for 1970–1984 were generally less than 50 percent of those for 1950–1959, with deficits during 1981–1984 equal to or exceeding those of the disastrous early 1970s' drought. The deficits continued into the 1990s. It has been suggested that the severe drought is related to weakening of the tropical easterly jet stream and limited northward penetration of the West African southwesterly monsoon flow. However,

Sharon Nicholson attributes the precipitation fluctuations to contraction and expansion of the Saharan arid core rather than to north–south shifts of the desert margin. In Australia, rainfall changes have been related to changes in the location and intensity of subtropical anticyclones and associated changes in atmospheric circulation. Winter rainfall decreased in southwestern Australia while summer rainfall increased in the southeast, particularly after 1950. Northeastern Australia shows decadal oscillations and large inter-annual variability.

Figure 13.12 illustrates winter and summer fluctuations in precipitation for England and Wales. There is wide interannual variability and some large decadal shifts are evident. There are also longer term changes. For example, winters have been wetter from about 1860 onward compared with the earlier part of the record. Changes also depend on season – while winter rainfall increased from 1960 to the end of the record, summer precipitation generally decreased over the same time. Records for individual stations show that even over relatively short distances there may be considerable differences in the magnitude of anomalies, especially in an east–west direction across the British Isles.

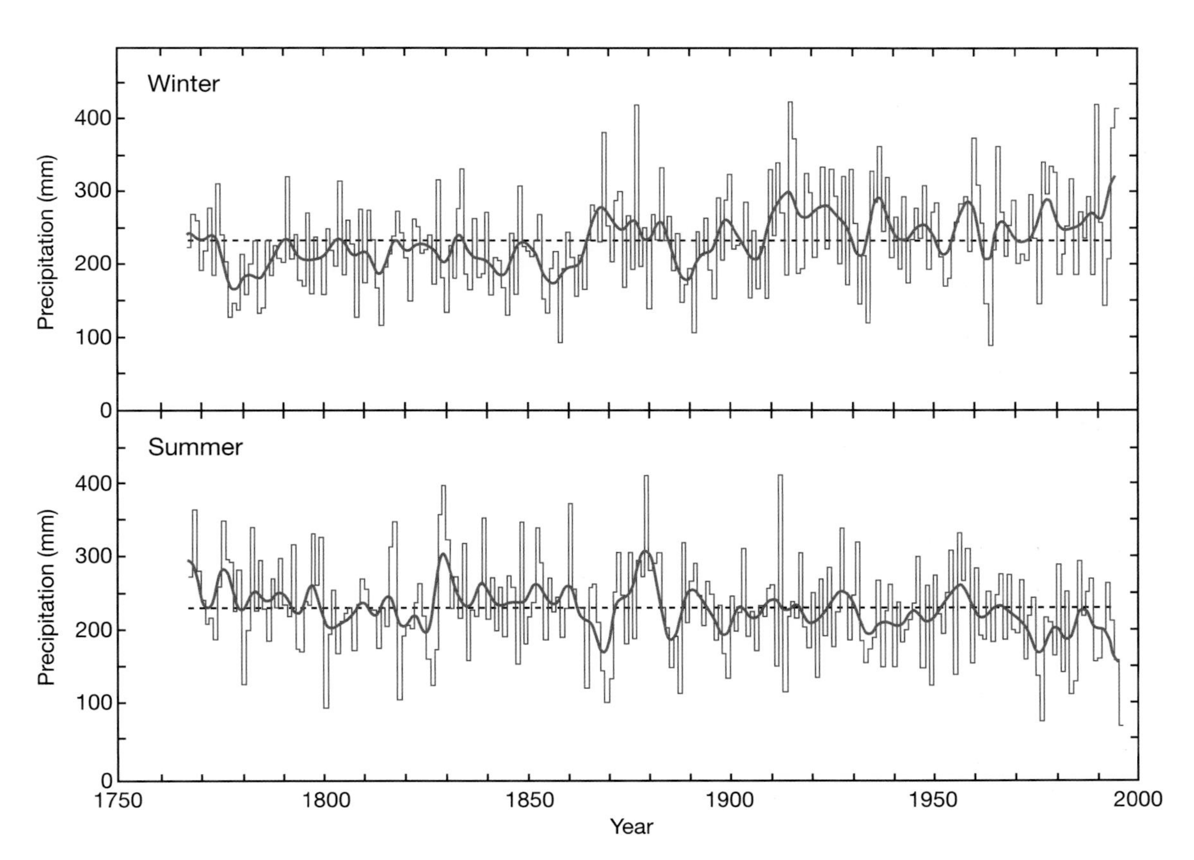

Figure 13.12 Time series of winter and summer precipitation (mm) for England and Wales, 1767–1995. The smooth line is a filter that suppresses variations of ≤10 yr. length.

Source: From P. Jones, D. Conway and K. Briffa (1997, p. 2004, fig. 10.5). By permission of Routledge, London.

The late twentieth and early twenty-first century has seen more frequent climatic extremes. For example, Britain has experienced several major droughts during this period (1976, 1984, 1989–1992 and 1995); seven severe winter cold spells occurred between 1978 and 1987 (compared with only three in the preceding 40 years); and several major windstorms (1987, 1989 and 1990) were recorded. The driest 28-month spell (1988–1992) recorded in Britain since 1850 was followed by the wettest 32-month interval of the twentieth century. Europe experienced unprecedented heatwaves in 2003 and 2008 (see Plate 13.2). In the United States, recent decades saw a marked increase in the inter-annual variability of mean winter temperatures and total precipitation. The year 1983 saw the most intense El Niño event for a century, followed by a comparable event in 1998. There is also some evidence of an increase in the frequency of intense hurricanes (Category 4 and 5).

D UNDERSTANDING RECENT CLIMATIC CHANGE

While the evidence is strong that much of the global warming over the past 100 years is a response to rising concentrations of atmospheric greenhouse gases, we have seen that the global temperature time series is characterized by fluctuations from inter-annual to decadal and even longer timescales (Figure 13.7). As just discussed, variability is in turn very pronounced at regional scales. Regional fluctuations and shorter term global fluctuations may be viewed as expressions of natural climate variability – a term which allows for the influence of non-anthropogenic radiative forcing. It is useful to review some of the causes of recent climate fluctuations embedded in the overall global warming trend, helping to set the stage for more focused discussion of anthropogenic-induced change.

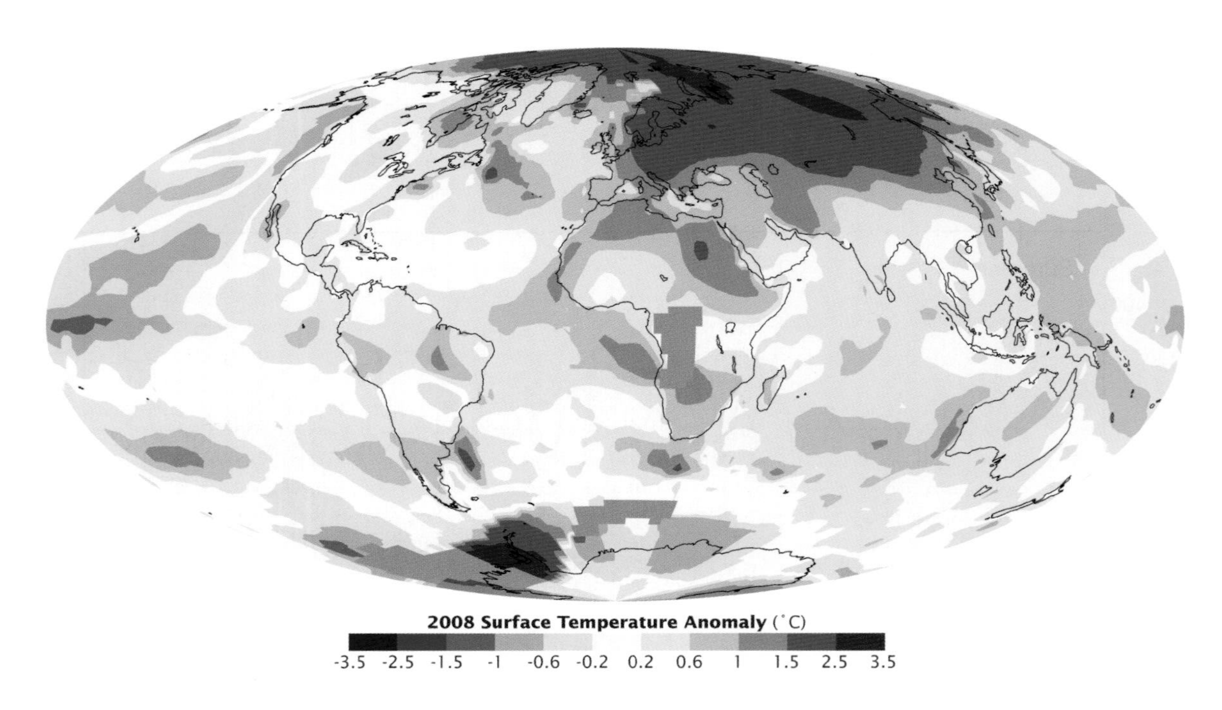

Plate 13.1 Global air temperatures for 2008.

Source: NASA image by Robert Simmon. http://earthobservatory.nasa.gov/IODT/view.php?id=06699.

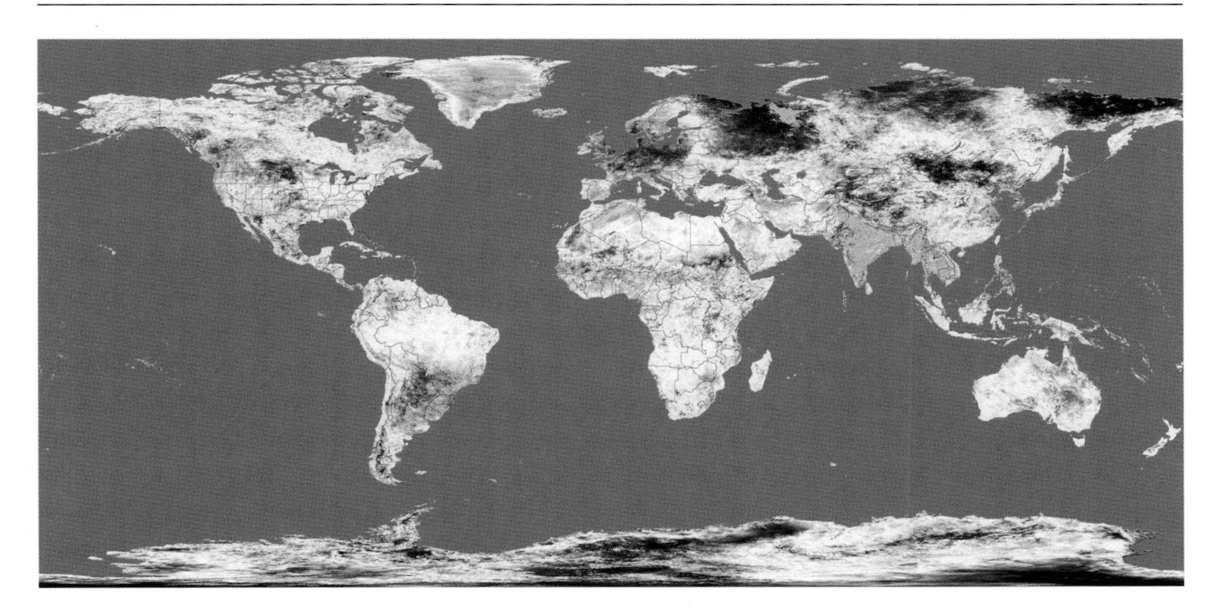

Plate 13.2 Heatwave over Western Europe shown by MODIS land surface temperature anomaly, 20–27 July 2006.
Source: http://earthobservatory.nasa.gov/IODT/view.php?id=M094.

1 Circulation changes

One immediate cause of climatic fluctuations is variability in the global and regional atmospheric circulation and associated heat transports. The first 30 years of the twentieth century saw a pronounced increase in the vigor of the westerlies over the North Atlantic, the northeast trades, the summer monsoon of South Asia and the Southern Hemisphere westerlies (in summer). Over the North Atlantic, these changes consisted of an increased pressure gradient between the Azores high and the Icelandic low as the latter deepened, and also between the Icelandic low and the Siberian high, which spread westward. These changes were accompanied by more northerly depression tracks, and this resulted in a significant increase in the frequency of mild southwesterly airflow over the British Isles between about 1900 and 1930, as reflected by the average annual frequency of Lamb's westerly airflow type (see Chapter 10A.3). For 1873–1897, 1898–1937, 1938–1961 and 1962–1995 the figures are 27, 38, 30 and 21 percent, respectively. Coinciding with the westerly decline, cyclonic and anticyclonic types increased substantially (Figure 13.13). The decrease in westerly airflow during the last 30-year interval, especially in winter, is linked with greater continentality in Europe. These regional indicators reflect a general decline in the overall strength of the mid-latitude circumpolar westerlies, accompanying an apparent expansion of the polar vortex.

Marked climatic fluctuations have occurred in the North Atlantic sector and Eurasia in association with the changing phase of the North Atlantic Oscillation (NAO). The winter NAO was mostly negative from the 1930s to 1970s (with a weak Icelandic Low and Azores High) but then exhibit a general upward trend to strong positive values in the mid-1990s (giving enhanced westerly flow). A sharp rise in winter temperatures over much of northern Eurasia from about 1970 through the mid-1990s can be linked to this shift in the NAO. From the late 1900s through 2007, the winter North Atlantic Oscillation has bounced between positive and negative phases.

Other coupled atmosphere–ocean anomalies may affect climatic trends on a global scale. For example, the occurrence and intensity of the warm

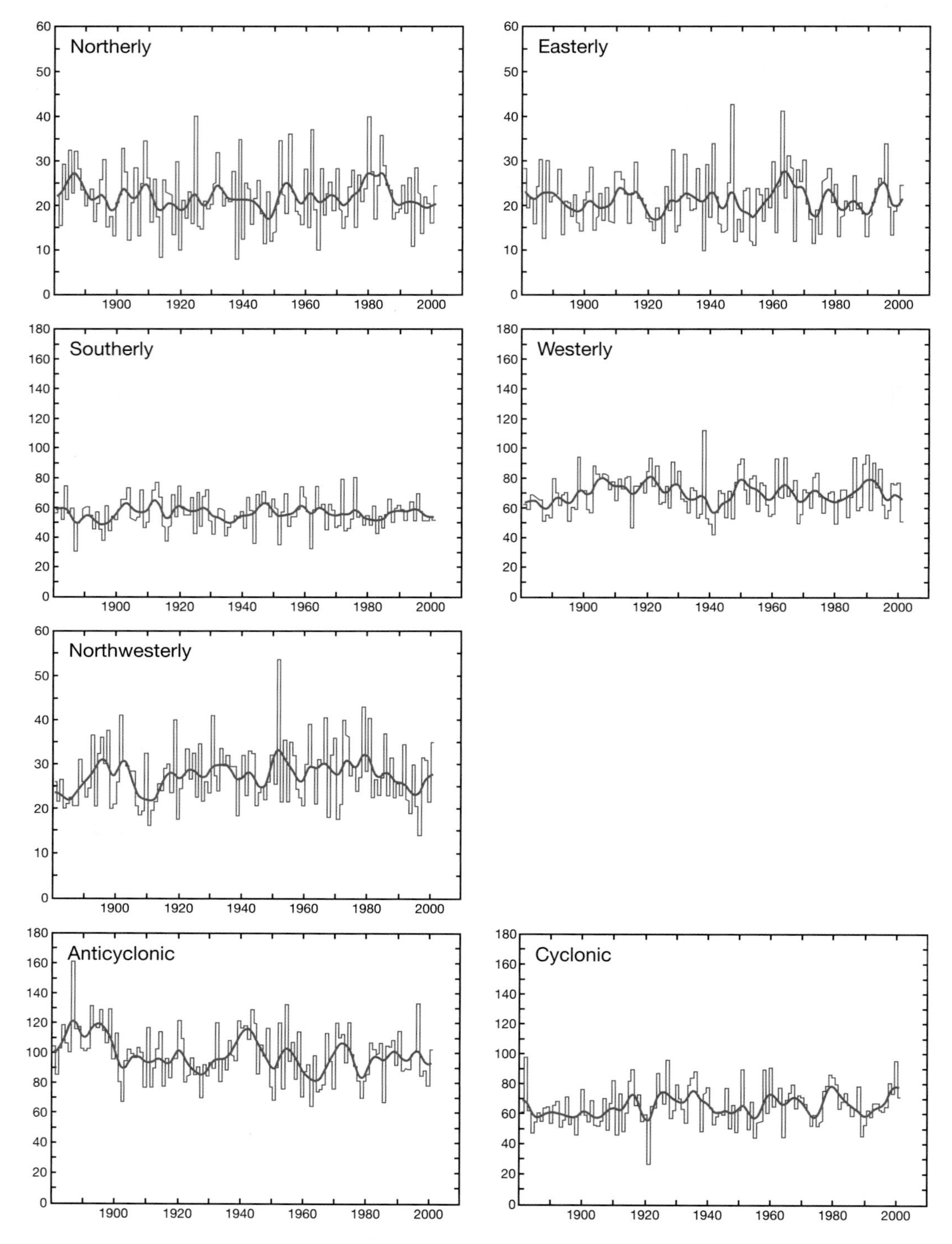

Figure 13.13 Annual totals and ten-year moving averages of the frequency (days) of the Lamb–Jenkinson circulation types over the British Isles, 1961–1999. Note scale changes.

Source: Lamb (1994). Reprinted from the Climate Monitor and from the Climatic Research Unit data with permission of the University of East Anglia.

phases of ENSO events are estimated to have increased the global mean temperature by about 0.06°C during 1950–1998. The very high global average temperature of 1998 can be linked to the strong El-Niño event of the same year.

A long-lived El-Niño-like pattern is the Pacific Decadal Oscillation (PDO) of North Pacific SST variability. It has undergone major 20–30 year cycles – cool (negative) from 1890–1924 and 1947–1976 (West Pacific warm, East Pacfic cool) and warm (positive) from 1925–1946 and 1977–mid-1990s (Wesst Pacific cool, East Pacific warm). Its causes are not yet known. The 15–30 year Interdecadal Pacific Oscillation similarly affects the northern and southern Pacific.

2 Solar variability

The ultimate driver of the climate system is of course the sun. The well-known solar cycle of approximately 11 years is usually measured with reference to the period between sunspot maxima and minima (see Figure 3.2). As shown in satellite records available since 1980 and as discussed earlier, irradiation varies by a modest 1W m^{-2} over the 11-year cycle (the radiation flux averaged across the top of the atmosphere is only 25 percent as large). As pointed out in Chapter 3A.1, the explanation is that sunspot darkening is accompanied by increased emission from faculae that is 1.5 times greater than the darkening effect. The 11-year cycle corresponds to global air temperature fluctuation of <0.1°C.

What of longer term variations? Strong global warming since 1980 cannot be attributed to solar activity, since the satellite data show no discernible trend. On the other hand, based on reconstructions, solar variability may account for perhaps half of the warming between 1860 and 1950. Variations in solar irradiance may also offer a partial explanation of the Little Ice Age.

It is suggested by David Rind of NASA that the direct solar forcing on climate may pale in comparison with the potential for solar forcing to trigger interactions involving a variety of feedback processes. There appear to be regional patterns in the temperature response to solar variability with the largest signals in low latitudes where there are large insolation totals and over oceans where the albedo is low. Hence, maximum responses are likely to occur over eastern tropical ocean areas. Climate model simulations suggest that enhanced solar irradiance during a sunspot maximum, with a corresponding increase of about 1.5 percent in column ozone, modifies the global circulation; the Hadley cells weaken and the subtropical jet streams and Ferrel cells shift poleward. A statistical relationship has also been found between the occurrence of droughts in the western United States over the past 300 years, determined from tree ring data and the approximately 22-year double (Hale) cycle of the reversal of the solar magnetic polarity. Drought areas are most extensive in the two to five years following a Hale sunspot minimum (i.e., alternate 11-year sunspot minima). A clear mechanism is not established, however.

3 Volcanic activity

Links between climate variability and volcanic activity are clear (see Box 13.3). As discussed, surface cooling driven by increased amounts of volcanic dust and sulfate aerosols in the stratosphere occurs one to two years after major explosive events. The effects of the eruption of Mt. Pinatubo in the Philippines in June 1991 (Plate 13.3) may be seen in Figure 13.7A as lower global average temperatures in 1992 and 1993 compared to surrounding years. Regionally the impacts were larger. Surface temperatures over the northern continents were up to 2°C below average in summer 1992 but, owing to impacts on atmospheric circulation patterns, up to 3°C above average in the winters of 1991–1992 and 1992–1993 (see Box 13.3). As noted earlier, given the short timescale of the forcing, prolonged cooling would require a chain of eruptions events; such a series of events may help to explain the 'Little Ice Age'. The period 1883–1912 also saw frequent eruptions (see Figure 2.12). Conversely, reduced volcanic activity after 1914 may have contributed in part to the early twentieth-century warming.

SIGNIFICANT 20TH-CENTURY ADVANCE

13.3 Volcanic eruptions and climate

The eruption of Krakatoa in Indonesia in 1883 demonstrated the global significance of large explosive events in andesitic volcanic cones. The eruption, which injected dust and sulfur gases into the stratosphere, was followed by cool conditions and dramatic red sunsets around the world. However, after the 1912 eruption of Katmai in the Aleutian Islands there was a lull in volcanic activity until Agung in Bali erupted in 1963. Plumes from equatorial eruptions can disperse into both hemispheres whereas those in middle and high latitudes cannot be transferred equatorward due to the upper circulation structure. Non-explosive eruptions of basaltic shield volcanoes of the Hawaiian type do not inject material into the stratosphere.

Volcanic aerosols are often measured in terms of a dust veil index (DVI), first proposed by H. H. Lamb, that takes account of the maximum depletion of monthly average direct incoming radiation measured in the mid-latitudes of the hemisphere concerned, the maximum spatial extent of the dust veil, and persistence of the dust veil. However, this cannot be directly calculated for historical events.

The largest DVI values are estimated for 1835 and 1815–1816. Volcanologists use a Volcanic Explosivity Index (VEI) to rank eruptions on a scale of 0–8. El Chichon (1982) and Agung (1963) are rated 4, but the index may not necessarily be a good indicator of climatic effects.

The surface cooling agents include both the transformation of sulfur dioxide (a gas) into sulfuric acid droplets (a reflective aerosol) and microparticles of dust that absorb solar radiation in the stratosphere (large particles rapidly settle out). Increased acidity in the snow falling on ice sheets can be measured by determining the signal of electrical conductivity in an ice core. This yields records of past eruptions.

Global average temperatures the year following a major eruption may be reduced by several tenths of a degree C, but impacts can be much larger at hemispheric and regional scales. Dramatic evidence of such effects was provided by the 'year without a summer' in 1816, following the

Plate 13.3 The first major eruption of Mt. Pinatubo on 12 June 1991. Mt. Pinatubo is located on the southwestern part of the island of Luzon in the Philippines. Prior to 1991, it had been dormant for more than 635 years.

Source: Photo credit, K. Jackson, U.S. Air Force. NOAA/NGDC.

eruption of Tambora in 1815. This seriously impacted upon societies in many parts of the world. However, it also followed a series of cold winters in Europe. The eruption of Krakatoa, Indonesia, in August 1883 was recorded by barographs around the world. Ash was propelled up to 80km altitude. Average global temperatures experienced a 1.2°C cooling in 1884 and the effects persisted for 3–4 years. It appears that repeated major eruptions are required in order for there to be long-term climatic effects. Ice core records provide long histories of volcanic eruptions through the late Pleistocene and do show episodes of more frequent eruptions.

4 Anthropogenic factors

As introduced earlier, the effects of human activities are best viewed in the framework of global radiative forcing, which refers to the amount by which a factor alters the radiation balance at the top of the atmosphere and is expressed in units of W m^{-2}.

Figure 13.14 summarizes the different components of radiative forcing in 2005 relative to 1750. Changes in atmospheric greenhouse gas concentrations associated with the explosive growth of world population, industry and technology have been described in Chapter 2A.2. The largest single positive radiative forcing is from the increased concentration of carbon dioxide (about 1.7W m^{-2}). This means that compared to 1750, the increase in carbon dioxide, considered by itself, would lead to a radiation imbalance of this amount. Methane (CH_4), nitrous oxide (N_2O) and halocarbons together contribute about another 1W m^{-2}. Hence the total radiative forcing from long-lived greenhouse gases (long-lived in that they have a long residence time in the atmosphere) is about 2.2W m^{-2}. Halocarbons is

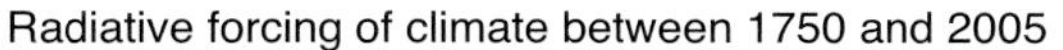

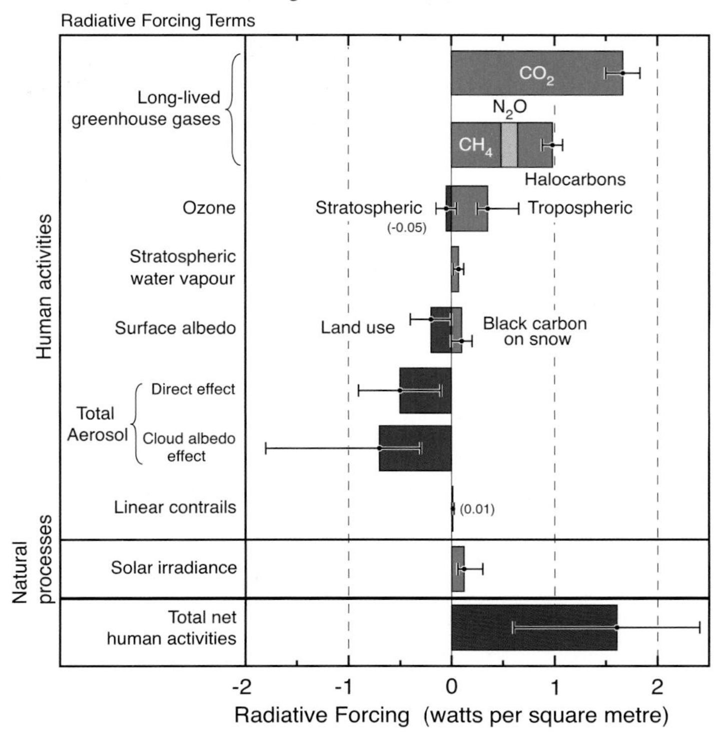

Figure 13.14 Components of global radiative forcing (Wm^{-2}) for the year 2005, expressed as relative to the year 1750. The error bars indicate uncertainty ranges.

Source: IPCC (2007). Reproduced by permission of the IPCC (ch. 2, Changes in atmospheric constituents and in radiative forcing, Report of WG1 1, IPCC, p. 136, FAQ 2.1, fig. 2).

a collective term for the group of partially halogenated organic species, and includes the chlorofluorocarbons (CFCs). Other smaller factors with a positive radiation forcing include tropospheric ozone, black carbon on snow (essentially soot from fossil fuel burning) and solar irradiance (which is of course not associated with human activities). These positive forcings are in part countered by negative forcings due to increased aerosol concentration and increased surface albedo associated with land use, yielding the estimated total forcing due to human activities of about 1.6W m^{-2}. The uncertainty in this value is largely due to uncertainty in the aerosol effects. Because of their highly episodic nature, Figure 13.4 does not include the effects of volcanic eruptions.

While CFCs (one of the halogens) have a positive radiative forcing, the student may be more familiar with the link between CFCs and the destruction of stratospheric ozone. Despite the Montreal Protocol that has helped to control the production and use of CFCs, CFCs are long-lived and are still impacting upon the ozone layer (see Chapter 2A.4). Emissions of H_2O and NOx by jet aircraft and by surface emissions of N_2O are contributing to the problem. Ozone circulates in the stratosphere from low to high latitudes and thus the occurrence of ozone in polar regions is diagnostic of its global concentration. In October 1984, an area of marked ozone depletion (termed the 'ozone hole') was observed in the lower stratosphere (i.e., 12–24km) centered on, but extending far beyond, the Antarctic continent. Ozone depletion is always greatest in the Antarctic spring, but in that year the ozone concentration was more than 40 percent lower than in October 1977. By 1990, Antarctic ozone concentrations had fallen to about 200 Dobson units in September to October (see Figure 2.9), compared with 400 units in the 1970s. In the extreme years (1993–1995), record minima of 116 DU have been recorded at the South Pole. It has been estimated that, owing to the slowness of the global circulation of CFCs and of their reaction with ozone, even a cut in CFC emissions to the level of that in 1970 would not eliminate the Antarctic ozone hole for at least 50 years. Winter ozone depletion also occurs in the Arctic stratosphere and was well marked in 1996 and 1997, but absent in 1998. Localized mini-holes are fairly common, but extensive holes are rare even in cold stratospheric winters. It seems that whereas the Antarctic vortex is isolated from the mid-latitude circulation, the Arctic vortex is more dynamic so that transport of ozone from lower latitudes makes up much of the loss.

Aerosol forcings are both direct and indirect. Together, they have an estimated radiative forcing of about –1.2W m^{-2}. The direct effects relate to how aerosols absorb and scatter both solar and longwave radiation; a variety of aerosol types, including fossil fuel organic carbons, fossil fuel black carbon, biomass burning and mineral dust and sulfate aerosols, exert a significant radiative forcing. The indirect effect relates to how aerosols alter clouds. A key issue is how effectively an aerosol particle can act as a cloud condensation nucleus, which depends on factors such as the chemical composition and size of the aerosol. The indirect aerosol effect includes impacts on cloud albedo (often termed the first indirect effect) and impacts on cloud liquid water, height and lifetime (the second indirect effect). Reducing the high uncertainty in the direct and indirect effects of aerosols is a key focus area of climate research.

Regarding land use, the basic issue is that increasing population pressures have led to overgrazing and forest clearance, acting to increase the planet's surface albedo. While the radiative forcing relative to 1750 is a modest – 0.2Wm^{-2}, human effects on vegetation cover have a long history. Burning of vegetation by Aborigines in Australia has been traced over the last 50,000 years, while significant deforestation began in Eurasia during Neolithic times (about 5000 years ago), as evidenced by the appearance of agricultural species and weeds. Deforestation expanded in these areas between about AD 700 and 1700 as populations slowly grew, but it did not take place in North America until the westward movement of settlement in the eighteenth and nineteenth centuries. During the past half-century extensive deforestation has occurred in the tropical rainforests of Southeast Asia, Africa and South America. Estimates of current tropical deforestation suggest losses of 105km^2/year out of a total tropical forest area of 9×10^6km^2. This annual figure is more than half the total land surface currently under irrigation. Forest destruction causes an increase in albedo of about 10 percent locally, with consequences for surface energy and moisture budgets.

It should be noted that deforestation is difficult to define and monitor. It can refer to loss of forest cover with complete clearance and conversion to a different land use, or species' impoverishment without major changes in physical structure. The term desertification, applied in semi-arid regions, creates similar difficulties. Desertification also contributes to an increase of wind-blown soil. The 'dust-bowl' years of the 1930s in the United States and the African Sahel drought since 1972 illustrate this, as well as dust transported from western China across the Pacific to Hawaii, and from the Sahara westward across the North Atlantic. The process of vegetation change and associated soil degradation is not solely attributable to human-induced changes, as it can be triggered by natural rainfall fluctuations leading to droughts.

Deforestation and associated biomass burning has also contributed to rising carbon dioxide concentrations. Forests store great amounts of carbon, and left alone buffer the carbon dioxide cycle in the atmosphere. The carbon retained in the vegetation of the Amazon basin alone is equivalent to at least 20 percent of the entire atmospheric carbon dioxide load. Deforestation and biomass burning in the Amazon and elsewhere is estimated to account for about 25 percent of the increase in atmospheric carbon dioxide since pre-industrial times.

E PROJECTIONS OF TEMPERATURE CHANGE THROUGH THE TWENTY-FIRST CENTURY

1 Applications of General Circulation Models

The most powerful tools for examining the emerging signatures of climate change and projecting changes through the twenty-first century are General Circulation Models (GCMs), the most sophisticated being those that are fully coupled to the ocean, known as Atmosphere–Ocean GCMs, or AOGCMs. As outlined in Chapter 8, such global models are based on detailed mathematical representations of the structure and operation of the earth–atmosphere–ocean system. The possible future (as well as past) states of the system can be simulated by applying assumed climate forcings, such as greenhouse gas concentrations, solar irradiance and (in the case of paleoclimate studies) ice sheet extent and topography. GCMs are very powerful but involve the need for a thorough understanding of the variables, states, feedbacks, transfers and forcings of the complex system, together with the laws of physics of the atmosphere and oceans on which they are based.

2 The IPCC simulations

GCMs have been developed by modeling groups all over the world. The Intergovernmental Panel on Climate Change (IPCC) has served as a key focal point for model development. As introduced in Chapter 1, a goal of the IPCC is to assess the impacts of projected increases in greenhouse gas concentrations and other anthropogenic climate forcings through the twenty-first century. The IPCC has issued four comprehensive reports, (in 1990, 1995 and 2001 and 2007), each making use of increasingly sophisticated models.

Models used in the First Assessment Report (1990) were primitive by today's standards. Only two models, from NCAR and GFDL (respectively, the National Center for Atmospheric Research and the Geophysical Fluid Dynamics Laboratory) included ocean coupling (i.e., could be classified as AOGCMs). Others employed a 'slab' mixed layer (top 50m or so) ocean and with various other simplifications, such as no horizontal ocean heat transport, prescribed horizontal ocean heat transport, and fixed zonally averaged cloud cover. Horizontal resolution (model grid cell size) was coarse, typically in the order of 500km. Simulations included climate responses to transient increases (1 percent per year) in CO_2 (with the NCAR and GFDL models) and equilibrium experiments for a doubling of CO_2 (in which models were run until an equilibrium climate state was reached). By the Second Assessment Report (1995), typical horizontal resolution had increased to about 250km, and ocean coupling had been improved. Additional refinements included treatment of the radiative effects of anthropogenic sulfate aerosols and volcanic eruptions. Eleven groups with 11 AOGCMs participated. Simulations included transient increases in CO_2 of 1 percent per year as well as other greenhouse gas change scenarios. By the Third Assessment Report (2001), horizontal resolutions had been further increased, with more robust treatment of the ocean (e.g., of overturning circulations) and land–surface interactions. Nineteen AOGCMs participated. Models used for the Fourth Assessment Report (2007) were even more mature, with some including atmospheric chemistry and interactive vegetation. A total of 23 AOGCMs were evaluated, representing the work of 16 modeling groups from 11 countries.

An important feature of the third and fourth Assessment Reports is that simulations with the different models used a range of greenhouse gas emission scenarios (contained in a Special Report on Emission Scenarios, or SRES) based on differing views of the global future. This was a major advance over simply assuming a 1 percent per year growth rate or a doubling of CO_2. One set of emissions scenarios (A1) assumes rapid economic growth, global population peaking in mid-century and then declining, and the introduction of more efficient technologies. Three variants are: A1F1, fossil fuel intensive; A1T, non-fossil energy sources; and A1B, a balance across all energy sources. Scenario A2 considers global heterogeneity, increasing population, and fragmented and slower technological change. A second set includes B1 where population trends are as in A1, but the global economy is service and information based, with clean, resource-efficient technologies. B2 envisages slower population increase, intermediate levels of economic development, and diverse, regionally oriented technological change. Of all these scenarios, A1B (often termed 'Business as Usual', or BAU) has been the most widely examined.

Figure 13.15 shows projected changes in the concentrations of CO_2, CH_4 and CFC-11 through the twenty-first century, based on four of the scenarios. Depending on the scenario, CO_2 concentrations are projected to rise to between 540 and 970ppm by 2100, corresponding to increases of 90 and 250 percent above the pre-industrial level. Methane concentration changes will range between – 190ppb and + 1970ppb above 1998 levels by 2100. In 1995 it was estimated that to stabilize the concentration of greenhouse gases at 1990 levels would require the following percentage reductions in emissions resulting from human activities: CO_2 >60 percent; CH_4 15–20 percent; N_2O 70–80 percent; CFCs 70–85 percent. The 2001 IPCC report notes that to stabilize CO_2 concentrations at 450 (650)ppm would require anthropogenic emissions to drop below 1990 levels within a few decades (about a century). Given the strong growth in emissions since 2001, even stronger reductions would be needed today.

The associated projected increases in anthropogenic radiative forcing (relative to pre-industrial conditions) corresponding to the SRES cases of Figure 13.15 are shown in Figure 13.16. The projected range is 4 to 9W m^{-2} by 2100. Aerosol impacts would reduce these numbers

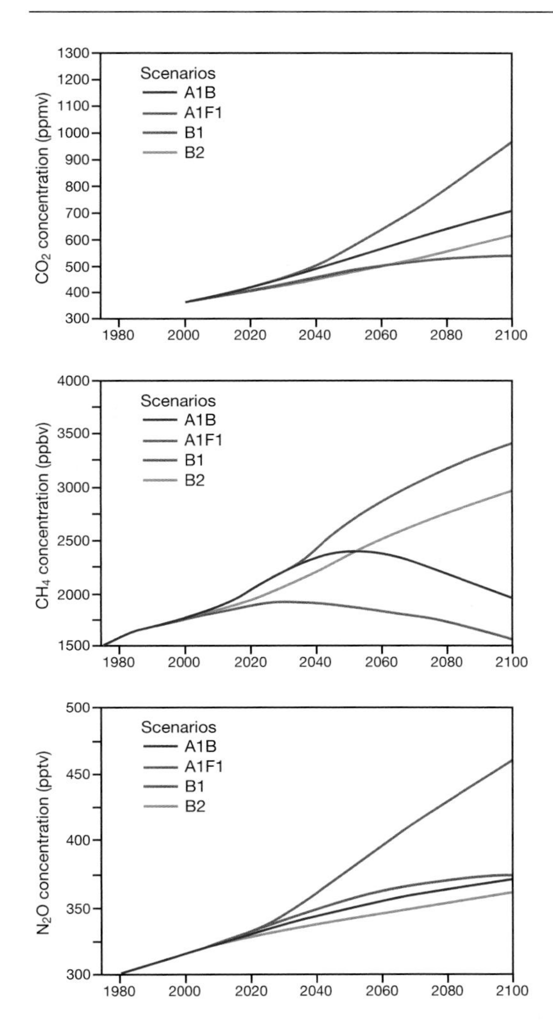

Figure 13.15 Predicted changes of CO_2, CH_4 and N_2O between 1980 and 2100 with scenarios from the Special Report on Emission Scenarios (SRES). A1F1, A1B and B1 (see text).

Source: Adapted from Houghton *et al.* (2001). Reproduced by permission of the IPCC (Summary for Policymakers, Report of WG 1, IPCC, p. 65, fig. 18).

Note: Units are in parts per million by volume (ppmv), parts per billion (ppbv) and parts per trillion (pptv), respectively.

somewhat. Recall that the estimated anthropogenic forcing in 2005 was 1.6W m^{-2}.

Figure 13.18 summarizes projected changes in global annual mean surface air temperature from 1900 through 2100 based on models participating in the Fourth Assessment Report. Through the end of the twentieth century, the simulations are driven with the best available estimates of observed radiative forcings (in particular, changes in greenhouse gas concentrations). Starting in the twenty-first century, the runs make use of forcings based on different SRES emissions scenarios, including a scenario of greenhouse gas concentrations held constant at levels observed for the year 2000. Results are shown for the multi-model mean (averaging outputs from the different models together) and for the ± 1 standard deviation of the individual model annual averages. Given that different models have different architectures, parameterizations and level of complexity, in turn contributing to differences in their climate sensitivity, use of the multi-model mean is viewed as giving a more robust projection of change rather than the output from any single model. The spread of projections from the different models (in the case of Figure 13.17 based on the ± 1 standard deviation) for each emissions scenario may be viewed as defining an envelope of uncertainty reflecting differences in model architectures and physics, which in turn impact on their climate sensitivity.

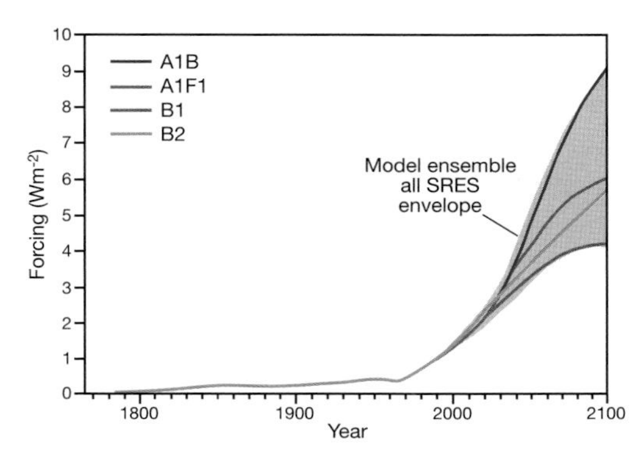

Figure 13.16 Model projected radiative forcing (Wm^{-2}) for the emission scenarios shown in Figure 13.17.

Source: Adapted from Houghton *et al.* (2001). Reproduced by permission of the IPCC. (Summary for Policymakers, Report of WG 1, IPCC, p. 66, fig. 19).

Based on the multi-model mean, and expressed with respect to the base period 1980–1999, the global mean temperature by the year 2100 is

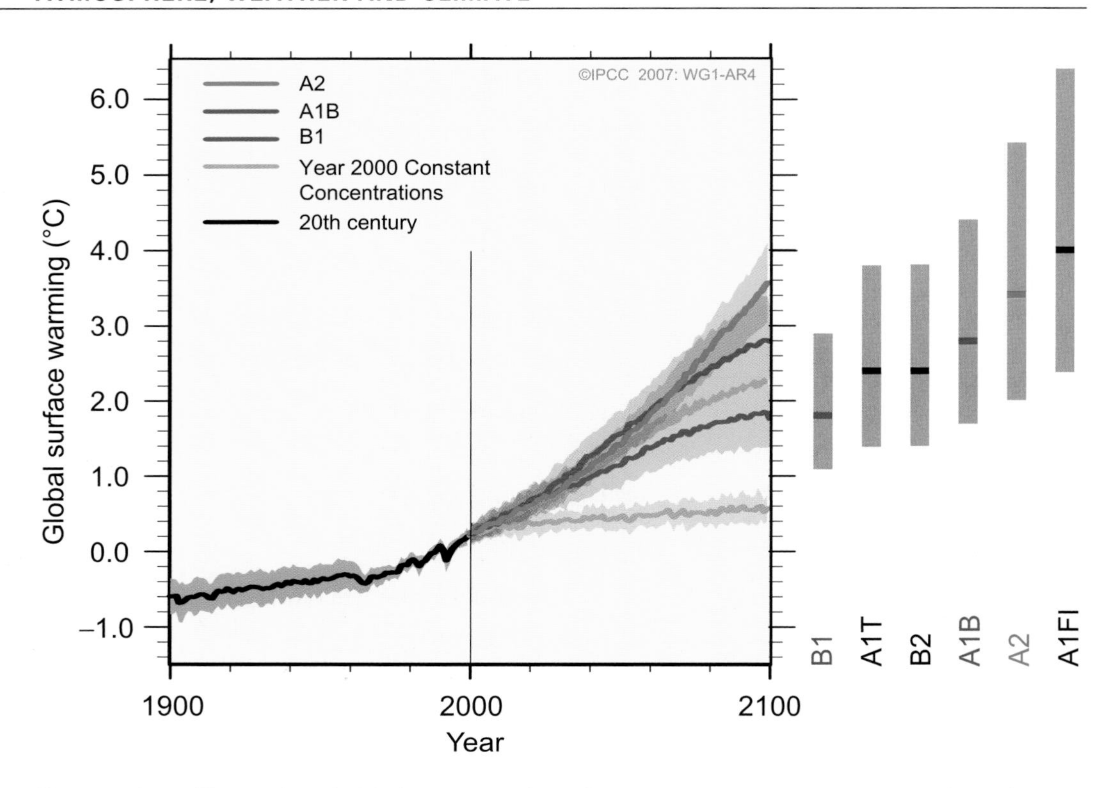

Figure 13.17 Time series of global average surface air temperature, expressed as anomalies relative to the base period 1980-1999, as simulated by global climate models participating in the IPCC Fourth Assessment Report. Results for the twentieth century are based on observed radiative forcings, projections for the twenty-first century employ different emissions scenarios. The solid lines represent the multi-model means, while the shading indicates the spread between different models based on the +/– 1 standard deviation.

Source: IPCC (2007). Reproduced by permission of the IPCC (Summary for Policymakers, Report of WG1 1, IPCC, p. 14, fig. SPM.5).

expected to have increased by 1.8°C (B1 scenario) to 4.1°C (A2 scenario). It is important to note that as time progresses, the uncertainty in greenhouse gas emissions (the range in projections from the different scenarios) starts to become of increasing importance relative to the range between simulations from different models for a given scenario. Phrased differently, uncertainty as to how much warmer it will be in 2100 is more a function of uncertainties in human behavior than uncertainties in how well the climate system can be modeled. If one assumed that greenhouse gas concentration could have been kept at levels for the year 2000, a small warming would occur over the next two decades. This warming is essentially the heat 'in the pipeline' that would ensue as the system comes into radiative equilibrium with the radiative forcing for the year 2000.

As is clear from Figure 13.18, the magnitude of projected surface warming has distinct spatial patterns which stay largely consistent through the twenty-first century. The expectation is for amplified warming, relative to the global mean, over the north polar region (a pattern already seen in observed trends: see Figure 13.9). Building on earlier discussion, this in large part reflects loss of the Arctic's sea ice cover. For most of the year, sea ice acts to insulate a relatively warm Arctic Ocean

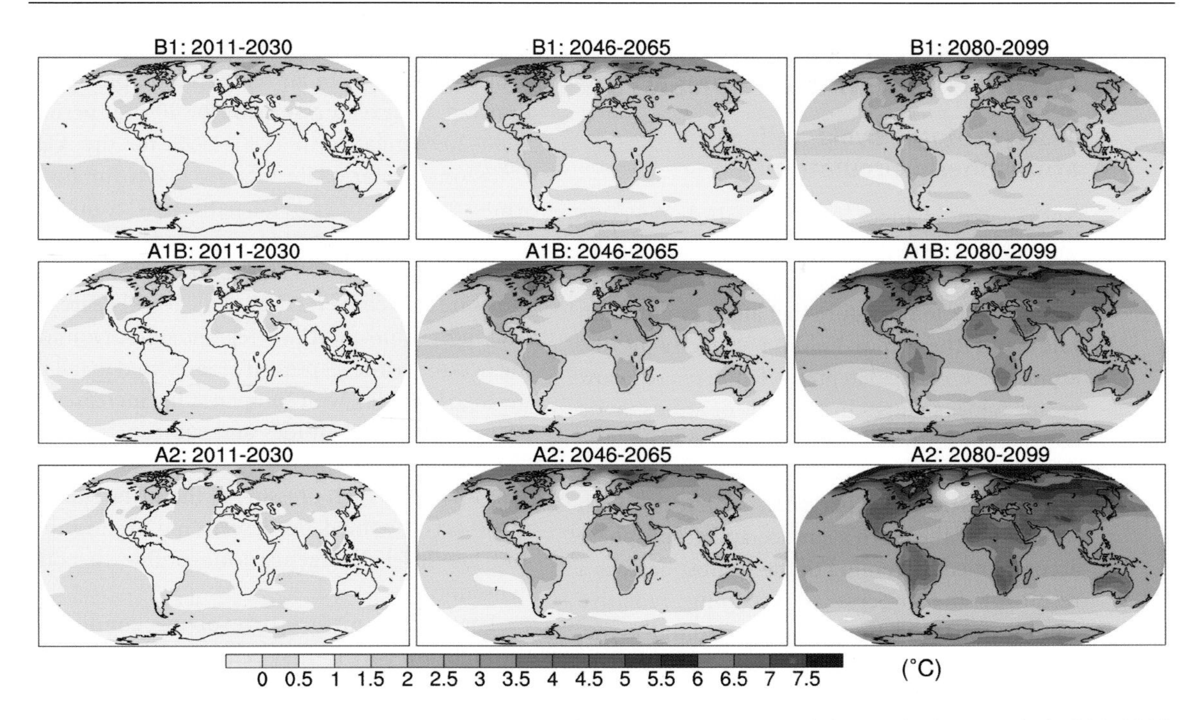

Figure 13.18 Projected changes in annual mean surface air temperature, relative to the base period 1980–1999, for the 20-year periods 2011–2030, 2046–2065 and 2080-2099. Results are given for the B1 (top), A1B (middle) and A2 (bottom) emissions scenarios based on global climate models participating in the IPCC Fourth Assessment Report. The maps represent the multi-model average.

Source: IPCC (2007). Reproduced by permission of the IPCC (ch. 10, Global climate projections, Report of WG1 1, IPCC, p. 766, fig. 10.8).

from a much colder atmosphere. However, as the climate warms, the summer melt season lengthens and intensifies, leading to less sea ice at summer's end. Summertime absorption of solar radiation in open-water areas increases, raising the heat content of the ocean mixed layer. Sea ice formation in autumn is hence delayed, and the ice that grows is thinner than it once was. This results in large upward heat fluxes from the ocean to the atmosphere during autumn and winter. A longer snow-free season over land (meaning stronger absorbtion of solar radiation and hence stronger heating of the lower atmosphere) contributes to the amplification effect. While the results in Figure 13.18 are for the multi-model average, Arctic amplification is a feature of all models. Note that projected warming in the Antarctic is not as great. This largely manifests the different nature of the ocean circulation in southern high latitudes. In the Arctic, the upper ocean is very stably stratified, such that ocean heat gained in summer stays near the surface to help melt sea ice (and delay autumn ice growth). By contrast, heat absorbed at the ocean surface in high southern latitudes is rapidly mixed to deeper ocean levels. Another interesting feature of Figure 13.18 is that, given the large thermal inertia of the ocean, there is a general pattern of smaller warming over the ocean than land. Finally, note the distinct region of only small warming over the northern North Atlantic in the projections for 2046–2065 and (even more clearly) for 2080–2099. This manifests projected slowing of poleward ocean heat transport via the Atlantic Meridional Overturning Circulation.

F PROJECTED CHANGE IN OTHER SYSTEM COMPONENTS

1 Hydrologic cycle and atmospheric circulation

Anticipated changes in the hydrologic cycle through the twenty-first century must consider complex interactions between rises in surface and tropospheric temperature that affect evaporation rates and the vapor-holding capacity of the atmosphere, changes in precipitation phase (snow versus rain), changes in patterns of atmospheric convection and shifts in circulation from the synoptic to global scale. Given anticipated changes in the vertical temperature structure (with warming of the earth's surface and troposphere accompanied by cooling of the stratosphere, as a result of the process of attaining radiative equilibrium) and the strong horizontal asymmetry in warming patterns such as shown in Figure 13.18, shifts in atmospheric circulation should come as no surprise. As an illustration of some of this complexity, Figure 13.19 summarizes projected changes in air temperature, precipitation and sea-level pressure for the 20-year period 2080–2099, relative to 1980–1999, for the A1B emissions scenario. Results are given for both winter and summer. The pattern is complex. Precipitation is expected to increase in high latitudes and along the ITCZ (pointing to stronger moisture flux convergence) while it will decrease over most subtropical land regions and surrounding oceans. Sea-level pressure is expected to fall in the high latitudes, with compensating increases in parts of the mid-latitudes and subtropics where precipitation amounts are expected to decline. This follows as high pressure at the surface tends to be accompanied by descending air motion and low-level divergence unfavorable to condensation.

The overall picture of evolving conditions through the twenty-first century, based on the IPCC models and results from other studies, includes:

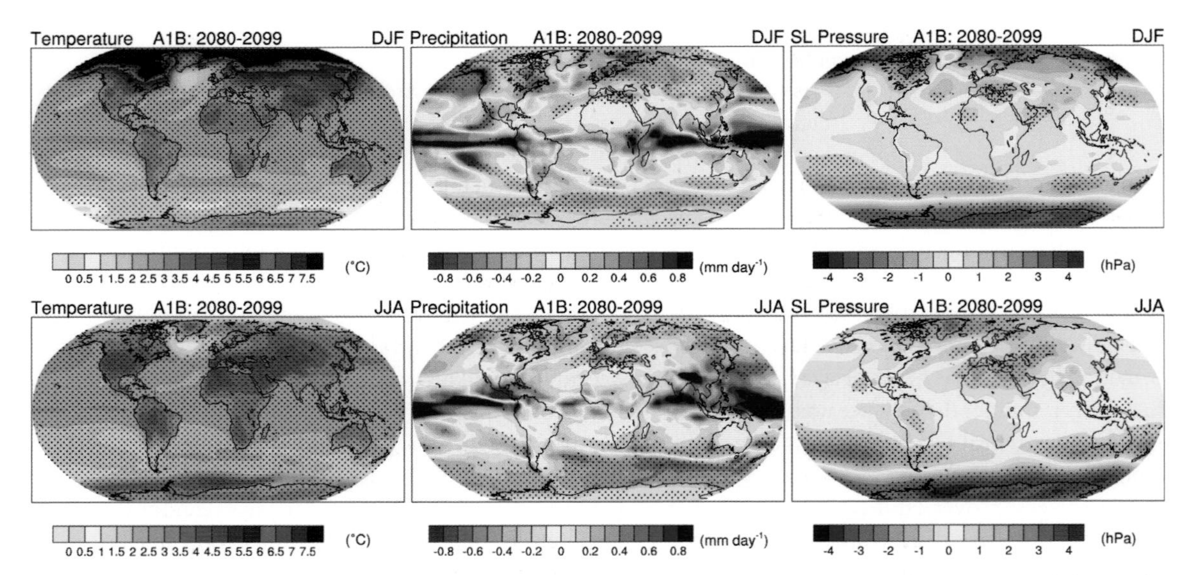

Figure 13.19 Projected changes in surface air temperature, precipitation and sea-level pressure, relative to the base period 1980–1999, for the 20-year period 2080–2099. Results are given for both winter and summer using the A1B emissions scenario, based on global climate models participating in the IPCC Fourth Assessment Report. The maps represent the multi-model average.

Source: IPCC (2007). Reproduced by permission of the IPCC (ch. 10, Global climate projections, Report of WG1 1, IPCC, p. 767, fig. 10.9).

1. A more vigorous global hydrological cycle overall.
2. More severe droughts and/or floods in some places and less severe ones in others.
3. An increase in precipitation intensities with possibly more extreme rainfall events.
4. Greater hydrological effects of climate change in drier areas than in wetter ones.
5. An overall increase in evaporation.
6. An increase in the variability of river discharges along with that of rainfall.
7. A shift of peak snow-melt runoff to earlier in spring as temperatures rise.
8. The greatest falls in lake water levels in dry regions with high evaporation.
9. More intense tropical cyclones (still controversial as of writing).

Note that projections of changes in the hydrologic cycle and atmospheric circulation are especially uncertain on the regional scale and the scales important to human affairs. The hydrological impacts of climate change may be greatest in currently arid or semi-arid regions, implying that the more severe runoff events there will particularly exacerbate soil erosion.

2 Global sea-level

The mechanisms influencing global sea-level are complex and operate over a broad spectrum of timescales. Over timescales of millions of years, one must consider issues such as plate tectonics which alter the shape and size of ocean basins, and the effects of erosion that slowly fill ocean basins with sediment. Moving to timescales of thousands to tens of thousands of years, we know that following the Last Glacial Maximum, global sea-level rose rapidly as the major ice sheets of North America and Northern Europe melted. By 6000 years ago, around the Holocene Thermal Maximum, sea-level had risen by about 120m from the glacial low stand. Sea-level then stabilized around 2000–3000 years ago, not changing significantly until the late nineteenth century, when, as climate warmed, it began to slowly rise. Based on the IPCC Fourth Assessment Report, the best estimate is that global sea-level rose at about 1.7mm/year over the twentieth century, but faster in recent decades. Satellite altimeter data point to a value of around 3mm/year since 1993.

The primary contributions to sea-level rise over the period 1961–2003 and from 1993–2003, based on available estimates compiled for the IPCC Fourth Assessment Report, are summarized in Figure 13.20. These include:

1. Thermal expansion of ocean waters. The upper ocean has warmed, and warmer water occupies a larger volume per unit mass than colder water.
2. The melting of glacier and ice caps, which is transferring water from terrestrial storage into the ocean.
3. Mass loss from the Greenland and Antarctic ice sheets, also transferring water from the land to the ocean For the Greenland ice sheet, this includes contributions from both liquid water runoff and the process of iceberg discharge (calving). For Antarctica, calving dominates.

All of the individual terms are larger over the second period, especially thermal expansion (0.42mm/year for 1961–2003 versus 1.6mm/year for 1993–2003) which is the largest contributor in the later period. For both periods, the observed sea-level rise exceeds the change assessed from adding estimates for the individual components. The causes of this discrepancy remain to be resolved. Effects of human impoundments offer no explanation, as impoundments have a negative impact on sea-level. Note that both the observations and estimates for the different components contain substantial uncertainty. Difficulties in estimating the magnitude of thermal expansion include the lack of knowledge of deep ocean temperature changes and the effects of oceanic circulations. Uncertainty in the ice sheet contributions include uncertainties in accumulation (by snowfall) and ice thickness at the grounding line where ice sheets float.

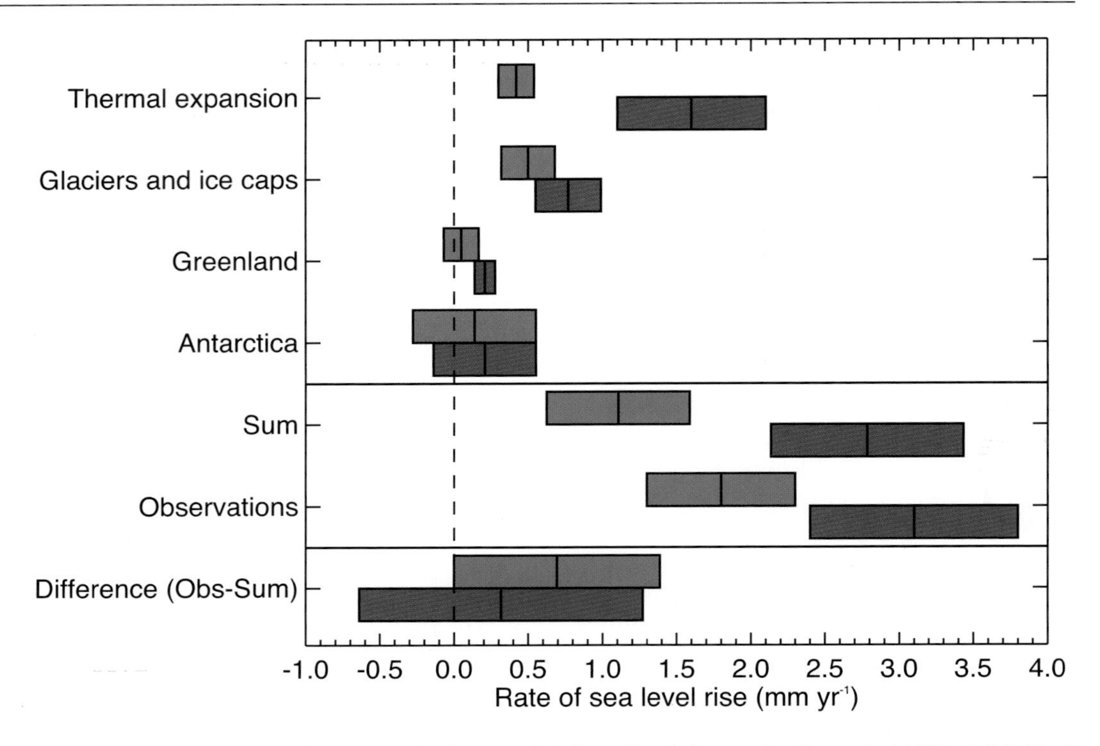

Figure 13.20 Estimates of the contributions to global sea-level change for the period 1961–2003 (blue) and 1993–2003 (red). Also given for each time period are the sum of the individual components, the observed sea-level change, and the difference between the sums and observations. The bars represent the 90 percent error range.

Source: IPCC (2007). Reproduced by permission of the IPCC (ch. 4, Observations: Oceanic climate change and sea level, Report of WG1 1, IPCC, p. 419, fig. 5.21).

Figure 13.21 shows the time series of sea-level in the past and projected through the twenty-first century from the IPCC models using the A1B emissions scenario. Relative to the 1989–1999 mean (the zero line on the y axis), sea-level by the year 2100 is expected to have risen by 200 to 500 mm. Many uncertainties remain. The key wild card is the behavior of the ice sheets. Recent finding suggest that the IPCC estimate of sea-level rise is too low because effects of changing ice dynamics in Greenland and Antarctic leading to accelerated iceberg discharge have not been taken into account. There is an outside possibility that a rise in sea-level might cause the West Antarctic ice sheet to be buoyed up and melt bodily (not just around the edges, as in the past) and cause a further catastrophic sea-level rise but spread over several hundred years.

3 Snow and ice

The effects of twentieth-century climate change on global snow and ice cover are apparent in many ways, but the responses differ widely as a result of the different factors and timescales involved. Snow cover is essentially seasonal, related to storm system precipitation and temperature levels. Sea ice is also a seasonal feature around much of the Antarctic continent (see Figure 10.35A), but the Arctic Ocean holds part of its sea ice cover throughout the year. This is known as multiyear ice as it has survived at least one summer melt

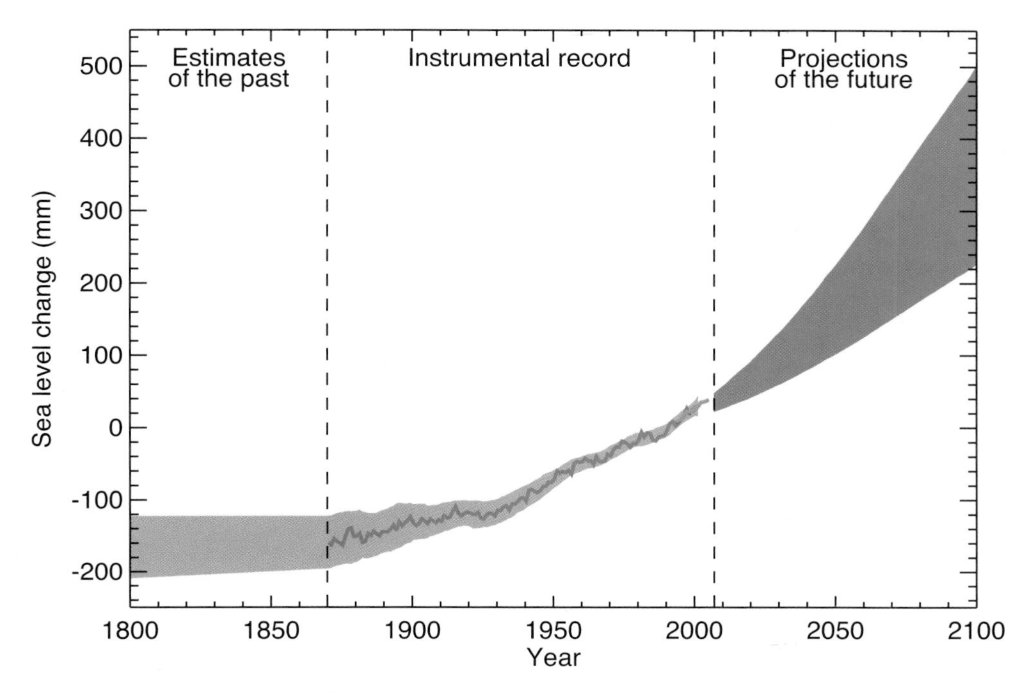

Figure 13.21 Time series of global mean sea level, expressed as anomalies with respect to the base period 1980–1999, for the period before instrumental records (grey shading, designating an estimated uncertainty in the estimated long-term rate of sea-level change) over the instrumental record and projected through the twenty-first century. The red shading represents results from tide gauges, the green line is based on satellite altimetry. The projections are from models participating in the IPCC Fourth Assessment Report with the A1B emissions scenario, the blue shading is the range in model projections.

Source: IPCC (2007). Reproduced by permission of the IPCC (ch. 5, Observations: Oceanic climate change and sea level, Report of WG1 1, IPCC, p. 409, FAQ 5.1, fig. 1).

season. It tends to be thicker than firstyear ice, which is ice that grows in a single season. Some of the Arctic's multiyear ice can be a decade old. Sea ice grows and melts in response to the heat budgets at top and bottom of the ice cover. In the Arctic, ice is also continuously transported into the North Atlantic via winds and ocean currents. Most of this export is in the form of the thicker multiyear ice. Glacier ice builds up from the net balance of snow accumulation and summer melt (ablation), but glacier flow transports ice towards the terminus, where it may melt or calve into water. In small glaciers, the ice may have a residence time of 10s–100s of years, but in ice caps and ice sheets this increases to 10^3–10^6 years.

The contribution of Greenland, Antarctica and the melt of glaciers and ice caps to recent sea-level rise has already been addressed. However, it must be stressed that glacier and ice cap retreat is very much a global phenomenon (Figure 13.22). This is consistent with a warming climate, acting to lengthen the melt season with a corresponding rise of the snowline. In the past 15–20 years, the freezing level in the troposphere has risen in the inner tropics by 100–150m, contributing to rapid ice loss on equatorial glaciers in East Africa and the northern Andes. While even a decade ago, some glaciers in Scandinavia were advancing due to increased precipitation, the pattern is now one of net mass loss. Of course one can always find some

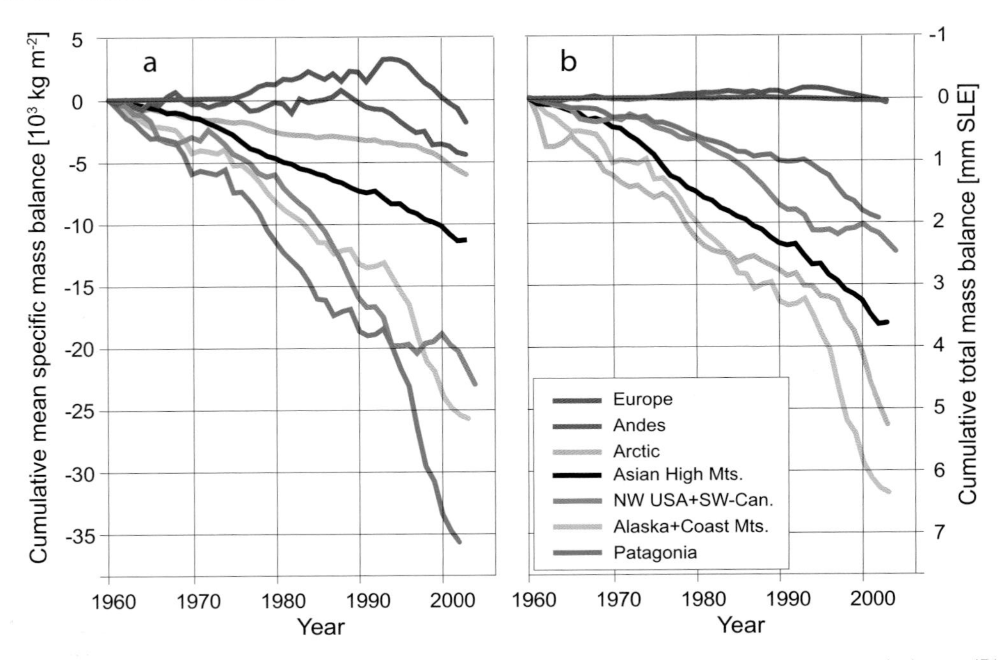

Figure 13.22 Cumulative mean specific mass balance (A) and cumulative total mass balance (B) of glaciers and ice caps calculated for large regions, based on the analysis of Dyurgerov and Meier (2005). Mean specific mass balance shows the strength of climate change in the particular regions. Total mass balance is the contribution from each region to sea-level rise.

Source: IPCC (2007). Reproduced by permission of the IPCC (ch. 4, Observations: Changes in snow, ice and frozen ground, Report of WG1 1, IPCC, p. 359, fig. 4.15).

advancing glaciers but the overall picture is clear. Projections for the year 2050 suggest that a quarter of the world's present glacier mass may disappear with critical and irreversible long-term consequences for water resources in alpine countries.

Another clear indicator of climate change is the Arctic Ocean's shrinking sea ice (plates 13.4 and 13.5). Over the modern satellite record, which begins in 1979, ice extent shows significant downward trends in all months, but largest in September (the end of the melt season) at about 10 percent per decade. The pace of summer ice loss appears to have accelerated since the turn of the twenty-first century. Figure 13.23 plots observed Arctic sea ice extent for September over an extended record spanning the years 1953 through 2006, along with simulated extent for the period 1900 through 2100 from a suite of the IPCC models. The simulations employ observed radiative forcings through the twentieth century and the A1B emissions scenario for the twenty-first century. Essentially all models indicate that sea ice extent should be declining over the period of observations. This consensus is strong evidence for a role of greenhouse gas loading on the observed decline. However, none of the simulations over the period 1953–2006 yield a downward trend as large as is observed. One explanation is that natural variability in the observed coupled system has been a very strong player. Changes in cloud cover, wind-driven alterations in sea ice circulation and ice thickness associated with the North Atlantic Oscillation and other patters of atmospheric variability and altered ocean heat transport have all been implicated in the observed retreat. The alternative explanation is that the IPCC models as a group under-represent the sensitivity of the sea ice cover

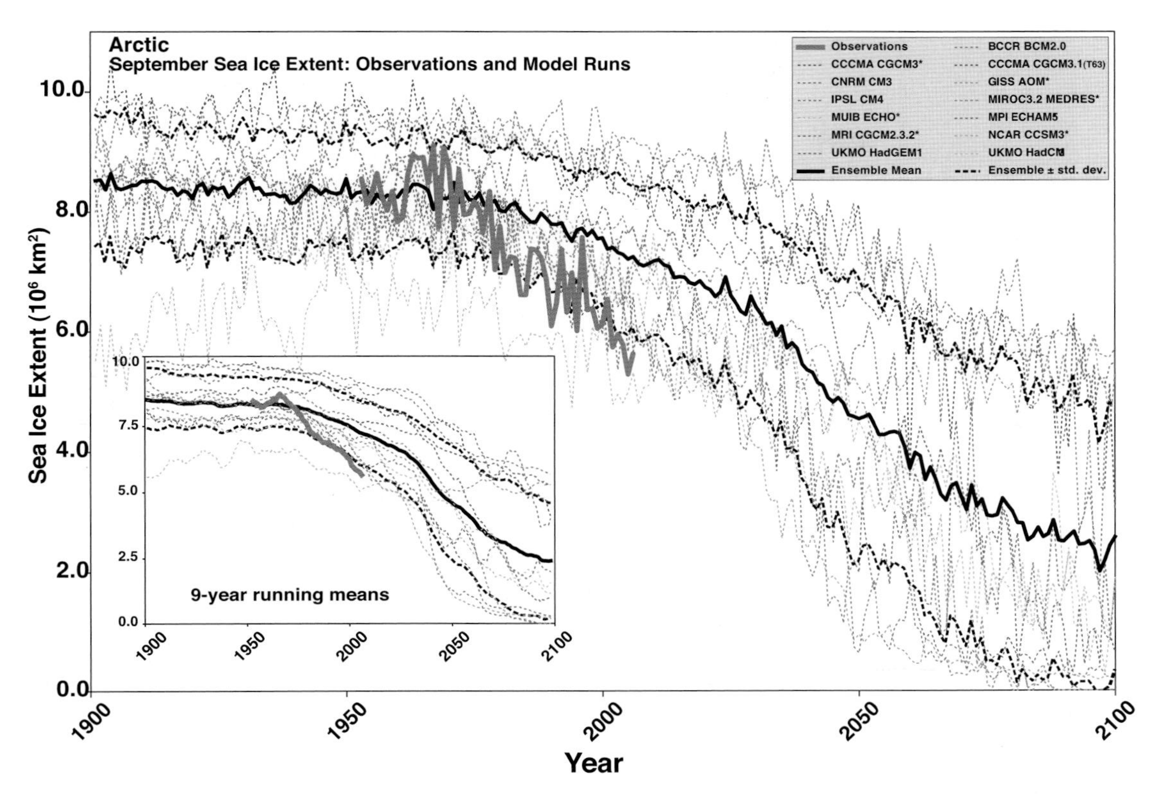

Figure 13.23 Arctic September sea ice extent from observations (thick red line, 1953-2006) and from 13 models participating in the IPCC Fourth Assessment Report, together with the multi-model mean (solid black line) and standard deviation (dotted black line). The inset shows nine-year running means.

Source: Stroeve *et al.* (2007, fig. 1). Courtesy of the American Geophysical Union.

to the effects of greenhouse gas loading. The IPCC models indicate that ice-free conditions in September might be realized any time from the year 2050 to well beyond 2100. Given the discrepancy between modeled and observed trends over the period of overlap, ice-free conditions may well occur much sooner.

Antarctic sea ice extent, which has been accurately monitored since 1979, actually shows small upward trends in most months (based on data through 2007). While perhaps counter-intuitive, this is largely in accord with projections from even very early generation GCMs of a much slower and delayed response of the Antarctic to greenhouse gas loading in comparison to the Arctic. Recall from earlier discussion (Section 13E.2) the very different nature of the ocean circulation in southern high latitudes, in which heat input to the ocean surface tends to be rapidly mixed to deeper ocean levels. The small upward trends that have been observed appear to reflect the especially persistent zonal circulation of the atmosphere that for several decades has characterized the region surrounding the ice sheet (a persistently positive Southern Annular Mode). This has helped to keep the region cool. The notable exception is the Antarctic peninsula, which has warmed by about 2.5°C in the past 50 years, (see also Figure 13.9). An interesting aspect of the Antarctic peninsula which seems broadly related to this warming is the major calving events that have occurred during the past 10–15 years along

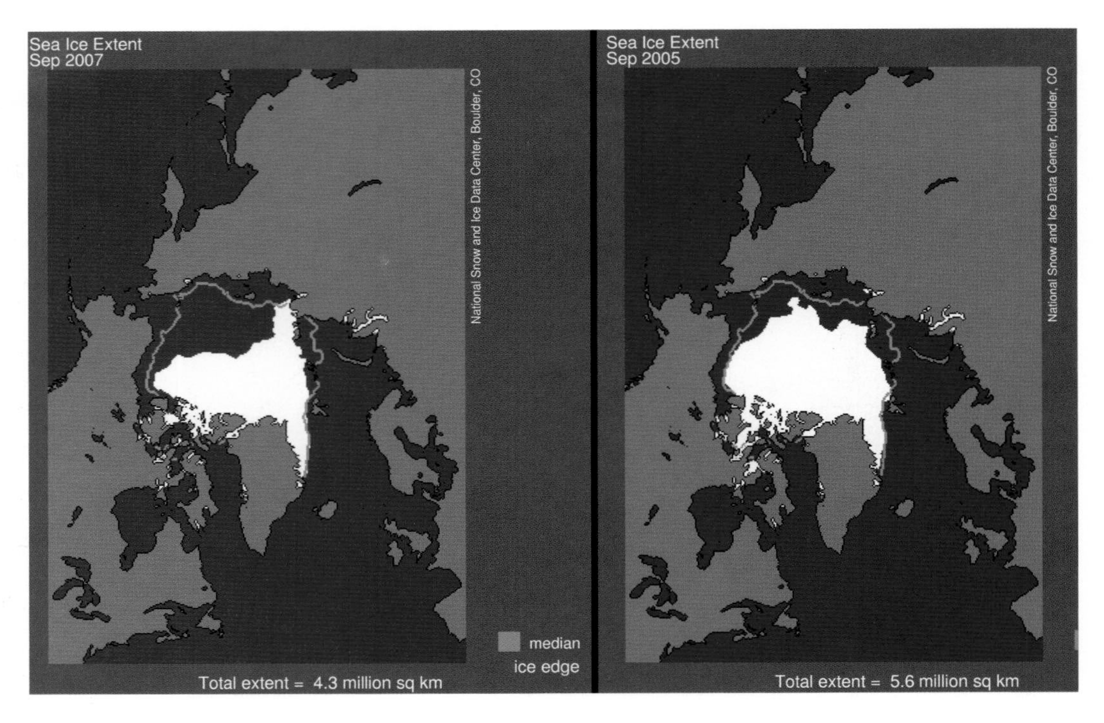

Plate 13.4 The record-breaking minimum Arctic sea ice extent of September 2007 compared with the previous minimum in 2005. The mean September limit for 1979–2007 is shown.
Source: Courtesy NSIDC.

its ice shelves. These include breakup of the Wordie shelf on the west side in the 1980s, the northern Larsen shelf on the east side between 1995 and March 2002, and the Wilkins ice shelf in 2008.

Snow cover extent also shows clear indication of a response to recent temperature trends. Northern Hemisphere snow cover has been mapped by visible satellite images since 1966. Compared with the 1970s to mid-1980s, annual snow cover since 1988 has shrunk by about 10 percent. The decrease is pronounced in spring and is well correlated with springtime warming (Figure 13.24). Winter snow extent shows little or no change. Nevertheless, annual snowfall in North America north of 55°N increased during 1950–1990. Scenarios for the mid-twenty-first century point to a shorter snow cover period in North America, with a decrease of 70 percent over the Great Plains. In alpine areas, snowlines will rise by 100–400m, depending on precipitation.

4 Vegetation

An increase in CO_2 up to around 1000ppmv is expected to enhance global plant growth, beyond which a saturation limit may be reached. However, deforestation could decrease the biosphere's capacity to act as a carbon sink. A sustained increase of only 1°C can cause considerable change in tree growth, regeneration and species extent. Species migrate only slowly but, eventually, extensive forested areas may change to new vegetation types, and it has been estimated that 33 percent of the present forest area could be affected, with up to 65 percent of the boreal zone being subject to change. Alpine tree lines appear to be quite resistant to climatic fluctuations. However, surveys of plant species on peaks in the European Alps indicate an upward migration of alpine plants by 1–4m per decade during the past century.

Plate 13.5 The Northwest Passage in the Canadian Arctic Archipelago largely free of sea ice, 15 September 2007.
Source: http://earthobservatory.nasa.gov/IODT/view.php?id=18964.

Tropical forests are likely to be affected more by human deforestation than by climate change. However, decreases in soil moisture are particularly damaging in hydrologically marginal areas. In the Amazon, climatic predictions support the idea of increased convection, and therefore of rainfall, in its western equatorial portion, where present rainfall is most abundant. Because of the particularly large temperature rises projected for the high northern latitudes, boreal forests are expected to be strongly affected by their advance northward into tundra regions. This may yield further regional warming owing to the lower albedo of forests during the snow season. Climate change over the twenty-first century is expected to have the least effect on temperate forests. The Arctic, by contrast, is already seeing areas of tundra being replaced by shrub vegetation. This trend is expected to continue. A large store of carbon is locked up in permafrost. A growing concern is that as permafrost thaws, this carbon could be released into the atmosphere (either as CO_2 or CH_4), representing a strong feedback that leads to further warming.

Wetlands currently cover 4–6 percent of the land surface, having been reduced by human activities by more than half over the past 100 years. Climate change will affect wetlands mainly by altering their hydrological regimes. It is believed that eastern China, the United States and southern Europe will suffer a natural decline in the area of wetlands during this century, decreasing the methane flux to the atmosphere.

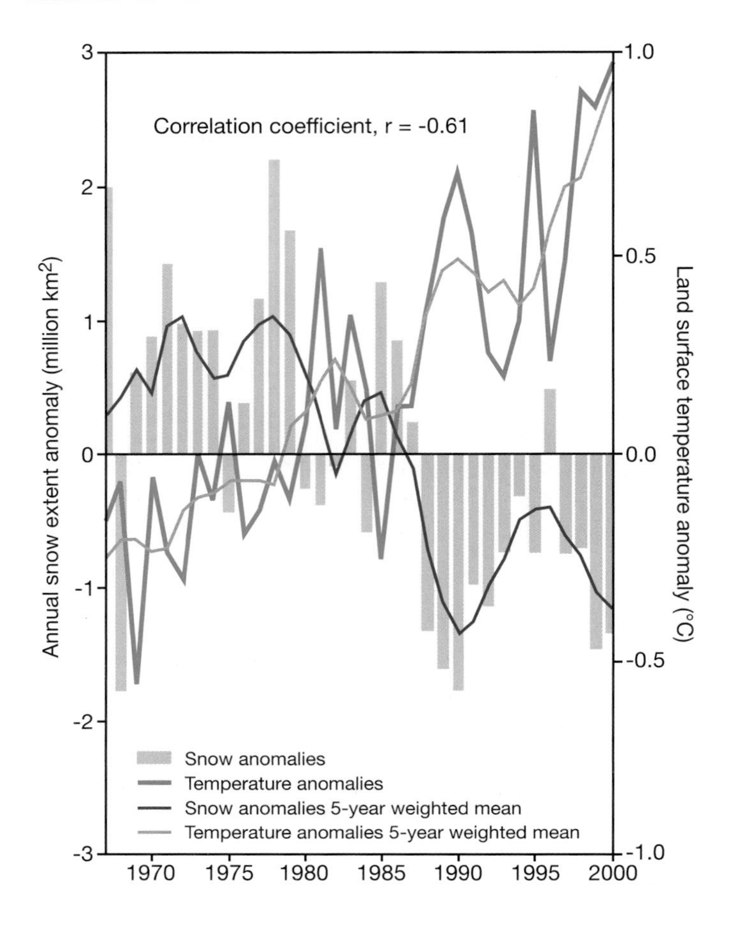

Figure 13.24 Time series of annual snow extent and land surface temperature anomalies. Annual anomalies are the sum of monthly anomalies, area-averaged over the region north of 20°N, for the snow hydrological year, October to September. The snow anomaly (in million km^2) is on the left vertical axis, the temperature anomaly (°C) is on the right vertical axis. Bar plot indicates snow anomalies, the fine line indicates temperature anomalies. The correlation coefficient, r, is –0.61. The thick curves are five-year weighted mean values. Snow cover calculations are based on the NOAA/NESDIS snow cover maps for 1967-2000. Temperature calculations are based on the Jones datasets; anomalies are with respect to the period 1960–1990.

Source: D. Robinson, Rutgers University, and A. Bamzai (NOAA/.OGP).

Dry lands are expected to be profoundly affected by climatic change. Dry rangelands (including grasslands, shrublands, savannas, hot and cold deserts) occupy 45 percent of the terrestrial land surface, contain one-third of the world's total carbon in their biomass, and support half of the world's livestock and one-sixth of the world's population. Low-latitude rangelands are most at risk both because an increase in CO_2 (increasing the carbon/nitrogen ratio) will decrease the nutrient value of forage and because the increasing frequency of extreme events will cause environmental degradation. Most deserts are likely to become hotter but not significantly wetter. Any increases in rainfall will tend to be associated with increased storm intensity. Greater wind speeds and evaporation may be expected to increase wind erosion, capillary rise and salinization of soils. Central Australia is one of the few places where desert conditions may improve.

A major consequence of global warming will be that desiccation and soil erosion will increase in semi-arid regions, rangelands and savannas adjacent to the world's deserts. This will accelerate the current rate of desertification, which is proceeding at six million hectares per year partly due to high rainfall variability and partly to unsuitable human agricultural activities such as overgrazing and over-intensive cultivation. Desertification was estimated to affect nearly 70 percent of the total dry land area in the 1990s.

G POSTSCRIPT

Our ability to understand and project climate change has increased considerably since the first IPCC Report appeared in 1990, but many problems and uncertainties remain. Key needs include (not in order of importance):

- The development of more refined forcing scenarios through a better understanding of impacts of economic growth, forest clearances, land-use changes, sulfate aerosols, carbonaceous aerosols generated by biomass burning, and radiative trace gases other than CO_2 (e.g., methane and ozone). Refined estimates of forcing by indirect aerosol effects merit particular attention.
- Incorporation of a realistic carbon cycle and ice sheet dynamics in AOGCMs.
- Better understanding of feedback processes, notably those involving clouds, water vapor, sea ice and the carbon cycle. Feedbacks involving polar cloud cover and carbon release due to thawing permafrost merit particular attention.
- Further increases in climate model resolution so that small-scale physical processes can be better represented (e.g., those relating to clouds).
- Greater understanding and modeling of ocean processes and atmosphere–ocean coupling that bear on the heat flux at the ocean surface, the ability of oceans to absorb CO_2, especially by biological processes, and their role in heat uptake that delays the climate system's response to radiative forcing.
- An improved ability to distinguish between anthropogenic climate change and natural variability, especially through the use of ensemble simulations.
- An improved understanding of threshold behaviors (sometimes termed 'tipping points') through which a warming climate may precondition key systems, such as ice sheets, sea ice and permafrost, to exhibit rapid decay.
- Continued, systematic collection of instrumental, proxy and space-based observations of climate variables. This requires a commitment by national governments to maintain surface observation networks and satellite remote sensing systems.

SUMMARY

The most fundamental measure of the earth's climate state is the global averaged surface air temperature. It is influenced by a variety of climate forcing factors operating on a suite of timescales. Climate variations over timescales of millions of years can be linked to plate tectonics. The great Ice Ages and interglacials that have characterized the past two million years can be linked to periodicities in the earth's orbit around the sun, influencing the seasonal distribution of solar radiation over different parts of the surface. The observed increase in global mean surface air temperature over the past 100 years may be attributed primarily to human-induced increases in atmospheric carbon dioxide and other greenhouse gases, partly compensated by the cooling effects of aerosol loading. These are known as radiative forcings in that they alter the globally averaged radiation budget at the top of the atmosphere. Solar variability, another radiative forcing, has played a minor role since the mid-twentieth century. The general rise in global mean surface temperature over the past 100 years contains inter-annual to multi-decadal variations. These reflect natural internal variability in the coupled atmosphere–ocean–land system as well as transient radiative forcings such as large volcanic eruptions (e.g., Mt. Pinatubo).

The magnitude of the response of global temperature to a radiative forcing of given magnitude, or set of forcings in combination, depends on the climate feedbacks. Positive feedbacks dominate, and hence act to amplify the temperature response to a forcing. In terms of human-induced climate change, the most important are the fast water vapor and ice-albedo feedbacks.

Climate projections through the twenty-first century, assuming a variety of emission scenarios for greenhouse gases and aerosols, indicate a mean global temperature increase in the range of 2–4°C by the year 2100, together with sea-level rises of 200–500mm. Given the rapid growth of greenhouse gas concentrations in recent decades, the effects of ice sheet dynamics and other wild cards in the system, these may be underestimates. The Arctic will eventually become free of sea ice in summer. Warming will also be accompanied by continued shrinking of glaciers, ice caps and permafrost, changes in the hydrologic cycle, atmospheric circulation, and vegetation.

Critical research needs include improved understanding of climate feedbacks, including feedbacks in the carbon cycle, and the role of the oceans in the uptake of heat and carbon dioxide.

DISCUSSION TOPICS

- Examine the figures showing climatic time series in Chapter 13 for evidence of changes in mean and variance and consider where step function changes have occurred and where trends can be detected.
- What are the different climate forcing factors at work on geological and historical timescales?
- What are the main advantages and limitations of different types of proxy records of paleoclimate? Consider the climatic variables that may be inferred and the temporal resolution of the information.
- What are the main reasons for uncertainties in projections of climate for the year 2100?
- What are some of the possible impacts of projected climate changes in your region and country?

REFERENCES AND FURTHER READING

Books

Adger, W. N. and Brown, K. (1995) *Land Use and the Causes of Global Warming*, John Wiley & Sons, New York, 282pp. [Greenhouse gas emissions from land use sources and the effects of land-use changes]

Bradley, R. S. (1999) *Quaternary Paleoclimatology: Reconstructing Climates of the Quaternary*, 2nd edn, Academic Press, San Diego, 683pp. [Details methods of paleoclimatic reconstruction, dating of evidence and modeling paleoclimates; numerous illustrations and references]

Bradley, R. S. and Jones, P. D. (eds) (1992) *Climate Since* A. D. *1500*, Routledge, London, 679pp. [Collected contributions focusing on the Little Ice Age and subsequent changes from proxy, historical and observational data]

Crowley, T. J. and North, G. R. (1991) *Paleoclimatology*, Oxford University Press, New York, 339pp. [Surveys the pre-Quaternary and Quaternary history of the earth's climate, presenting observational evidence and modeling results]

Dyurgerov, M. and Meier, M. F. (2005) *Glaciers and the Changing Earth System: A 2004 Snapshot*, Occasional Paper 58, Institute of Arctic and Alpine Research, University of Colorado, Boulder, CO, 118pp. [An evaluation of recent global changes in glacier mass balance]

Fris-Christensen, E., Froehlich, C., Haigh, J. D., Schluesser, M. and von Steiger, R. (eds) (2001) *Solar Variability and Climate*, Kluwer, Dordrecht, 440pp. [Contributions on solar variations, solar influences on climate, climate observations and the role of the sun by leading specialists]

Goodess, C. M., Palutikof, J. P. and Davies, T. D. (eds) (1992) *The Nature and Causes of Climate Change: Assessing the Long-term Future*, Belhaven Press, London, 248pp. [Contributions on natural and anthropogenic forcing, proxy records, Pleistocene reconstructions and model studies]

Grove, J. M. (2004) *Little Ice Ages, Ancient and Modern*, Routledge, London, 2 vols, 718pp. [Detailed account of the Little Ice Age in terms of the response of glaciers around the world]

Harvey, L. D. D. (1998) *Global Warming: The Hard Science,* Prentice Hall/Pearson Education, Harlow, 336pp. [Detailed discussion of global warming – processes, feedbacks and environmental impacts]

Houghton, J. T. *et al.* (eds) (1996) *Climate Change 1995: The Science of Climate Change*, Cambridge University Press, Cambridge, 572pp.

Houghton, J. T. *et al.* (2001) *Climate Change 2001: The Scientific Basis*, Cambridge University Press, Cambridge, 881pp.

Hughes, M. K. and Diaz, H. F. (eds) (1994) *The Medieval Warm Period*, Kluwer Academic Publishers, Dordrecht 342pp. [Contributions on the proxy and historical evidence concerning climates around the world during about AD 900–1300]

Hughes, M. K., Kelly, P. M., Pilcher, J. R. and La Marche, V. (eds) (1981) *Climate from Tree Rings*, Cambridge University Press, Cambridge (400pp.).

Imbrie, J. and Imbrie, K. P. (1979) *Ice Ages: Solving the Mystery*, Macmillan, London, 224pp. [Readable account of the identification of the role of astronomical forcing – the Milankovitch effect – in Ice Age cycles, by one of the paleoclimatologists involved]

IPCC (2007) *Climate Change 2007: The Physical Science Basis. Contribution of Working Group I to the Fourth Assessment Report of the Intergovernmental Panel on Climate Change*, Solomon, S., Qin, D., Manning, M., Chen, Z., Marquis, M., Avery, K. B., Tignor, M. and Miller, H. L., (eds), Cambridge University Press, Cambridge, UK and New York, USA, 996pp [The Fourth Assessment Report of the IPCC on observed climate changes, the physical basis of climate change, and projections from climate models]

Jones, P., Conway, D. and Briffa, K. (1997) *Climate of The British Isles. Present, Past and Future*, Routledge, London.

Lamb, H. H. (1977) *Climate: Present, Past and Future, 2: Climatic History and the Future*, Methuen, London, 835pp. [Classic synthesis by a renowned climate historian]

Serreze, M. C. and Barry, R. G. (2005) *The Arctic Climate System*, Cambridge University Press, 385pp. [The climate system of the Arctic region, Arctic paleoclimates, modeling and projected future states]

Williams, J. (ed.) (1978) *Carbon Dioxide, Climate and Society*, Pergamon, Oxford, 332pp. [Proceedings of one of the first wide-ranging conferences on the effects of increasing carbon dioxide on climate and the environment, and societal consequences]

Articles

Anderson, D. E. (1997), Younger Dryas research and its implications for understanding abrupt climatic change. *Progr. Phys. Geog.* 21(2), 230–49.

Beer, J., Mende, W. and Stellmacher, W. (2000) The role of the sun in climate forcing. *Quatern. Sci. Rev.* 19, 403–16.

Broecker, W. S. and Denton, G. S. (1990) What drives glacial cycles? *Sci. American* 262, 48–56.

Broecker, W. S. and Van Donk, J. (1970) Insolation changes, ice volumes and the O^{18} record in deep sea cores. *Rev. Geophys.* 8, 169–96.

Chu, P-S., Yu, Z-P. and Hastenrath, S. (1994) Detecting climate change concurrent with deforestation in the Amazon Basin. *Bull. Amer. Met. Soc.* 75(4), 579–83.

Davidson, G. (1992) Icy prospects for a warmer world. *New Scientist* 135(1833), 23–6.

Diaz, H. F. and Kiladis, G. N. (1995) Climatic variability on decadal to century time-scales, in Henderson-Sellers, A. (ed.) *Future Climates of the World: A Modelling Perspective*, World Surveys of Climatology 16, Elsevier, Amsterdam, 191–244.

Douglas, B. C. and Peltier, W. R. (2002) The puzzle of global sea-level rise. *Physics Today* 55 (3), 35–41.

French, J. R., Spencer, T. and Reed, D. J. (1995) Editorial – Geomorphic response to sea-level rise: existing evidence and future impacts. *Earth Surface Processes and Landforms* 20, 1–6.

Grootes, P. (1995) Ice cores as archives of decade-to-century scale climate variability, in *Natural Climate Variability on Decade-to-century Time Scales*, National Academy Press, Washington, DC, 544–54.

Haeberli, W. (1995) Glacier fluctuations and climate change detection. *Geogr. Fis. Dinam. Quat.* 18, 191–9.

Hansen, J. E. and Lacis, A. A. (1990) Sun and dust versus greenhouse gases: an assessment of their relative roles in global climate change. *Nature* 346, 713–19.

Hansen, J. E. and Sato, M. (2001) Trends of measured climate forcing agents. *Proceedings, National Academy of Science* 98(26), 14778–83.

Hansen, J. E. *et al.* (2006) Global temperature change. *Proceedings, National Academy of Science* 103, 14288–93.

Hansen, J., Nazarenko, L., Reudy, R., Sato, M., Willis, J., Del Genio, A., Koch, D., Lacis, A., Lo, K., Menon, S., Novakov, T., Perlwitz, J., Russell, G., Schmidt , G.A. and Tausnev, N. (2005) Earth's energy imbalance: confirmation and implications, *Science* 308, 1431–35.

Hare, F. K. (1979) Climatic variation and variability: empirical evidence from meteorological and other sources, in *Proceedings of the World Climate Conference*, WMO Publication No. 537, WMO, Geneva, 51–87.

Henderson-Sellers, A. and Wilson, M. F. (1983) Surface albedo data for climate modelling. *Rev. Geophys. Space Phys.* 21, 1743–8.

Hulme, M. (1992) Rainfall changes in Africa: 1931–60 to 1961–90. *Int. J. Climatol.* 12, 685–99.

Jäger, J. and Barry, R. G. (1991) Climate, in Turner, B. L. II (ed.) *The Earth as Transformed by Human Actions*, Cambridge University Press, Cambridge, 335–51.

Jones, P. D. and Bradley, R. S. (1992) Climatic variations in the longest instrumental records, in Bradley, R. S. and Jones, P. D. (eds) *Climate Since A.D. 1500*, Routledge, London, 246–68.

Jones, P. D., Wigley, T. M. L. and Farmer, G. (1991) Marine and land temperature data sets: a comparison and a look at recent trends, in Schlesinger, M. E. (ed.) *Greenhouse-gas-induced Climatic Change*, Elsevier, Amsterdam, 153–72.

Kukla, G. J. and Kukla, H. J. (1974) Increased surface albedo in the Northern Hemisphere. *Science* 183, 709–14

Kutzbach, J. E. and Street-Perrott, A. (1985) Milankovitch forcings of fluctuations in the level of tropical lakes from 18 to 0 kyr BP. *Nature* 317, 130–9.

Lamb, H. H. (1970) Volcanic dust in the atmosphere; with a chronology and an assessment of its meteorological significance. *Phil. Trans. Roy. Soc.* A 266, 425–533.

Lamb, H. H. (1994) British Isles daily wind and weather patterns 1588, 1781–86, 1972–1991 and shorter early sequences. *Climate Monitor* 20, 47–71.

Lean, J. and Rind, D. (2001) Sun–climate connections: earth's response to a variable star. *Science* 252(5515), 234–6.

Lean, J., Beer, J. and Bradley, R. (1995) Reconstruction of solar irradiance since 1610: implications for climate change. *Geophys. Res. Lett.* 22(23), 3195–8.

Levitus, S., *et al.* (2001) Anthropogenic warming of the Earth's climate system. *Science* 292(5515), 267–70.

Manley, G. (1958) Temperature trends in England, 1698–1957. *Archiv. Met. Geophy. Biokl.* (Vienna) B 9, 413–33.

Mann, M. E. and Jones, P. D. (2003) Global surface temperatures over the past two millennia. *Geophys. Res. Lett.* 30(15), 1820.

Mann, M. E., Park, J. and Bradley, R. S. (1995) Global interdecadal and century-scale climate oscillations during the past five centuries. *Nature* 378(6554), 266–70.

Mann, M. E., Bradley, R. S. and Hughes, M. K. (1999) Northern Hemisphere temperatures during the past millennium: inferences, uncertainties and limitations. *Geophys. Res. Lett.* 26, 759–62.

Mann, M. E., Zhang, Z., Hughes, M. K., Bradley, R. S., Miller, S. K., Rutherford, S. and Ni, F. (2008) Proxy-based reconstructions of hemispheric and global surface temperature variations over the past two millennia. *Proc. Natl. Acad. Sci.* 105, 13252–7.

Mather, J. R. and Sdasyuk, G. V. (eds) (1991) *Global Change: Geographical Approaches* (Sections 3.2.2, 3.2.3). University of Arizona Press, Tucson.

Meehl, G. A. and Washington, W. M. (1990) CO_2 climate sensitivity and snow–sea-ice albedo parameterization in an atmospheric GCM coupled to a mixed-layer ocean model. *Climatic Change* 16, 283–306.

Meehl, G. A. *et al.* (2000) Anthropogenic forcing and decadal climate variability in sensitivity experiments of twentieth- and twenty-first century climate. *J. Climate* 13, 3728–44.

Meier, M. F. and Wahr, J. M. (2002) Sea level is rising: do we know why. *Proc. Nat. Acad. Sci.* 99(10), 6524–6.

Mitchell, J. F. B., Johns, T. C., Gregory, J. M. and Tett, S. F. B. (1995) Climate response to increasing levels of greenhouse gases and sulphate aerosols. *Nature* 376, 501–4.

Mitchell, T. D. and Hulme, M. (2002) Length of the growing season. *Weather* 57(5), 196–8.

Nicholson, S. E. (1980) The nature of rainfall fluctuations in subtropical West Africa. *Monthly Weather Review* 108, 473–87.

Nicholson, S. E., Some, B. and Kone, B. (2000) An analysis of recent rainfall conditions in West Africa, including the rainy seasons of the 1997 El Niño and the 1998 La Niña years. *J. Climate* 13(14), 2628–40.

Parker, D. E., Horton, E. B., Cullum, D. P. N. and Folland, C. K. (1996) Global and regional climate in 1995. *Weather* 51(6), 202–10.

Pfister, C. (1985) Snow cover, snow lines and glaciers in central Europe since the 16th century, in Tooley, M. J. and Sheail, G. M. (eds) *The Climatic Scene*, Allen & Unwin, London, 154–74.

Quinn, W. H. and Neal, V. T. (1992) The historical record of El Niño events. In Bradley, R. S. and Jones, P. D. (eds) *Climate since AD 1500*, Routledge, London, 623–48.

Rind, D. (2002) The sun's role in climate variations. *Science* 296 (5569), 673–7.

Serreze, M. C. and Francis, J. A. (2006) The Arctic amplification debate, *Climatic Change* 76, 241–64.

Sioli, H. (1985) The effects of deforestation in Amazonia. *Geog. J.* 151, 197–203.

Sokolik, I. N. and Toon, B. (1996) Direct radiative forcing by anthropogenic airborne mineral aerosols. *Nature* 381, 501–4.

Solomon, S. (1999) Stratospheric ozone depletion: a review of concepts and history. *Rev. Geophys.* 37(3), 275–316.

Stark, P. (1994) Climatic warming in the central Antarctic Peninsula area. *Weather* 49(6), 215–20.

Street, F. A. (1981) Tropical palaeoenvironments. *Prog. Phys. Geog.* 5, 157–85.

Stroeve, J., Holland, M. M., Meier, W., Scambos, T. and Serreze, M. (2007) Arctic sea ice decline: faster than forecast. *Geophysical Research Letters* 34, L09501, doi: 10.1029/2007GL029703.

Thompson, R. D. (1989) Short-term climatic change: evidence, causes, environmental consequences and strategies for action. *Prog. Phys. Geog.* 13(3), 315–47.

Wild, M. (2009) Global dimming and brightening. A review, *J. Geophys. Res.*, 114: DOOD16, 31pp.

APPENDIX 1

Climate classification

The purpose of any classification system is to obtain an efficient arrangement of information in a simplified and generalized form. Climate statistics can be organized in order to describe and delimit the major types of climate in quantitative terms. Obviously, any single classification can serve only a few purposes satisfactorily and many different schemes have therefore been developed. Many climatic classifications are concerned with the relationships between climate and vegetation or soils and rather few attempt to address the direct effects of climate on humans.

Only the basic principles of the four groups of the most widely known classification systems are summarized here. Further information may be found in the listed references.

A GENERIC CLASSIFICATIONS RELATED TO PLANT GROWTH OR VEGETATION

Numerous schemes have been suggested for relating climate limits to plant growth or vegetation groups. They rely on two basic criteria – the degree of aridity and of warmth.

Aridity is not simply a matter of low precipitation, but of the 'effective precipitation' (i.e., precipitation minus evaporation). The ratio of rainfall/temperature is often used as an index of precipitation effectiveness, since higher temperatures increase evaporation. W. Köppen developed the pre-eminent example of such a classification. Between 1900 and 1936, he devised several classification schemes that involve considerable complexity in their full detail. The system has been used extensively in geographical teaching. The key features of Köppen's approach are temperature criteria and aridity criteria.

Temperature criteria

Five of the six major climate types are based on monthly mean temperature values.

1. Tropical rainy climate: coldest month >18°C.
2. Dry climates.
3. Warm temperate rainy climates: coldest month between –3° and +18°C, warmest month >10°C.
4. Cold boreal forest climates: coldest month <–3°, warmest month >10°C. Note that many American workers use a modified version with 0°C as the C/D boundary.
5. Tundra climate: warmest month 0–10°C.
6. Perpetual frost climate: warmest month <0°C.

The arbitrary temperature limits stem from a variety of criteria. These are as follows: the 10°C summer isotherm correlates with the poleward limit of tree growth; the 18°C winter isotherm is critical for certain tropical plants; and the –3°C isotherm indicates a few weeks of snow cover. However, these correlations are far from precise! De Candolle determined the criteria in 1874 from the study of vegetation groups defined on a physiological basis (i.e., according to the internal functions of plant organs).

Aridity criteria

The criteria imply that, with winter precipitation, arid (desert) conditions occur where $r/T < 1$, semi-arid conditions where $1 < r/T < 2$. If the rain falls in summer, a larger amount is required to offset evaporation and maintain an equivalent effective precipitation.

Subdivisions of each major category are made with reference, first, to the seasonal distribution of precipitation. The most common of these are: *f* = no dry season; *m* = monsoonal, with a short, dry season and heavy rains during the rest of the year; *s* = summer dry season; *w* = winter dry season. Second, there are further temperature criteria based on seasonality. Twenty-seven subtypes are recognized, of which 23 occur in Asia. The ten major Köppen types each have distinct annual energy budget regimes, as illustrated in Figure A1.1.

Figure A1.2A illustrates the distribution of the major Köppen climate types on a hypothetical continent of low and uniform elevation. Experiments using GCMs with and without orography show that, in fact, the poleward orientation of BS/BW climatic zones inland from the West Coast is largely determined by the western cordilleras. It would not be found on a low, uniform-elevation continent.

A new analysis of world climate data has been used by Peel *et al.* (2007) to map the distribution of Köppen-Geiger climate types for each continent and the world. The land areas covered by the major classes, (in percent) are as follows: arid (B) 30.2; cold (D) 24.6; tropical (A) 19.0; temperate (C) 13.4; and polar (E) 12.8.percent.

The Köppen climatic classification has proved useful in evaluating the accuracy of GCMs in simulating current climatic patterns, as a convenient index of recent climatic change, and for climate scenarios projected for CO_2 doubling.

C. W. Thornthwaite introduced a further empirical classification in 1931. An expression for *precipitation efficiency* was obtained by relating measurements of pan evaporation to temperature and precipitation. The second element of the classification is an index of *thermal efficiency*, expressed by the positive departure of monthly mean temperatures from freezing point. Distribution for these climatic provinces in North America and over the world have been published, but the classification is now largely of historical interest.

B ENERGY AND MOISTURE BUDGET CLASSIFICATIONS

Thornthwaite's most important contribution was his 1948 classification, based on the concept of potential evapotranspiration and the moisture budget (see Chapters 4C and 10B.3c). Potential evapotranspiration (*PE*) is calculated from the

Table A1.1 Thornthwaite's climatic classification

		PE		
Im (1955 system)*		**cm**	**in**	**Climatic type**
>100	Perhumid (*A*)	>114	>44.9	Megathermal (*A'*)
20 to 100	Humid (B_1 to B_4)	57 to 114	22.4 to 44.9	Mesothermal (B'_1 to B'_4)
0 to 20	Moist subhumid (C_2)	28.5 to 57	11.2 to 22.4	Microthermal (C'_1 to C'_2)
–33 to 0	Dry subhumid (C_1)	14.2 to 28.5	5.6 to 11.2	Tundra (*D'*)
–67 to –33	Semi–arid (*D*)	<14.2	<5.6	Frost (*E'*)
–100 to –67	Arid (*E*)			

Note: $*Im = 100(S - D)/PE$ is equivalent to $100(r/PE - 1)$, where: *r* = annual precipitation (cm); *T* = mean annual temperature (°C).

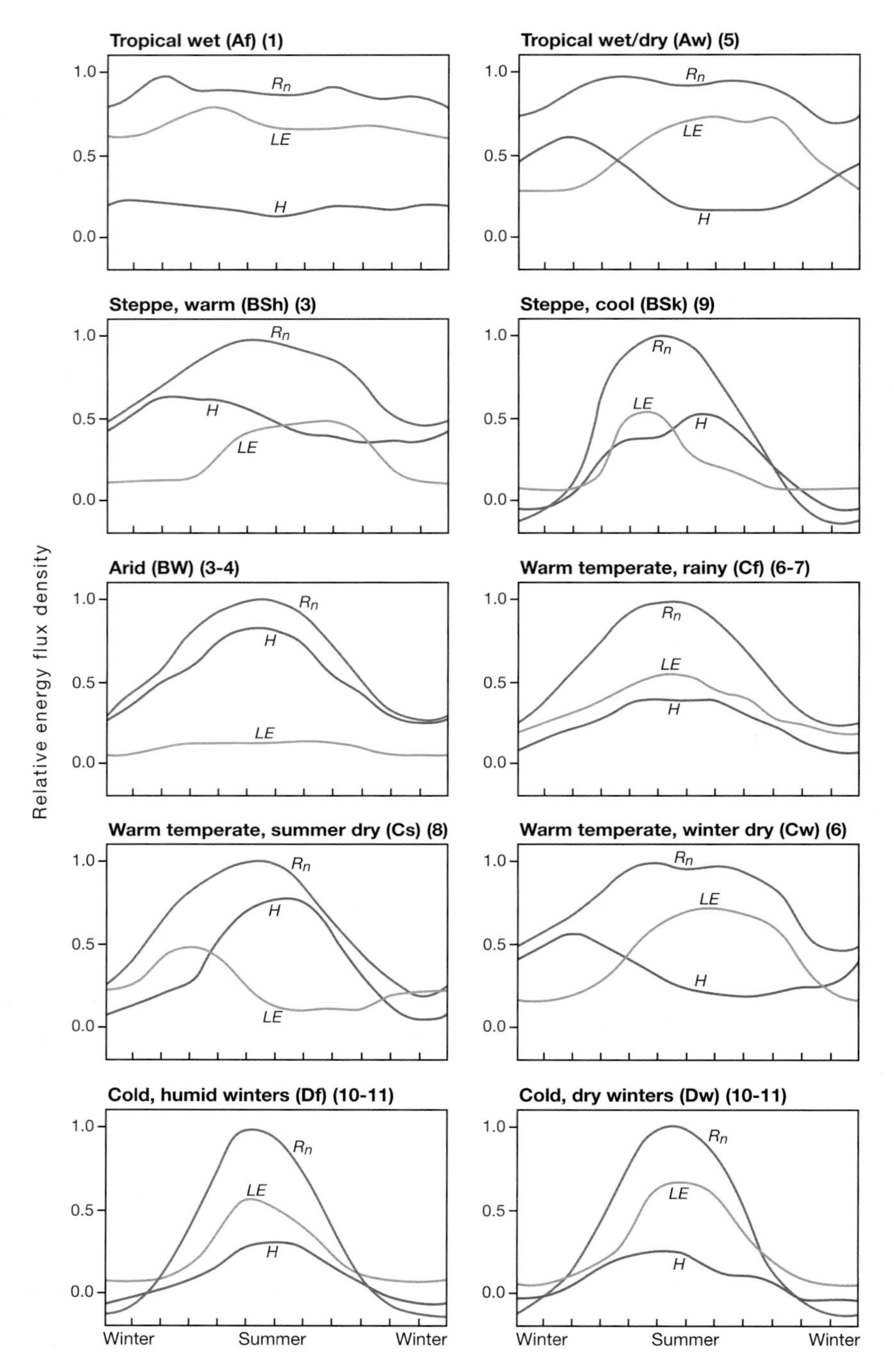

Figure A1.1 Characteristic annual energy balances for ten different climatic types (Köppen symbols and Strahler classification numbers shown). The ordinate shows energy flux density normalized with the maximum monthly net all-wavelength radiation (Rn), normalized with the maximum monthly value as unity. The abscissa intervals indicate the months of the year with summer in the center. H = turbulent flux of sensible heat, LE = turbulent flux of latent heat to the atmosphere.

Source: From Kraus and Alkhalaf (1995). Copyright © John Wiley & Sons Ltd. Reproduced with permission.

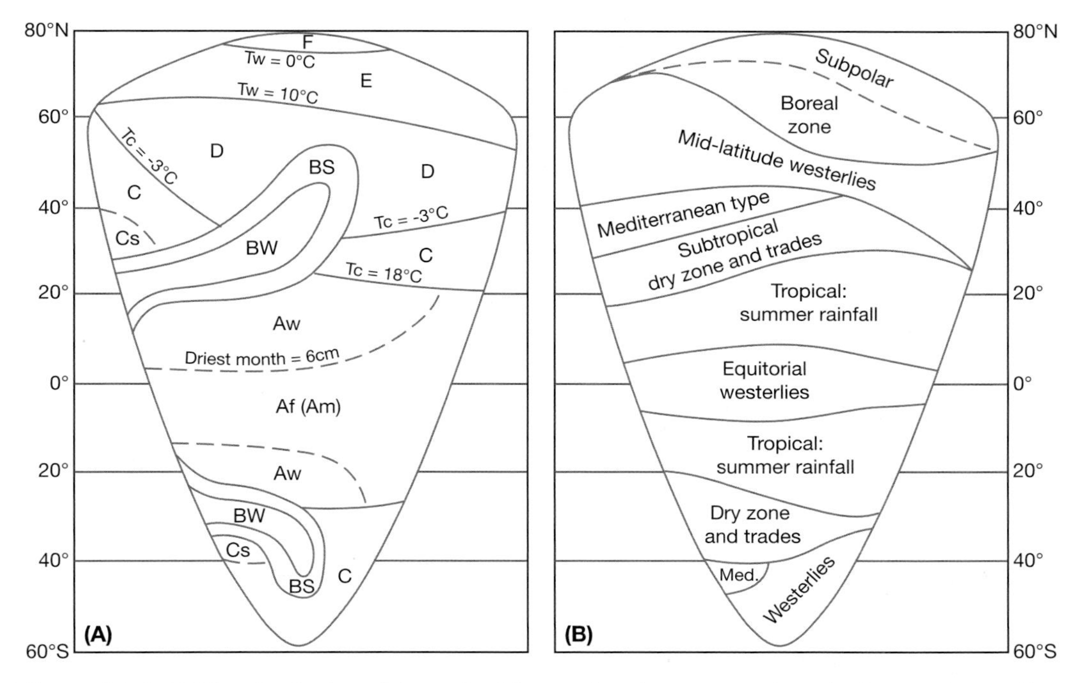

Figure A1.2 (A) The distribution of the major Köppen climatic types on a hypothetical continent of low and uniform elevation. Tw = mean temperature of the warmest month, Tc = mean temperature of the coldest month. (B) The distribution of Flohn's climatic types on a hypothetical continent of low and uniform elevation.

Source: From Flohn (1950). Copyright © *Erdkunde*. Published by permission.

mean monthly temperature (in °C), with corrections for day length. For a 30-day month (12-hour days):

$$PE \text{ (in cm)} = 1.6(10t/I)^a$$

where: I = the sum for 12 months of $(t/5)^{1.514}$
a = a further complex function of I.

Tables have been prepared for the easy computation of these factors.

The monthly water surplus (S) or deficit (D) is determined from a moisture budget assessment, taking into account stored soil moisture (Thornthwaite and Mather 1955; Mather 1985). A moisture index (Im) is given by:

$$Im = 100(S - D)/PE.$$

This allows for variable soil moisture storage according to vegetation cover and soil type, and permits the evaporation rate to vary with the actual soil moisture content. The average water balance is calculated through a bookkeeping procedure. The mean values of the following variables are determined for each month: PE, potential evapotranspiration, precipitation minus PE; and Ws, soil water storage (a value assumed appropriate for that soil type at field capacity). Ws is reduced as the soil dries (DWs). AE is actual evapotranspiration. There are two cases: $AE = PE$, when Ws is at field capacity, or $(P - PE) > 0$; otherwise $AE = P + DWs$. The monthly moisture deficit, D, or surplus, S, is determined from $D = (PE - AE)$, or $S = (P - PE) > 0$, when $Ws <$ field capacity. Monthly deficits or surpluses are carried forward to the subsequent month.

A novel feature of the system is that the thermal efficiency is derived from the PE value, which itself is a function of temperature. The climate types defined by these two factors are shown in Table

A1.1; both elements are subdivided according to the season of moisture deficit or surplus and the seasonal concentration of thermal efficiency. A revised Thornthwaite-type classification is proposed by J. Feddema (2005).

The system has been applied to many regions, but no world map has been published. Unlike the Köppen approach, vegetation boundaries are not used to determine climatic ones. In eastern North America, vegetation boundaries do coincide reasonably closely with patterns of *PE*, but in tropical and semi-arid areas the method is less satisfactory. Shifts in climatic boundaries according to Thornthwaite's classification have been evaluated for the past 111 years for the United States (Grundstein, 2008).

M. I. Budyko developed a more fundamental approach using net radiation instead of temperature (see Chapter 4A). He related the net radiation available for evaporation from a wet surface (*Ro*) to the heat required to evaporate the mean annual precipitation (*Lr*). This ratio *Ro*/*Lr* (where *L* = latent heat of evaporation) is called the *radiational index of dryness*. It has a value of less than unity in humid areas and greater than unity in dry areas. Boundary values of *Ro*/*Lr* are: Desert (>3.0); Semi-desert (2.0–3.0); Steppe (1.0–2.0); Forest (0.33–1.0); Tundra (<0.33). In comparison with the revised Thornthwaite index ($Im = 100(r/PE - 1)$), note that $Im = 100(Lr/Ro - 1)$ if all the net radiation is used for evaporation from a wet surface (i.e., no energy is transferred into the ground by conduction or into the air as sensible heat). A general world map of *Ro*/*Lr* has appeared, but there are few measurements of net radiation over large parts of the earth.

Energy fluxes were used by Terjung and Louie (1972) to categorize the magnitude of energy input (net radiation and advection) and outputs (sensible heat and latent heat), and their seasonal range. On this basis, 62 climatic types were distinguished (in six broad groups), and a world map was provided. Smith *et al.* (2002) determine net shortwave and net longwave radiation criteria for a climate classification with nine global types that is similar to Köppen's, developed by Trewartha and Horn (1980). Table A1.2 summarizes their criteria.

Table A1.2 Radiation budget criteria for major climatic types (from Smith *et al.*, 2002). Units W m^{-2}

Climatic type	Annual shortwave	Annual longwave	Annual range of shortwave
LAND			
Tropical:	> 140	<70	<100
Wet	> 140	<50	<100
Wet/dry	>140	>50	<100
Desert		>90	
Steppe		$70 < L_N < 90$	
Subtropical	>140	<70	>100
Temperate	$100 < S_N < 140$	<70	
Boreal	$50 < S_N < 100$	<70	
Polar	$0 < S_N < 50$	<50	
OCEANS			
Tropical	>210		<140
Convergence and stratus	$170 < S_N < 210$		<140
Subtropical	> 150		>140
Temperate	$80 < S_N < 150$		
Polar	$0 < S_N < 80$		

W. Lauer *et al.* (1996) have prepared a new classification and map of world climate types based on thermal and hygric thresholds for both natural vegetation and crops. The limits of the four primary zones (tropical, subtropical, mid-latitude and polar regions) are determined from a radiation index (duration of daily sunshine hours). Climate types are then based on a thermal index (temperature sums) and a moisture index, which takes account of the difference between monthly precipitation and potential evaporation.

C GENETIC CLASSIFICATIONS

The genetic basis of large-scale (macro-) climates is the atmospheric circulation, and this can be related to regional climatology in terms of wind regimes or air masses.

H. Flohn proposed one system in 1950. The major categories are based on the global wind belts and precipitation seasonality, as follows:

1 Equatorial westerly zone: constantly wet.
2 Tropical zone, winter trades: summer rainfall.
3 Subtropical dry zone (trades or subtropical high pressure): dry conditions prevail.
4 Subtropical winter-rain zone (Mediterranean type): winter rainfall.
5 Extra-tropical westerly zone: precipitation throughout the year.
6 Subpolar zone: limited precipitation throughout the year.
6a Boreal, continental subtype: summer rainfall; limited winter snowfall.
7 High polar zone: meagre precipitation; summer rainfall, early winter snowfall.

Temperature does not appear explicitly in the scheme. Figure A1.2B shows the distribution of these types on a hypothetical continent. Rough agreement between these types and those of Köppen's scheme is apparent. Note that the boreal subtype is restricted to the Northern Hemisphere and that the subtropical zones do not occur on the eastern side of a land mass. Flohn's approach has value as an introductory teaching outline.

A. N. Strahler (1969) proposed a simple but effective genetic classification of world climates, based on the fundamental planetary mechanisms. Following a tripartite division by latitude (low, middle and high), regions are grouped according to the relative influence of the ITCZ, the subtropical high pressure cells, cyclonic storms, high-latitude frontal zones and Polar/Arctic air sources. This gives 14 classes and a separate category of Highland Climates. Briefly, these are as follows:

1 Low-latitude climates controlled by equatorial and tropical air masses.
- Wet equatorial climate (10°N–10°S; Asia 10°S–20°N) – converging equatorial and mT air masses produce heavy convectional rains; uniform temperature.
- Trade Wind littoral climate (10°–25°N and S) – high-sun Trade Winds alternate seasonally with subtropical high pressure; strong seasonality of rainfall, high temperatures.
- Tropical desert and steppe (15°–35°N and S) – dominance of subtropical high pressure gives low rainfall and high maximum temperatures with moderate annual range.
- West Coast desert climate (15°–30°N and S) – dominance of subtropical high pressure. Cool seas maintain low rainfall with fog and small annual temperature range.
- Tropical wet–dry climate (5°–15°N and S) – high-sun wet season, low-sun dry season; small annual temperature range.

2 Mid-latitude climates controlled by both tropical and polar air masses.
- Humid subtropical climate (20°–25°N and S) – high-sun moist mT air and low-sun cyclones give well-distributed annual rainfall with moderate temperature regime.
- West Coast marine climate (40°–60°N and S) – windward coasts with cyclones all year. Cloudy; well-distributed rainfall with low-sun maximum.
- Mediterranean climate (30°–45°N and S). Hot, dry summers associated with the

subtropical highs alternate with winter cyclones bringing ample rain.
- Mid-latitude continental desert and steppe (35°–50°N and S). Summer cT air alternates with winter cP air. Hot summers and cold winters give a large annual temperature range.
- Humid continental climate (35°–60°N). Central and eastern continental locations. Frontal cyclones. Cold winters, warm to hot summers, large annual temperature range. Well distributed precipitation.

3 High-latitude climates controlled by polar and Arctic air masses
- Continental sub-Arctic climates (50°–70°N). Source region for cP air. Very cold winters, short, cool summers, extreme annual temperature range. Year-round cyclonic precipitation.
- Marine sub-Arctic climate (50°–60°N and 45°–60°S). Dominated by the winter arctic frontal zone. Cold, moist winters, cool summers; small annual temperature range.
- Polar tundra climates (north of 55°–60°N and south of 60°S). Arctic coastal margins dominated by cyclonic storms. Humid and cold, moderated somewhat by maritime influences in winter.
- Ice sheet climates (Greenland and Antarctica). Source regions of Arctic and Antarctic air. Perpetual frost, low snowfall except near coasts.

4 Highland climates – localized and varied in character.

D CLASSIFICATIONS OF CLIMATIC COMFORT

The body's thermal equilibrium is determined by metabolic rate, heat storage in body tissues, radiative and convective exchanges with the surroundings, and evaporative heat loss by sweating. In indoor conditions, about 60 percent of body heat is lost by radiation and 25 percent by evaporation from the lungs and skin. Outdoors, additional heat is lost by convective transfer due to the wind. Human comfort depends primarily on air temperature, relative humidity and wind speed (Buettner, 1962). Comfort indices have been developed by physiological experiments in test chambers. They include measures of heat stress and windchill.

Windchill describes the cooling effect of low temperature and wind on bare skin. It is commonly expressed via a windchill equivalent temperature. For example, a 15 m s^{-1} wind with an air temperature of –10°C has a windchill equivalent of –25°C. A windchill of –30°C denotes a high risk of frostbite and corresponds to a heat loss of approximately 1600 W m^{-2}. Nomograms to determine windchill have been proposed, as well as other formulae that include the protective effect of clothing.

Heat discomfort is assessed from measurements of air temperature and relative humidity. The U.S. National Weather Service uses a Heat Index based on a measure of *apparent temperature* developed by R. G. Steadman for normally clothed individuals. The value of apparent temperature in the shade (TAPP) is approximately:

$$\text{TAPP} = -2.7 + 1.04\,T_A + 2.0e - 0.65\,V_{10}$$

where T_A = midday temperature (°C), e = vapour pressure (mb), V_{10} = 10m wind speed (m s^{-1}). Warnings are issued in the United States when the apparent temperature reaches 40.5°C for more than three hours/day on two consecutive days.

Another approach measures the thermal insulation provided by clothing. One '*clo*' unit maintains a seated/resting person comfortable in surroundings of 21°C, relative humidity below 50 percent and air movement of 10 cm s^{-1}. For example, the clo values of representative clothing are: tropical wear <0.25, light summer clothes 0.5, typical male/female day wear, 1.0, winter wear with hat and over- coat 2.0–2.5, woollen winter sportswear 3.0, and polar clothing 3.6–4.5. The clo units correlate closely with windchill and inversely with the heat index.

A bioclimatic classification that incorporates estimates of comfort using temperature, relative humidity, sunshine and wind speed data has been proposed for the United States by W. H. Terjung (1966).

BIBLIOGRAPHY

Bailey, H. P. (1960) A method for determining the temperateness of climate. *Geografiska Annaler* 42, 1–16.

Budyko, M. I. (1956) *The Heat Balance of the Earth's Surface* (trans. by N. I. Stepanova), US Weather Bureau, Washington, DC.

Budyko, M. I. (1974) *Climate and Life* (trans. D. H. Miller), Academic Press, New York, 508pp.

Buettner, K. J. (1962) Human aspects of bioclimatological classification, in Tromp, S. W. and Weihe, W. H. (eds) *Biometeorology*, Pergamon, Oxford and London, 128–40.

Carter, D. B. (1954) Climates of Africa and India according to Thornthwaite's 1948 classification, in *Publications in Climatology* 7(4), Laboratory of Climatology, Centerton, NJ.

Chang, J-H. (1959) An evaluation of the 1948 Thornthwaite classification. *Ann. Assn. Amer. Geog.* 49, 24–30.

Dixon, J. C. and Prior, M. J. (1987) Wind-chill indices – a review. *Met. Mag.* 116, 1–17.

Essenwanger, O. (2001) Classification of climates, in *General Climatology, Vol.1C. World Survey of Climatology*, Elsevier, Amsterdam, 1–102.

Feddema, J. (2005) A revised Thornthwaite-type global climate classification. *Phys. Geog.* 26, 442–66.

Flohn, H. (1950) Neue Anschauungen über die allgemeine Zirkulation der Atmosphäre und ihre klimatische Bedeutung. *Erdkunde* 4, 141–62.

Flohn, H. (1957) Zur Frage der Einteilung der Klimazonen. *Erdkunde* 11, 161–75.

Gentilli, J. (1958) *A Geography of Climate*, University of Western Australia Press, 120–66.

Gregory, S. (1954) Climatic classification and climatic change. *Erdkunde* 8, 246–52.

Grundstein, A. (2008), Assessing climate change in the contiguous United States using a modified Thornthwaite climate classification scheme. *Prof. Geog.* 60, 398–412.

Kraus, H. and Alkhalaf, A. (1995) Characteristic surface energy budgets for different climate types. *Internat. J. Climatol.* 15, 275–84.

Lauer, W., Rafiqpoor, M. D. and Frankenberg, P. (1996) Die Klimate der Erde. Eine Klassifikation auf ökophysiologicher Grundlage auf der realen Vegetation. *Erdkunde.* 50(4), 275–300.

Lohmann, U. *et al.* (1993) The Köppen climate classification as a diagnostic tool for general circulation models. *Climate Res.* 3, 277–94.

Mather, J. R. (1985) The water budget and the distribution of climates, vegetation and soils, in *Publications in Climatology* 38(2), Center for Climatic Research, University of Delaware, Newark 36pp.

Oliver, J. E. (1970) A genetic approach to climatic classification. *Ann. Assn. Amer. Geog.* 60, 615–37. (Commentary, see 61, 815–20.)

Oliver, J. E. and Wilson, L. (1987) Climatic classification, in Oliver, J. E. and Fairbridge, R. W. (eds) *The Encyclopedia of Climatology*, Van Nostrand Reinhold, New York, 231–6.

Osczevski, R. and Bluestein, M. (2005) The new windchill equivalent temperature chart. *Bull. Amer. Met. Soc.*, 86, 1453–58.

Peel, M. C., Finlayson, B. L. and McMahon, T. A. (2007) Updated world map of the Köppen-Geiger climate classification. *Hydrol. Earth Syst. Sci.* 11, 1633–44.

Rees, W. G. (1993) New wind-chill nomogram. *Polar Rec.* 29(170), 229–34.

Salmoud, J. and Smith, C. G. (1996). Back to basics: world climatic types. *Weather* 51, 11–18.

Sanderson, M. E. (1999) The classification of climates from Pythagoras to Koeppen. *Bull. Amer. Met. Soc.* 669–73.

Smith, G. L. *et al.* (2002) Surface radiation budget and climate classification. *J. Climate* 15(10), 1175–88.

Steadman, R. G. (1984) A universal scale of apparent temperature. *J. Clim. Appl. Met.* 23(12), 1674–87.

Steadman, R. G., Osczewski, R. J. and Schwerdt, R. W. (1995) Comments on 'Wind chill errors'. *Bull. Amer. Met. Soc.* 76(9), 1628–37.

Strahler, A. N. (1969) *Physical Geography* (3rd edn), Wiley, New York, 733pp.

Terjung, W. H. (1966) Physiologic climates of the conterminous United States: a bioclimatological classification based on man. *Ann. Assn. Amer. Geog. 56*, 141–79.

Terjung, W. H. and Louie, S. S-F. (1972) Energy input–output climates of the world. *Archiv. Met. Geophys. Biokl.* B, 20, 127–66.

Thornthwaite, C. W. (1948) An approach towards a rational classification of climate. *Geog. Rev.* 38, 55–94.

Thornthwaite, C. W. and Mather, J. R. (1955) The water balance, in *Publications in Climatology* 8(1), Laboratory of Climatology, Centerton, NJ, 104pp.

Thornthwaite, C. W. and Mather, J. R. (1957) Instructions and tables for computing potential evapotranspiration and the water balance, *Publications in Climatology* 10(3), Laboratory of Climatology, Centerton, NJ, 129pp.

Trewartha, G. E. and Horn, L. H. (1980) *An Introduction to Climate*, McGraw-Hill, New York, 416pp.

Troll, C. (1958) Climatic seasons and climatic classification. *Oriental Geographer* 2, 141–65.

Wang, M. and Overland, J. E. (2004) Detecting climate change using Köppen climate classification scheme. *Climaic Change* 67, 1573–80.

Yan, Y. Y. and Oliver, J. E. (1996) The clo: a utilitarian unit to measure weather/climate comfort. *Int. J. Climatol.* 16(9), 1045–56.

APPENDIX 2

Système International (SI) units

The *basic SI units* are meter, kilogram, second (m, kg, s):

1mm	= 0.03937in	1in	= 25.4mm
1m	= 3.2808 feet	1ft	= 0.3048m
1km	= 0.6214 miles	1mi	= 1.6090km
1kg	= 2.2046lb	1lb	= 0.4536kg
1m s^{-1}	= 2.2400mi hr^{-1}	1mi hr^{-1}	= 0.4460m s^{-1}
1m^2	= 10.7640ft^2	1 ft^2	= 0.0929m^2
1km^2	= 0.3861mi^2	1mi^2	= 0.5900km^2
1°C	= 1.8°F	1°F	= 0.555°C

Temperature conversions can be determined by noting that:

$$\frac{T(°C)}{5} = \frac{T(°F) - 32}{9}$$

Density units = kg m^{-2}

Pressure units = N m^{-2} (= Pa) ; 100 Pa (= hPa) = 1mb.

Mean sea-level pressure = 1013mb (= 1013hPa)

Radius of the sun = 7 × 10^8m
Radius of the earth = 6.37 × 10^6m
Mean earth–sun distance = 1.495 × 10^{11}m

Energy conversion factors:

4.1868J = 1 calorie
J cm^{-2} = 0.2388cal cm^{-2}
Watt = J s^{-1}
W m^{-2} = 1.433 × 10^{-8}cal^{-2}min^{-1}
697.8W m^{-2} = 1 cal cm^{-2}min^{-1}

For time sums:

Day: 1W m^{-2} = 8.64J $cm^{-2}dy^{-1}$
= 2.064cal $cm^{-2}dy^{-1}$
Day: 1W m^{-2} = 8.64 x 10^4J $m^{-2}dy^{-1}$
Month: 1W m^{-2} = 2.592M J $m^{-2}(30\ dy)^{-1}$
= 61.91cal $cm^{-2}(30\ dy)^{-1}$
Year: 1W m^{-2} = 31.536M J $m^{-2}yr^{-1}$
= 753.4cal $cm^{-2}yr^{-1}$

Gravitational acceleration (g) = 9.81m s^{-2}

Latent heat of vaporization (288K) = 2.47 × 10^6Jkg^{-1}
Latent heat of fusion (273K) = 3.33 × 10^5Jkg^{-1}

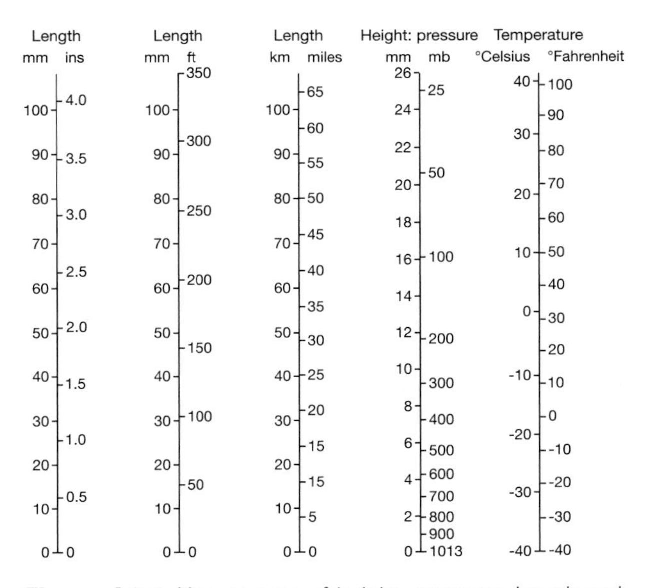

Figure A2.1 Nomograms of height, pressure, length and temperature.

APPENDIX 3

Synoptic weather maps

Table A3.1 Synoptic code (World Meteorological Organization, January 1982)

Symbol	Key	Example	Comments
yy	Day of the month (GMT)	05	All groups are in blocks
GG	Time (GMT) to nearest hour	06	{ of 5 digits
i_w	Indicator for type of wind speed observation and units	4	Measured by anemometer (knots)
Iliii	International index number of station		
i_R	Indicator: precipitation data included/omitted (code)	3	Data omitted
i_X	Indicator: station type + ww W1 W2 included/ omitted (code)	1	Manned station with ww W1 W2 included
h	Height of lowest cloud (code)	3	
vv	Visibility (code)	66	
N	Total cloud amount (oktas)	7	
dd	Wind direction (tens of degrees)	32	
ff	Wind speed (knots, orm s-1)	20	Knots
I	Header	1	
S_n	Sign of temperature (code)	0	Positive value
TTT	Temperature (0.1°C), plotted rounded to nearest 1°C	203	(1 = negative value)
2	Header	2	
S_n	Sign of temperature (code)	0	
$T_dT_dT_d$	Dewpoint temperature (as TTT)	138	
4	Header	4	
PPPP	Mean sea-level pressure (tenths of mb, omitting thousands)	0105	
5	Header	5	
a	Characteristic of pressure tendency (coded symbol)	3	
ppp	3-hour pressure tendency (tenths of mb)	005	
7	Header	7	
ww	Present weather (coded symbol)	80	
W_1	Past weather (coded symbol)	9	(W_1 must be greater
W_2	Past weather (coded symbol)	8 }	than W_2)
8	Header	8	
N_h	Amount of low cloud (oktas)	4	
C_L	Low cloud type (coded symbol)	2	
C_M	Medium cloud type (coded symbol)	5	
C_H	High cloud type (coded symbol)	2	

Note: Group 3 is for a report of surface pressure and group 6 for precipitation data.

MODEL (enlarged)

CH
TT CM PPP
VV ww N ppa
ff dd CL W1 W2
TdTd Nh/h

EXAMPLE

20 105
66 05
14
4/3

	KEY	EXAMPLE
N	Total cloud (oktas)[1]	7
dd	Wind direction (tens of degrees)	32
ff	Wind speed (knots)	20
VV	Visibility (code)	66
ww	Present weather (coded symbol)	80
W_1	Past weather (coded symbol)	9
W_2	Past weather (coded symbol)	8
PPP	Sea-level pressure (mb)[2]	105
TT	Temperature (°C)[4]	20
N_h	Low cloud (oktas)[1]	4
C_L	Low cloud type (coded symbol)	2
h	Height of C_L (code)	3
C_M	Medium cloud type (coded symbol)	5
C_H	High cloud type (coded symbol)	2
TdTd	Dew-point temperature (°C)[4]	14
a	Barograph trace (coded symbol)	3
pp	3-hour pressure change (mb)[3]	05

[1]okta = eighth
[2]Pressure in tens, units and tenths mb: omitting initial 9 or 10 i.e. 105 = 1010.5
[3]Pressure change in units and tenths mb
[4]Rounded to nearest °C

Figure A3.1 Basic station model for plotting weather data. The key and examples are tabulated in the internationally agreed sequence for teletype messages. These data would be preceded by an identifying station number, date and time.

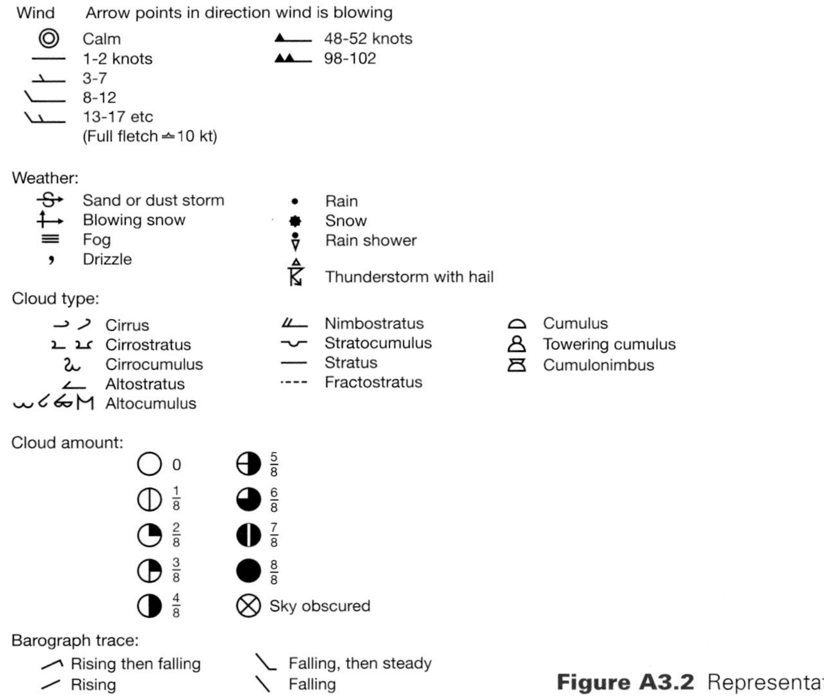

Figure A3.2 Representative synoptic symbols.

APPENDIX 4

Data sources

A DAILY WEATHER MAPS AND DATA

Western Europe/North Atlantic: *Daily Weather Summary* (synoptic chart, data for the UK). London Weather Center, 284 High Holborn, London WC1V 7HX, England.

Western Europe/North Atlantic: *Monthly Weather Report* (published about 15 months in arrears; tables for approximately 600 stations in the UK). London Weather Center.

Europe – Eastern/North Atlantic: *European Daily Weather Report* (synopic chart). Deutsche Wetterdienst, Zentralamt D6050, Offenbach, Germany.

Europe – Eastern/North Atlantic: *Weather Log* (daily synopic chart, supplement to *Weather* magazine). Royal Meteorological Society, Bracknell, Berkshire, England.

North America: *Daily Weather Reports* (weekly publication). National Environmental Satellite Data and Information Service, NOAA, US Government Printing Office, Washington, DC 20402, USA.

B SATELLITE DATA

NOAA operational satellites (imagery, digital data). Satellite Data Services NCDC, NOAA/NESDIS, Asheville, NC 28801–5001, USA.

Defense Meteorological Satellite Program (data). National Geophysical Data Center, NOAA/NESDIS, 325 Broadway, Boulder, CO 80303, USA.

Metsat (imagery, digital data). ESOC, Robert-Bosil Str. 5, D-6100 Darmstadt, Federal Republic of Germany.

NASA research satellites (digital data). National Space Science Data Center, Goddard Space Flight Center, Greenbelt, MD 20771, USA.

UK Direct readout data from NOAA and Meteosat satellites are received at Dundee, Scotland. Dr P. E. Baylis, Department of Electrical Engineering, University of Dundee, Scotland, UK.

C CLIMATIC DATA

Canadian Climate Center: *Climatic Perspectives* (197 weekly and monthly summary charts). Atmospheric Environment Service, 4905 Dufferin Street, Downsview, Ontario, Canada M3H 5T4.

Carbon Dioxide Information Center (CDIC): Data holdings and publications on climate-related variables and indices. Carbon Dioxide Information Center, Oak Ridge National Laboratory, Oak Ridge TN 37931, USA.

National Center for Atmospheric Research, Boulder, CO 80307–3000, USA. Archives most global analyses and many global climate records.

National Climatic Center, NOAA/NESDIS: *Local Climatological Data* (1948–) (monthly tabulations, charts); *Monthly Climatic Data for the World* (May 1948–). National Climatic Data Center, Federal Building, Asheville NC 28801, USA.

Climate Analysis Center, NOAA/NESDIS: *Climatic Diagnostics Bulletin* (1983–) (monthly summaries of selected diagnostic product from NMC analyses). Climate Analysis Center, NMC NOAA/NWS, World Weather Building, Washington, DC 20233, USA.

Climatic Research Unit, University of East Anglia: *Climate Monitor* (1976–) (monthly summaries, global and UK). Climatic Research Unit, University of East Anglia, Norwich NR4 7TJ, England.

World Climate Data Programme: *Climate System Monitoring Bulletin* (1984–) (monthly). World Climate Data Programme, WMO Secretariat, CP5, Geneva 20 CH-1211, Switzerland.

World Meteorological Center, Melbourne: *Climate Monitoring Bulletin for the Southern Hemisphere* (1986–). Bureau of Meteorology, GPO Box 1289 K, Melbourne, Victoria 3001, Australia.

D SELECTED SOURCES OF INFORMATION ON THE WORLD WIDE WEB

World Meteorological Organization, Geneva, Switzerland http://www.wmo.ch

National Oceanic and Atmospheric Administration, Washington, DC, USA http://www.noaa.gov

National Climate Data Center, Asheville, NC, USA hhtp://www.ncdc.noaa.gov/ncdc.html

Climate Analysis Branch, ESRL, NOAA, Boulder, CO, USA http://www.cdc.noaa.gov/

National Snow and Ice Data Center, Boulder, CO, USA http://nsidc.org

National Center for Atmospheric Research http://www.rap.ucar.edu/weather/

Environment Canada http://www.weatheroffice.gc.ca

European Center for Medium-range Weather Forecasting http://www.ecmwf.int

Climate Diagnostics Bulletin (US) http://nic.fb4.noaa.gov

UK Meteorological Office http://www.meto.gov.uk

US National Severe Storms Laboratory http://www.nssl.uoknor.edu

Windchill http://www.crh.noaa.gov/den/windchill.html#definitions

BIBLIOGRAPHY

Ahlquist, J. (1993) Free software and information via the computer network. *Bull. Amer. Met. Soc.* 74 (3), 377–86.

Brugge, R. (1994) Computer networks and meteorological information. *Weather* 49(9), 298–306.

Carleton, A. M. (1991) *Satellite Remote Sensing in Climatology*, Belhaven Press, London, and CRC Press, Boca Raton, FL, 291pp.

European Space Agency (1978a) *Introduction to the Metsat System*, European Space Operations Center, Darmstadt, 54pp.

European Space Agency (1978b) *Atlas of Meteosat Imagery, Atlas Meteosat*, ESA-SP-1030, ESTEC, Nordwijk, Netherlands, 494pp.

Finger, F. G., Laver, J. D., Bergman, K. H. and Patterson, V. L. (1958) The Climate Analysis Center's user information service. *Bull. Amer. Met. Soc.* 66, 413–20.

Hastings, D. A., Emery, W. J., Weaver, R. L., Fisher, W. J. and Ramsey, J. W. (1987) *Proceedings North American NOAA Polar Orbiter User Group First Meeting*, NOAA/NESDIS, US Department of Commerce, Boulder, CO, National Geophysical Data Center, 273pp.

Hattemer-Frey, H. A., Karl, T. R. and Quinlan, F. T. (1986) *An Annotated Inventory of Climatic Indices and Data Sets*, DOE/NBB-0080, Office of Energy Research, US Deptartment of Energy, Washington, DC, 195pp.

Jenne, R. L. and McKee, T. B. (1985) Data; in Houghton, D. D. (ed.) *Handbook of Applied Meteorology*, Wiley, New York, 1175–281.

Meteorological Office (1958) *Tables of Temperature, Relative Humidity and Precipitation for the World*, HMSO, London.

Singleton, F. (1985) Weather data for schools, *Weather* 40, 310–13.

Stull, A. and Griffin, D. (1996) *Life on the Internet – Geosciences*, Prentice Hall, Upper Saddle River, NJ.

US Department of Commerce (1983) *NOAA Satellite Programs Briefing*, National Oceanic and Atmospheric Administration, Washington, DC, 203pp.

US Department of Commerce (1984) *North American Climate Data Catalog. Part 1*, National Environmental Data Referral Service, Publication NEDRES-1, National Oceanic and Atmospheric Administration, Washington, DC, 614pp.

World Meteorological Organization (1965) *Catalogue of Meteorological Data for Research*, WMO No. 174. TP-86, World Meteorological Organization, Geneva.

Notes

2 Atmospheric composition, mass and structure

1 Mixing ratio = ratio of number of molecules of ozone to molecules of air (parts per million by volume, ppm(v)). Concentration = mass per unit volume of air (molecules cubic meter).

2 K = degrees Kelvin (or Absolute). The degree symbol is omitted.
°C = degrees Celsius
°C = K – 273
Conversions for °C and °F are given in Appendix 2.

3 Joule = 0.2388cal. The units of the International Metric System are given in Appendix 2. At present the data in many references are still in calories; a calorie is the heat required to raise the temperature of 1g of water from 14.5°C to 15.5°C. In the United States, another unit in common use is the Langley (ly) (ly min^{-1} = 1cal $cm^{-2}min^{-1}$).

4 The equation for the so-called 'reduction' (actually the adjusted value is normally greater!) of station pressure (p_h) to sea-level pressure (p_0) is written:

$$p_0 = p_h \exp [g_0 Z_h / R_d T_v]$$

where R_d = gas content for dry air; g_0 = global average of gravitational acceleration ($9.8ms^{-2}$); Z_h = geopotential height of the station (≃ geometric height in the lowest kilometer or so); T_v = mean virtual temperature. This is a fictitious temperature used in the ideal gas equation to compensate for the fact that the gas constant of moist air exceeds that of dry air. Even for hot moist air, T_v is only a few degrees greater than the air temperature.

5 The official definition is the lowest level at which the lapse rate decreases to less than, or equal to, 2°C/km (provided that the average lapse rate of the 2km layer does not exceed 2°C/km).

3 Solar radiation and global energy budget

1 The radiation flux (per unit area) received normal to the beam at the top of the earth's atmosphere is calculated from the total solar output weighted by $1/(4\pi D^2)$, where the solar distance $D = 1.5 \times 10^{11}$m, since the surface area of a sphere of radius r (here equivalent to D) is $4\pi r^2$ – i.e., the radiation flux is (6.24 ´× 10^7Wm^{-2}) ($61.58 \times 10^{23}m^2$)/4π (2.235×10^{22}) = $1367Wm^{-2}$.

2 The albedos refer to the solar radiation received on each given surface: thus the incident radiation is different for planet earth, the global surface and global cloud cover, as well as between any of these and the individual cloud types or surfaces.

5 Atmospheric instability, cloud formation and precipitation processes

1. It is significant that some storms also occur downwind of other and plateau regions in Mexico, the Iberian peninsula and West Africa.

6 Atmospheric motion: principles

1. The centrifugal 'force' is equal in magnitude and opposite in sign to the centripetal acceleration. It is an apparent force that arises through inertia.
2. Apparent gravity, $g = 9.78 m\ s^{-2}$ at the equator, $9.83 m\ s^{-2}$ at the poles.
3. The vorticity is a vector measure of the local rotation, or spin, in a fluid flow. It is given by the product of the rotation on its boundary (vR) and the circumference ($2\pi R$) where R = radius of the fluid disc. The vorticity is then $2v\pi R^2$, or 2v per unit area. It comprises the sum of the (anti-)cyclonic shear across a flow and the (anti-)cyclonic curvature of the flow. Cyclonic vorticity is defined as positive. The relative vertical vorticity is that around a local vertical axis on the earth's surface. The absolute vorticity is the sum of the relative vorticity and the earth's vorticity, which is the Coriolis parameter, *f*.

7 Planetary-scale motions in the atmosphere and ocean

1. The geostrophic wind concept is equally applicable to contour charts. Heights on these charts are given in geopotential meters (gpm) or dekameters (gpdkm).
2. The World Meteorological Organization recommends an arbitrary lower limit of $30 m\ s^{-1}$.
3. Equatorial speed of rotation is $465 m\ s^{-1}$.
4. Note that, at the equator, an east/west wind of $5 m\ s^{-1}$ represents an absolute motion of $460/470 m\ s^{-1}$ towards the east.

9 Mid-latitude synoptic and mesoscale systems

1. Resultant wind is the vector average of all wind directions and speeds.
2. This latter term tends to be restricted to the tropical (hurricane) variety.

10 Weather and climate in middle and high latitudes

1. Standard indices of continentality developed by Gorcynski (see p. 270), Conrad and others are based on the annual range of temperature, scaled by the sine of the latitude angle as a reciprocal in the expression. This index is unsatisfactory for several reasons. The small amplitude of annual temperature range in humid tropical climates renders it unworkable for low latitudes. The latitude weighting is intended to compensate for summer–winter differences in solar radiation and thus temperatures, which were thought to increase uniformly with latitude. For North America, the differences peak about 55°N. The index used in Figure 10.20 is based on *departures* from the regression line of annual temperature range versus latitude. The specific regression constants will differ between continents. It should be noted that indices of the Gorcynski type are appropriate for regions of limited latitudinal extent as shown in Figure 10.2.

11 Tropical weather and climate

1. This periodicity is the Madden–Julian Oscillation.

13 Climate change

1. Statistics commonly reported for climatic data are: the *arithmetic mean*,

$$\bar{x} = \Sigma \frac{x_i}{n}$$

where Σ = sum of all values for i = 1 to n
x_i = an individual value
n = number of cases

and the *standard deviation*, *s* (pronounced sigma).

$$s = \Sigma \frac{S\,(x_i - \bar{x})^2}{n}$$

which expresses the variability of observations.

For precipitation data, the *coefficient of variation*, CV is often used:

$$CV = \frac{s}{\bar{x}} \times 100 \text{ (percent)}$$

For a *normal* (or Gaussian) bell-shaped symmetrical frequency distribution, the arithmetic mean is the central value; 68.3 percent of the distribution of values are within ± 1 s of the mean and 94.5 percent within ± 2 s of the mean.

The frequency distribution of mean daily temperatures is usually approximately normal. However, the frequency distribution of annual (or monthly) totals of rainfall over a period of years may be 'skewed' with some years (months) having very large totals whereas most years (months) have low amounts. For such distributions the *median* is a more representative average statistic; the median is the middle value of a set of data ranked according to magnitude. Fifty percent of the frequency distribution is above the median and 50 percent below it. The variability may be represented by the 25 and 75 percentile values in the ranked distribution.

A third measure of central tendency is the *mode* – the value which occurs with greatest frequency. In a normal distribution the mean, median and mode are identical.

Frequency distributions for cloud amounts are commonly bimodal with more observations having small or large amounts of cloud cover than are in the middle range.

General Bibliography

Ahrens, C. D. (2003) *Meteorology Today: An Introduction to Weather, Climate and the Environment*, 7th edn, Brooks/Cole (Thomson Learning, 624pp. [Basic introduction to meteorology, including weather forecasts, air pollution, global climate and climatic change; accompanying CD]

Anthes, R. A. (1997) *Meteorology*, 7th edn, Prentice-Hall, Upper Saddle River, NJ, 214pp. [Introduction to meteorology, weather and climate]

Atkinson, B. W. (1981a) *Meso-scale Atmospheric Circulations*, Academic Press, London, 496pp. [Discusses the theoretical ideas and current understanding of the major mesoscale circulation features – sea/land breezes, mountain/valley winds, convective systems]

Atkinson, B. W. (ed.) (1981b) *Dynamical Meteorology*, Methuen, London, 250pp. [Collected papers, most originally published in *Weather*, introducing readers to the basic dynamical concepts of meteorology]

Barry, R. G. (2008) *Mountain Weather and Climate*, 3rd edn, Cambridge University Press, Cambridge 506pp. [Details the effects of altitude and orography on climatic elements, orographic effects on synoptic systems and airflow, the climatic characteristics of selected mountains, and climate change in mountains]

Barry, R. G. and Carleton, A. M. (2001) *Dynamic and Synoptic Climatology*, Routledge, London 620pp. [Graduate-level text on the global circulation and its major elements – planetary waves, blocking, and teleconnection patterns – as well as synoptic systems of middle and low latitudes and approaches to synoptic classification and their applications; there is also a chapter on climate data, including remote sensing data. and their analysis; extensive bibliographies]

Berry, F. A., Bollay, E. and Beers, N. R. (eds) (1945) *Handbook of Meteorology*, McGraw-Hill, New York, 1068pp. [A classic handbook covering many topics]

Bigg, G. (1996) *The Oceans and Climate*, Cambridge University Press, Cambridge, 266pp. [Undergraduate text dealing with the physical and chemical interactions of the oceans and climate, air–sea interaction and the role of the oceans in climate variability and change]

Blüthgen, J. (1966) *Allgemeine Klimageographie*, 2nd edn, W. de Gruyter, Berlin, 720pp. [A classic German work on climatology with extensive references]

Bradley, R. S., Ahern, L. G. and Keimig, F. T. (1994) A computer-based atlas of global instrumental climate data. *Bull. Amer. Met. Soc.* 75(1), 35–41.

Bruce, J. P. and Clark, R. H. (1966) *Introduction to Hydrometeorology*, Pergamon, Oxford, 319pp. [Valuable introductory text with no modern equivalent]

Carleton, A. M. (1991) *Satellite Remote Sensing in Climatology*, Belhaven Press, London, 291pp. [A monograph on basic techniques and their climatological application in the study of clouds, cloud systems, atmospheric moisture and the energy budget]

Crowe, P. R. (1971) *Concepts in Climatology*, Longman, London, 589pp. [A geographical climatology that covers processes non-mathematically, the general circulation, air masses and frontal systems, and local climates]

Geiger, R., Aron, R. and Todhunter, P. (2003) *The Climate Near the Ground*, 6th edn, Rowman & Littlefield, Lanham, MD, 584pp. [Classic text on local, topo- and microclimates; extensive references to European research]

Glickmann, T. S. (ed.) (2000) *Glossary of Meteorology*, 2nd edn, American Meteorological

Society, Boston, MA, 855pp. [Indispensable guide to terms and concepts in meteorology and related fields]

Gordon, A., Grace, W., Schwerdtfeger, P. and Byron-Scott, R. (1995) *Dynamic Meteorology: A Basic Course*, Arnold, London, 325pp. [Explains the basic thermodynamics and dynamics of the atmosphere, with key equations; synoptic analysis and the tropical cyclone]

Haltiner, G. J. and Martin, F. L. (1957) *Dynamical and Physical Meteorology*, McGraw-Hill, New York, 470pp. [Comprehensive account of the fundamentals of atmospheric dynamics and physical processes]

Hartmann, D. L. (1994) *Global Physical Climatology*, Academic Press, New York, 408pp. [Covers the physical bases of climate – the energy and water balances, the atmospheric and ocean circulations, the physics of climate change, climate sensitivity and climate models]

Henderson-Sellers, A. and Robinson, P. J. (1999) *Contemporary Climatology*, 2nd edn, Pearson Education, London, 342pp. [A non-mathematical treatment of physical climatology, the general circulation, selected regional and local climates and climate change and modeling]

Hess, S. L. (1959) *Introduction to Theoretical Meteorology*, Henry Holt, New York, 362pp. [A clear and readable introduction to meteorological principles and processes]

Houghton, D. D. (1985) *Handbook of Applied Meteorology*, Wiley, New York, 1461pp. [A valuable reference source on measurements, a wide range of applications, societal impacts, resources includng data]

Houghton, H. G. (1985) *Physical Meteorology*, MIT Press, Cambridge, MA, 442pp. [Advanced undergraduate-graduate text and reference work in atmospheric science; treats atmospheric aerosols, radiative transfer, cloud physics, optical phenomena and atmospheric electricity]

Houghton, J. T. (ed.) (1984) *The Global Climate*, Cambridge University Press, Cambridge, 233pp. [Contributions by experts relating to the World Climate Research Programme including climate variability, GCMs, the role of clouds, land surface, deserts, the cryosphere, the upper ocean and its circulation, biogeochemistry and carbon dioxide]

Kendrew, W. G. (1961) *The Climates of the Continents*, 5th edn, Oxford University Press, London, 608pp. [Classic climatography with many regional details, figures and tables]

Lamb, H. H. (1972) *Climate: Present, Past and Future 1: Fundamentals and Climate Now*, Methuen, London, 613pp. [Detailed presentation on the mechanisms of global climate and climatic variations; useful supplementary tables; second part summarizes world climatic conditions with extensive data tables; numerous references]

List, R. J. (1951) *Smithsonian Meteorological Tables*, 6th edn, Smithsonian Institution, Washington, 527pp. [A unique collection of atmospheric reference data]

Lockwood, J. G. (1974) *World Climatology: An Environmental Approach*, Arnold, London, 330pp. [Undergraduate climatology text; following an overview of climatic processes and the general circulation, five chapters treat low-latitude climatic regions and five chapters cover mid- and high-latitude climates; clear diagrams and chapter references]

Lockwood, J. G. (1979) *Causes of Climate*, Arnold, London, 260pp. [Covers the physical components of the climate system, atmospheric circulation, glacial and interglacial climates and model projections of the future]

Lutgens, F. K. and Tarbuck, E. J. (1995) *The Atmosphere: An Introduction to Meteorology*, 6th edn, Prentice-Hall, Englewood Cliffs, NJ, 462pp. [Descriptive introduction to atmospheric processes, weather, including forecasting, and world climates]

McIlveen, R. (1992) *Fundamnetals of Weather and Climate*, Chapman and Hall, London, 497pp. [Mainly quantitative introduction to atmospheric processes, regional weather systems and the general circulation; mathematical derivations in appendices; numerical problems]

McIntosh, D. H. and Thom, A. S. (1972) *Essentials of Meteorology*, Wykeham Publications, London, 239pp. [Introductory text covering main concepts in dynamics and thermodynamics of the atmosphere]

Malone, T. F. (ed.) (1951) *Compendium of Meteorology*, American Meteorological Society, Boston, MA, 1334pp. [Status reports on all areas of meteorology in 1950]

Martyn, D. (1992) *Climates of the World*, Elsevier, Amsterdam, 435pp. [A brief outline of climatic factors and elements followed by a climatography of land and ocean areas. Sources are mostly from the 1960s–1970s; few tables]

Moran, J. M. and Morgan, M. D. (1997) *Meteorology: The Atmosphere and Science of Weather*, 5th edn, Prentice-Hall, Upper Saddle

River, NJ, 550pp. [introduction to meteorology, weather and climate for non-science majors]

Musk, L. F. (1988) *Weather Systems*, Cambridge University Press, Cambridge, 160pp. [Basic introduction to weather systems]

Oliver, J. E. and Fairbridge, R. W. (eds) (1987) *The Encyclopedia of Climatology*, Van Nostrand Reinhold, New York, 986pp. [Comprehensive, illustrated articles and many references]

Palmén, E. and Newton, C. W. (1969) *Atmosphere Circulation Systems: Their Structure and Physical Interpretation*, Academic Press, New York, 603pp. [A work that still lacks a modern equal in its overview of the global circulation and its major component elements]

Pedgley, D. E. (1962) *A Course of Elementary Meteorology*, HMSO, London, 189pp. [A useful concise introduction]

Peixoto, J. P. and Oort, A. H. (1992) *Physics of Climate*, American Institute of Physics, New York, 520pp. [Advanced description of the mean states of the atmosphere and oceans and their variability, the characteristics of momentum, energy and moisture transports, the global energy cycle, and climate simulation]

Petterssen, S. (1956) *Weather Analysis and Forecasting* (2 vols), McGraw-Hill, New York, 428 and 266pp. [Classical standard work of the 1950s; useful basic information on weather systems and fronts]

Petterssen, S. (1969) *Introduction to Meteorology*, 3rd edn, McGraw-Hill, New York, 333pp. [Classic introductory text including world climates]

Pickard, G. L. and Emery, W. J. (1990) *Descriptive Physical Oceanography*, 5th edn, Pergamon Press, Oxford. [Readable introduction to physical properties, processes and circulation in the world's oceans]

Reiter, E. R. (1963) *Jet Stream Meteorology*, University of Chicago Press, Chicago, IL, 515pp. [Classic text on jet streams, their mechanisms and associated weather]

Rex, D. F. (ed.) (1969) *Climate of the Free Atmosphere. World Survey of Climatology Vol. 4*, Elsevier, Amsterdam, 450pp. [Details the structure of the atmosphere, cloud systems the tropospheric circulation and jet streams, ozone and ultraviolet radiation, and the dynamics of the stratosphere]

Rohli, R. V. and Vega, A. J. (2008) *Climatology*, Jones & Bartlett, Sudbury, MA, 467pp. [Upper-level text covering climatological principles, global water balance, atmospheric circulation and climate classification]

Schaefer, V. J. and Day, J. A. (1981) *A Field Guide to the Atmosphere*, Houghton Mifflin, Boston, MA, 359pp. [Popular text with good photographs of atmospheric phenomena]

Singh, H. B. (ed.) (1992) *Composition, Chemistry, and Climate of the Atmosphere*, Van Nostrand Reinhold, New York, 527pp.

Strahler, A. N. (1965) *Introduction to Physical Geography*, Wiley, New York, 455pp. [Broad survey that includes climatic characteristics, controls and distribution of climate types]

Strangeways, I. (2003) *Measuring the Natural Environment,* 2nd edn, Cambridge University Press, Cambridge, 548pp. [Well-illustrated book on instruments for observing all surface weather and climate elements, together with practical information and many references]

Stringer, E. T. (1972a) *Foundations of Climatology, An Introduction to Physical, Dynamic, Synoptic and Geographical Climatology*, W. H. Freeman, San Francisco, CA, 586pp. [Text for climatology students; detailed coverage of atmospheric properties and processes, the general circulation, elementary dynamics and thermodynamics and synoptic methods; appendices with notes and formulae]

Stringer, E. T. (1972b) *Techniques of Climatology*, W. H. Freeman, San Francisco, CA, 539pp. [Companion volume to 1972a dealing with weather observations and their analysis, applications to radiation, temperature and clouds, and the study of regional climates]

Sverdrup, H. V. (1945) *Oceanography for Meteorologists*, Allen & Unwin, London, 235pp. [Classic text on the meterologically-relevant aspects of oceanography]

Sverdrup, H. V., Johnson, M. W. and Fleming, R. H. (1942) *The Oceans: Their Physics, Chemistry and General Biology*, Prentice-Hall, New York, 1087pp. [Classic reference work on the oceans]

Trewartha, G. T. (1981) *The Earth's Problem Climates* (2nd edn), University of Wisconsin Press, Madison (371pp.). [Focuses on climatic regimes that do not fit into the usual global climatic classifications; circulation controls are emphasized]

Trewartha, G. T. and Horne, L. H. (1980) *An Introduction to Climate*, 5th edn, McGraw-Hill, New York, 416pp. [A traditional climatology text]

van Loon, H. (ed.) (1984) *Climates of the Oceans: World Survey of Climatology* 15, Elsevier, Amsterdam, 716pp. [Substantive chapters on the climate of each of the world's ocean basins]

Wallace, J. M. and Hobbs, P. V. (2006) *Atmospheric Science: An Introductory Survey*, 2nd edn, Academic Press, New York, 467pp. [Comprehensive introductory text for meteorologists; atmospheric thermodynamics and dynamics, synoptic disturbances, global energy balance and the general circulation; numerical qualitative problems]

World Meteorological Organization (1962) *Climatological Normals (CLINO) for CLIMAT and CLIMAT SHIP Stations for the Period 1931–60*, World Meteorological Organization, Geneva.

Index

A World of Online Content!

Did you know that Taylor & Francis has over 20,000 books available electronically?

What's more, they are all available for browsing and individual purchase on the Taylor & Francis eBookstore.

www.ebookstore.tandf.co.uk

eBooks for libraries

Free trials available

Choose from annual subscription or outright purchase and select from a range of bespoke subject packages, or tailor make your own.

www.ebooksubscriptions.com

For more information, email
online.sales@tandf.co.uk

ROUTLEDGE
Revivals

Are there some elusive titles you've been searching for but thought you'd never be able to find?

Well this may be the end of your quest. We now offer a fantastic opportunity to discover past brilliance and purchase previously out of print and unavailable titles by some of the greatest academic scholars of the last 120 years.

Routledge Revivals is an exciting new programme whereby key titles from the distinguished and extensive backlists of the many acclaimed imprints associated with Routledge are re-issued.

The programme draws upon the backlists of Kegan Paul, Trench & Trubner, Routledge & Kegan Paul, Methuen, Allen & Unwin and Routledge itself.

Routledge Revivals spans the whole of the Humanities and Social Sciences, and includes works by scholars such as Emile Durkheim, Max Weber, Simone Weil and Martin Buber.

FOR MORE INFORMATION

Please email us at **reference@routledge.com** or visit:
www.routledge.com/books/series/Routledge_Revivals

www.routledge.com

LIBRARY, UNIVERSITY OF CHESTER